CHROMOSOMES AND CANCER

This is the first volume in
CHROMOSOMES: A WILEY BIOMEDICAL-HEALTH SERIES

ADVISORS TO THE EDITOR

John H. Edwards, *Birmingham*
C. E. Ford, *Oxford*
Jean de Grouchy, *Paris*
Patricia A. Jacobs, *Honolulu*
Jérôme Lejeune, *Paris*
Orlando J. Miller, *New York*
Klaus Patau, *Madison*
Curt Stern, *Berkeley*

CONSULTING EDITOR

Judith R. Tennant, *New York*

CHROMOSOMES AND CANCER

JAMES GERMAN, Editor

The New York Blood Center

New York, New York

1974

A WILEY BIOMEDICAL-HEALTH PUBLICATION

JOHN WILEY & SONS, New York • London • Sydney • Toronto

Copyright © 1974, by John Wiley & Sons, Inc.

All rights reserved. Published simultaneously in Canada.

No part of this book may be reproduced by any means, nor transmitted, nor translated into a machine language without the written permission of the publisher.

Library of Congress Cataloging in Publication Data:

German, James.
   Chromosomes and cancer.

   (Chromosomes series) (A Wiley biomedical-health publication)
   Includes bibliographical references.
   1. Cancer— Genetic aspects. 2. Human chromosome abnormalities. 3. Cancer cells. I. Title.
II. Series. [DNLM: 1. Chromosome abnormalities.
2. Cytogenetics. 3. Neoplasms—Etiology. QZ202
G373c 1974]

RC262.G42   616.9'94'071   74-897

ISBN 0-471-29682-1

Printed in the United States of America

10 9 8 7 6 5 4 3 2 1

CONTENTS

THE CHROMOSOMES SERIES vii
INTRODUCTION TO CHROMOSOMES AND CANCER ix
SPECIAL RECOGNITION: PROFESSOR CURT STERN xii
A GENETICIST'S JOURNEY, *Curt Stern* xiii

BACKGROUND

THEODOR BOVERI AND HIS BOOK ON THE PROBLEM OF THE ORIGIN OF MALIGNANT TUMORS, *Ulrich Wolf* 3
THE BIOLOGY OF CANCER, *Sir Macfarlane Burnet* 21

DISTURBANCES OF THE GENETIC MATERIAL

CHROMOSOME INTEGRATION OF VIRAL DNA: THE OPEN-REPLICON HYPOTHESIS OF CARCINOGENESIS, *H. L. K. Whitehouse* 41
ANEUPLOIDY AS A POSSIBLE MEANS EMPLOYED BY MALIGNANT CELLS TO EXPRESS RECESSIVE PHENOTYPES, *Susumu Ohno* 77
WHAT IS A CHROMOSOME BREAK? *David E. Comings* 95
THE APPLICATION OF BANDING TECHNIQUES TO TUMOR CHROMOSOMES, *Margery W. Shaw and T. R. Chen* 135
VIRUSES, CHROMOSOMES, AND TUMORS: THE INTERACTION BETWEEN VIRUSES AND CHROMOSOMES, *D. G. Harnden* 151
EFFECTS OF IONIZING RADIATION ON MAMMALIAN CHROMOSOMES, *H. J. Evans* 191
MITOTIC ABNORMALITIES AND CANCER, *Tarvo Oksala and Eeva Therman* 239

CANCER AS A CLONE

CHROMOSOME CHANGES AND THE CLONAL EVOLUTION OF CANCER, *Peter C. Nowell* 267
CLONAL EVOLUTION IN THE MYELOID LEUKEMIAS, *Jean de Grouchy and Catherine Turleau* 287

UTILIZATION OF MOSAIC SYSTEMS IN THE STUDY OF THE ORIGIN AND PROGRESSION OF TUMORS, *Stanley M. Gartler* 313

CYTOGENETICS OF CANINE VENEREAL TUMORS: WORLDWIDE DISTRIBUTION AND A COMMON KARYOTYPE, *S. Makino* 335

CYTOGENETICS OF CERTAIN SPECIFIC CANCERS

CHROMOSOMES IN HUMAN MALIGNANT TUMORS: A REVIEW AND ASSESSMENT, *N. B. Atkin* 375

CYTOGENETICS OF CANCER AND PRECANCEROUS STATES OF THE CERVIX UTERI, *A. I. Spriggs* 423

CHROMOSOMES AND LEUKEMIA, *S. Muldal and L. G. Lajtha* . . . 451

CHROMOSOME PATTERNS IN BENIGN AND MALIGNANT TUMORS IN THE HUMAN NERVOUS SYSTEM, *Joachim Mark* 481

THE HUMAN MENINGIOMA: A BENIGN TUMOR WITH SPECIFIC CHROMOSOME CHARACTERISTICS, *Joachim Mark* 497

SPECIAL APPROACHES

CELL HYBRIDIZATION IN THE STUDY OF THE MALIGNANT PROCESS, INCLUDING CYTOGENETIC ASPECTS, *Orlando J. Miller* 521

EPSTEIN-BARR VIRUS INFECTION OF LYMPHOID CELLS AND THE CYTOGENETICS OF ESTABLISHED HUMAN LYMPHOCYTE CELL LINES, *Arthur D. Bloom, Jeanne A. McNeil, and Frank T. Nakamura* . . 565

BLOOM'S SYNDROME. II. THE PROTOTYPE OF GENETIC DISORDERS PREDISPOSING TO CHROMOSOME INSTABILITY AND CANCER, *James German* 601

ATAXIA TELANGIECTASIA SYNDROME: CYTOGENETIC AND CANCER ASPECTS, *D. G. Harnden* 619

CYTOGENETIC AND ONCOGENIC EFFECTS OF THE IONIZING RADIATIONS OF THE ATOMIC BOMBS, *Akio A. Awa* 637

THE ROUS SARCOMA VIRUS STORY: CYTOGENETICS OF TUMORS INDUCED BY RSV, *Felix Mitelman* 675

AUTHOR INDEX 695
SUBJECT INDEX 739

THE CHROMOSOMES SERIES

Each volume in the Chromosomes series is devoted to cytogenetic aspects of one broad subject, or to some aspect of cytogenetics itself. The chapters within a volume review various areas of knowledge pertaining to the given subject. Each chapter is authoritative and definitive, and each is written by an eminent scientist who himself has made important experimental contributions.

Authors have been given a free hand in style of writing and in approach. However, because the review paper plays an important role in contemporary science, a major objective in the preparation of this series is that the articles be comprehensible, not only to those in cytogenetics or in the specific discipline of an author but also to those working in other branches of science.

I acknowledge with gratitude my Advisors, whose valuable views aid me both in the choice of major subject areas and possible chapter topics for review and in the selection of individual contributors.—J. G.

INTRODUCTION TO *CHROMOSOMES AND CANCER*

Since the last decade of the nineteenth century, chromosome disturbances have been known to be associated with cancer. Although interest in the cytogenetics of cancer has waxed and waned, innumerable papers have appeared on this and closely related topics. Fifty years ago, Theodor Boveri published a book entitled *Zur Frage der Enstehung maligner Tumoren** in which he advanced the theory that mutation in the genetic constitution of somatic cells, specifically in the chromosome complement, explains the change from normal to malignant growth; rightfully, frequent reference has been made to this book by many authors writing scientific papers on various aspects of cancer cytogenetics.

It is my impression, however, that few working today either in oncology or in cytogenetics itself have a close familiarity with this vast and complex literature. One "easy way out" for many has been to adopt the view that chromosome changes of many kinds do occur in cancer but that they are a feature of established cancer itself, unimportant in its inception. This may be shown eventually to be the true situation, but, as I have written elsewhere [1], to accept this view at present seems unwise because it very possibly is incorrect, at least for many human cancers. A more acceptable view to me is that chromosome changes *of various types* may be associated with cancer, and that they may be important in *different ways* in the predisposition to and the inception of cancer as well as in its progression and response to therapy. At any rate, cytogenetic changes have been observed at all these stages of cancer, and, until they are proven to be unimportant (or even relatively so), they deserve to be known by all those interested in cancer—whether in its etiology, its nature, or its management.

I have attempted to include reviews of the major areas investigated during the half century since publication of Boveri's book, by authors who have themselves developed hypotheses, devised original experiments, and made important observations or surveys about some aspect of the cytogenetics of cancer. The contributors comprise a distinguished group, and their exceptional abilities are elegantly displayed in their papers here, so

---

* Professor Ulrich Wolf's "book review" of this 1914 work of Boveri constitutes a first chapter for *Chromosomes and Cancer*.

that I feel the book's first objective has, with a few exceptions to be mentioned below, been accomplished. Together, these papers present and critically evaluate under one cover much that has been learned about the cytogenetics of neoplasia. The contents of the volume suggest that chromosome change is of importance, though very possibly in ways quite different from the one of which Boveri wrote!

Four additional chapters might well have been included in the volume, and the reader may wish to consider papers on these subjects as supplements to the volume: (1) C. E. Ford made observations that contributed toward our present concept of the single-cell origin of cancer, and of cancer as an evolving clone of mutated cells. His experiments on the cytogenetics of radiation-induced lymphoreticular neoplasia in rodents are of major importance in cancer cytogenetics, and I list a reference to this work [*2*]. (2) A review of the interrelationships between oncogenic, mutagenic, and cytogenetic effects of various classes of chemicals is missing from this volume. Most chemicals known to be oncogenic can be shown to be mutagenic as well, and they normally can also be shown to be capable of breaking and rearranging chromosomes. (3) Disturbances of immunity are known to enter the cancer picture in several places, and these sometimes appear to be genetically determined. The interplay between genetically determined disturbances of immunity, cytogenetic change, and cancer is so poorly understood at present that no author I approached was yet ready to undertake a comprehensive review of it. (4) The cytogenetics of the Burkitt lymphoma is currently being clarified by Dr. George Manolov in Sofia, in collaboration with his Swedish colleagues. His important and extensive observations concerning characteristic marker chromosomes as well as characteristic patterns of numerical and structural change in this virus-associated and possibly vector-transmitted neoplasm are of importance because they demonstrate the complexity of the cytogenetics of cancer. They eventually will be reported elsewhere.

It is worth pointing out for the reader who studies this volume in years hence that it was written at the beginning of the era of cytogenetic "banding" techniques [*3,4*]. Most of the investigations reviewed in *Chromosomes and Cancer* were made with cytogenetic techniques in conventional use before 1971. Because many of the questions discussed in the volume will be re-opened during coming years by application of banding techniques, this seemed to be a highly suitable time to review previously accumulated knowledge.

I am particularly grateful to the authors who contributed to *Chromosomes and Cancer*, not just for preparing such splendid and comprehensive papers but for working patiently with me in achieving an additional objective, that of making the papers comprehensible to those working in most areas of science, not just in the area of the author nor just in cytogenetics. This has been a painful and time-consuming procedure—for both editor and author—but I believe the results justify the effort and will speak for themselves.—J. G.

LITERATURE CITED

1. German J: Genes which increase chromosomal instability in somatic cells and predispose to cancer. *In* Progress in Medical Genetics (Steinberg AG, Bearn AG, eds) vol 8, New York, Grune and Stratton, 61–101, 1972.
2. Ford CE, Clarke CM: Cytogenetic evidence of clonal proliferation in primary reticular neoplasms. *In* Canadian Cancer Conference, New York, Academic Press, 129–146, 1963.
3. Caspersson T, Zech L: Chromosome Identification, New York, Academic Press, 1973.
4. German J: An Advance in Cytogenetics (Book Review, ref. 3) *Science* 183: 647–648, 1974.

Professor Curt Stern, Berkeley, 1950

Editor's Note: *In each volume of* Chromosomes *an outstanding cytogeneticist will be honored and invited to compose an autobiographical sketch. It is with pleasure and a deep personal respect that I select Professor Curt Stern for acknowledgement in this first volume of the series.*—J.G.

A GENETICIST'S JOURNEY

I was born in Hamburg, Germany, on the thirtieth of August, 1902. When I grew up my parents influenced my scientific career in diverse ways. My father had only eight years of schooling, but he had early learned to appreciate the artistic values of antiques for which he was a dealer in his youth, before building up a dental supply business. He did not understand the role of pure science but neither did he lay hindrances in my way of becoming a university student with a major in zoology. He taught me accuracy in observation and patience with refractory problems —attitudes which served me well as a scientist. My mother came of generations of teachers. She looked up to the achievements of learned men and women and often regretted that she had not become such a woman herself. She fostered in me a desire to join the ranks of scientists. Her ambition spilled over to me. It was a time in which ambition was regarded by many as an unadulterated virtue, witness for example Clark's admonition to the students of his agricultural school in Japan, "Boys be ambitious" (1876), which still stands on a monument on the Hokkaido University campus.

My own attitude to ambition has varied. I felt that from an absolute point of view ambition should play no role in the achievements of scientists but also that it was often a prerequisite for bringing to conclusion valuable work which otherwise would have remained fragmentary. Many times I have wondered whether the selfish motives of a scientist tend to degrade his status; whether a complete divestment from subjective elements is necessary to act as a truly "pure" scientist. I have come to the belief that it is impossible in science as in other fields to separate the subjective human component from the objective component. The ideal of *Dienst am Werke*—dedication to the task—is approached in different degrees, in specific cases ranging from overwhelming self-centeredness to nearly saintly abrogation of human nature. Much of my awareness of the varieties of elements involved in attitudes to research I owe to my wife Evelyn.

"Ontogeny recapitulates phylogeny." The child or youth who grows up to be a biologist will often pass through successive stages representing steps in the development of his science. As a happy naturalist I captured lizards to observe and feed them at home, collected the skulls of wild rabbits, and became acquainted with many of the microscopic protozoa and protophytes in the water of ponds and in hay infusions. Only when I became a predoctoral student did the pleasures of recognizing and classifying a variety of species make way for more "modern" occupations. In the early twenties I attended a lecture course in general biology offered by Max Hartmann at the University of Berlin. It was a course which drew together most branches of biology: the cell as the basis of life, metabolism, reproduction, sexuality, genetics, developmental biology, evolution, and neurophysiology and all this crowned by philosophical considerations involving such problems as that of the body-mind relation, the mechanism-vitalism controversy, teleology, and the epistemological foundations of biology. I enjoyed the course tremendously even though I understood only a fraction of the material delivered by the arm-swinging, enthusiastic lecturer. I determined to master many of the areas outlined by him.

Professor Hartmann accepted me as a predoctoral student. After several unsuccessful starts a topic was chosen for my thesis research, a study of mitosis in a protozoan of the order *Heliozoa*. It possesses a centrosome-like "central body" whose role in cell division had not been clarified. This thesis work was a good training-ground for my later cytological analyses in *Drosophila*.

When my thesis was completed I was deeply shocked by the fact that I did not know where to go next. Further work on the theme of the thesis did not promise essential new insights, and the same was true for various ideas from other branches of biology which had occurred to me. I began to read in the literature of genetics and came across a paper by Goldschmidt on crossing over in *Drosophila* which explained in a new manner the varying frequencies of recombination between linked genes. Instead of accepting the theories of the Morgan School which assumed a breakage and reunion mechanism of chromosomal segments to be the basis of recombination of linked genes, Goldschmidt suggested that crossing over is the consequence of the genes leaving and rejoining a nongenic chromosomal skeleton. I analyzed Goldschmidt's hypothesis and came to the conclusion that it did not account satisfactorily for the known facts of linkage. I wrote a short manuscript and with great hesitation handed it to Professor Goldschmidt in whose department I occupied a very temporary position. I never heard anything about this, my first attempt in genetics, but some time after the stillbirth of the manuscript the Rockefeller Foundation established a few fellowships for postdoctoral biologists and, on the recommendation of Goldschmidt, I was awarded one of these fellowships with the understanding that I would spend a period in the laboratory of Morgan, Bridges, and Sturtevant at Columbia University in order to become a *Drosophila* worker who would bring back to Germany his personal acquaintance with the famous fly genetics. I do not know whether my unpublished paper had anything to do with Goldschmidt's recommendation.

At Columbia University I began with some elementary analyses of genetic phenomena. Soon, however, I had my first opportunity to combine genetic and cytological techniques in the discovery that the Y chromosome of *Drosophila*, instead of being "empty" of genes as it had been believed, carries the normal allele for bobbed bristles, as does also the X chromosome (1927). This normal allele is dominant over its mutant allele. Females were found which were homozygous for the mutant yet had nonmutant bristles. Genetic analysis showed that the nonmutant, normal phenotype was caused by a genetic element which recombined freely with genes in all other chromosomes, the X as well as the second, third, and fourth chromosomes. Cytological preparations showed the normal females to have two X chromosomes like females in general and in addition to have a Y chromosome. When such an XXY female was bred she produced normal-appearing and mutant offspring in a 1:1 ratio. Twelve normal and twelve mutant females, all sisters, were selected and their cytological constitution determined. The twelve normal flies were XXY, the twelve mutant flies were XX.

This work with the Y chromosome led to the finding of Y-chromosomal segments which had been derived from spontaneous translocations of parts of the Y to the X chromosome. The segments permitted further analyses of the genetic structure of the Y chromosome. Males without a Y chromosome are sterile, as had been known earlier. Now it could be shown that single fragments of the Y are not sufficient to produce fertility in XY-fragment males but that different fragments could complement each other so that they led to restoration of fertility.

Among the Y-chromosomal fragments was one consisting mostly of the long arm of the Y chromosome. It had been translocated to one end of the X chromosome. It was at this stage of study that I remembered a passage in a protozoological paper which I had read several years earlier. Its author, Karl Belar, had pointed out that genetic crossing over between two identical homologous chromosomes would not be recognizable in cytological preparations nor would it be recognizable in cases in which the two homologous chromosomes were heteromorphic at some one place. If, however, the two homologous chromosomes were heteromorphic at two places, crossing over would result in new chromosomes. My findings of a chromosome which had a Y arm attached to one of its ends could be used to produce such new chromosomes, provided that still another kind of X chromosome could be found, for instance one which was heteromorphic at the other end. Such a chromosome had been produced by Muller in his famous, then very recent, experiments with X-ray induced chromosomal aberrations. He generously made it available to me. Essentially, this enabled me to construct females which had the chromosomal constitution, "long X with long Y-arm/short X without Y-arm." Crossing over in such females should produce two new kinds of chromosomes, in addition to the original two kinds "long X without Y-arm" and "short X with Y-arm." By means of crosses these chromosomes were marked with certain genes which enabled me to determine the genotypes of the offspring by simple inspection as having been derived from noncrossover or single crossover events. Flies of the different phenotypes were selected and their cytological chromosome constitutions determined. It was found that genetic noncrossover flies were also cytological noncrossovers and that genetic crossovers were also cytological crossovers (1931).

The same type of experiments used in the *Drosophila* work had been reported nearly simultaneously by Creighton and McClintock, using corn. Together the experiments represented "the cytological proof" of crossing over (1931). They seemed to prove the existence of a breakage-fusion mechanism of recombination and neglected to recognize the alternative possibility of a copy-choice mechanism. Only many years later was the regular occurrence of this latter possibility excluded.

As an unexpected outgrowth of some of the work described above I should like to refer to a study which I carried out long ago and which has again become of interest in recent times. It concerns the *bobbed* (*bb*) gene which, as seen earlier, has loci in both the X and the Y chromosome. This fact makes it possible to build up flies with multiple doses of mutant *bobbed* alleles and to observe their effects as expressed in length of the bristles located on the surface of the flies (1929). Restricting our attention to females and simplifying the discussion somewhat, the following genotypes were experimentally obtained: $X^{bb}X^{bb}$; $X^{bb}X^{bb}Y^{bb}$; $X^{bb}X^{bb}Y^{bb}Y^{bb}$. The first of these constitutions had short bristles, as expected from a recessive homozygote. It was also expected that the two other constitutions would produce short bristles, presumably even shorter than in the first type, but this expectation turned out to have been wrong. The second and third genotypes had bristles of normal, wild-type length! Why should an allele for short bristles make long bristles when accumulated? The answer was one of semantics: *bobbed* was not a gene for short bristles but one for not-long-enough bristles and the effects of different doses of the mutant *bobbed* alleles were adding up toward an approach to normal bristle length. At present, molecular geneticists are deepening our knowledge of the nature of the *bobbed* locus, which is intimately related to the nucleolar organizer and its function as a multiple ribosomal DNA region.

The cumulative effects of the *bobbed* alleles early raised a question of regulation of gene activity. It had been known that most *Drosophila* females homozygous for a mutant X-linked gene express that gene's effect to the same degree as do males. Since males have only one X chromosome while females have two, one would expect that the single dose of X-linked loci would have a lesser effect than the double dose at the same locus. This was indeed true in the case of *bobbed*, the single dose of this gene in an X male producing much shorter bristles than the double dose in an XX female. The equality of the effects of the other X-linked genes in males and females had to be attributed to a process of regulation of gene activity which Muller later termed dosage compensation.

After two years of study with the triumvirate Morgan, Sturtevant, and Bridges, I had returned to Germany where I had been given an appointment as an investigator (1926). Six years later, I went back to the United States in order to attend the Sixth International Congress of Genetics in Ithaca. I had intended to spend a year in America before going back to Germany, but the events in Germany's political history made this impossible, so I was glad to have had an offer from the University of Rochester. I stayed there until 1947, when I moved to the University of California at Berkeley (where I became Professor Emeritus in 1970).

For several years I was occupied with an attempt to solve what at first seemed to be a minor problem. In 1925 Bridges had discovered a strange effect of a dominant X-linked gene for fine bristles and slow development, *Minute-n*. He dealt with females in one of whose X chromosomes there was the dominant gene for not-yellow as well as *Minute-n* and in whose other X chromosome was present the recessive allele for yellow and that for not-Minute. Such flies are not-yellow and Minute. Unexpectedly, however, many of them had somewhere an area of yellow not-Minute phenotype. From his analysis of numerous such "spots" on females of the stated or of related genotypes, Bridges concluded that *Minute-n* has the property of sometimes eliminating the chromosome on which it is located, thus resulting in spots in which only the X chromosome occupied by *yellow* and *not-Minute* was left. Such losses of an X chromosome were not unknown. They accounted for the origin of many gynanders, which are flies composed of a mixture of large female and male areas. Elimination of an X chromosome from one of an XX zygote's nuclei had occurred during early cleavage, resulting in equal or similar numbers of XX and X nuclei. The new feature of Bridges' spot mosaics was the apparent origin of the new genotype late in development as well as the specific influence of *Minute-n* on the postulated elimination of an X chromosome. Why and how did *Minute-n* lead to loss of chromosomes? This puzzle led to a variety of experimental approaches. Ultimately the answer was that actually no chromosome loss occurred at all. The decisive experiments on which I stumbled involved the finding (1) that the presence of *Minute-n* was not a necessity and (2) that more different kinds of spots occurred on females of a given genotype than could be accounted for by simple loss of X chromosomes. To illustrate the second point, consider an experiment in which the female had one X chromosome marked by the recessive gene for yellow body color and the dominant gene for not-singed bristle shape, while the other X chromosome was marked by the dominant gene for not-yellow color and the recessive gene for singed bristle shape: yellow not-singed/not-yellow singed. Loss of one or the other X chromosome would result in yellow not-singed and not-yellow singed spots. Actually these two types of spots were found, but in addition there were twin spots consisting of a yellow not-singed area adjacent to a not-yellow singed. In another experiment one X chromosome carried the recessive alleles for yellow and singed, the other X chromosome the dominant alleles not-yellow not-singed: yellow singed/not-yellow not-singed. Loss of one or the other X chromosome would result in yellow singed spots (and not-yellow not-singed spots which phenotypically would be like the normal background). Yellow singed spots were found indeed, but in addition there were yellow not-singed and not-yellow singed spots. How to account for these results? It turned out that the overall solution was based on the unexpected existence of "somatic crossing over," not on chromosomal loss (1936). Recombination of linked genes is not restricted to meiosis but occurs occasionally during mitosis with subsequent mitotic distribution of recombined chromosomes. The analysis of the mitotic spots was an early example of somatic cell genetics.

The discovery of mitotic recombination is a good example of a long gap between observation and explanation. Investigators living in an earlier century might well have observed flies of normal phenotype except for the presence of a single yellow bristle somewhere, but it would have been impossible to account for such mosaic individuals. In order to understand the origin and presence of the yellow bristle the whole fields of Mendelism and Morganism had to have developed.

In my student days the two outstanding branches of biology were classical genetics and experimental embryology. The two branches had developed side by side without close association. I had been under the influence of Goldschmidt who stressed a physiological dynamic approach to genetics in opposition to the Morgan School whose members he felt continued in a static consideration of genetic mechanisms. While I did not fully share Goldschmidt's views, I agreed with him that it was desirable to fuse the two fields in a new entity to be called developmental genetics. I now took up work in this new field. I chose to investigate a trait in *Drosophila* which under the influence of alternative genotypes leads to variations in the shape of the testis. In adult males of *D. melanogaster* this organ is a long tube which is wound up so as to form a spiral. Dobzhansky had observed that the spiral form of the testis developed only when the sperm ducts which embryologically arise independently of the gonads become attached to them; otherwise the testis remains a slightly elongated, pear-shaped vesicle. My own later work made use of the recessive, X-linked mutant gene *sexcombless* which when present in males causes an abnormal morphology of the genital apparatus. In agreement with Dobzhansky's observations it was shown that the variations in the morphology of the gonads depend on variations in the development of the duct system.

Further insight into this relation made use of the fact that the normal males of some species of the genus *Drosophila* regularly have a spiral testis, e.g., *D. azteca*, while the normal males of other species, e.g., *D. pseudoobscura*, have a short vesicle-like nonspiral testis. What was responsible for the different shapes of the testes? There were two different interpretations of these facts. Either the testes owed their shape to their own genetic constitution independently of the morphology of the duct system or the testes become spiralized under the influence of the male ducts. A decision on this alternative was possible with the use of the method of transplantation of *Drosophila* tissues and organs, a method which had just been worked out by Beadle and Ephrussi. Would a vesicle-like larval testis of *D. azteca* if transplanted into a larval *D. pseudoobscura* host spiralize according to its own genotype or would it remain unspiralized according to its host's genotype? And, reciprocally, would a larval testis of *D. pseudoobscura* if transplanted into a larval *D. azteca* host remain unspiralized according to its own genotype or become spiralized according to its host's genotype? The answers to these questions in all cases signified nonautonomy of testis shape. It could be shown that a sperm duct when attached to a testis of *D. azteca* induced an asymmetrical growth of the gonad resulting in spiral elongation while the limited nonspiral elongation of the testis of *D. pseudoobscura* resulted from symmetrical-growth induction (1941).

Having answered the question of the determination of testis shape by means of the sperm ducts, I chose a new topic of investigation, that of the position effect of genes. Pioneered by Sturtevant in 1925, the work had not moved well enough as measured by earlier expectations. My own work did not fare better, and I must agree with Sturtevant's statement in his "History of Genetics": "The position effect of the gene cubitus interruptus (ci) has been studied in great detail, especially by Dubinin and Stern and their coworkers. There are many interesting observations, some of which are rather puzzling, but it does not seem (to me, at least) that they have led to any close insight into the nature of such cases. . . ."

One side-result of these studies was the discovery that "the" normal allele of the mutant ci exists in at least two different allelic forms. They were distinguishable by means of special crosses. It was proposed to call such different normal alleles "isoalleles," a term which later has been found useful in the analyses of the innumerable cases in which biochemical polymorphism has been recognized as depending on similarly acting genes (1943).

Simultaneously with the fly work I began to think of a reorganization of my teaching program. For a good number of years I had taught a course in general genetics. Its principles were derived from the work on flies and corn, moths and birds, rabbits and peas, and other experimental organisms. Human genetics was not totally omitted, but it often consisted of no more than an appendix to the main discussion: "and it is the same in man." I knew that a large number of my students desired to enter a medical career. Was it then not time I presented them with a special course in human genetics?

I have always regarded fundamental research as a proper task for its own sake but have also rejoiced when it brought with it applications to human needs and understanding. Here then was the challenge to construct a course in human genetics based as much as possible on man. I started my own preparation for such a course by offering a graduate seminar on human genetics. It was followed by a course primarily for undergraduates, subsequently given at Berkeley. I wondered what textbook to recommend to the students. The usual texts on genetics did not address themselves specifically to man, and those which did so—often under the label "eugenics"—had not developed with the times which now demanded a much more rigorous approach to the field. As a result I wrote my own book, "Principles of Human Genetics." It found a favorable reception. It came at a period when many medical schools were considering the introduction of some genetics into their curricula. Many American biomedical teachers and investigators trained themselves in human genetics by studying the "Principles." They in turn taught classes and seminars to undergraduate and graduate students. Foreign interest was likewise awakened, as seen in the fact that the book in its three editions (1949, 1960, and 1973) was translated into German, Japanese, Polish, Russian, and Spanish. My mastery of various aspects of the content originally had been very limited. I had to work hard to understand some of the algebraic problems such as those concerned with ascertainment, consanguinity, and mutation rates. Thus I was aware of the difficulties in understanding by the uninitiated, and my own struggles with the material resulted in a clearer presentation than that which an expert in specific branches of human genetics might have provided.

From 1949 on, my scientific personality was a split one. I spent years in preparing the enlarged and revised successive editions of the book, tried to follow the literature on human genetics, and did some limited research in this field. At the same time I did not want to abandon work in general developmental genetics as studied in *Drosophila*. *Ars longa, vita brevis*!

My original contribution to human genetics consisted primarily in a revision of the role of the human Y chromosome (1957) apart from its sex-determining action which became well established independently. A survey of all alleged cases of Y-chromosomal inheritance in man eliminated each one, except possibly a single gene, from being located on the Y chromosome. The following two cases represent examples of such elimination. The famous porcupine skin of the Lambert family in England, which had been assigned to a gene in the Y chromosome, had to yield to a joint attack by Penrose and myself. The former obtained evidence against Y linkage from parish records, while I found genealogical information not compatible with Y linkage, but rather with dominant, probably autosomal, inheritance. The much cited pedigree of webbed toes in the Schofield family when considered together with many other pedigrees of this condition was shown to be consistent with autosomal rather than Y-chromosomal inheritance. A third case, that of hairy ear rims, represents the residual situation in which there is the possibility or rather the probability that the gene or genes causing this trait may indeed be carried by the Y chromosome.

The hairy-ear-rim problem gave me the opportunity of making surveys of population samples, particularly in India in which a high frequency of the trait occurs. I still cherish the recollection of an early morning in Vellore where several hundred cadets and officer candidates of the Police College had assembled in military formation at a parade field after they had voted to support my research. There I went from one man to the next examining with a reading glass first the left, then the right, ear and dictating to Dr. Centerwall, with whom this study was undertaken, a code number designating the phenotype of the hairy rim condition. The case for Y linkage was strong for some time, but then difficulties of classification and of penetrance arose so that I am still reserving my judgment. It amuses me that the same person who found the localization on the Y chromosome of the bobbed gene in *Drosophila* now is skeptical about the localization on the Y chromosome of a gene in man.

Two nomenclatural proposals also came from my interest in human genetics. One is the replacement of the designation Hardy's Law by Hardy-Weinberg's Law following the finding that Weinberg had priority in discovering the population formula. My proposal to change the name was immediately adopted by the scientific community. My other proposal, supported by an international group of human geneticists, was intended to replace the designation "mongolian idiocy" by some less painful and misleading term. "Down's syndrome" has largely replaced the old designation.

It was natural for me to become involved in human genetics counseling. Students and other university members came to ask me for information and advice concerning their genetic problems, and physicians sent me their patients. Over the years, hundreds of such counseling cases came before me. I enjoyed this application of basic research to specific personal problems, the more so since a favorable prognosis for healthy offspring can be made more frequently than one might have expected. More recently I have usually suggested to persons in need of counseling that a medical school with its teams of physicians and other skilled staff is a better place to go for advice than a single person.

During World War II I was asked to organize a study in *Drosophila* of the effect on mutation of very low doses of ionizing radiation. The work was carried out by Spencer, Caspari, and Uphoff. Information of a quantitative nature was needed concerning the exposure to radiation of persons working in nuclear reactor installations so that necessary protective measures could be taken. *Drosophila* seemed an appropriate organism to use for a model investigation. Since then many similar studies by various investigators with *Drosophila*, mice, and other organisms have greatly enlarged our knowledge of radiation genetics.

Still other work carried out more recently consisted of analyses of the distribution of pigmentation types in the American black population. Three or four pairs of pigmentation genes gave a reasonable fit to the observed distributions of phenotypes. Also it was shown theoretically that consanguinity, which usually results in lowering of the viability of the offspring, has the opposite effect in regard to the Rh factor.

I have greatly enjoyed teaching large classes and small groups of students in seminars, even though I found teaching tasks very demanding. I also gave many lectures before physicians and other groups. However, in recent years I have never seen some of "my" students but they see me, on color film. It has been interesting to organize in this medium a whole introductory course of genetics under the auspices of AIBS, the American Institute of Biological Sciences.

In later years, supported by students and outstanding co-workers including Drs. Aloha Hannah and Chiyoko Tokunaga, I returned once more to the study of developmental genetics in *Drosophila*. Genetic mosaicism was employed as a tool available to study differentiation in the fly. Two main methods of producing mosaics are known: loss of a chromosome during development and mitotic recombination. Loss of an X chromosome in an originally XX embryo results in sexual mosaicism (gynandrism), XX and X. Loss of the large autosomal chromosomes is lethal. Here genetic mosaicism can be produced by means of mitotic recombination which leads to homozygosity of recessive autosomal genes which were carried heterozygously in the fertilized egg.

The first experiment in developmental genetics, using mosaics, dealt with the sex comb (1950). This row of an average of about 11 very heavy bristles is present on male forelegs but absent on female forelegs. Why, it was asked, does maleness not express itself in the differentiation of sex comb teeth all over the body? Why, also, does the female have no teeth anywhere? Genetic mosaics of male and female tissues might throw some light on these problems. Flies were obtained which once carried a ring X chromosome and which had lost this ring from a nucleus at an early cleavage stage. These then develop into gynanders. Dependent on the variable cell lineage from one gynander to the next, different parts of the embryo were male or female in genotype. Many gynanders would form nonmosaic forelegs. A male foreleg would have a normal sex comb but a female one would lack this organ. These were of little interest. Some of the gynanders would have mosaic forelegs. These were the specimens on which the analysis was focused. The experiment required one more refinement. If the rod chromosome carried a recessive gene such as that for singed bristle shape, and the ring chromosome carried the dominant normal allele, then each bristle on the body including the sex comb teeth could be assigned its gender. All normally shaped bristles were female in genotype; all singed bristles were male. What was to be expected regarding the appearance or nonappearance of sex comb teeth? It seemed to me that the differences between male and female forelegs in the arrangement of the many bristles which cover the relevant segment of the foreleg and the presence or absence of the sex comb would be due to an overall difference of organization of male and female forelegs. If this were so it was expected that a mosaic foreleg which was predominantly male would stimulate the differentiation of a sex comb regardless of the

XX or X genotype of the cells. Inversely, I believed that a predominantly female foreleg would result in suppression of the innate male genotype of some of its cells and thus result in the absence of a sex comb. These expectations turned out to be wrong. Instead of two different overall organizations of the forelegs which would endow them with the power to override the genetic constitution it was found that a sex comb had differentiated wherever the male genotype was present in cells lying in a specific region of the leg. Depending on the area of maleness the sex comb would be as large as in nonmosaic forelegs or as small as to comprise a single tooth only. If, on the other hand, female tissue intruded into an area which was mainly male, no sex comb teeth were differentiated but often a single, typical female bristle appeared within a gap in a male sex comb. The answer to the question "why does maleness not express itself in the differentiation of teeth all over the body?" is answered by stating that the experiments suggest that the genetic basis of sex comb differentiation requires (1) a male constitution and in addition (2) the presence of a region differing from all others which permits the differentiation of the sex comb. A mosaic foreleg without sex comb does possess the same potentially sex-comb-forming singular region as any male, but a female genotype does not respond to this singularity.

A similar situation has been found for various individual genes concerned with the presence or absence of bristles at specific positions on the head or thorax. For instance the recessive mutant *achaete* does not form a bristle at a specific site, in contrast to the dominant allele which leads to bristle differentiation at this site. Flies heterozygous for *achaete* may possess spots homozygous for *achaete* due to mitotic recombination. Such mosaic individuals demonstrated that the presence of the bristle at its specific location is determined solely by the genotype of the spot itself. This and similar results with different genes show that the formation of surface features of the adult flies occurs as a mosaic of separable developmental events. Most mutant genes tested exert their effects by having a reduced ability to respond to whatever is the stimulus of the specific singularities to which the normal allele is tuned. At least one example is known where it is rather the singularity which is changed than the responding tissue. These findings were first reported under the title "Two or Three Bristles" at a Sigma Xi lecture tour which sent me to 19 different campuses. The findings brought up-to-date were also presented in a small book *Genetic Mosaics and other Essays* (1968).

I have always admired those scientists who are able to make long-range plans for the solution of problems far in the future. I have counted myself as belonging to another kind of scientist who does not believe that he can forecast successfully decades ahead what steps to take to approach a distant goal. This second group of scientists "makes its living" from having an open mind. It occupies itself with minor tasks and keeps its eyes open for unexpected exceptions. In my own work there has been no straight line which would represent a single central idea. On several occasions I started new experiments which were no more than minor variations of older experiments which had remained inconclusive. Then, unexpectedly, a solution of the old problem would present itself. Looking at a mountain from one angle may not be sufficient to grasp its configuration. Looking at it from several diverse angles may suddenly reveal what had been missed before.

To have been a researcher for a lifetime is a great privilege. It means periods of distress when the work seems not to progress and the phenomena studied will not yield to analysis. But it also means times of exhilaration when one's own mind or that of someone else, or the group mind of one's contemporaries, has successfully reached a new level of understanding.

Curt Stern
University of California
Berkeley, California
November, 1973

CHROMOSOMES AND CANCER

This is the first volume in
CHROMOSOMES: A WILEY BIOMEDICAL-HEALTH SERIES

BACKGROUND

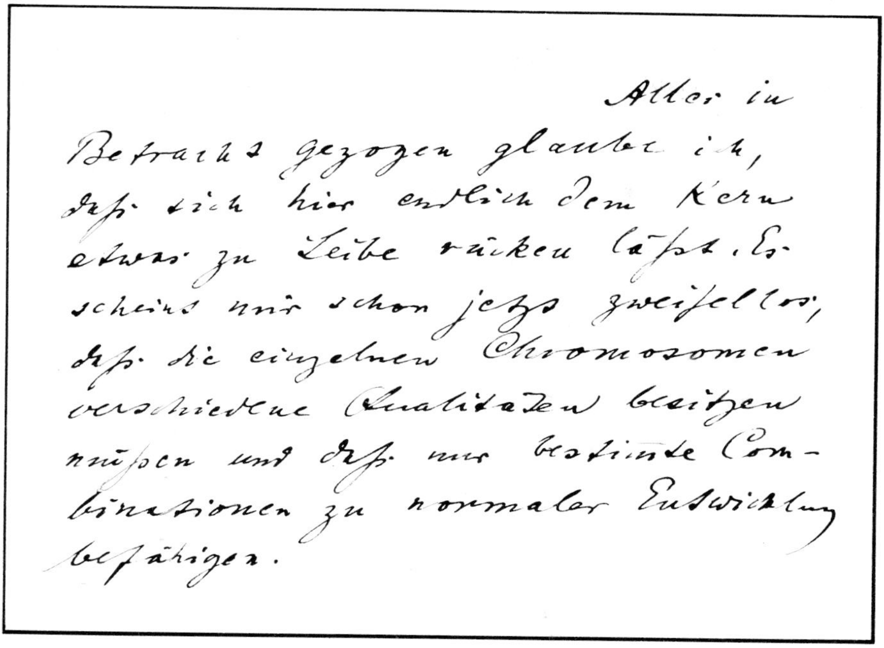

Alles in
Betracht gezogen glaube ich,
daß sich hier endlich dem Kern
etwas zu Leibe rücken läßt. Es
scheint mir schon jetzt zweifellos,
daß die einzelnen Chromosomen
verschiedene Qualitäten besitzen
müssen und daß nur bestimmte Com-
binationen zu normaler Entwicklung
befähigen.

Figure 2. Excerpt from a letter to Spemann, dated "Neapel, 1. Dez. 1901." (Courtesy of Dr. Margret Boveri) Translation: "Taking everything into consideration, I believe that the essential point can finally be approached. I feel beyond any doubt that the individual chromosomes must be endowed with different qualities and that only certain combinations permit normal development."

ogy), Otto Warburg (biochemistry), Richard Goldschmidt (genetics), and Max Hartmann (protozoology), who were still young people then. For various personal reasons, but especially because of his poor health, Boveri finally refused this honorable call in 1913. But the department heads mentioned kept their offices when the botanist C. Correns, one of the rediscoverers of the Mendelian laws, was appointed director of the institute after Boveri's refusal.

The book on malignant tumors is the last publication by Boveri. He died in the year 1915, at the age of 53, after a long illness, the nature of which is not exactly known. Unfortunately, the Zoological Institute in Würzburg, which had kept important parts of his estate, was destroyed in World War II during an air raid. Boveri's house was lost in the same way.

The personality of Boveri has been honored by his students and friends in numerous publications. Here, I only wish to point out that he took a very extensive interest in many fields, especially in art. His scientific designs do not only give evidence of his outstanding gift of observation but of his artistic talent as well. Often he used to paint in his free time. It is said that he was an excellent pianist, and indeed music played an important role in his family. In keeping with this artistic gift, he advocated high standards in the writing of scientific papers in which construction and style are exemplary. Wilson [15] wrote in his memories of Boveri:

> The work of Theodor Boveri, not less than his life, recalls to mind that saying of the ancient scriptures that man does not live by bread alone. For, as the interests of his life reached far beyond the limits of the laboratory—he was, for instance, a skilled amateur of painting and of music—so his work was remarkable not alone for what he did but also in the manner of its doing. That work was in high degree original, logical, accurate, thorough. It enriched biological science with some of the most interesting discoveries and fruitful new conceptions of our time. But beyond all this it is distinguished by a fine quality of constructive imagination, by a sureness of grasp and an elegance of demonstration, that make it almost as much a work of art as of science. In this respect, as I think, Boveri stood without a rival among the biologists of his generation; and his writings will long endure as classical models of conception, execution and exposition.

This mental attitude guided Boveri's research, as best expressed in the introduction to one of his works, in which he writes [6]:

> Everyone who has followed the development of our field during the past 20 years has to acknowledge, willingly, and with gratitude, how valuable, perhaps irreplaceable in this particular instance, the approach has been to connect the sparse facts by persistent hypothesizing and construct a unifying vision, the true nature of which we may hope to know only in the far distant future. Building theoretical castles in the air can provide the stimulus which is essential for carrying out painstaking experimental work. However, in order to assess the true progress we have made, we have to realize just how far observation and experimentation by themselves are able to carry us today. To demarcate this is the purpose of my treatise. However, by confining myself to this task I do not want to deprive myself of hypotheses entirely, without which a mere body of factual data must remain lifeless. (Translation from the German.)

BOVERI'S RESEARCH UNDERLYING HIS THEORY OF THE ORIGIN OF MALIGNANCY

Boveri's field of research would be named nowadays "developmental genetics." This implied at that time that the chromosomes had to be recognized first as the carriers of the genetic properties before chromosome aberrations could come into question as the cause of developmental disturbances. Boveri was among the first who considered the chromosomes to be the matter of heredity, and he may have been the first one who presented a clear experimental proof for this assumption. In a series of brilliant experiments, he first demonstrated the individuality of chromosomes as such, and subsequently, by means of the correlations between chromosomal aberrations and developmental anomalies, he showed that they were endowed with differing qualities. It is amazing that, on the basis of these discoveries, he did not make himself a proponent of a one-sided chromosome theory; but he was just as intensely engaged in the role of the cytoplasm, and he was the first to demonstrate its influence on differentiation in early development. However, these experiments, as well as some other outstanding work carried out by Boveri, cannot be included in this article.

The following statement illustrates the same tendency to consider alternative possibilities just after having gained new insight: though a morphologist, he ascribes to biochemistry the greater explanatory potential when he writes in 1904 [6]: "Even the morphologist could not have a better imagination than that the morphologic analysis is brought to the point when its final elements are defined as chemical units." Therefore, Goldschmidt [10] is right in stating that Boveri is the last cytologist of the classical period and, at the same time, the founder of modern cytology.

In the following summary, it is intended to give an outline of some of Boveri's chromosome studies which provided the basis for his theory on the origin of malignant tumors. The organisms preferentially employed by him for various studies were *Ascaris megalocephala* and several species of sea urchin, in particular *Paracentrotus*. He called a famous series of publications "Zellenstudien" (studies on cells). In the second Zellenstudie of 1888 [2], he demonstrated, in early cleavage stages of *Ascaris*, that the chromosomes always emerged at mitosis with the same morphology and number (four) as they had exhibited before their disappearance in the interphase nucleus after the previous mitosis. In this paper he created the hypothesis of the individuality of chromosomes, according to which the chromosomes preserve their independence also within the interphase nucleus. Since this hypothesis was exposed to various attacks, Boveri dedicated another study to this problem in 1909 by employing a variant form of *Ascaris* endowed with only two chromosomes [8]. After this, even his last opponent was convinced.

A striking proof of the assumption that the cell nucleus contains the matter of heredity was provided by Boveri's studies with the sea urchin egg. He fragmented sea urchin eggs, using the method of O. and R. Hertwig [12], by which shaking resulted in some cell fragments that included the nucleus and others that did not [3]. After fertilization, the fragments without the egg nucleus developed into normal pluteus larvae, although they contained only the paternal haploid chromosome set. Occasionally an unfertilized egg also developed into a pluteus which then possessed the maternal haploid chromosome set only. With this experiment, it was shown that the gametic nuclei of both parents contained homologous information.

Since it could be considered as settled that the genetic material is included in the nucleus and that the chromosomes preserve their individuality from one cell generation to the other, the conclusion was at hand that the chromosomes themselves are the matter of heredity. That this was indeed the case, and that they were functionally distinct elements, was shown by Boveri in a number of experiments on multipolar mitoses and their products.

Already mentioned in his second Zellenstudie [2] were Boveri's observations of multipolar mitoses in *Ascaris*, which sometimes occurred spontaneously, and he realized that the distribution of chromosomes to the daughter cells was irregular, since each chromosome had a bilateral structure allowing a connection with two of the poles only. In the subsequent experiments with the sea urchin, this observation became the starting point for the demonstration of qualitative differences between the chromosomes, and it also constituted one of the fundamentals of his theory of malignancy.

If a sea urchin egg is fertilized with two sperm, a tetrapolar spindle is formed during the first cleavage division. Boveri had discovered earlier that the centrosome was introduced into the egg cytoplasm by the sperm [2]. Thus, after double fertilization, two centrosomes which divided and gave rise to the tetrapolar spindle apparatus were combined with three haploid chromosome complements. As a consequence, the chromosomes were distributed unequally to the daughter cells in almost all instances. The majority of the embryos developed normally, indeed until gastrulation, but at this stage they died off.

In another experiment, Boveri produced embryos with a tripolar spindle apparatus. As discovered by Morgan [14], the division of one of the centrosomes formed after double fertilization can be suppressed by shaking. In this case, a regular distribution of the chromosomes into the three daughter cells should occur more frequently than in the tetraster cases. The resulting embryos proceed to various stages of development; in addition, a number of abnormalities occur.

The sea urchin species employed in these experiments has 18 chromosomes in the haploid complement. Under the assumption that each chromosome had different properties, Boveri performed a model experiment with 3 × 18 = 54 balls, designated by the numbers 1 to 18 in sets of three, which were distributed at random on a circular plane. Then he divided the plane into three or four equal sectors, and in a large sample of distributions he ascertained how frequently a complete series of balls occurred in all of the respective three or four sectors. The approximate expectation values thus obtained agreed surprisingly well with the frequencies of normal pluteus larvae occurring in the experiments with the embryos. But also the abnormally developed embryos were illuminating. In many cases, the abnormalities were restricted to certain segments of the embryo, while other segments were not affected. Similarly, in the model experiment, sectors containing a complete series, and others with an irregular number of balls, occurred within the same distribution.

From the experiments with the dispermic sea urchin egg, Boveri concluded that the individual chromosomes are endowed with different qualities. Thus the chromosome theory of inheritance was settled. But even after these brilliant and far-reaching results, he states [5]: "The more our insight grows, the more we feel that the morphologic aspect of this problem is only a foundation for what we finally want to know: what physiological importance these chromatin elements may have which suffer such strange fates."

At about the same time that Boveri was engaged in these classic experiments, the Mendelian laws were rediscovered (in 1900). Boveri was among the first to realize that the Mendelian modes of inheritance and the cytologic events could be reconciled [4]. In 1904 he states [6]: "The probability that the traits examined in the Mendelian experiments are linked to individual chromosomes is extraordinarily high." In the same paper he predicts that linkage of traits must be based on their localization at the same chromosome; also the possibility of genetic recombination by exchange of chromosome pieces is mentioned here.

With these ideas Boveri brings about the unification of genetics and cytology and the new field of cytogenetics emerged.

"ZUR FRAGE DER ENTSTEHUNG MALIGNER TUMOREN"

In his first publication on dispermic sea urchin eggs in 1902 [4], Boveri had conceived of the idea that malignant tumors could be due to an abnormal chromosome constitution, originating, for example, as the consequence of a multipolar mitosis. Since the possible connection between abnormal mitoses and malignant tumors was discussed now and then in the literature, but time and again rejected, he decided to deal with the problem in a separate treatise. Apparently, this problem had a strong attraction for Boveri, since it was contrary to his habits to discuss hypotheses in detail without contributing to them any proper observations.

For his justification, he states in the introduction that, though his own experiments had not been with malignant tumors, he felt some competence since "the tumor problem is a cell problem." Some properties of the cell would have to be considered which could not be derived from the study of the tumors themselves.

As the only other author who had uttered similar ideas, Boveri points to David von Hansemann [11], who was at that time forgotten, however, possibly because his arguments did not seem to be satisfactory.

Boveri first makes some general statements on the nature of tumors, including the following:

(i) Malignant cells can be derived from normal tissue cells.

(ii) The cause of the abnormal behavior lies within the tumor cell itself, not in its environment.

(iii) There is a clear difference between benign and malignant tumors. The transition from a benign into a malignant growth is to be regarded as a process similar to the transition from normal tissue to malignant growth.

(iv) The malignant cell has acquired a secondary indifference by loss of properties that were present before. Thus it is not a cell that remained undifferentiated histologically (incomplete tissue maturity), but rather a defective cell as compared with the normal cell (von Hansemann's term "anaplasia" applies here).

(v) During regular tissue differentiation, the single cell is integrated into the requirements of the respective tissues. The behavior of the cell becomes "organotypical" (R. Hertwig). The malignant cell, in contrast, has returned to the "cytotypical" behavior of unlimited proliferation which is characteristic of unicellular organisms.

These statements lead to the central problem: how can a property of a cell be taken away from it? Boveri discusses the experiments showing that loss of parts of the protoplasma would be regenerated under the influence of the nucleus, that portions of the protoplasma have the qualities of the whole, and that partial loss does not result in a permanent disturbance (if the cell survives). In contrast, the situation with the nucleus is quite different. To illustrate this, Boveri reviews his experiments with the dispermic sea urchin eggs in which multipolar mitoses occur. The significance of cells having three or more poles, rather than the normal two, is that the bipartite chromosome can align itself between only two poles; the distribution of chromosomes to the daughter cells necessarily is uneven, some daughters lacking chromosomes that others possess. The distribution of chromosomes to daughter cells is variable.

Boveri considers the following mechanisms to be responsible for the occurrence of multipolar spindles: (1) In anomalous multiple division of the centrosome in a diploid cell, there result four poles for two haploid sets of chromosomes. (2) In double fertilization of an ovum, each sperm introduces one centrosome which divides; the result is four poles for three haploid sets. (3) In suppression of anaphase, four poles and four haploid sets will be present following abortive mitosis.

In the illustration taken from Boveri's book (Fig. 3), daughter nuclei of cells with multiple spindles are shown to be formed with missing elements. It can be observed that there are three consequences of these anomalous distributions: (1) Only embryos containing the complete chromosome number develop normally. (2) Missing chromosomes or even pieces of chromosomes cannot be regenerated; the defect is transmitted to subsequent cell generations, and the embryo exhibits anomalous development. (3) The great majority of defective cells die off.

From these experiments, put together in his sixth Zellenstudie of 1907 [7], it follows that the individual chromosomes are endowed with different qualities. If these insights are extended to the problem of malignancy the essential hypothesis, as already formulated by von Hansemann, is: *The cell of the malignant tumor is a cell with a certain abnormal chromatin content.* In support of this hypothesis, Boveri uses the following arguments. On each of the respective 23 or 24 chromosomes of man [16] a large number of genetic factors, or "units of heredity," must

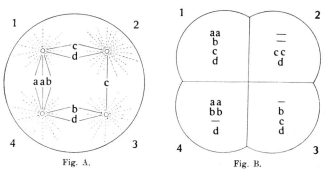

Figure 3. The sole illustration in Boveri's book *Zur Frage der Entstehung maligner Tumoren*. Unequal distribution of chromosomes (a b, c, d) to daughter cells is shown to be the inevitable consequence of a tetrapolar mitosis in a diploid cell.

be located one after another. As an example, to illustrate the attachment of a genetic quality to an individual chromosome, he mentions color blindness and the sex chromosome. He further explains that gross morphologic traits as well as constituents of the cytoplasm, like a pigment or a secretion, will come about by the separate or combined function of certain chromatin elements. And also, different parts of the chromatin will have different significance. If the normal cell function is dependent on the interaction of a number of these elements, it may well happen that a certain disturbance within this equilibrium results in the capacity of unlimited proliferation characteristic of tumor cells. Here Boveri points to tissue culture cells: if normal cells are isolated from their tissue, they start dividing again. Similarly, the malignant cell apparently does not respond to some inhibitory factor normally effective within the tissue assemblage, and this may be due to some disturbance of the chromatin equilibrium. In both benign and malignant cells this inhibition is not effective; the difference, therefore, must be sought in additional chromatin changes of other kinds within the malignant cell.

The essence of Boveri's hypothesis, therefore, is not the abnormal mitosis itself, but "a certain abnormal chromatin constitution, the way in which it originates having no significance. Each process which brings about this chromatin constitution, would result in the origin of a malignant tumor."

The remainder of the book discusses the explanatory power of the hypothesis. Since this is surprisingly comprehensive, and since Boveri includes a number of predictions which are of great interest in view of recent cytogenetic results obtained on malignant tumors, some of his arguments may be briefly reported here.

(i) Since tumors can originate from the most diverse tissues, the induction of malignancy cannot in each case be caused by certain continuously acting exogenous influences (which in special cases may of course be responsible). An irregular distribution of chromosomes, however, is always possible if only the respective cell population undergoes mitoses.

(ii) Each type of tumor has a characteristic cytology which is expressed also in metastases and transplants. However, tumors come into existence not suddenly and diffusely but from imperceptible initial stages. From this, Boveri concludes that "typically each tumor takes its origin from one and a single cell." This original cell must have acquired an abnormal chromosome constitution, and it must be endowed already with all the properties of the tumor.

(iii) It is well known that a tumor cell has an abnormal metabolism. If the individual chromosomes have different qualities, chromosome aberrations will result in deviant metabolic functions. If, therefore, certain chromosomes are missing and others are present in abundance, certain substances will be produced also in abundance, and there will be a deficiency in others.

(iv) Various types of tumors may originate from one and the same organ or tissue. If the tumor type is dependent on which particular chromosome has been lost or added, the greater the deviation of the chromosome set from the normal complement, the more the tumor will differ from the mother tissue.

(v) If a certain type of tumor shows multiple occurrence, it is to be assumed that one and the same inducing factor causes the same chromosome aberration each time. Boveri suggests that future research should try to eliminate individual chromosomes experimentally.

(vi) On the problem of tumor inheritance, he assumes the existence of a corresponding predisposition. This may have its cause in a variable capacity to suppress the primary tumor cell, in a tendency of the centrosome to undergo multiple division leading to an increased occurrence of multipolar mitoses, or in a predisposition to non-disjunction. Indeed, Boveri discusses the possibility that malignancy may originate through an asymmetrical mitosis, and the concept of non-disjunction is clearly conceived in his mind. Under the assumption of such predisposing factors, inbreeding should be of significance for the risk to develop a malignant tumor.

(vii) If several different secondary tumors arise from a primary tumor, as is true in some transplantable tumors, a secondary anomalous distribution of chromosomes can be assumed. In principle, as many different types of tumors could emerge from a heterogeneous malignant tissue as there are different cell types present.

(viii) In particular cases, the origin of a tumor could be dependent on a defect carried by one chromosome which, however, is compensated in the diploid state. After loss of the intact homologous chromosome, the defect would become manifest.

(ix) If the idea of differential activity of chromosomes, which is assumed to be responsible for the process of differentiation, is applied to tumors, it could be imagined that some change in differential activity leads to a type of cell that differs from the mother tissue and develops malignant properties.

(x) As had been shown experimentally already at that time, certain exogenous influences give rise to irregular mitoses. This was demonstrated with certain chemical agents [12] as well as Röntgen rays (results by Boveri himself on *Ascaris* embryos). The time interval often passing between the time of the insult and the origin of a tumor may be explained by the assumption that the agent first interferes with the mitotic process, giving rise to an incomplete mitosis; in a second step, the cell must be stimulated to divide further, and then a multipolar mitosis will occur. As an example, von Hansemann studied pigmented nevi of the skin which sometimes develop into a carcinoma [11].

(xi) In heavily proliferating tissues, the risk of the occurrence of a tumor is increased. A further increase is expected if some chronic irritation is added. This corresponds well with the increasing risk of abnormal mitosis under these conditions.

(xii) The risk of tumors also increases with age. The assumption that abnormal mitoses are involved here is plausible, since in aging cells the mitotic process is more frequently disturbed. Boveri points to the analogous situation in aging oocytes; he is already familiar with the phenomenon of developmental errors caused by delayed fertilization.

(xiii) In malignant tissues, mitotic errors, pronounced variability in nuclear size, and cell degeneration can be observed. These phenomena are interpreted as different steps of the same process, i.e., as the consequences of a primary chromatin defect.

(xiv) It is quite compatible with this theory for a tumor to be composed of uniform cells of normal size, and even to have a normal number of chromosomes. In this case, it is expected that the cells are pseudodiploid.

(xv) It may be mentioned that Boveri also includes in his considerations the leukemias, which may originate from leukocytes, just as solid tumors originate from tissue cells.

Boveri summarizes his conception in the following way. The tendency of continuous proliferation is an original property of the cell which is inhibited only secondarily in the multicellular organism. The sensitivity toward this inhibitory effect can be lost. Then the cell escapes the tissue assemblage. Changes in the constitution of the nucleus are responsible for the loss of reactivity toward the conditions of the normal environment.

From future research, Boveri anticipates improvements in methods of chromosome analysis. It should then be possible to determine whether tumors with little chromosome variation are more uniform histologically and show relatively less degeneration than those with widely varying chromosome numbers. Also, when the chromosome numbers are normal, the combinations of chromosomes may turn out to be aberrant. Boveri's book closes with an appeal to future scientists to consider their results in view of the hypothesis he has put forward.

BOVERI'S HYPOTHESIS IN VIEW OF RECENT RESULTS

Boveri's hypothesis on the origin of malignant tumors represents the first unifying concept which tries to explain causally the manifold and heterogeneous characteristics of malignant cells and tissues. In order to be able to develop this concept, Boveri has traversed a long way, at the beginning of which, 30 years earlier, the nature of the chromosomes was still largely obscure, and the most bizarre ideas existed concerning the functions of the nucleus, the protoplasma, and the chromosomes. On his way, through truly inquisitive experiments, he, step by step, recognized the nature of the chromosomes as the matter of heredity. He already knew that the "hereditary units" are located on the chromosomes and that an equilibrium among them is the prerequisite that guarantees the normal behavior of the cell and the entire organism. He succeeded in producing developmental anomalies on the basis of the experimental induction of chromosome aberrations. He was quite aware that such anomalies can be the consequence of gene defects as well, though he did not live to see the rise of *Drosophila* genetics which lay immediately ahead and which made possible an exact definition of the gene. Since a significant portion of Boveri's work was dedicated to chromosomes, it is not surprising that the mechanisms he explored also serve him as models for his tumor theory. It should not, however, be overlooked that, while demonstrating chromosome aberrations as a cause of genetic imbalance, he also includes as possible causes of malignancy events that we call gene mutations nowadays.

It was confirmed in the meantime not only that "the problem of malignancy is a cell problem" but also that the transition to the malignant cell presupposes some change within the genome. Even today there is little evidence that a chromosome aberration is the primary defect indeed, but in the case of the Philadelphia chromosome, and in tumors induced by radiation or some chemical agents, this possibility is still under discussion. However, the problem of whether chromosome aberrations are the *primum movens* or not is without relevance if the characteristics of a tumor are dependent on its chromosome constitution. Thus the chromosomal situation may be responsible for the stability and flexibility of a tumor. In this sense, Koller [13] wrote: "The configuration of anomalous chromosome patterns is a decisive factor which determines whether the aneuploid cell has proliferative capacity or not."

If one looks over the catalog of arguments which Boveri has put together in order to test the evidence in favor of his hypothesis, it is really amazing to what extent his predictions were confirmed, or are still under discussion, without remembering him. To me, it appears particularly remarkable that Boveri's ideas were mostly derived in a deductive way, proceeding from the just-emerging chromosome theory of inheritance.

In reading the extracts from the above-mentioned catalog, the reader will already have drawn comparisons to what we know today about the cytogenetics of malignant tumors. Nevertheless, a few striking examples may be given which can be considered as confirming Boveri's ideas.

It is most exciting to read that Boveri, some three-quarters of a century ago, had already postulated the monoclonal origin of malignancy, which in the meantime has been demonstrated to be true for several tumors.

It has been confirmed that certain tumors can be characterized by their chromosome constitution; there, the chromosome numbers vary around a modal number. Primary tumors in particular have been shown to have a rather stable chromosome constitution. The subsequent heterogeneity is secondary and should be traced back to abnormal mitoses, as was assumed by Boveri.

It has also been confirmed that different tumors possess considerably different physiological and biochemical characteristics as expressed by the activities of several enzymes, hormone dependency, resistance to chemicals, or immunological behavior. These differences without doubt reflect genetical differences, and some correlation between these parameters and the chromosome condition has been shown. There are a number of arguments in favor of the assumption that the specific properties of certain tumors are based on the respective chromosome constitutions.

In some induced tumors it is obvious that the nature of the respective agent does not determine which type of tumor will develop. Rather, the same agent can be used to produce tumors of different cytology. Here again, the assumption is at hand that the inducing agent first produces chromosome aberrations or mitotic disturbances at random, while the properties of the tumors finally coming into existence are determined by its chromosome constitution.

Furthermore, it has been confirmed that tumors of normal chromosome number can be pseudodiploid. In virus-induced primary tumors, the majority of cells show a normal chromosome complement. Thus the tumor characteristics cannot be demonstrated to reflect chromosome aberrations. Here Boveri's concept of a change in the differential genetic activity of the cell leading to malignancy might come into one's mind, the change in this case being the consequence of the virus infection.

Genetic predisposition to malignant disease has been supported in particular by the detection of the chromosomal breakage syndromes [9]. The origin of tumors could depend in such cases on the general tendency to chromosome aberrations. The finding that the risk of malignancy is enhanced in certain constitutional chromosome diseases—such as trisomy 21 and perhaps others—may be interpreted, in the sense of Boveri, as genetic predisposition to non-disjunction.

It seems needless to increase the number of examples here, since reading of the chapters in the present volume suggests comparisons of results in modern cancer cytogenetics with Boveri's hypothesis. My point was to illustrate that his hypothesis is frequently too simply interpreted, as when attention is focused only on the single concept that mitotic disturbances may result in malignant deviation. In his book, Boveri considers this to be only one possibility, though it is the one he discusses almost exclusively. And indeed, it is true that the nearly unlimited adaptability of malignant cells (which may be the most typical characteristic of tumors) is dependent on the extraordinary variability of the chromosome constitution, and that this again is the consequence of the frequent occurrence of abnormal mitosis. In principle, however, Boveri only postulates that the genetic equilibrium must be disturbed in some special way to make a cell a malignant cell. With this postulate, he has mapped a perspective for future research which still remains valid for the problem of the origin of malignant tumors.

LITERATURE CITED

1. BALTZER FRITZ: Theodor Boveri. Leben und Werk. Grosse Naturforscher (Degen H, ed) vol 25, Stuttgart, Wissenschaftliche Verlagsgesellschaft, 1962.
2. BOVERI THEODOR: Zellenstudien II. Die Befruchtung und Teilung des Eies von Ascaris magalocephala. Z Naturwiss 22:685–882, 1888.
3. BOVERI THEODOR: Ein geschlechtlich erzeugter Organisms ohne mütterliche Eigenschaften. Sitzungsber Ges Morphol Physiol (München) 5, 1889.
4. BOVERI THEODOR: Über mehrpolige Mitosen als Mittel zur Analyse des Zellkerns. Verhandl Phys-med Ges (Würzburg), 1902.
5. BOVERI THEODOR: Über die Konstitution der chromatischen Kernsubstanz. Verhandl Deut Zool Ges 13. Vers (Würzburg), 1903.
6. BOVERI THEODOR: Ergebnisse über die Konstitution der chromatischen Substanz des Zellkerns. Jena, G. Fischer, 1904.
7. BOVERI THEODOR: Zellenstudien VI. Die Entwicklung dispermer Seeigeleier. Ein Beitrag zur Befruchtungslehre und zur Theorie des Kerns. Z Naturwiss 43:1–292, 1907.
8. BOVERI THEODOR: Die Blastomerenkerne von Ascaris megalocephala und die Theorie der Chromosomenindividualität. Arch Zellforsch 3:181–268, 1909.
9. GERMAN J: Genes which increase chromosomal instability in somatic cells and predispose to cancer. In Progress in Medical Genetics (Steinberg AG, Bearn AG, eds) vol 8, New York and London: Grune & Stratton, 1972, 61–101.
10. GOLDSCHMIDT R: Portraits from Memory. Seattle, University of Washington Press, 1955.
11. HANSEMANN D VON: Über asymmetrische Zellteilung in Epithelkrebsen und deren biologische Bedeutung. Virchow's Arch Path Anat 119:299–326, 1890.

12. HERTWIG O, HERTWIG R: Über den Befruchtungs- und Teilungsvorgang des tierischen Eies unter dem Einfluss äusserer Agentien. Jena, 1887.
13. KOLLER PC: The Role of Chromosomes in Cancer Biology. Berlin-Heidelberg-New York, Springer-Verlag, 1972.
14. MORGAN TH: A study of variation in cleavage. *Rous Arch Entwickl Mechan Organismen* 2, 1895.
15. RÖNTGEN WC (ed): Erinnerungen an Theodor Boveri. Including contributions from: W. Boveri, H. Beeg, H. Spemann, F. Baltzer, E. B. Wilson, A. Leiber, W. Wien, W. C. Röntgen. Tübingen, J. C. B. Mohr, 1918.
16. WINIWARTER H VON: Etudes sur la spermatogenèse humaine. *Arch Biol* 27:91–189, 1912.

THE BIOLOGY OF CANCER

SIR MACFARLANE BURNET
School of Microbiology, University of Melbourne, Parkville, Victoria, Australia

---

Monoclonal Character of Cancer 23
Impact of Carcinogenic Chemicals or Viruses on the Cell 25
    Effects of Chemical Carcinogens 26
    Effects of Oncogenic Viruses 28
Ecology and Evolution of Cancer Cells in the Body 29
Age-Specific Incidence of Cancer 31
Conclusions 34
Literature Cited 35

In 1957 I published a fairly long discussion of the biology of cancer in two consecutive numbers of the *British Medical Journal*, and in the following year I provided a more compact version of the same theme at the Third Canadian Cancer Conference at Honey Harbor [10, 11]. On rereading these papers I find very little that needs to be altered despite the enormous amount of experimental work that has been published in the intervening years.

My attitude then, as now, was that cancer is basically a very simple phenomenon to understand. It needs only one postulate: that all somatic cells are continually subject to random somatic mutation (some mutations arising spontaneously, others resulting from chemical or genetic intrusion into the genetic mechanism), producing changes that still leave the cell viable. It is axiomatic that only those changes in the genome which leave the cell capable of mitosis are relevant. The term somatic "mutation" is used, I believe legitimately, to cover any type of inheritable change occurring in the genome, including point mutation, deletion, gene duplication, or recombination, and even gross chromosomal changes. Except for the last, it is normally impossible in a mammalian cell to analyze the nature of the inheritable change at a strictly genetic level. All the evidence suggests that when the relevant internal and environmental conditions are constant, somatic mutations occur at random in relation both to time and to their informational content, and at a rate of the same order as that of mutations in germ-line cells.

In any tissue, the proportion of cells undergoing mutation in a given period (a month, say) will probably be of the order of 10^{-5} or less. So few cells will be involved that any gross immediate effect will be quite undetectable. If the mutation is lethal, the cell involved will autolyze and disappear; if a functional alteration leaves the cell still viable, at most an undetectable weakening of functional efficiency can be expected. In a long-lived animal like man, there must be a steady accumulation of deleterious mutations compatible with continued cell survival, some occurring sequentially on previously mutated cells. This has two important consequences. A progressive increase in the proportion of more or less inadequate somatic cells will diminish the general efficiency of the body and will play, perhaps, the chief role in the increased vulnerability to all damaging impacts of the environment that is typical of old age. A second inevitable consequence will be a progressively increasing likelihood that a cell line will arise that is capable of proliferating beyond the normal limits of local control imposed upon its congeners. This means, in general, benign or malignant tumor production.

With insignificant exceptions, mutation in a somatic cell only becomes detectable when the cell proliferates sufficiently to give rise to an aberrant mass of cells or to liberate detectable amounts of some specific cell product (an active hormone, for example). In reverse, this implies that every neoplasm is initiated from a single cell that has undergone a specific somatic mutation or equivalent type of inheritable change. The final change giving rise to the malignant clone may be only the last step in a sequence of deviations from the genetic norm. These may have occurred as germ-line mutations or as somatic mutations at any point in the embryogenesis or adult life of the animal. The view will be adopted that, in general, any mutation that initiates the process of carcinogenesis will give a proliferative advantage to the mutant.* In the mutant population so built up, another mutation can give rise to a subclone with further advantage. This process can, in principle, continue until malignant growth occurs

---

* A computer simulation of such a process by Williams and Bjerknes [51] gives patterns of replacement by "malignant" cells in a basal layer of epithelium which resemble clinical findings.

and further steps of the same basic process become evident as "progression" of the established cancer.

Essentially, the interpretation is that cancer results from a series of random processes which arise as an inevitable consequence of the cellular structure of higher organisms. Its appearance can, however, be accelerated by any agent—physical, chemical, or biological—that can increase the probability of a significant primary mutation.

A modern analysis of this general hypothesis of cancer can be divided into four principal themes: (1) The monoclonal nature of cancer. (2) The mutagenic impact of carcinogenic chemicals and viruses on the cell. (3) The ecology and evolution of malignant cells in the body. (4) The age-specific incidence of different types of cancer.

MONOCLONAL CHARACTER OF CANCER

There are two requirements for the establishment of any condition characterized by the presence of many abnormal cells as monoclonal. First, there must be certain determinable characteristics by which a cell type can be divided into two or more subgroups with some representatives of each always present in the body. Second, tumors of a given type *may* have the character of any of the subgroups, but any particular tumor is of *one* specific type. The simplest example comes from immunology. All individuals have some immunoglobulins with light chains carrying the antigen κ and another, smaller component with antigen λ. The plasma cells that produce the immunoglobulins are similarly divisible unequivocally into κ and λ producers. With only about 1% of exceptions, the characteristic plasma cell tumor, multiple myeloma, produces either κ chains or λ chains, never both [26]. Each myeloma protein (the immunoglobulin characteristically produced by the malignant plasma cells) in fact is unique and homogeneous in composition and in literally hundreds of characteristics establishes the strictly monoclonal character of the condition. The whole of modern immunochemistry only became possible because multiple myeloma was a monoclonal tumor.

Most tumors, however, derive from less tractable material. In a given individual, one thinks of all the cells of the fixed tissues as carrying the same genetic information but expressed to varying degrees according to the differentiation required. This holds equally for homozygous or heterozygous qualities, but there is one major exception. The Lyon phenomenon [32] involves the functional inhibition of one X chromosome in all female somatic cells. In the first divisions of the zygote both X chromosomes (the paternally derived Xpat and the maternal Xmat) are equivalent, but at an early stage of embryonic life, probably about 12 to 13 days in man, each cell makes a random choice of one or the other (Xpat or Xmat) for expression. From that time onward every somatic cell is either Xpat or Xmat functionally; the other chromosome persists and divides regularly but is inert and is visible in interphase as a chromatin mass, the Barr body. Every female mammal is therefore a mosaic of Xpat and Xmat cells. There are about 50% of each, but it is not a wholly random mosaic. Depending on the details of cell proliferation and morphogenesis, since the "decisions" were made early in embryonic life, there tend to be domains of a few hundred cells of uniform character. If a female is heterozygous for a quality carried by chromosome X, and methods are available for recognizing the different products of the allelic gene, it is possible, in principle, to tell whether any common type of tumor is or is not monoclonal. There will always be at least a theoretical possibility that some mutational process will cause a retrogression to a cellular phase when both X's are expressed or when the restriction to one is unstable. It must also be kept in mind that any genetic or environmental quality that significantly increases the likelihood of somatic mutation will ipso facto increase the possibility that two separately initiated cancer lines may fuse and show both alleles within the tumor substance. The net result of these qualifications is to make only the finding of a monoclonal situation significant. Evidence of activity of more than one allele carried on the X chromosome does not disprove the general hypothesis of origin from an individual cell.

Most of the actual investigations using this approach have been concerned with women of African ancestry in which mutation at the G6PD locus has occurred, so that in heterozygous females some cells produce the A form of glucose-6-phosphate dehydrogenase and some the B form. Prima facie evidence of monoclonal origin has been well established for uterine leiomyomas [31], for chronic myelocytic leukemia [16], for the Burkitt lymphoma [17], and for common warts [37]. No positive evidence for monoclonal origin was obtained in chronic lymphatic leukemia.

In view of the widely held view that the Burkitt lymphoma *must* be caused by a virus, the demonstration must be taken seriously that in seven heterozygous patients one tumor was A and six B, with none mixed. The monoclonal character of the Burkitt lymphoma was equally well supported by the recent, carefully substantiated finding that the lymphoma cells from an African boy showed both a female karyotype and a consistent, minor autosomal anomaly. It is immaterial whether the tumor was derived from a maternal cell that passed the placenta or whether it had some other origin, as Manolov et al. suggested [35]. Such findings are completely opposed to any view by which virus passing "horizontally" from cell to cell provokes all or a significant proportion of those cells it infects to malignant activity.

IMPACT OF CARCINOGENIC CHEMICALS OR VIRUSES ON THE CELL

There are very many synthetic organic chemicals which are demonstrably carcinogenic, and they include representatives of a wide diversity of chemical types. They differ almost as widely in the requirements for dosage and route of administration that are needed for consistent effect. The highly active carcinogens share no obvious common feature, apart from the fact that all are structures quite alien to the body. Many are known to be reactive with DNA, and it will be our working hypothesis that wherever there is demonstrable carcinogenicity, the agent must directly or indirectly (e.g., via histones) react in such a fashion as to modify informationally significant DNA structures, without, of course, affecting proliferative capacity and all that this entails.

I have recently published a speculative review on the general topic of chemical carcinogenesis in relation to the diversity of new antigens manifested amongst tumors invoked in genetically homogeneous strains of mice or guinea pigs [13]. This should be consulted for details, and only an outline will be attempted here. There is near agreement that the cell surface, and particularly the specific proteins of the cell membrane, are of special importance in the interactions of the cell with the rest of the body [50]. This would hold for the specific antibody-type receptors of immunocytes, for the histocompatibility antigens expressed on all cells, and for other antigens manifesting tissue or cell-type specificity. Most of the hormone and drug receptors are there too, and it is well known that in the change from normal to malignant activity there are often striking changes in the reactivity of surfaces.

Certainly for the specific immune receptors (antibody), and probably for all other specific proteins incorporated into the cell membrane, there is a highly complex genetic mechanism to ensure that the right pattern of protein is produced for each surface component at the right stage of differentiation or in accord with a current functional need. I have used Smithies's [45] concept of a switch gear to call this or that alternative pattern into action as a means of visualizing the delicacy and complexity of the situation. By hypothesis, it is by intrusion into this switch gear that a chemical capable of reacting with informational DNA may induce distortions of pattern that are responsible for the development of malignant change and for any other abnormal characteristics shown by the malignant line, such as specific differences in immunogenicity. Since there is suggestive evidence that malignancy may be essentially a manifestation of change in cell membrane reactivity, it is possible that on occasion a single DNA lesion, by changing the amino acid sequence of a single species of cell membrane protein, may be responsible directly for both the antigenic change and the malignant character. This, however, need not always be the case. It must be emphasized that the only type of mutation that, as it were, selects itself for recognition is one that enforces proliferation of the mutant cell; any associated inheritable traits are irrelevant. When we know a great deal more about the genetic control of cell surface structure, it may become possible to show some meaningful association between the variable antigenic change and the development of malignancy.

At the present time there is active interest in the appearance of fetal antigens in cancer cells, and there are even suggestions that the new antigens found in experimental tumors, whether induced by viruses or by chemical carcinogens, may all be fetal antigens [2]. Current evidence seems to favor the view that while fetal antigens frequently appear in tumor cells, they are distinct from the well defined tumor-specific transplantation antigens (TSTAs) and probably play relatively little part in immunological surveillance. The fact that nonmalignant processes (e.g., acute viral hepatitis) can also allow the expression of fetal antigens suggests that their emergence is a casual phenomenon, not one directly responsible for the malignant change.

The present somatic genetic approach implies that many other forms of distortion of DNA pattern will occur under mutagenic impact. Some will leave the cell nonviable, others may allow multiplication with loss or change of some function; but unless there is a capacity for gross proliferation as well, none of these will be demonstrable.

Effects of Chemical Carcinogens

Most of the relevant experimental data come from the use of a standard carcinogen—often methylcholanthrene—in inbred strains of animals, most frequently mice. The general finding [4, 5, *18*, *39*, *52*, *53*] is that there are considerable differences amongst the tumors produced in regard to the incubation period before a standard-sized tumor appears and the ease with which syngeneic animals can be immunized with tumor cells; some tumors are not detectably immunogenic. When antigenic tumors are cross-tested, the usual result is that each is antigenically distinct and will immunize only against its own cancer-cell strain.

Findings of this sort have some important theoretical implications. (1) They indicate that there is no standard determinative process by which a malignant clone of cells is initiated. Apart from the central fact that the emergent clone *is* malignant, there is no uniformity in any of the other characteristics that have been recorded. One could also venture to predict that suitably detailed study would show types of heterogeneity other than antigenic differences, when malignant cells arising in different individuals were compared. Such variability is, of course, completely in line with the hypothesis of somatic mutation. (2) The fact that many tumor lines show clear-cut antigenic individuality implies that at least these are monoclonal. (3) The existence of apparently nonimmunogenic strains of malignant cells [42] suggests that the immunogenicity of cancer cells plays no significant part in the process of carcinogenesis.

It is accepted that homograft rejection, whether of normal or malignant tissue, is a cell-mediated (T-system) response, and this almost certainly holds as well for the rejection of tumor implants in specifically immunized syngeneic animals. This can be shown most clearly in pure-line guinea pigs [8, 14, 29, 30] in which typical delayed hypersensitivity reactions to tumor cells injected intradermally in immunized animals have been shown to have the same specificity as the immune reactions responsible for preventing tumor development. Essentially similar findings for mouse tumors induced by methylcholanthrene have been reported [23].

The suggestion that the diverse TSTAs are reemergent fetal antigens was mentioned earlier. I have found only one experimental result to suggest that this may in some sense be correct. In work described by Alexander [1] at the Oak Ridge Conference, rats treated with benzpyrene (implanted as a pellet) developed fibrosarcomas, each with its characteristic, specific antigen. In testing antigenicity it was found necessary to remove the entire tumor surgically and to allow a week to elapse before testing with the autologous tumor cells, because as long as the animal carried the tumor, challenge with a fragment from the same tumor induced a secondary tumor. However, another type of immune response shown in rats from which a benzpyrene tumor had been removed was less specific. Delayed hypersensitivity induced in the footpad occurred after intradermal injection of extracts of *any* benzpyrene rat tumors or of normal rat embryo extract but was not evident in rats still carrying a primary tumor. The suggestion here is that a surface protein of fetal character may have undergone a variety of minor changes (as a result, perhaps, of point mutations in the gene concerned) sufficient to produce a diversity of TSTA immunogenicities, each also retaining a common specificity with the others, as tested in the footpad. It is well known that two *Salmonella* flagellins may be quite specific when tested serologically but may cross-react extensively in tolerance and delayed hypersensitivity tests [38, 40]. The analogy is not exact since there is no evidence that B cells are involved in tumor rejection responses; but the hypothesis should be amenable to test.

If we allow for minor qualifications of this sort, all these characteristics of tumors induced by a single carcinogen in syngeneic animals are consistent with the hypothesis outlined at the beginning of the chapter. Probably it would be more correct to say that our ignorance of the detailed genetic control of cell membrane proteins precludes any useful speculation beyond the rather superficial and obvious approach I have adopted.

Effects of Oncogenic Viruses

The influence of the standard DNA oncogenic viruses in inducing malignant change can be interpreted along similar lines [22]. Here, however, the intrusion into the cell genome is not of a substance which by biological accident can react with DNA but of a form of DNA itself which has evolved a capacity to suborn the nuclear processes of the cell for its own replication. There is no real biological significance in the fact that a small proportion of the cells infected by oncogenic viruses develop malignant characteristics. The process has no survival value for virus or host species. Anyone who has followed the literature on genetic interactions in bacteria will be well aware of the innumerable ways by which the intrusion of DNA moieties from other bacteria, or bacterial viruses, can modify the inheritable functions of the organism involved. It is not always fully realized, however, that all such intrusions are rare accidents which lend themselves to recognition and investigation because of the ease with which highly selective environments can be provided for the isolation of these rare modified forms. The essential point, however, is that it is the nature of DNA chains to interact with each other and to be prone to accident.

The viral carcinogens are the liberated DNA of the virus, or, in the case of RNA viruses, DNA formed from the viral RNA pattern by the recently defined enzyme present in some or all these viruses [6, 47]. Again, it is impossible to define or even speculate on the physicochemical or informational details of the intrusion. Everything suggests that details will vary from one invaded nucleus to another, the only common feature being that those cells that develop malignant change regularly express one or more new antigens; these are common to all tumors induced in this fashion by a given virus and are based on information carried in the viral genome. In addition, however, refined experimentation has shown that other antigens [36, 48], with the same sort of diversity as those present in methylcholanthrene-induced tumors, are also often present. On this reading, the only major difference between virus-induced tumors and those initiated by chemical carcinogens is the possible incorporation of a functional segment of viral DNA in the former instance, and the random distortion of host DNA pattern leading to the individual changes in antigenic determinants in the second instance. Fetal antigens can also be detected in such tumors, and some suspicion is being voiced that the "virus-coded" antigens may in fact be of host (fetal) pattern [24]. For the present, however, I would judge the classical interpretation—that the main new antigen is coded for by the viral genome—is more likely.

At one period I was interested in the possibility that there were pre-existent antigenic differences amongst body cells arising by some diversification process analogous to that responsible for antibody and immune receptor diversity. It is not unreasonable that such differences should only become immunologically significant when a substantial clone of cells, all with the same distinctive antigen, develops. This hypothesis could and should be tested experimentally, but it is hard to see any survival advantage in such diversification, and the whole concept now seems unattractive.

ECOLOGY AND EVOLUTION OF CANCER CELLS IN THE BODY

If malignant disease arises by somatic mutation (probably in most instances by a sequence of two or more), it would be surprising if the process ceased with the establishment of definite malignancy. There is in fact a very large literature on "progression" of tumors, as judged by clinical evidence in man and as shown experimentally in animals [20]. In general terms, once a large population of proliferating cancer cells has developed, many new genetic anomalies will develop in individual cells. When the phenotypic effect of any such mutations gives increased opportunity for short-term proliferation, the new subclone will prosper. At the clinical level this can often be followed by karyotype studies because aneuploid mutants very commonly occur [19]. It is common, too, to find that a conditioned malignancy of breast or prostate can be controlled for a period by appropriate hormonal manipulation but that eventually a refractory mutant emerges. The "escape" of childhood leukemia after a period of control by cytotoxic drugs is basically a similar phenomenon. There may be various explanations for the common phenomenon of a long-standing tumor suddenly taking on a lethal anaplastic form, but "progressive" mutation is much the simplest and most likely.

Spontaneous retrogression of malignant foci and the phenomena of immunological surveillance [12] are only marginally relevant to the theme of somatic mutation and cancer, but a few comments may be of interest. The childhood cancers (neuroblastoma, retinoblastoma, and the Wilms tumor) do not arise once the embryonic cell lines from which they spring have matured. Neuroblastoma is specially prone to retrogression [7, 44], which has been ascribed by some to maturation of the malignant cells, by others to immunological factors.

The frequent but not invariable immunogenicity of tumor cells, both in the primary host and on transfer to syngeneic animals, is now well established [25]. This provides a basis for the concept of immunological surveillance as the main protection against the development of malignant disease. There is some doubt as to how effective this immunological protection is at the clinical level. In many ways the most convincing evidence is to be found in the recent reports of an undue appearance of malignant tumors in patients on prolonged immunosuppressive therapy after kidney transplantation [34, 41, 43, 46]. The most serious has been the rare occurrence of malignant lymphomas with an unusual predilection for the brain. Superficial carcinoma of skin and cervix uteri are also more frequent among transplant patients than in other people of comparable age. One of the most interesting findings comes from Sydney, where 7 of 51 transplant patients developed a total of 19 epithelial skin tumors. As is well known, basal cell carcinoma associated with long exposure to direct sunshine is particularly common in Australia, and hyperkeratoses on the back of the hands or other exposed areas of skin are even commoner. Walder et al. [49] found most of their cases were squamous cell carcinomas, often arising on the site of a preexistent hyperkeratosis. This is an unusual spontaneous change, and further analysis of the process by which immunosuppression allows its appearance might be fruitful.

Another aspect of somatic genetic change in human cancer is seen in the ectopic production of hormones by several types of tumors, notably bronchial carcinoma. This is presumably a manifestation of the accepted dogma that every nucleus in the body possesses the full genetic information of the zygote, but with most of it repressed. Pathological derepression of structural genes concerned with hormone synthesis could be associated with the variety of gross chromosome abnormalities to be seen in most fully developed tumors. Perhaps the most important point to be made is that if a small mass of tumor tissue, and perhaps only a subclone within the mass, is producing a protein inappropriate to a bronchial epithelium cell, the only likely circumstances under which this will be recognized are (1) if the protein is a highly active hormone or (2) if the protein is one that can produce an autoimmune-type response with symptoms. A probable example of that last process is the almost regular association of dermatomyositis with a malignant tumor somewhere in the body [15]. Fetal antigens can reasonably be placed in the same category as these ectopic proteins. Their significance has been discussed earlier.

Finally, a little may be said about the process of metastasis—in particular, blood-borne metastases, with their characteristic localization in the lung. There is sufficient evidence to show that many cancer cells can be released into the blood stream without giving rise to metastases, and it has been suggested that in general a malignant embolism must lodge in the pulmonary circulation for a metastasis to form. Presumably, discrete cancer cells are removed from the circulation by phagocytosis in the sinusoids of liver, spleen, or bone marrow. There is some limited evidence that immunological factors play a part in inhibiting metastasis [21, 27] in experimental animals and that, on occasion, metastases from an immunogenic tumor are composed of non-antigenic cells.

The overall picture of the natural history of cancer, once it has been initiated, is characteristically Darwinian. Malignant cells are genetically highly labile; chromosomal abnormalities may be undetectable in early tumors, but they appear almost invariably at later stages and after metastasis or repeated experimental transfer. Antigenic surface components may be lost or new ones expressed by derepression of genetic potentialities. All these inheritable changes will have some influence on extinction, survival, or proliferation of the clones concerned in the bodily environment, where hormonal, immunological, and probably other controlling factors, still to be defined, may also be acting against the malignant cells. In addition to intrinsic proliferative vigor in the cells, the malignant process may be aided by immunological factors (enhancing antibody) and probably by a variety of local microenvironmental factors.

AGE–SPECIFIC INCIDENCE OF CANCER

The uniformity in the age-specific incidence of particular cancers is well known and has been extensively studied (see ref. [9]). To a reasonable approximation, when the (age-specific) incidence of any definable form of cancer—excluding leukemia and lymphomatous conditions—is plotted against age, using logarithmic scales for both, a straight-line curve is obtained. If the tenfold distances on ordinates and abscissae are the same, the line has a slope of 5 to 6. The exceptions to the rule are: (1) In cancers influenced by hormonal factors in the female, such as uterine cancer, the slope decreases sharply after age 40. (2) In general, the slope diminishes slightly in old age. (3) The slope for lung cancer in males is significantly steeper than that for females, and the slope for cancer of the prostate is about twice as steep as for most other cancers with a value around 10 to 11.

As Doll has suggested informally in Melbourne, one can bring more regularity into the results if one assumes that a slope of 5 to 6 is the characteristic result of a random process acting at the same intensity over the whole period to be considered in the graph. If the process begins to act at birth (or conception) and persists throughout life, and the age scale runs from 0 to 80 according to actual age, the slope will be 5 to 6. Its position (i.e., the age at which a standard incidence is reached) will depend on the intensity and effectiveness of the random process. One can test this approach in relation to lung cancer using data from McKenzie ([33]; Fig. 1). One can reasonably assume that females of this cohort included hardly any cigarette smokers; males, however, were beginning to smoke extensively by 1900. It is also reasonable to assume that the exposure responsible for the vastly greater overall incidence of lung cancer in the males of this cohort began about the age of 15, and wholly obscured a background incidence due to something active over the whole of life and shown equally in both males and females. If, therefore, one uses an age scale with actual age X reduced to $X - 15$ and then plots logarithmically, the slope for males becomes parallel to that for females. Nothing is known about the etiology of the prostatic carcinoma, but it may be significant that if the figures for the same cohort are similarly changed so that $X - 25$ is used instead of X, a straight line of standard slope results.

Taking the primary lung cancer figures as a basis for discussion [33], the character of the curve seems to require both the occurrence in a cell of some random and rare change which has an indefinitely persisting effect and a means by which accumulation of such changes eventually leads to recognizable malignant disease. The persisting effect must be of somatic mutational character, since there is no other known biological phenomenon with the necessary properties. There is no satisfactory explanation for the regularity of the slope in these curves. Of the various hypotheses that have been suggested, the most likely seems to be that of Armitage and Doll [3]. They suggested that it resulted from the occurrence of two (or more) sequential mutations, each of which allowed a proliferative advantage over unmutated cells of the same population. This seems to be in general accord with what we know of the pathology of malignant disease. The absolute sex difference in lung cancer would be ascribed to the number of primary mutations. We assume that, on

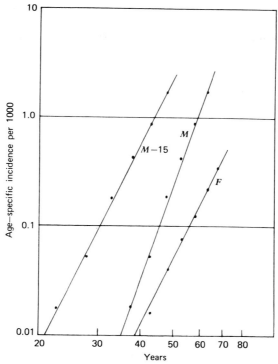

Figure 1. Age incidence of lung cancer deaths in persons born 1880–1885 in England and Wales. Both age and age-specific incidence are on logarithmic scales: F = female; M = male, with normal age scale; M—15 = males with age 15 taken as zero, on the assumption that the impact of the major carcinogenic agent (cigarette smoking) began at that age in males. It will be seen that the slope of the line is now parallel with the female graph. (See text.)

the average, one significant mutation occurs per unit of time in both males and females as a result of some lifelong influence, such as thermal molecular agitation or background ionizing radiation. The rate is raised by cigarette smoking, beginning at, say, age 15, to perhaps 10 significant mutations in the same time unit.

Of all human malignant tumors, retinoblastoma is the most interesting from the genetic angle. Knudson [28] has recently analyzed the age incidence of this tumor: it is seen only in young children, it is relatively amenable to surgical treatment, and it is known to be on occasion inherited as a dominant gene with variable penetration. He finds that the results are consistent with the occurrence of two sequential mutations *a* and *b*. Mutation *a* can occur in the germ line, and in such children all retinoblasts, while they exist, are subject to somatic mutation *b*. This means that multiple tumors are possible, and all bilateral tumors are accepted as inherited, mutation *a* having been incorporated in the germ line. Genetically normal children can also develop retinoblastomas by the occurrence of both *a* and *b* as somatic cell mutations. As would be expected, such tumors are always unilateral and, on the average, appear later. Approximate calculations suggest that mutation *a* occurs with about the same frequency in germ-line cells as in somatic cells. The age distribution of incidence in a large series indicates that bilateral tumors have a "one-hit" distribution, but unilateral ones a "multiple-hit" distribution.

CONCLUSIONS

It is inevitable that general hypotheses about the nature of malignant change will be influenced by the field of experimental study with which each writer is most familiar. Experimental work on cancer tends to be concentrated on (1) the viral etiology of tumors, (2) biochemical and immunological changes associated with the development of malignancy, and (3) the relationship of fetal antigens to those present in tumors. Each has given rise to corresponding hypotheses. Unfortunately there is still no effective *genetic* approach at the experimental level to the differences between malignant and normal cells, and a general hypothesis based on somatic mutation is therefore less attractive to most workers.

The available facts, however, seem to be fully in accord with the hypothesis that cancer is a somatic genetic condition in which the result of a sequence of mutational changes is expressed in the capacity of a cell line to proliferate without experiencing normal morphogenetic control. My contention would be that this is the only logical basis broad enough to accommodate and coordinate experimental findings as they emerge. If effective methods of analyzing somatic genetic changes in terms of standard genetic concepts are developed, it may become possible to devise hypotheses that give a more closely defined role to chemical or viral carcinogens.

The complexity of all the relevant factors—the structures of the cell surface and of the chromosomal apparatus, the mechanism by which mitosis is initiated, and the morphogenetic control of cells in the mammalian organism—is intimidating, and it is far from certain that a more detailed interpretation of cancer than the one on which this paper is

DISTURBANCES OF THE GENETIC MATERIAL

CHROMOSOME INTEGRATION OF VIRAL DNA: THE OPEN–REPLICON HYPOTHESIS OF CARCINOGENESIS

H. L. K. WHITEHOUSE

Reader in Genetic Recombination, University of Cambridge, Cambridge, England

---

Introduction 42
Why Do Many Viruses Exist in a Proviral State Integrated with the Host DNA? . . 42
    The Mechanism of Integration of Bacterial Virus Genomes 42
    The Evidence for Integration of Animal Virus Genomes 46
    The Selective Advantages of Integration and Immunity 50
Why are Animal Proviruses Apparently Often Associated with Cancer? 51
    Cell Transformation and Viral Genes 52
        Temperature-sensitive Viral Mutants and Transformation 52
        Reversion and Gene Loss 53
        Diversity of Transforming Viruses 54
    Cell Transformation and DNA Synthesis 55
    DNA Synthesis, Recombination, and Viral Genome Integration 57
    Chromosome Replication, Crossing-over, and Replicons 59
    Viral Genome Integration and the Open-Replicon Hypothesis of Carcinogenesis . 61
Why are Ionizing Radiations, Many Chemicals, and Certain Genetic Defects Carcinogenic? 64
    Mutation at Virus-suppressor Loci 64
    Known Genetic Defects, Cellular Transformation, and Increased Risk of Cancer . 65
Summary: The Open-Replicon Hypothesis of Carcinogenesis 68
Acknowledgements 68
Literature Cited 68

INTRODUCTION

I wish to discuss three related questions that seem to me to be of central importance for an understanding of the cause of cancer. First, why do many viruses exist in a proviral state integrated in the host DNA? This is established for a number of bacterial viruses and seems likely to be true also for many animal viruses. Second, when apparently integrated in this way, why are so many animal viruses associated with cancer? At least 25 viruses, in diverse viral families, are carcinogenic. Third, if viruses cause cancer, how is it that ionizing radiations, many chemicals, and certain genetic defects are also carcinogenic?

WHY DO MANY VIRUSES EXIST IN A PROVIRAL STATE INTEGRATED WITH THE HOST DNA?

Before attempting to answer the first question, I will describe the best-known examples of integration—virus lambda (λ) of *Escherichia coli*, polyoma virus of *Mus musculus* (Mouse), and simian virus 40 of *Macaca mulatta* (Rhesus Monkey) and *Cercopithecus aethiops* (African Green Monkey)—making brief references to other examples, as well.

The Mechanism of Integration of Bacterial Virus Genomes

A detailed account of λ integration, with references to original sources, has been given by Echols [34], and the following paragraphs summarize our current knowledge.

Strains of *E. coli* carrying the λ genome integrated into the host chromosome are immune to infection by λ. The integration and the immunity are determined separately, although they normally occur together. Lambda is a relatively large virus, having about 17 μm of DNA, equivalent to about 54,000 nucleotide pairs. This is enough to code for about 80 polypeptides containing an average of 200 amino acid residues each.

Integration is brought about through the action of a λ gene called *int*. The product of this gene is believed to be an enzyme which can recognize a specific nucleotide sequence at the integration site in the *E. coli* chromosome, and a specific nucleotide sequence at the integration site in the λ chromosome. It is far from certain that the two sequences are identical, because Davis and Parkinson [25] could find no evidence of homology between them. They studied electron micrographs of DNA molecules obtained by annealing complementary nucleotide chains of various deletion mutants and normal λ, and inferred that the host and viral integration sites were unlikely to share a sequence of more than 12 nucleotides. The specificity of integration thus seems to reside in the virus-coded enzyme. The enzyme would need to cut both chains of the DNA of both chromosomes and rejoin them as follows: bacterial to viral, and viral to bacterial. The λ chromosome is known to be circular at the time of integration, as postulated by Campbell [19]; thus the outcome of the integration process is that the circular λ DNA is opened at a specific point and inserted as a linear molecule at a specific point in the host chromosome (Fig. 1).

When the λ genome is released from the host chromosome, a similar reciprocal exchange takes place between the DNA at the two ends of the provirus. The *int* gene product is again involved and also the product of a λ gene called *xis*, the activity of which is peculiar to the excision process. The function of *xis* is not known.

The immunity to infection by λ conferred on *E. coli* cells when they contain the λ provirus is now understood in some detail. A viral regulatory gene called *c1* codes for a protein molecule called the λ repressor. This protein binds to λ DNA at two specific sites, called the left-hand and right-hand operators, respectively. The left-hand operator is the regulator site for gene *N* and the right-hand operator for an operon containing regulatory genes *cro* and *c2* and genes *O* and *P* concerned with viral replication. These five genes are the first to function when the virus infects a cell, and the initiation of the activity of all the other λ genes is dependent on their activity. The binding of the λ repressor to its specific operators thus prevents the initiation of activity throughout the λ provirus and also throughout any λ infecting the cell. The repressor thus confers on such cells an immunity to λ. Quite a sophisticated control system is involved, however, to bring the *c1* gene into action and so bring about repression.

When λ infects a cell and causes its destruction with liberation of progeny λ particles, the viral genes function in a precise sequence. Three steps may be recognized (Fig. 2). First, there is the so-called "immediate early" stage involving, as already mentioned, genes *N* and *c2*, among others. Second, there is the "delayed early" stage, which includes *int* and two regulatory genes, *c3* and *Q*. Stage 3, the "late" stage, involves numerous genes activated by the product of gene *Q*, and also *c1* activated by the combined products of *c2* and *c3*. Whether infection of an *E. coli* cell lacking λ provirus leads on the one hand to integration and immunity or on the other hand to cell destruction therefore depends on the action of a second stage gene (*int*) and also on which of the third stage genes gains the upper hand: the repressor gene (*c1*) or the genes

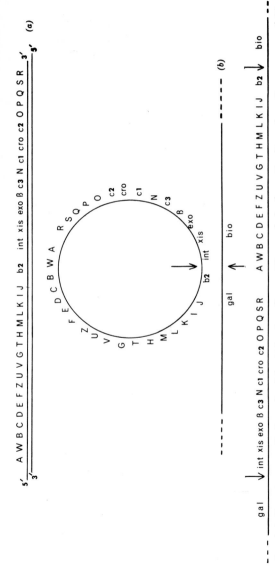

Figure 1. (a) Genetic map of phage lambda (λ) when linear. The symbols above the lines show the sequence of 33 of the viral genes that have been mapped. The two lines represent the two nucleotide chains. DNA is synthesized from 5'-nucleotides, that is, from nucleotides with the phosphate group attached to the 5' carbon atom of the deoxyribose, the 3' carbon atom carrying a hydroxyl radical. The synthesis involves the covalent joining of the 3'-hydroxyl group of one nucleotide to the phosphate group of the next, with the result that the chain begins with 5'-phosphate and ends with 3'-hydroxyl. At both ends of λ, the 5'-phosphate terminus extends 12 nucleotides beyond the 3'-hydroxyl terminus of the complementary chain, and the nucleotide sequences of these projecting ends are complementary to one another. (b) Genetic map of phage λ when circular. Base pairing has occurred between the complementary sequences of nucleotides in the projecting 5'-terminal chains, followed by joining of the phosphodiester axes to give a circular, double-stranded molecule, shown as a single circle. The straight line below the circle represents part of the chromosome of Escherichia coli, gal and bio indicating particular groups of genes. The sites of attachment of the bacterial and viral chromosomes, when integration takes place, are indicated by arrows. (c) Prophage map, following integration as a result of an exchange between the DNA of the host chromosome and that of the virus at the attachment sites. The λ gene sequence is now a circular permutation of that in (a). Arrows again indicate the attachment sites.

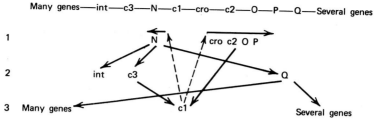

Figure 2. The regulation of gene activity in phage λ of *Escherichia coli*. The horizontal axis shows the relative position of the genes in the DNA of the virus and the vertical axis the stage at which they function: (1) immediate early, (2) delayed early, (3) late. The horizontal arrows show the direction of transcription from the two operators to which the λ repressor—that is, the product of gene *c1*—binds. Oblique broken and unbroken arrows indicate repression and activation, respectively.

activated by *Q*. The *c1* product activates the synthesis of its own messenger. Such behavior of the λ repressor—namely, the activation of its own structural gene—explains how the repressor can continue to be produced, once formed, even though all the other viral genes are shut down.

In addition to the basic control system just outlined, there are several cross controls. Thus, the *c3* product inhibits genes *O* and *P*, the *c1* product prevents the products of genes *N*, *O*, and *P* from functioning, and the *cro* product is an inhibitor of *c1*. This elaborate control system must, among other things, enable the λ repressor to be inactivated when the provirus is excised from the host chromosome and starts the production of progeny virus particles, but the details of this release from repression are not understood.

Since integration and immunity are distinct phenomena determined by different genes, it might be thought that each would occur alone. Mutants with the single properties are in fact known. Those defective in repression have clear plaques because immune host cells cannot arise: the turbid appearance of normal plaques is caused by the survival of immune cells. The mutants lacking repressor—that is, defective in gene *c1* or its regulators *c2* and *c3* (the *c* stands for clear)—will always produce progeny virus particles. Whether or not a viral genome was integrated into the host chromosome is then of no consequence, since the host cell will not survive. Conversely, mutants defective in integration will be able to establish repression in the normal way; but since this will prevent viral replication, the *int* mutants, not being integrated in the host chromosome, will soon be diluted out of the *E. coli* cells as these multiply, and the repressor protein will then likewise be diluted out. It is evident that integration alone, or immunity alone, has only a temporary effect. Both phenomena must occur together if a stable alternative to cell destruction by the virus is to be provided.

In virus P2 of *Shigella* and in a number of viruses of *E. coli* the proviral state is not maintained by a repressor of the kind found in λ (see ref. [54]). Nevertheless, bacteria carrying any of these proviruses are immune to infection by the corresponding virus, although the mechanism by which this immunity is conferred is unknown. In P2 it has been shown that, as with λ, immunity and integration are separate functions. The *E. coli* sex factor *F*, which seems to be a virus that has developed the conjugation mechanism as a means for its own dissemination [54], also confers immunity to further infection of the cell by *F* factors.

Hayes [54] pointed out that certain mutations in a provirus allow the viral genome to survive only in permanent association with the host chromosome. For example, mutants of the *O* or *P* genes of λ are defective in viral DNA replication, but in the proviral state replication is brought about by the host; thus such mutants could survive as provirus. The *cI* gene of λ would then appear to be a host gene conferring immunity to infection by λ. Hayes also drew attention to the frequency with which viruses incorporate host genes into their structure—the basis of the phenomenon of transduction. He concluded that in bacteria it is not possible always to draw a clear distinction between host and viral genes.

The Evidence for Integration of Animal Virus Genomes

Polyoma virus and simian virus 40 (SV40) particles contain a circular DNA molecule about 1.6 μm long, which is sufficient to code for about 7 genes of 600 nucleotide pairs. Polyoma virus normally kills cultured cells of *Mus musculus* (Mouse), which are therefore described as permissive; similar cells of another rodent, *Mesocricetus auratus* (Syrian Golden Hamster), are said to be nonpermissive, because they do not support its multiplication. On the other hand, the virus does cause changes in these cells, which are said to be transformed by it. Transformed cells are recognized by their continued growth under conditions where normal cells would no longer multiply, and by their random rather than polarized orientation. Surprisingly, these transformed *Mesocricetus* cells contained no infectious virus particles. Nevertheless, it was demonstrated unequivocally that polyoma virus was responsible for the transformation by purifying its DNA and showing that the DNA would bring about transformation [30].

The presence of viral DNA in transformed cells was demonstrated with the *Cercopithecus* virus SV40, which will transform, but does not multiply in, *Mus* cells. Watkins and Dulbecco [128] fused uninfected (permissive) *Cercopithecus* and transformed (nonpermissive) *Mus* cells, and found that infectious SV40 particles were produced. Fusion was brought about through the presence of inactivated Sendai virus, the protein envelope of which is known to promote cell fusion [51]. When the experiment was repeated using single cells, it became evident that each transformed *Mus* cell contained at least one complete SV40 genome.

Nucleic acid hybridization has shown that the SV40 genome is present in the host genome of transformed cells. Gelb et al. [42] used an improved technique involving measurement of the DNA reassociation kinetics on hydroxyapatite and concluded that one to three copies of the SV40 genome were present per diploid mammalian cell: four different cultures of *Mus* cells transformed by SV40 gave values of 1.0, 1.4, 1.6, and 3.9, while an SV40 tumor in *Mesocricetus* contained 2.1 copies per cell. These values are considerably lower than previous estimates. Control experiments using untransformed cells of *Mus* and *Cercopithecus* revealed about half the SV40 nucleotide sequence per diploid cell. Sambrook et al. [97] showed that the SV40 genome was covalently bound to the *Mus* DNA in transformed cells, because RNA synthesized in vitro with SV40 DNA as template was found to hybridize with high-molecular-weight DNA extracted from the transformed cells.

Hybridization with SV40 DNA has shown that RNA complementary to at least part of the SV40 DNA is present in SV40-transformed *Mus* cells. Lindberg and Darnell [77] found high-molecular-weight RNA containing virus-specific sequences in such cells, and Wall and Darnell [126] showed by hybridization studies that this RNA also contained host-specific sequences covalently linked to the viral ones. On the other hand, the smaller SV40-specific RNA molecules associated with ribosomes contained little or no RNA complementary to *Mus* DNA. Wall and Darnell suggested that virus-specific messenger RNA was formed by selective cleavage of high-molecular-weight RNA containing host and viral sequences. These experiments confirm that the viral DNA is integrated into a host chromosome with covalent joining of host and viral DNA.

In cells which have been transformed by polyoma virus or SV40, the activity of much of the viral genome is evidently blocked, since viral progeny are not produced. The mechanism of this inactivation is not understood. With nonpermissive cells, as Butel, Tevethia, and Melnick [17] pointed out, it is meaningless to say that transformed cells are resistant to further infection, because it is not known whether the production of progeny virus is prevented at an earlier point in the pathway than the block implied by nonpermissiveness. With permissive cells, there are differences of opinion about the immunity of transformed cells to further infection. Resistance is necessary for transformation to be revealed in such cells, but the rarity of transformants means that selection for resistance may have occurred. Hahn and Sauer [50] concluded that resistance to SV40 in *Cercopithecus* cells transformed by this virus is a response of the host cells and is induced by the virus. Butel et al. [17] found that some permissive cell lines were susceptible to further infection and others were resistant to varying degrees. Basilico and Wang [7] found that hybrid cells arising from fusion of polyoma-transformed *Mesocricetus* cells and uninfected *Mus* cells were susceptible to infection by polyoma virus. They concluded that, contrary to some earlier claims, the transformed cells do not synthesize a dominant diffusible viral repressor.

Papilloma viruses such as those responsible for Shope papilloma in *Oryctolagus cuniculus* (Rabbit) and verruca vulgaris (wart) in man contain DNA for about 12 genes of 600 nucleotide pairs. Owing to difficulties in their culture, papilloma viruses have been less studied than polyoma virus and SV40. The complete viral genome is known to persist, however, in cells transformed by papilloma viruses, since infectious particles appear in vivo in cornified cells [24].

Adenoviruses have enough DNA for about 55 genes and so approach virus λ of *E. coli* in size. Doerfler [28, 29] showed by differential labeling experiments that DNA of human adenovirus type 12 is integrated into the DNA of *Mesocricetus* cells when these are infected by the virus. The virus does not replicate in these cells but can bring about transformation. Schlesinger [98], in discussing tumorigenesis by adenoviruses, favored the idea that the viral genes that act late in the development of progeny virus were repressed in transformed cells, but enough viral genes were active to permit transformation. According to him, the response of adenovirus-induced tumor cells to superinfection with other adenoviruses strongly suggests repressive control. Williams, however, finds no evidence for repression in some type 5 transformed cells, and he believes that the repressor hypothesis is open to question (J. F. Williams, personal communication).

Herpesviruses have large genomes—for example, in herpes simplex, sufficient for about 200 genes. Nucleic acid hybridization studies have shown [72] that human lymphoblastoid cell lines contain DNA homologous with that of Epstein-Barr herpesvirus, while this viral DNA has not been found in other human cell lines. Marek's disease herpesvirus causes tumors in lymphoid tissues of *Gallus domesticus* (Chicken) from which no infectious virus has been isolated. Cultivation with susceptible cells reveals the virus, however, and as with papilloma viruses, infectious virus is produced in keratinizing cells—in this instance, the feather follicle. Herpesvirus saimiri of *Saimiri sciureus* (Squirrel Monkey) causes lymphoma in *Aotus trivirgatus* (Owl Monkey), *Cebus albifrons* (Cinnamon Ringtail Monkey), and *Callithrix sp.* (Marmoset). Virus particles or viral antigens cannot be demonstrated in the tumor cells but, again, virus can be recovered by cultivating the cells in association with susceptible cell lines. The Lucké herpesvirus is almost certainly the cause of tumors in the kidneys of *Rana pipiens* (Frog). A remarkable feature of this tumor is that growth occurs only at high temperatures, and then no viral particles are present, while progeny virus particles are produced only at low temperatures [72].

All the RNA tumor viruses contain single-stranded RNA sufficient for about 40 genes. Baltimore [5] and Temin and Mizutani [118] discovered an RNA-dependent DNA polymerase in particles of Rous sarcoma virus of *Gallus domesticus*, and Baltimore found such an enzyme in Rauscher leukemia virus of *Mus musculus*. The existence of this reverse transcriptase provided strong support for Temin's hypothesis that the RNA of Rous sarcoma virus is synthesized on a DNA template. Spiegelman et al. [104, 105] found reverse transcriptase, and also DNA-dependent DNA polymerase, in a number of other RNA tumor viruses. The presence of reverse transcriptase in all the RNA tumor viruses tested and its absence from five nontumorigenic RNA viruses looked significant. Subsequently, the enzyme was detected in normal human and *Mus* cells in culture [99]. The discovery of reverse transcriptase in RNA tumor viruses implied not only the likely existence of DNA copies of the genome but also the possibility that this DNA might become integrated into a host chromosome. The presence of virus-specific RNA in cells transformed by murine sarcoma viruses [49] gave support to this possibility.

Unlike the DNA tumor viruses, the production of progeny particles of RNA tumor viruses is not associated with death of the host cell. In consequence, cells transformed by RNA tumor viruses may also produce progeny virus. Indeed, Temin [117] believes that progeny viral RNA is formed directly from the provirus as template. In some instances, transformed cells do not produce progeny particles (for example, *Mus* cells transformed by avian sarcoma virus); but just as with DNA viruses, it is meaningless to describe such nonpermissive cells as immune to the virus when nonpermissivity is not understood in molecular terms. No immunity of the kind known with virus λ of *E. coli* has been found with RNA tumor viruses, but there is virus-specified interference in cells already infected with an RNA tumor virus of the same subgroup [117].

From the evidence outlined above, it seems possible that the genome of all cancer viruses can exist as DNA and can become integrated into that of the host. The clearest evidence for this integration has been obtained with SV40 but, as shown, data consistent with this hypothesis are available for all the other groups of cancer viruses.

The Selective Advantages of Integration and Immunity

I will now consider why many virus genomes can apparently exist integrated into a host chromosome, whether prokaryotic or eukaryotic. Watson [129] raised this question and pointed out that it was hard to believe that integration was a matter of chance, but he offered no explanation for it. Dulbecco [31] argued that integration was unlikely to be the result of selection for the sake of the virus since the virus does not benefit appreciably from it. He pointed out that in fact the integrated viral DNA of polyoma or SV40 only exceptionally initiates a viral replication cycle. Dulbecco, however, seems to have taken a narrow view of the selective forces acting on parasitic organisms. With obligate parasites such as viruses, selection will not favor the maximum degree of virulence, since this may lead to extinction of the host. This is exemplified by the introduction in 1950 of a highly virulent strain of the myxoma virus into the Australian population of *Oryctolagus cuniculus* (Rabbit), where the virus previously had been unknown. Within one year less virulent strains appeared, and these became predominant throughout the country within three or four years, replacing the original highly lethal virus [38]. Some sort of balance between host cell destruction and host cell immunity would seem to offer the maximum potentiality for virus multiplication.

An analogous situation applies to parasitic fungi. I have suggested [130] that the occurrence in such fungi of a mutation causing increased virulence is likely to be followed by selection favoring a mutation in the host for increased resistance. This gene-for-gene hypothesis [40] has now gained wide acceptance [39]. Soon after the myxoma virus had been introduced into the Australian rabbit population, strains of the host appeared which were genetically more resistant to the virus [38]. The development by a host of increased resistance to infection by a specific parasite is clearly of selective value to the host. It is likely also, though less obviously, to be of long-term selective value to the parasite, since it allows the host to survive. Increased virulence by the parasite is of course of immediate selective value to it, but would ultimately be disadvantageous to the parasite if the hosts were eliminated or greatly reduced in number.

Extending these ideas, it would seem that integration to give the proviral condition, in conjunction with increased immunity of the host to the specific virus, would be of selective value in the long term to the virus, as well as being of obvious value to the host. It was pointed out in discussing virus λ of *E. coli* that an immunity mechanism determined by a viral gene would be ineffective unless accompanied by integration,* since the immunity would otherwise soon be lost. Immunity and integration, in association with one another, will then have selective value. According to this hypothesis, an unstable equilibrium will exist, virulence having immediate advantages to the virus, but host resistance associated with provirus having long-term benefits for the virus—provided the immunity mechanism is not so efficient that it can never be broken down by the virus.

A selective value to both host and virus for the occurrence of provirus, such as this hypothesis postulates, will explain the evidence for virus λ of *E. coli* that both organisms participate in integration. Thus there are specific integration sites in the DNA of both host and virus. It is likely that a similar situation occurs with animal viruses. McDougall et al. [79], who made *in situ* hybridization studies using ^{3}H-labeled adenovirus type 12 and human kidney cells in culture, found preferential labeling of chromosome No. 1 on autoradiographs. This suggests that there is a specific integration site on that chromosome.

Spiegelman [103] argued that the advantage to a virus of integration into a host chromosome is in allowing the viral genome to be duplicated by the host at every mitotic cycle, and Smith [101] suggested that integration assures the survival of the viral DNA. They assumed, no doubt, that sooner or later integration would be followed by excision and the production of progeny virus particles. A dormant phase, such as integration would give, might well be of selective value to the virus, but by itself it would seem to bring only disadvantages to the host.

It is difficult to understand the apparently widespread occurrence of integration and the likely participation of the host in the process, unless there are advantages to both partners. It is for this reason that I favor the combined effect of integration and immunity as the primary source of the selective advantage, even though in particular instances, such as the RNA tumor viruses, there is little evidence for immunity associated with integration.

WHY ARE ANIMAL PROVIRUSES APPARENTLY OFTEN ASSOCIATED WITH CANCER?

The general properties of cancer cells were discussed by Dulbecco [30] and may be summarized as follows. First, cancer cells do not respond to the regulatory influences that control the multiplication of normal cells;

---

* Regulated replication of the viral genome would be an alternative to integration, provided the replication kept pace with that of the genome of the host cell, as exemplified by mitochondrial DNA.

second, they show altered neighbor relationships and so disrupt the normal histological cell pattern; and third, they transmit the cancerous character to the daughter cells. Several authors [24, 30] have pointed out that this third property shows that cancer is caused by an alteration of a hereditary cellular component—most probably, therefore, a change in its DNA.

A phenomenon called cell transformation occurs in tissue culture [125]. Transformed cells have been found to express one or several of the characteristics of cancer cells: they may continue to grow under conditions that would limit the multiplication of normal cells, their orientation toward one another as they proliferate may be random, and in some cases such cells may grow into a tumor when injected into the host of origin. Investigations of this transformation in vitro should greatly facilitate the study of carcinogenesis. For several reasons, however, transformed cells will not necessarily give rise to cancer in vivo. Foremost are the immune responses of the host, which evidently keep most cancers in check [2, 47, 71]. Dulbecco [31] discussed a further factor: the occurrence of changes through mutation and selection subsequent to the initial change of growth behavior. The key step, however, in understanding the cause of cancer is the initial change, which is evidently common to transformation in vitro and to carcinogenesis in vivo.

Cell Transformation and Viral Genes

There is good evidence that cell transformation by tumor viruses involves the activity of viral genes. In the first place, cells transformed in vitro by one virus often differ strikingly in appearance from those transformed by another [31]. Secondly, cells transformed by one virus differ antigenically from those transformed by another, while those transformed by the same virus express antigens that are cross-reactive [71]. Third, investigation by hybridization techniques has revealed the presence of SV40 messenger RNA in *Mus* cells transformed by this virus; these studies have suggested that not more than one-third of the viral genome is essential for the maintenance of transformation [31]. Fourth, cell culture experiments with temperature-sensitive viral mutants indicate that consistent alterations have occurred in the qualities of the cells, consequent to infection by the specific mutant.

Temperature-sensitive Viral Mutants and Transformation. Study of temperature-sensitive mutants of polyoma virus has shown that a mutant called *Ts-a* prevents the stable transformation of *Mesocricetus* cells at 38.5°, though not at 31.5°. The mutant also fails at the higher temperature to accumulate viral DNA in *Mus* cells, while giving rise to progeny virus particles at the lower temperature. However, the mutant-infected *Mesocricetus* cells can maintain the transformed state at 38.5°, once it has been induced, and can also undergo an abortive transformation (a temporary change in growth characteristics) at that temperature [*109*]. This shows that the *Ts-a* gene, though necessary for the establishment of transformation, is not involved in the changes of cell physiology characterizing the transformed state. Study of other temperature-sensitive mutants of polyoma virus has shown that several of the viral genes play no part in transformation.

Temperature-sensitive mutants affecting the maintenance of transformation have been identified in Rous sarcoma virus, an RNA tumor virus of *Gallus domesticus* [68, 80]. Experiments with these mutants strongly suggest that a virus-coded protein is essential for the maintenance of the cell in the transformed state but is not required for the production of progeny viral particles. This shows, as Green [48] has emphasized, that the viral genes active in transformation do not necessarily also function when productive infection occurs.

Reversion and Gene Loss. Defendi et al. [27] made somatic hybrids between polyoma-transformed and normal *Mus* cells and found that the hybrid cells retained the transformed properties and gave rise to tumors when inoculated into the host. Cell fusion was achieved by growing the mixed cultures at low temperature (29°). Contrary to these results, Harris et al. [52] found that *Mus* tumor cells lost their cancerous properties on fusion with certain nonmalignant cells. The tumor cells used were from tumors of three kinds: one of spontaneous origin, one induced by polyoma virus, and a third induced by a chemical carcinogen. Inactivated Sendai virus was used to bring about cell fusion, and the behavior of the hybrid cells was tested by injection into the host. Harris et al. found that malignant segregants from the hybrid cell lines were associated with the loss of chromosomes. Later studies [51, 135] supported their earlier work and led them to conclude that malignancy was recessive. The evidence indicated that the generation of malignant segregants from the hybrid cells requires the elimination of one or more specific chromosomes derived from the nonmalignant parent cell. They favored the idea that cancerous growth results from a specific genetic loss.

Pollack et al. [85] found that, following exposure of *Mus* cells transformed by polyoma virus or SV40 to 5-fluorodeoxyuridine, it was possible by selection to obtain cells sensitive to contact inhibition, that is, showing loss of transformation. Similar revertants of cells transformed by polyoma virus were obtained spontaneously by Rabinowitz and Sachs [87]. The viral genome had not been lost from these cells, because the nuclear tumor antigen specific to polyoma virus was still present. Evidently the expression of the polyoma virus transformation can be repressed.

Hitotsumachi et al. [58] studied reversion to normal growth in *Mesocricetus* cells transformed by polyoma virus and concluded that certain chromosomes carried genes determining the expression of malignancy and other chromosomes had genes for its suppression. They found that a reversion of transformation could be associated with either an increase or a decrease in chromosome number, and they concluded that transformation and its reversion depended on a balance between the numbers of chromosomes with opposing effects. It is evident that cell transformation by a tumor virus does not simply depend on the activity of one or more viral genes. Host genes evidently can act, either directly or indirectly, to suppress the viral gene activity.

Harris's conclusion that transformation is due to genetic loss can be reconciled with the evidence from tumor virus studies that one or more proviral genes are actively involved, by postulating that the genetic loss involves host repressor loci. Defendi et al. [27] pointed out that the inheritance of polyoma-induced transformation as a dominant character in their cell-hybridization experiments was incompatible with hypotheses attributing cancer to deletion of structural information or to virus-induced recessive mutations of structural genes. Crawford [24] drew attention to the fact that if viral cancer arose through the induction by the virus of mutation or a deletion in host DNA, the changes induced would be expected to be rather nonspecific, and, furthermore, to segregate normal progeny for the first few cell divisions. These predictions do not agree with what is observed. Hitotsumachi et al. [58] favored an explanation similar to their own to account for the results obtained by Harris and associates. To explain the lack of transformed properties in the immediate hybrids, they suggested that the nonmalignant parent cells in Harris's experiments had an excess of suppressor genes. A balance, such as they favor, between one or more active proviral genes and a number of host suppressor loci, seems in general to fit the viral tumor data better than the deletion hypothesis.

Diversity of Transforming Viruses. One of the most remarkable features of cell transformation is the diversity of viruses that will bring it about. As already indicated, they belong to five different groups—four with duplex DNA and one with single-stranded RNA as the genetic material. Their sizes range from about 7 genes (polyoma virus and SV40) to about 200 (herpesviruses). Furthermore, there is good evidence that they are of diverse origin. Subak-Sharpe et al. [*110, 111, 113*] analyzed the nearest-neighbor base sequences of the DNA of a number of viruses and found that polyoma virus, SV40, and human and Shope papilloma viruses had similar shortages of the CpG doublet as the polypeptide-coding part of the nuclear DNA of their hosts, while large viruses like herpes simplex had a doublet pattern quite unlike the host.

These studies have been extended by Spiegelman (see ref. [111]) to avian myeloblastosis virus. This RNA tumor virus was also found to have the vertebrate type of doublet pattern. Subak-Sharpe [110, 111] concluded that the DNA of polyoma virus, SV40, and the papilloma viruses, and the RNA of avian myeloblastosis virus, have originated within the polypeptide-specifying stretches of the ancestral host DNA, or at least are of vertebrate origin, while the DNAs of large viruses like herpes simplex are derived from distant, possibly bacterial, lines of descent. A similar dichotomy of origin—some internal, some external—was found for non-carcinogenic mammalian viruses. That viruses of such diversity of origin should have developed the capacity to liberate host cells from growth control suggests that this ability is associated with some basic requirement of considerable selective value.

In discussing possible selective advantages of tumor formation for a virus, Subak-Sharpe [112] argued that with DNA tumor viruses, where the transformed cell is immune to further infection by the same virus, selection must have acted to bring about this immunity. He concluded that "tumor development may already be a stage past some really important cryptic situation which has some selective advantage", and he suggested that integration of the viral genome without tumor formation might be an earlier stage with survival advantage to the virus.

Cell Transformation and DNA Synthesis

It was discovered in 1965 that infection of *Mus* cells with polyoma virus leads to a marked increase in the activity of the host enzymes involved in DNA synthesis and, furthermore, that host DNA synthesis is stimulated (see review: ref. [136]). It has been shown that a viral function is involved in this stimulation because it is sensitive to interferon [33, 116] and to inactivation of the virus by nitrous acid or ultraviolet light. Host DNA synthesis can be stimulated in the absence of viral DNA synthesis, and experiments reveal that polyoma virus infection can overcome blocks in cell DNA synthesis imposed by a great diversity of factors, including contact inhibition, high temperature, X-rays, and various chemical inhibitors such as 5-fluorodeoxyuridine. Winocour [136] suggested that a primary switch in the cells' regulation of DNA synthesis is triggered. Stimulation of host DNA synthesis, both in viral multiplication and in transformation, has been demonstrated for polyoma virus and SV40. It has also been shown to occur with adenovirus multiplication and in transformation by RNA tumor viruses [117].

The temperature-sensitive *Ts-a* mutant of polyoma virus which, as already described, is defective in the establishment but not in the maintenance of transformation, is able to stimulate host cell DNA synthesis at the restrictive temperature, 38.5° [41, 109], suggesting that the *Ts-a* gene is concerned with some other step in the establishment of transformation. Benjamin and Burger [9] found that host-range mutants of polyoma virus (mutants blocked in their growth in normal cells but still able to grow in cells previously transformed by the virus) were capable of inducing cellular DNA synthesis under the restrictive conditions—that is, when normal cells were infected—but were unable to cause transformation under these conditions. The existence of these mutants, and of the *Ts-a* mutant, indicates that induction of host DNA synthesis is not sufficient to bring about transformation.

Dulbecco and Eckhart [32] obtained a temperature-sensitive mutant, *ts-3*, of polyoma virus, defective at the restrictive temperature in the ability to stimulate DNA synthesis in *Mus* cells. Eckhart et al. [36] found that the *ts-3* mutant was also defective in the alteration that polyoma virus infection normally causes in the cell membrane. Unlike normal virus, the *ts-3* mutant failed to cause increased binding to the cell membrane of an agglutinin obtained from *Triticum aestivum* (Wheat). It is not known whether the lack of stimulation of host DNA synthesis is a direct result of the absence of the normal *ts-3* gene product or an indirect result of the absence of a cell surface alteration [35].

In addition to triggering synthesis of the nuclear DNA of the host, it has been found that infection by polyoma virus or SV40 also induces the replication of the DNA of the mitochondria [75, 124]. This mitochondrial DNA synthesis takes place after the nuclear DNA replication [70], thus it is probably a secondary effect of the latter. On the other hand, Kára et al. [66] discovered that Rous sarcoma virus replicates within the mitochondria of *Gallus domesticus* cells, Richert and Hare [89] showed that viral replication and cell transformation by this virus were sensitive to inhibitors of mitochondrial function, and Kára et al. [65] detected viral reverse transcriptase in particles containing the complete viral genome which they isolated from within the mitochondria of tumor cells. These studies suggest that the mitochondria are directly involved in Rous sarcoma virus replication and tumor induction.

Sachs and associates [46, 95] discovered that the transforming activity of polyoma virus is inactivated by treatment with nitrous acid at the same rate as the DNA-inducing capacity of the virus. This led them to suggest that the cell transformation and the induction of cellular DNA synthesis are expressions of the same function of the viral genome. Todaro and Green [120], from study of the initiation of transformation in *Mus* cells by SV40, concluded that fixation of the virus-induced transformed state as a heritable property requires a process associated with cell duplication. This led them to surmise that stimulation of cell DNA synthesis is the event required to fix transformation, since most other macromolecular biosyntheses occur in resting cells. Dulbecco [31] suggested that induction of host DNA synthesis was related to transformation, because both phenomena imply loss of the normal response to factors limiting cell multiplication. Furthermore, induction of host DNA synthesis seems not to be found with viruses that do not cause transformation. Dulbecco therefore suggested that the viral gene which directly or indirectly initiates host DNA synthesis may be the main agent of transformation. Lehman and Defendi [73], from study of SV40 infection of *Cricetulus griseus* (Chinese Hamster) cells, likewise suggested that the initial event relevant to transformation might occur at the level of the control of cellular DNA synthesis. In agreement with these views, it was recently stated that "although nobody has yet suggested a good reason why it should be so, many tumor virologists maintain that the transformation of cells by both RNA and DNA tumor viruses depends on the cell going through at least one round of division, a quantal mitosis, after it has been infected" [4].

DNA Synthesis, Recombination, and Viral Genome Integration

In attempting to understand the significance of cellular DNA synthesis, it has been suggested that integration might somehow be linked to the phenomenon.

Dulbecco [31], in reviewing work on polyoma virus and SV40, suggested that the enzymes necessary for DNA replication could also, under restricted circumstances, bring about recombination, and this might account for the occurrence of integration. In this view, integration would be a side effect of the virus's ability to promote host DNA synthesis. Conversely, Kit et al. [70] argued that a requirement for the switching on of cellular DNA synthesis during transformation might be expected if recombination, assumed to be required for integration, were coupled to host cell DNA synthesis. Kit et al. referred to Boon and Zinder's [13, 14] break-and-copy recombination model, which invokes replication in the recombination process. Boon and Zinder put forward this hypothesis, however, to account for their experimental results with the single-stranded

DNA phage *f1* of *E. coli*, and it seems unlikely that the model is applicable to eukaryotes.

Kit et al. [70] stated that "recombination through chromosome breakage and reunion does not involve DNA synthesis, so that transformation might also occur without concomitant cellular DNA synthesis." Although it is true, as in Holliday's model [59], that recombination by breakage and reunion need not involve DNA synthesis, nevertheless one of the peculiarities of recombination in eukaryotes is its occurrence after the chromosomes have duplicated (Fig. 3a). This was first established about 50 years ago for crossing-over at meiosis in *Drosophila melanogaster* and subsequently was shown to apply to all eukaryotes tested (reviews: refs. [37, 132]). The most straightforward evidence has come from organisms

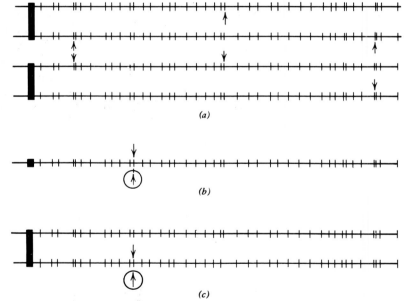

Figure 3. A representation of one of the chromosomes of a higher organism, illustrating (a) the process of crossing-over in eukaryotes, whether meiotic or somatic; (b) and (c) the proposed mechanism of integration of a viral genome into a eukaryotic chromosome (or chromatid): either (b)—before replication, or (c)—after replication. The black square or rectangle represents the centromere, the horizontal lines indicate chromosomes (or chromatids), and the circles represent viral genomes. The short vertical lines indicate the points at which recombination can be initiated; it is suggested that these are the replicon junctions or, in the viral genome, the replication origin. Arrows show where crossing-over might be initiated in a particular cell.

in which the four products of meiosis are retained together, as in fungi belonging to the Ascomycetes. For example, in the fungus *Sordaria brevicollis* linked mutants are known, one giving yellowish brown (buff-colored) and the other yellow spores instead of the normal black spores. When these mutants are crossed, most of the asci will have four buff and four yellow spores. Whenever a crossover occurs between the buff and yellow spore color genes, however, an ascus results in which the four pairs of spores (corresponding to the products of meiosis) are buff, yellow, colorless (double mutant), and black (wild-type), respectively (see Fig. 4). The regular occurrence of a spore-pair of each parental genotype in crossover asci provides direct evidence that crossing-over occurs between chromatids. Stern [107], in his classic investigation of somatic recombination in *Drosophila*, showed that here also the exchanges took place at the four-chromatid stage, that is after chromosome duplication. The evidence came from an analysis of the types and frequencies of twin and single spots on the bodies of flies heterozygous for genes affecting body color or hair morphology. Pontecorvo and associates [86] found that mitotic recombination in the fungus *Aspergillus nidulans* likewise occurred after chromosome replication (review: ref. [37]), and German [43] found that mitotic recombination in human cells in culture is similar (see below).

Chromosome Replication, Crossing-over, and Replicons

Other features of chromosome structure and organization enable an explanation to be offered for this relationship between chromosome replication and crossing-over. Chromosomes of eukaryotes have been shown to replicate in a large number of discrete segments or replicons. This was first demonstrated by Taylor [115]. The replicons are believed to correspond to the chromomeres, on the basis primarily of studies of replication in the giant salivary gland chromosomes of *Diptera* [84]. There is considerable evidence, notably that of Judd et al. ([64], review: ref. [119]), that the chromomeres correspond to individual genes, although the amount of DNA in a chromomere is much greater than that required for one gene. The excess of DNA can be explained either by Beermann's packaging hypothesis [8] or by Callan's master-and-slaves hypothesis [18]. These alternatives have been discussed by Hess and Meyer [55].

Recombination in eukaryotes seems always to be polarized. This phenomenon was discovered in fungi (reviews: refs. [37, 134]) and has now

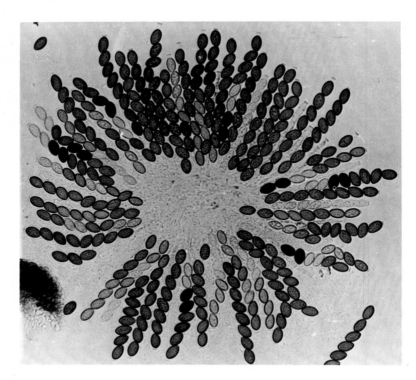

Figure 4. A cluster of asci from a cross between buff and yellow spore color mutants of the fungus *Sordaria brevicollis*. Most of the asci have four buff and four yellow spores, both colors appearing grey in the photograph. The buff spores, however, often have a less mottled appearance than the yellow ones, and they are almost colorless in immature asci. Whenever a crossover has occurred in the interval of the chromosome between the loci of the two spore-color genes, it has resulted in asci containing a pair of wild-type (black) and a pair of double mutant (colorless) spores. The regular occurrence of two buff and two yellow spores in these crossover asci provides direct evidence that crossing-over takes place after the chromosomes have replicated and that it involves only one of the two chromatids from each parent. The two lowest spores (buff) of the crossover ascus at 4 o'clock in the photograph are separated from the other six and lie beyond the lowest spore of an adjoining ascus. The top three spores of the latter ascus lie on the far side of a third ascus. None of the other asci shown in the photograph is entangled with another.

been demonstrated in *Drosophila* [20]. Hastings and I [53] suggested that polarity arose through the existence of fixed points in the chromosome where the recombination process is initiated. This hypothesis has received considerable support from the discovery by Catcheside and associates ([3, 21], review: ref. [134]) of dominant repressors of recombination in specific regions of the chromosomes of *Neurospora crassa*. These repressors affect the polarity as well as the frequency of recombination in the particular regions. The data are in keeping with the hypothesis that the repressors block the opening of specific initiation points for recombination.

There is evidence that the opening points for recombination are situated between the genes and not within them, because discontinuity in the polarity gradient is known between neighboring genes but not within them ([82], review: ref. [134]). Furthermore, the intragenic polarity gradient is often steep—for example, in the data (illustrated in ref. [134]) of Pees [83] for the *lysine-51* gene of *Aspergillus nidulans*. Such steepness implies that the opening points for recombination are closely spaced, possibly at one or both ends of every gene.

In the light of the evidence that the eukaryotic chromosome consists of a tandem array of replicons, that the replicon corresponds to the chromomere, and the chromomere to the gene, I suggested that the fixed intergenic opening points for recombination are at replicon junctions [131]. Recombination in eukaryotes might then be initiated through a failure to join nucleotide chains after the previous replication. This would explain why crossing-over occurs at the four-chromatid stage. The idea that the premeiotic DNA replication is incomplete is supported by the discovery by H. Stern and associates [60, 63, 108] that in *Lilium* anthers a rapidly annealing guanine-cytosine-rich DNA fraction amounting to about 0.3% of the total DNA is synthesized at zygotene and represents a delayed replication. Synthesis of this fraction, which is distributed at numerous points along the chromosomes, is not delayed before mitosis.

Viral Genome Integration and The Open-Replicon Hypothesis of Carcinogenesis

If recombination in eukaryotes is initiated at replicon junctions that have not been closed after DNA replication, integration of a viral genome is likely also to require an open replicon junction to initiate the crossover process. This would explain the need for cellular DNA synthesis if integration is to occur, thus accounting for the association between host cell transformation and the proviral state.

The *open-replicon hypothesis of carcinogenesis* is based, therefore, on the following assumptions: (1) integration of a viral genome has strong selective advantages, (2) integration requires the initiation of host DNA synthesis in order to provide the necessary open replicon junctions, and (3) transformation is a consequence of the triggering of host DNA replication, with the breakdown of replication control that this entails. According to this hypothesis, carcinogenesis is a subsidiary effect of integration and need not be regarded as conferring any selective advantages to the virus. This is in keeping with the conclusions of Subak-Sharpe [112], already discussed, and also with those of Klein [71], who emphasized that tumor induction is not part of viral strategy, conveys no survival advantage on the virus, and must be regarded as an incidental consequence of other features of its life history.

It seems possible that a hypothesis similar to that proposed here for carcinogenesis might also explain cell differentiation. Bischoff and Holtzer [11] drew attention to the need for a critical mitosis—or quantal mitosis, as they called it—for differentiation of many tissues. An *open-replicon hypothesis of differentiation*, according to which an episome with regulatory functions has to be integrated into a host chromosome to activate the genes involved in differentiation, would explain the need for mitosis if, as already suggested, episomal integration requires unsealed replicon ends. The idea of controlling episomes is of course not new [26, 78, 102]. In cell differentiation, unlike carcinogenesis, DNA synthesis is often suppressed following the critical mitosis.

Bischoff and Holtzer [12] discovered that 5-bromodeoxyuridine, which is incorporated into DNA in place of thymidine, inhibits the synthesis of tissue-specific components in differentiating cells but does not inhibit the synthesis of molecules required for cell growth and division. Bischoff and Holtzer found this effect paradoxical, since they saw no reason to believe that the activity of DNA is different in controlling different cellular processes. Their discovery is understandable, however, by the open-replicon hypothesis if the bromodeoxyuridine inhibits the integration of a plasmid required to bring about differentiation.

If the integration of episomes into chromosomes forms part of the controlling mechanism of cell differentiation, this behavior might explain how viruses have evolved. They could have originated from controlling episomes by the acquisition of a more independent existence. Their ability to integrate into a host chromosome would then be an ancestral character.

In bacteria, following integration of a phage genome, new characters may appear (see review: ref. [106]), for example, a new antigen on the surface of *Salmonella anatum* cells when infected by phage ε-15; and the diphtheria toxin when a specific prophage in *Corynebacterium diphtheriae* gives rise to viral particles and destroys the bacterial cells. With phages, however, there is no evidence of host cell transformation in the sense of uncontrolled bacterial cell division triggered by the virus. This is in keeping with the open-replicon hypothesis, because in bacteria there is only a single replicon to the genome and no evidence that recombination is polarized or that it occurs only after replication in the way it does in eukaryotes (see review: ref. [*133*]).

When the DNA is organized in a linear array of replicons, the initial step in replication is likely to be the cutting of one nucleotide chain, presumably by an endonuclease, at each replication origin. Such nicking of the duplex DNA is also required at each replication terminus, to allow rotation of the double helical molecule about the helix axis to permit strand separation. If recombination is initiated only at replicon ends, there seem to be two possibilities: the process might occur after the initial nicking but before any DNA synthesis has occurred (Fig. 3*b*), or it might occur after replication has taken place but before the strands are sealed, presumably by a ligase, across the replicon junctions (Fig. 3*c*). The latter alternative would fit the eukaryotic recombination data. The former is ruled out for reasons already given (p. 58), and because the homologous chromosomes do not associate until after replication has occurred [*91*]. With viral integration, however, no such restriction would seem to apply.

Doerfler [*29*] discovered that integration of the DNA of adenovirus type 12 into the *Mesocricetus* chromosomes was not affected by D-arabinosyl cytosine (cytosine arabinoside), which is an inhibitor of DNA synthesis. He concluded that replication of cellular DNA was not required for integration. Similarly, Hirai, Lehman, and Defendi [*57*] found that SV40 DNA could be integrated into *Cricetulus* DNA in the presence of cytosine arabinoside. They concluded that although integration occurs during the induction of cellular DNA synthesis, it is not dependent on it. They argued, however, that although cellular or viral DNA replication is not required for integration, nevertheless, some intermediary changes of the cellular DNA may be necessary, such as availability or activation of an endonuclease. The open-replicon hypothesis predicts that at least the initial nicking of nucleotide chains at replicon ends (and presumably at the replication origin in the virus) is required for integration. This possibility does not seem to have been ruled out by these experiments. Indeed the good chronological as well as quantitative correlation which they found between integration and induction of cellular DNA synthesis is in agreement with the hypothesis. Later experiments [*56*] showed that integration of SV40 DNA into the DNA of permissive *Cercopithecus* cells was likewise not inhibited by cytosine arabinoside.

To test the open-replicon hypothesis would seem to require an inhibitor, not of DNA synthesis, but of the action of the host endonuclease or endonucleases presumed to be involved in the initiation of replication. The hypothesis predicts that in the absence of such endonucleolytic activity, integration would not occur. Conversely, an inhibitor of the ligase required to join phosphodiester axes after replication should promote integration.

WHY ARE IONIZING RADIATIONS, MANY CHEMICALS, AND CERTAIN GENETIC DEFECTS CARCINOGENIC?

If it is true that viral integration into a host chromosome has selective value for both partners, it is to be expected that host chromosomes will contain proviruses permanently integrated. This might come about through mutation in proviral genes concerned with excision of the viral genome from the host chromosome, with the result that detachment became impossible. Another possibility, discussed earlier in connection with bacterial proviruses, is mutation in proviral genes controlling replication of the viral nucleic acid.

Mutation at Virus-suppressor Loci

Huebner and Todaro [62] suggested that the cells of vertebrates contained DNA transcripts of the genomes of RNA tumor viruses and that this DNA was inherited through the germ line. Considerable support for this hypothesis has since been obtained [122]. The evidence includes the finding of the group-specific antigen characteristic of RNA tumor viruses in one or more of the tissues in embryo mice of all strains tested, the induction of infectious virus from normal mice by X-rays or chemicals such as 5-bromodeoxyuridine [92], and the demonstration of the widespread presence in *Gallus* of DNA complementary to the RNA of tumor viruses [6, 90].

Todaro and Huebner [122] argued that radiation and chemical carcinogens act by destroying the repressors of proviral genes that induce tumor formation. This idea is supported by the discovery that cancer induction by radiation or chemicals is lessened or even inhibited by interferon [76, 96]. The mechanism of action of interferon in preventing tumor formation is not understood, but it probably interferes with the translation of viral RNA into protein or with the functioning of viral proteins. The interferon might therefore inhibit the activity of a tumor virus released from a host chromosome by the action of the carcinogen, or it might suppress the action of a proviral gene activated by the carcinogen. Furthermore, experiments, for example, by Rhim, Creasy, and Huebner [88], in which mammalian cells are treated with both a cancer virus and a chemical carcinogen, have shown a much higher frequency of transformation by the two agents together than by either alone. This can be interpreted as activation by the chemical carcinogen of the viral genes responsible for transformation, following integration of the viral genome. A rather similar explanation might account for the discovery by Watkins [127] that the number of cells producing infective virus, following fusion of *Mus* cells transformed by SV40 with normal *Cercopithecus* cells, is greatly increased by treatment of the transformed cells with iododeoxyuridine or 8-azaguanine before fusion. He favored the idea that these base analogs, through substitution for normal bases at the DNA or RNA level, inactivated a repressor and so promoted viral release. It might then be expected, however, that RNA tumor viruses would also be released, but there was no evidence for this.

The idea that ionizing radiations and chemical carcinogens cause cancer by activating proviruses is in agreement with the hypothesis discussed earlier of Hitotsumachi et al. [58] that transformation and its reversion depend on a balance between proviral genes that bring about the cell transformation and host genes capable of suppressing the viral ones. Physical and chemical carcinogens, many—perhaps all—of which are mutagens, would then act by causing mutation of the suppressor loci and so inactivating them. There is now considerable support for the belief that at least some chemical cancer-inducing agents are carcinogenic because they are able to cause mutation [22, 61].

An alternative hypothesis to the suppressor-inactivation theory is to postulate that physical and chemical carcinogens induce cancer through mutation of cellular genes that cause loss of control of cell division. The difficulty with this hypothesis, as Sachs and associates [10, 15] pointed out, is that the frequency of induction of transformation (for example, in their experiments, by treating *Mesocricetus* cells with 3-methylcholanthrene or with X-rays) is much greater than that expected from random mutation. The high frequency would be accounted for, however, if mammalian chromosomes contain numerous suppressor loci that keep proviral genes inactive and if mutation of any of these suppressor genes is liable to tip the balance, allowing a transforming gene in one of the proviruses to become active.

Known Genetic Defects, Cellular Transformation, and Increased Risk of Cancer

Patients with primary immunodeficiency diseases have a high cancer risk, but Kersey et al. [69] found that cells of all but one of 15 such patients had normal susceptibility to transformation by SV40. This is in contrast to those with Fanconi's anemia [121] or Down's syndrome [123], who also carry a high risk of malignancy and whose cells are more readily transformed by SV40. Fanconi's anemia is inherited as an autosomal recessive, but heterozygotes as well as homozygotes seem to have an increased risk of developing cancer [44, 114]; thus it is of particular interest that cells from heterozygotes were found to be 6 to 8 times as susceptible to transformation by SV40 as normal cells, and homozygotes 12 to 24 times as susceptible [121]. The frequent occurrence of malignancy in patients with primary immunodeficiency is likely to be due to abnormalities of the immunologic surveillance mechanism, whereas in Fanconi's anemia and Down's syndrome there seems to be an intrinsically greater likelihood of transformation occurring. Fanconi's anemia is associated with chromatid breaks or gaps in cultured cells [44]. The breaks might directly favor viral integration, or the loss of chromosome segments caused by the breaks might provide a situation comparable to that found by Harris [51] in which genetic loss promotes cancerous growth. As already discussed, such loss could cause cancer through loss of hypothetical suppressor loci of proviral genes for transformation. The high transformation frequency and cancer risk in Down's syndrome can also be explained on the suppressor-inactivation hypothesis if trisomy for chromosome No. 21 alters the balance between suppressor loci and proviral transformation genes in favor of the latter.

Unlike Fanconi's anemia, cells from patients with another rare genetic defect, xeroderma pigmentosum, show no abnormally high transformation frequency with SV40 [1]. This disease seems to be caused by a deficiency in a UV-specific endonuclease presumed to initiate the excision of pyrimidine dimers from the DNA following exposure to UV light [100]. As expected with this defect, xeroderma pigmentosum patients have an unusually high sensitivity to sunlight. There is also a greatly increased susceptibility to malignant skin tumors [44], which would be accounted for if the deficiency in excision repair of DNA implies greater dependence on a recombinational repair pathway such as Rupp and Howard-Flanders [93, 94] have demonstrated in *E. coli*. Such repair is known to be associated with mutation [81, 137]. The occurrence of reciprocal translocations in chromosome preparations derived from cultures of skin cells from a xeroderma pigmentosum patient [44] might be the result of defective recombinational repair. On the other hand, Lehmann [74] showed that the gaps formed opposite pyrimidine dimers, when the DNA of UV-irradiated *Mus* cells underwent replication, were filled with newly synthesized DNA and not with old DNA such as a recombinational repair process would insert. Bridges and Huckle [16] found that the mutation rate of cultured cells of *Cricetulus griseus* to 8-azaguanine resistance after ultraviolet and X-irradiation was much higher than in *E. coli*. A possible source of the ultraviolet-induced mutations is a mistake-prone postreplicational repair of gaps formed opposite pyrimidine dimers. If proviral genes for transformation are held in check by host suppressor loci, mutations in these suppressor genes resulting from such postreplicational repair would be expected to cause skin tumors.

German [43–45] found that quadriradial chromosome configurations occur frequently at mitosis in cell cultures derived from patients with Bloom's syndrome, another rare genetic defect. This disease is associated, among other features, with retarded growth, a disturbed immunity system, and a high cancer risk [44]. The quadriradial chromosomes evidently arise through the occurrence of somatic crossing-over between homologous chromosomes. As with other examples of mitotic crossing-over already discussed, the exchanges occur after replication and involve only two of the four chromatids. With Bloom's syndrome the crossovers seem to be largely restricted to the centromere regions of particular chromosomes. This suggests derepression of specific opening points for recombination, a situation that might be expected to favor viral integration at these points. This could conceivably contribute to the high cancer risk of these patients, though the immunoglobulin abnormality is likely to be an important factor.

Chromosomal aberrations of many kinds occur in cultured cells from patients with Fanconi's anemia or Bloom's syndrome. These aberrations are not necessarily directly related to the high cancer risk but, as German [44] suggested, may provide a predisposing background. Chromosomal aberrations can be induced in many ways, including treatment with carcinogenic chemicals [67], with viruses which cause tumors, and with those which do not [23]. It is particularly interesting that McDougall et al. [79], using in situ hybridization of ^{3}H-labeled adenovirus type 12 DNA with human chromosomes, found that the sites of chromosomal damage by the virus were not necessarily those of major viral DNA accumulation.

In conclusion, the data on carcinogenesis by physical and chemical agents, or in association with genetic defects, seem to be in agreement with the hypothesis that cancer induction is primarily the result of inactivation of suppressor loci that normally block the action of proviral genes inducing transformation. Other explanations do not seem to fit the data so well.

SUMMARY: THE OPEN-REPLICON HYPOTHESIS OF CARCINOGENESIS

Evidence is presented for the open-replicon hypothesis of carcinogenesis, the main features of which are the following:

1. Integration of a viral genome into a host chromosome has strong selective advantages to both partners. The primary factor of selective value is immunity of the host to the particular virus. This immunity will not be effective without integration. Survival of the host, which the immunity promotes, has selective value for the virus.

2. Recombination in eukaryotes is postreplicational, probably because it is initiated at the ends of the replicons before rejoining has occurred after DNA synthesis. Integration of a viral genome requires the initiation of host DNA synthesis to provide the necessary open replicon junctions. Viral carcinogenesis is a consequence of this triggering of host DNA synthesis, with the breakdown of replication control that this entails.

3. Physical and chemical carcinogens inactivate, by mutation, host genes whose products normally keep in check the proviral genes that can initiate host DNA synthesis. Genetic defects associated with a high cancer risk may affect the immunologic responses of the host, or they may alter the balance between proviral genes and their suppressors, either directly or through mutation.

Acknowledgements.—I am indebted to Professor J. H. Subak-Sharpe for his critical reading of this article in manuscript and for most helpful discussion. I am grateful to Dr. J. F. Williams for permission to refer to his unpublished work. I thank Dr. B. A. Bridges for information about mammalian DNA repair.

LITERATURE CITED

1. AARONSON SA, LYTLE CD: Decreased host cell reactivation of irradiated SV40 virus in xeroderma pigmentosum. *Nature* 228:359–361, 1970.
2. AMBROSE KR, ANDERSON NG, COGGIN JH: Cytostatic antibody and SV40 tumour immunity in hamsters. *Nature* 233:321–324, 1971.
3. ANGEL T, AUSTIN B, CATCHESIDE DG: Regulation of recombination at the *his*-3 locus in *Neurospora crassa*. *Aust J Biol Sci* 23:1229–1240, 1970.
4. ANONYMOUS: Mitosis required. Comment on *Tumour Virology* by Cell Biology Correspondent. *Nature* 235:13–14, 1972.
5. BALTIMORE D: RNA-dependent DNA polymerase in virions of RNA tumour viruses. *Nature* 226:1209–1211, 1970.
6. BALUDA MA: Widespread presence, in chickens, of DNA complementary to the

RNA genome of avian leukosis viruses. *Proc Nat Acad Sci (US)* 69:576–580, 1972.

7. BASILICO C, WANG R: Susceptibility to superinfection of hybrids between polyoma "transformed" BHK and "normal" 3T3 cells. *Nature New Biol* 230: 105–107, 1971.
8. BEERMANN W: Gene action at the level of the chromosome. *In* Heritage from Mendel (Brink RA, ed) Madison, University of Wisconsin Press, 179–201, 1967.
9. BENJAMIN TL, BURGER MM: Absence of a cell membrane alteration function in non-transforming mutants of polyoma virus. *Proc Nat Acad Sci (US)* 67:929–934, 1970.
10. BERWALD Y, SACHS L: *In vitro* transformation of normal cells to tumor cells by carcinogenic hydrocarbons. *J Nat Cancer Inst* 35:641–661, 1965.
11. BISCHOFF R, HOLTZER H: Mitosis and the processes of differentiation of myogenic cells in vitro. *J Cell Biol* 41:188–200, 1969.
12. BISCHOFF R, HOLTZER H: Inhibition of myoblast fusion after one round of DNA synthesis in 5-bromodeoxyuridine. *J Cell Biol* 44:134–150, 1970.
13. BOON T, ZINDER ND: A mechanism for genetic recombination generating one parent and one recombinant. *Proc Nat Acad Sci (US)* 64:573–577, 1969.
14. BOON T, ZINDER ND: Genotypes produced by individual recombination events involving bacteriophage f1, *J Mol Biol* 58:133–151, 1971.
15. BOREK C, SACHS L: *In vitro* cell transformation by X-irradiation. *Nature* 210: 276–278, 1966.
16. BRIDGES BA, HUCKLE J: Mutagenesis of cultured mammalian cells by X-radiation and ultraviolet light. *Mutat Res* 10:141–151, 1970.
17. BUTEL JS, TEVETHIA SS, MELNICK JL: Oncogenicity and cell transformation by papovavirus SV40: the role of the viral genome. *Adv Cancer Res* 15:1–55, 1972.
18. CALLAN HG: On the organization of genetic units in chromosomes. *J Cell Sci* 2:1–7, 1967.
19. CAMPBELL AM: Episomes. *Adv Genet* 11:101–145, 1962.
20. CARLSON PS: A genetic analysis of the *rudimentary* locus of *Drosophila melanogaster*. *Genet Res* 17:53–81, 1971.
21. CATCHESIDE DG, AUSTIN B: Common regulation of recombination at the *amination-1* and *histidine-2* loci in *Neurospora crassa*. *Aust J Biol Sci* 24:107–115, 1971.
22. COOKSON MJ, SIMS P, GROVER PL: Mutagenicity of epoxides of polycyclic hydrocarbons correlates with carcinogenicity of parent hydrocarbons. *Nature New Biol* 234:186–187, 1971.
23. COOPER JEK, YOHN DS, STICH HF: Viruses and mammalian chromosomes. X. Comparative studies of the chromosome damage induced by human and simian adenoviruses. *Exp Cell Res* 53:225–240, 1968.
24. CRAWFORD LV: Nucleic acids of tumour viruses. *Adv Virus Res* 14:89–152, 1969.
25. DAVIS RW, PARKINSON JS: Deletion mutants of bacteriophage lambda. III. Physical structure of *att*$^\phi$. *J Mol Biol* 56:403–423, 1971.

26. DAWSON GWP, SMITH-KEARY PF: Episomic control of mutation in *Salmonella typhimurium*. *Heredity* 18:1–20, 1963.
27. DEFENDI V, EPHRUSSI B, KOPROWSKI H, YOSHIDA MC: Properties of hybrids between polyoma-transformed and normal mouse cells. *Proc Nat Acad Sci (US)* 57:299–305, 1967.
28. DOERFLER W: The fate of the DNA of adenovirus type 12 in baby hamster kidney cells. *Proc Nat Acad Sci (US)* 60:636–643, 1968.
29. DOERFLER W: Integration of the deoxyribonucleic acid of adenovirus type 12 into the deoxyribonucleic acid of baby hamster kidney cells. *J Virol* 6:652–666, 1970.
30. DULBECCO R: Cell transformation by the small DNA-containing viruses. *Harvey Lect* 63:33–46, 1969.
31. DULBECCO R: Cell transformation by viruses. *Science* 166:962–968, 1969.
32. DULBECCO R, ECKHART W: Temperature-dependent properties of cells transformed by a thermosensitive mutant of polyoma virus. *Proc Nat Acad Sci (US)* 67:1775–1781, 1970.
33. DULBECCO R, JOHNSON T: Interferon-sensitivity of the enhanced incorporation of thymidine into cellular DNA induced by polyoma virus. *Virology* 42:368–374, 1970.
34. ECHOLS H: Lysogeny: viral repression and site-specific recombination. *Ann Rev Biochem* 40:827–854, 1971.
35. ECKHART W: Polyoma gene functions required for cell transformation. *In* Strategy of the Viral Genome (Wolstenholme GEW, O'Connor M, eds), Ciba Foundation Symposium, Edinburgh and London, Churchill Livingstone, 267–274, 1971.
36. ECKHART W, DULBECCO R, BURGER MM: Temperature-dependent surface changes in cells infected or transformed by a thermosensitive mutant of polyoma virus. *Proc Nat Acad Sci (US)* 68:238–286, 1971.
37. EMERSON S: Linkage and recombination at the chromosome level. *In* Genetic Organization (Caspari EW, Ravin AW, eds), vol 1, New York and London, Academic Press, 267–360, 1969.
38. FENNER F: The Biology of Animal Viruses. New York and London, Academic Press, 845 p, 1968.
39. FINCHAM JRS, DAY PR: Fungal Genetics, 2nd ed, Oxford, Blackwell, 326 p, 1965.
40. FLOR HH: The complementary genic systems in flax and flax rust. *Adv Genet* 8:29–54, 1956.
41. FRIED M: Characterization of a temperature-sensitive mutant of polyoma virus. *Virology* 40:605–617, 1970.
42. GELB LD, KOHNE DE, MARTIN MA: Quantitation of simian virus 40 sequences in African green monkey, mouse and virus-transformed cell genomes. *J Mol Biol* 57:129–145, 1971.
43. GERMAN J: Cytological evidence for crossing-over in vitro in human lymphoid cells. *Science* 144:298–301, 1964.

44. GERMAN J: Genes which increase chromosomal instability in somatic cells and predispose to cancer. *Progr Med Genet* 8:61–101, 1972.
45. GERMAN J, ARCHIBALD R, BLOOM D: Chromosomal breakage in a rare and probably genetically determined syndrome of man. *Science* 148:506–507, 1965.
46. GERSHON D, HAUSEN P, SACHS L, WINOCOUR E: On the mechanism of polyoma virus-induced synthesis of cellular DNA. *Proc Nat Acad Sci (US)* 54:1584–1592, 1965.
47. GOOD RA: Relations between immunity and malignancy. *Proc Nat Acad Sci (US)* 69:1026–1032, 1972.
48. GREEN M: Oncogenic viruses. *Ann Rev Biochem* 39:701–756, 1970.
49. GREEN M, ROKUTANDA H, ROKUTANDA M: Virus specific RNA in cells transformed by RNA tumour viruses. *Nature New Biol* 230:229–232, 1971.
50. HAHN EC, SAUER G: Initial stage of transformation of permissive cells by simian virus 40: development of resistance to productive infection. *J Virol* 8:7–16, 1971.
51. HARRIS H: Cell fusion and the analysis of malignancy. *Proc Roy Soc (London) B* 179:1–20, 1971.
52. HARRIS H, MILLER OJ, KLEIN G, WORST P, TACHIBANA T: Suppression of malignancy by cell fusion. *Nature* 223:363–368, 1969.
53. HASTINGS PJ, WHITEHOUSE HLK: A polaron model of genetic recombination by the formation of hybrid deoxyribonucleic acid. *Nature* 201:1052–1054, 1964.
54. HAYES W: The Genetics of Bacteria and their Viruses, 2nd ed, Oxford, Blackwell, 925 p, 1968.
55. HESS O, MEYER GF: Genetic activities of the Y chromosome in *Drosophila* during spermatogenesis. *Adv Genet* 14:171–223, 1968.
56. HIRAI K, DEFENDI V: Integration of simian virus 40 deoxyribonucleic acid into the deoxyribonucleic acid of permissive monkey kidney cells. *J Virol* 9:705–707, 1972.
57. HIRAI K, LEHMAN J, DEFENDI V: Integration of simian virus 40 deoxyribonucleic acid into the deoxyribonucleic acid of primary infected Chinese hamster cells. *J Virol* 8:705–715, 1971.
58. HITOTSUMACHI S, RABINOWITZ Z, SACHS L: Chromosomal control of reversion in transformed cells. *Nature* 231:511–514, 1971.
59. HOLLIDAY R: A mechanism for gene conversion in fungi. *Genet Res* 5:282–304, 1964.
60. HOTTA Y, STERN H: Analysis of DNA synthesis during meiotic prophase in *Lilium*. *J Mol Biol* 55:337–355, 1971.
61. HUBERMAN E, ASPIRAS L, HEIDELBERGER C, GROVER PL, SIMS P: Mutagenicity to mammalian cells of epoxides and other derivatives of polycyclic hydrocarbons. *Proc Nat Acad Sci (US)* 68:3195–3199, 1971.
62. HUEBNER RJ, TODARO GJ: Oncogenes of RNA tumor viruses as determinants of cancer. *Proc Nat Acad Sci (US)* 64:1087–1094, 1969.
63. ITO M, HOTTA Y, STERN H: Studies of meiosis *in vitro*. II. Effect of inhibiting

DNA synthesis during meiotic prophase on chromosome structure and behavior. *Dev Biol* 16:54–77, 1967.
64. JUDD BH, SHEN MW, KAUFMAN TC: The anatomy and function of a segment of the X chromosome of *Drosophila melanogaster*. *Genetics* 71:139–156, 1972.
65. KÁRA J, DVOŘÁK M, ČERNÁ H: Presence of viral RNA-instructed DNA polymerase in the oncogenic subviral particles (virosomes) isolated from the mitochondria of Rous sarcoma cells. *FEBS Lett* 25:33–37, 1972.
66. KÁRA J, MACH O, ČERNÁ H: Replication of Rous sarcoma virus and the biosynthesis of the oncogenic subviral ribonucleoprotein particles ("virosomes") in the mitochondria isolated from Rous sarcoma tissue. *Biochem Biophys Res Commun* 44: 162–170, 1971.
67. KATO R: Chromosome breakage induced by a carcinogenic hydrocarbon in Chinese hamster cells and human leukocytes in vitro. *Hereditas* 59:120–141, 1968.
68. KAWAI S, HANAFUSA H: The effects of reciprocal changes in temperature on the transformed state of cells infected with a Rous sarcoma virus mutant. *Virology* 46:470–479, 1971.
69. KERSEY JH, GATTI RA, GOOD RA, AARONSON SA, TODARO GJ: Susceptibility of cells from patients with primary immunodeficiency diseases to transformation by simian virus 40. *Proc Nat Acad Sci (US)* 69:980–982, 1972.
70. KIT S, DUBBS DR, SOMERS K: Strategy of simian virus 40. *In* Strategy of the Viral Genome (Wolstenholme GEW, O'Connor M, eds), Ciba Foundation Symposium, Edinburgh and London, Churchill Livingstone, 229–265, 1971.
71. KLEIN G: Virus-induced, tumour-associated antigens. *In* Strategy of the Viral Genome (Wolstenholme GEW, O'Connor M, eds), Ciba Foundation Symposium, Edinburgh and London, Churchill Livingstone, 295–315, 1971.
72. KLEIN G: Herpesviruses and oncogenesis. *Proc Nat Acad Sci (US)* 69:1056–1064, 1972.
73. LEHMAN JM, DEFENDI V: Changes in deoxyribonucleic acid synthesis regulation in Chinese hamster cells infected with simian virus 40: *J Virol* 6:738–749, 1970.
74. LEHMANN AR: Postreplication repair of DNA in ultraviolet-irradiated mammalian cells. *J Mol Biol* 66:319–373, 1972.
75. LEVINE AJ: Induction of mitochondrial DNA synthesis in monkey cells infected by simian virus 40 and (or) treated with calf serum. *Proc Nat Acad Sci (US)* 68:717–720, 1971.
76. LIEBERMAN M, MERIGAN TC, KAPLAN HS: Inhibition of radiogenic lymphoma development in mice by interferon. *Proc Soc Exp Biol Med* 138:575–578, 1971.
77. LINDBERG U, DARNELL JE: SV40-specific RNA in the nucleus and polyribosomes of transformed cells. *Proc Nat Acad Sci (US)* 65:1089–1096, 1970.
78. MCCLINTOCK B: The control of gene action in maize. *Brookhaven Symp Biol* 18:162–184, 1965.
79. MCDOUGALL JK, DUNN AR, JONES KW: *In situ* hybridization of adenovirus RNA and DNA. *Nature* 236:346–348, 1972.

80. MARTIN GS: Rous sarcoma virus: a function required for the maintenance of the transformed state. *Nature* 227:1021–1023, 1970.
81. MIURA A, TOMIZAWA J: Studies on radiation-sensitive mutants of *E. coli*. III. Participation of the Rec system in induction of mutation by ultraviolet irradiation. *Mol Gen Genet* 103:1–10, 1968.
82. MURRAY NE: Recombination events that span sites within neighbouring gene loci of *Neurospora*. *Genet Res* 15:109–121, 1970.
83. PEES E: Genetic fine structure and polarized negative interference at the lys-51 (FL) locus of *Aspergillus nidulans*. *Genetica* 38:275–304, 1967.
84. PELLING C: A replicative and synthetic chromosomal unit—the modern concept of the chromomere. *Proc Roy Soc (London) B* 164:279–289, 1966.
85. POLLACK RE, GREEN H, TODARO GJ: Growth control in cultured cells: selection of sublines with increased sensitivity to contact inhibition and decreased tumor-producing ability. *Proc Nat Acad Sci (US)* 60:126–133, 1968.
86. PONTECORVO G, KÄFER E: Genetic analysis based on mitotic recombination. *Adv Genet* 9:71–104, 1958.
87. RABINOWITZ Z, SACHS L: Reversion of properties in cells transformed by polyoma virus. *Nature* 220:1203–1206, 1968.
88. RHIM JS, CREASY B, HUEBNER RJ: Production of altered cell foci by 3-methylcholanthrene in mouse cells infected with AKR leukemia virus. *Proc Nat Acad Sci (US)* 68:2212–2216, 1971.
89. RICHERT NJ, HARE JD: Distinctive effects of inhibitors of mitochondrial function on Rous sarcoma virus replication and malignant transformation. *Biochem Biophys Res Commun* 46:5–10, 1972.
90. ROSENTHAL PN, ROBINSON HL, ROBINSON WS, HANAFUSA T, HANAFUSA H: DNA in uninfected and virus-infected cells complementary to avian tumour virus RNA. *Proc Nat Acad Sci (US)* 68:2336–2340, 1971.
91. ROSSEN JM, WESTERGAARD M: Studies on the mechanism of crossing over. II. Meiosis and the time of meiotic chromosome replication in the ascomycete *Neottiella rutilans* (Fr.) Dennis. *CR Trav Lab Carlsberg* 35:233–260, 1966.
92. ROWE WP, LOWY DR, TEICH N, HARTLEY JW: Some implications of the activation of murine leukemia virus by halogenated pyrimidines. *Proc Nat Acad Sci (US)* 69:1033–1035, 1972.
93. RUPP WD, HOWARD-FLANDERS P: Discontinuities in the DNA synthesized in the excision-defective strain of *Escherichia coli* following ultraviolet irradiation. *J Mol Biol* 31:291–304, 1968.
94. RUPP WD, WILDE CE, RENO DL, HOWARD-FLANDERS P: Exchanges between DNA strands in ultraviolet-irradiated *Escherichia coli*. *J Mol Biol* 61:25–444, 1971.
95. SACHS L: A theory on the mechanism of carcinogenesis by small deoxyribonucleic acid tumour viruses. *Nature* 207:1272–1274, 1965.
96. SALERNO RA, WHITMIRE CE, GARCIA IM, HUEBNER RJ: Chemical carcinogenesis in mice inhibited by interferon. *Nature New Biol* 239:31–32, 1972.
97. SAMBROOK J, WESTPHAL H, SRINIVASAN PR, DULBECCO R: The integrated state

of viral DNA in SV40-transformed cells. *Proc Nat Acad Sci (US)* 60:1288–1295, 1968.

98. SCHLESINGER RW: Adenoviruses: the nature of the virion and of controlling factors in productive or abortive infection and tumorigenesis. *Adv Virus Res* 14:1–61, 1969.

99. SCOLNICK EM, AARONSON SA, TODARO GJ, PARKS WP: RNA dependent DNA polymerase activity in mammalian cells. *Nature* 229:318–321, 1971.

100. SETLOW RB, REGAN JD, GERMAN J, CARRIER WL: Evidence that xeroderma pigmentosum cells do not perform the first step in the repair of ultraviolet damage to their DNA. *Proc Nat Acad Sci (US)* 64:1035–1041, 1969.

101. SMITH JD: In discussion on advantages of oncogenicity. *In* Strategy of the Viral Genome (Wolstenholme GEW, O'Connor M, eds), Ciba Foundation Symposium, Edinburgh and London, Churchill Livingstone, 380–382, 1971.

102. SMITH-KEARY PF, DAWSON GWP: Episomic suppression of phenotype in *Salmonella*. *Genet Res* 5:269–281, 1964.

103. SPIEGELMAN S: In discussion on advantages of oncogenicity. *In* Strategy of the Viral Genome (Wolstenholme, GEW, O'Connor M, eds), Ciba Foundation Symposium, Edinburgh and London, Churchill Livingstone, 380–382, 1971.

104. SPIEGELMAN S, BURNY A, DAS MR, KEYDAR J, SCHLOM J, TRAVNICEK M, WATSON K: Characterization of the products of RNA-directed DNA polymerases in oncogenic RNA viruses. *Nature* 227:563–567, 1970.

105. SPIEGELMAN S, BURNY A, DAS MR, KEYDAR J, SCHLOM J, TRAVNICEK M, WATSON K: DNA-directed DNA polymerase activity in oncogenic RNA viruses. *Nature* 227:1029–1031, 1970.

106. STENT GS: Molecular Biology of Bacterial Viruses. San Francisco and London, W. H. Freeman, 474 p, 1963.

107. STERN C: Somatic crossing over and segregation in *Drosophila melanogaster*. *Genetics* 21:625–730, 1936.

108. STERN H, HOTTA Y: Chromosome behavior during development of meiotic tissue. *In* The Control of Nuclear Activity (Goldstein L, ed), Englewood Cliffs, N.J., Prentice-Hall, 47–76, 1967.

109. STOKER M, DULBECCO R: Abortive transformation by the Tsa mutant of polyoma virus. *Nature* 223:397–398, 1969.

110. SUBAK-SHARPE JH: Base doublet frequency patterns in the nucleic acid and evolution of viruses. *Brit Med Bull* 23:161–168, 1967.

111. SUBAK-SHARPE JH: In discussion on doublet analysis and the possible origin of viruses. *In* Strategy of the Viral Genome (Wolstenholme GEW, O'Connor M, eds), Ciba Foundation Symposium, Edinburgh and London, Churchill Livingstone, 129–133, 1971.

112. SUBAK-SHARPE JH: In discussion on advantages of oncogenicity. *In* Strategy of the Viral Genome (Wolstenholme GEW, O'Connor M, eds), Ciba Foundation Symposium, Edinburgh and London, Churchill Livingstone, 380–382, 1971.

113. SUBAK-SHARPE JH, BÜRK RR, CRAWFORD LV, MORRISON JM, HAY J, KEIR HM: An approach to evolutionary relationships of mammalian DNA viruses through analysis of the pattern of nearest neighbor base sequences. *Cold Spring Harbor Symp Quant Biol* 31:737–748, 1967.

114. SWIFT M: Fanconi's anaemia in the genetics of neoplasia. *Nature* 230:370–373, 1971.
115. TAYLOR JH: Asynchronous duplication of chromosomes in cultured cells of Chinese hamster. *J Biophys Biochem Cytol* 7:455–463, 1960.
116. TAYLOR-PAPADIMITRIOU J, STOKER M: Effect of interferon on some aspects of transformation by polyoma virus. *Nature New Biol* 230:114–117, 1971.
117. TEMIN HM: Mechanism of cell transformation by RNA tumor viruses. *Ann Rev Microbiol* 25:609–648, 1971.
118. TEMIN HM, MIZUTANI S: RNA-dependent DNA polymerase in virions of Rous sarcoma virus. *Nature* 226:1211–1213, 1970.
119. THOMAS CA: The genetic organization of chromosomes. *Ann Rev Genet* 5:237–256, 1971.
120. TODARO GJ, GREEN H: Cell growth and the initiation of transformation by SV40. *Proc Nat Acad Sci (US)* 55:302–308, 1966.
121. TODARO GJ, GREEN H, SWIFT MR: Susceptibility of human diploid fibroblast strains to transformation by SV40 virus. *Science* 153:1252–1254, 1966.
122. TODARO GJ, HUEBNER RJ: The viral oncogene hypothesis: new evidence. *Proc Nat Acad Sci (US)* 69:1009–1015, 1972.
123. TODARO GJ, MARTIN GM: Increased susceptibility of Down's syndrome fibroblasts to transformation by SV40. *Proc Soc Exp Biol Med* 124:1232–1236, 1967.
124. VESCO C, BASILICO C: Induction of mitochondrial DNA synthesis by polyoma virus. *Nature* 229:336–338, 1971.
125. VOGT M, DULBECCO R: Virus-cell interaction with a tumor-producing virus. *Proc Nat Acad Sci (US)* 46:365–370, 1960.
126. WALL R, DARNELL JE: Presence of cell and virus specific sequences in the same molecules of nuclear RNA from virus transformed cells. *Nature New Biol* 232:73–76, 1971.
127. WATKINS JF: The effects of some metabolic inhibitors on the ability of SV40 virus in transformed cells to be detected by cell fusion. *J Cell Sci* 6:721–737, 1970.
128. WATKINS JF, DULBECCO R: Production of SV40 virus in heterokaryons of transformed and susceptible cells. *Proc Nat Acad Sci (US)* 58:1396–1403, 1967.
129. WATSON JD: Molecular Biology of the Gene, 2nd ed, New York, W. A. Benjamin, 662 p, 1970.
130. WHITEHOUSE HLK: Genetics of parasitic fungi. Lecture to Agricultural Research Council Subject Meeting on Disease Resistance in Crop Plants, Cambridge, England, March 24–26, 1953.
131. WHITEHOUSE HLK: A cycloid model for the chromosome. *J Cell Sci* 2:9–22, 1967.
132. WHITEHOUSE HLK: Towards an Understanding of the Mechanism of Heredity, 3rd ed, London, Arnold, and New York, St. Martin's Press, 528 p, 1973.
133. WHITEHOUSE HLK: The mechanism of genetic recombination. *Biol Rev* 45:265–315, 1970.

134. WHITEHOUSE HLK: Chromosomes and recombination. *Brookhaven Symp Biol* 23:293–325, 1972.
135. WIENER F, KLEIN G, HARRIS H: The analysis of malignancy by cell fusion. III. Hybrids between diploid fibroblasts and other tumour cells. *J Cell Sci* 8: 681–692, 1971.
136. WINOCOUR E: Some aspects of the interaction between polyoma virus and cell DNA. *Adv Virus Res* 14:153–200, 1969.
137. WITKIN EM: The mutability toward ultraviolet light of recombination-deficient strains of *Escherichia coli*. *Mutat Res* 8:9–14, 1969.

ANEUPLOIDY AS A POSSIBLE MEANS EMPLOYED BY MALIGNANT CELLS TO EXPRESS RECESSIVE PHENOTYPES

SUSUMO OHNO

Department of Biology, City of Hope Medical Center, Duarte, California

---

Introduction 77
Viruses as Oncogens 79
Malignancy as a Recessive Trait 81
A Hemizygous State as a Prelude to Malignant Transformation 83
Recessive Traits That Occur in Aneuploid Cells at the Frequency of Dominants . . 84
A Few Conceivable Mechanisms That Might be Responsible for the
 Ploidy-independent Expression of Recessive Phenotypes 87
 Somatic Recombination Due to Crossing-over 88
 Somatic Segregation 88
Direct Effect of Ploidy on Genetic Regulatory Systems 90
Summary 91
Acknowledgements 91
Literature Cited 92

INTRODUCTION

While bacteria and viruses that cause disease are external agents alien to the host body, malignant cells are descendants of one's own somatic cells gone astray. One can hardly call them alien. The idea that malignant cells are mutant clones that have acquired a genetically unbalanced chromosome complement (aneuploidy) is an old and honorable one dating back to Boveri [4] and other great and classical biologists. By 1952 this notion was fully confirmed in the Yoshida ascites tumor of the rat [30] and in various other experimental tumors of the mouse [26].

Human malignant tumors studied in situ have also been shown to be aneuploid. Atkin and his colleagues [1] reported that carcinomata of the uterine cervix are either hyperdiploid or hypertetraploid, while carcinomata of the corpus uteri tend to be strictly hyperdiploid. Cells of acute leukemia are often aneuploid [47], as are other tumors of diverse types such as the 52 studied by Makino et al. [32].

Meanwhile, evidence has accumulated that chromosomal instability, either inherited or induced, makes normal somatic cells more prone to malignant transformation. Chromosome breaks occur at higher frequency in humans with two particular recessively inherited disorders: Bloom's syndrome and Fanconi's anemia. Such individuals are exceptionally prone to develop malignant tumors [48], as are individuals affected by a third disorder, the Louis-Bar syndrome, which is also associated with a high frequency of chromosome breaks [14, 19]. In individuals who are homozygous for the xeroderma pigmentosum mutation, the skin is severely affected by exposure to sun, and skin cancers (sometimes of multiple types) occur and often cause death before age 30 [12]. The defect lies in the repair process of DNA damaged by ultraviolet rays [5]. It appears that a mutation affects the gene locus that specifies the UV-specific endonuclease, so that abnormal thymine-thymine dimers in a DNA strand formed by ultraviolet cannot be removed and replaced by the proper bases [50].

A careful study was made in Britain of a large group of patients with ankylosing spondylitis treated by X-radiation. About six years after treatment, the incidence of chronic granulocytic leukemia in this group increased to nearly 13 times that of the general population [7, 8]. There is a great deal of other evidence that chronic exposure to chromosome-damaging agents causes increased incidence of malignant tumors.

It is tempting to be satisfied with a sweeping statement such as "genetic imbalance (aneuploidy) is either the prerequisite or the cause of malignancy," inasmuch as conditions that enhance production of aneuploid somatic cells appear also to predispose to the development of malignant tumors, and since most, if not all, malignant cells are aneuploid. The trouble with such a statement is that neither the reader nor the maker of the statement has any idea of what it means precisely.

Normal somatic cells are obedient servants that slave for the good of the body as a whole. Just as individual members of the species are mortal, so are the somatic cells that constitute the body of an individual programmed to have a finite life span [18]. Malignant cells, on the other hand, are rebellious and behave as a population of unicellular organisms with a capacity for self-perpetuation. In 1951 Gey [13] isolated a cell line from an endocarcinoma of the uterine cervix of a 31-year-old woman. Descendants of this HeLa cell line are thriving still, in countless tissue culture laboratories of the world. The Ehrlich ascites tumor that originated from a solid tumor that arose in a mouse about 1938 [17] is kept multiplying either in vitro or by mouse-to-mouse transplantation in many laboratories. While a 2-year-old mouse is a very old mouse indeed, this 34-year-old tumor cell population is in splendid shape.

How does a malignant cell population perpetuate itself without benefit of a sexual cycle? The adaptability of a population to changing environments is dependent normally upon the degree of diversity in heritable traits exhibited by its individual members. Diploid organisms express their full range of phenotypes only through the sexual cycle, because newly acquired traits are more often recessive than dominant. However, somatic cells do not go through a haploid phase and mating, and simple fusion between diploid cells (so-called somatic cell mating) per se does not lead to expression of recessive phenotypes. In this paper, I will try to develop the argument that as a substitute for the sexual cycle, malignant cells do go through a functionally haploid phase, and that aneuploidy is merely an overt consequence of the mechanism through which malignant cells express their recessive phenotype in high frequency.

VIRUSES AS ONCOGENS

The existence of DNA as well as RNA oncogenic viruses in mammals has now been firmly established. The only question remaining appears to be whether or not there are other direct causes of cancer. Oncogenic viruses behave in one of two ways, as do bacteriophages. In permissive cells, they proliferate utilizing the host's machinery to produce more infectious progeny, which may kill the host cells by lysis in the process. In nonpermissive cells when mature virus particles are not produced, however, a small segment (one to five genes) of the viral genome may become integrated into the host genome and thus establish a stable lysogenic state. The host cells are not killed by this process, but instead undergo transformation and acquire characteristics suggestive of malignant cells. Subsequent competition among transformed cells may yield a truly malignant clone.

While it is easy to imagine the genomic integration of DNA viruses such as polyoma and SV 40, that of RNA viruses such as murine leukemia virus (MuLV) would require that a DNA copy of the viral genome first be made. The existence of an enzyme for this task (reverse transcriptase) was first postulated by Temin [53] and has now been amply confirmed [3, 54].

It might be expected that a viral genome that could live in the stable lysogenic state would sooner or later become a truly integral part of the host genome and would then be transmitted from one generation of individuals to the next. This indeed seems to happen with MuLV (Gross) in mice of strains with a high incidence of spontaneous leukemia (e.g., AKR and C58). Aside from internal group-specific (gs) antigens, and type-specific components of the virion envelope, the MuLV genome also almost certainly induces specific cellular antigens, such as a cell surface antigen (GCSA), an internal, soluble antigen (GSA), and another cell surface antigen (GIX) that is expressed only in lymphoid cells, especially thymocytes from which leukemia originates.

In a high leukemia strain, a gene for GIX has been localized on the linkage group IX autosome of the mouse and is loosely linked to the H-2 locus. Nonlinkage between the genes determining the GIX and gs antigens has been shown [52]. Nonlinkage of integrated viral genes can now be explained, for it was found that the oncogenic RNA viral genome actually consists of three different subunits. DNA copies of individual subunits apparently prefer different sites of the same chromosome or different chromosomes for integration.

If viral genes can become part of the cell's genome, those genes, in turn, must be able to specify the viral genome. Thus the question of "chicken or egg" becomes legitimate. The idea that the genome of mammalian and other vertebrate species contains genetic information required to specify RNA tumor viruses was first mooted by Payne and Chubb [42] and others, but Huebner and Todaro [21] converted this idea into the now familiar theory of oncogenes and virogenes. The theory is deceptively simple; if all animal genomes contain oncogenes, the transformation of any particular cell to malignancy depends on whether its oncogenes are switched on. Likewise, the liberation of an RNA tumor virus from the cell depends on whether its virogenes are expressed.

This hypothesis could account for the transformation of any and every cell by any and every carcinogen. The cells that have active oncogenes but inactive virogenes would be malignant without containing recognizable tumor viruses, which would explain why most malignant cells do not disseminate infectious units of oncogenic viruses. Evidence has begun to accumulate in favor of this hypothesis, but none against it. For example, it has been known for some time that whether or not normal chick cells express an antigen indistinguishable from the group specific antigen of the avian tumor viruses is genetically determined: gs^+ versus gs⁻. Does the gs⁻ genome lack oncogenes and virogenes? Using various chemical carcinogens and mutagens, Weiss et al. [55] have now shown that gs⁻ cells can be induced to release a virus that very closely resembles, and may even be identical to RAV.0 virus, which is the virus spontaneously released by gs⁺ cells.

MALIGNANCY AS A RECESSIVE TRAIT

The foregoing discussion suggests that transformation of normal somatic cells to malignant cells does not require the acquisition of a new set of genes but only that normally silent oncogenes become expressed. In this view, understanding malignant transformation apparently depends on understanding a more or less irreversible change occurring in the genetic regulatory systems (e.g., a derepression or activation). Such a change might or might not require a mutation affecting a regulatory gene or genes. Nevertheless, with regard to the conversion of diploid cells to malignant cells, the most pertinent question is whether such a pivotal change behaves as a dominant or recessive trait.

Either a repressor or an activator can control the regulatory system in which the expression of regulated structural genes depends upon the presence or absence of an inducer (peptide and steroid hormones, etc.). A repressor or activator molecule has to have binding affinity to an inducer, on the one hand, and to the operator base sequence of structural genes it controls, on the other. Furthermore, the inducer-binding should modify a regulatory molecule's binding affinity for the operator. Thus if inducer-binding decreases the affinity for the operator, a regulatory molecule functions as a repressor, whereas if the affinity is increased as a result of the inducer-binding, the regulatory molecule functions as an activator. Although it has often been said that mammalian genetic regulatory systems must follow a principle entirely different from that of prokaryotes as described mainly by Jacob and Monod [22], I have pointed out that there can be no regulatory system which lacks any of the three essential components: (1) an inducer, (2) a regulatory gene product, or (3) the operator base sequence in the regulated structural genes [41].

In a repressive system, a switch to the noninducible state (no production even in the presence of an inducer) is dominant, whereas a switch to the constitutive state (production in the absence of an inducer) is recessive, to the wild-type inducible state (production only in the presence of an inducer). In an activating system, the situation is reversed.

Is it a dominant or a recessive regulatory switch that causes malignant transformation? For the sake of argument, let us assume that a switch requires a mutational change of a regulatory locus and that such mutation occurs among somatic cells at the spontaneous rate of 10^{-7} per cell generation. If the mutation behaves as a dominant trait, 10^{-7} is the frequency of malignant transformation, whereas if it is a recessive trait, the frequency is reduced to 10^{-14}. The point I wish to make, therefore, is that (depending upon the nature of regulatory switch) there can occur a ten-million-fold difference in the frequency of malignant transformation. It would appear likely that malignant transformation does not depend on a dominant switch. If such were so, the incidence of malignant tumors would become too high and would endanger the survival of the species.

Needless to say, a regulatory switch need not involve the regulatory locus per se. For example, it has been shown that the action of peptide hormones is mediated by an internal inducer such as cyclic AMP [46]. Thus a switch might actually be a change either in the cell-surface peptide hormone-receptor or the adenyl cyclase. It has recently been shown that cyclic AMP levels are lower in transformed cells than in normal cells and that so long as intracellular cyclic AMP levels of transformed cells are artificially raised, either by the direct administration of dibutyl cyclic AMP or by the administration of ubiquitous hormone prostaglandins (PGE_1 or PGE_2), they behave like normal cells.

In the case of mouse neuroblastoma cells, they even differentiate to look like neuronal cells [43, 51]. It follows then that either the decreased production of prostaglandins or the decreased binding affinity of the cell-surface receptor toward prostaglandins might be involved in a regulatory switch that causes malignant transformation. Both of these changes are likely to behave as recessive traits. Unlike other hormones, prostaglandins are not produced by specialized endocrine cells. Thus there is a possibility that they are produced by differentiating cells themselves.

While the predisposition toward malignant tumor development is clearly inherited in a Mendelian manner from one generation of individuals to the next, the malignant change itself occurs only in somatic cells. Thus although classical progeny analysis cannot be used to test for whether or not malignancy behaves as a recessive trait among somatic cells, the technique of somatic cell hybridization has come to the rescue. Initial studies by Ephrussi's group indicated that malignant–nonmalignant hybrid cells almost invariably behaved as malignant. Thus malignancy was thought of as a dominant trait [10]. However, Harris's group later showed that the apparent dominance of malignancy observed by Ephrussi's group was due to unintended and inapparent selection of hybrid clones that had regained malignancy after shedding certain chromosomes. Their experiment suggested that so long as certain chromosomes introduced from the nonmalignant cell are kept, they complement the defect sustained by a parental malignant cell and suppress the manifestation of the malignant phenotype in hybrid cells [15]. Malignancy is apparently due to deficiency of certain gene products. This deficiency can be remedied by the introduction of certain wild-type chromosomes, which indicates that malignancy must be a recessive trait.

If oncogenes and virogenes exist as wild-type components of the vertebrate genome, as recent experiments seem to indicate, it must be that their existence has never endangered the survival of species in the past. Indeed, development of malignant tumors is relatively harmless to the species, since the individuals already past the reproductive age are those primarily affected. Such individuals are dead genetically, at any rate. Relatively late occurrence of malignant tumors, in turn, must mean a low rate of spontaneous malignant transformation among somatic cells. This low rate can best be explained by concluding that malignancy is a recessive trait.

A HEMIZYGOUS STATE AS A PRELUDE TO MALIGNANT TRANSFORMATION

If the rate of spontaneous malignant transformation among normal diploid cells is 10^{-14} per cell generation, the rate would be 10^{-7} for abnormal monosomic cells that have become hemizygous with regard to the gene loci involved in malignant transformation of that particular type. Since there exists a ten-million-fold greater likelihood of malignant transformation among such overtly or covertly (small deletion, etc.) monosomic cells than among normal cells, one could deduce that a particular type of malignant tumor would almost invariably arise from a certain type of monosomic cell.

Indeed, the only known specific chromosomal alteration that could be viewed as directly associated with the cause of a particular type of malignancy is the Philadelphia chromosome seen in cells of human chronic myeloid leukemia [2, 37]. The cell carrying a Philadelphia chromosome is hemizygous with regard to about 50% of the long arm of chromosome No. 22. My interpretation is that the malignant transformation of normal human myeloid stem cells to leukemic cells requires a recessive and heritable change including mutation of a gene locus, or loci, carried by the long arm of chromosome No. 22. It is for this reason that chronic myeloid leukemia almost invariably develops from a cell that has sustained a deletion in this region of chromosome No. 22 [40]. I believe that any chromosomal alteration associated with initial malignant transformation must necessarily be a deletion, and that there is often such an association merely because the rate of spontaneous deletion is much higher than the rate of spontaneous point mutation, and because malignancy is a recessive trait. The chance of recognizing such a deletion in other types of malignancy, however, must be very small, since a large and easily recognizable deletion is apparently deleterious not only to normal somatic cells but also to malignant cells. No instance of monosomy 22 has been found in human chronic myeloid leukemia. Thus it appears that total monosomy 22 places both normal and leukemic myeloid cells at a competitive disadvantage against the diploid myeloid stem cells.

The transformation of a normal somatic cell to a malignant cell is by no means the whole story of tumors; it is merely the beginning. For a newly emerged malignant clone that is extremely small, the formidable task of overwhelming vast numbers of normal counterparts lies ahead. Malignant cells are up to this task because they are now capable of behaving as a population of unicellular organisms. Since the viability of a population depends upon the heritable diversity of phenotypes exhibited by its individual members, and since heritable traits are more often recessive than dominant, how can an asexual population that is not haploid generate recessive phenotypes with significant frequency?

RECESSIVE TRAITS THAT OCCUR IN ANEUPLOID CELLS AT THE FREQUENCY OF DOMINANTS

Resistance to a poisonous analog of thymidine, BUdR, is conferred upon a cell by the loss of thymidine kinase activity (TK$^-$ cells), for a cell that cannot incorporate a poison cannot be killed by that poison. Since BUdR resistance is dependent upon an enzyme deficiency, the trait should and does indeed behave as a recessive. When a TK$^-$ cell is fused with a wild-type TK$^+$ cell, the resultant hybrid clone becomes TK$^+$ and loses BUdR resistance [28]. If we assume the spontaneous deficiency mutation rate to be 10^{-7} per thymidine kinase locus per cell generation, we would expect the incidence of TK$^-$ mutant cells arising in a diploid cell population to be 10^{-14}, and in a tetraploid cell population to be 10^{-28}. For some years it has been clear to many of us that the incidence of spontaneously appearing, apparently recessive traits in a somatic cell population in vitro does not conform to the expectation. Resistance to a poisonous purine analog, 8-azaguanine, is acquired by a cell that has lost inosinic acid pyrophosphorylase, or HGPRT (hypoxanthine-guanine phosphoribosyl transferase), activity [27]. Since this enzyme locus has been shown to be X-linked in man [49], and in view of the apparent conservation of X-linkage in toto by eutherian as well as by marsupial mammals [6, 39, 44], it would be reasonable to expect X-linkage of this locus in all other mammalian species. Due to X-inactivation [29], the expression of X-linked genes is effectively haploid in diploid cells and diploid in tetraploid cells; therefore, the incidence of HGPRT⁻ cells should be much greater than that of TK⁻ cells: 10^{-7} v. 10^{-14} in a diploid population, and 10^{-14} v. 10^{-28} in a tetraploid population. In fact, however, most observers seem to agree that the two traits occur at about the same frequency. Another disturbing fact about somatic cell mutants generated in vitro has been the unexpectedly high rate of spontaneous reversion to the wild type, a frequency of 10^{-4} being not uncommon. While a revertant mutation should be dominant over a deficiency mutation so that ploidy should not affect the revertant mutation rate, it should be the rarest of all mutations, for a second mutation has to restore the function that was lost as a result of the first mutation.

During the last several months, suspicions long held concerning the true nature of so-called somatic mutations have been confirmed by more exacting experiments. Morgan Harris [16] cloned the Chinese hamster $V_{-79}122$ D_1 cell line according to ploidy and examined the spontaneous occurrence of 8-azaguanine resistant cells. The estimated mutation frequency was about 10^{-5} not only in diploid but also in tetraploid and octaploid populations. An enzyme deficiency was apparently behaving as a completely dominant trait so far as the mutation rate was concerned. Yet 8-azaguanine resistance is clearly a recessive trait, for a hybrid cell clone produced by fusion between a HGPRT⁻ cell and a wild-type HGPRT⁺ cell manifests a wild-type phenotype that has lost 8-azaguanine resistance [28].

Mezger-Freed [35] compared haploid and pseudodiploid cell lines of the frog *Rana pipiens* in response to various mutagens for the occurrence of BUdR resistant cells. Some haploids showed more mutants than appeared among the diploids, and some less, but at any rate nothing to compare with the million to ten-million-fold differences to be expected if the BUdR resistance were the result of recessive mutations in the TK structural gene locus.

It was argued that BUdR resistance was not due to reduction in thymidine kinase activity per se, but rather to a heritable change in the membrane that inhibits the entrance to the cell of BUdR. But a deficiency of permease should also behave as a recessive trait. Both Harris and Mezger-Freed argued that so-called somatic cell mutations are sensus stricto not mutations but that they represent, instead, some sort of epigenetic change which is stable and heritable from generation to generation. Although what is meant by a stable epigenetic change is not clear, I will conjecture that the authors meant the types of change that accompany somatic cell differentiation during embryonic development. If we think of the variety of proteins they synthesize, the phenotypic differences between liver cells and lymphocytes of the same body are far greater than the different species of unicellular organisms. Yet very few would contend that changes in genes (mutations) cause differentiation. There is little doubt that a great phenotypic change that is heritable can be brought about by a more or less stable change in the state of genetic *regulatory* systems.

Whatever the cause of apparent somatic mutations, however, we should not forget that somatic hybridization experiments have clearly shown that the resistance acquired in vitro to BUdR, or to 8-azaguanine, behaves as an indisputably recessive trait. Thus the most amazing thing about all this is not the doubtful nature of apparent somatic mutations but the ease with which diploid, tetraploid—and even octaploid—tissue-cultured cell populations generate variants that manifest recessive phenotypes. Somatic cells that have been selected for continuous growth in vitro are almost invariably aneuploids, and when transplanted to animals they often reveal themselves to be malignant. Thus the idea that ploidy-independent expression of recessive phenotypes is a characteristic of aneuploid populations appears to find a measure of support.

There is no evidence that variants manifesting recessive phenotypes occur with appreciable frequency among normal somatic cells in vivo. If they occur, they would not have eluded detection this long, since an ample variety of coat-color and other pertinent recessive marker genes have been available for some years in the mouse. However, it might be argued that they do occur but that they have no chance of becoming clones of recognizable size, since normal somatic cells are programmed to go through only a finite number of cell generations.

I believe that natural selection has made certain that expression of the initial transformation to malignancy is a rare event in the wild-type populations. A true recessive mutation or mutations is probably required, and it is for this reason that I believe transformation primarily affects an already abnormal somatic cell that has become hemizygous for a pertinent gene or genes by a mitotic error—that is, deletion, in effect. Subsequent manifestation of recessive phenotypes by established malignant cells, however, would be frequent events and apparently ploidy-independent. Ploidy-independent expression of recessive phenotypes would be a source of tremendous advantage to a malignant cell population and a great source of trouble to the host.

A FEW CONCEIVABLE MECHANISMS THAT MIGHT BE RESPONSIBLE FOR THE PLOIDY-INDEPENDENT EXPRESSION OF RECESSIVE PHENOTYPES

A large volume of information on chromosome constitutions of malignant tumors has been published. It shows that tumors can be near-diploid, near-triploid, near-tetraploid, and even near-octaploid in chromosome numbers. Yet neither a haploid nor substantially hypodiploid tumor cell line has ever been found. Atkin and his colleagues [1] showed that even though the stemline chromosome number of certain human tumors may be below the normal diploid number of 46, the DNA content of the stemline is almost invariably above the normal diploid value (hyperdiploid). The normal diploid chromosome number of the dog is 78. Of these, only the X and the Y are metacentrics, all 76 autosomes being acrocentrics. The venereal lymphosarcoma cells of this species [33], on the other hand, invariably contain only 59 chromosomes, but 17 of these are metacentrics. Thus the normal diploid complement is made up of 80 chromosome arms, and the lymphoma complement of 76. The DNA content of the lymphoma cell, however, is slightly higher than the normal diploid value [31]. It would appear that the monosomic condition, in general, is just as disadvantageous to malignant cells as it is to normal somatic cells. Witness the aforementioned observation that while the half-deletion of the long arm of the chromosome No. 22 (Philadelphia chromosome) invariably accompanies the development of human chronic myeloid leukemia, no monosomy 22 (total loss) leukemic line has ever been found. In the case of malignant cells, too, the only truly dispensable chromosome of the diploid complement might be the redundant second X of the female and the Y of the male.

Aneuploidy that characterizes a malignant cell population must be essentially polysomic in nature. However, my view that an initial chromosomal change that predisposes an abnormal somatic cell to malignant transformation must necessarily be hemizygous (a deletion) is not in conflict with the apparently polysomic nature of an established malignant cell population, since whatever has been lost initially can hardly be regained. If malignant transformation is the consequence of a recessive mutation, the cell that became hemizygous for a pertinent gene locus, or loci, by a prior mitotic error would primarily be affected. Once transformed, however, a chromosome carrying a mutated gene or genes could become polysomic, and the cell would still be homozygous for mutated genes.

Needless to say, the gene loci involved in the malignant transformation of one somatic cell type need not be involved in the transformation of another cell type. For example, the Philadelphia chromosome is seen in association with chronic myeloid leukemia but not with other human malignant tumors. Furthermore, Henry Harris's group has shown that when the genomes of two particular malignant mouse lines (YAC and SEWA) were brought together by somatic fusion, mutual suppression of each other's malignancy occurred [15]. Therefore, the kind of chromosome involved in initial deletion is likely to be different from one kind of malignant tumor to the next. In the same manner, the type of chromosomes primarily involved in polysomy (trisomy in diploid, pentasomy in tetraploid, etc.) is not likely to be the same in different kinds of established malignant cell populations.

How might the frequent manifestation of recessive phenotypes be permitted in a cell population that is predominantly polysomic? I have no ready answer but, instead, I shall suggest two possible mechanisms for readers' consideration.

Somatic Recombination Due to Crossing-over

Normal mammalian somatic cells have actually been caught under the microscope in the act of crossing-over between presumptive homologues [11]. If homologous chromosomes involved in a somatic crossing-over were heterozygous at any of the gene loci situated distal to an exchange point away from the centromere, there is a 25% chance that one of the daughter cells would be homozygous for a recessive gene or genes as shown in Fig. 1. Evidence of somatic recombination involving D and K regions of the H-2 histocompatibility locus has actually been obtained in F_1-mouse tumors [23, 24], and there is some evidence that somatic crossing-over occurs in normal lymphocytes of the rabbit among variable-region genes and constant-region genes of immunoglobulin heavy chains [9]. By means of somatic crossing-over, a diploid cell population could generate a variant expressing a recessive phenotype, without having either to sustain independent mutations on both copies of the same locus or to go through monosomic phase. This alone, however, would be quite ineffective for a population that is trisomic or tetrasomic.

Somatic Segregation

In certain individuals heterozygous for chromosomal rearrangements (Robertsonian fusions in the rainbow trout and pericentric inversions in the deer mouse), we have observed polymorphism among somatic cells

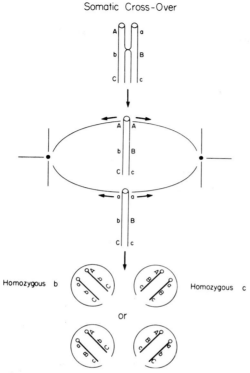

Figure 1. Consequences of a somatic crossing-over are illustrated on a pair of homologous chromosomes carrying three marker gene loci: A, B, C represent the dominant alleles, and a, b, c, the recessive.

of the same individual. It appeared as though some of the somatic cells had become homozygous for some of the rearrangements, and we defined this phenomenon as "somatic segregation" [38]. The same or a similar phenomenon has been observed by others in the same and related species [20, 25, 45] and also in human cells [34]. Although there are a number of mechanisms that conceivably can cause somatic segregation [38], the simplest would be through two successive nondisjunctions and monosomy, as illustrated in Fig. 2.

It has been shown that a tetraploid cell line periodically throws off diploid progeny [34]. In a similar manner, a near-diploid cell line that is trisomic for a particular chromosome might periodically produce a variant that is monosomic for that chromosome; some of such variants might remain viable by reverting to the disomic, or even trisomic, state by

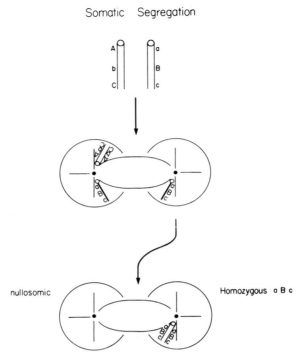

Figure 2. One means of achieving somatic segregation. Production of a recessive homozygous cell from a heterozygous cell by successive nondisjunctions is illustrated. Alternative means are discussed in the text.

successive nondisjunctions. Although such a process involves a great deal of waste, not only would most of the monosomic cells be eliminated, but also, in its reversion to the disomic and then trisomic state, nonviable nullisomic cells would be produced. A malignant cell population can afford to and does sustain a large loss of individual members.

If the frequency with which aneuploid cells go through a transitory monosomic state is high enough, it would permit frequent production of variants manifesting recessive phenotypes.

DIRECT EFFECT OF PLOIDY ON GENETIC REGULATORY SYSTEMS

As already mentioned, a regulatory gene product has to have two different binding sites: one for an inducer, which is more often a hormone or cyclic AMP than a substrate in the case of mammals, and the other for the operator base sequence, which is a part of each of a number of the structural genes it controls. Furthermore, the binding with an inducer should either decrease or increase the binding affinity of a regulatory gene product to the operator region. In the case of decrease, a regulator exercises negative (repressive) control, and, in the case of increase, positive (activating) control.

Inasmuch as the above is the basis of genetic regulatory mechanisms, a system can function in a finely tuned manner only if the intracellular concentration of a regulatory gene product is kept at a very low level. Indeed, the wild-type *E. coli* contains only 10 molecules of a repressor of the *lac*-operon per cell [36].

When the regulator concentration becomes abnormally high, a regulatory gene product would begin to bind not only with the proper operator but also with other, similar base sequences in the genome, and even an inducer-bound repressor or inducer-unbound activator may stay bound to the operator. Thus an increased regulator concentration should result in a mutant-like, noninducible phenotype in the case of repressive control, and in a mutant-like, constitutive phenotype in the case of activating control.

Some of the mutant-like phenotypes exhibited by aneuploid cells might simply be due to polysomy of a chromosome carrying a regulatory gene or genes. Such phenotypes should be stable and heritable so long as the high regulatory-gene dosage due to polysomy is maintained, but the loss of polysomic state would result in reversion to the wild-type phenotype.

SUMMARY

The relative rarity of malignant tumors, as well as the behavior of malignant–nonmalignant somatic hybrids, points to the recessive nature of malignant transformation. Consequently, there may be a million- to ten-million–fold difference in the probability of malignant transformation between a normal diploid somatic cell and an abnormal cell that has become hemizygous for pertinent gene loci. I contend, therefore, that an initial chromosome change that accompanies malignant transformation must necessarily be a deletion, as exemplified by the Philadelphia chromosome seen in human chronic myeloid leukemia.

Once established, an aneuploid cell population generates variants manifesting recessive phenotypes with unexpectedly high frequency, and the frequency is independent of ploidy. Thus aneuploid cells that are polysomic rather than monosomic behave as though haploids. Three possible mechanisms that may contribute to this peculiar behavior were discussed.

Acknowledgments.–This work was supported in part by a grant (CA 05138) from The National Cancer Institute, U.S. Public Health Service.

LITERATURE CITED

1. ATKIN NB, RICHARDS BM, Ross AJ: The deoxyribonucleic acid content of carcinoma of the uterus: an assessment of its possible significance in relation to histopathology and clinical course based on data from 165 cases. Brit J Cancer 13:773–787, 1959.
2. BAIKIE AG, COURT BROWN WM, BUCKTON KE, HARNDEN DG, JACOBS PA, TOUGH IM: A possible specific chromosome abnormality in human chronic myeloid leukemia. Nature 188:1165–1166, 1960.
3. BALTIMORE D: RNA-dependent DNA polymerase in virions of RNA tumour viruses. Nature 226:1209–1211, 1970.
4. BOVERI T: Zur Frage der Enstehung maligner Tumoren. Jena, Gustav Fischer, 1914.
5. CLEAVER JE: Defective repair replication of DNA in xeroderma pigmentosum. Nature 218:652–656, 1968.
6. COOPER DW: Directed genetic change model for X chromosome inactivation in Eutherian mammals. Nature 230:292–294, 1971.
7. COURT BROWN WM, ABBATT JD: The incidence of leukemia in ankylosing spondylitis treated with X-rays. Lancet 1:1283, 1955.
8. COURT BROWN WM, DOLL R: Adult leukemia. Brit Med J 1:1753, 1960.
9. DUBISKI S: Does antibody synthesis involve somatic recombination? In Protides of the Biological Fluids (Peeters H, ed), Oxford, Pergamon Press, 117–123, 1970.
10. EPHRUSSI B: Hybridization of somatic cells and phenotypic expression. In Developmental and Metabolic Control Mechanism and Neoplasia. Baltimore, Williams & Wilkins Company, 496–503, 1965.
11. GERMAN J: Cytological evidence for crossing-over *in vitro* in human lymphoid cells. Science 144:298–301, 1964.
12. GERMAN J: Genes which increase chromosomal instability in somatic cells and predispose to cancer. In Progress in Medical Genetics (Steinberg AG, Bearn AG, eds), vol 8, New York, Grune & Stratton, 61–101, 1972.
13. GEY GO: Some aspects of the constitution and behavior of normal and malignant cells maintained in continuous culture. The Harvey Lectures, Series L, New York, Academic Press, 154–229, 1954–1955.
14. GROPP A, FLATZ G: Chromosome breakage and blastic transformation of lymphocytes in ataxia-telangiectasia. Humangenetik 5:77–79, 1967.
15. HARRIS H, MILLER OJ, KLEIN G, WORST P, TACHIBANA T: Suppression of malignancy by cell fusion. Nature 223:363–368, 1969.
16. HARRIS M: Mutation rates in cells at different ploidy levels J Cell Physiol 78:177–184, 1971.
17. HAUSCHKA TS, LEVAN A: Cytologic and functional characterization of single cell clones isolated from the Krebs-2 Ehrlich ascites tumors. J Nat Cancer Inst 21:77–135, 1958.
18. HAYFLICK L, MOORHEAD PS: The serial cultivation of human diploid cell line. Exp Cell Res 25:585–621, 1961.

19. HECHT F, KOLER RD, RIGAS DA, DAHNKE GS, CASE MP, TISDALE V, MILLER RW: Leukemia and lymphocytes in ataxia-telangiectasia. *Lancet* 2:1193, 1966.
20. HECKMAN JR, ALLENDORF FW, WRIGHT JE: Trout leukocytes: growth in oxygenated cultures. *Science* 173:246–247, 1971.
21. HUEBNER RJ, TODARO GJ: Oncogenes of RNA tumor viruses as determinants of cancer. *Proc Nat Acad Sci (US)* 64:1087–1094, 1969.
22. JACOB F, MONOD J: Genetic repression, allosteric inhibition, and cellular differentiation. In Cytodifferentiation and Macromolecular Synthesis (Locke M, ed), New York, Academic Press, 30–64, 1963.
23. KLEIN E: Studies on the mechanism of isoantigenic variant formation in heterozygous mouse tumors. I. Behavior of H-2 antigens D and K: quantitative absorption tests on mouse sarcomas. *J Nat Cancer Inst* 27:1069–1093, 1961.
24. KLEIN E, KLEIN G: Studies on the mechanism of isoantigenic variant formation in heterozygous mouse tumors. III. Behavior of H-2 antigens D and K when located in the *trans* position. *J Nat Cancer Inst* 32:569–578, 1964.
25. KREIZINGER JD, SHAW MW: Chromosomes of *Peromyscus* (Rodentia, Cricetidae) II. The Y chromosome of *Peromyscus maniculatus*. *Cytogenetics* 9:52–70, 1970.
26. LEVAN A, HAUSCHKA TS: Chromosome numbers of three mouse ascites tumors. *Hereditas* 38:251–255, 1952.
27. LITTLEFIELD JW: The inosinic acid pyrophosphorylase activity of mouse fibroblasts partially resistant to 8-azaguanine. *Proc Nat Acad Sci (US)* 50:568–576, 1963.
28. LITTLEFIELD JW: Selection of hybrids from matings of fibroblasts *in vitro* and their presumed recombinants. *Science* 145:709, 1964.
29. LYON MF: Sex chromatin and gene action in the mammalian X-chromosome. *Am J Human Genet* 14:135–148, 1962.
30. MAKINO S: A cytological study of the Yoshida sarcoma, an ascites tumor of white rats. *Chromosoma* 4:649–674, 1952.
31. MAKINO S: Some epidemiologic aspects of venereal tumors of dogs as revealed by chromosome and DNA studies. *Ann NY Acad Sci* 108:1106–1122, 1963.
32. MAKINO S, SASAKI MS, TONOMURA A: Cytological studies of tumors. XI. Chromosome studies in fifty-two human tumors. *J Nat Cancer Inst* 32:741–777, 1964.
33. MAKINO S: Cytogenetics of canine venereal tumors. This volume, 335–372, 1974.
34. MARTIN GM, SPRAGUE CA: Parasexual cycle in cultivated human somatic cells. *Science* 166:761–763, 1969.
35. MEZGER-FREED L: Effect of ploidy and mutagens on bromodeoxyuridine resistance in haploid and diploid frog cells. *Nature New Biol* 235:245–246, 1972.
36. MULLER-HILL B: *Lac* repressor. *Angew Chem Int Ed* 10:160–172, 1971.
37. NOWELL PC, HUNGERFORD DA: Chromosome studies on normal and leukemic human leukocytes. *J Nat Cancer Inst* 25:85–109, 1960.
38. OHNO S: Cytologic and genetic evidence of somatic segregation in mammals, birds and fishes. In Phenotypic Expression, *In Vitro* vol 2, Publication of the Tissue Culture Association, Baltimore, Williams & Wilkins Company, 1966.
39. OHNO S: Sex Chromosomes and *Sex-linked Genes* (Labhart A, Mann T, Samuels

LT, eds), vol 1, Monograph on Endocrinology. Heidelberg, Springer-Verlag, 1967.
40. OHNO S: Genetic implications of karyological instability of malignant somatic cells. *Physiol Rev* 51:496–526, 1971.
41. OHNO S: Gene duplication, mutation load and mammalian genetic regulatory systems. *J Med Genet* (in press).
42. PAYNE LN, CHUBB R: Studies on the nature and genetic control of an antigen in normal chick embryos which reacts in the COFAL test. *J Gen Virol* 3:379–391, 1968.
43. PRASAD KN: Morphological differentiation induced by prostaglandin in mouse neuroblastoma cells in culture. *Nature New Biol* 236:49–52, 1972.
44. RICHARDSON BJ, CZUPPON AB, SHARMAN GB: Inheritance of glucose-6-phosphate dehydrogenase variation in kangaroos. *Nature New Biol* 230:154–155, 1971.
45. ROBERTS FL: Chromosomal polymorphism in North American landlocked *Salmo salar*. *Can J Genet Cytol* 10:865–875, 1968.
46. ROBINSON GA, BUTCHER RW, SUTHERLAND EW: Cyclic AMP. *Ann Rev Biochem* 37:149–174, 1968.
47. SANDBERG AA, ISHIHARA T, CROSSWHITE LH, HAUSCHKA TS: Comparison of chromosome constitution in chronic myelocytic leukemia and other myeloproliferative disorders. *Blood* 20:393–423, 1962.
48. SAWITSKY A, BLOOM D, GERMAN J: Chromosomal breakage and acute leukemia in congenital telangiectatic erythema and stunted growth. *Ann Int Med* 65:487–495, 1966.
49. SEEGMILLER JE, ROSENBLOOM FM, KELLEY WN: Enzyme defect associated with a sex-linked human neurological disorder and excessive purine synthesis. *Sciene* 155:1682–1684, 1967.
50. SETLOW RB, REGAN JD, GERMAN J, CARRIER WL: Evidence that xeroderma pigmentosum cells do not perform the first step in the repair of ultraviolet damage to their DNA. *Proc Nat Acad Sci (US)* 64:1035–1041, 1969.
51. SHEPPARD JR: Difference in the cyclic adenosine 3',5'-monophosphate levels in normal and transformed cells. *Nature New Biol* 236:14–16, 1972.
52. STOCKERT E, OLD LJ, BOYSE EA: The G_{IX} system: a cell surface alloantigen associated with murine leukemia virus; implications regarding chromosomal integration of the viral genome. *J Exp Med* 133:1334–1355, 1971.
53. TEMIN HM: Nature of the provirus of Rous sarcoma. *Nat Cancer Inst Monogr* 17:557–570, 1964.
54. TEMIN HM, MIZUTANI S: RNA-dependent DNA polymerase in virions of Rous sarcoma virus. *Nature* 226:1211–1213, 1970.
55. WEISS RA, FRIIS RR, KATZ E, VOGT PK: Inductions of avian tumor viruses in normal cells by physical and chemical carcinogens. *Virology* 46:920–938, 1971.

WHAT IS A CHROMOSOME BREAK?

DAVID E. COMINGS

Department of Medical Genetics, City of Hope National Medical Center,
Duarte, California

Introduction 96

Types of Chromosome Aberrations 96
   Gaps vs. Breaks 97
   Chromatid vs. Chromosome Lesions 97
   Breaks vs. Exchanges 98
   U vs. X Exchanges 98
   Complete vs. Incomplete Exchanges 99
   Restitution 99
   Isodiametric Deletions 99
   Intrachange vs. Interchange 99

The Breakage-First and Exchange Hypotheses 100
   The Breakage-First Hypothesis 100
   The Exchange Hypothesis 100
   Evidence Favoring the Exchange Hypothesis 104
   Objections to the Exchange Hypothesis 105
   Some Breaks Are Due to Exchange 106
   Exchanges and the Breakage-First Hypothesis 107
   A (Dose)² Component for Breaks 108
   Increased Breaks in Heterochromatin 108
   Chromosome Breakage at the Molecular Level 108

Chromatid Gaps vs. Breaks in DNA 109
   Fine Structure of Gaps 109
   Evidence that Gaps Are Not True Breaks 109
   Evidence that Gaps May Be True Breaks 111
   Influence of Chromatin Fiber Folding 113
   Acentric Chromosomes Show Fewer Breaks 118

Breaks in G_1 and Chromosome Strandedness 119
   Transition from Chromosome to Chromatid Aberrations 119
   Relevance of a G_1 Split 120

Do Half-Chromatids and Half-Chromatid Aberrations Really Exist? 122

The Site-Limitation Hypothesis 123
An Alternative: Real vs. Effective Limitation 125
Concluding Comments 126
Acknowledgements 126
Literature Cited 126

INTRODUCTION

Since the demonstration in 1927 by Muller [75] that X-rays can cause rearrangements in the linear order of genes, induced chromosome breaks have been used to study many aspects of chromosome structure. For example, the first clue that chromosome duplication did not occur just before mitosis came from the studies by Sax [99] on the effects of X-rays on *Tradescantia* microspores. He noted that the transition from chromosome to chromatid breaks occurred in interphase rather than during prophase. Breaks have also been used to study the problem of chromosome strandedness, and even the mechanism by which breaks arise is a matter of considerable controversy.

This discussion will speak to the following questions. (1) Is the mechanism of breaks best described by the "breakage-first" hypothesis or by the "exchange" hypothesis? (2) Do gaps represent true breaks in the continuity of the DNA molecule? (3) Does the observation that chromosomes appear to be double in response to X-ray breakage (i.e., appear to give chromatid breaks) *prior to* the S period mean that chromosomes are double-stranded? (4) Are half-chromatids and half-chromatid exchanges real? (5) Is there a real or only an effective limitation in the number of exchange sites in the nucleus?

A major point I would like to emphasize is that both the manner of folding of the chromatin fiber and certain basic observations about mechanisms of DNA replication are important in the interpretation of various aspects of chromosome breakage.

TYPES OF CHROMOSOME ABERRATIONS

When a chromatid is broken, a number of different types of aberrations may arise, depending on the stage in the cell cycle at which the break has occurred and whether the broken ends reunite with other chromatids or other chromosomes. Only the variations that are pertinent to the present discussion will be illustrated. A more detailed discussion of the various types of chromosome aberrations can be found elsewhere [37, 68, 89, 100].

Gaps vs. Breaks. Exposure to various agents may result in a severe attenuation of the chromatid, producing an achromatic lesion or gap.

When there is a displacement of the distal fragment, the lesion can be positively identified as a true break.

Chromatid vs. Chromosome Lesions. Chromatid breaks occur in only one of two chromatids. They are usually the result of breakage that has taken place after DNA replication (post-split breaks). When a break is present at the same site on both chromatids, it is termed a chromosome break, isochromatid break, or isolocus break.

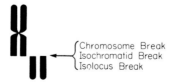

The same terminology holds for chromosome gaps.

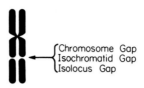

Chromosome breaks are the result of a lesion that has occurred before DNA replication (pre-split breaks.) At replication, it is reproduced in both chromatids. Similar appearing lesions called isochromatid breaks are due to the chance occurrence of two independent chromatid breaks at the same site. Although chromosome breaks and isochromatid breaks are morphologically indistinguishable, they can be distinguished on the basis of the amount of time that has elapsed between exposure to the breaking agent and the mitosis at which they are observed. Thus if the G_2 period is 4 hours long, and 2 hours after exposure to a breaking agent an isolocus break is found at mitosis, it would be an isochromatid break. The same lesion observed at mitosis after exposure to a breaking agent during the G_1 period would be a chromosome break. The noncommittal term, not implying either mechanism, is isolocus break.

Breaks vs. Exchanges. A simple break results in a chromosome deletion and a chromosome fragment. If two breaks occur in reasonable proximity to each other, an exchange or reunion may take place.

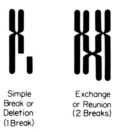

U vs. X Exchanges. If the fusion occurs between both proximal ends or both distal ends, a U-type exchange occurs. If fusion occurs from the proximal end onto the other end, an X-type exchange occurs. The latter is the type seen during recombination in meiosis.

WHAT IS A CHROMOSOME BREAK?

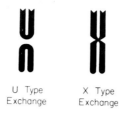

U Type Exchange   X Type Exchange

Complete vs. Incomplete Exchange. If all four ends involved in a potential exchange are reunited, the exchange is termed complete. If none or only two are reunited, it is incomplete.

In the U type of exchange, these may be divided into incomplete proximal (NUp) and incomplete distal (NUd). If there is no exchange, it is incomplete proximal and distal (NUpd).

Complete   Incomplete Proximal   Incomplete Distal

Restitution. Not all breaks become visible at the next mitosis. This is particularly true if the broken ends are reunited to reproduce the original configuration, a process called restitution. Lea [68] has estimated that more than 90% of breaks produced in *Tradescantia* by X-rays undergo restitution.

Isodiametric Deletions. Minute interstitial deletions that appear at mitosis as tiny rings are called isodiametric deletions [46, 79, 80, 91].

Isodiametric Deletion

Intrachange vs. Interchange. An interchromosomal exchange involving reunion between two different chromosomes is termed an interchange.

An intrachromosomal exchange involving reunion between different parts of the same chromosome is an intrachange.

THE BREAKAGE–FIRST AND EXCHANGE HYPOTHESES

Formulation of the Breakage-First Hypothesis

According to the breakage-first hypothesis [*100, 109*], the immediate effect of X-irradiation is chromosome breakage, and after this three things can occur: restitution can take place, the breaks can remain open, or the breaks can interact to produce an exchange. The breaks are single-hit events, and exchanges are the result of a reunion between two different breaks. A large part of the evidence for this hypothesis comes from the following observations.

(1) Breaks increase linearly with the dose of radiation. Thus, in the equation $Y = a + kD^n$, where Y is the total yield of breaks, a is the frequency of spontaneous breaks, k is a constant, and n is the power of the dose D of radiation, n is usually found to be 1.0 or close to it [*5, 16, 41, 57, 69, 91, 93, 112, 122*]. This linear relationship is also seen when DNA breaks are assayed by alkaline-sucrose gradients [*55, 56, 70*]. In this equation the D^n component defines whether there is one-hit kinetics, (dose)^1, or two-hit kinetics, (dose)^2.

(2) Exchanges increase with the square of dose. Since exchanges involve interaction between two different and usually independent breaks, their frequency should increase with the square of the dose of radiation; this has been repeatedly observed [*3–5, 9, 14, 28, 36, 48, 57, 65, 67, 77, 81, 82, 90, 100*]. In this relationship n is equal, or approximately equal, to 2.0.

(3) The number of exchanges decreases as the rest period increases. This is shown by studies in which a dose of radiation is followed by a rest period, followed by a second dose of radiation [*72, 100, 120, 125*]. It was interpreted to suggest that if the breaks produced by the first dose undergo restitution, they will be unable to interact with the breaks produced by the second dose, and thus no exchanges can be formed.

Formulation of the Exchange Hypothesis

In contrast to the idea of breakage followed by subsequent exchange, Revell [*89, 90*] suggested that the immediate effect of X-irradiation was the production of a "primary event" that was not a true chromosome break. When left to themselves, the primary events would soon heal. However, if another primary event is available in a nearby chromatid, the two can interact and stabilize each other and can lead to a second stage termed "exchange initiation," during which exchange between the two chromatids can take place. If the exchange involves a different chromosome or different parts of the same chromosome, a typical interchange results, the same as with the breakage-first hypothesis. However, if the exchange involves sister chromatids (during the S or G_2 period) and is incomplete, the end result will have the same appearance as a simple chromatid break. Since this hypothesis suggests that "single breaks" are actually the result of two breaks involved in an incomplete exchange, it has been termed the exchange hypothesis.

The production of breaks requires that sister chromatids be intimately associated with each other throughout the G_2 period and that they form small loops. Each line or dotted line in the following diagrams represents one double DNA helix.

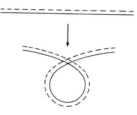

Two types of breaks could then occur. The first would be at two places in the DNA of only one chromatid.

The second would involve one break in the DNA of each chromatid.

Each of these could rejoin in two ways. The double break in one chromatid would rejoin like this

or like this,

and a single break in both chromatids could rejoin like this

or like this.

The rejoining could be complete or incomplete as shown in Fig. 1. The consequences of this as far as the configuration of the metaphase chromosomes is concerned are shown in Fig. 2.

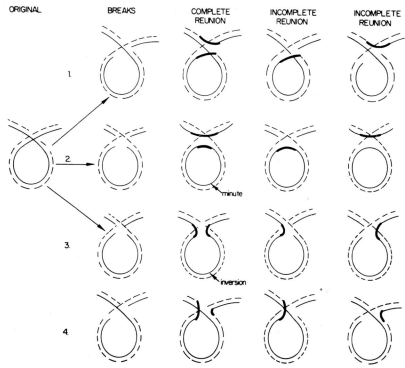

Figure 1. Diagram of the complete and incomplete reunions for the four types of break–reunion that could take place between paired loops of the sister chromatids.

To test the exchange hypothesis it was necessary to make two assumptions. First, it was assumed that there was an equal likelihood of either chromatid participating in an exchange so that the four types of interchange would occur with the same frequency—and second, it was assumed that the likelihood of incompleteness was the same for all exchanges so that the eight incomplete forms would occur with equal frequency. Since it would not be possible to score partially or totally the complete form of types 1, 2, or 3 intrachanges (Fig. 2), the hypothesis had to be tested on the basis of the incomplete forms. From these it would predict a ratio of five chromatid breaks to two isochromatid, incomplete, sister nonunions (Fig. 2). A second assumption is that the frequency of minutes (type 2) would be equal to the frequency of isochromatid lesions (type 4). Since not all the minutes could be scored, these would actually be fewer in number than the isochromatid lesions.

In some extensive experiments on *Vicia faba*, Revell (89) found the

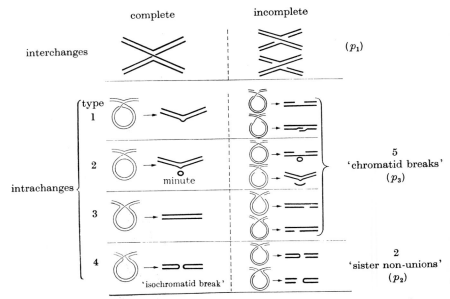

Figure 2. Diagram of the exchange hypothesis. The complete forms of exchanges are shown on the left and the incomplete forms on the right. The interchanges are above. The four types of intrachanges are below, along with a diagram of their early states with the chromatids completely paired and their later states as they would appear at metaphase (from Revell, 1959, by permission; see ref. [89]).

ratio of "chromatid breaks" to "sister non-unions" to be 3:2–1 and 2.52–1 (close to the predicted ratio of 2:5–1), and the frequency of minutes was less than the total frequency of isochromatid breaks. Others have also found similar ratios (98). In all previous studies, the ratio of the chromatid breaks to isochromatid breaks or interchanges was much higher than this as would be anticipated on the classical assumption that breaks are single-hit events and reunions are double-hit events. Most of these studies, however, scored gaps as breaks, and one of Revell's major points was the assumption that gaps do not represent true breaks. Only when this assumption is made, does the ratio of breaks to exchanges become low enough to be compatible with the proposal of the exchange hypothesis that simple breaks and exchanges are both two-break events.

Evidence Favoring the Exchange Hypothesis

Since its formulation, there have been a number of additional observations that favor the exchange hypothesis.

(1) If breaks are the result of exchanges between two "primary lesions," they should show (dose)^2 kinetics. Such kinetics have been reported by several investigators [9, 44, 90, 107]. Revell suggests that the reports of (dose)^1 kinetics for breaks are due to the inclusion of gaps [89, 90].

(2) The exchange hypothesis would allow a significant delay between a hit and a break. As such, it would provide an explanation of the prolonged interval between radiation and reunion that Lane [65, 66] and others [38, 50] have observed.

(3) Fox [43] has suggested that modifications of the exchange hypothesis provide a reasonable explanation for some of the aberrations observed in irradiated locust embryos.

(4) Autoradiographic studies, discussed below, indicate that sister chromatid exchanges frequently occur at the site of some chromatid breaks [52].

Objections to the Exchange Hypothesis

There are a number of points about the exchange hypothesis that may be raised as mild or serious objections, depending on one's prejudices.

(1) The hypothesis makes the assumption that sister chromatids are in extremely intimate contact with each other at all times after DNA replication. Several observations about the chromosome structure lead one to doubt this. For example, whole-mount electron microscopy of prophase chromosomes shows that each chromatid has its own complex manner of folding and its own multiple set of attachment points to the nuclear membrane [23–25].

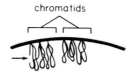

To claim, as the exchange hypothesis does, that a hit in the lateral portion of one chromatid (arrow) is usually going to involve an exchange with DNA in the other chromatid seems unlikely.

(2) The exchange hypothesis holds that both chromatid exchanges and chromatid breaks are the result of a complete or incomplete exchange involving two breaks. This, however, is inconsistent with the observation that chromatid breaks usually show (dose)^1 kinetics, while exchanges show (dose)^2 kinetics. Although there have been several reports of a

(dose)^2 kinetics for simple breaks [44, 90, 107], most observers report (dose)^1 kinetics. The presence of two-hit kinetics could be explained on the basis of the tertiary structure of the chromosome, which makes it necessary to have two breaks before complete separation of the chromatid is produced (see "Evidence that Gaps May Be True Breaks"). This possibility seems to be enhanced by the observation that the (dose)^2 relationship for breaks is most prominent when cells are radiated in the G_2 period [107]. Thus, as chromosomes begin to condense in G_2 two breaks may be required to produce the discontinuity. On the other hand, there are some reports of a (dose)^2 component to breaks induced during the G_1 period [9].

(3) If both isochromatid and chromatid breaks are the result of an exchange, both should show the same dose kinetics. This, however, is generally not the case. Chromatid and chromosome aberrations vary with (dose)^1 kinetics, while isochromatid breaks vary with (dose)$^{1.5}$ kinetics [8, 16, 17, 28, 36, 37, 62, 76, 77, 90, 91, 111–113]. The difference between these two events is further indicated by the observation that the ratio of isochromatid breaks to chromatid breaks increases with increasing linear energy transfer [30, 111]; this would be expected with the classical breakage-first hypothesis, while both should increase together on the basis of the exchange hypothesis.

(4) In some reports, the ratio of chromatid breaks to incomplete isochromatid breaks has differed significantly from that predicted by the exchange hypothesis [28, 107].

(5) In some reports, the frequency of carefully scored gaps is actually smaller than the frequency of complete chromatid breaks [28].

(6) With the type of interlooping required by the exchange hypothesis, interlocking aberrations should be frequently seen but are not [37].

(7) The exchange hypothesis proposes that ionizing radiation does not break chromosomes directly, but instead, it produces a "primary lesion," and that the interaction of two primary lesions results in exchange initiation. If the exchange is incomplete, a break results. There is molecular evidence both for and against this. Sucrose-gradient studies show that radiation produces many single-stranded breaks that could represent the "primary lesion." The same evidence, however, also shows an immediate production of double-stranded breaks [19, 56, 58].

An Autoradiographic Test:
Some Breaks Are Due to Exchange, Some Are Not

If the breakage-first hypothesis is correct, single-chromatid breaks should not be associated with an exchange between sister chromatids. If the exchange hypothesis is correct, 40 to 50% of the chromatid breaks should be associated with an exchange between sister chromatids [52, 53].

If a cell is exposed to tritiated thymidine during one cell cycle, then allowed to pass through a second cell cycle in the absence of tritiated thymidine, one chromatid will contain radioactive DNA while the other will not. If a sister chromatid exchange takes place it can be detected by autoradiography. In an experiment of this type, Heddle and Bodycote [52] observed that, depending on the X-ray dose and other conditions, 6 to 40% of single breaks were associated with an exchange between sister chromatids. This indicated that some breaks are the result of an exchange and some are not.

Exchanges Can Occur with the Breakage-First Hypothesis

The observation that some breaks are associated with a sister-chromatid exchange does not prove that such breaks result from the mechanism proposed by the exchange hypothesis. For example, irradiation during the G_2 period could result in an isochromatid break. It has long been known that such breaks may undergo complete or partial U-type reunion to result in SU, NUp, or NUd lesions.

It is, however, just as likely that they could undergo partial X-type reunion and thus could appear as single breaks but actually be the remnants of isochromatid breaks.

Since isochromatid breaks have a (dose)^2 component, this could explain the presence of a (dose)^2 component for single breaks produced during the G_2 [107], and the entire lesion would still be the result of breaks followed later by reunions à la the breakage-first hypothesis. It would not explain the (dose)^2 component occasionally reported for G_1 breaks [9].

Alternative Explanations for a (Dose)² Component For Breaks

Since the exchange hypothesis requires a (dose)² component for single breaks, there is a subtle implication that if such a component is found the hypothesis is proven. There are, however, several possible explanations for the presence of a (dose)² component for single breaks. These are (1) the exchange hypothesis; (2) the above suggestion that some breaks may arise from an incomplete X-type reunion or an isochromatid break; (3) the extensive and complex manner of folding of the chromatin during both interphase and metaphase may make it necessary for a chromosome to sustain two breaks before a complete separation is obtained; and (4) some small interstitial deletions, in which the two-break points are moderately well separated, show a (dose)² component [79, 80, 91]. The end result is a long fragment and a small isodiametric deletion. If the latter is not taken into consideration, the lesion may be scored as a simple break rather than a two-break interstitial deletion.

The Exchange Hypothesis Could Explain the Increased Breaks in Heterochromatin

Under varying circumstances, both ionizing radiation and chemicals can produce selective chromosome breakage in heterochromatic regions [18, 42, 61, 73, 108]. Since heterochromatin is highly condensed during interphase, there should be a much greater opportunity for interaction between two primary lesions in heterochromatin than in euchromatin. Thus, if at least some breaks are the result of the mechanism suggested by the exchange hypothesis, this could explain the increased breakability in heterochromatin.

Chromosome Breakage at the Molecular Level

The use of alkaline or neutral sucrose-gradient centrifugation of DNA has shown that both single- and double-stranded breaks can occur immediately after irradiation, and that single-stranded breaks predominate 10 to 1 [116]. This has in it all the elements of a compromise. One could say that the immediate production of double-stranded breaks proves that the breakage-first hypothesis is sometimes correct and that all those single-stranded breaks represent the primary lesions of the exchange hypothesis which, if not immediately repaired, could interact to produce exchange breaks.

CHROMATID GAPS VS. BREAKS IN DNA

Fine Structure of Gaps

When chromosomes are exposed to various forms of ionizing radiation or to certain chemicals and viruses, the lesions induced may vary from slight attenuation of the chromatin, to attenuated secondary constrictions, to "achromatic regions" or gaps, to complete breaks with displacement of the fragments. Fine-structure studies show that the gaps come in two classes: those with chromatid fibers passing across the achromatic region, and those which are true discontinuities (Fig. 3) *[10, 11, 105, 106]*. The latter can be true breaks, or the result of breakage of fibers during the fixation *[11]*. Breaks that remain aligned but appear by light microscopy to be complete discontinuities may occasionally be seen by electron microscopy to possess chromatin fibers bringing the discontinuity *[10]*. These obviously cause some problems in scoring if one is trying to distinguish between simple achromatic lesions without discontinuity and true breaks with discontinuity. However, for the purposes of the present discussion, a gap will be taken to mean an achromatic region traversed by one or more chromatin fibers.

Opinions on what gaps represent are varied and include the suggestions that they are (1) breaks that are not complete due to chromosome matrix or other factors [99]; (2) half-chromatid or subchromatid breaks ("half-chromatid" or "subchromatid" breaks are breaks, or breaks with reunions, which appear to involve less than the whole width of the chromatid; see p. 122) [87]; (3) primary lesions or events that will heal (see p. 100); (4) incomplete chromatid exchanges [43, 44, 52]; or (5) defects in chromosome coiling *[11]*. These basically boil down to two alternatives: (1) gaps actually represent a break somewhere in the continuity of the DNA, despite the chromatin fibers passing between the fragment and the rest of the chromosome; (2) gaps do not represent a break in the DNA.

Evidence that Gaps Are Not True Breaks

Many of the early studies on the frequency of chromatid breaks following irradiation showed significant differences between laboratories in the relative frequency of chromatid breaks compared with exchanges. The studies of Revell [89] helped to emphasize that much of this inconsistency was due to variation in scoring; some laboratories score gaps as breaks, while others score only displaced fragments as breaks. Because of this problem, most studies now score gaps and breaks separately,

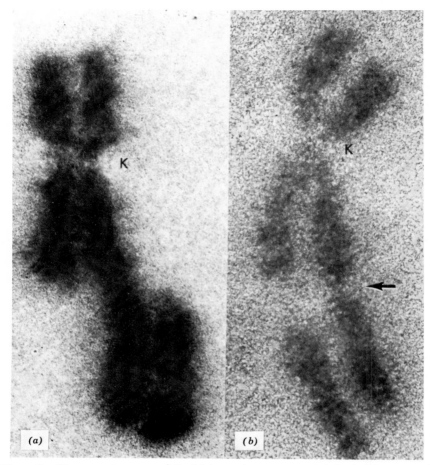

Figure 3. Electron microscopy of fixed human chromosomes: (a) displaced chromatid break, (b) break without displacement in one chromatid and gap in the other chromatid (Brinkley BR, Shaw MW, 1970; ref. [10]).

and many have suggested that gaps do not represent true breaks. Some of the reasons for this view follow. (1) By light microscopy the gaps are not displaced, and they frequently appear only as hypochromatic regions rather than true discontinuities. (2) Ultrastructural studies frequently show the presence of some chromatid strands passing across the gap regions [10, 11, 105, 106]. (3) Gaps do not separate into centric and acentric fragments at anaphase [28, 37, 89]. (4) Gaps do not show up as breaks in the second mitosis (M_2) [37, 89], whereas subchromatid breaks do. These reasons have usually been sufficient to convince most workers that gaps are not merely unresolved breaks. There are, however, some reasons to suspect that gaps might be true breaks after all.

Evidence that Gaps May Be True Breaks

Evidence in at least a half-dozen areas can be cited to support the position that gaps may be true breaks.

(1) Gaps occur after exposure to all agents that cause chromosome breakage, including X-rays, other forms of ionizing radiation, and chemicals. (2) Gaps show a linear increase with dose [44, 77, 90, 107], although in one report [107] a (dose)^2 component was observed. (3) Autoradiographic studies [52] show that there is frequently an exchange of label at the site of gaps, which indicates that some gaps represent breaks resulting from a sister-chromatid exchange. (4) The frequency of gaps is increased when irradiation is carried out in the presence of oxygen [77], just as the frequency of chromatid breaks is increased. (5) The distribution of X-ray-induced gaps along the chromosome arms in *Vicia faba* [104] is the same as the distribution of breaks [39, 92]. (6) The time of occurrence in the cell cycle following X-irradiation fits well with the probability that subchromatid breaks, gaps, and chromatid breaks are all different manifestations of a break in the continuity of the chromatin fiber.

This last point is made particularly well in Revell's own data: *Vicia faba* root tips were exposed to 50 R of X-rays at multiple intervals, from 10 minutes to 5 hours before mitosis. The results in terms of frequency per anaphase cell of subchromatid bridges, gaps, and breaks (chromatid and isochromatid) are shown in Fig. 4. Irradiation at prophase, which occurred during the 0- to 2-hour interval, produced mostly subchromatid bridges and, as discussed elsewhere, these are actually chromatid breaks (p. 122). At 2 hours before mitosis, there was a sudden increase in the frequency of both gaps and breaks [89, 107]. They both reached a peak at 3 hours but dropped in frequency at 4 hours, and both began to rise again at 5 hours. When subchromatid lesions were added to the gaps and breaks, to give a total curve, the initial asymmetry in the gap and break curves disappeared, and the curves became symmetrical. This, plus the manner in which the frequency curve for the gaps almost exactly mimics the curve for the breaks, and the fact that subchromatid bridges are also chromatid breaks, suggest that actually all three lesions are breaks, each manifested in a different way, and that the way is dictated by the tertiary structure or manner of folding of the chromatin fiber into the chromatid.

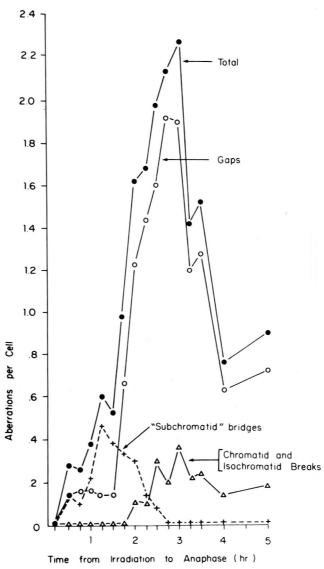

Figure 4. Diagram of the data of Revell [89] on the frequency of subchromatid bridges, gaps, and chromatid and isochromatid breaks in *Vicia faba* at various times after exposure of the root meristems to 50 r of X-rays.

Influence of Chromatin Fiber Folding on How Breaks Present Themselves

How is it possible for a single event—namely, a break—to be manifested in three different ways depending on the manner of folding of the chromatin fiber? Many details of the tertiary structure of chromosomes have been presented and reviewed elsewhere and will not be repeated here [20, 22–27]. It is sufficient to say that the concept of the chromatin fiber being completely dispersed during interphase,

then folding into a chromatid at mitosis,

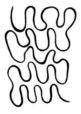

is an oversimplification. It fits well with the proposal that a break in the interphase chromatin fiber

should present as a break in the metaphase chromatid.

Whole-mount electron microscopy, however, suggests a more complex situation in which the interphase chromatin, instead of being freely dispersed in the interphase nucleus, is actually attached at many sites to a localized portion of the nuclear membrane (arrows) [20, 23–25]

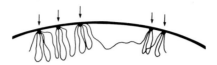

Studies of water-spread interphase chromatin suggest it clusters together at points (arrow),

and that these points are sites of attachment to the nuclear membrane [23–25, 35, 64]. Similar studies of metaphase chromosomes suggest that even though the chromatin pulls away from the nuclear membrance at mitosis, these cluster points may persist and lend a certain degree of stability to the folding pattern of metaphase chromatin [24]. Thus, the interphase chromatin (without the nuclear membrane) may be organized in a manner like this,

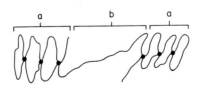

with clusters of closely knit chromatin *a* connected by an unclustered chromatin fiber *b*.

I should hasten to emphasize that the major point of what follows is that the tertiary structure of the chromosome may be playing a very important role in determining the way in which breaks are manifested.

Since our knowledge of this tertiary structure is still incomplete, the subsequent diagrams present just one of several possible variations on a theme. For the present, the details of the folding patterns are less important than the general concept.

The theme will be developed that a break in the *a* region might present as a gap, while a break in the *b* region might present as a break, and since much more of the chromatin is in the *a* configuration, gaps will be more prevalent than breaks. The third variable comes with the onset of chromosome condensation at prophase where the *b* region may be obliterated.

Now let us carry this proposal to its logical conclusion and first examine chromatid breaks. During interphase these may represent breaks in the *b* regions.

At prophase the proximal and distal chromatin would contract independently, and at metaphase a complete break would result.

A chromatid gap, on the other hand, may represent a break in the *a* region.

During prophase this region would partially separate, but cluster points would prevent complete separation,

and at metaphase this would present as a gap but not a complete discontinuity.

At the next interphase the original configuration would occur, with a good possibility for repair of the break.

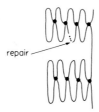

Thus at the following mitosis, unlike Humpty Dumpty, everything would be put back together again,

and gaps would not show up as chromatid breaks at M_2. Such repair could also explain the observation that gaps produced by irradiation during the S phase show dose kinetics with an n less than 1 [*107*].

Finally, with subchromatid breaks the irradiation is not administered until prophase condensation has occurred, and this would prevent a break in the b region from producing a complete chromatid break.

At mitosis a subchromatid bridge would occur. By the next interphase, however, this protective effect would be lost,

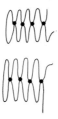

and following DNA replication a chromosome break would be detected [*21*].

If the prophase irradiation caused a break in the *a* region, it might be totally protected and repaired and thus never show up as a chromatid break. Thus the "gaps" of prophase irradiation would never be seen, and this might explain the low frequency of "breaks" during the first 2 hours prior to mitosis. This would also suggest that the frequency of subchromatid bridges would be similar to the frequency of chromatid breaks occurring from the 2 to 4 hours before mitosis and, as seen in Fig. 4, this is the case. In fact, if the frequencies of subchromatid bridges, and chromatid and isochromatid breaks, are combined, almost a flat plateau is produced from 1 to 4 hours, and the variation of breaks during the G_2 period is not all that striking.

Whether this interpretation is correct or not, it does provide an alternative way of looking at these lesions and shows that it is possible that gaps may merely be one of three ways in which a chromatid break may be manifested. Even if gaps are breaks, they should still be scored separately, because "true gaps" and secondary constrictions induced in regions of heterochromatin may not always be distinguishable.

Acentric Chromosomes Show Fewer Breaks

When cells are irradiated in the G_1 period, chromosome breaks result in acentric fragments. A second irradiation in the G_2 period will now add chromatid breaks to the picture. At mitosis, the frequency with which such chromatid breaks show up is one-ninth as great in the acentric fragments as in the centric chromosomes [*101*]. This indicates that the strains placed on a chromosome by traction on the centromere during chromosome movement are an important factor in changing latent breaks into visible breaks. This is further substantiated by the observation that cells exposed to colchicine, which inhibits chromosome movement, show fewer breaks than control cells [*13*]. Stress produced by centrifugation after X-irradiation accentuates breaks, while centrifugation alone does not cause breakage [*102*]. These observations further emphasize the fact that simple scissions of the chromatid fiber alone do not automatically result in a visible break. The tendency for the tertiary structure of the chromosome to hold the broken ends together must still be overcome if a clean separation is to result.

BREAKS IN G_1 AND CHROMOSOME STRANDEDNESS
Transition from Chromosome to Chromatid Aberrations

In his early studies of chromosome breakage in *Tradescantia*, Sax [100] recorded the types of aberrations occurring in cells fixed at hourly intervals after irradiation. The first breaks to appear at metaphase were of the chromatid type. At 26 to 30 hours before mitosis, chromosome aberrations began to predominate. He interpreted this to indicate that before this transition the chromosomes had not yet divided, and after the transition they had divided. In terms of the double DNA helix, chromosome aberrations can be represented by

and the chromatid aberrations by

These and other, similar studies [6, 15, 16] provided the first evidence that chromosome splitting occurred in mid-interphase rather than at the beginning of mitosis, as originally believed on cytological grounds.

To provide precise evidence on the relationship between the chromosome-to-chromatid break transition and DNA synthesis, it was necessary to have an independent measure of the time of DNA replication. When irradiation studies were done in combination with exposure of cells to tritiated thymidine, a very interesting phenomenon was noted. Instead of a gradual transition extending throughout the S period, during which some cells showed both chromosome and chromatid breaks, the transition was actually rather abrupt, with only occasional cells showing both types. The transition occurred either in very early S or in late G_1 [12, 33, 40, 49, 54, 57, 74, 124, 126]. Various aspects of these studies have strongly suggested that the transition period was actually occurring in late G_1, before DNA synthesis.

Relevance of a G_1 Split to Chromosome Strandedness

How can a chromosome appear to be "split" by X- and γ-rays, in the absence of either DNA or protein synthesis? Wolff [123, 124] has suggested that this is an indication that chromosomes are actually double or multistranded. The reasoning behind this is as follows. If an early G_1 chromosome contains two closely knit DNA helices instead of one, a single hit will produce chromosome lesions.

If two DNA helices separate from each other at late G_1, they will form two independent targets, and the result will be a chromatid lesion before the actual onset of DNA synthesis.

There is, however, a flaw in this line of reasoning in that with DNA replication the daughter molecules do not segregate as shown in the diagram. In this diagram the G_1 chromatid is composed of two DNA helices *a* and *b*, the daughter strands of *a* are passing to the same side of the centromere to give an *aa'* chromatid, and both *b* daughter strands are passing to the other side to give a bb' chromatid. This violates a basic tenet of DNA replication—that one daughter strand passes to one chromatid and the other passes to the other chromatid. When this rule is followed, a break in late G_1 involving separated DNA molecules should show up as two half-chromatid lesions at mitosis.

Since half-chromatid lesions are not seen at mitosis, and since most evidence favors a single-stranded model of chromosome structure [22], we have to look elsewhere for an explanation for this sudden transition in late G_1.

The reason may become obvious when the mechanism of DNA replication in eukaryotes is more thoroughly understood. One possibility, for example, may be that during late G_1, DNA strands at the initiating points of all replicons separate or replicate a few bases in the form of either a DNA or an RNA primer. If these are more susceptible to X-rays than unreplicated or completely replicated DNA, then on the basis of X-ray breakage the chromosomes would all seem to have suddenly split. The occurrence of fragmentation of Chinese hamster cells when exposed to FuDR at early S [83] may be a different manifestation of the same phenomenon.

That there is much to be learned about the mechanism of the chromosome-to-chromatid break transition is indicated by the effect of Colcemid on Chinese hamster tissue culture cells. When these cells were synchronized by mitotic selection without using Colcemid, and irradiated in mitosis or G_1, 89 to 94% of the aberrations were of the chromosome type [34]. However, when cells were synchronized by mitotic selection with Colcemid (where the mitotic cells were accumulated by exposure to Colcemid for 2 hours) and then irradiated in mitosis or G_1, chromatid aberrations were just as frequent as chromosome aberrations. This was due to an absolute increase in the frequency of chromatid aberrations. The relative frequency of chromosomal aberrations remained unchanged. Thus the simple presence of Colcemid resulted in the appearance of chromatid breaks in cells irradiated in G_1.

DO HALF-CHROMATIDS AND HALF-CHROMATID ABERRATIONS REALLY EXIST?

For many years, cytogeneticists have observed the presence of half-chromatids in the coiled arms of both plant and animal chromosomes [47, 71, 78, 114]. These are not artifacts of fixation, since they can also be observed in living material [2]. Their appearance is enhanced by treatment of the chromosomes with trypsin [115, 119], and they have also been observed by whole-mount electron microscopy of large *Vicia faba* chromosomes [119]. When chromosomes are irradiated during prophase, aberrations that have been termed half-chromatid exchanges are seen at the following anaphase [31, 32, 63, 86, 103, 110, 117, 118]. These two observations have frequently been used as evidence that the chromosomes are at least binemic; that is, they contain two longitudinal DNA helices per chromatid. This conclusion, however, was challenged by Ostergren and Wakonig [85]. They reasoned that if a true half-chromatid aberration occurred at the first anaphase A_1, then full-chromatid aberrations should be present at the next mitosis M_2; and yet, when they tested this using coumarin as a breaking agent, they found chromosome breaks predominantly at M_2. This is the lesion that would be expected if the M_1 event were actually a full-chromatid break involving a single-stranded chromatid. Kihlman and Hartley [60], using *Vicia faba* cells, confirmed this observation of chromosome breaks at M_2 after prophase irradiation. These authors' conclusion that half-chromatid breaks were actually chromosome breaks was challenged by Heddle [51], who found 8 to 17% chromatid aberrations at M_2 in irradiated *Vicia faba* cells.

This controversy again touches on the question of how DNA daughter molecules segregate following DNA replication. According to the multi-stranded model, these aberrations are the result of a half-chromatid union or exchange (Fig. 5). They may also, however, be the result of a break and reunion in a single-stranded chromosome, as shown in Fig. 6. Comings [21] has pointed out that when the mode of segregation of the DNA molecules is taken into consideration, one daughter strand passing to each side of the centromere, the lesion that should really be expected at M_2 is the presence of two half-chromatid breaks. On the other hand, if the subchromatid lesion is actually an incompletely resolved, double-stranded DNA break, at M_2 it should show up as a chromosome break. All observers, regardless of their prior convictions, have noted that chromosome breaks are by far the predominant lesion at M_2, and no one has found any evidence of double half-chromatid lesions. In extensive whole-mount electron microscopy studies of chromosomes from many different species of mammals and birds, Comings

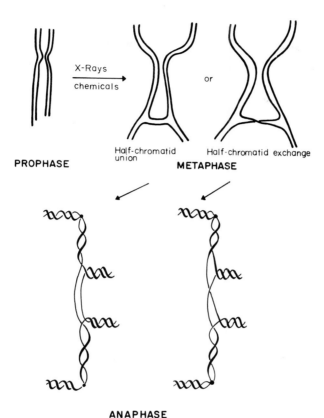

Figure 5. Classical interpretation of half-chromatid exchanges based on a multistranded model of chromosome structure (from [21]).

and Okada [26] could find no evidence for any subdivision of the chromatid other than the 100- to 250-Å chromatid fiber, and they suggested that the half-chromatids so frequently seen by light microscopy are an optical illusion resulting from the coiling of the chromatid. This is consistent with numerous other lines of evidence suggesting that chromosomes are single-stranded [22].

THE SITE–LIMITATION HYPOTHESIS

Although Lea [68] and others [67] observed that the frequency of nuclei with zero, one, two, or more aberrations after treatment followed a Poisson distribution, Atwood and Wolff [1, 121] found that in *Tradescan-*

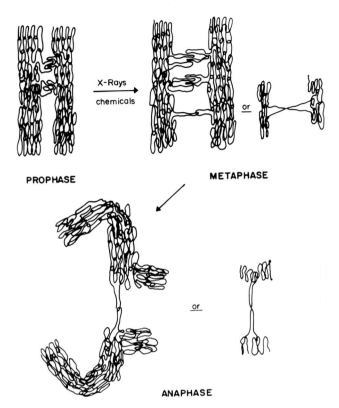

Figure 6. Half-chromatid exchanges on the basis of a single-stranded model of chromosome structure (from [21]).

tia, Vicia, and barley the number of dicentrics and rings did not follow a Poisson distribution. There was an excess of nuclei with one aberration and a deficiency of nuclei with two or more [1, 94, 95, 121]. Selective loss of nuclei with multiple aberrations was ruled out by the observation that other aberrations, such as isodiametric deletions, did follow a Poisson distribution. When the dosage of radiation was increased in an attempt to produce multiple aberrations, the dose–response curve for rings, di- and multicentrics reached a saturation point at which higher doses did not produce increased aberrations [29, 82, 96, 122]. This led Wolff and Atwood [123, 127] to propose that there was a severe restriction on the number of sites in the nucleus where two chromatid strands come close enough to allow an exchange to occur.

As an example, in *Tradescantia* cells irradiated with neutrons, the effective number of sites (N) at saturation was 2.7 per nucleus [96]. With *Tradescantia* irradiated by X-rays, Wolff [122] calculated an N of 5.35, while Savage [96] found no evidence for saturation, indicating that the number of sites differs with different types of radiation.

In *Vicia faba* irradiated in G_1, the number of sites was estimated at two per nucleus [45, 121]. By contrast, the chromatid aberrations produced by irradiation during the G_2 period conformed to the Poisson distribution [45]. This, it was suggested, indicated that DNA replication resulted in more than doubling of the number of exchange sites [45]. This increase in the number of sites has been advanced as the explanation for the increased number of exchanges produced when cells are irradiated during the G_2 period [7, 45].

Additional evidence that can be taken as support for the site hypothesis comes from the observation that the frequency of sister-chromatid exchanges in endoreduplicated cells is about 100-fold greater than in control cells. This could be explained on the basis of intimate pairing between the chromatids in endoreduplicated cells, leading to an increase in the effective number of sites [84].

ALTERNATIVE TO THE SITE HYPOTHESIS: REAL VS. EFFECTIVE LIMITATION

One needs only to look at an electron microscopic cross section of the nucleus and see the highly compacted quality of the chromatin lining the nuclear membrane, and the nucleolus to realize that there are thousands of sites where the chromatin fibers are less than 0.1 μm apart, which is estimated to be the maximum distance between two chromatids that will still allow an exchange to occur [121, 127]. Other than considerably reducing the 0.1 μm figure [94], how can this be reconciled with the above hypothesis which suggests that there are fewer than 10 such sites per nucleus? The resolution revolves around the question of whether there is an actual limitation of sites or only an effective limitation.

The following are some factors which would lead to an effective limitation of rings or dicentrics. (1) Higher doses of radiation may tend to inhibit rejoining [65, 66, 90, 107]. (2) Since the rejoining is estimated largely on the basis of dicentrics, the total number of centromeres present would put an absolute limit on the number of possible dicentrics that can be formed [82, 89]. (3) Inherent distortions in the binomial distribution of exchanges could explain some of the results [97]. (4) Sax [101] found that once a chromosome was broken, it was difficult to break it again. He attributed this to release of stresses required to actuate a break. (5) The site hypothesis is difficult to reconcile with data describing the production of two-break aberrations by α-particles in *Vicia faba* [88].

In summary, it is clear that with increasing doses of radiation there is a saturation level beyond which no further increases in exchanges are produced. While this can be due to either an absolute or an effective limitation in the number of potential exchange sites, the latter seems more likely.

CONCLUDING COMMENTS

As one peruses the literature on radiation and chemically induced breaks, the following statements appear: (1) Gaps are not true breaks. (2) The fact that chromosomes respond to X-rays as though they are split in the G_1 period indicates that chromosomes are multistranded. (3) Subchromatid aberrations represent breaks and exchanges in half-chromatids. (4) There are a limited number of sites at which exchange can occur in the nucleus.

When the tertiary structure of the chromatin fibers (both in the interphase nucleus and in the mitotic chromosomes) and the manner of segregation of newly replicated DNA are taken into consideration, the following conclusions may be drawn. (1) Gaps may in fact represent true chromatid breaks that do not become visible as such because of the manner in which the chromatin fibers are arranged in the mitotic chromosome. (2) The G_1 split does not provide any evidence that chromosomes are multistranded. (3) Subchromatid aberrations are actually chromatid aberrations. (4) There is probably an effective rather than an absolute limitation of exchange sites in the nucleus.

Acknowledgments.–This work was supported in part by National Institutes of Health Grant GM 15886.

LITERATURE CITED

1. ATWOOD KC: Numbers of nuclear sites for aberration formation and the distribution of aberrations. *In* Radiation Induced Chromosome Aberrations (Wolff S, ed), New York, Columbia University Press, 73–86, 1963.
2. BAJER A: Subchromatid structure of chromosomes in the living state. *Chromosoma* 17:291–302, 1965.
3. BELL AG, BAKER DG: Irradiation-induced chromosome aberrations in normal human leukocytes in culture. *Can J Genet Cytol* 4:340–351, 1962.
4. BENDER MA, BARCINSKI MA: Kinetics of two-break aberration production by X-rays in human leukocytes. *Cytogenetics* 8:214–246, 1969.

5. BENDER MA, GOOCH PC: Types and rates of X-ray induced chromosome aberrations in human blood irradiated *in vitro*. *Proc Nat Acad Sci (US)* 48:522–532, 1962.
6. BISHOP CJ: Differential X-ray sensitivity of *Tradescantia* chromosomes during the mitotic cycle. *Genetics* 35:175–187, 1950.
7. BREWEN JG: Studies on the frequencies of chromatid aberrations induced by X-rays at different times of the cell cycle of *Vicia faba*. *Genetics* 50:101–107, 1964.
8. BREWEN JG: Kinetics of X-ray induced chromatid aberrations in *Vicia faba* and studies relating aberration frequencies to the cell cycle. *Mutat Res* 1:400–408, 1964.
9. BREWEN, JG, BROCK RD: The exchange hypothesis and chromosome-type aberrations. *Mutat Res* 6:245–255, 1968.
10. BRINKLEY BR, SHAW MW: Ultrastructural aspects of chromosome damage. *In* Genetic Aspects and Neoplasia, Twenty-third Annual Symposium on Fundamental Cancer Research, M. D. Anderson Hospital, Baltimore, Williams & Wilkins Co., 313–345, 1970.
11. BROGGER A: Apparently spontaneous chromosome damage in human leukocytes and the nature of chromatid gaps. *Humangenetik* 13:1–14, 1971.
12. BROOKS AL, PETERS RF, ROLLAG MD: Metaphase chromosome aberrations in Chinese hamster liver cells *in vivo* after single acute ^{60}Co exposure. *Radiat Res* 45:191–201, 1971.
13. BRUMFIELD RT: Effect of colchicine pretreatment on the frequency of chromosomal aberrations induced by X-irradiation. *Proc Nat Acad Sci (US)* 29:190–193, 1943.
14. BUCKTON KE, LANGLANDS AO, SMITH PG: Chromosome aberrations following partial- and whole body X-irradiation in man. Dose–response relationships. *In* Human Radiation Cytogenetics (Evans, J, Court Brown, WM, McLean, AS, eds), Amsterdam, North Holland Publishing Co., 122–135, 1967.
15. CATCHESIDE DG: Genetic effects of radiations. *Adv Genet* 2:271–358, 1948.
16. CATCHESIDE DG, LEA DE, THODAY JM: The production of chromosome structural changes in *Tradescantia* microspores in relation to dosage, intensity and temperature. *J Genet* 47:137–149, 1946.
17. CHU EHY, GILES NH, PASSANO K: Types and frequencies of human chromosome aberrations induced by X-rays. *Proc Nat Acad Sci (US)* 47:830–839, 1961.
18. COHEN MM, SHAW MW: Effects of mitomycin C on human chromosomes. *J Cell Biol* 23:386–395, 1964.
19. COLE A, CORRY PM, LANGLEY R: Effects of radiation and other agents on the molecular structure and organization of the chromosome. *In* Genetic Concepts and Neoplasia, Twenty-third Annual Symposium on Fundamental Cancer Research, M. D. Anderson Hospital, 1969, Baltimore, Williams & Wilkins Co., 346–379, 1970.
20. COMINGS, DE: The rationale for an ordered arrangement of chromatin in the interphase nucleus. *Am J Hum Genet* 20:440–460, 1968.
21. COMINGS DE: Half-chromatid aberrations and chromosome strandedness. *Can J Genet Cytol* 12:960–964, 1970.

22. COMINGS DE: The structure and function of chromatin. *In* Advances in Human Genetics (Harris H, Hirschhorn, K, eds), New York, Plenum Press, 237–431, 1972.
23. COMINGS DE, OKADA TA: The association of nuclear membrane fragments with metaphase and anaphase chromosomes as observed by whole mount electron microscopy. *Exp Cell Res* 63:62–68, 1970.
24. COMINGS DE, OKADA TA: The association of chromatin fibers with the annuli of the nuclear membrane. *Exp Cell Res* 62:293–302, 1970.
25. COMINGS DE, OKADA TA: The condensation of prophase chromosomes onto the nuclear membrane. *Exp Cell Res* 63:471–473, 1970.
26. COMINGS DE, OKADA TA: Do half-chromatids exist? *Cytogenetics* 9:450–459, 1970.
27. COMINGS DE, OKADA TA: Electron microscopy of chromosomes. *In* Perspectives in Cytogenetics (Wright SW, Crandall BF, Boyer L, eds), Springfield, Ill., Charles C. Thomas, 223–250, 1972.
28. CONGER AD: Real chromatid deletions versus gaps. *Mutat Res* 4:449–459, 1967.
29. CONGER AD, CURTIS HJ: Abnormal anaphases in regenerating mouse livers. *Radiat Res* 33:150–161, 1968.
30. CONGER AD, RANDOLPH ML, SHEPPARD CW, LUIPPOLD HJ: Quantitative relation of RBE in *Tradescantia* and average LET of gamma x-rays, and 1.3-2.5-, and 14.1- Mev fast neutron. *Radiat Res* 9:525–547, 1958.
31. CROUSE HV: X-ray breakage of lily chromosomes at first meiotic metaphase. *Science* 119:485, 1954.
32. CROUSE HV: Irradiation of condensed meiotic chromosomes in lilium longiflorum. *Chromosoma* 12:190–214, 1961.
33. DEWEY WC, HUMPHREY RM, SEDITA BA: Cell cycle kinetics and radiation induced chromosomal aberrations studied with C^{14} and H^3 labels. *Biophys J* 6: 247–260, 1966.
34. DEWEY WC, MILLER HH: X-ray induction of chromatid exchanges in mitotic and G_1 Chinese hamster cells pretreated with Colcemid. *Exp Cell Res* 57:63–70, 1969.
35. DUPRAW EJ: The organization of nuclei and chromosomes in honeybee embryonic cells. *Proc Nat Acad Sci (US)* 53:161–168, 1965.
36. EVANS HJ: Chromatid aberrations induced by gamma irradiation. I. The structure and frequency of chromatid interchanges in diploid and tetraploid cells of *Vicia faba*. *Genetics* 46:257–275, 1961.
37. EVANS HJ: Chromosome aberrations induced by ionizing radiations. *Int Rev Cytol* 13:221–321, 1962.
38. EVANS HJ: Repair and recovery at chromosome and cellular levels: similarities and differences. *Brookhaven Symp Biol* 20:111–133, 1968.
39. EVANS HJ, BIGGER TRL: Chromatid aberrations induced by gamma irradiation. II. Non-randomness in the distribution of chromatid aberrations in relation to chromosome length in *Vicia faba* root-tip cells. *Genetics* 46:277–289, 1961.
40. EVANS HJ, SAVAGE JRK: The relation between DNA synthesis and chromosome structure as resolved by X-ray damage. *J Cell Biol* 18:525–540, 1963.

41. FABERGE AC: An experiment on chromosome fragmentation in *Tradescantia* by X-rays. *J Genet* 39:229–248, 1940.
42. FORD CE: Chromosome breakage in *Vicia faba* root tip cells. *Proc 8th Int Congr Genet, Hereditas Suppl* 570–571, 1949.
43. Fox DP: The effects of X-rays on the chromosomes of locust embryos. III. The chromatid aberration types. *Chromosoma* 20:386–412, 1967.
44. Fox DP: The effects of X-rays on the chromosomes of locust embryos. IV. Dose–response and variation in sensitivity of the cell cycle for the induction of chromatid aberrations. *Chromosoma* 20:413–441, 1967.
45. GARCIA-BENITEZ C, WOLFF S: On the increase of sites for chromosome exchange formation after chromosome duplication. *Science* 135:438–439, 1962.
46. GILES NH: Comparative studies of the cytological effects of neutrons and X-rays. *Genetics* 28:398–418, 1943.
47. GIMEMEZ-MARTIN G, LOPEZ-SAEZ JF, GONZALEZ-FERNANDEZ A: Somatic structure (observations with the light microscope). *Cytologia* 28:381–389, 1963.
48. GOOCH PC, BEDNER MA, RANDOLPH ML: Chromosome aberrations induced in human somatic cells by neutrons. *In* Biological Effects of Neutron and Proton Irradiation, vol 1, Vienna, International Atomic Energy Agency, 325–342, 1964.
49. GRANT CJ: Chromosome aberrations and the mitotic cycle in Trillium root tips after X-irradiation. *Mutat Res* 2:247–262, 1965.
50. HAQUE: The fractionation effect in *Tradescantia*. *Heredity* 6:Suppl, 35–40, 1953.
51. HEDDLE JA: The strandedness of chromosomes: Evidence from chromosomal aberrations. *Can J Genet Cytol* 11:783–793, 1969.
52. HEDDLE JA, BODYCOTE J: On the formation of chromosomal aberrations. *Mutat Res* 9:117–126, 1970.
53. HEDDLE JA, WHISSEL D, BODYCOTE DJ: Changes in chromosome structure induced by radiations: a test of the two chief hypotheses. *Nature* 221:1158–1160, 1969.
54. HEDDLE JA, TROSKO JE: Is the transition from chromosome to chromatid aberrations the result of the formation of single-stranded DNA? *Exp Cell Res* 42:171–177, 1966.
55. HORIKAWA M, FUKUHARA M, SUZUKI F, NIKAIDO O, SUGAHARA T: Comparative studies on induction and rejoining of DNA single-strand breaks by radiation and chemical carcinogen in mammalian cells *in vitro*. *Exp Cell Res* 70:349–359, 1972.
56. HORIKAWA M, NIKAIDO O, TANAKA T, NAGATA H, SUGAHARA T: Comparative studies on rejoining of DNA-strand breaks induced by X-irradiation in mammalian cell lines *in vitro*. *Exp Cell Res* 63:325–332, 1970.
57. HSU TC, DEWEY WC, HUMPHREY RM: Radiosensitivity of cells of Chinese hamster *in vitro* in relation to the cell cycle. *Exp Cell Res* 27:441–452, 1962.
58. KAPLAN HS: DNA-strand scission and loss of viability after X-radiation of normal and sensitized bacterial cells. *Proc Nat Acad Sci (US)* 55:1442–1448, 1966.

59. KELLY S, BROWN CD: Chromosome aberrations as a biological dosimeter. Am J Pub Health 55:1419–1429, 1965.
60. KIHLMAN BA, HARTLEY B: "Sub-chromatid" exchanges and the "folded-fiber" model of chromosome structure. Hereditas 57:289–294, 1967.
61. KIHLMAN BA, LEVAN A: Localized chromosome breakage in Vicia faba. Hereditas 37:382–388, 1951.
62. KIRBY-SMITH JS, DANIELS DS: The relative effects of X-rays, gamma rays and beta rays on chromosomal breakage in Tradescantia. Genetics 38:375–388, 1953.
63. LaCOUR LF, RUTISHAUSER A: X-ray breakage experiments with endosperm. I. Sub-chromatid breakage. Chromosoma 6:696–709, 1954.
64. LAMPERT F: Attachment of human chromatin fibers to the nuclear membrane, as seen by electron microscopy. Humangenetik 13:285–295, 1971.
65. LANE GR: X-ray fractionation and chromosome breakage. Heredity 5:1–35, 1951.
66. LANE GR: Interpretation in X-ray chromosome breakage experiments. Heredity 6:Suppl, 23–34, 1953.
67. LANGLANDS AO, SMITH PG, BUCKTON KE, WOODCOCK GE, McLELLAND J: Chromosome damage induced by radiation. Nature 218:1133–1135, 1968.
68. LEA DE: Actions of Radiations on Living Cells, New York, Macmillan, 1st ed 1947, 2nd ed 1955.
69. LEA DE, CATCHESIDE DG: The mechanism of the induction by radiation of chromosome aberrations in Tradescantia. J Genet 44:216–245, 1942.
70. LETT JT, CALDWELL I, DEAN CJ, ALEXANDER P: Rejoining of X-ray induced breaks in the DNA of leukemia cells. Nature 214:790–792, 1967.
71. MANTON I: New evidence on the telophase split in Todea barbara. Am J Bot 32:342–348, 1945.
72. MARINELLI LD, NEBEL BR, GILES NH, CHARLES DR: Chromosomal effects of low X-ray doses on five-day Tradescantia microspores. Am J Bot 29:866–874, 1942.
73. McLEISH J: The action of maleic hydrazide in Vicia. Heredity 6:Suppl, 125–147, 1953.
74. MONESI V, CRIPPA M, ZITO-BIGNAMI R: The stage of chromosome duplication as revealed by X-ray breakage and H^3-thymidine labeling. Chromosoma 21:369–386, 1967.
75. MULLER HJ: Artificial transmutation of the gene. Science 66:84–87, 1927.
76. NEARY GJ: The relation between the exponent of dose response for chromosome aberrations and the relative contribution of "two-track" and "one-track" processes. Mutat Res 2:242–246, 1965.
77. NEARY GJ, EVANS HJ: Chromatid breakage by irradiation and the oxygen effect. Nature 182:890–891, 1958.
78. NEBEL BR: Chromosome structure in Tradescantiae. I. Methods and morphology. Z Zellforsch Mikr Anat 16:251–284, 1932.
79. NEWCOMBE HB: Effects of X-rays on chromosomes. I. The chromosome variable. J Genet 43:145–171, 1942.
80. NEWCOMBE HB: The action of X-rays in the cell. II. The external variable. J Genet 43:237–248, 1942.

81. NORMAN A: Multihit aberrations. *In* Human Radiation Cytogenetics (Evans HJ, Court Brown WM, McLean AS, eds), Amsterdam North Holland Publishing Co., 53–57, 1967.
82. NORMAN A, SASAKI MS: Chromosome exchange aberrations in human lymphocytes. *Int J Radiat Biol* 11:321–328, 1966.
83. OCKEY CH, HSU TC, RICHARDSON LC: Chromosome damage induced by 5-fluoro-2'-deoxyuridine in relation to the cell cycle of the Chinese hamster. *J Nat Cancer Inst* 40:465–475, 1968.
84. OLIVIERI G, BREWEN JG: Evidence for nonrandom rejoining of chromatid breaks and its relation to the origin of sister-chromatid exchanges. *Mutat Res* 3:237–248, 1966.
85. OSTERGREN G, WAKONIG T: True or apparent sub-chromatid breakage and the induction of labile states in cytological chromosome loci. *Bot Notis* 4:357–375, 1955.
86. PEACOCK WJ: Sub-chromatid structure and chromosome duplication in *Vicia faba*. *Nature* 191:832, 1961.
87. READ J: Radiation Biology of *Vicia faba* in Relation to the General Problem. Oxford, Blackwell, 1959.
88. READ J: The induction of chromosome exchange aberrations by ionizing radiations: the "site concept." *Int J Radiol* 9:53–65, 1965.
89. REVELL SH: The accurate estimation of chromatid breakage, and its relevance to a new interpretation of chromatid aberrations induced by ionizing radiations. *Proc Roy Soc London B,* 150:563–589, 1959.
90. REVELL SH: Evidence for a dose-squared term in the dose–response curve for real chromatid discontinuities induced by X-rays, and some theoretical consequences thereof. *Mutat Res* 3:34–53, 1966.
91. RICK CM: On the nature of X-ray induced deletions in *Tradescantia* chromosomes. *Genetics* 25:467–482, 1940.
92. RIEGER R, MICHAELIS A: Vergleichende untersuchungen zur Verteilung durch verschiedene Mutagene induzierter bruche über den Chromosomensatz von *Vicia faba* L. *Chromosoma* 10:163–178, 1959.
93. RILEY HP, GILES NH, BETTY AV: The effect of oxygen on the induction of chromatid aberrations in *Tradescantia* microspores by X-irradiation. *Am J Bot* 39:592–597, 1952.
94. SAVAGE JRK: Chromosome-exchange sites in *Tradescantia paludosa*. *Int J Radiat Biol* 9:81–86, 1965.
95. SAVAGE JRK: Non-interaction of radiation-induced chromosome lesions in *Tradescantia* microspores. I. Fractionated X-day dose studies. *Int J Radiat Biol* 11:287–300, 1966.
96. SAVAGE JRK: Saturation of the dose–response curve for chromosome exchanges in *Tradescantia* microspores. *Mutat Res* 4:295–306, 1967.
97. SAVAGE JRK, PAPWORTH DG: Distortion hypothesis: an alternative to a limited number of sites for radiation-induced chromosome exchange. *J Theor Biol* 22:493–514, 1969.
98. SAVAGE JRK, PRESTON RJ, NEARY WL: Chromatid aberrations in *Tradescantia bracteata* and a further test of Revell's hypothesis. *Mutat Res* 5:47–56, 1968.

99. Sax K: Chromosome aberrations induced by X-rays. *Genetics* 23:494–516, 1938.
100. Sax K: An analysis of X-ray induced chromosomal aberrations in *Tradescantia*. *Genetics* 25:42–68, 1940.
101. Sax K: Mechanism of X-ray effects on cells. *J Gen Physiol* 25:533–537, 1942.
102. Sax K: Effect of centrifuging upon production of chromosomal aberrations by X-rays. *Proc Nat Acad Sci (US)* 29:18–21, 1943.
103. Sax K, King ED: An X-ray analysis of chromosome duplication. *Proc Nat Acad Sci (US)* 41:150–155, 1955.
104. Scheid W, Traut H: Non-random distribution of X-ray induced achromatic lesions ("gaps") in the chromosomes of *Vicia faba*. *Mutat Res* 6:481–483, 1968.
105. Scheid W, Traut H: Ultraviolet-microscopical studies on achromatic lesions ("gaps") induced by X-rays in the chromosomes of *Vicia faba*. *Mutat Res* 10:159–161, 1970.
106. Scheid W, Traut H: Visualization by scanning electron microscopy of achromatic lesions ("gaps") induced by X-rays in chromosomes of *Vicia faba*. *Mutat Res* 11:253–255, 1971.
107. Scott D, Evans HJ: X-ray induced chromosomal aberrations in *Vicia faba*: Changes in response during the cell cycle. *Mutat Res* 4:579–599, 1964.
108. Sparrow AH: Radiation sensitivity of cells during mitotic and meiotic cycles with emphasis on possible cytochemical changes. *Ann NY Acad Sci* 51:1508–1540, 1951.
109. Stadler LJ: The experimental modification of heredity in crop plants. I. Induced chromosomal irregularities. *Sci Agr* 11:557–572, 1931.
110. Swanson CP: X-ray and ultraviolet studies on pollen tube chromosomes. II. The quadripartite structure of the prophase chromosomes of *Tradescantia*. *Proc Nat Acad Sci (US)* 33:229–232, 1947.
111. Swanson CP: Relative effects of qualitatively different ionizing radiations on the production of chromatid aberrations in air and in nitrogen. *Genetics* 40:193–203, 1955.
112. Thoday JM: The effect of ionizing radiations on the broad bean root. Part IX. Chromosome breakage and the lethality of ionizing radiations to the root meristem. *Brit J Radiol (NS)* 24:572–576, 1951.
113. Thoday JM: Sister-union isolocus breaks in irradiated *Vicia faba*: the target theory and physiological variation. *Heredity* 6:Suppl, 299–309, 1953.
114. Trosko JE, Brewen JG: Cytological observations on the strandedness of mammalian metaphase chromosomes. *Cytologia* 31:208–212, 1966.
115. Trosko JE, Wolff S: Strandedness of *Vicia faba* chromosomes as revealed by enzyme digestion studies. *J Cell Biol* 26:125–135, 1965.
116. Veatch W, Okada S: Radiation-induced breaks of DNA in cultured mammalian cells. *Biochem J* 9:330–346, 1969.
117. Wilson GB, Sparrow AH: Configurations resulting from isochromatid and isosubchromatid unions after meiotic and mitotic prophase irradiation. *Chromosoma* 11:229–244, 1960.

118. Wilson GB, Sparrow AH, Pond V: Sub-chromatid arrangements in *Trillium erectum*. I. Origin and nature of configurations induced by ionizing radiation. *Am J Bot* 46:309–316, 1959.
119. Wolfe SL, Martin PG: The ultrastructure and strandedness of chromosomes from two species of *Vicia*. *Exp Cell Res* 50:140–150, 1968.
120. Wolff S: Delay of chromosome rejoining in *Vicia faba* induced by irradiation. *Nature* 173:501, 1954.
121. Wolff S: Interpretation of induced chromosome breakage and rejoining. *Radiat Res Suppl* 1:453–462, 1959.
122. Wolff S: The kinetics for two-break chromosome exchanges. *J Theor Biol* 3:304–314, 1962.
123. Wolff S: The splitting of human chromosomes into chromatids in the absence of either DNA or protein synthesis. *Mutat Res* 8:207–214, 1969.
124. Wolff S: On the "tertiary" structure of chromosomes. *Mutat Res* 10:405–414, 1970.
125. Wolff S, Luippold HE: Metabolism and chromosome break rejoining. *Science* 122:231–232, 1955.
126. Wolff S, Luippold HE: Chromosome splitting as revealed by combined X-ray and labeling experiments. *Exp Cell Res* 34:548–556, 1964.
127. Wolff S, Atwood KC, Randolph ML, Luippold HE: Factors limiting the number of radiation-induced chromosome exchanges. I. Distance: evidence from non-interaction of X-ray and neutron-induced breaks. *J Biophys Biochem Cytol* 4:365–372, 1958.

THE APPLICATION OF BANDING TECHNIQUES TO TUMOR CHROMOSOMES

MARGERY W. SHAW and T. R. CHEN

The University of Texas, Graduate School of Biomedical Sciences,
Medical Genetics Center, Houston, Texas

Introduction 135
Hematologic and Lymphatic Disorders 136
Childhood Tumors 137
Long-term Cell Lines 138
Human Fibroblast Cell Strains 139
Newly Established Solid-tumor Cell Strains 140
Concluding Comments 147
Acknowledgements 148
Literature Cited 149

INTRODUCTION

The horizons of the cytogeneticist have been considerably widened by the development of the modern staining techniques that allow differentiation of specific regions along the chromosome arms [20, 29]. As it has become apparent that all of the chromosomes of man thus could be defined and so identified, there has been a flurry of activity to apply these methods to the many problems awaiting just such a technologic breakthrough [32]. Included among these are: the discovery of new chromosomal polymorphisms not associated with genetic disease or defect [27, 9], the reinvestigation of patients with known structural rearrangements [11, 14], the search for small chromosomal mutations not detected by conventional staining methods [16], and the comparison of banding patterns in studies of the evolution of species [33]. Detailed descriptions of Q-bands, G-bands, C-bands, and R-bands have already been published elsewhere [27] and so will not be repeated here. Discussions of the mechanism of chromosome banding and the relationship of bands to chromosome structure and function also are given elsewhere [7, 8]. This review is confined to the applications made of these techniques to the banding of chromosomes in tumor cells.

HEMATOLOGIC AND LYMPHATIC DISORDERS

Soon after the quinacrine fluorescence staining method was described by Caspersson et al. [1], it was unexpectedly discovered that the Philadelphia (Ph^1) chromosome of chronic myelocytic leukemia (CML) is autosome No. 22 instead of No. 21 [2, 26]. Previously it had been assumed, based on the fact that leukocyte alkaline phosphatase levels are elevated in trisomy 21 and reduced in CML, that the chromosome that is trisomic in Down's syndrome (No. 21) is the same autosome that is deleted in the Ph^1 chromosome. However, through application of Q-, G-, R-, and C-staining methods, there is now no doubt that the banding of the Ph^1 chromosome matches better with that of a No. 22.

More recently, Rowley has studied the complete Q-band and G-band karyotypes of nine patients with CML who are Ph^1-positive. She finds, added to the end of the long arm of chromosome No. 9, a small, weakly fluorescent segment that is approximately the same length as the region missing from the long arm of chromosome No. 22 [30]. If this finding is confirmed, it suggests that the primary lesion in CML is a reciprocal translocation rather than a chromosomal deletion.

Ever since the discovery of the close association of the Ph^1 chromosome with CML, there has been a search for other specific associations between chromosomes and malignancy. Because the Ph^1 deletion (or translocation) has been found to involve such a small segment of autosomal material, it seems possible that other specific rearrangements could have gone undetected by conventional staining techniques. There is now evidence that this may be true, at least in some disorders, as illustrated by the recent findings in the karyotypes of the Burkitt lymphoma by Manalov and Manalova, who have reported a specific banding abnormality in this disease [23]. Among 12 individuals studied, 10 were shown to have an added band at the end of the long arm of one chromosome No. 14. This occurred in biopsy material and persisted during subsequent culture. It could not be determined whether the extra band originated as a translocation of a segment from another chromosome or as a duplication of a segment within No. 14. However, the former interpretation appeared to be more likely.

Another example of a specific chromosomal rearrangement in lymphoid tumors has been described by Fleischmann and coworkers [12, 13]. They used banding techniques to analyze lymph-node cell karyotypes from five patients with the following diagnoses: lymphosarcoma, Hodgkin's disease, undifferentiated malignant lymphoma, and lymphoblastic sarcoma. The modal chromosome number varied in the different tumors from the hypodiploid to the hyperdiploid range. However, among several markers present in each tumor strain, one marker is unique in that it appeared in all five tumors! It was described as lying between chromosomes No. 12 and No. 13 in total length but as having a strictly median centromere and an evenly bright fluorescence distributed equally in the two arms. It is quite likely an isochromosome, but its origin remains undetermined.

Two individuals with hematologic disorders associated with ineffective erythropoiesis were found to have No. 8 trisomy confined to the bone marrow [10]. Among five other patients with trisomy 8 and multiple developmental defects, three were mosaics [18, 3]. It will be important to follow these cases to determine whether they have an increased tendency to develop leukemia as a consequence of their underlying chromosomal imbalance.

Although not tumors of hematopoietic origin, the human meningiomas are characterized by either monosomy for a group G chromosome or deletion of a group G chromosome indistinguishable from the Ph[1] chromosome found in chronic myelocytic leukemia [36]. Q-banding studies have now shown the deletion to have occurred in a No. 22 chromosome [37, 22]. In an established cell line from a meningioma that had been cultured for 6 weeks, frozen for 28 months, and repropagated for 3 months, G-banding revealed the loss of a No. 22 chromosome in all hypodiploid cells and the loss of two or three Nos. 22 in hypotetraploid cells [28].

CHILDHOOD TUMORS

Even before banding methods were available, development of retinoblastoma had been associated with a group D chromosome deletion in eight patients with severe congenital malformations (see first eight citations under ref. [35]). More recently, Wilson et al. [35] described the banding patterns in a patient with chromosome No. 13q − and retinoblastoma. The interstitial deletion of the long arm of chromosome No. 13 appeared to involve the fourth large band distal to the centromere. Wilson's summary of the findings in four other individuals, in whom retinoblastoma and 13q − or 13r coexisted, suggests that this deletion may be a specific lesion related to this malignancy. Since retinoblastoma ordinarily is a dominantly inherited disorder, it is tempting to speculate that the "retinoblastoma gene" resides in the long arm of chromosome No. 13. Retinoblastoma may be familial or sporadic, that is, it may be caused by a deleterious gene either transmitted through the germ line or arising de novo through mutation in a somatic cell. If it exists in the germ line, a chromosomal lesion would be expected to be present in all tissues of the body. Thus banding studies of patients with familial, bilateral retinoblastoma may prove rewarding.

Bilateral aniridia, which follows a Mendelian dominant pattern of inheritance, is a condition in which the irises of the eyes are absent. Miller first reported an association of sporadic aniridia with Wilms' tumor, a childhood cancer of the kidney [25]. Littlefield (personal communication), investigating the chromosomes of a patient with aniridia and Wilms' tumor, found a reciprocal translocation: 46, XY, t(11q −; 8p+). Knudson et al. have suggested that perhaps all types of human cancers are inherited in some families and sporadic in others, postulating that a mutation (either in the germ line or in a somatic cell) must be followed by another "event," in order to give rise to the cancer phenotype [21]. The examples of retinoblastoma and Wilms' tumor raise the possibility that other chromosomal rearrangements will be found for other inherited cancers.

LONG-TERM CELL LINES

There are two classes of long-term "established" human cell lines in culture: those derived from tumor tissue and maintained in vitro over many years, and those derived from circulating lymphocytes and continued as long-term lymphoblastoid cultures. The tumor lines are characterized by abnormal chromosome numbers and many marker chromosomes, whereas the lymphoblastoid lines are predominantly diploid and with no structural rearrangements detectable by conventional analysis.

Q-banding techniques have been applied to several established tumor cell lines by Miller et al. [24]. The HeLa line, originated in 1951 from a carcinoma of the uterine cervix, has been analyzed in detail. Sublines of HeLa were obtained from three separate laboratories; each subline was found to contain different chromosome numbers, but six of the seven markers detected were present in either two or all three sublines. Another interesting observation was made on the D98/AG line; this line was reportedly of independent origin from normal sternal bone marrow of an adult Caucasian male, but it was found by Gartler [15] to be a HeLa contaminant, based on isozyme similarities with HeLa. In this line, three of the seven HeLa markers were present as well as two unique markers.

Sinha and Pathak [31] compared HeLa and HEp-2 cell lines derived from a female and a male donor, respectively. Using the C-band staining technique, they were able to demonstrate the Y chromosome in 4.5% of the HEp-2 cells. Both lines had modal numbers near 75 and wide variation about the mode. Both had many unique markers; however, one long subtelocentric marker was common to both lines. They concluded that HEp-2 was not a contaminant of HeLa in spite of the presence of one common marker. If certain segments are predisposed to rearrangement, then it is conceivable that two similar marker chromosomes could arise independently. G-band or Q-band studies would shed further light on this marker.

Miller's group also reported on an established lymphoblastoid line WI-L2 [24], derived from the spleen of a male patient having hereditary spherocytic anemia. They found the karyotype to be 46,XY, but pseudodiploid: quinacrine-fluorescence studies revealed monosomy for No. 21 and the presence of a small submetacentric marker derived partially from a No. 21 chromosome. In contrast, Chen (unpublished observations) examined four established lymphoblastoid lines from patients with infectious hepatitis and was unable to detect any consistent banding aberrations, whereas when he examined two independent lymphoblastoid lines, not known to have originated from persons with a viral infection, he found subtle banding differences. It may prove rewarding to follow those lymphoblastoid lines that are "apparently diploid" over a long period of time to determine whether minute, sometimes undetectable, chromosomal changes will give rise to further rearrangements which can then be identified by banding techniques.

HUMAN FIBROBLAST CELL STRAINS

In contrast to the long-term cell lines, human fibroblasts established in vitro have a finite life span of approximately 40 to 50 cell generations. They then undergo senescence and eventual cell death [19]. Q-band analysis of the WI-38 fibroblast strain revealed a 46,XX complement with no alterations in the normal banding patterns. More detailed studies of this strain by Chen and Ruddle [4] showed the presence of minor populations of three stemlines; these were found at various passages but particularly during the senescent phase (i.e., after 40 passages). The first stemline (46,XX,4p−) remained relatively stable and comprised about 3% of the total cell population. The second stemline (46,XX,1p−) was detectable for about 10 generations and then disappeared. The third stemline [46,XX,t(17p+,11p−)] increased to constitute approximately 35% of the senescent cell population. Although it is known that chromosomal abnormalities are observed more frequently in primary cell populations in vitro than in comparable tissues in vivo, it is not possible to distinguish between two alternative explanations for this difference. Is there truly a higher chromosomal mutation rate in vitro, or is there stronger selection against chromosomal mutants in vivo? It is tempting to speculate that there is a fundamental relationship between the greater probability of chromosomal rearrangements in senescing fibroblasts and the increased risk of malignancy with age.

NEWLY ESTABLISHED SOLID-TUMOR CELL STRAINS

We have had the opportunity to obtain biopsy material from 39 malignant tumors through the cooperation of the Department of Surgery at the M. D. Anderson Hospital and Tumor Institute. Of these, 35 have been established in culture and banded karyotypes obtained. In some cases, only normal fibroblasts with 46 chromosomes were recovered; in others, growth was slow and the cultures were discarded. We will report here eight tumor cell strains that have been extensively analyzed cytologically. In these eight strains there were normal fibroblast cells present in the early stages of growth from the original explants, but also morphologically abnormal cells were present. The latter types tended to show a disoriented, disturbed growth pattern with loss of contact inhibition and multiple layers of cells which were presumed to be the cancer cells. Indeed, karyotype analysis of early subcultures often revealed a mixture of normal diploid and aneuploid metaphases. After manipulation of the culture conditions such as growth in serum-free medium or shaking off the loose cells for further subculture, the tumor cells could be rescued at the expense of the normal fibroblasts. Karyotype banding analysis of both the diploid and aneuploid cells was accomplished by staining with quinacrine mustard to produce Q-bands, followed by C-band staining on the same slides. Thus double karyotypes could be produced on each of the cells, demonstrating both Q- and C-bands.

All of the eight cell strains were derived from solid tumors. Five were malignant melanomas, and there was one each of fibrosarcoma, gastric adenocarcinoma, and mammary adenocarcinoma. In three cases (C26, C32, and C48), karyotypes were recovered within 4 hours after the biopsy was taken and thus reflected the in vivo constitution of the tumor cells. All strains were monitored for chromosome constitution over a period of several weeks to 2 years.

Two of the eight strains had a hypodiploid modal chromosome number, two were hypotetraploid, and four were between diploid and triploid. Within each cell strain there was a rather narrow range of variation of chromosome counts above and below the mode. As demonstrable by analysis of the banding patterns, all strains had chromosome "markers," some of which would have escaped detection by conventional analysis. Thus the reshuffling of the genome in tumor cells is perhaps greater than has been suspected previously. Our findings in each of the tumor strains are summarized in Table 1. A Q-band karyotype, illustrating both intact chromosomes and markers, is shown in Fig. 1.

An analysis of the apparently intact chromosomes, as evidenced by their normal banding pattern, reveals a remarkable consistency. Table 2 gives the modal number of intact chromosomes in each strain. All of the six strains with a modal number greater than 46 had no more than four homologs for each chromosome, with two exceptions. In strain C10 there were five No. 20 chromosomes, and in strain SH1 there were five No. 2 chromosomes. This suggests that these six strains could have been derived from a tetraploid ancestral cell with subsequent loss of chromosomes and rearrangements of chromosome segments with nondisjunction occurring very rarely. In strain C9, the possibility should be entertained that the ancestral cell was diploid even though the chromosome number is hyperdiploid, since only one chromosome is represented in the tetrasomic state and none is trisomic.

Nullisomy for whole autosomes is not a common condition, even in the hypodiploid state. When nullisomy for intact chromosomes did occur (see Table 2), a marker chromosome nearly always could be shown by banding techniques to carry segments of the missing chromosome. A case in point is chromosome No. 9, in which there is nullisomy in three strains (C9, C26, and C27). In each instance, the long arm of the No. 9 could be identified in marker chromosomes because of its characteristic and easily recognized C-band in combination with the Q-bands. Perhaps true nullisomy is quite rare in tumor cells and may in fact be lethal. However, the possibility should not be dismissed that nullisomy for a segment may be, in fact, the *sine qua non* of malignancy. The present state of the art of banding analysis is not sufficiently refined to answer this question.

Four of the eight cell strains were quite stable, showing similar karyo-

Table 1. Cytogenetic Findings in Cell Strains Derived from Eight Human Cancers

Strain Identity and Sex of Donor	Type of Tumor	Cells Analyzed	Mode (and Range) of Chromosomes per Cell			Number of Different Markers Observed	Time in Culture	Stability†
			Total Chromosomes	Intact Chromosomes*	Marker Chromosomes			
C31, male	Fibrosarcoma	10	43 (42–45)	30 (29–35)	10 (10–14)	26	3 Months	No
C32, male	Melanoma	50	45 (41–47)	39 (36–41)	6 (5–7)	7	1–3.5 Months	Yes
C9, male	Melanoma	50	52 (48–53)	40 (38–43)	10 (10–12)	12	1–10 Months	Yes
C26, male	Melanoma	7	58 (57–60)	45 (44–47)	13 (10–14)	27	2 Weeks	No
C27, male	Melanoma	10	63 (61–67)	45 (42–47)	18 (15–21)	45	3 Months	No
C10, male	Gastric adenocarcinoma	10	67 (64–68)	51 (49–55)	13 (10–17)	44	5 Months	No
C48, male	Melanoma	10	81 (81–84)	68 (65–72)	13 (12–17)	16	2 Weeks	Yes
SH1, female	Mammary adenocarcinoma	22	89 (83–90)	80 (77–83)	6 (4–7)	9	24 Months	Yes

* Those with apparently normal banding patterns.
† Stability refers to similar karyotypes from cell to cell within a culture observed over a period of months; instability is evidenced by wide variability in karyotypes, particularly among the marker chromosomes.

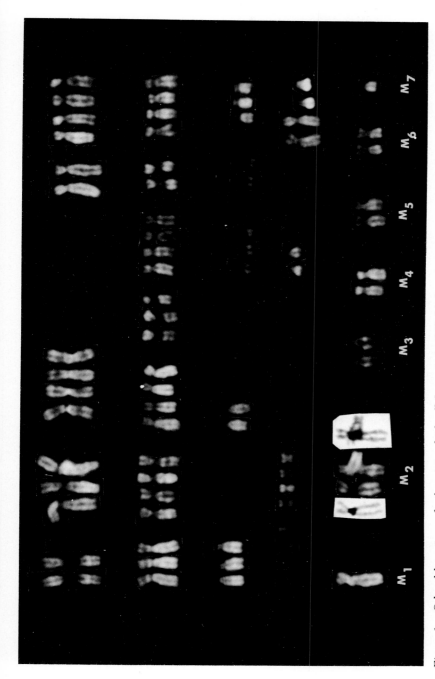

Figure 1. Q-band karyotype of a hypotetraploid cell from strain C48. Marker chromosomes are shown in the bottom row (M_1–M_7). Five of the seven markers occurred in pairs. The two M_2 chromosomes with white background have been added from a photograph of the same cell stained for C-bands

Table 2. Distribution of Modal Number of Intact Chromosomes in Eight Tumor Cell Strains*

Cell Strain Number	Chromosome Modal Number	Chromosome Group and Number†									
		A			B					C	
		1	2	3	4	5	6	7	8	9	10
Hyperdiploid											
C31	43	1	2	2	1	2	1	1	1(2)	1(2)	0(1)
C32	45	1	1	2	2	2	2	2	1	1	1
Hyperdiploid-Hypotriploid											
C9	52	1	2	2	2	2	2	4	2	0	2
C26	58	3	2	2(3)	3	3	3	3	1(2)	0(1)	3
C27	63	1	3	4	3	3	3	4	2(3)	0	2(3)
C10	67	2	3	2	1	3	2(3)	2	2	2	3(4)
Hypotetraploid											
C48	81	2	4	4	2	4	4	4	4	3	4
SH1	89	3	5	2(3)	3	4	4	3	4	2	4

* "Intact chromosomes" refers to those with normal banding patterns.
† Hypermodal numbers are given in parentheses.

types from cell to cell (C9, C32, C48, and SH), at various cell culture generations. These had been in continuous cultivation over many months, demonstrating that little or no in vitro selection had occurred after the strains had been established. Although selection very early in the culture period cannot be ruled out in every case, in strains C32 and C48 the karyotypes recovered after 4 hours of culture matched well with karyotypes obtained months later. One of these strains (C32) has been reported in detail [5]; the remaining four strains displayed a marked degree of instability from cell to cell within cultures. This instability, however, is not attributable to variability in the numbers of the intact chromosomes but rather to the number of, and features of, the markers that were present in the lines; these markers seemed to be continually appearing and disappearing, increasing and decreasing in numbers, and changing in their forms. Secondary rearrangements of marker chromosomes occasionally appeared, suggesting a continual reshuffling of certain chromosomal segments but specifically of those which already had participated in rearrangements. This was in contrast to the stability of the intact (i.e., apparently never rearranged) chromosomes. *Forty-five different markers were observed in strain C27, and 44 markers in C10.* Figure 2 illustrates several marker chromosomes derived from the various strains.

An example of clonal evolution involving marker chromosomes was observed in strain C9. After 4 months in culture this strain became stable (i.e., it showed little variation in chromosome complement from cell to cell). Soon after the strain was established in vitro there was a modal

Table 2. (continued)

		D			E			F		G			
11	12	13	14	15	16	17	18	19	20	21	22	X	Y
2	2	1	0(1)	1	2	2	2	2	2(3)	2	1(2)	0(1)	1
1	2	1	2	2	2	2	2	2	2	2	2	1	1
1	2	2	2	1	2	2	2	0	1	2	2	1	1
1	3(4)	2	3	0	2	2	1	1	2	2	2	1	0
0	4	0	2(3)	3	2	3(4)	1(2)	2	3(4)	1	0(1)	1(2)	0
4	4	2(3)	2	3(4)	2	2	2	2(3)	5	2	0	1(2)	0
2(3)	4	4	0	2	3	2	4	4	4	2	1	2	2
4	4	2	2	4	4	4	4	4	4	4	4	4	—

count of 52 chromosomes, including among its markers two representatives of a certain translocation t(19,11q) and one representative of another t(13q,14q). Two separate cultures derived from this strain and grown in parallel fashion have shown, for the past 10 months, chromosome numbers of 51 and 50 chromosomes, respectively. In the former, one t(19,11q) was lost, while in the latter, one t(19,11q) and the t(13q,14q) were lost. During this time, however, tetrasomy for the intact chromosome No. 7 remained stable, and the normal Nos. 13 and 14 did not undergo loss. This illustrates again the stability of the intact chromosomes and the variability of the markers.

Some of the chromosomes appear to be less likely or more likely than others to become rearranged, to be lost, or both. This statement is made with caution, because strains from only eight tumors have been examined in detail. However, in this limited sample, the Nos. 2, 5, 12, and 20 appear to be quite stable, unlikely to be lost or to engage in rearrangement. Conversely, the Nos. 1, 9, 13, 14, and 22 seem to be more susceptible to change and loss. All of these more vulnerable chromosomes have large blocks of constitutive heterochromatin (C-bands), chromatin that probably is dispensable and is quite likely to have an increased tendency to breakage. Curiously, chromosomes Nos. 15, 16, and 21, which also have extra C-band material, do not seem to be involved commonly in loss or rearrangement. Is it merely coincidence that Nos. 9, 13, 14, and 22 are the very chromosomes involved in rearrangements in chronic myelocytic leukemia, the Burkitt lymphoma, meningioma, and retinoblastoma, discussed earlier in this chapter?

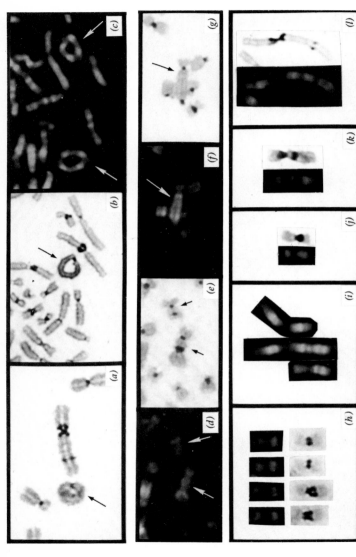

Figure 2. Examples of marker chromosomes. (a) Ring chromosome, derived from a No. 2, from strain C32; (b) same marker as in (a) but from a different cell; (c) two dicentric ring chromosomes, derivatives of No. 11, in a cell from strain C26; (d) Q-bands of two dicentric chromosomes, from strain C10 [cf. (e)]; (e) C-band of the two dicentrics in (d); (f) Dicentric chromosome, derived from a No. 10 and a No. 14, from strain C10 [Q-banding; cf. (g)]; (g) same dicentric as in (f) (C-banding); (h) No. 16 chromosomes from hypotetraploid cell showing homolog asymmetry of the constitutive heterochromatin in strain SH1; (i) translocation chromosome (center), showing (from top of the chromosome down) material of 1p, 1q, and 7q, intact No. 7 on left, intact No. 1 on right; (j) pericentric inversion in a No. 9, from strain C31; (k) marker chromosome with interstitial C-band, showing (from the top down) material of 1q, 9p, and 9q, from strain C48; (l) marker chromosome with an interstitial C-band, showing (from the top down) material of 1q, 9q, 9q, and 10q, from strain C32. [Note: C-band staining in (a), (b), (e), (g), (h) (lower), (i), and (i)–(l) (right); Q-band staining in (c), (d), (f), (h) (upper), (i), and (i)–(l) (left).]

If loss of whole chromosomes is nonrandom in tumor cells, as indicated by our observations, is there any correlation between those human chromosomes more apt to be lost in malignant cells and those eliminated in interspecific somatic cell hybrids? Not enough data have been accumulated to answer this question, but it is worth pointing out that No. 9 is one of the first chromosomes to be lost in human/mouse hybrid cells and is the least represented among intact chromosomes in Table 2.

CONCLUDING COMMENTS

German [17] has made a compelling argument in favor of the theory that chromosomal rearrangement occurs in the ancestral cell that originally converted to cancer and, furthermore, that the mutation is an integral part of that conversion. He has pointed out that individuals who carry certain genes that predispose to chromosomal instability as well as individuals exposed to radiation, chemical carcinogens, and perhaps even viruses, all have in common an underlying disturbance in the normal chromosomal phenotypes, predisposing them to an increased risk of malignancy. Knudson et al. [21] argue strongly for an underlying mutation, whether somatic or germinal, as the first step in oncogenesis. This "mutation" could, in fact, be a small chromosomal rearrangement. Studies are in progress in our laboratory to determine whether, in dominantly inherited childhood tumors (e.g., retinoblastoma, Wilms' tumor), a small deletion, duplication, translocation, or inversion can be demonstrated by refined banding analysis. If this occurs, it may be visible in all of the cells of a patient who had an affected parent, whereas in sporadic cases it would more likely be confined to the tumor tissue itself.

The banding analysis of benign tumor cells should be another fertile field of investigation. Many benign tumors have a strong tendency to undergo malignant transformation, suggesting the possibility that an underlying chromosomal mutation may be found in all of the cells of the benign tumor which could be looked upon as a "clone."

Assuming the hypothesis to be true which German has so eloquently argued (i.e., that the chromosomal rearrangement is the critical first event in oncogenesis), two related questions remain unanswered. First, *why* does an abnormal sequence of genes in the genome predispose to further chromosomal instability? Second, what are the events that must take place subsequently to give the unstable, highly disorganized cell priority over its less disorganized sister cells?

To address the first question, we would like to suggest a hypothesis about chromosome changes in cancer cells which is partly derived from a discussion of normal cells by Comings [6], which may be paraphrased to state: "There is a place for everything in the nucleus and everything is in its place." In other words, in the normal interphase nucleus there is an orderly spatial organization of all of the components (euchromatin, heterochromatin, telomeres, kinetochores, nucleolar organizers, etc.), and these all have a specific relationship to each other and to all points on the nuclear membrane. If this is true, then it is not far-fetched to presume that when critical molecules in this complex spatial pattern are transposed, they may play havoc with the orderly sequence of cellular events, not only during interphase but also during mitosis. Observations of mitotic abnormalities in cancer cells are legion. One could almost envision a game of musical chairs taking place in the nucleus when one aberration is present, leading to a plethora of subsequent changes. Suggestive evidence for this hypothesis is the observation that trisomic cells may be more easily transformed by SV40 virus [34], and that many families are described in which both translocations and trisomies have occurred. The drastic karyotypic changes seen in malignant cells are far removed from the first basic misstep in the nuclear dance.

The second question reveals our ignorance not only of the disorganized state of malignancy but also of the organized and regulated state of nonmalignancy. Because malignancies themselves vary tremendously in invasiveness, tendency to metastasize, and degree of aberrant cell types, the selective advantages of tumor cells may be random rather than specific. Once a cell is committed (i.e., malignant), new forms of daughter cells may arise wholly by random events taking place in the disorganized nucleus. There may be specific sites or regions of genetic material which, when lost, or mutated, or moved about, may confer upon the cell any of a number of advantages that are detrimental to the normal cells surrounding it. Careful banding studies of small changes in clones of cells from many different tumors may shed light on the critical chromosomal regions involved in different malignant characteristics.

One of the beauties of the theory of chromosomal origin of cancer is that it does not exclude other theories as a prerequisite. Chemical oncogenesis, viral oncogenesis, and even immuno-oncogenesis may all be encompassed in the chromosomal theory. Future studies, utilizing the rapidly evolving techniques for differential staining of chromosomes, may elucidate some of these vexing problems.

Acknowledgements.–We are particularly grateful to Dr. Marvin Romsdahl, Department of Surgery, M. D. Anderson Hospital and Tumor Institute, who provided the biopsy material for the studies reported under the section on newly established solid tumor cell strains, and also for the technical assistance of Mss. I. Mendiola, J. Pointer, Y. Chen, and A. Craig-Holmes, and the secretarial help of N. Carter. Support for this study was obtained from U.S. Public Health Service grant GM 19513.

LITERATURE CITED

1. CASPERSSON T, FARBER S, FOLEY GE, KUDYNOWSKI J, MODEST EJ, SIMONSSON E, WAGH U, ZECH L: Chemical differentiation along metaphase chromosomes. Exp Cell Res 49:219–222, 1968.
2. CASPERSSON T, GAHRTON G, LINDSTEN J, ZECH L: Identification of the Philadelphia chromosome as no. 22 by quinacrine mustard fluorescence analysis. Exp Cell Res 63:238–240, 1970.
3. CASPERSSON T, LINDSTEN J, ZECH L, BUCKTON KE, PRICE WH: Four patients with trisomy 8 identified by the fluorescence and Giemsa techniques. J Med Genet 9:1–7, 1972.
4. CHEN TR, RUDDLE FH: Subtle chromosome changes revealed by Q-band staining method during cell senescence. (In preparation, 1973).
5. CHEN TR, SHAW MW: Stable chromosome changes in a human malignant melanoma. Cancer Res 33:2042–2047, 1973.
6. COMINGS DE: The rationale for an ordered arrangement of chromatin in the interphase nucleus. Am J Hum Genet 20:440–460, 1968.
7. COMINGS DE: The structure and function of chromatin. In Advances in Human Genetics (Harris H, Hirschhorn K, eds), vol 3, New York, Plenum Press, 237–431, 1972.
8. COMINGS DE, AVELINO E, OKADA TA, WYANDT E: The mechanism of C- and G-banding of chromosomes. Exp Cell Res 77:469–493, 1973.
9. CRAIG-HOLMES AP, MOORE FB, SHAW MW: Polymorphism of human C-band heterochromatin: 1. Frequency of variants. Am J Hum Genet 25:181–192, 1973.
10. DE LA CHAPPELLE LA, SCHRÖDER J, VUOPIO P: 8-trisomy in the bone marrow. Report of two cases. Clin Genet 3:470–476, 1972.
11. EVANS HJ, BUCKTON KE, SUMNER AT: Cytological mapping of human chromosomes: results obtained with quinacrine fluorescence and the acetic-saline-Giemsa techniques. Chromosoma 35:310–325, 1971.
12. FLEISCHMANN T, HAKANSSON CH, LEVAN A: Fluorescent marker chromosomes in malignant lymphomas. Hereditas 69:311–314, 1971.
13. FLEISCHMANN T, HAKANSSON CH, LEVAN A, MOLLER T: Multiple chromosome aberrations in a lymphosarcomatous tumor. Hereditas 70:243–258, 1972.
14. FRANCKE U: Quinacrine mustard fluorescence of human chromosomes: characterization of unusual translocations. Am J Hum Genet 24:189–213, 1972.
15. GARTLER SM: Apparent HeLa cell contamination of human heteroploid cell lines. Nature 217:750–751, 1967.
16. GEORGE KP, POLANI PE: Y heterochromatin and XX males. Nature 228:1215–1216, 1970.
17. GERMAN J: Genes which increase chromosomal instability in somatic cells and predispose to cancer. In Progress in Medical Genetics (Steinberg AG, Bearn AG, eds), vol 8, New York, Grune & Stratton, 61–101, 1972.
18. GROUCHY J DE, TURLEAU C, LEONARD C: Étude en fluorescence d'une trisomie C mosaique probablement 8:46, XY/47, YX, ?8+. Ann Génét 14:69–72, 1971.
19. HAYFLICK L, MOORHEAD PS: The serial cultivation of human diploid cell strains. Exp Cell Res 25:585–621, 1961.

20. Hsu TC: Longitudinal differentiation of chromosomes. *Ann Rev Genet* (in press, 1973).
21. Knudson AG, Strong LC, Anderson DE: Heredity and cancer in man. In Progress in Medical Genetics (Steinberg AG, Bearn AG, eds), vol 9, New York, Grune & Stratton, 113–158, 1973.
22. Mark J, Levan G, Mitelman F: Identification by fluorescence of the G chromosome lost in human meningiomas. *Hereditas* 71:163–168, 1972.
23. Manalov G, Manalova Y: Marker band in one chromosome 14 from Burkitt's lymphoma. *Nature* 237:33–34, 1972.
24. Miller OJ, Miller DA, Allderdice PW, Dev VG, Grewal MS: Quinacrine fluorescent karyotypes of human diploid and heteroploid cell lines. *Cytogenetics* 10:338–346, 1971.
25. Miller RW, Fraumeni JF, Manning MD: Association of Wilms' tumor with aniridia, hemihypertrophy and other congenital malformations. *New Engl J Med* 270:922–927, 1964.
26. O'Riordan ML, Robinson JA, Buckton KE, Evans HT: Distinguishing between the chromosomes involved in Down's syndrome (trisomy 21) and chronic myeloid leukemia (Ph[1]) by fluorescence. *Nature* 230:167–168, 1971.
27. Paris Conference 1971: Standardization in human cytogenetics (Bergsma D, ed), New York, National Foundation, Original Article Ser, 1973.
28. Paul G, Porter IH: Giemsa banding of an established line of a human malignant meningioma. *Humangenetik* 18:185–187, 1973.
29. Pearson P: The use of new staining techniques for human chromosome identification. *J Med Genet* 9:264–275, 1972.
30. Rowley JD: A new consistent chromosomal abnormality in chronic myelogenous leukemia identified by quinacrine fluorescence and Giemsa staining. *Nature* 243:290–293, 1973.
31. Sinha AK, Pathak S: Distribution of constitutive heterochromatin in HeLa and HEp-2 cell lines. *Humangenetik* 18:47–54, 1973.
32. Shaw MW: Uses of the banding techniques for the identification of human diseases of cytogenetic origin. *Environmental Health Perspectives* (in press 1973).
33. Stock AD, Hsu TC: Evolutionary conservatism in arrangement of genetic material: A comparative analysis of chromosome banding between the Rhesus Macaque ($2n=42$, 84 arms) and the African green monkey ($2n=60$, 120 arms). *Chromosoma* 43:211–224, 1973.
34. Todaro GJ, Martin GM: Increased susceptibility of Down's syndrome fibroblasts to transformation by SV40. *Proc Soc Exp Biol Med* 124:1232–1236, 1967.
35. Wilson MG, Towner JW, Fujimoto A: Retinoblastoma and D-chromosome deletions. *Am J Hum Genet* 25:57–61, 1973.
36. Zankl H, Zang KD: Cytological and cytogenetical studies on brain tumors. III. Ph[1]-like chromosomes in human meningiomas. *Humangenetik* 12:42–49, 1971.
37. Zankl H, Zang KD: Cytological and cytogenetical studies on brain tumors. IV. Identification of the missing G chromosome in human meningiomas as No. 22 by fluorescence technique. *Humangenetik* 14:167–169, 1972.

VIRUSES, CHROMOSOMES, AND TUMORS: THE INTERACTION BETWEEN VIRUSES AND CHROMOSOMES

D. G. HARNDEN

Department of Cancer Studies, University of Birmingham,
Birmingham, England

Introduction	152
Cell–virus Interactions	153
Effects of DNA Viruses	154
Adenoviruses	154
Adenovirus Type 12	154
Other Adenoviruses	158
Herpesviruses	158
Herpes Simplex Virus Type 1	159
Herpes Simplex Virus Type 2	160
Cytomegalovirus	161
Herpes Zoster Virus	161
Marek's Disease Virus	161
Epstein-Barr (EB) Virus	161
Papovaviruses	163
Papilloma Virus	163
Polyoma Virus	164
Simian Virus 40 (SV40)	164
Poxviruses	167
Effects of RNA Viruses	168
Paramyxoviruses	168
Measles	168
Mumps	169
Sendai Virus	169
Rabies	169
Myxoviruses	169
Arboviruses, Picornaviruses, and Unclassified Viruses	170
Yellow Fever Virus	170
Poliovirus	170

Hepatitis Virus	170
Rubella	170
Oncornaviruses	170
Avian Leukemia–Sarcoma Complex	170
Murine Leukemia–Sarcoma Complex	172
General Considerations	172
Chromosome Changes Occurring Early after Virus Infection	174
Chromosome Changes Occurring Late after Virus Infection	176
Acknowledgements	180
Literature Cited	180

INTRODUCTION

The significance of the chromosome changes observed in tumor cells is still not understood in relation to the initiation of the malignant process and its subsequent progression. The chromosomal hypothesis of cancer suggests that these changes in the chromosomes are visible manifestations of the primary event in neoplasia. Since some tumor cells have normal chromosomes, this is at best an oversimplification, and at worst wrong; but the observed association between chromosomal abnormalities and neoplasia is so strong that it is clearly necessary to try to elucidate the nature of the relationship.

The observed facts are undisputed: many environmental carcinogenic agents cause chromosome damage, though not all mutagens are carcinogens; many, but not all, cancer cells have visible chromosome abnormalities; these chromosome abnormalities may undergo an evolution during the growth of a tumor. Controversy arises only in the interpretation: (1) chromosome changes caused by environmental agents may predispose a cell to undergo malignant change; (2) the initial event in carcinogenesis may be the visible change in the chromosomes; (3) chromosome changes may play an important role in the progression of a tumor; or (4) visible changes in the chromosomes may be coincidental and may merely accompany the significant changes that are themselves not visualized by ordinary light microscopy. This last does not, of course, rule out the possibility that alterations in the chromosomes may play an important part in the development of neoplasia, since quite sizable deletions, duplications, and reciprocal exchanges would not be visualized using conventional techniques. Even by using quinacrine or Giemsa banding techniques, it would not be possible to detect some chromosomal changes that at the molecular level would amount to massive alterations in the genotype.

Since the specific biological characteristics of tumor cells are passed on from one cell generation to the next, it must be assumed that there has been an alteration either in the structure of the tumor cell genome or in its regulation. The separation of chromosomal events into those that can be visualized with a light microscope and those that cannot is artificial. It must be remembered that visible changes are the consequence of molecular events and that no discussion of chromosome changes can be meaningful unless these are related to underlying molecular events. In discussion of a chromosomal hypothesis of cancer, therefore, the basic question to consider is whether the change in a tumor cell is a regulatory change or a mutational change, rather than whether the change is chromosomal or not chromosomal. Mutational and regulatory changes are not, of course, mutually exclusive, especially when it is borne in mind that current hypotheses favor the idea that much of the chromosomal DNA may function in regulation rather than in transcription [33, 126].

CELL-VIRUS INTERACTIONS

Recognition that viruses can cause visible damage to chromosomes of mammalian cells is relatively recent [50]. There is no major body of literature on viruses as mutagens as there is for radiation and chemicals, although viruses have been implicated in mutagenesis in a small number of instances [14, 24]. In considering the relationship between environmental agents and chromosomes, we know that many different types of chromosome aberration can apparently be induced by many of the agents. In the case of viruses, however, one must also consider the possibility of the addition of new genetic material to the cell, either by integration into the chromosomes or by close association with them. Such events would not of course be visible but would lead to consequences that must therefore concern us here. It is also necessary to consider the precise nature of the virus cycle in a particular cell type. Many cytogenetic studies on virus-infected cells give only scant information on interaction. This often makes interpretation of results difficult. The cell–virus interactions range from a productive infection leading to the death of the cell to situations in which the cell survives, either because it is nonpermissive for the particular virus or because the productive infection does not lead to cell death but to the transformation of cells or to tumor production.

Clearly, the subsequent fate of the cell is relevant to any association between virus-induced chromosome change and neoplasia. It is necessary to consider not only the nature of the particular cell–virus system but also the conditions of infection, particularly the multiplicity of infection and the time relationship between the infection and the cytogenetic observation. Infection with different multiplicities of a virus in the same cell type can give completely different results. Similarly, observations made a short time after infection may be quite different from those made some time later. Purification of the virus is also important, since damage could be caused by other substances in a crude virus preparation.

During the following discussion it will be useful to consider the various chromosome changes inducible by viruses. Chromosome breakage may be caused directly by the input virus or more indirectly by events occurring during synthesis of new virus. Breakage may or may not be repaired, and in either case the damage may be incompatible with survivial of the cell. If the cell survives, persistence of the virus genome within the cell may lead to continuing instability of the karyotype. This instability, or indeed the primary damage, may play a part in events leading up to the transformation of the cell. Further chromosome changes may either accompany transformation of the cell by the virus or arise as a consequence of transformation. The role of the virus in inducing such changes must be considered, as well as the importance of such changes or instabilities of the karyotype to the further progression of the tumor.

EFFECTS OF DNA VIRUSES

Adenoviruses

The adenoviruses illustrate many of the complexities encountered in studying the cytogenetics of virus-infected cells. There are 31 different serotypes of human adenoviruses and many other adenoviruses of nonhuman origin. Various cell–virus combinations give the whole range of possible interactions from nonproductive infection to tumor production. It seems probable, however, that all members of the adenovirus group have the potential for transformation in some cell type [79, 80], though attempts to demonstrate that adenoviruses are involved in naturally occurring neoplasms have so far proved unsuccessful [78, 133]. The human adenoviruses are divided into three groups according to their degree of oncogenicity (high, intermediate, or low) in hamsters [62]. While this classification does not necessarily reflect the oncogenicity of these viruses in other systems, it is a valid classification in that members of each of the three groups also share other properties (e.g., antigenicity, GC content). Similarly, the animal adenoviruses also appear to differ in range of oncogenic potential in different cell systems.

Adenovirus Type 12. Most of the cytogenetic studies on cells infected with adenoviruses have been made using adenovirus 12, probably because of its high oncogenicity. In *Mastomys* cells, Huang [60] reported the induction of gaps and breaks but also of small numbers of exchanges and dicentrics at 2 days after adenovirus 12 infection. In cells of a Chinese hamster cell strain, mainly chromatid exchanges, rings, and possibly dicentric chromosomes, as well as gaps and breaks, were seen 48 hours after infection [151]. In Syrian hamster cells, infection with adenovirus 12 led to the production of a high incidence of chromatid and isochromatid breaks, fragments, endoreduplication, and pulverization [32, 84, 153, 178]. Occasional chromatid exchanges were seen; but no rings, dicentrics, or abnormal chromosomes were reported. The amount of damage was roughly proportional to virus input and, since there was no evidence of virus DNA synthesis or structural–antigen synthesis, the damage must have been caused by an early event in infection. This conclusion is supported by the finding that UV-irradiated virus still retained the capacity to damage hamster cell chromosomes [153]. Zur Hausen [176] showed, however, that UV-irradiation of the virus did reduce its capacity to induce chromosome damage, but that, with increasing doses of radiation, the rate of reduction was four times less than the rate of inactivation of infectivity in human cells. Later Zur Hausen [178] concluded that the chromosome damage seen in hamster cells was associated with cell killing even though these cells produced no virus (i.e., were nonpermissive), since the amount of chromosome damage was inversely related to the plating efficiency of the cells; surviving cells were cytogenetically normal [151, 178].

Similarly, in phytohemagglutinin (PHA)-stimulated human lymphocytes, which are also nonpermissive, breaks and gaps were induced by adenovirus 12 [120]. Nichols found that the break points were distributed randomly throughout the karyotype, but McDougall (personal communication, 1972) has found specific damage to chromosomes Nos. 1 and 17 in PHA-stimulated human lymphocytes as well as in fibroblasts (see below). In the human amnion cell line AV3, which supports adenovirus-12 replication (i.e., is permissive), MacKinnon et al. [84] found no significant increase in chromosome aberrations 24 to 72 hours after infection at a low multiplicity, and later Stich and Yohn [153] reported that chromosome aberrations in adenovirus-infected human cell lines were seen only in those cells capable of entering division before the onset of viral replication and mitotic inhibition. Other authors, however, using primary human-embryo kidney cells that are also permissive for adenovirus 12, found not only extensive random chromosome damage, but also specific lesions (Fig. 1a) of chromosomes Nos. 1 and 17 [83, 176, 179]. Random and specific chromosome damage could also be demonstrated in both normal and genetically abnormal human dermal fibroblasts that supported adenovirus-12 growth rather poorly [81, 82]. All these aberrations were chro-

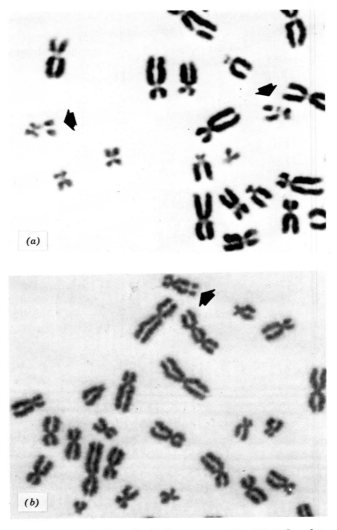

Figure 1. Specific lesions. (a) On both chromosomes No. 17, induced in a human-embryo kidney cell by adenovirus type 12. (b) Of a group C chromosome in a cell from the GOR lymphoid cell line carrying the herpes-type EB virus.

matid and chromosome gaps and breaks, with a few chromatid interchanges. In no instance were dicentrics or abnormal chromosomes observed. The nature of the specific lesions on chromosomes Nos. 1 and 17 has been further investigated. Zur Hausen [179], using radioactive virus, failed to demonstrate an association between input virus and the sites of the specific lesions; he did, however, find some evidence of specific association with other sites. McDougall has shown that, while specific lesions were most easily demonstrated at low multiplicity, a larger input of virus increased the amount of random damage, but that the specific breakage of chromosome No. 17 could still be demonstrated. Since the locus for thymidine kinase is known to be on chromosome No. 17 [99], and since adenoviruses that do not code for thymidine kinase [69] are known to stimulate synthesis of this enzyme, McDougall [83] has suggested that the site of the lesions on chromosome No. 17 may be at that locus.

There have been claims that fibroblasts from patients with Fanconi's anemia [159] or from foreskin of normal subjects [141] can be transformed by adenovirus 12, but these claims seem doubtful, since no stable cell line of altered morphology was produced in either case. Cells from patients with Fanconi's anemia do, however, survive for several weeks after virus infection, and virus-specific chromosome damage can be demonstrated throughout this period of survival [82]. However, there is no evidence at present to suggest that either the random damage, or the specific damage associated with adenovirus infection of human cells, is in any way connected with cellular transformation, tumor production, or even with integration of the virus genome into the cell.

If, however, one returns to consideration of the situation in hamster cells, a different picture emerges. Studies of Chinese hamster cells transformed by adenovirus 12 [23] and of cells from tumors induced in hamsters by adenovirus 12 [11, 152, 155, 162] showed a wide range of chromosome abnormalities persisting in the cells for a very long time after virus infection. Brailovsky et al. [23] found chromosomes in many cells to be normal, but also found a wide range of abnormal chromosomes which, they suggested, were independent of malignant conversion. Similar observations have been made on a line of Gecko cells transformed by adenovirus 12 [28].

The other studies, on Syrian hamster tumor cells in vivo and in vitro, were characterized by a pronounced instability of the karyotype. Chromosome breakage and rearrangements of both stable and unstable types were observed through many generations both in vivo and in cultured tumor cells. Although minor clones were not uncommon in some tumors, no predominant cell line was observed, and the authors argued that this was evidence against virus-induced changes playing a part in tumor progression. In one instance, however, a major stemline was reported in a transplanted tumor [155]. These observations led Stolz and his colleagues to suggest that the DNA of the virus persisted in these cells and led to a karyotypic instability. The integration of the genome of adenovirus 12 into the DNA of hamster cells has since been demonstrated by Doerfler [38], but the precise role of the virus genome has not been elucidated. Fjelde et al. [43] have found most cells to be diploid in primary and transplanted adenovirus 12-induced tumor in C3H mice, but they did report a low incidence of chromosome and chromatid aberrations and a tendency to polyploidy.

Other Adenoviruses. Little work has been reported on chromosome changes in cells infected with the other adenoviruses. Stich and Yohn [153] and Cooper et al. [31] reported the occurrence of chromosome breakage and other chromosome changes, such as coiling deficiencies, in a Syrian hamster cell line (BHK 21) infected with human adenoviruses 2, 4, 7, and 18 and simian adenoviruses SA7, SV15, and SV20. Adenoviruses SV20 and SA7 and adenovirus 18 are all highly oncogenic in hamsters; adenovirus 7 is weakly oncogenic, while adenoviruses 2, 4, and SV15 do not produce tumors at all in hamsters. It can be concluded, therefore, that the capacity to induce chromosome damage early after infection is not associated with the oncogenic potential of these viruses in hamsters. However, McDougall [83] has shown that while human adenovirus types 2, 7, 18, and 31 all induced random chromosome damage in human-embryo kidney cells at low multiplicity of infection, 18 and 31 also caused a specific lesion on chromosome No. 17 similar to that induced by adenovirus 12. It is of interest that these three viruses (12, 18, and 31) constitute Huebner's highly oncogenic group [62]. As with adenovirus 12, it is probable that the induction of chromosome damage is associated with an early event after infection, since UV-impaired virus can still cause chromosome damage [149].

Few studies have been carried out with other species, but Gallimore [46] has shown that adenovirus type 2, which transformed rat embryo cells in vitro [45], also produced chromosomal instability in the transformed cells.

Herpesviruses

Several members of the herpes group of viruses have been associated with the occurrence of neoplasms, although in some instances the association is tenuous [71]. A relationship has been suggested between herpes simplex virus (HSV) and lip carcinoma in man [76] but this association has not been confirmed, nor have there been any reports of cellular transformation by HSV type 1. A seroepidemiological relationship has been demonstrated between HSV type 2 and carcinoma of the uterine cervix [145], and transformation of hamster cells by HSV type 2 has been reported [40]. Cytomegalovirus has frequently been isolated from patients with malignant disease, but it seems probable that the presence of the virus is fortuitous and not causally related to the induction of the neoplasms [53]. It is accepted by many workers that the association between the Epstein-Barr (EB) virus and African lymphoma may be a causal relationship [70], and there seems to be little doubt that the herpesvirus associated wth Marek's disease in chickens is the causative organism [17]. The Lucké virus, first demonstrated many years ago, is now known to be a herpesvirus and is probably the causative agent of renal adenocarcinoma in frogs [101]. Herpesvirus saimiri causes tumors in several different species of primate [63, 107]. There is good justification, therefore, for considering as relevant to tumorigenesis the chromosome damage caused by all members of the herpesvirus group even though some, such as herpesvirus zoster, have not been associated with malignancies.

Herpes Simplex Virus Type 1. The first demonstration that viruses could cause damage to chromosomes was the report by Hampar and Ellison [50] that HSV caused multiple chromosome gaps and breaks in an established strain of Chinese hamster cells. They also found chromatid exchanges and possibly dicentric chromosomes, and a tendency for breakage to occur at the centromeric and secondary constriction regions of chromosomes Nos. 1 and 2. In a later paper [51], they suggested that the more grossly damaged cells were those in the process of being killed by the virus, and they found evidence of chromosomal rearrangements in cloned survivor cells. Mazzone and Yerganian [97] and Rapp and Hsu [131] have reported chromatid and chromcsome breakage caused by HSV in primary Chinese hamster cells infected at low multiplicity. Similar results of HSV infection have been reported for primary monkey kidney cells [156] and for the established human cell lines, HeLa and Hep-2 [134].

In contrast to the results with adenovirus, both Hamper and Ellison [51] and Rapp and Hsu [131] found that inactivated herpesvirus did not cause chromosome damage; Rapp and Hsu concluded that virus DNA synthesis was required for the production of chromosome damage. However, this is difficult to reconcile with the report by Mazzone and Yerganian that damage appears within three hours of virus infection [97]. Similarly, in monkey kidney cells, chromatid gaps, breaks, and diffuse lesions due to partial despiralization of the chromosomes have appeared within 2 to 3 hours of infection with HSV [19], which is 2 hours before the onset of virus DNA synthesis; UV-irradiated virus had no effect in this system. However, in a very careful study using hamster (BHK 21) cells, Waubke et al. [168] showed very convincingly that chromosome lesions caused by herpes simplex virus developed prior to, and were independent of, viral DNA synthesis. They found a reduction in chromosome damage by UV-treated virus but found that the rate of reduction was four times slower than the rate of reduction in infectivity with increasing doses of UV irradiation. They suggested that the chromosome-damaging effect might be due to the action of an early enzyme under the control of the virus genome. This suggestion is supported by the finding of Donner and Gonczol [39] that chromosome damage by HSV was inhibited by actinomycin D or puromycin, but not by cytosine arabinoside added immediately after infection.

In human embryo lung (HEL) cells, Stich et al. [150] found a very high incidence of chromatid gaps, breaks (35%), and pulverization (32%) following HSV infection at low multiplicity. O'Neill and Miles [122], however, found only a small increase in breaks and gaps shortly after HSV type I infection of human-embryo kidney (HEK) cells and failed to find any damage in HEL cells. They did report, however, a striking accentuation of the secondary constriction regions of chromosomes Nos. 1, 9, and 16 in HEK cells. Although O'Neill and Miles find no effect on PHA-stimulated peripheral blood leukocytes, Aya et al. [10] and Makino and Aya [87] have reported a large excess of chromatid aberrations from 12 hours after HSV infection (10 plaque-forming units, or PFU, per cell). A few dicentrics were found, and there was said to be a specificity of breakage at the midpoints of a group D and a group B chromosome, at the middle and proximal segment of the long arm of No. 2, and near the middle of the long arm of a group C chromosome.

Specificity of break points induced by HSV was also suggested by Stich et al. [150], who found increased breakage at specific locations on chromosomes Nos. 1 and X of Chinese hamster cells. They regarded this specificity of breakage to be similar to that induced by BUDR as reported by Hsu and Somers [59]. As with the adenoviruses, the amount of damage was found to be proportional to the virus input [134, 156, 168].

Herpes Simplex Virus Type 2. Relatively few observations have been made with HSV type 2. O'Neill and Miles [122], however, have found that it causes breaks and attenuation of the secondary constriction regions in HEL cells, HEK cells, and cultured peripheral blood leukocytes. There were slightly more aberrations in the HEK cells. A small number of pulverized cells were observed in lymphocyte cultures. O'Neill and Rapp [123], also using HEL cells, found considerable numbers of chromatid aberrations 4 hours after infection with plaque-purified HSV type 2 (5 PFU/cell). They found a dramatic synergistic effect for the virus with cytosine arabinoside and, since cytosine arabinoside inhibits DNA synthesis, this observation suggested that in this system also, neither cell nor virus DNA synthesis is required for the production of chromosome aberrations. The importance of this synergistic effect in carcinogenesis is stressed because of the possible association between HSV type 2 and carcinoma of the cervix. It is surprising that no other observations have been made and, in particular, that no attempt has been made to examine cultures of cervical cells infected with HSV type 2.

Cytomegalovirus. In the only reported study, Demidova et al. [37] found no chromosomal abnormalities in human embryo fibroblast cultures infected with cytomegalovirus or in peripheral blood cultures of children with cytomegalovirus infection.

Herpes Zoster Virus. There have been a number of reports of chromosome breakage in cultured leukocytes from patients with chickenpox [4, 5, 49]. Other authors have failed to find such an effect, however [27, 36, 52]. From the report by Benyesh-Melnick et al. [16], there can be no doubt that zoster causes chromosome gaps and breaks in cultured HEL cells. They also reported that overcontraction of the chromosomes occurred and that this was sometimes accompanied by apparent fragmentation. This curious effect is found in several other virus infections (Fig. 4).

Marek's Disease Virus. I can find only one report of cytogenetic observations on Marek's disease virus. Bloom [18] reported normal chromosomes in the cells of resistant and sensitive chickens and also in tumor cells from birds with Marek's disease.

Epstein-Barr (EB) Virus. Since the establishment of proliferating cultures of lymphoid cells takes a considerable time after exposure of the target cells to EB virus, it is difficult to be dogmatic about the role of EB virus in causing chromosome damage. The work of Zur Hausen and Schulte-Holthausen [182] and Zur Hausen et al. [180] using molecular hybridization techniques, however, leaves no doubt that the virus genome persists even in lymphoid cell lines in which virus particles cannot be demonstrated either by immunofluorescence or by electron microscopy. There are as yet no studies on early infection, but it is relevant to consider briefly the role of virus in the cytogenetic abnormalities that have been observed in the lymphoid cell lines derived from patients with malignant disease or from normal individuals. Early after establishment of such lines, many have essentially normal chromosomes [147], and it can be argued that EB virus causes no cytogenetic damage immediately after infection. However, since the lines probably originated from a small number of cells, this is not necessarily true; but the results do suggest that visible chromosome damage is not a prerequisite for the establishment of the lines.

Cell lines derived from normal individuals or from patients with nonmalignant diseases are apparently more likely to remain cytogenetically normal for some time than are those derived from patients with malignant diseases [61, 103, 148], although Steel et al. suggested that the age of the donor may also be an important factor. Most lines do develop chromosomal aberrations at some stage. Chromosome exchanges leading to the formation of marker chromosomes, pseudodiploidy, and an occasional dicentric chromosome are the most common, but gaps and breaks are occasionally seen. Later in the evolution of the cultures, the cells tend to become polyploid, and at this stage marker chromosomes become more common [160]. Marker chromosomes similar in appearance have been seen in several different cell lines, but the most recent evidence [146] suggests that this is a morphological similarity only and that the apparently similar markers are not derived from identical rearrangements. However, Manolov and Manolova [88] have described an extra band at the end of the long arm of chromosome No. 14 in cultures and biopsies from Burkitt lymphoma cases, and a strong impression does remain that the exchange points are not random.

There is strong evidence of clonal evolution within the Burkitt lines [146] and of divergence when a line is maintained in different laboratories [160]. The features of the lymphoid lines suggest that they are "transformed" lines; the evolution of these karyotypically abnormal clones is of interest and closely resembles the evolution of lines of cells found in vivo in acute leukemias. The general picture, therefore, is one of karyotypic instability, with selection nevertheless in favor of particular rearrangements. The role of the persistent virus genome in inducing chromosome abnormalities is not known.

One interesting aspect is the occurrence of a near-terminal constriction on the long arms of a group C chromosome (Fig. 1b) in some of these lines of cells [72] and the role of the virus in inducing this aberration. The elusive nature of this abnormality has led some authors to deny its existence [171], but to those who have observed a line of cells containing the marker, there can be no doubt of its reality, since it may be present in a very high proportion of cells [100, 148, 160]. It was originally suggested [72] that this marker appeared on chromosome No. 10, but more recent evidence suggests that it is on chromosome No. 7 [146]. It has been observed mainly in lines of cells derived from patients with the Burkitt tumor, but it has been reported also in lines established from leukemic patients [100, 177] and from patients with infectious mononucleosis [61]. The proportion of cells showing the specific marker is low in lines from infectious mononucleosis patients, and many workers have failed to find evidence of it at all. Similarly, lines from normal individuals often fail to show the marker [147]. Perhaps the most interesting observation comes from Henle et al. [57], who were able to demonstrate the marker chromosome in lines of cells established from normal cells by co-cultivation with irradiated virus-containing cells, which could suggest that the presence of the virus is closely associated with the appearance of the marker chromosome. Steel [147] was not able to confirm this observation.

Papovaviruses

The papova group of viruses includes the papilloma viruses, polyoma virus, and simian vacuolating virus 40 (SV40). The papilloma viruses are of course associated with the production of benign tumors in several species, including man, while polyoma and SV40 are capable of transforming cells from several species in vitro and of producing tumors in some of these animals. SV40 can transform human cells in vitro [73, 142], and there is one claim that polyoma virus can transform human cells [143]. So far neither SV40 nor polyoma virus has been shown to be oncogenic in its natural host in the wild, however. Recently, papovaviruses other than wart viruses have been isolated from human sources—one from patients with renal transplants [47], and others from patients with progressive multifocal leukoencephalopathy [124], a disease which is often associated with leukemia.

The role of all members of this group of viruses in inducing chromosome damage is therefore of interest. I am not aware of cytogenetic studies on cells infected with human wart virus or with the new papovaviruses. Reports on chromosome studies using the papilloma viruses are few; reports on polyoma virus are surprisingly few, and are also apparently contradictory.

Papilloma Virus. Prunerias et al. [129] found an increased incidence of chromosome breakage and aneuploidy in cultured rabbit cells infected with Shope papilloma virus, but their work was complicated by a high incidence of aberrations in control cultures. Bayreuther [15] has reported normal chromosomes in five benign papillomas induced by Shope papilloma virus; however, after malignant transformation, all five had abnormal and hyperdiploid, but different, chromosome complements. Similarly, Palmer [125] found early papillomas to be normal, but many abnormalities in chromosome number appeared in transplantable carcinomas arising from them. In contrast, McMichael et al. [85] in an extensive study found chromosome abnormalities at all stages in the development of papillomatous tumors. Evidence of chromosome instability was not marked, but a few dicentrics were noted. The incidence of aberrations increased with time and with conversion to malignancy, but it is of interest that stemlines were noted even at the earliest benign stage. Prunerias et al. also recorded chromosome abnormalities in benign tumors.

Polyoma Virus. Both Ford et al. [44] and Yerganian et al. [174] reported no increase in chromosome aberrations following infection of Chinese hamster cells with polyoma virus. Yerganian et al. further reported that the transformed cells remained euploid over many generations in vitro and through several transplant generations in vivo. The only evidence of any disturbance of the chromosomes was the appearance, after many weeks in culture, of a specific lesion at the secondary constriction of the X chromosome in both males and females.

On the other hand, Vogt and Dulbecco [163] found a very high incidence of chromatid breaks during the period shortly after infection of primary Syrian hamster cells with polyoma virus. The possibility of difference in the response of cells of different species is of interest. McPherson [86] reported accumulation of tetraploid cells in a clone of polyoma-transformed hamster cells, BHK 21 clone 13. He suggested that, although chromosome abnormality may not necessarily accompany transformation, there might be an increased tendency to mitotic errors following transformation. Defendi and Lehman [34] found no evidence of virus-induced chromosome aberrations in Syrian hamster cells shortly after infection with polyoma virus. Aneuploidy and abnormal marker chromosomes appeared only following transformation. The findings differed in different lines. Some highly aneuploid cell lines showed no evidence of stemline formation, but in one line a clear stemline emerged which was also present in a tumor induced by inoculation of these cells. Breaks were not common, but dicentric and ring chromosomes of abnormal morphology occurred regularly in all lines. Studies on tumors induced by polyoma virus in Syrian hamsters in vivo [35] gave similar results, though no stemlines were recognized; the number of abnormal cells increased with repeated passage of the cells in vitro. The picture in the mouse is, however, quite different. Hellstrom et al. [54, 55] found 5 of 20 polyoma-induced mouse tumors to be diploid, but the majority were aneuploid, containing marker chromosomes; the latter showed clear evidence of stemline proliferation that remained constant through several transplant generations. Some breakage did occur in these tumors.

Simian Virus 40 (SV40). Only a few studies have been carried out on the cytogenetics of human cells in the period immediately after infection with SV40. Wolman et al. [173] studied adult skin fibroblasts that were at a high passage level though still cytogenetically normal and found that gaps and breaks appeared very quickly after infection. Within a few days, dicentric chromosomes and chromatid interchanges were also seen. There was some enhancement of secondary constrictions, but otherwise the chromosome and chromatid aberrations seemed to be random. One of the X chromosomes frequently appeared to be lost from such cells, however. The telomeric associations and chromosome stickiness suggested to the authors a possible involvement of heterochromatic areas. Moorhead and Saksela [106] also studied the early changes and found that the time of appearance of aberrations was different in two human embryo fibroblast cell lines of different ages (eleventh and thirtieth passages). In the older line, their observations were similar to those of Wolman et al. in that chromatid breaks increased in frequency about a week after infection, but dicentrics and other abnormal chromosomes did not increase till 6 weeks after infection. In the younger culture a similar wave of chromatid breaks appeared at 10 weeks after infection, while dicentrics and abnormal chromosomes became common after the fourteenth week. Furthermore, tetraploidy, which increased in frequency from the first week in the older culture, did not increase till after the tenth week in the younger culture. It is interesting to note that morphological transformations first became apparent in the two cultures at 5 and 10 weeks post-infection, respectively.

It therefore appears that there is a definite sequence of chromosome changes in human cell cultures infected with SV40 and that these changes are related in time to the appearance of transformation. Attempts to relate more precisely the chromosomal changes to the transformation event illustrate the difficulty of making this point. Girardi et al. [48] found that the appearance of chromosomally abnormal cells was closely related to the loss of contact inhibition, but Weinstein and Moorhead [170] concluded that even after loss of contact inhibition has occurred, at least some of the cells are still karyotypically normal. They argued that this suggests that the visible chromosomal events are not essential for transformation. It has been found, however, that cultures of human fibroblasts that have become aneuploid [105, 137], and cells from patients with a genetically determined syndrome that shows chromosome instability [1], are more readily transformed by SV40 virus.

When fully transformed human cells were studied, chromosome changes were found in almost all the cells [73, 105, 106, 108, 135, 142, 169, 170, 175]. There is good agreement about the finding that there is a tendency to polyploidy in transformed cells. Some chromosome and chromatid breakage occurred, and abnormal chromosomes, including dicentrics, were common (Fig. 2). Such instability of the karyotype did not usually lead to a selection of stable clones of karyotypically marked cells, though transient clones were sometimes found; the heteroploidy and chromo-

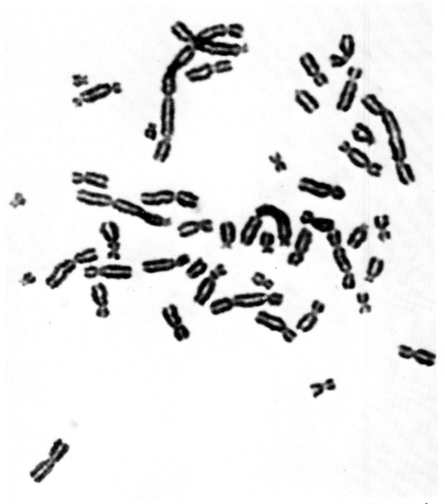

Figure 2. A cell from a culture of human skin fibroblasts transformed in vitro by SV40 virus. Note the remarkable frequency of exchanges leading to the formation of dicentric and multicentric chromosomes.

somal instability persisted after the crisis period through which many of these SV40 transformed lines passed [48]. Some evidence of specificity in the chromosome aberrations has been reported—in particular, the specific loss of a group G chromosome [73, 105, 142], of a group D chromosome, and, less frequently, of a group E chromosome [105]. An association of breakage points with the secondary constriction regions on chromosome Nos. 1, 9, and 16 has also been noted [48, 105].

Studies on the cells of other species are less numerous. Cooper and Black [29] found gross chromosomal instability, with both chromosome and chromatid aberrations, persisting for a long period after transformation of Syrian hamster cells by SV40. It was later shown [30] that the chromosomal instability persisted after clonal selection, and it was concluded that this was in some way related to the continued presence of the virus genome. More recently Nachtigal et al. [110] have found some differences among different tissues of the Syrian hamster in their susceptibility to SV40 virus-induced chromosome damage, and also some correlation between dicentric formation and the appearance of T antigen. Furthermore, cells transformed by the more oncogenic SV40 (PARA) variants, which have an SV40 genome but an adenovirus capsid, showed more chromosome damage. This suggests that in this system also there is a strong correlation between early events in the virus infection, chromosomal instability, and transformation. May et al. [96] suggested that in SV40-infected mouse kidney cells, following transcription of early virus messenger and synthesis of T-antigen, there was a virus-induced replication of the chromosomes followed by a normal or abnormal mitosis. Most cells underwent only one or two cycles, but the abortive infection conferred on a small fraction of the cells a continuous mitotic stimulus and pronounced cytogenetic instability. Again, the close relationship between chromosomal instability, T-antigen production, and transformation in this system is in line with the report by Aaronson and Todaro [1] that there is a correlation between the susceptibility of human cells to transformation by SV40 virus and the percentage of cells that show T-antigen early after infection. Very few chromosome studies have been carried out on tumors induced by SV40, but both Nachtigal et al. [109] and Lehman and Defendi [77] have reported numerous abnormalities in tumors induced in Syrian hamsters.

Poxviruses

Most members of the poxvirus group cause an initial stimulation of cell division shortly after infection even if subsequently a lytic cycle occurs. The poxvirus group also contains several members that produce benign tumors (e.g., molluscum contagiosum virus of man and Yaba virus of the monkey). Another member, Shope fibroma virus of the rabbit, normally produces benign lesions; but if inoculated into young animals or immunosuppressed hosts, it can produce malignant lesions. Bayreuther [15] has reported both fibromas and fibrosarcomas induced by the Shope fibroma virus to have normal chromosomes, but other cytogenetic observations on cells infected with the poxviruses are few.

Zur Hausen and Lanz [181] found some evidence of chromosome

breakage by vaccinia virus in mouse L cells in culture, while Schuler et al. [140] found an increased incidence of breakage in lymphocytes from two out of eight children after vaccination with vaccinia virus. In vitro infection of lymphocytes, however, resulted in complete disintegration of the chromosomes, presumably something akin to the "pulverization" seen with many RNA viruses. Matsaniotis et al. [94] reported normal chromosomes in the bone marrow in four boys and eight girls between the ninth and twenty-fourth day following successful vaccination with vaccinia virus. Koziorowska et al. [74] claimed to have transformed mouse embryo cells with vaccinia virus. The "transformed" cells are highly aneuploid and show considerable instability of the karyotype. One cannot reach any conclusions on these few observations, and it does seem that further observations should be carried out on the poxviruses.

EFFECTS OF RNA VIRUSES

With the important exception of the C-type viruses of the leukemia-sarcoma complex and the mammary tumor viruses, there is no clear-cut evidence that any RNA virus is involved in the transformation of cells or the production of malignant tumors in any species. For this reason my consideration of chromosome damage by these groups will be brief; but it is necessary to compare the effects of the DNA viruses with those of the oncogenic RNA viruses. Reviews by Boué and Boué [20], Nichols [113, 114], Moorhead [104], and Stich and Yohn [154] give more details.

Paramyxoviruses

Measles. There are many reports of chromosome breakage by measles virus in vitro and in vivo. The first report from Nichols et al. [115] showed that the incidence of chromosome and chromatid breaks was higher in cultured peripheral blood lymphocytes from patients with measles than in controls. Similar results have been obtained by Grippenberg [49] and by Aula [5], though others have failed to find damage [27, 36, 52, 157]. The differences seem to be real but are unexplained. Patients receiving living attenuated measles virus vaccine also showed chromosome damage in the lymphocytes [111]. Further studies using in vitro systems showed clearly that measles virus causes extensive damage to human cells [25, 116, 118, 130]. The damage was essentially of two types: simple chromatid and chromosome breakage, and multiple fragmentation of the chromosomes—a phenomenon that has come to be known as "pulverization." There is no evidence of either unstable or stable chromosome rearrangements following infection with measles virus. It seems, therefore, that directly or indirectly measles virus induces chromosome breakage that is not repaired. The role of the virus in inducing this damage may be due, at least in part, to its action in fusing cells to form syncytia. Sandberg et al. [138] have shown that chromosome pulverization may follow the fusion of asynchronous cells.

While the biological phenomena underlying chromosome damage induced by measles virus are undoubtedly of interest, it is difficult to envisage any role for such damage in the induction or progression of neoplastic disease, since it seems probable that the affected cells are killed. However, there has been some suggestion that the break points may be nonrandom [5], and a study of such nonrandom change may indicate the nature of the cell–virus interaction. One interesting report by Mauler and Hennesen [95] showed that chromosome breakage may also be induced in cells nonproductively infected with measles virus. In their studies with normal rabbit kidney cells they also reported the induction of dicentric chromosomes. This is one of the very few reports of a chromosomal rearrangement induced by a non-oncogenic RNA virus.

Mumps. Reports very similar to those for measles have suggested that the incidence of chromosome breakage is high in lymphocytes of patients with mumps [5, 49]. Chun et al. [27], however, failed to find any damage. Chromosome pulverization by mumps virus in HeLa cells has also been reported [25].

Sendai Virus. Chromosome pulverization has appeared early after infection of HeLa cells with Sendai (parainfluenza type 1) virus [7, 25, 136]. As with other systems, the amount of damage was related to the multiplicity of infection, and the chromosome-damaging action of the virus was less sensitive to UV-inactivation than were the infectivity and the hemolytic activity of the virus.

Rabies Virus. No abnormalities were reported in leukocytes from the one case of rabies studied by Marquez-Monter et al. [92].

Myxoviruses

Cantell et al. [25] found pulverization with Newcastle disease virus but not with influenza viruses (A/PR8 and B/LEE), although Ter Meulen and Love [158] did report that influenza virus induced clumping of chromosomes in HeLa cells. Homma et al. [58] claimed a highly specific loss of a group G chromosome and an X chromosome from a surprisingly stable HeLa S3 line carrying hemadsorption type 2 virus, as well as in the same line newly infected with this virus.

Arboviruses, Picornaviruses, and Unclassified Viruses

Yellow Fever Virus. Harnden [52] and de Grouchy et al. [36] have noted increased chromosome breakage and pulverization in cultured lymphocytes from patients inoculated with live, attenuated, yellow fever virus. The results were inconsistent from one patient to another and, though no rearrangements were reported, de Grouchy did find some evidence of increased aneuploidy.

Poliovirus. Chromosome breakage in human cell strains and in an established cell line has been reported to occur within 6 hours of infection with a high multiplicity of poliovirus [12]. Again, there was no evidence of chromosome rearrangements.

Hepatitis Virus. There are several reports of chromosome breakage in leukocytes cultured from hepatitis patients [8, 9, 41, 42, 87]; there are also reports of chromosome damage in bone marrow cells [42, 93]. These are probably the only reports of virus-induced chromosome damage in vivo in man.

Rubella. Chromosome breakage in lymphocytes of patients with congenital rubella has been reported [75, 87, 121, 172], although several authors have failed to find any evidence of damage [13, 98, 167]. In cells cultured from fetuses aborted because of maternal rubella infection [26], and in human cells infected with rubella virus in vitro [21, 22, 26], an increased incidence of breakage was reported in cases where an adequate number of cells were studied. In a study of over 2000 metaphases, Chang et al. [26] found no evidence of chromosome rearrangement, however, and none has been reported in any of the other studies.

Oncornaviruses

Avian Leukemia–Sarcoma Complex. The complications of the avian sarcoma virus system and the interrelationship of these viruses with the avian leukosis viruses has been dealt with in detail by Mitelman [102]. Briefly, this large group of morphologically identical (C-type) RNA viruses can be subdivided according to biological and serological characteristics; but a distinction between oncogenic members and non-oncogenic members cannot be made in any meaningful manner, since even viruses of proven oncogenicity may behave in a non-oncogenic manner in certain cell systems.

The replication of these viruses requires a DNA synthesis stage, and each carries within the virion an RNA-dependent DNA polymerase, and other enzymes that enable the virus to integrate a DNA copy into the host cell chromosomes. This intimate association has led to the suggestion that the viral genome may function as a cellular gene and that the control of the behavior of this locus is closely related to the control of the neoplastic properties of the cell. The genotype and phenotype of the virus and the genotype of the cell are all important in determining the nature of the relationship between the virus and the cell, and the interpretation of cytogenetic results may be difficult, since details of the cell–virus interactions are not always given.

Aberration analysis on normal chicken chromosomes is difficult because of the large number of microchromosomes. Ponten [127], however, in a careful study, found essentially normal chromosomes in a Rous sarcoma virus (RSV)-induced tumor and in a virus-induced erythroleukemia. Most of the other work with RSV has concerned the effect of the avian C-type viruses in mammalian cell systems. Ponten and Lithner [128]—using Engelbreth-Holm, Bryan, and Schmidt-Ruppin (SR) strains of RSV—found no excess of chromosome aberrations in cultured bovine lung fibroblasts, although there was some morphological alteration of the cells. However, Nichols [111], using cultured rat fibroblasts, found breaks and a small number of rearrangements at the second division following infection with RSV–SR. In Chinese hamster cells Kato [66] found increased breakage with RSV–SR and the Mill Hill strain of RSV. Most aberrations were simple breaks, but a few chromatid exchanges and pulverizations were noted.

Kato [67] followed up this study by reporting that the chromosomes of primary sarcomas induced in the Chinese hamster by RSV–SR were sometimes normal, and he concluded, therefore, that visible chromosome changes are not necessary for tumor induction. More often, however, the tumors had abnormal chromosomes, since there was an inverse correlation between the number of normal cells and the time elapsed since virus inoculation. Most aberrations were simple aneuploidy and polyploidy, but marker chromosomes and, more rarely, dicentric chromosomes and chromosome breakage were seen. Stemline evolution was clear in many instances. Similarly, Ahlström et al. [2] found stable stemline progression in two sarcomas induced in the Chinese hamster by RSV–SR. Tumors induced in rats with RSV–SR were found to have largely normal or near-diploid chromosome complements [111], but cytogenetically marked stemlines emerged at a very early stage, and thereafter the tumor progressed by stemline evolution [*111, 112*]. This was also true for rat tumors induced by RSV–Prague strain [*164, 165*]. The tumor cells appeared to be very stable, with few rearrangements, but these increased when the tumor cells were cultured in vitro [*111*]. Results with RSV–B77 avian sarcoma virus [*144*] yielded similar results, except that the karyotype of the tumor cells was markedly unstable. Mark [*89, 91*] found the majority of RSV-induced tumors in mice to have predominantly diploid karyotypes, although many also had aneuploid secondary lines. Pseudodiploid and triploid and tetraploid tumors also occurred, and many of them had marker chromosomes, indicating clonal evolution of the tumor. There was some evidence of karyotype instability, and in particular of the formation of numerous double-minute fragments [*90*]. It was suggested that the karyotype of the tumors was normal early in their development, but that the presence of the virus caused instability in the karyotype leading to the development of tumors that were chromosomally abnormal.

Nichols et al. [*117, 119*] have described chromosome breakage induced by RSV–SR in cultured human leukocytes. The aberrations were mainly of the delayed isolocus type of breakage, and reunions were rare. Nichols et al. also commented on the unusual degree of contraction of the chromosomes. RSV–Bryan, on the other hand, had no effect on the chromosomes. However, Schendler and Harris [*139*] were unable to confirm these results.

Murine Leukemia–Sarcoma Complex. The murine leukemia and sarcoma viruses share many characteristics with the avian C-type viruses, and the intimate association between the virus and the host cell chromosomes is probably similar. Surprisingly, few cytogenetic observations have been made with murine viruses, and most of these studies were done on fully developed leukemias. Rich et al. [*132*], using their own isolate of murine leukemia virus, found no evidence of cytogenetic change in the preleukemic phase of virus-inoculated mice. They suggested that the abnormalities in chromosome number that appeared in the leukemic phase were a consequence of the neoplastic transformation. Transplantable Moloney virus-induced leukemia was found to be cytogenetically abnormal, with good evidence of clonal proliferation [*64*]. A similar result was obtained for Friend virus-induced leukemia [*166*]. However, Tsuchida and Rich [*161*] found chromosome abnormalities in Rauscher leukemia and Friend leukemia but reported no evidence of clone formation.

GENERAL CONSIDERATIONS

In examining the significance of chromosome changes in virus-infected cells, it is important to consider the fate of the virus in a particular cell type. Events may be considered in three groups. The *early* events include uncoating of the virus nucleic acid, transcription of early virus messenger RNA, and in some cases inhibition of cellular DNA synthesis or the stimulation of the production of a host-coded enzyme. Then follows a period during which new virus nucleic acid is synthesized. *Late* events include transcription of late virus messenger RNA, synthesis of virion proteins, and assembly and release of virus.

Whether or not such a full productive cycle occurs will depend on the particular cell–virus interaction. A lytic cycle is more likely to occur in cells of the natural host, but productive infections in other species are not uncommon. The differentiated state of the cell is very important, and a virus may undergo a lytic cycle in cells of one tissue but fail to complete the cycle in cells of a different tissue from the same organism. The point at which the cycle is aborted varies, but it is more common for the early events to take place before the cycle breaks down. Normally, a productive cycle results in the death of the cell; but in some instances, when the virus is released by budding from the cell membrane, the cell may survive for a considerable time, in which case it is obvious that the virus genome persists within the cell. When the virus cycle is abortive, the cell may be killed or may survive the initial infection. In these survivor cells, at least part of the virus genome may persist and may be detected by the presence of virus-specified antigens. In some instances, it is possible to recover the virus by induction, or by co-cultivation or fusion with susceptible cells, but in others all attempts to recover the virus have failed. These various states may reflect differences in the relationship between the persistent virus genome and the host cell genome, or may mean that only part of the virus genome is present.

Those factors that determine the nature of the interaction between the virus and the genetic material of the cell will also be important in determining the visible chromosomal consequences. For example, immediate inhibition of mitosis by the virus could make visualization of chromosome damage impossible even if it did occur.

A most important distinction to make, however, would be between those chromosome aberrations due to input virus itself or to the early events in virus synthesis, and those changes occurring later, when damage could be due either to persistence of the virus genome or to instability in the cell genome induced by the virus. However, this distinction may be difficult to make, especially with low-input multiplicities, since a slow or highly asynchronous infection may be established and new cells continuously be infected.

From a consideration of the data so far presented, it is clear that there are many important gaps in our knowledge of virus–chromosome interactions. Therefore, any attempt to bring together the information currently available could possibly lead to erroneous conclusions. Nevertheless, it is probably of value to attempt to discern an overall pattern in the hope that this will highlight some of the areas in which our knowledge is slight and areas in which further investigation might be most profitable.

Chromosome Changes Occurring Early after Virus Infection

It is clear that many DNA and RNA viruses cause chromosome damage soon after infection of the cell. This is true of viruses that replicate their nucleic acid in the nucleus and those that replicate it in the cytoplasm. It is not necessary for the RNA viruses to have a DNA intermediate step, and there is no consistent difference between oncogenic and non-oncogenic viruses at this stage. The damage that may appear within a few hours of infection is dependent on the input multiplicity and can often be induced by UV-inactivated virus. This suggests that virus replication is not a prerequisite for the induction of early damage. The early damage is largely comprised of chromatid and chromosome gaps and breaks (Fig. 3). There are only occasional reports of chromatid or chromosome exchanges, but some of these (e.g., adenovirus 12 in Chinese hamster cells [151]) are very dramatic. Other aberrations common at this stage are "pulverization," overcontraction of the chromosomes (C-mitosis), and mitotic abnormalities such as clumping of the chromosomes and formation of micronuclei.

The mechanisms by which these aberrations are induced are not known, but there may be several different causes. Pulverization, for example, may be caused by agents other than virus [68] and also by the fusion of two cells whose DNA synthesis is asynchronous [138]. Since many viruses cause cellular fusion, the "pulverization" may be a consequence [56] of this effect, but this could not explain pulverization in mononucleate cells. Chromosome gaps and breaks may be due similarly to direct and indirect effects. In the case of the RNA viruses that do not interact with the cellular DNA, the effect must be indirect. One possible indirect effect may be that lysosomal enzymes are released following virus infection [3, 6]. Another possibility is that enzymes carried in or on the virion, or coded by early virus mRNA, may play a part in inducing damage.

There is good evidence that even in productive infections virus DNA may become covalently linked to the cellular DNA. In these instances the creation of integration sites by enzyme action might be visualized as chromosome damage. That rearrangements are rarely present could suggest either that chromosome damage of this type is not readily repaired

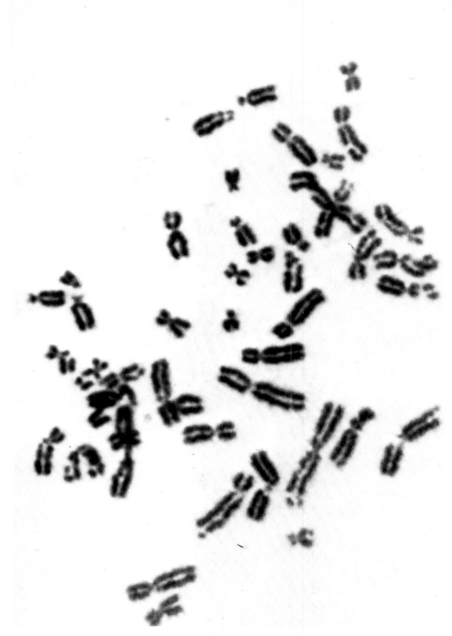

Figure 3. Chromatid and chromosome lesions induced in a human-embryo kidney cell 24 hours after infection with adenovirus type 12.

or that the necessary repair enzymes are not functioning during this phase of virus infection. The occasional appearance of rearrangements could be explained by the functioning of repair enzymes only for a short period of time during the course of virus infection. It is also possible that some of the lesions occurring at this stage are not true breaks. Overcontraction of the chromosomes is interesting in that it suggests a disturbance in the configuration of the entire cellular nucleoprotein following virus infection. It may be relevant to note that similar overcontraction is often a feature of animal tumor cells (Fig. 4) and human tumor cells. For example, in a leukemic bone marrow, the chromosomes of diploid cells may be normal, whereas adjacent, cytogenetically abnormal leukemic cells have overcontracted chromosomes. Whether this is a true correlation is, however, a matter of speculation.

Many of the cells that exhibit early chromosome damage, even if they do not produce virus, are killed by the infection, and, while cell death may be of importance in the pathogenesis of virus diseases and in teratogenesis, this early visible damage is likely to be irrelevant as far as carcinogenesis is concerned.

There is some evidence that the breaks or gap points are not randomly distributed throughout the karyotype. In the case of the non-oncogenic RNA viruses, such as measles virus, this could be due to the specific action of a particular enzyme or to the inhibition of host cell DNA synthesis at a particular point. In the case of the DNA viruses, however, nonrandom damage could suggest the association of the virus with a particular locus or loci or, more probably, with a particular type of DNA. It is not necessary to postulate that a virus integrates at the locus of a particular structural gene. Indeed, in those cases in which a highly specific aberration is found there is no evidence that the input virus is located at that point [179]. It seems much more probable that the virus, in order to regulate its own synthesis, associates with regions of the DNA that have a specific regulatory function; for this reason it will be of interest to determine the relationship between virus-induced chromosome breaks and the banding patterns of chromosomes, and also their relationship to the distribution of the various subfractions of host cell DNA that are clearly located at particular sites in the genome [65].

Chromosome Changes Occurring Late after Virus Infection

If the cell survives the virus infection, a number of different cell–virus interactions may result. Rarely, the surviving cells may persist untransformed. In such a situation, Hampar and Ellison [51] have shown by

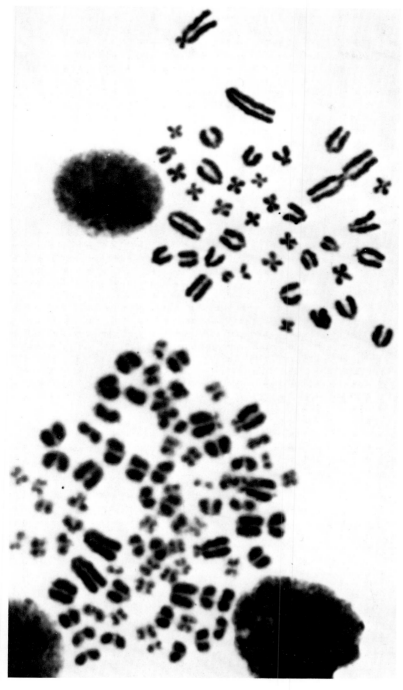

Figure 4. Two cells from a direct preparation of an adenovirus-12 induced rat tumor. The cell on the right contains a fragment and at least two abnormal chromosomes. The cell on the left illustrates the overcontraction phenomenon observed in many virus-induced cells and cells from virus-induced tumors. Such cells resemble cells that have been exposed to colchicine for prolonged periods.

cloning techniques that chromosome rearrangements may be demonstrated in Chinese hamster cells some time after herpes simplex virus infection. These cells were not virus-producing, and the situation was probably quite different from late effects reported for measles [118] and adenovirus 12 [82], in which continuing chromosome breakage was probably due to the infection of previously uninfected cells. In other instances, cells surviving adenovirus 12 infection have been found to have normal chromosomes [151, 178].

Most other observations made late after virus infection have been made of cells that were transformed or were undergoing transformation. Visible chromosome changes are clearly not a prerequisite for transformation, but frequently accompany it, so that they could be attributable either to the persistence of the virus genome or to the effects of the transformation event itself. At present it is not possible to distinguish between these two possibilities, and indeed, since the two are likely to be closely interlinked, such a distinction may not be possible. The kinds of chromosome aberrations appearing at this time are different from those occurring earlier in that the main features are chromosome and chromatid exchanges, aneuploidy, and a tendency to polyploidy (Fig. 5). Chromosome gaps and breaks do occur, but they are normally repaired.

This chromosomal instability is a feature of transformed cells, but the degree of instability varies with the system. In the in vitro systems, the karyotypic instability persists for many generations and, while minor clones may develop, a stemline with a relatively stable abnormal karyotype does not usually evolve. Such instability has been very well documented for adenovirus in hamster cells [155], SV40 in human cells [48], and SV40 in hamster cells [30]. Specific attempts to evolve stemlines by repeated cloning have failed [30]. However, there are exceptions. Defendi and Lehman [34] have described the evolution in vitro of a clone in polyoma-transformed Syrian hamster cells. In the case of EB herpesvirus-induced lymphoid cell lines, there was also clear-cut evidence of clonal proliferation in vitro [147]. Some evidence suggests, however, that it is the particular cell–virus combination that determines the stability of the karyotype. It should also be noted that instability of the karyotype is also characteristic of permanent cell lines in vitro that have not originated from transformation by virus.

The situation in vivo appears to be different. The evolution of stemlines is the rule rather than the exception. When a tumor is induced by direct inoculation of virus into the animal, stemlines may be present from a very early stage. McMichael et al. [85] found stemline evolution even at the premalignant stage of rabbit papillomas. In the case of Rous sarcoma virus, stemlines evolved in a number of different mammalian

Figure 5. The chromosome of a rat embryo cell transformed by adenovirus type 2 showing polyploidy, aneuploidy, and stable and unstable chromosome rearrangements.

hosts [67, 89, 112, 165]. The observations are too incomplete to allow us to be too sure that this distinction between the in vivo and in vitro situation is real. They could, however, suggest that the chromosomal events underlying transformation in vitro are different from those underlying tumor production in vivo. It is more probable, however, that the relationship is similar, if not the same, and that the differences are due to more rigorous selection of a particular cell type in vivo. This is supported by the observation of Nichols [111] that RSV rat tumors become markedly more unstable when cultured in vitro, following a period of growth in vivo.

This does not necessarily imply that the chromosomal events are an important part of transformation, or that they are responsible for the progression of the tumor. It is clear, however, that the appearance of visible chromosome changes is very closely associated with the transformation event, and that tumor progression is frequently accompanied by the selection of chromosomally abnormal lines of cells from an unstable cell population. It is most important, therefore, to attempt to understand the mechanisms underlying these associations. While the critical changes may be occurring at the molecular level, a more careful study of the visible chromosomal changes could be one of the more productive ways of determining the nature of these molecular events.

Acknowledgements.–I would like to express my thanks to J. K. McDougall and P. H. Gallimore for their comments on this manuscript and for providing photographs; and also the Cancer Research Campaign for their financial support.

LITERATURE CITED

1. AARONSON SA, TODARO GJ: SV40 T antigen and transformation in human fibroblast cultures. *Virology* 36:254–261, 1968.
2. AHLSTRÖM CG, KATO R, LEVAN A: Rous sarcoma in Chinese hamsters. *Science* 144:1232–1233, 1964.
3. ALLISON AC, PATON GR: Chromosome damage in human diploid cells following activation of lysosomal enzymes. *Nature* 207:1170–1173, 1965.
4. AULA P: Chromosome breaks in leucocytes of chicken pox patients. *Hereditas* 49:451–453, 1963.
5. AULA P: Virus associated chromosome breakage. A cytogenetic study of chicken pox, measles and mumps patients and of cell cultures infected with measles virus. *Ann Acad Sci Fenn, Ser A, IV, Biol* 89:1–78, 1965.
6. AULA P, NICHOLS WW: Lysosomes and virus induced chromosome breakage. *Exp Cell Res* 51:595–601, 1968.
7. AULA P, SAKSELA E: Early morphology of the chromosome damage induced by Sendai virus. *Hereditas* 55:362–366, 1966.
8. AYA T, MAKINO S: Notes on chromosome abnormalities in cultured leucocytes from serum hepatitis patients. *Proc Jap Acad* 42:648–653, 1966.
9. AYA T, MAKINO S, HIRAYAMA A: Some chromosomal studies on patients with infectious hepatitis. *Proc Jap Acad* 42:1088–1093, 1966.
10. AYA T, MAKINO S, YAMADA M: Chromosome aberrations induced in cultured human leucocytes by herpes simplex virus infection. *Proc Jap Acad* 43:239–244, 1967.
11. BARSKI G, CORNEFERT F: Caractéristiques caryologiques des tumeurs pulmonaire des hamsters produit par l'adenovirus 12. *Ann Inst Pasteur* 107:114–120, 1964.
12. BARTSCH HD, HABERMEHL KO, DIEFENTHAL W: Chromosomal damage after infection with poliomyelitis virus. *Exp Cell Res* 48:671–675, 1967.
13. BAUGHMAN F, HIRSCH B, BENDA CE: Chromosomes in Gregg's syndrome. *Lancet* 2:101, 1964.
14. BAUMILLER RC: Virus induced point mutation. *Nature* 214:806–807, 1967.
15. BAYREUTHER K: Chromosomes in primary neoplastic growth. *Nature* 186:6–9, 1960.

16. BENYESH-MELNICK M, STICH HF, RAPP F, HSU TC: Viruses and mammalian chromosomes. III. Effect of herpes zoster virus on human embryonal lung cultures. *Proc Soc Exp Biol Med* 117:546, 1964.
17. BIGGS PM, PAYNE LN: Studies on Marek's disease. I. Experimental transmission. *J Nat Cancer Inst* 39:267–280, 1967.
18. BLOOM SE: Marek's disease: chromosome studies of resistant and susceptible strains. *Avian Dis* 14:478–490, 1970.
19. BOIRON M, TANZER J, THOMAS M, HAMPE A: Early diffuse chromosome alterations in monkey kidney cells infected *in vitro* with herpes simplex virus. *Nature* 209:737–738, 1966.
20. BOUÉ A, BOUÉ JG: Virus et chromosomes humaines. *Pathol Biol* 16:677–690, 1968.
21. BOUÉ JG, BOUÉ A, LAZAR P: Chromosomal alterations induced in human diploid cell cultures *in vitro* by rubella virus and by measles virus. *Pathol Biol* 15:997–1007, 1967.
22. BOUÉ JG, BOUÉ A, MOORHEAD PS, PLOTKIN SA: Alterations chromosomique induite par le virus rubéole dans les cellules embryonaire diploid humaine cultivé *in vitro*. *CR Acad Sci Paris* 259:687–690, 1964.
23. BRAILOVSKY C, WICKER R, SUAREZ H, CASSINGENA R: Transformation *in vitro* de cellules de hamster chinois par l'adenovirus 12: étude cytogénétique. *Int J Cancer* 2:133–142, 1967.
24. BURDETTE WJ, YOON JS: Mutations, chromosomal aberrations, and tumors in insects treated with oncogenic virus. *Science* 155:340–341, 1967.
25. CANTELL K, SAKSELA E, AULA P: Virological studies on chromosome damage of HeLa cells induced by myxoviruses. *Ann Med Exp Fenn* 44:255–259, 1966.
26. CHANG TH, MOORHEAD PS, BOUE JG, PLOTKIN SA, HOSKINS JM: Chromosome studies of human cells infected *in utero* and *in vitro* with rubella virus. *Proc Soc Exp Biol Med* 122:236–243, 1966.
27. CHUN T, ALEXANDER DS, BRYANS AM, HAUST MD: Chromosomal studies in children with mumps, chicken pox, measles and measles vaccination. *Can Med Assoc J* 94:126–129, 1966.
28. COHEN MM, CLARK HF, JENSEN F: Effects of SV40 virus on the chromosomes of poikilothermic cells (*Gekko Gecko*) culivated at different temperatures. *Int J Cancer* 9:618–625, 1972.
29. COOPER HL: BLACK PH: Cytogenetic studies of hamster kidney cell antigens transformed by the simian vacuolating virus (SV40). *J Nat Cancer Inst* 30:1015–1025, 1963.
30. COOPER HL, BLACK PH: Cytogenetic studies of three clones derived from a permanent line of hamster cells transformed by SV40. *J Cell Comp Physiol* 64:201–219, 1964.
31. COOPER JEK, STICH HF, YOHN DS: Viruses and mammalian chromosomes. VIII. Dose–response studies with human adenovirus types 18 and 4. *Virology* 33:533–541, 1967.
32. COOPER JEK, YOHN DS, STICH HF: Viruses and mammalian chromosomes. X. Comparative studies of the chromosome damage induced by human and simian adenoviruses. *Exp Cell Res* 53:225–240, 1968.

33. CRICK FH: General model for the chromosomes of higher organisms. *Nature* 234:25–27, 1971.
34. DEFENDI V, LEHMAN JM: Transformation of hamster cells *in vitro* by polyoma virus in morphological, karyological, immunological and transplantation characteristics. *J Cell Comp Physiol* 66:351–409, 1965.
35. DEFENDI V, LEHMAN JM: Biological characteristics of primary tumours induced by polyoma virus in hamsters. *Int J Cancer* 1:525–540, 1966.
36. DE GROUCHY J, TUDELA V, FEINGOLD J: Etudes cytogénétiques *in vivo* et *in vitro* après infections virales et après vaccination anti-amarile. *Pathol Biol* 15:879–885, 1967.
37. DEMIDOVA SA, STONOVA NS, SELEZNIOVA TG, GAVRILOV VI: Morphological and cytogenetical studies of the effect of the cytomegalovirus in cultures of human cells. *Genetika* 4:126–132, 1968.
38. DOERFLER W: Integration of the DNA of adenovirus type 12 into the DNA of baby hamster kidney cells. *J Virol* 6:652–666, 1970.
39. DONNER L, GONCZOL E: The influence of inhibitors of macromolecular synthesis on capacity of herpes simplex virus to induce chromosomal damage. *J Gen Virol* 10:243–250, 1971.
40. DUFF R, RAPP F: Oncogenic transformation of hamster cells after exposure to herpes simplex virus type 2. *Nature New Biol* 233:48–49, 1971.
41. EL ALFI OS, SMITH PM, BIESELE JJ: Chromosomal breaks in human leucocyte cultures induced by an agent in the plasma of infectious hepatitis patients. *Hereditas* 52:285–294, 1965.
42. EMERIT I, EMERIT J: Virus induced chromosome breakage. *Lancet* 2:494, 1972.
43. FJELDE A, TRENTIN JJ, BRYAN E: The chromosomes of tumors induced in mice by human adenovirus 12. *Proc Soc Exp Biol Med* 116:1102–1105, 1964.
44. FORD DK, BUGUSZEWSKI C, AURSPERG N: Chinese hamster cell strains *in vitro* spontaneous chromosome changes and latent polyoma virus infection. *J Nat Cancer Inst* 26:691–706, 1961.
45. FREEMAN AE, BLACK PH, VANDERPOOL EA, HENRY PH, AUSTIN JB, HUEBNER RJ: Transformation of primary rat embryo cells by adenovirus type 2. *Proc Nat Acad Sci (US)* 58:1205–1212, 1967.
46. GALLIMORE PH: Studies with adenovirus type 2; cellular transformation and oncogenicity. M.Sc. Thesis, University of Birmingham, 1972.
47. GARDNER SD, FIELD AF, COLEMAN DV, BULME B: New human papovirus (B.K.) isolated from urine after renal transplantation. *Lancet* 1:1253–1257, 1971.
48. GIRARDI AJ, WEINSTEIN D, MOORHEAD PS: SV40 transformation of human diploid cells. A parallel study of viral and karyologic parameters. *Ann Med Exp Fenn* 44:242–254, 1966.
49. GRIPPENBERG U: Chromosome studies on some virus infections. *Hereditas* 54:1–18, 1965.
50. HAMPAR B, ELLISON SA: Chromosomal aberrations induced by animal virus. *Nature* 192:145–147, 1961.
51. HAMPAR B, ELLISON SA: Cellular alterations in the M.C.H. line of Chinese hamster cells following infection with herpes simplex virus. *Proc Nat Acad Sci (US)* 49:474–480, 1963.

52. HARNDEN DG: Cytogenetic studies on patients with virus infections and subjects vaccinated against yellow fever. *Am J Hum Genet* 16:204–213, 1964.
53. HARNDEN DG, ELSDALE TR, YOUNG DE, Ross A: The isolation of cytomegalovirus from peripheral blood. *Blood* 30:120–125, 1967.
54. HELLSTRÖM KE, HELLSTRÖM I, SJÖGREN HO: Karyologic studies on polyoma virus induced mouse tumours. *Exp Cell Res* 26:434–457, 1962.
55. HELLSTRÖM KE, HELLSTRÖM I, SJÖGREN HO: Karyotype and polyoma virus sensitivity in clones isolated from a polyoma-induced mouse tumour. *J Nat Cancer Inst* 32:635–643, 1964.
56. HENEEN WK, NICHOLS WW, LEVAN A, NORRBY E: Polykaryocytosis and mitosis in a human cell line after treatment with measles virus. *Hereditas* 64:53–84, 1970.
57. HENLE W, DIEHL V, KOHN G, ZUR HAUSEN H, HENLE G: Herpes-type virus and chromosome marker in normal leukocytes after growth with irradiated Burkitt cells. *Science* 157:1064–1065, 1967.
58. HOMMA M, OHIRA M, ISHIDA N: Specific chromosome aberrations in cells persistently infected with type 2 hemadsorption virus. *Virology* 34:60–68, 1968.
59. HSU TC, SOMERS CE: Effect of 5-bromodeoxyuridine on mammalian chromosomes. *Proc Nat Acad Sci (US)* 47:396–403, 1961.
60. HUANG CC: Induction of a high incidence of damage to the X chromosomes of *Rattus (Mastomys) natalensis* by base analogues, viruses, and carcinogens. *Chromosoma* 23:162–179, 1967.
61. HUANG CC, MINOWADA J, SMITH RT, OSUNKOYA BO: Reevaluation of relationship between C chromosome marker and Epstein-Barr virus: chromosome and immunofluorescence analyses of 16 human hematopoietic cells lines. *J Nat Cancer Inst* 45:815–823, 1970.
62. HUEBNER RJ: Adenovirus-directed tumor and T antigens. In Perspectives in Virology (Pollard M, ed), vol 5, New York and London, Academic Press, 147–166, 1967.
63. HUNT RD, MELENDEZ LV, KING NW, GILMORE CE, DANIEL MD, WILLIAMSON ME, JONES TC: Morphology of disease with features of malignant lymphoma in marmosets and owl monkeys inoculated with herpesvirus saimiri. *J Nat Cancer Inst* 44:447–465, 1970.
64. IDA N, OHBA Y, FUKUHARA A, KOHNO M: Chromosome conditions of hypodiploid Moloney virus-induced mouse leukaemia, its transplantability and inductivity. *J Nat Cancer Inst* 40:97–110, 1968.
65. JONES KW, CORNEO G: Location of satellite and homogenous DNA sequences. *Nature New Biol* 233:268–271, 1971.
66. KATO R: Localization of "spontaneous" and Rous sarcoma virus-induced breakage in specific regions of the chromosomes of the Chinese hamster. *Hereditas* 58:221–247, 1967.
67. KATO R: The chromosomes of forty-two primary Rous sarcomas of the Chinese hamster. *Hereditas* 59:63–120, 1968.
68. KATO H, SANDBERG AA: Chromosome pulverisation in human binucleate cells following Colcemid treatment. *J Cell Biol* 34:33–45, 1967.

69. KIT S, NAKAJIMA K, DUBBS DR: Origin of thymidine kinase in adenovirus-infected human cell lines. *J Virol* 5:446–450, 1970.
70. KLEIN G: Immunological studies on Burkitt's lymphoma. *Postgrad Med J* 47:141–155, 1971.
71. KLEIN G: Herpesviruses and oncogenesis. *Proc Nat Acad Sci (US)* 69:1056–1064, 1972.
72. KOHN G, MELLMAN WJ, MOORHEAD PS, LOFTUS J, HENLE G: Involvement of C group chromosomes in five Burkitt lymphoma cell lines. *J Nat Cancer Inst* 38:209–222, 1967.
73. KOPROWSKI H, PONTEN JA, JENSEN F, RAVDIN RG, MOORHEAD P, SAKSELA E: Transformation of cultures of human tissue infected with simian virus SV40. *J Cell Comp Physiol* 59:281–292, 1962.
74. KOZIOROWSKA J, WLODARSKI K, MAZUROWA N: Transformation of mouse embryo cells by Vaccinia virus. *J Nat Cancer Inst* 46:225–241, 1971.
75. KUROKI Y, MAKINO S, AYA T, NAGAYAMA T: Chromosome abnormalities in cultured leucocytes from rubella patients. *Jap J Hum Genet* 11:17–23, 1966.
76. KVASNICKA A: Relationship between herpes simplex and lip carcinoma. III. *Neoplasia* 10:199–203, 1963.
77. LEHMAN JM, DEFENDI V: Cytogenetic studies of primary tumours indirect by SV40, adeno 7 and SV40 adeno 7 hybrid. Abstract in *Mammalian Chromosome Newsl* 9:35; Sixth Conference on Mammalian Cytology and Somatic Cell Genetics, 1968.
78. McALLISTER RM, GILDEN RV, GREEN M: Adenoviruses in human cancer. *Lancet* 1:831–833, 1972.
79. McALLISTER RM, NICHOLSON MO, LEWIS AM, MACPHERSON I, HUEBNER RJ: Transformation of rat embryo cells by adenovirus type 1. *J Gen Virol* 4:29–36, 1969.
80. McALLISTER RM, NICHOLSON MO, REED G, KERN J, GILDEN RV, HUEBNER RJ: Transformation of rodent cells by adenovirus 19 and other group D virus. *J Nat Cancer Inst* 43:917–924, 1969.
81. McDOUGALL JK: Effects of adenoviruses on the chromosomes of normal human cells and cells trisomic for an E chromosome. *Nature* 225:456–458, 1970.
82. McDOUGALL JK: Spontaneous and adenovirus type 12 induced chromosome aberrations—Fanconi's anemia fibroblasts. *Int J Cancer* 7:526–534, 1971.
83. McDOUGALL JK: Adenovirus-induced chromosome aberrations in human cells. *J Gen Virol* 12:43–51, 1971.
84. MACKINNON E, KALNINS VI, STICH HF, YOHN DS: Viruses and mammalian chromosomes. VI. Comparative karyological and immunofluorescent studies on Syrian hamster and human amnion cells infected with human adenovirus type 12. *Cancer Res* 26:612–618, 1966.
85. MCMICHAEL H, WAGNER PC, NOWELL PC, HUNGERFORD DA: Chromosome studies of virus induced rabbit papillomas and derived carcinomas. *J Nat Cancer Inst* 31:1197–1215, 1963.
86. McPHERSON IA: Characteristics of a hamster cell clone transformed by polyoma virus. *J Nat Cancer Inst* 30:795–816, 1963.
87. MAKINO S, AYA T: Cytogenetic studies in leukocyte cultures from patients with

some viral diseases and in those infected with HSV. *Cytologica* 33:370–396, 1968.
88. MANOLOV G, MANOLOVA Y: Marker band in one chromosome 14 from Burkitt lymphomas. *Nature* 237:33–34, 1972.
89. MARK J: Chromosomal analysis of ninety-one primary Rous sarcomas in the mouse. *Hereditas* 57:23–62, 1967.
90. MARK J: Double minutes—a chromosomal aberration in Rous sarcomas in mice. *Hereditas* 57:1–22, 1967.
91. MARK J: Chromosomal analyses of Rous sarcomas in mice—comparison between the findings in the tumor and in material explanted *in vitro*. *Acta Pathol Microbiol Scand* 70:37–52, 1967.
92. MARQUEZ-MONTER H, HIGUERA-BALLESTEROS FJ, ALFARO-KOFMAN SA, MAYOR EGM: Chromosomes in human rabies. *Lancet* 2:102, 1967.
93. MATSANIOTIS N, KIOSSOGLOU KA, MAOUNIS F, ANAGNOSTAKIS DE: Chromosomes in infectious hepatitis. *Lancet* 2:1421, 1966.
94. MATSANIOTIS N, MAOUNIS F, KIOSSOGLOU KA, ANAGNOSTAKIS DE: Chromosomes after smallpox vaccination. *Lancet* 1:978, 1968.
95. MAULER R., HENNESSEN W: Virus induced alterations of chromosomes. *Arch Ges Virusforsch* 16:175–181, 1964.
96. MAY E, MAY P, WEIL R: Analysis of the events leading to SV40 induced chromosome replication and mitosis in primary mouse kidney cell cultures. *Proc Nat Acad Sci (US)* 68:1208–1211, 1971.
97. MAZZONE HM, YERGANIAN G: Gross and chromosomal cytology of virus infected Chinese hamster cells. *Exp Cell Res* 30:591–592, 1963.
98. MELLMAN WJ, PLOTKIN SA, MOORHEAD PS, HARTNETT EM: Rubella infection of human leukocytes. *Am J Dis Child* 110:473–476, 1965.
99. MIGEON BR, MILLER CS: Human–mouse somatic cell hybrids with single human chromosomes (group 6): link with thymidine kinase activity. *Science* 162:1005–1006, 1968.
100. MILES CP, O'NEILL F, ARMSTRONG D, CLARKSON B, KEANE J: Chromosome patterns of human leukocyte established cell lines. *Cancer Res* 28:481–485, 1968.
101. MINZELL M, TOPLIN I, ISAACS JJ: Tumour induction in developing frog kidneys by a zonal centrifuge purified fraction of the frog herpes-type virus. *Science* 165:1134–1137, 1969.
102. MITTELMAN F: The Rous virus story: cytogenetics of tumors induced by RSV. This volume, 675–693, 1974.
103. MOORE GE, MINOWADA J: Human haemopoietic cell lines: a progress report. *In* Hemic Cells In Vitro (Farnes P, ed), Baltimore, Williams & Wilkins Company, 100–114, 1969.
104. MOORHEAD PS: Virus effects on host chromosomes. *In* Genetic Concepts and Neoplasia, Twenty-third Annual Symposium on Fundamental Cancer Research, M. D. Anderson Hospital and Tumor Institute, University of Texas, 281–312, 1970.
105. MOORHEAD PS, SAKSELA E: Nonrandom chromosomal aberrations—SV40 transformed human cells. *J Cell Comp Physiol* 62:57–84, 1963.

106. MOORHEAD PS, SAKSELA E: The sequence of chromosome aberrations during SV40 transformation of a human diploid cell strain. *Hereditas* 52:271–284, 1965.
107. MORGAN DG, EPSTEIN MA, ACHONG BG, MELENDEZ LV: Morphological confirmation of the herpes nature of a carcinogenic virus of primates (herpes saimiri). *Nature* 228:170–172, 1970.
108. MUSTAFINA AN: Changes in the karyotype of human cells *in vitro* induced by the simian virus 40. *Cytology* 8:249–258, 1966.
109. NACHTIGAL M, LUNGEANU A, NACHTIGAL S, ADERCA I: Cytogenetic study of two tumour lines induced in the golden hamster by SV40. *Rev Roum Inframicrobiol* 5:119–129, 1968.
110. NACHTIGAL M, MELNICK JL, BUTEL JS: Chromosomal changes in Syrian hamster cells transformed by simian virus 40 (SV40) and variants of defective SV40 (PARA). *J Nat Cancer Inst* 47:35–45, 1971.
111. NICHOLS WW: Relationship of viruses, chromosomes and carcinogenesis. *Hereditas* 50:53–80, 1963.
112. NICHOLS WW: Studies on the role of viruses in somatic mutation. *Hereditas* 55:1–27, 1966.
113. NICHOLS WW: Interactions between viruses and chromosomes. In Handbook of Molecular Cytology, (Lima De Faria A, ed), Amsterdam and London, North Holland Publishing Company, 732–750, 1969.
114. NICHOLS WW: Virus induced chromosome abnormalities. *Ann Rev Microbiol* 24:479-500, 1970.
115. NICHOLS WW, LEVAN A, HALL B, ÖSTERGREN G: Measles associated chromosome breakage. Preliminary communication. *Hereditas* 48:367–370, 1962.
116. NICHOLS WW, LEVAN A, AULA P, NORRBY E: Extreme chromosome breakage induced by measles virus in different *in vitro* systems. Preliminary communication. *Hereditas* 51:380–382, 1964.
117. NICHOLS WW, LEVAN A, GOLDNER H, CORIELL LL, AHLSTROM CG: Chromosome abnormalities *in vitro* in human leukocytes associated with Schmidt-Ruppin Rous sarcoma virus. *Science* 146:248–250, 1964.
118. NICHOLS WW, LEVAN A, AULA P, NORRBY E: Chromosome damage associated with the measles virus *in vitro*. *Hereditas* 54:101–118, 1965.
119. NICHOLS WW, LEVAN A, HENEEN W, PELUSE M: Synergism of the Schmidt-Ruppin strain of the Rous sarcoma virus and cytidine triphosphate in the induction of chromosome breaks in human cultured leukocytes. *Hereditas* 54:213–236, 1965.
120. NICHOLS WW, PELUSE M, GOODHEART C, MCALLISTER R, BRADT C: Autoradiographic studies on nuclei and chromosomes of cultured leukocytes after infection with tritium labelled adenovirus type 12. *Virology* 34:303–311, 1968.
121. NUSBACHER J, HIRSCHHORN K, COOPER LZ: Chromosomal abnormalities in congenital rubella. *New Eng J Med* 276:1409–1413, 1967.
122. O'NEILL FJ, MILES CP: Chromosome changes in human cells induced by herpes simplex, types 1 and 2. *Nature* 223:851–852, 1969.
123. O'NEILL FJ, RAPP F: Synergistic effect of herpes simplex virus and cystosine arabinoside on human chromosomes. *J Virol* 7:692–695, 1971.

124. PADGETT BL, WALKER DL, ZURHEIN GM, ECKROADE RJ, DESSEL BH: Cultivation of papova-like virus from human brain with progressive multifocal leucoencephalopathy. *Lancet* 1:1257–1260, 1971.
125. PALMER CG: The cytology of rabbit papillomas and derived carcinomas. *J Nat Cancer Inst* 23:241–248, 1959.
126. PAUL J: General theory of chromosome structure and gene activation in eukaryotes. *Nature* 238:444–446, 1972.
127. PONTEN J: Chromosome analysis of three virus associated chicken tumors: Rous sarcoma, erythroleukemia and RPL 12 lymphoid tumors. *J Nat Cancer Inst* 30:897–921, 1963.
128. PONTEN J, LITHNER F: Absence of specific chromosome alterations in bovine lung fibroblasts exposed to Rous sarcoma virus. *Int J Cancer* 1:589–598, 1966.
129. PRUNERIAS M, JACQUEMONT M, CHARDONNET Y, GAZZOLO L: Études sur les rapports virus chromosomes. VI. Étude caryotypique du papillome de Shope. *Ann Inst Pasteur* 110:145–174, 1966.
130. RADSEL-MEDVESCEK A, BLATNIK D: Chromosome abnormalities in measles patients and in children given living attenuated measles virus vaccine. *Zdrav Vestn* 36:179–181, 1967.
131. RAPP F, HSU TC: Viruses and mammalian chromosomes. IV. Replication of herpes simplex virus in diploid Chinese hamster cells. *Virology* 25:401–411, 1965.
132. RICH MA, TSUCHIDA R, SIEGLER R: Chromosome aberrations: their role in the etiology of murine leukemia. *Science* 146:252–253, 1964.
133. SABIN AB: Viral carcinogenesis: phenomena of special significance in the search for a viral etiology in human cancers. *Cancer Res* 28:1849–1858, 1968.
134. SABLINA OV, BOCHAROV EF: Chromosomal aberrations in the HeLa and Hep-2 tissue culture cells induced by herpes simplex virus. *Genetika* 4:128–134, 1968.
135. SAKSELA E: Chromosomal changes in human cells after infection with simian virus 40. *Hereditas* 52:250–251, 1964.
136. SAKSELA E, AULA P, CANGELL K: Chromosomal damage of human cells induced by Sendai virus. *Ann Med Exp Fenn* 43:132–136, 1965.
137. SAKSELA E, MOORHEAD PS: Aneuploidy in the degeneration phase of serial cultivation of human cell strains. *Proc Nat Acad Sci (US)* 50:390–395, 1963.
138. SANDBERG AA, AYA T, IKEUCHI T, WEINFELD H: Definition and morphologic features of chromosome pulverization: a hypothesis to explain the phenomenon. *J Nat Cancer Inst* 45:615–623, 1970.
139. SCHENDLER S, HARRIS RJC: The effect of Rous sarcoma virus (Schmidt-Ruppin strain) on the chromosomes of human leucocytes *in vitro*. *Int J. Cancer* 2:109–115, 1967.
140. SCHULER D, HERVEI S, GACS G, KIRCHNER M, SZATHMARY J: The influence of vaccinia viruses on human chromosomes *in vivo* and *in vitro*. *Humangenetik* 6:55–60, 1968.
141. SHAW GJ, MCFARLANE ES: Adenovirus 12 transformation of human foreskin cells. *Can J Microbiol* 15:811–812, 1969.
142. SHEIN HM, ENDERS JF: Transformation induced by simian virus 40 in human

renal cell cultures. 1. Morphology and growth characteristics. *Proc Nat Acad Sci (US)* 48:1164–1172, 1962.

143. SHEVLIAGHYN VJ, KARAZAS NV: Transformation of human cells by polyoma and Rous sarcoma viruses mediated by inactivated Sendai virus. *Int J Cancer* 6:234–244, 1970.

144. ŠIMKOVIČ D, POPOVIC M, SVEC J, GROFOVA M, VALENTOVA N: Continuous production of avian sarcoma virus B77 by rat tumour cells in tissue culture. *Int J Cancer* 4:80–85, 1969.

145. SKINNER GRB, THOULESS ME, JORDAN JA: Antibodies to types 1 and 2 herpes virus in women with abnormal cervical cytology. *J Obstet Gynaec Brit Commonw* 78:1031–1038, 1971.

146. STEEL CM: Non-identity of apparently similar chromosome aberrations in human lymphoblastoid cell lines. *Nature* 233:555–556, 1971.

147. STEEL CM: Human lymphoblastoid cell lines. III. Cocultivation technique for the establishment of new lines. *J Nat Cancer Inst* 48:623–628, 1972.

148. STEEL CM, MCBEATH S, O'RIORDAN ML: Human lymphoblastoid cell lines. II. Cytogenetic studies. *J Nat Cancer Inst* 47:1203–1214, 1971.

149. STICH HF, AVILA L, YOHN DS: Viruses and mammalian chromosomes. IX. The capacity of UV-impaired adenovirus type 18 to induce chromosome aberrations. *Exp Cell Res* 53:44–54, 1968.

150. STICH HF, HSU TC, RAPP F: Viruses and mammalian chromosomes. 1. Localization of chromosome aberrations after infection with herpes simplex virus. *Virology* 22:439–445, 1964.

151. STICH HF, VAN HOOSIER GL, TRENTIN JJ: Viruses and mammalian chromosomes. II. Chromosome aberrations by human adenovirus type 12. *Exp Cell Res* 34:400–403, 1964.

152. STICH HF, YOHN DS: Viruses and mammalian chromosomes. V. Chromosome aberrations in tumors of Syrian hamsters induced by adenovirus type 12. *J Nat Cancer Inst* 35:603–615, 1965.

153. STICH HF, YOHN DS: Mutagenic capacity of adenovirus for mammalian cells. *Nature* 216:1292–1294, 1967.

154. STICH HF, YOHN DS: Viruses and chromosomes. *Progr Med Virol (Karger)* 12:78–118, 1970.

155. STOLTZ DB, STICH HF, YOHN DS: Viruses in mammalian chromosomes. VII. The persistence of a chromosomal instability in regenerating, transplanted and cultured neoplasms induced by human adenovirus type 12 in Syrian hamsters. *Cancer Res* 27:587–592, 1967.

156. TANZER J, STOITCHKOVY Y, BOIRON M, BERNARD J: Alterations chromosomiques observées dans des cellules de rein de singe infectées *in vitro* par le virus de l'herpes. *Ann Inst Pasteur* 107:366–373, 1964.

157. TANZER J, STOITCHKOVY Y, HAREL P, BOIRON M: Chromosomal abnormalities in measles. *Lancet* 2:1070–1071, 1963.

158. TER MEULEN V, LOVE R: Virological, immunochemical and cytochemical studies on four HeLa cell lines infected with two strains of influenza virus. *J Virol* 1:626–639, 1967.

159. TODARO GT, AARONSON SA: Human cell strains susceptible to focus formation by human adenovirus type 12. *Proc Nat Acad Sci (US)* 61:1272–1278, 1968.
160. TOUGH IM, HARNDEN DG, EPSTEIN MA: Chromosome markers in cultured cells from Burkitt's lymphoma. *Eur J Cancer* 4:637–646, 1968.
161. TSUCHIDA R, RICH MA: Chromosomal aberrations in viral leukemogenesis. 1. Friend and Rauscher leukemia. *J Nat Cancer Inst* 33:33–47, 1964.
162. UTSUMI KR, KITAMUMA I, TRENTIN JJ: Karyologic studies of normal cells and of adenovirus type 12-induced tumor cells of the Syrian hamster. *J Nat Cancer Inst* 35:759–769, 1965.
163. VOGT M, DULBECCO R: Steps in the neoplastic transformation of hamster embryo cells by polyoma virus. *Proc Nat Acad Sci (US)* 49:171–179, 1963.
164. VRBA M, DONNER L: Chromosome numbers and karyotypes of two rat tumours induced by Rous sarcoma *in vitro*. *Folia Biol, Praha* 10:373–380, 1964.
165. VRBA M, DONNER L: Summary of findings on the rat karyotypes and on the chromosome characteristics of rat tumours induced by Rous sarcoma virus *in vitro*. *Neoplasia* 12:265–284, 1965.
166. WAKONIG-VAARTAJA R: Chromosomes in leukaemias induced by S37 and Friend viruses. *Brit J Cancer* 15:120–122, 1960.
167. WALKER M, ARMSTRONG G, NAHMIAS A: Chromosomal analyses of infants with the rubella syndrome. *Am J Dis Child* 111:110, 1966.
168. WAUBKE R, ZUR HAUSEN H, HENLE W: Chromosomal and autoradiographic studies of cells infected with herpes simplex virus. *J Virol* 2:1047–1054, 1968.
169. WEINSTEIN D, MOORHEAD PS: Karyology of permanent human cell line, W-18VA2, originated by SV40 transformation. *J Cell Comp Physiol* 65:85–92, 1965.
170. WEINSTEIN D, MOORHEAD PS: The relation of karyotypic change to loss of contact inhibition of division in human diploid cells after SV40 infection. *J Cell Physiol* 69:367–376, 1967.
171. WHANG-PENG J, GERBER P, KNUTSEN T: So-called C-marker chromosome and Epstein-Barr virus. *J Nat Cancer Inst* 45:831–839, 1970.
172. WIEDEMAN HR: Chromosomes in Gregg's syndrome. *Lancet* 1:721,1964.
173. WOLMAN SR, HIRSCHHORN K, TODARO GJ: Early chromosomal changes in SV40 infected human fibroblast cultures. *Cytogenetics* 3:45–61, 1964.
174. YERGANIAN G, HO T, CHO SS: Retention of euploidy and mutagenicity and heterochromatin in culture. *In* Cytogenetics of Cells in Culture (Harris RJC, ed), New York, Academic Press 79–96, 1964.
175. YERGANIAN G, SHEIN HM, ENDERS JF: Chromosomal disturbances observed in human fetal renal cells transformed in vitro by simian virus 40 and carried in culture. *Cytogenetics* 1:314–325, 1962.
176. ZUR HAUSEN H: Induction of specific chromosomal aberrations by adenovirus type 12 in human embryonic kidney cells. *J Virol* 1:1174–1185, 1967.
177. ZUR HAUSEN H: Chromosomal changes of similar nature in seven established cell lines derived from the peripheral blood of patients with leukemia. *J Nat Cancer Inst* 38:683–696, 1967.

178. Zur Hausen H: Chromosomal aberrations and cloning efficiency in adenovirus type 12-infected hamster cells. *J Virol* 2:915–917, 1968.
179. Zur Hausen H: Association of adenovirus type 12 DNA with host cell chromosomes. *J Virol* 2:218–223, 1968.
180. Zur Hausen H, Diehl V, Wolf H, Schulte-Holthausen H, Schneider U: Occurrence of Epstein-Barr virus genomes in human lymphoblastoid cell lines. *Nature New Biol* 237:189–190, 1972.
181. Zur Hausen H, Lanz E: Chromosomale Aberrationen bei L-Zellen nach Vaccinia-Virus-Infektion. *Z Med Nikrobiol Immunol* 152:60–65, 1966.
182. Zur Hausen H, Schulte-Holthausen H: Presence of EB virus nucleic acid homology in a "virus free" line of Burkitt tumour cells. *Nature* 227:245–248, 1970.

EFFECTS OF IONIZING RADIATION ON MAMMALIAN CHROMOSOMES

H. J. EVANS

MRC Clinical and Population Cytogenics Unit, Western General Hospital, Edinburgh, Scotland

---

Introduction	192
Types of Aberration Produced and Changes in Response throughout the Cell Cycle	193
Chromosome-type Aberrations	195
Chromatid-type Aberrations	198
Sub-chromatid-type Aberrations	200
"Derived" Chromosome-type Aberrations	201
Quantitative Changes during the Cell Cycle	201
Mechanisms of Formation of Aberrations and Dose–Response Kinetics	203
The Classical Theory of Aberration Formation	203
The Exchange Hypothesis	206
The Primary Events Involved in the Formation of Aberrations and the Notion of Misrepair	207
Some Physical and Biological Factors Influencing Aberration Yield	211
Dose, Dose-rate, Exposure Time, and Radiation Quality	211
Oxygen, Temperature, and Associated Variables	213
"Endogenous" Radiation	217
In Vitro vs. In Vivo Responses	219
Chromosome Organization and Other Influences of Genotype and Phenotype	220
The Use of Aberrations in Biological Dosimetry	223
Aberrations in Germ Cells	225
Concluding Comments	227
Literature Cited	228

INTRODUCTION

Interest in the effects of ionizing radiations on chromosomes dates back to within a decade of the discovery of X-rays by Röntgen in 1895 and of radium by the Curies in 1898. In 1905 Koernicke [89] studied the effects of radium rays on pollen cells in lily flower buds and provided the first clear demonstrations that ionizing radiations could fragment chromosomes and that broken ends from different fractures could fuse and thus result in rearrangements in chromosome structure. It was not until the early 1930s, however, that detailed studies on the types and frequencies of chromosome changes produced by radiations were undertaken. In 1927 Muller [104] demonstrated that X-rays were mutagenic in *Drosophila*, and in the following 30 years a great deal of information on the actions of radiations on chromosomes was accumulated, and theories of the mechanisms whereby chromosome aberrations were produced were developed and tested (refer to reviews by Wolff [173] and by Evans [51]). Most of these early studies were carried out using species of plants (such as *Tradescantia* and *Vicia faba*) with a small number of relatively large chromosomes, or fruit flies and their larvae (particularly *Drosophila melanogaster*) with their large polytene chromosomes and their suitability for large-scale breeding experiments. Thus, although many genetic studies were carried out on mice exposed to radiations, the fact that mammals were known to have a large number of relatively small chromosomes dissuaded most cytogeneticists from studying irradiated mammalian materials.

In the late 1950s routine cytogenetic studies on mammalian chromosomes became possible as a consequence of the development of reliable methods of culturing mammalian cells in vitro and of the application of cytological techniques similar to those used earlier in plant cytogenetics. Two of the many significant advances that at last enabled human chromosomes to be easily accessible for radiation studies were the development by Moorhead et al. [103] of a simple and reliable technique for obtaining mitotic preparations from cultured peripheral blood lymphocytes, and the introduction by Rothfels and Siminovitch [130] of an air-drying technique for producing flattened, well-spread chromosomes. As a consequence of these and other technical advances, the past decade has seen a great deal of work devoted to the effects of radiations on mammalian chromosomes. Particular attention has been devoted to human somatic cells and to the germ cells of the mouse. In the present article we shall be concerned with the kinds of chromosome damage induced in mammalian cells exposed to ionizing radiation, their frequencies, the influence of some important modifying factors, and the mechanisms believed to be involved in the formation of aberrations. Particular attention will be devoted to some of the findings with human chromosomes and to some of the more recent results.

THE TYPES OF ABERRATION PRODUCED AND THE CHANGES IN RESPONSE THROUGHOUT THE CELL CYCLE

The types of chromosome aberrations induced in mammalian cells are identical in structure and behavior to those induced in other eukaryotic plant and animal cells with similarly organized chromosomes. These aberrations are usually considered to be of two basic forms: (1) the simple *deletion*, which may be the result of a single break in the chromosome thread, and (2) the *exchange*, which involves at least two breaks and an exchange of parts. If the exchange occurs between different chromosomes, then the resulting aberration is described as an *interchange*; if the exchange takes place within a single chromosome, then the aberration is described as being an *intrachange*. These chromosome structural changes which appear as aberrations are of course observed only at mitosis (or at meiosis in the gonads). However, since proliferating cells usually spend most of their lifetimes in interphase and pass rapidly through the division processes, then the majority of the aberrations produced are a consequence of damage sustained in interphase.

Exposure of mammalian interphase cells to X- or γ-rays results in the production of single-strand and, to a lesser extent, double-strand breaks in the DNA polynucleotide chains which are fairly rapidly repaired. Other types of DNA damage are also induced in nuclei irradiated at interphase (see below), and these are also repaired, but the processes of reparation themselves involve breakage and rejoining of the DNA chains. It is usually presumed, therefore, that the chromosome aberrations later seen in dividing cells are a consequence of interactions between breaks in the DNA chains induced in interphase nuclei. (Refer to the "Mechanisms of Formation of Aberrations and Dose–Response Kinetics" section.)

Since the aberrations observed in dividing cells are only the visible manifestations of radiation damage sustained at an earlier point in time, then a multitude of cellular processes may intervene between the initial radiation exposure and the final development of an aberration. Thus, for a given cell type, radiation dose, quality, etc., the type and yield of aberrations may be modified by a variety of physiological as well as physical factors. The first factor of some importance in this context is that relating to the state of multiplicity of the DNA in the interphase nucleus (i.e., whether the DNA—and the chromosome—is in a replicated or unreplicated state) and the unit of breakage.

The types of aberration induced following radiation exposure fall into three categories, according to the unit of breakage or exchange involved. Aberrations involving both chromatids of the chromosome at identical loci are generally referred to as *chromosome-type* aberrations, whereas those in which the unit of aberration formation is the half chromosome, or chromatid, are termed *chromatid-type* aberrations. The third category of aberrations, the *sub-chromatid-type* changes, have been considered to involve a sub-unit of the chromatid, although they may be more plausibly interpreted as a particular category of chromatid-type changes (e.g., see [88]). Which of the three basic types of chromosome aberration is observed at mitosis (or at meiosis) depends upon the stage of development of the cell at the time of radiation.

In a mitotically proliferating cell, the interphase period of the cell cycle can be partitioned into three phases; the pre-DNA synthesis or G_1 phase of early interphase, the DNA synthesis or S phase, and the post-DNA synthesis or G_2 phase of late interphase [79]. Exposure of cells to ionizing radiations in the G_1 phase results in production of chromosome-type aberrations. However, at the very end of the G_1 phase, there is a transition from the chromosome-type to the chromatid-type aberration [60, 176], and this transitional phase extends from late G_1 to early S (Fig. 1). Thus most of the cells irradiated whilst in S, and all the cells exposed whilst in G_2, yield chromatid-type aberrations. Sub-chromatid-type aberrations are only produced in cells exposed to radiation whilst in the early prophase of mitosis or meiosis, but cells exposed to radiation at the metaphase, or at later stages in division, yield chromosome-type changes that develop in the subsequent interphase and so are observed only at the second mitosis after radiation (see Fig. 1).

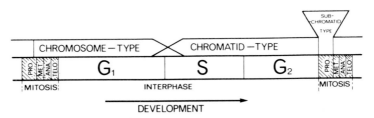

Figure 1. Type of aberration induced by radiation and the stage in the cell cycle at the time of irradiation.

Chromosome-type Aberrations

Chromosome-type changes, in which both chromatids of the chromosome are broken or exchanged at the same locus and in an identical fashion, are the aberrations that have been most frequently studied in mammalian cells. This is partly because much of the work has been carried out using human peripheral blood lymphocytes exposed to a mitotic stimulant whilst in culture; these cells are in G_1 phase whilst in the peripheral blood and are only stimulated to pass into an S phase after exposure to the stimulant. Examination of irradiated G_1 cells at their first mitosis in culture reveals that seven different kinds of chromosome-type aberration may be distinguished (Fig. 2).

Intrachanges. The five intrachange types include terminal and interstitial (intercalary) deletions, the latter clearly requiring two breakages within the single chromosome arm. A terminal deletion will result in the shortening of a chromosome and the formation of an *acentric fragment* (Fig. 2). A similar appearance would result if two breaks within a chromosome arm yielded an intercalary-derived acentric fragment, in which the ends of the fragment remained unjoined. On the other hand, union of the ends of such an excised fragment results in an *acentric ring*

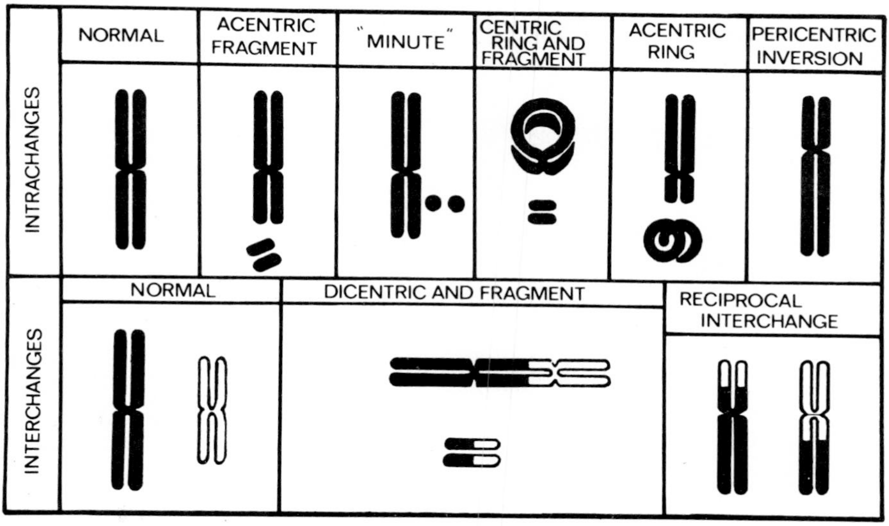

Figure 2. The seven principal types of chromosome-type aberrations distinguishable at mitosis (see text for details). The term "minute" in the figure is an alternative name for a small intercalary deletion.

structure. If the excised intercalary region is short, the ring aberration has the appearance of a pair of "dots," which may be equal to, or smaller than, the diameter of a chromatid; these are often referred to as *"minutes"* or dot deletions (Fig. 2). If the breaks occur in the two arms of a bi-armed chromosome, then the intra-arm intrachange can result in the formation of either a *centric ring and a fragment* or, alternatively, in the inversion (*pericentric inversion*) of that centric region (see Fig. 2). Two breaks occurring within the chromosome arm and resulting in inversion (paracentric inversion) cannot be detected morphologically in conventionally stained preparations of mitotic cells.

Interchanges. The interchange aberrations are basically of two sorts; those that involve a symmetrical exchange of parts between two chromosomes giving a *reciprocal interchange,* and those in which the exchange between the broken ends of two chromosomes is asymmetrical and results in the formation of a *dicentric* chromosome and its accompanying fragment. These are illustrated in Fig. 2 (see also Fig. 3).

The seven kinds of aberrations diagrammed in Fig. 2 cannot be scored with equal efficiency. For instance, in the case of the terminal and interstitial deletions, the amount of material that may be deleted from the chromosome may be very small, and in a contracted metaphase cell the fragments that are present may be too small to be resolved. Similarly, if the two breaks giving a pericentric inversion (i.e., an inversion that results from an exchange between breaks on either side of the centromere) are equidistant from the centromere, then the inversion may not be detected using conventional methods, since there is no overall change in the morphology of the chromosome. In the cases of the interchanges, dicentric aberrations are usually easily recognized and can be stored with high efficiency. On the other hand, symmetrical interchanges present a problem of recognition, since if equal amounts of material are exchanged between chromosomes, then there is no overall change in chromosome morphology. For this reason, the scoring of symmetrical interchanges is much less efficient. Studies on plant chromosomes [74] suggest that the frequency of dicentric interchanges is approximately equal to the frequency of symmetrical interchanges, so that the scoring of only dicentric aberrations can give a reliable guide to the total amount of interchange events present in the irradiated cell population. However, with the recent introduction of the banding techniques that permit all the chromosomes in the human complement to be independently recognized and some 300 different chromosome bands to be distinguished [120], the efficiency of scoring both the symmetrical interchanges and the intrachange events that do not alter overall chromosome morphology, should be increased.

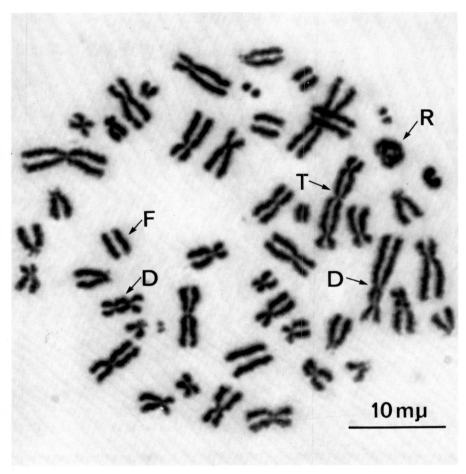

Figure 3. X-ray-induced chromosome-type aberrations in a human peripheral blood lymphocyte. The cell shows 10 pairs of acentric fragments (F), two dicentrics (D), one tricentric (T), and one ring (R) chromosome.

Interchange events are not restricted to two chromosomes but may involve three or more, and these may give rise to tricentric or polycentric derivatives. We should also note that a proportion of exchanges may be *incomplete*, only two of the four broken ends completing a joining process, and this incompleteness may occur on either the proximal (centromeric) or the distal side of the exchange. For example, distal incompleteness will result in the presence of two acentric fragments associated with a dicentric or a ring chromosome.

It should be emphasized here that most radiation cytogeneticists classify aberrations in the purely descriptive manner outlined in Fig. 2.

Categorizing the changes in these ways does not involve making assumptions as to the precise mechanism of formation of any particular aberration, does not group together aberrations that are structurally different but may have similar mechanical consequences at mitosis, and is not based on the biological importance of particular types. Thus, although the symmetrical aberrations that result in little or no change in chromosome morphology are difficult to score accurately, they nevertheless may constitute a very important category of change, since they do not result in cell lethality. On the other hand, the two centromeres in the easily recognized dicentric aberrations have, on the average, a 0.5 probability of moving to opposite poles at anaphase and forming a bridge, which will prevent the normal processes of chromosome separation and cell cleavage and may lead to polyploidy, cell death, or both.

Chromatid-type Aberrations

Chromatid-type aberrations occur if damage is sustained either at the time of, or following, chromosome splitting and replication in late G_1 and S phases of the cell cycle. In the case of the unsplit G_1 chromosome, any radiation damage sustained in the G_1 state is itself replicated when the cells proceed through S, so that the whole chromosome (both chromatids) is involved in the chromosome-type change. Chromatid-type aberrations are therefore distinguished solely by the fact that the unit of breakage is the single chromatid. The kinds of chromatid-type changes produced are essentially similar to their chromosome-type counterparts. In other words, there are intrachanges and interchanges. However, the intrachange events can now involve the subunits of the chromosome independently, so that there are both intra- and interchromatid changes, giving a greater variety of possible aberrations. These include *terminal* and *intercalary deletions, duplications* (an insertion of a segment of one chromatid into its sister), *inversions,* and an important aberration known as the *isochromatid aberration.* The isochromatid aberration involves an asymmetrical exchange between sister chromatids at roughly the same locus and results in the formation of a dicentric chromatid and an acentric fragment. As with chromosome-type changes, a proportion (usually less than 10%) of the exchanges are incomplete giving even further variety. A diagram illustrating some of these changes is shown in Fig. 4.

In the case of interchanges, symmetrical and asymmetrical events again occur; but in contrast to chromosome-type interchanges, both types are easily recognized at mitosis. This follows because the chromatids of each chromosome remain paired, so that both chromosomes in a symmetrical (as well as asymmetrical) interchange remain associated at the point of

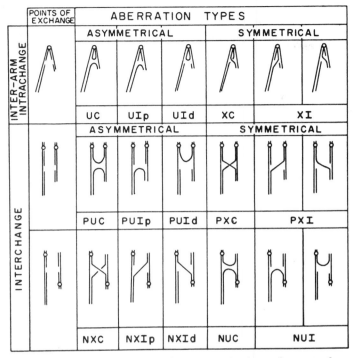

Figure 4. Types of chromatid-type aberrations resulting from single exchanges between chromosomes and chromosome arms. The symbols P and N represent polarized (centromeres oriented in the same direction) and nonpolarized (centromeres oriented in opposite directions) chromosomes. U and X are types of exchange that are complete (C) or incomplete proximally (Ip) or distally (Id) (see Evans, ref. [51]).

exchange. Such interchanges therefore have the appearance of four-armed structures and are often generally referred to as *quadriradials* (Fig. 5). More complex interchanges involving three or more chromosomes may also be induced in which all the chromosomes involved remain in association, because of chromatid pairing.

Finally to be considered is the *achromatic lesion* or *gap*, an aberration type that is not usually considered in itself to be a structural change but may in fact be associated with obvious structural changes. Gaps appear as unstained constricted regions in the chromatid; they may be single, affecting one locus on one chromatid, or paired, affecting the same locus on sister chromatids. Gaps do not represent complete fractures across a chromatid and are often observed at a point where exchange has occurred and resulted in a recognizable inter- or intrachange, although the majority do not appear to be associated with obvious exchange events [52, 54]. Recent studies of gaps in plant chromosomes induced in the second cell

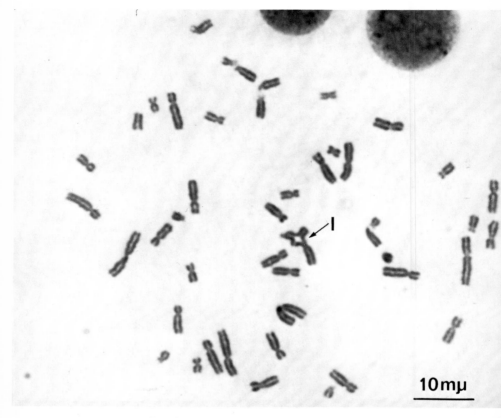

Figure 5. An X-irradiated human peripheral blood lymphocyte in metaphase showing a chromatid interchange (I) between a metacentric and an acrocentric chromosome.

cycle after prelabeling of the chromosomes with ^{3}H-thymidine [75] have shown that a proportion of those gaps not associated with a visible exchange nevertheless occur at a point at which there is a switch of ^{3}H label between chromatids (i.e., at sites of sister-chromatid exchange). It should be noted that gaps are extremely frequent aberrations in cells irradiated in G_2 and are particularly common in cells exposed to chemical agents and viruses.

Sub-chromatid-type Aberrations

We have already pointed out that sub-chromatid-type changes may well be a category of chromatid-type effects that are produced in early prophase nuclei, and it is sufficient here simply to note that sub-chromatid events that are observed all involve exchanges either between chromosome arms or between chromosomes.

"Derived" Chromosome-type Aberrations

In addition to the three basic types of aberrations, a fourth category of changes that may develop from these basic forms should be considered, and such changes are referred to as "derived" chromosome-type aberrations [58]. Symmetrical chromatid intrachanges (including duplications, deficiencies, and pericentric inversions) and symmetrical chromatid interchanges will all result in the transfer of an abnormal monocentric chromosome to one (or possibly to both in the case of interchange) of the daughter cells produced as a result of mitosis. If these daughter cells proceed through a second interphase, these abnormal chromatids will be replicated and, as a consequence of replication, will appear as derived symmetrical *chromosome-type* aberrations at the second mitosis following their induction. Similarly, if asymmetrical chromatid interchanges and chromatid fragments are included in the daughter nuclei, these will also be duplicated in the second cell cycle after their formation and again will result in derived chromosome-type aberrations appearing at the second mitosis (see Fig. 6). Asymmetrical chromatid aberrations, for example, dicentric chromatids, have a probability of 0.5 of being transferred intact to one of the two daughter nuclei resulting from a mitosis, so that one half of these aberrations will inevitably end up as derived chromosome-type changes. Since it is, in many cases, difficult or even impossible to distinguish between the derived chromosome change resulting from an original chromatid-type aberration and a real chromosome-type aberration, it is obviously of utmost importance in quantitative studies to ensure that cells that are being analyzed for chromosome aberrations are analyzed at their first mitosis after irradiation.

Quantitative Changes during the Cell Cycle

In addition to influencing the type of aberration induced following radiation exposure, cell developmental stage may also have a profound effect upon the number of aberrations produced. On a simplistic target-theory approach [92], it might be expected that since the amount of chromosome material present in a cell doubles as it passes from a G_1 into a G_2 state, then at a given level of exposure, the amount of radiation energy absorbed per nucleus is doubled and the aberration yield per cell should increase. Added to this is the fact that the duplicated chromatids in S and G_2 nuclei are very closely associated and therefore offer a greater oppor-

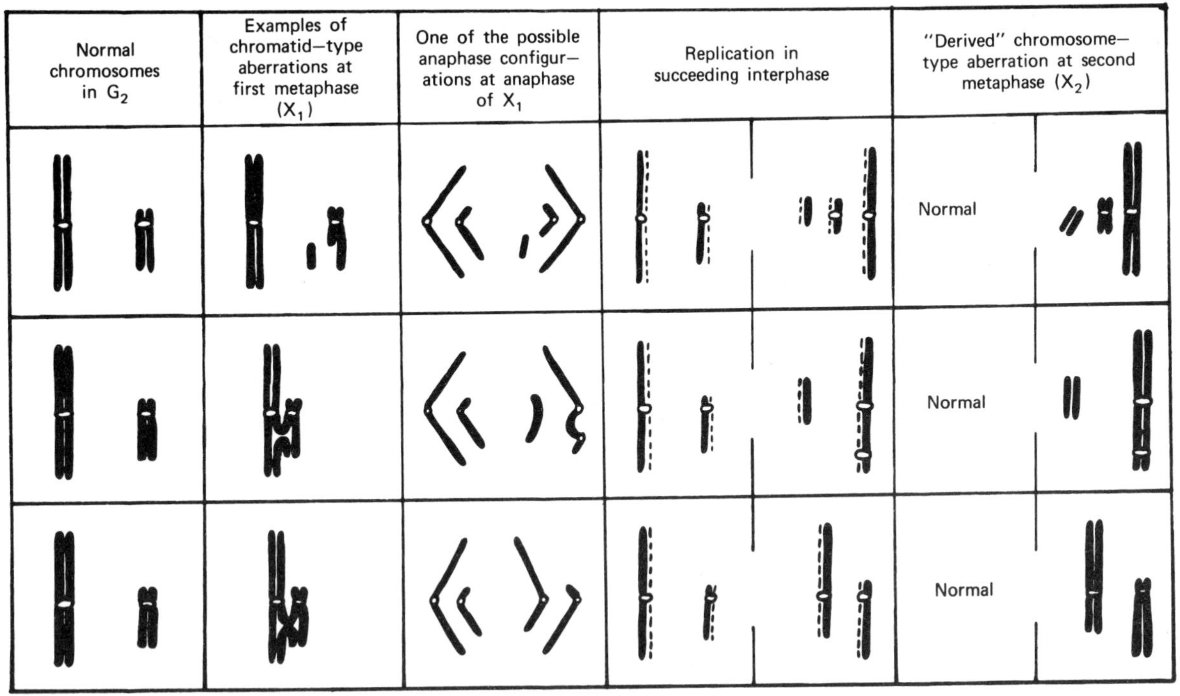

Figure 6. Examples of "derived" chromosome-type aberrations, observed at the second mitosis (X_2) after exposure to radiation. The aberrations pictured at the first division (X_1) are of the chromatid type, and only a limited number of the possible anaphase configurations are shown. It should be noted that in many instances acentric fragments may be lost and will not appear at the X_2 division with the chromosomes from which they were derived.

tunity for the interaction of damaged sites necessary to produce an aberration.

A number of studies on plant materials have shown that on a per cell basis the number of chromatid aberrations produced in S and G_2 cells is indeed greater than the number of chromosome-type aberrations induced in cells from the same population exposed whilst in the G_1 phase. Where detailed studies have been carried out, it has been shown that the yield of chromatid aberrations increases as cells progress through S and reaches a peak in mid-G_2 [148]. At dose levels giving around one aberration per cell in G_2 cells, there is only one aberration per 10 cells in cells irradiated in G_1. However, variations in response occur within each of the three interphase periods.

Similar findings have been reported in mammalian cells. For example, in Chinese hamster fibroblasts receiving 250 rads of ^{60}Co γ-rays in vitro, the yield of chromatid-type aberrations has been shown to be some three times higher in cells exposed whilst in G_2 than in cells exposed whilst in S [80]. Similarly, in earlier studies [7] on bone-marrow cells from Chinese hamsters exposed to 100 rads of X-rays or ^{60}Co γ-rays in vivo, the G_2 phase was also shown to be the most sensitive for aberration induction. In vitro studies on a marsupial (*Potorous*) cell line [11] and also on monkey (*Macaca mulatta*) peripheral blood leukocytes [84] have shown a similar correlation in number of aberrations with cell phase. Other data, on X- or γ-irradiated mouse ascites tumor and fibroblast cells, have suggested that the S phase is the most sensitive to aberration induction [42]; but the experimental procedure and methods of scoring used in this work have been criticized [39]. In the case of human cells, a number of investigations on fibroblasts [28, 87] and peripheral blood leukocytes [1, 9, 16, 85] have shown that chromatid-type aberrations (and chromatid deletions in particular) are in general more frequent in G_2 cells than in cells irradiated whilst in S. In leukocytes, chromatid-type aberrations are only slightly more frequent than the chromosome-type changes induced in G_1 cells, and there is now clear evidence that there is some variation in response not only in S and G_2 populations but also in the G_1 population [6, 168].

MECHANISMS OF FORMATION OF ABERRATIONS AND DOSE–RESPONSE KINETICS

The Classical Theory of Aberration Formation

The first critical evaluation of radiation-induced chromosome aberrations was made by Sax [144–146], who studied the effects of X-irradiation on the chromosomes in pollen grain cells of the plant *Tradescantia*. Sax studied the influence of a number of physical variables and in particular obtained information on the influence of dose level, on the shape of the dose–response curve for different aberration types, and on the influence of exposure time and dose rate; he also examined the effect of fractionated exposures separated by varying time intervals. From these detailed studies Sax found that the relation between yield (y) and X-ray dose (D) for his data on dicentric and ring chromosome-type aberrations could be adequately described by the relation $y = D^{1.8}$. In other words, the yields of these aberrations appeared to increase approximately in proportion to the square of the X-ray dose. On the other hand, terminal deletions showed an approximately linear response and fitted the expression $y = D^{1.1}$ or $y = D^{1.2}$. A somewhat better expression for these relationships would be $y = c + aD^n$, where c is the aberration yield in unirradiated control cells and a is a constant.

Sax also showed that the dose exponent n for dicentrics decreased if the exposures were delivered at a constant, but relatively low, dose rate (giving different exposure times for each dose), as compared with the value of n obtained when all doses were given in the same short exposure time, that is, at different dose rates. He further demonstrated that the yield of dicentrics at a given dose level was decreased if the dose was split into two fractions separated by a time interval, but no further decrease occurred if the rest time between fractions was more than one hour. Varying exposure time and fractionation intervals appeared to have little effect on the dose response and yields of terminal deletions.

These results obtained by Sax were interpreted as support for the "breakage-first" hypothesis originally proposed by Stadler [161]. This hypothesis considered that a physical break was produced in a continuous interphase chromatid as a consequence of the passage through it of a single ionizing particle. It was presumed that the broken ends of a fractured chromosome then: (1) restituted to give back the original configuration, that is, a complete repair; (2) rejoined with other broken ends to form an exchange that was then seen as an intra- or interchange at mitosis; or (3) remained open to give a simple terminal deletion. Deletions were therefore thought to be the survivors of a proportion of the primary breakage events. Since these primary events would be expected to increase as a linear function of dose, then the frequency of deletions should also be linearly proportional to dose, and Sax's results showed that this appeared to be approximately true. Support for this interpretation was also obtained from the results of the experiments in which fractionated doses and varying exposure times were used, and it was concluded that broken ends must be closely associated spatially for rejoining to occur and that they only remained available for rejoining with other broken ends for a limited time period of less than one hour. Since simple deletions were assumed to be the consequence of one-hit events, they have been referred to as one-hit, or more properly one-track, aberrations, in contrast to the intensity-dependent exchange, or two-track, aberrations such as the chromosome dicentrics and rings.

This general theory of aberration formation has been referred to as the "classical theory," and it received considerable support from the later work of Lea and Catcheside (see Lea, [92]), again using *Tradescantia* microspores. These authors pointed out that since the yield of a proportion of the X-ray-induced exchange aberrations was independent of exposure time and dose rate, then a more appropriate expression relating dose to yield for the exchange aberrations would be: $y = c + aD + bD^2$. In this equation, a is the coefficient for the exchange aberrations produced by single electron tracks (i.e., one-hit types) and b the corresponding coefficient for the two-hit exchanges [93]. In the case of densely ionizing radiations (e.g., fast neutrons, Giles [71]), all aberrations, including exchanges, were found to increase linearly with dose. Thus, provided two chromosomes or chromosome regions were sufficiently close together in space, then both chromosomes would be, more often than not, broken by the energy imparted by a single densely ionizing radiation track passing in their near vicinity. More recent studies on human and other mammalian cells (see below) have shown that the quadratic equation defines adequately most dose response curves and have confirmed the linear relationship between aberration yield and dose with densely ionizing radiations (radiations of high linear energy transfer, or LET).

It was realized early, of course, that for breaks to interact they must not only be formed simultaneously but must also be very closely associated in space, since exchanges between parts of the same chromosome occurred much more frequently than expected on the basis of a random interaction of breaks throughout the interphase nucleus. On the basis of the number of electrons from a dose of X-rays, or protons from a dose of neutrons, which will be found in the nucleus at a given exchange yield, Lea [92] concluded that the distance over which breaks may interact (the so-called rejoining distance) was about 1 μm. More refined methods of estimating "rejoining distance" [109, 177], including the use of ultrasoft X-rays of quantum energies below 3 keV (where the extended electron track lengths in tissue are less than 0.25 μm), suggested that exchanges take place only between breaks separated by not more than around 0.2 μm [108]. Although these studies were carried out on plant chromosomes, these conclusions are almost certainly applicable to mammalian cells.

The Exchange Hypothesis

The classical theory of Sax received no challenge until in 1959 Revell [127] published an alternative hypothesis, called the exchange hypothesis, which was based on data obtained from irradiated plant root-tip cells. According to the exchange hypothesis, all aberrations, including the so-called simple chromatid deletions, were believed to be a consequence of exchange. Such exchanges occurred at regions of the chromosome that were in close spatial proximity due to the coiled and looped nature of the interphase chromatid thread. A diagram of the possible intrachange types is shown in Fig. 7, and it may be seen from this figure that the deletions are believed to be a consequence of an *incomplete* exchange between two regions within a chromosome, so that the deletion is associated with an inversion or duplication of a short length of a chromosome at the point of "failed union." Thus, by this hypothesis, a proportion of simple deletions could result from interaction of the effects of two separate electron tracks and could show either a negligible or a significant "dose-squared" component in their rate of increase with increasing X-ray dose. *An important feature of the exchange hypothesis is that all aberra-*

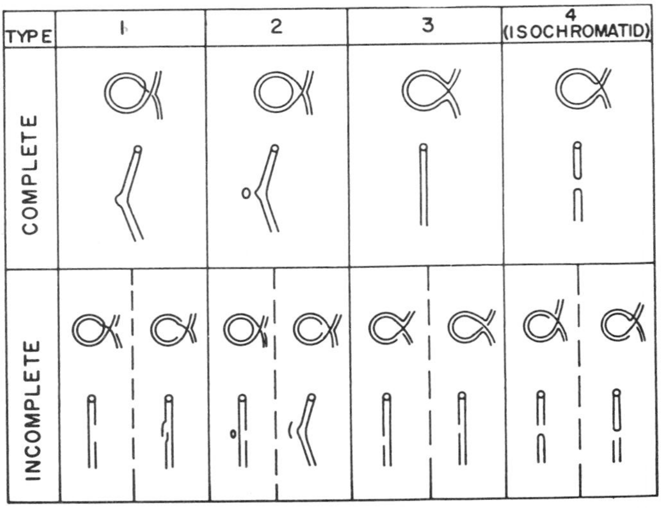

Figure 7. Types of chromatid intrachange resulting from a single exchange within a chromosome arm (after Revell, ref. [127]). The upper figures in each set depict the suggested nature of the exchange in interphase, and the lower figures the observed aberrations at metaphase.

tions must be a consequence of the interaction between two chromosomes or chromosome regions that have been damaged by radiation, but this damage need not be a complete fracture across the chromatid.

Revell's data [127, 129], and his interpretation, were soon supported by other work on plant chromosomes [51, 52, 55, 61, 128, 142, 148]. More recently, a number of authors have published results of experiments with mammalian cells which are also in accord with the exchange hypothesis. Studies on dose–response curves for chromosome-type aberrations induced in peripheral blood chromosomes of man ([57], and Bender and Brewen, quoted by Brewen and Brock [17]), wallaby [17], and Chinese hamster fibroblasts [15] have shown that terminal deletions increase in a nonlinear fashion with increasing X-ray dose, indicating the presence of "two-hit" type events.

Heddle et al. [75, 77] have recently carried out a more direct test of the exchange hypothesis using X-irradiated cultures of fibroblasts from the rat kangaroo (*Potorous tridactylis*). In this study, chromosomes were labeled with tritiated thymidine, X-irradiated one cell cycle later, and then examined at the second mitosis after labeling. Autoradiography revealed a switch of label from one chromatid to the other in many of the chromosomes at the point of an assumed breakage associated with a deletion. This implied that the terminal deletions involved an exchange between units of the chromosome and did not reflect simply a straightforward fracture across the chromatid. It is clear therefore that many terminal deletions are not unrejoined single breaks but, rather, are incomplete exchanges.

Details of the results obtained in the rat kangaroo cell cultures, in human peripheral blood leukocyte cells (Adams and Evans, unpublished), and in plant root tips [148] indicate that neither the classical "breakage-first" nor the exchange hypothesis can account for all the results and that both mechanisms may be operative in the formation of aberrations. It should be pointed out that by either hypothesis it would be expected that with high LET radiations all aberration types would increase linearly with increasing dose; this has certainly been shown to be the case in all experiments carried out so far in mammals (see below).

The Primary Events Involved in the Formation of Aberrations and the Notion of Misrepair

Although plausible suggestions for the formation of aberrations based on interaction between damaged sites have been proposed, there is a dearth of information on the nature of the basic lesions. Since the longitudinal integrity of the chromosome is a function of its DNA (for discussion see Comings [36]), then a breakage and a rejoining of DNA must eventually be involved in the formation of all aberrations. Studies on pollen-tube chromosomes in plants [109] have shown that with very soft X-rays one energy-loss event is sufficient to induce a chromatid interchange. One such energy-loss event amounts to around 70 eV, and this is sufficient to produce at least one single-strand breakage of the DNA [38, 69], although in the formation of an exchange, two such breakages must eventually occur. However, chain breakage is not the only kind of damage induced in irradiated DNA, and other changes (including base damage, cross-linkage, and sugar–phosphate bond cleavage) are also found [86].

In 1966, McGrath and Williams [107] developed a method that permitted examination of the DNA of whole cells for breaks. Using this technique, Lett et al. [97] showed that one single-strand break was produced for every 70 eV in X-irradiated murine lymphoma cells. It was further shown that irradiation under anoxia gave fewer single-strand breaks (one-third to one-half the number observed in the presence of oxygen, see Palcic and Skarsgard [119]); it is of course well known that the killing effects of radiation and the incidence of chromosome aberrations induced by X-rays are reduced if the radiation is given in the absence of oxygen [39]. On the other hand, a number of pieces of evidence [2, 68, 96, 97, 110] have shown that oxygen does not significantly increase the effectiveness of radiation in inducing double-strand breakage in bacterial and phage DNA. At this point we should note that single-strand breaks are about 10 to 20 times more frequent than double-strand breaks [38, 69, 110] and that about 10 single-strand breaks per rad are produced in an X-irradiated mammalian nucleus.

Lett et al. [97] were the first to observe, in mammalian cells, that such single-strand breaks underwent a rejoining and that this rejoining occurred very rapidly after exposure to radiation. Since this original demonstration, a number of authors have used a modified alkaline sucrose-gradient technique to study the induction and rejoining of single-strand breaks in the DNA of human [98], murine [119, 165], and Chinese hamster cells [48, 82]. In synchronously developing X-irradiated Chinese hamster cells, it was found that the amount of breakage induced in DNA was independent of cell-cycle stage at the time of irradiation [82], although this did not appear to be the case in synchronized human kidney cells [98]. In both cell types, however, rejoining of breaks occurred at all stages of the cell cycle, with some evidence in human cells for fluctuation (in the capacity for rejoining) with cell stage. There is some dispute over the question of whether double-strand breaks are rejoined ([143], but see Painter [116]), but it is clear that the rejoining of single-strand breaks is a rapid process occurring in the first 30 minutes or so after exposure to radiation.

The time factor involved in the rejoining of DNA chain breaks appears to be similar to the time factor involved in the formation of chromosome aberrations. This does not mean that aberrations are a direct consequence of direct breakages in the DNA, however; and it may be significant that if certain inhibitors of protein synthesis are present at the time of (or shortly after) irradiation, they may have profound effects on the induction of aberrations [172] but little or no effect on the rejoining of single-strand DNA breaks [143, 165].

Nearly all the breaks induced by the X-irradiation of DNA in vitro are accompanied by loss of base or sugar or both (see Painter [116] for review), so that some synthesis is required before a complete rejoining can occur. Evidence for synthesis and for the insertion of new bases into DNA outside the normal replication or S phase was first provided in mammalian cells by Rasmussen and Painter [125, 126]. In autoradiographic studies on HeLa cells incubated in ^{3}H-thymidine after exposure to X-rays or UV, these authors reported that in irradiated cultures all cells were labeled over the nuclei, whereas in control cultures only cells in S phase were labeled. This phenomenon of DNA synthesis outside the normal S phase has been referred to as "unscheduled DNA synthesis" [46], and in many cases it has been shown to involve a nonsemiconservative "repair replication" [30, 117, 118]. Moreover, such synthesis could be induced in a wide variety of mammalian cells, including human lymphocytes exposed to a wide variety of physical and chemical mutagens [30, 64].

"Repair replication" in mammalian cells appears to be analogous to "excision repair" in bacteria, in which damaged single-strand regions of DNA are excised (degraded) and resynthesized using the complementary DNA strand as template [13, 122, 155]. It would seem that "repair replication" may be involved in the rejoining of perhaps the bulk of the single-strand breaks induced in DNA by X-rays and is a major repair pathway for the restoration of other types of damaged sites in DNA. The process of excision of small, damaged DNA segments followed by a subsequent localized polymerization of new bases into the excised gap, thus renewing the linear integrity of the DNA, is a strong candidate for the kind of mechanism necessary for the formation of chromosome aberrations. Such a mechanism would imply that cross-polymerization would occur, at the time of repair, between two damaged sites that were in close proximity. The aberrations would therefore be a consequence of a "misrepair" of two lesions closely associated in space and time, and suggestions as to how such a mechanism might operate have been published elsewhere [53–55].

In 1968, as a consequence of their work on bacterial cells, Rupp and Howard-Flanders [131] described a further repair mechanism, referred to as "recombination repair," in which gaps on newly replicated DNA, made on radiation-damaged templates, are sealed by a process involving recombination events between daughter strands. Such a repair mechanism could be of considerable importance in those mammalian cells that show only a limited amount of the DNA excision associated with "repair replication," and it is of interest to note that "recombination repair" has now been demonstrated in Chinese hamster [34] and mouse L cells [132]. In this context it may be pertinent to note that chromatid aberrations induced by many chemical mutagens are actually formed only during the S phase of the cell cycle [62], and one may speculate that "recombination repair" events may be important in the development of such aberrations. Indeed, the similarity between the normal exchanges giving chiasmata at meiosis and the mutagen-induced exchanges in somatic cells has of course long been recognized, and it has been proposed that normal recombination and induced exchange make use of common pathways in the cell [53–55].

Indirect evidence in support of the possibility that aberrations produced during S may involve "recombinational repair" mechanisms comes from studies on the action of UV light on mammalian cells. Studies on cultured Chinese hamster cells have shown that by far the majority of aberrations induced in these cells are chromatid-type changes induced in the S phase of the cycle [27, 81]. Exposure to UV results in the formation of thymine dimers in the DNA of the hamster chromosomes, but loss of dimers from the DNA by an excision repair process has not been detected [164]. In the case of human cells, Cleaver [12, 31] showed that fibroblasts from patients with xeroderma pigmentosa were unable to undergo "repair replication" after exposure to UV light, although such repair takes place if the cells are exposed to mutagens that produce DNA strand breaks [32]. It is now clear that xeroderma pigmentosum cells are unable to perform the first step in the excision repair process [32, 156] and, unlike other human cells, can not excise UV-induced pyrimidine dimers [33]. It is of significance, therefore, that exposure of xeroderma cells to UV light results in the formation of chromatid aberrations at a frequency considerably increased over that of normal cells [121].

SOME PHYSICAL AND BIOLOGICAL FACTORS INFLUENCING ABERRATION YIELD

Dose, Dose-rate, Exposure Time, and Radiation Quality

In our discussion on mechanisms and theories of aberration formation in the previous section, we considered, in general terms, dose–response kinetics and the influence of dose-rate, exposure time, and radiation quality on aberration yield. Over the past few years, a considerable amount of information on the influence of these factors on aberrations induced in human cells has been accumulated, and it may be pertinent to consider some of these data here. Much of this information has been obtained from experiments on peripheral blood lymphocytes that were withdrawn from the body and irradiated in vitro, before culturing by standard blood-culture techniques [103].

In early studies on human lymphocytes exposed to X-rays in vitro, there were marked differences among laboratories in the yields of aberrations obtained at given dose levels and in the shapes of the dose–response curves. It has emerged (see UN report [165]) that, in addition to possible differences in the methods used, and in the efficiency of scoring, these differences were largely due to three factors, enumerated below.

(1) Differences in the timing and methods of exposure. For example, in some cases, lymphocytes were irradiated while in culture after exposure to the mitotic stimulant phytohemagglutinin (PHA), whereas in others, whole blood was irradiated before culture. A number of groups have reported low-dose exponents ($n=1.2$) in the dose–response curves for dicentric aberrations induced by X-rays when cells were exposed to radiation early in their culture history ([4, 56, 167]; but see Bender and Brewen [6]). In these experiments, the aberration yields at low doses (below 200 rads) were generally higher than the yields obtained from cells exposed to radiation before culture [59]. However, when cells were exposed to X-rays before culture, with certain other variables controlled (see below), then there was good agreement between results obtained from a variety of laboratories. With use of conventional X-rays (180 to 300 keV), the yield of exchange aberrations increased as the 1.6 power of dose. Typical results obtained from four different laboratories are shown in Fig. 8.

(2) Differences in culture conditions and the time of sampling of cells after exposure. The incidence of aberrations in a cell population decreases as the cells proceed through successive mitoses. Reduced aberration yields are therefore to be observed in cultures containing cells in their second, or later, postirradiation division. For this reason, under good culture con-

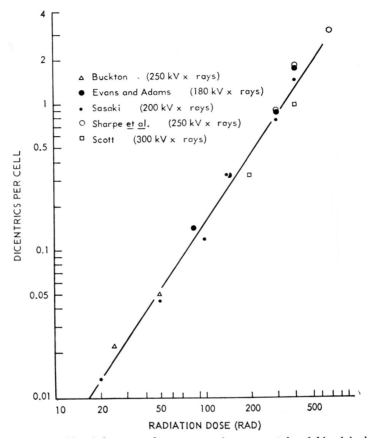

Figure 8. The yields of dicentric aberrations in human peripheral blood leukocytes in relation to X-ray dose (180 to 300 kVp) from results of in vitro experiments in four different laboratories: triangle, Buckton (unpublished); large solid circle, Evans and Adams [59]; small solid circle, Sasaki [136]; open circle, Sharpe et al. [160]; square, Scott (unpublished).

ditions, most workers culture lymphocytes for 48 to 55 hours; after 60 hours, a considerable proportion of the dividing population consists of cells in their second mitosis in culture [76, 138]. Differences have also been reported between cells cultured in different ways [24, 157], and in this context it is important to note that the lymphocyte population is heterogeneous and that there is evidence that "fast-developing" and "slow-developing" cells show different frequencies of aberrations at their first mitosis in culture [6, 158, 168].

(3) Differences in radiation quality. A number of laboratories have studied the response of lymphocyte chromosomes to 2.0-MeV X-rays and have shown that the form of the dose–response curve for dicentric and ring aberrations with these radiations approximates more nearly the dose-squared law with a dose exponent of $n = 2$ [24, 111, 136]. Moreover, the efficiency of these radiations relative to conventional 180 to 300 keV X-rays (the relative biological efficiency, or RBE) is around 0.8 [166]. We pointed out above that with conventional X-rays a dose of exponent of around 1.6 was obtained for exchange aberrations, an exponent similar to that found for exchange aberrations in *Drosophila* [105]. In the case of ^{60}Co γ-rays, an intermediate dose exponent of $n = 1.8$ has been reported [136, 157], and these data are summarized in Fig. 9. ^{60}Co γ-rays are less efficient than conventional X-rays, with an RBE of around 0.85 [150, 166], so that there is a parallel between increasing dose exponent and decreasing relative efficiency of these radiations. A summary showing a comparison between results obtained with these three radiations, and with 14.1-MeV fast neutrons, in the same laboratory [136], is shown in Fig. 10.

A number of studies with fast neutrons including 2.5-MeV DD, 14.1-MeV DT, and reactor fast neutrons with a mean energy of 0.7 MeV, show that, in contrast to results obtained with radiations of low LET, the yield of all aberrations induced by these neutrons increases linearly with dose, that is, $n = 1$ [136, 149, 150, 166]. Again in contrast with X-rays [19, 53, 123], ^{60}Co γ-rays [150], and ^{137}Cs γ-rays [19], in which the yields of aberrations are markedly reduced if exposures are fractionated or protracted over long time periods, the aberration yields with fast neutrons are independent of dose-rate and exposure times. Since the shapes of the dose–response curves for aberrations induced by fast neutrons and X-rays are very different, then the relative biological efficiency of neutrons relative to X-rays will vary according to the level of damage considered, and there will be no single RBE value. All the fast neutrons studied are more efficient than X-rays in inducing aberrations; in the case of the 0.7-MeV neutrons, when dicentric aberration yields are up to about 2 per cell, RBE values (relative to 250-keV X-rays) of between 3 and 4 are obtained [149].

Oxygen, Temperature, and Associated Variables

The fact that the presence or absence of oxygen has a profound effect on the yield of X-ray-induced chromosome aberrations was first noted by Thoday and Read [163] in their work on plant cells. These authors also

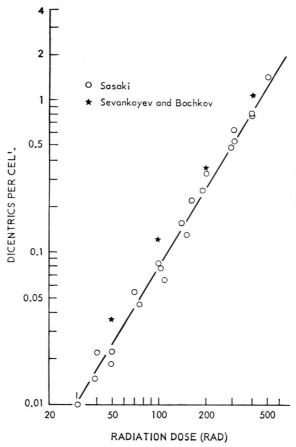

Figure 9. The yields of dicentric aberrations in human peripheral blood leukocytes following acute in vitro exposure to ^{60}Co γ-rays: star, data of Sevenkayev and Bochkov [157]; circle, data of Sasaki, [136].

showed that oxygen did not influence the yield of aberrations induced by densely ionizing particles such as α-rays. This work led to a number of detailed studies on plant chromosomes and later to studies on mammalian chromosomes (mouse ascites tumor cells), and essentially similar results were obtained [37, 41].

The "oxygen effect," as it became known as shortly after its discovery, is one of the most important, and most widely studied, modifying influences on radiation damage, and it will not be considered in any detail here. In general, the presence of oxygen effectively modifies the dose,

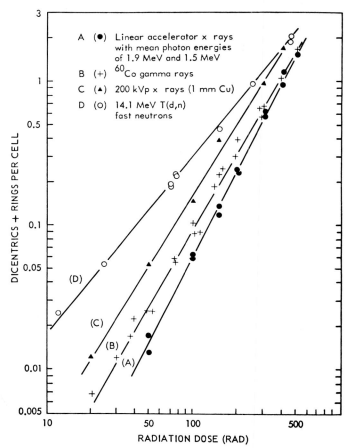

Figure 10. Comparisons between the yields of dicentric and ring aberrations induced in human peripheral blood leukocytes following exposure in vitro to radiations of different quality (from Sasaki, [136]).

enhancing the action of sparsely ionizing radiations such that in many systems the ratio of doses given in air and in nitrogen in order to yield the same damage (the oxygen-enhancement ratio, or OER) may differ by up to a factor of around 3. With radiations of increasing LET, there is a decreasing oxygen-enhancement factor, and little or no effect is apparent with very densely ionizing radiations [51].

Two recent studies have described the influence of the presence of oxygen during irradiation on X-ray-induced chromosome damage in human cells. Bushong et al. [26] examined the effect of oxygen on induced aberrations in HeLa cells, and Watson and Gillies [168] examined the effects of oxygen on aberration yields in X-irradiated human lymphocytes. In the first study, results were in accord with earlier work: oxygen enhanced the radiation effect and acted as a dose modifier; the OER was 3 : 1. In the second study, using lymphocytes, an increase in chromosome damage in the presence of oxygen was also observed (see Fig. 11), but the oxygen did not appear to act strictly as a dose modifier, and the OER values were from 3 to 9 : 1. It seems almost certain, however, that the somewhat anomalous results obtained with the lymphocytes reflected complications of the cell system itself [168].

As with oxygen, the influence of temperature on the frequency of radiation-induced chromosome damage was first reported in studies on plant chromosomes [65, 147]. Since the temperature influences the solu-

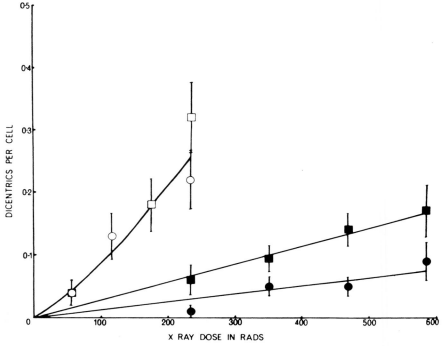

Figure 11. Dose–response curves for the yields of dicentric chromosomes for X-irradiated human lymphocytes under oxygenated and anoxic experimental conditions: open squares, oxygenated, irradiated lymphocytes cultured for 60 hours; open circles, oxygenated irradiated lymphocytes cultured for 50 hours; solid squares, anoxic irradiated cells cultured for 60 hours; solid circles, anoxic irradiated cells cultured for 50 hours (from Watson and Gillies, ref. [168]).

bility of oxygen, then an indirect effect due to oxygen solubility might be expected. However, Deschner and Gray [41] reported increases in X-ray-induced aberration yields with increasing temperatures when mouse ascites cells were maintained under carefully controlled conditions of oxygenation. Bajerska and Liniecki [4] examined the effect of temperature at the time of X-irradiation on aberrations induced in human lymphocytes; they reported a twofold increase in response in cells irradiated at 37°C compared with those irradiated at 18 to 20°C. More recently, Dewey et al. [44], using synchronously dividing Chinese hamster cells, have studied the effects of temperature during and shortly after X-irradiation. These authors carried out experiments in which an X-ray dose was given as two half-fractions separated by varying time intervals of up to 60 minutes. As expected, the aberration yield for a given total dose was shown to decrease as the rest interval between fractions increased. In other words, with increasing time between fractions, an increasing proportion of the lesions induced by the first fraction was unavailable for interaction with lesions induced by the second fraction. The rate of decline in aberration yield with increasing fractionation interval indicated that the half-time for the interaction of lesions to give aberrations was around 12 minutes. Furthermore, the loss (repair without interaction) of these lesions was not influenced by protein synthesis, since the presence of cyclohexamide had no effect. If cells were held at a low temperature (0°C) between fractions, then increasing the rest interval did not decrease the aberration yield; full interaction occurred between half-doses separated by 60 minutes—a result previously obtained with plant chromosomes [141, 175]. If cells were exposed to a single dose and held at 20°C for varying time intervals before returning to 37°C, it was found that the longer the cells were held at 20°C (up to 60 minutes), the lower the aberration yield (Fig. 12). These data point to an effect of temperature on the mechanisms involved in the formation of aberrations postirradiation (i.e., in the repair or misrepair of the initial lesions) and serve to demonstrate that the postirradiation conditions that have been shown to influence aberration yields in plant cells [173] may also be important in modifying the response of mammalian cells.

"Endogenous" Radiation

Here we are concerned with the effects of radiations emitted within a cell on the chromosomes of that cell or with the effects on chromosomes of cells exposed to radiations from isotopes deposited in nearby cells or tissues.

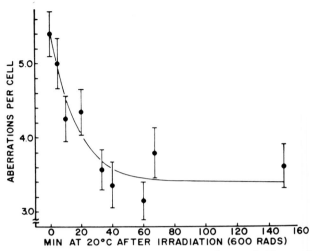

Figure 12. Yield of chromosome aberrations in synchronized Chinese hamster cells exposed to 600 rads of X-rays and held for various time intervals at 20°C after irradiation, prior to returning to 37°C (from Dewey et al. ref. [44]).

In the first category we are concerned with radiations of limited penetration and, very largely, with the effects of tritium β-rays. The use of ^{3}H-labeled nucleosides in cell biology studies has led to numerous reports that the radiations from tritium deposited in the DNA of the chromosomes gives rise to chromosome aberrations. Tritium incorporated as ^{3}H-thymidine results in chronic exposure, the level of exposure decreasing with successive DNA replications in the absence of label. Various studies on Chinese hamster cells have related the incidence of aberrations induced by β-particles from ^{3}H-thymidine to "dose," by determining grain counts in autoradiographs and relating these to observed effects (e.g., ref. [20]). Comparisons have also been made on differences in response with cell stage [45], the influence of ^{3}H in ^{3}H$_2$O and on the relative efficiency of tritium β-rays as compared with ^{60}Co γ-rays (e.g., ref. [43]). One feature of interest that has aroused controversy [174] is the question of whether the sister-strand chromatid exchanges, detectable only by autoradiography of cells at their second mitosis after exposure to a pulse label with ^{3}H-thymidine [162], represent radiation-induced exchanges or reflect a "normal" and frequent process of exchange [101]. Recent studies on a line of rat kangaroo cells by Gibson and Prescott [70] have now shown that the frequency of sister-chromatid exchange does indeed increase with increasing ^{3}H-thymidine incorporation, and their data also indicate that all of the exchanges observed are radiation induced and that none are of spontaneous origin.

The increasing utilization of radioisotopes in medicine has led to many studies of the effects of radiations from various nuclides in inducing chromosome aberrations in human peripheral blood leukocytes exposed in vivo (see UN report [166] for review). Perhaps the most detailed of these have been carried out on patients who had been given Thorotrast, a colloidal suspension of ^{232}thorium dioxide which is deposited in liver, spleen, and, to a lesser extent, bone marrow and lymph nodes. Only minute quantities of Thorotrast are excreted, so that nearby tissues are affected by continuous exposure to densely ionizing α-particles. A number of groups have reported significant yields of aberrations in blood leukocytes of persons given Thorotrast injections up to 30 years or so before study [23, 66, 83]. However, in view of difficulties with regard to dosimetry, there is little that can be made of the data in terms of relating aberration yield to dose.

In addition to studies on patients treated with isotopes, various individuals and populations of individuals have been accidentally, or occupationally, exposed to a variety of radioactive substances and have been subjected to cytogenetic study. Perhaps the best known group of exposed individuals coming under the category are workers in the luminizing industry, many of whom have high body contents of ^{226}Ra and show significant yields of chromosome damage in their peripheral blood lymphocytes (e.g., ref. [14]). Many of the results obtained from individuals who had ingested radioisotopes have been recently reviewed elsewhere [166].

In Vitro vs. In Vivo Responses

A problem that is of particular concern in relation to the possible use of radiation-induced chromosome aberrations for dosimetric purposes (refer to next section) is the question of whether the aberration yield induced by a given dose, under specific conditions in vitro, is the same as the yield observed in the same cell type exposed in vivo. It has been tacitly assumed that the response of human lymphocytes exposed in vitro can be used as an indication of their response in vivo. This assumption has seemed perfectly reasonable, particularly when the in vitro studies involved exposure of freshly drawn whole blood before its centrifugation or introduction into culture media. However, it is only recently that a certain amount of confirmatory evidence has become available.

Clemenger and Scott [35] have presented results obtained from peripheral blood cells of rabbits given whole-body exposure to ^{60}Co γ-rays simultaneously with irradiation of a sample of blood removed immediately before and kept at the appropriate temperature. The results showed that over a range of doses there was no significant difference in the yields of dicentric and ring aberrations induced under the two conditions of irradiation. More recently, Preston et al. [124] and Brewen and Genozian [18], using 250-keV X-rays, have carried out similar experiments with Chinese hamsters and their lymphocytes, and with marmosets. In these latter studies the aberration yields obtained from in vitro irradiation were identical with those obtained from cells exposed in vivo.

Similar, but necessarily more limited, comparisons have been carried out in man. Buckton et al. [24] compared aberrations induced by 2-MeV X-rays in human peripheral blood leukocytes exposed in vitro and those induced in cells obtained from the donors (cancer patients) *after* their whole-body exposure. Over the limited small range of doses used, the results showed no differences between the in vivo and in vitro responses. Somewhat similar results have been obtained by Winkelstein et al. [170] and by Sharpe et al. [159] on human lymphocytes exposed in an extracorporeal irradiation device. It is clear from these results, therefore, that at least with lymphocytes under appropriate conditions, the yield of chromosome aberrations obtained in in vitro exposures is identical with the yield obtained with in vivo exposures at the same level of absorbed dose.

Chromosome Organization and Other Influences of Genotype and Phenotype

In our summary of possible mechanisms involved in the formation of aberrations, it was stated that the original lesions induced following radiation exposure would be distributed randomly along the length of a chromosome of uniform density and that the interaction of two lesions is necessary for the formation of aberrations. The opportunity for interaction between two lesions may differ between chromosomes and chromosome regions. For example, it has long been known that condensed heterochromatic regions in interphase nuclei of many plant and invertebrate species are involved in aberration formation more frequently than expected on the basis of the relative lengths of these regions as measured at metaphase. This nonrandom distribution of aberrations within chromosomes is also evident in comparisons between chromosomes and has been commented on by a number of authors [51].

One of the main problems in studying the distribution of aberrations within and between chromosomes in mammalian nuclei is the fact that in many species only a few of the chromosomes in a complement can be individually distinguished. The recent development of new techniques that reveal distinctive bands across mammalian somatic chromosomes means that all the chromosomes in most complements can be readily distinguished and, in many cases, specific chromosome segments can be unambiguously identified [120]. In man, these techniques have already been used by many to identify points of breakage and rearrangement in structurally abnormal chromosomes in individuals with a constitutional chromosome anomaly. The methods are now being applied in studies on radiation and chemically induced chromosome damage in human cells. The results of one of the first such studies have been recently reported by Seabright [151], who has found a marked degree of nonrandomness in the distribution of exchange points in X-irradiated human leukocytes. In this study it was found that exchanges were localized in the pale-staining regions between the darkly stained G-bands resolved with the new Giemsa techniques.

In addition to a nonrandom distribution of aberrations within cells, there is considerable evidence from studies on plant chromosomes that there may be nonrandom distribution of dicentric and ring aberrations between cells. This nonrandom distribution between cells has been interpreted by the "site hypothesis" [171] to be due to a limitation in the nucleus of the numbers of sites at which exchange can take place. Savage [140] has recently reviewed thoroughly the "site hypothesis" concept and it seems clear that a limitation of sites, which would result in a nonrandom distribution of exchange aberrations between cells, may only be applicable to species with a relatively small number of chromosomes. Site limitations, therefore, may not be an important factor in influencing the frequency of radiation-induced aberrations in mammalian cells.

Changes in the spatial distribution of chromosomes during interphase, and in their degree of coiling and compaction, are certainly responsible for at least part of the changes in total aberration frequencies and in the relative proportions of various aberration types observed at different stages of the cell cycle [16, 21, 61, 87, 148]. Similarly, differences among species in chromosome numbers and in their organization (e.g., in terms of amount, degree of compaction, and distribution of heterochromatin) may also be important in influencing aberration yield. Since the number of primary lesions in a nucleus will be a function of its chromatin content, and as most mammalian species have very similar amounts of DNA, then it might be expected that similar amounts of initial damage would be present in the nuclei of various species at a given dose level. Mammals, however, show a considerable variation in chromosome number and organization, so that differences in the frequencies of various aberration types are to be expected. For example, in the laboratory mouse the centromeres occupy an almost terminal position in the chromosomes, so that the possibility of the formation of centric ring aberrations is remote. The opportunity for involvement in dicentric formation is increased in metacentric relative to acrocentric chromosomes and will of course be influenced by chromosome (centromere) and arm number [57, 63]. Differences in response are therefore to be expected among species, particularly in relation to interchange and large intrachange aberrations relative to small deletions. Information on X-ray-induced aberrations yields in peripheral blood cells of a variety of species (man, mouse, pig, rabbit, marmosets, hamsters, wallaby, etc., *loc. cit.*) is now available, and comparative studies could prove rewarding.

In addition to differences in the organization of chromosomes between species, differences in chromosome organization within species may also result in modifying the radiation response. Changes in chromosome number will certainly influence the yield of aberrations induced by radiations. A striking example of such an effect is illustrated in the case of the response of chromosomes of lymphocytes from patients with Down's syndrome (trisomy 21). An increased sensitivity of Down's syndrome cells was indicated in some early work by Dekaban et al. [40] and by Chudina et al. [29] in their studies on X-ray-induced chromatid-type aberrations. More recent studies [59, 91, 139] have confirmed and extended these studies to lymphocytes irradiated in G_1. The studies in G_1 cells show that the incidence of dicentric and ring aberrations in X- or γ-irradiated lymphocytes with an additional chromosome No. 21 is increased by 50 to 100% over the incidence of the same aberrations in chromosomally normal cells from mosaic patients (see Fig. 13). However, no difference in the frequency of deletions was observed between the different cell types. The increased aberration yield in the trisomic cells is not simply due to an increased involvement of the three No. 21 chromosomes but is a consequence of an increased participation of all the chromosomes in the complement in aberration formation.

Sasaki and Tonomura [139] and Evans and Adams [59] have also examined the influence of the age of the donor on the response of peripheral blood leukocyte chromosomes to radiation damage. The data on cells from the Japanese population show an increased sensitivity in cells from individuals below the age of 6 months or so, with little change over the age range from 1 to 40 years (Fig. 13). No information was available on cells from the Scottish population from individuals between birth and 6 years of age, but in this work a further increase in response

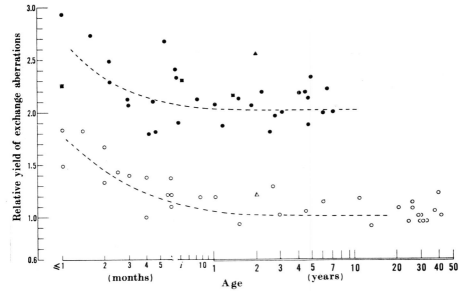

Figure 13. Relative incidence of ^{60}Co-γ-ray-induced aberrations in human peripheral blood cells taken from individuals of different ages and chromosome constitutions and irradiated in vitro (from Sasaki and Tonomura, ref. [139]). Solid circles and squares: results from cells from patients with Down's syndrome; open circles: results from patients with a normal chromosome constitution; triangles: normal (open) and trisomic (solid) cells in a mosaic Down's patient with a 46,XY/47,XY,+21 chromosome constitution.

with age occurred over the age range 40 to 101 years. We have already commented on the heterogeneity of response in lymphocyte cell populations, and the age-associated changes referred to here may well reflect changes in composition of the cell populations sampled.

Finally, it should be remembered that the processes involved in the formation of aberrations may require the interaction of the products of a number of gene loci, so that mutations of the sort exemplified by the xeroderma pigmentosum condition in man could have profound effects on the response of the chromosomes to radiation exposure.

THE USE OF ABERRATIONS IN BIOLOGICAL DOSIMETRY

The possibility of estimating the absorbed radiation dose received by individuals, through an analysis of the frequency of induced aberrations detectable in their peripheral blood lymphocytes, has obvious potential in the case of accidental or indeed occupational exposure to ionizing radiations. There are now a very large number of reported cases in which attempts have been made to relate chromosome damage to physical dose received accidentally, occupationally, or for therapeutic purposes. Because these data have been reviewed in extenso elsewhere [166], this work will not be considered in any detail here.

In principle, provided an individual has received an acute, uniform, whole-body exposure to radiation of known quality, then the dose received may be calculated by relating the aberration yield observed in cells from that individual to an equivalent yield obtained in lymphoctyes exposed in vitro to a known dose of that radiation. We have already indicated that the response to chromosome damage in lymphocytes in vivo is closely similar to the response in vitro, and various laboratories have produced "in vitro calibration dose–response curves" for different qualities of radiation [47]. Although the principles are clear and have indeed been successfully applied in a number of radiation accident cases (e.g., refs. [10, 22]), there are a number of problems.

A major problem is that in most cases of accidental irradiation the exposure is often to only part of the body and is nonuniform. In such cases the lymphocytes sampled for chromosome analysis will consist of unirradiated cells and of cells that were subjected to a variety of different exposures. It is to be anticipated that there will be some selection against the irradiated, relative to the unexposed, cells both in vivo prior to the withdrawal of blood samples and in vitro in the culture vessel. Sharpe [158] has recently investigated the problem of cell selection in vitro through studying aberration yields in mixed populations of irradiated and unirradiated cells and has shown that there is considerable selection against the radiation-damaged cells, which results in a reduction in the total aberration yield scored. This factor clearly complicates attempts at extrapolating from in vitro dose–response curves for partial body exposure.

Lymphocytes exchange rapidly from the peripheral blood into a large pool distributed throughout the body, and a proportion of these cells are long-lived and are not often involved in proliferation cycles. As a consequence, aberrations can be detected in peripheral blood cells of individuals many years after an original exposure to ionizing radiations; however, the number of aberrations observed declines with increasing time after irradiation [10, 25, 112]. The classic examples are the survivors of the atomic bombings of Hiroshima and Nagasaki (see ref. [3]). In these individuals it has been possible to relate aberration yield to dose in lymphocytes examined 22 years after the original exposure [137]. However, for practical dosimetry purposes it is necessary to sample cells from irradiated individuals shortly after their exposure.

ABERRATIONS IN GERM CELLS

There is virtually no information on the response of the chromosomes in the germ cells of man to radiation exposure, but a considerable amount of information has accumulated from the extensive genetic and cytogenetic studies that have been carried out on the laboratory mouse. Irradiation of gonial cells results in the production of aberrations identical to those produced in somatic cells. In the case of meiocytes, since in both plants [78] and mammals [90] the replication of all but a minute amount of DNA occurs before the entry of cells into meiotic prophase, irradiation of cells in meiotic prophase results in the formation of chromatid-type changes, later prophase stages yielding sub-chromatid-type aberrations [169].

After the original study of Oakberg and DiMinno [114], reporting changes in aberration yields throughout meiosis in X-irradiated male mice, there have been few detailed cytological analyses of meiotic cells themselves after low-dose exposures. Recently, however, Wennstrom [169] has described aberrations occurring in male spermatocytes after exposure to X-rays. His studies covered a period from 2.5 hours to 7 days, spanning the time required for cells to proceed from pachytene to diakinesis [113]; further studies were also carried out at later time intervals. He found that chromatid deletions and chromatid interchanges were present at diakinesis throughout the 2.5-hour to 7-day period in animals exposed to 250 R of X-rays. The highest frequency of deletions was seen at around one day after exposure, implying that the most sensitive period for the induction of chromatid deletions was at diplotene or early diakinesis—a finding in accord with earlier studies on plant materials (e.g., ref. [102]) and also in accord with the data obtained through anaphase scoring by Oakberg and DiMinno [114]. Of equal interest was the observation that these deletions increased in proportion to the square of the X-ray dose, a result that agrees with results obtained in somatic cells of other species referred to earlier, and with the results obtained by Wennstrom on somatic cells from bone marrow and corneal epithelium in the mouse.

The majority of cytogenetic studies on induced aberrations in the irradiated mouse have been largely concerned with translocations induced in spermatogonial cells analyzed at the succeeding diakinesis or metaphase I stages. As a consequence of chromosome pairing to form bivalents, reciprocal translocations are easily recognized (although not with 100% efficiency) as multivalent associations. It has long been known that translocations transmitted to offspring result in semisterility in the mouse [72, 73], but detailed studies on these translocations in the meiotic cells of irradiated animals, as opposed to the somatic and meiotic cells of their offspring, were only made possible by the development of an air-drying technique by E. P. Evans et al. [49], which was based on the technique developed earlier for somatic cells. Since radiation exposure results in a considerable degree of cell killing, and an associated period of sterility, cytological analyses in studies on the relationship between dose and effect have usually been made between 70 and 200 days after the initial exposure of the gonia.

The data obtained by a number of groups using X-ray doses of from 25 to 600 R are consistent in that the yields of translocations in irradiated spermatogonia increase linearly with dose; however, the slopes of the dose–response curves differ among laboratories and possibly among host strains [50, 95, 99, 106]. At dose levels above 600 R, the yield of translocations declines; and over the dose range 25 to 1250 R, the dose–response curve is humpbacked, with reduced frequencies at the high dose levels. Similar results, but with different critical dose levels, are also obtained in irradiated spermatogonial cells of other species, including rabbits and guinea pigs [100].

It might be expected that the yield of translocations should increase in proportion to the square of the X-ray dose, and it seems to be generally agreed that the observed linear response with X-rays and ^{60}Co γ-rays [152] over the early part of the dose–response curve is probably a secondary consequence of complicating factors intervening between the time of formation of aberrations and time of sampling some 70 to 200 days later. Since there is considerable cell killing and cell proliferation, and in view of the humpbacked nature of the dose–response curve, cell selection is clearly important. A selection against heavily damaged cells would tend to flatten the dose–response curve, reducing the dose-squared component to give a spurious linear response [115]. That cell proliferation and cell selection does occur is indicated by studies that have revealed the existence of clones of spermatocytes with multiple translocations [99, 153]. It should be noted that a similar linear response is obtained with 0.7-MeV fission neutrons [154], again giving a humpbacked dose–response curve with a peak in this case at 100 rads.

In most of the work referred to above, information on the effects of dose rate, exposure time, and fractionation on the yields of reciprocal translocations has also been examined. Generally, and especially at lower dose levels, it has been found that there may be a marked reduction in yield with increasing exposure time or with split doses. However, interpretation in all these studies is complicated by changes in the structure of the irradiated population during the irradiation procedure, changes which during fractionated or protracted exposures would lead to differential cell selection. That selection may occur at later stages is also evident from the fact that the frequency of translocation heterozygotes in the progeny of irradiated male mice is reduced to about one-half of what would be expected from the frequencies of multivalent configurations observed in the spermatocytes of their fathers [67].

Prior to the development of adequate cytological methods, virtually all of the information on the induction of chromosome structural changes in the mouse, and other mammalian species, had been largely derived from studies of dominant lethality expressed during embryonic development (e.g., ref. [5]) and from genetic analyses of viable offspring (e.g., ref. [133]). Indeed, such approaches still offer the only useful method of studying effects on the later germ cell stages and have enabled the relative sensitivities of all stages of spermatogenesis, including sperm maturation, to be compared [5, 135]. Breeding studies also offer a realistic approach to quantitative analysis of the actions of radiations on the chromosomes of female germ cells, and much work has been done in this area [100, 134]. Dominant lethal effects may of course result from a variety of structural chromosome and other genetic changes, and it is pertinent to note that Ford et al. [67] have recently presented data indicating that in the mouse all dominant lethals induced as a consequence of the irradiation of spermatogonial cells are derived from translocations. In this context we should also remember that the gonial cells that are now being studied so extensively at meiosis with cytogenetic techniques had previously been considered to be relatively free from chromosome aberrations after exposure to radiation as judged by studies on dominant lethality and viable offspring [94].

Finally, although in this article we have been primarily concerned with structural chromosome changes induced by radiation exposure, it should be noted that changes in chromosome number are also induced (e.g., ref. [133]) and that certain categories of chromosome loss, in particular those involving the sex chromosomes, may be viable and tolerated in germ cells and resultant offspring.

CONCLUDING COMMENTS

Studies on the effects of ionizing radiations on mammalian chromosomes, although initiated early in this century, have only blossomed as a consequence of the utilization of new techniques for the preparation of chromosomes that were developed in the late 1950s and early 1960s. Major advances in biology and medicine are invariably presaged by the development of new technical approaches and the development and application of new methods for studying the effects of radiation on macromolecules, and the recent major cytogenetic advances that have been described in the Paris Conference on Standardization in Human Genetics [120] must certainly herald further waves of advance in our knowledge on the actions of radiations on the mammalian genome.

LITERATURE CITED

1. ADAMS A: The response of chromosomes of human peripheral blood lymphocytes to X-rays. Ph.D. Thesis, University of Aberdeen, 1970.
2. ALEXANDER P, LETT JT, KOPP P, ITZHAKI R: Degradation of dry deoxyribonucleic acid by polonium alpha-particles. Radiat Res 14:363–373, 1961.
3. AWA AA: This volume, 637–674, 1974.
4. BAJERSKA A, LINIECKI J: The influence of temperature at irradiation in vitro on the yield of chromosomal aberrations in peripheral blood lymphocytes. Int J Rad Biol 16:483–493, 1969.
5. BATEMAN AJ: Mutagenic sensitivity of maturing germ cells in the male mouse. Heredity 12:213–232, 1958.
6. BENDER MA, BREWEN JG: Factors influencing chromosome aberration yields in the human peripheral system. Mutat Res 8:383–399, 1969.
7. BENDER MA, GOOCH PC: Spontaneous and X-ray-induced somatic–chromosome aberrations in the Chinese hamster. Int J Rad Biol 4:175–184, 1961.
8. BENDER MA, GOOCH PC: Types and rates of X-ray induced chromosome aberrations in human blood irradiated in vitro. Proc Nat Acad Sci (US) 48:522–532, 1962.
9. BENDER MA, GOOCH PC: Chromatid-type aberrations induced by X-rays in human leukocyte cultures. Cytogenetics, 2:107–116, 1963.
10. BENDER MA, GOOCH PC: Somatic chromosome aberrations induced by human whole-body irradiation: the "Recuplex" criticality accident. Radiat Res 29:568–582, 1966.
11. BICK YAE, BROWN JK: Variations in radiosensitivity during the cell cycle in a marsupial cell line. Mutat Res 8:613–622, 1969.
12. BOOTSMA D, MULDER MP, POT F, COHEN JA: Different inherited levels of DNA repair replication in Xeroderma pigmentosum cell strains after exposure to ultraviolet irradiation. Mutat Res 9:507–516, 1970.
13. BOYCE RP, HOWARD-FLANDERS P: Release of ultraviolet light-induced thymine dimers from DNA in E. coli K-12. Proc Nat Acad Sci (US) 51:293–300, 1964.
14. BOYD JT, COURT BROWN WM, WOODCOCK GE, VENNART J: Relationship between external radiation exposure and chromosome aberrations among luminous dial painters. In Human Radiation Cytogenetics (Evans HJ, Court Brown WM, McLean AS, eds), Amsterdam, North Holland Publishing Company, 208–214, 1967.
15. BREWEN JG: Dependence of frequency of X-ray-induced chromosome aberrations on dose rate in the Chinese hamster. Proc Nat Acad Sci (US) 50:322–329, 1963.

16. BREWEN JG: Cell-cycle and radiosensitivity of the chromosomes of human leukocytes. *Int J Rad Biol* 9:391–397, 1965.
17. BREWEN JG, BROCK RD: The exchange hypothesis and chromosome-type aberrations. *Mutat Res* 6:245–255, 1968.
18. BREWEN JG, GENOZIAN N: Radiation-induced human chromosome aberrations. II. Human *in vitro* irradiation compared to *in vitro* and *in vivo* irradiation of marmoset leukocytes. *Mutat Res* 13:383–391, 1971.
19. BREWEN JG, LUIPPOLD HE: Radiation-induced human chromosome aberrations: *in vitro* dose rate studies. *Mutat Res* 12:305–314, 1971.
20. BREWEN JG, OLIVIERI G: The kinetics of chromatid aberrations induced in Chinese hamster cells by tritium-labeled thymidine. *Radiat Res* 28:779–792, 1966.
21. BREWEN JG, OLIVIERI G, LUIPPOLD HE, PEARSON FG: Nonrandom rejoining in the formation of chromatid interchanges: variations through the cell cycle and the effect of chromosome pairing. *Mutat Res* 8:401–408, 1969.
22. BREWEN JG, PRESTON RJ, LITTLEFIELD LG: Radiation-induced human chromosome aberration yields following an accidental whole-body exposure to ^{60}Co γ-rays. *Radiat Res* 49:647–656, 1972.
23. BUCKTON KE, LANGLANDS AO, WOODCOCK GE: Cytogenetic changes following Thorotrast administration. *Int J Rad Biol* 12:565–577, 1967.
24. BUCKTON KE, LANGLANDS AO, SMITH PG, WOODCOCK GE, LOOBY PC: Further studies on chromosome aberration production after whole-body irradiation in man. *Int J Rad Biol* 19:369–378, 1971.
25. BUCKTON KE, SMITH PG, COURT BROWN WM: The estimation of lymphocyte lifespan from studies on males treated with X-rays for ankylosing spondylitis. *In* Human Radiation Cytogenetics (Evans HJ, Court Brown WM, McLean AS, eds), Amsterdam, North Holland Publishing Company, 106–114, 1967.
26. BUSHONG SC, WATSON JA, WALD N: The oxygen effect based on chromosome damage in HeLa cells. *Int J Rad Biol* 13:249–260, 1967.
27. CHU EHY: Effects of ultraviolet radiation on mammalian cells. I. Induction of chromosome aberrations. *Mutat Res* 2:75–94, 1965.
28. CHU EHY, GILES NH, PASSANO K: Types and frequencies of human chromosome aberrations induced by X-rays. *Proc Nat Acad Sci (US)* 47:830–839, 1961.
29. CHUDINA AP, MALUTINA TS, POGOSIANZ HE: Comparative radiosensitivity of chromosomes in the cultured peripheral blood leukocytes of normal donors and patients with Down's syndrome. *Genetika* 4:50–63, 1966.
30. CLARKSON JM, EVANS HJ: Unscheduled DNA synthesis in human leucocytes after exposure to UV light, γ-rays and chemical mutagens. *Mutat Res* 14:413–430, 1972.
31. CLEAVER JE: Defective repair replication of DNA in *Xeroderma pigmentosum*. *Nature* 218:652–656, 1968.
32. CLEAVER JE: *Xeroderma pigmentosum:* a human disease in which an initial stage of DNA repair is defective. *Proc Nat Acad Sci (US)* 63:428–435, 1969.
33. CLEAVER JE, TROSKO JE: Absence of excision of ultraviolet-induced cyclobutane dimers in Xeroderma pigmentosum. *Photochem Photobiol* 11:547–550, 1970.
34. CLEAVER JE, THOMAS B: Single stranded interruptions in DNA and the effects

of caffeine in Chinese hamster cells irradiated with ultraviolet light. *Biochem Biophys Res Comm*, 36:203–208, 1969.

35. CLEMENGER JF, SCOTT D: In vitro and in vivo sensitivity of cultured blood lymphocytes to radiation induction of chromosome aberrations. *Nature New Biol* 234:154, 1971.
36. COMINGS DE: The structure and function of chromatin. *In* Advances in Human Genetics (Harris H, Hirschhorn K, eds), vol 3, New York, Plenum Press, 237–431, 1972.
37. CONGER AD: The effect of oxygen on the radiosensitivity of mammalian cells. *Radiology* 66:63–69, 1956.
38. CORRY PM, COLE A: Radiation-induced double-strand scission of the DNA of mammalian metaphase chromosomes. *Radiat Res* 36:528–543, 1968.
39. DAVIES DR, EVANS HJ: The role of genetic damage in radiation-induced cell lethality. *In* Advances in Radiation Biology (Augenstein LG, Mason R, Zelle MR, eds), vol 2, New York, Academic Press, 243–353, 1966.
40. DEBAKAN AS, THRON R, STEUSING J: Chromosomal aberrations in irradiated blood and blood cultures of normal subjects and of selected patients with chromosomal abnormality. *Radiat Res* 27:50–63, 1966.
41. DESCHNER EE, GRAY LH: Influence of oxygen tension on X-ray-induced chromosomal damage in Ehrlich ascites tumor cells irradiated *in vitro* and *in vivo*. *Radiat Res* 11:115–146, 1959.
42. DEWEY WC, HUMPHREY RM: Relative radiosensitivity of different phases in the life cycle of L-P59 mouse fibroblasts and ascites tumor cells. *Radiat Res* 16:503–530, 1962.
43. DEWEY WC, HUMPHREY RM, JONES BA: Comparisons of tritiated thymidine, tritiated water and cobalt-60 gamma rays in inducing chromosomal aberrations. *Radiat Res* 24:214–238, 1965.
44. DEWEY WC, MILLER HH, LEEPER DB: Chromosomal aberrations and mortality of X-irradiated mammalian cells: emphasis on repair. *Proc Nat Acad Sci (US)* 68:667–671, 1971.
45. DEWEY WC, SEDITA BA, HUMPHREY RM: Chromosomal aberrations induced by tritiated thymidine during the S and G_2 phases of Chinese hamster cells. *Int J Rad Biol* 12:597–600, 1967.
46. DJORDEVIC B, TOLMACH JL: Responses of synchronous population of HeLa cells to ultraviolet irradiation at selected stages of the generation cycle. *Radiat Res* 32:327–346, 1967.
47. DOLPHIN GW, PURROT RJ: Use of radiation-induced chromosome aberrations in human lymphocytes for dosimetry. *In* Advances in Physical and Biological Radiation Detectors. Vienna, International Atomic Energy Agency, 611–622, 1971.
48. ELKIND MM: Sedimentation of DNA released from Chinese hamster cells. *Biophys J* 11:502–520, 1971.
49. EVANS EP, BRECKON G, FORD CE: An air-drying method for meiotic preparations from mammalian testes. *Cytogenetics* 3:289–294, 1964.
50. EVANS EP, FORD CE, SEARLE AG, WEST BJ: Studies on the induction of translocations in mouse spermatogonia. II. Effects of X-irradiation. *Mutat Res* 9:501–506, 1970.

51. EVANS HJ: Chromosome aberrations induced by ionizing radiations. *Int Rev Cytol* 13:221–321, 1962.
52. EVANS HJ: Chromosome aberrations and target theory. In Radiation-Induced Chromosome Aberrations (Wolff S, ed), New York, Columbia University Press, 8–40, 1963.
53. EVANS HJ: Repair and recovery from chromosome damage after fractionated X-ray dosage. In Genetical Aspects of Radiosensitivity: Mechanisms of Repair. Vienna, International Atomic Energy Agency, 31–48, 1966.
54. EVANS HJ: Repair and recovery at chromosome and cellular levels: similarities and differences. *Brookhaven Symp Biol* 20:111–133, 1967.
55. EVANS HJ: Repair and recovery from chromosome damage induced by fractionated X-ray exposures. In Radiation Research. Amsterdam, North Holland Publishing Company, 482–501, 1967.
56. EVANS HJ: Dose-response relations from *in vitro* studies. In Human Radiation Cytogenetics (Evans HJ, Court Brown WM, McLean AS, eds), Amsterdam, North Holland Publishing Company, 20–36, 1967.
57. EVANS HJ: The response of human chromosomes to X-irradiation: *in vitro* studies. *Proc Roy Soc Edin B* 70:132–151, 1968.
58. EVANS HJ: Population cytogenetics and environmental factors. In Symposium on Human Population Cytogenetics. Pfizer Medical Monographs 5, Edinburgh, Edinburgh University Press, 192–216, 1970.
59. EVANS HJ, ADAMS A: X-ray-induced chromosome aberrations in human lymphocytes irradiated *in vitro*: the influence of exposure conditions, genotype and age on aberration yields. In Advances in Radiation Research, Proceedings of the Fourth International Congress of Radiation Research (1970) (Duplan JF, Chapiro A, eds), New York, Gordon & Breach, 1973.
60. EVANS HJ, SAVAGE JRK: The relation between DNA synthesis and chromosome structure as resolved by X-ray damage. *J Cell Biol* 18:525–540, 1963.
61. EVANS HJ, SCOTT D: Influence of DNA synthesis on the production of chromatid aberrations by X-rays and maleic hydrazide in *Vicia faba*. *Genetics* 49:17–38, 1964.
62. EVANS HJ, SCOTT D: The induction of chromosome aberrations by nitrogen mustard and its dependence on DNA synthesis. *Proc Roy Soc (London) B* 173:491–512, 1969.
63. EVANS HJ, SPARROW AH: Nuclear factors affecting radiosensitivity. Parts I and II. *Brookhaven Symp Biol* 14:76–127, 1961.
64. EVANS RG, NORMAN A: Unscheduled incorporation of thymidine in ultraviolet irradiated human lymphocytes. *Radiat Res* 36:287–298, 1968.
65. FABERGE AC: An experiment on chromosome fragmentation in *Tradescantia* by X-rays. *J Genet* 29:229–248, 1940.
66. FISCHER P, GOLOB E, KUNZE-MÜHL E, BEN HAIM A, DUDLEY RA, MÜLLNER T, PARR RM, VETTER H: Chromosome aberrations in peripheral blood cells in man following chronic irradiation from internal deposits of Thorotrast. *Radiat Res* 29:505–515, 1966.
67. FORD CE, SEARLE AG, EVANS EP, WEST BJ: Differential transmission of translocations induced in spermatogonia of mice by irradiation. *Cytogenetics* 8:447–470, 1969.

68. FREIFELDER D: Mechanism of inactivation of coliphage T7 by X-rays. *Proc Nat Acad Sci (US)* 54:128–134, 1965.
69. FREIFELDER D: DNA strand breakage by X-irradiation. *Radiat Res* 29:329–338, 1966.
70. GIBSON DA, PRESCOTT DM: Induction of sister chromatid exchanges in chromosomes of rat kangaroo cells by tritium incorporated into DNA. *Exp Cell Res* 74:397–402, 1972.
71. GILES NH: The effect of fast neutrons on the chromosomes of *Tradescantia*. *Proc Nat Acad Sci (US)* 26:567–575, 1940.
72. GRIFFEN AB: Occurrence of chromosomal aberrations in pre-spermatocytic cells of irradiated male mice. *Proc Nat Acad Sci (US)* 44:691–694, 1958.
73. GRIFFEN AB, BUNKER MC: The occurrence of chromosomal aberrations in pre-spermatocytic cells of irradiated male mice. III. Sterility and semisterility in the offspring of male mice irradiated in the pre-meiotic and post-meiotic stages of spermatogenesis. *Can J Genet Cytol* 9:163–254, 1967.
74. HEDDLE JA: Randomness in the formation of radiation-induced chromosome aberrations. *Genetics* 52:1329–1334, 1965.
75. HEDDLE JA, BODYCOTE DJ: On the formation of chromosomal aberrations. *Mutat Res* 9:117–126, 1970.
76. HEDDLE JA, EVANS HJ, SCOTT D: Sampling time and the complexity of the human leucocyte system. *In* Human Radiation Cytogenetics (Evans HJ, Court Brown WM, McLean AS, eds), Amsterdam, North Holland Publishing Company, 6–19, 1967.
77. HEDDLE JA, WHISELL D, BODYCOTE DJ: Changes in chromosome structure induced by radiations: a test of the two chief hypotheses. *Nature* 221:1158–1160, 1969.
78. HOTTA Y, ITO M, STERN H: Synthesis of DNA during meiosis. *Proc Nat Acad Sci (US)* 56:1184–1191, 1966.
79. HOWARD A, PELC SR: Synthesis of desoxyribonucleic acid in normal and irradiated cells and its relation to chromosome breakage. *Heredity Suppl* 6:261–273, 1953.
80. HSU TC, DEWEY WC, HUMPHREY RM: Radiosensitivity of cells of Chinese hamster *in vitro* in relation to the cell cycle. *Exp Cell Res* 27:441–452, 1962.
81. HUMPHREY RM, DEWEY WC, CORK A: Relative ultraviolet sensitivity of different phases in the cell cycle of Chinese hamster cells grown *in vitro*. *Radiat Res* 19:247–260, 1963.
82. HUMPHREY RM, STEWARD DL, SEDITA BA: DNA-strand breaks and rejoining following exposure of synchronized Chinese hamster cells to ionizing radiation. *Mutat Res* 6:459–465, 1968.
83. ISIHARA T, KUMATORI T: Chromosome aberrations in human leukocytes irradiated *in vivo* and *in vitro*. *Acta Haematol Jap* 28:291–307, 1965.
84. JEMILEV ZA: Radiosensitivity of the chromosomes of monkey (*Macaca mulatta*) peripheral-blood leucocytes at the different stages of the mitotic cycle. *Genetika* 6:3, 155, 1970.
85. JEMILEV ZA: Radiosensitivity of the chromosomes of human peripheral blood leucocytes at the different stages of the cell cycle. *Genetika* 3(5):67, 1967.
86. KANAZIR DT: Radiation-induced alterations in the structure of deoxyribonucleic

acid and their biological consequences. *In* Progress in Nucleic Acid Research and Molecular Biology (Davidson JN, Cohn WE, eds), vol 9, New York, Academic Press, 117–222, 1969.

87. KANG YS, PARK SD: Studies on stage radiosensitivity and DNA synthesis of chromosomes in cultured human cells. *Radiat Res* 37:371–380, 1969.

88. KIHLMAN BA, HARTLEY B: "Sub-chromatid" exchanges and the "folded fibre" model of chromosome structure. *Hereditas* 57:289–295, 1967.

89. KOERNICKE M: Über die Wirkung von Röntgen- und Radium Strahlen auf pflanzliche Gewebe und Zellen. *Ber Deut Bot Ges* 23:405–415, 1905.

90. KOFMAN-ALFARO S, CHANDLEY AC: Radiation-initiated DNA synthesis in spermatogenic cells of the mouse. *Exp Cell Res* 69:33–44, 1971.

91. KUCEROVA M: Comparison of radiation effects *in vitro* upon chromosomes of human subjects. *Acta Radiol* 6:441–448, 1967.

92. LEA DE: Actions of Radiations on Living Cells, 2nd ed, Cambridge, Cambridge University Press, 416 pp, 1955.

93. LEA DE, CATCHESIDE DE: The mechanism of the induction by radiation of chromosome aberrations in *Tradescantia*. *J Genet* 44:216–245, 1942.

94. LEONARD A: Radiation-induced translocations in spermatogonia of mice. *Mutat Res* 11:71–88, 1971.

95. LEONARD A, DEKNUDT G: The sensitivity of various germ-cell stages of the male mouse to radiation induced translocations. *Can J Genet Cytol* 10:495–507, 1968.

96. LETT JT, ALEXANDER P: Crosslinking and degradation of deoxyribonucleic acid gels with varying water contents when irradiated with electrons. *Radiat Res* 15:159–173, 1961.

97. LETT JT, CALDWELL I, DEAN CJ, ALEXANDER P: Rejoining of X-ray induced breaks in the DNA of leukaemia cells. *Nature* 214:790–792, 1967.

98. LOHMAN PHM: Induction and rejoining of breaks in the deoxyribonucleic acid of human cells irradiated at various phases of the cell cycle. *Mutat Res* 6:449–458, 1968.

99. LYON M, MORRIS T: Gene and chromosome mutation after large fractionated or unfractionated radiation doses to mouse spermatogonia. *Mutat Res* 8:191–198, 1969.

100. LYON MF, SMITH BD: Species comparisons concerning radiation-induced dominant lethals and chromosome aberrations. *Mutat Res* 11:45–58, 1971.

101. MARIN G, PRESCOTT DM: The frequency of sister chromatid exchanges following exposure to varying doses of H^3-thymidine or X-rays. *J Cell Biol* 21:159–167, 1964.

102. MITRA S: Effects of X-rays on chromosomes of *Lilium longiflorum* during meiosis. *Genetics* 43:771–789, 1958.

103. MOORHEAD PS, NOWELL PC, MELLMAN WJ, BATTIPS DM, HUNGERFORD DA: Chromosome preparations of leukocytes cultured from human peripheral blood. *Exp Cell Res* 20:613–616, 1960.

104. MULLER HJ: Artificial transmutation of the gene. *Science* 66:84–87, 1927.

105. MULLER HJ: The nature of the genetic effects produced by radiation. *In* Radiation Biology (Hollander A, ed), vol 1, New York, McGraw-Hill Publishing Company, 351–474, 1954.

106. MURATMASU S, NAKAMURA W, ITO H: Radiation-induced translocations in mouse spermatogonia. Jap J Genet 46:281–283, 1971.
107. McGRATH RA, WILLIAMS RW: Reconstruction in vivo of irradiated Escherichia coli deoxyribonucleic acid; the rejoining of broken pieces. Nature 212:534–535, 1966.
108. NEARY GJ, PRESTON RJ, SAVAGE JRK: Chromosome aberrations and the theory of RBE. III. Evidence from experiments with soft X-rays and a consideration of the effects of hard X-rays. Int J Rad Biol 12:317–345, 1967.
109. NEARY GJ, SAVAGE JRK, EVANS HJ: Chromatid aberrations in Tradescantia pollen tubes induced by monochromatic X-rays of quantum energy 3 and 1.5 keV. Int J Rad Biol 8:1–19, 1964.
110. NEARY GJ, SIMPSON-GILDEMEISTER VF, PEACOCKE AR: The influence of radiation quality and oxygen on strand breakage in dry DNA. Int J Rad Biol 18:25–40, 1970.
111. NORMAN A, SASAKI MS: Chromosome-exchange aberrations in human lymphocytes. Int J Rad Biol 11:321–328, 1966.
112. NORMAN A, SASAKI MS, OTTOMAN RE, FINGERHUT AG: Elimination of chromosome aberrations from human lymphocytes. Blood 27:706–714, 1966.
113. OAKBERG EF: Duration of spermatogenesis in the mouse. Nature 180:1137–1138, 1957.
114. OAKBERG EF, DIMINNO RL: X-ray sensitivity of primary spermatocytes of the mouse. Int J Rad Biol 2:196–209, 1960.
115. OFTEDAL P: A theoretical study of mutant yield and cell killing after treatment of heterogeneous cell populations. Hereditas 60:177–210, 1968.
116. PAINTER RB: Repair of DNA in mammalian cells. In Current Topics in Radiation Research Quarterly (Ebert M, Howard A, eds), vol 7, Amsterdam, North Holland Publishing Company, 45–70, 1970.
117. PAINTER RB, CLEAVER JE: Repair replication in HeLa cells after large doses of X-irradiation. Nature 216:369–370, 1967.
118. PAINTER RB, CLEAVER JE: Repair replication, unscheduled DNA synthesis and the repair of mammalian DNA. Radiat Res 37:451–466, 1969.
119. PALCIC B, SKARSGARD LD: The effect of oxygen on DNA single-strand breaks produced by ionizing radiation in mammalian cells. Int J Rad Biol 21:417–433, 1972.
120. Paris Conference, Standardization in Human Cytogenetics, Birth Defects: Original Article Ser. vol 8, no 7, The National Foundation, 1–46, 1972.
121. PARRINGTON JM, DELHANTY JDA, BADEN HP: Unscheduled DNA synthesis u.v.-induced chromosome aberrations and SV_{40} transformation in cultured cells from Xeroderma pigmentosum. Ann Hum Genet 35:149–160, 1971.
122. PETTIJOHN D, HANAWALT PC: Evidence for repair-replication of ultraviolet damaged DNA in bacteria. J Mol Biol 9:395–410, 1964.
123. PREMPREE T, MERZ T: Radiosensitivity and repair time: the repair time of chromosome breaks produced during the different stages of the cell cycle. Mutat Res 7:441–451, 1969.
124. PRESTON RJ, BREWEN JG, JONES KP: Radiation-induced chromosome aberrations in Chinese hamster leukocytes: a comparison of in vivo and in vitro exposures. Int J Rad Biol 21:397–400, 1972.

125. RASMUSSEN RE, PAINTER RB: Evidence for repair of ultraviolet damaged deoxyribonucleic acid in cultured mammalian cells. *Nature* 203:1360–1362, 1964.
126. RASMUSSEN RE, PAINTER RB: Radiation-stimulated DNA synthesis in cultured mammalian cells. *J Cell Biol* 29:11–19, 1966.
127. REVELL SH: The accurate estimation of chromatid breakage, and its relevance to a new interpretation of chromatid aberrations induced by ionizing radiations. *Proc Roy Soc London B* 150:563–589, 1959.
128. REVELL SH: An attempt at continuous metaphase estimation of chromatid and choromosome aberration frequencies in broad bean root meristem cells in the period 2–23 h. after 50r or X-rays. *In* Effects of Ionizing Radiation on Seeds. Vienna, International Atomic Energy Agency, 229–242, 1961.
129. REVELL SH: Evidence for a dose squared term in the dose-response curve for real chromatid discontinuities induced by X-rays, and some theoretical consequences thereof. *Mutat Res* 3:34–54, 1966.
130. ROTHFELS KH, SIMINOVITCH L: An air drying technique for flattening chromosomes in mammalian cells grown *in vitro*. *Stain Technol* 33:73–77, 1962.
131. RUPP WD, HOWARD-FLANDERS P: Discontinuities in the DNA synthesized in an excision-defective strain of *Escherichia coli* following ultraviolet irradiation. *J Mol Biol* 31:291–304, 1968.
132. RUPP WD, ZIPSER E, VON ESSEN C, RENO D, PROSNITZ L, HOWARD-FLANDERS P: *In* Time and Dose Relationship in Radiation Biology as Applied to Radiotherapy. New York, Brookhaven Monograph (quoted by PAINTER RB, 1970).
133. RUSSELL LB: The use of X-chromosome anomalies for measuring radiation effects in different germ cell stages of the mouse. *In* Effects of Radiation on Meiotic Systems. Vienna, International Atomic Energy Agency, 27–41, 1968.
134. RUSSELL LB, RUSSELL WL: The sensitivity of different stages in oogenesis to the radiation induction of dominant lethals and other changes in the mouse. *In* Progress in Radiobiology (Mitchell JS, Holmes BE, Smith CL, eds), London, Oliver and Boyd, 187–195, 1955.
135. RUSSELL LB, SAYLORS CL: The relative sensitivity of various germ-cell stages of the mouse to radiation-induced nondisjunction chromosome losses and deficiencies. *In* Repair from Genetic Radiation (Sobels FH, ed), Oxford, Pergamon Press, 313–342, 1963.
136. SASAKI MS: Radiation-induced chromosome aberrations in lymphocytes: possible biological dosimeter in man. *In* Biological Aspects of Radiation Protection (Sugahara T, Hug O, eds), Tokyo, Igaku Shoin, 81–91, 1970.
137. SASAKI MS, MIYATA H: Biological dosimetry in atomic bomb survivors. *Nature* 220:1189–1193, 1968.
138. SASAKI MS, NORMAN A: Proliferation of human lymphocytes in culture. *Nature* 210:913–914, 1966.
139. SASAKI MS, TONOMURA A: Chromosomal radiosensitivity in Down's syndrome. *Jap J Hum Genet* 14:81–92, 1969.
140. SAVAGE JRK: Sites of radiation induced chromosome exchanges. *In* Current Topics in Radiation Research (Ebert M, Howard A, eds), Amsterdam, vol 6, North Holland Publishing Company, 130–194, 1970.
141. SAVAGE JRK, NEARY GJ, EVANS HJ: The rejoining time of chromatid breaks

induced by gamma radiation in *Vicia faba* root tips at 3°C. *J Biophys Biochem Cytol* 7:79–85, 1960.

142. SAVAGE JRK, PRESTON RJ, NEARY GJ: Chromatid aberrations in *Tradescantia bracteata* and a further test of Revell's hypothesis. *Mutat Res* 5:47–56, 1968.
143. SAWADA S, OKADA S: Rejoining of single-strand breaks of DNA in cultured mammalian cells. *Radiat Res* 41:145–162, 1970.
144. SAX K: The time factor in X-ray production of chromosome aberrations. *Proc Nat Acad Sci (US)* 25:225–233, 1939.
145. SAX K: An analysis of X-ray-induced chromosomal aberrations in *Tradescantia*. *Genetics* 25:41–68, 1940.
146. SAX K: Types and frequencies of chromosomal aberrations induced by X-rays. *Cold Spring Harbor Symp Quant Biol* 9:93–103, 1941.
147. SAX K, ENZMAN BV: The effect of temperature on X-ray-induced chromosome aberrations. *Proc Nat Acad Sci (US)* 25:397–405, 1939.
148. SCOTT D, EVANS HJ: X-ray-induced chromosomal aberrations in *Vicia faba*: changes in response during the cell cycle. *Mutat Res* 4:579–599, 1967.
149. SCOTT D, SHARPE H, BATCHELOR AL, EVANS HJ, PAPWORTH DG: Radiation-induced chromosome damage in human peripheral blood lymphocytes *in vitro*. I. RBE and dose–rate studies with fast neutrons. *Mutat Res* 8:367–381, 1969.
150. SCOTT D, SHARPE H, BATCHELOR AL, EVANS HJ, PAPWORTH DG: Radiation-induced chromosome damage in human peripheral blood lymphocytes *in vitro*. I. RBE and dose–rate studies with ^{60}Co γ- and X-rays. *Mutat Res* 9:225–260, 1970.
151. SEABRIGHT M: High resolution studies on the pattern of induced exchanges in the human karyotype. *Chromosoma* 40:333–346, 1973.
152. SEARLE AG, BEECHEY CV, EVANS EP, FORD CE, PAPWORTH DG: Studies on the induction of translocations in mouse spermatogonia. IV. Effects of acute gamma-irradiation. *Mutat Res* 12:411–416, 1971.
153. SEARLE AG, EVANS EP, FORD CE, WEST BJ: Studies on the induction of translocations in mouse spermatogonia. I. The effect of dose–rate. *Mutat Res* 6:427–436, 1968.
154. SEARLE AG, EVANS EP, WEST BJ: Studies on the induction of translocations in mouse spermatogonia. II. The effects of fast neutron irradiation. *Mutat Res* 7:235–240, 1969.
155. SETLOW RB, CARRIER WL: The disappearance of thymine dimers from DNA: an error-correcting mechanism. *Proc Nat Acad Sci (US)* 51:226–231, 1964.
156. SETLOW RB, REGAN JD, GERMAN J, CARRIER WL: Evidence that *Xeroderma pigmentosum* cells do not perform the first step in the repair of ultraviolet damage to their DNA. *Proc Nat Acad Sci (US)* 64:1035–1041, 1969.
157. SEVENKAYEV AV, BOCHKOV NP: The effect of gamma-rays on human chromosomes. I. The dependence of chromosome aberration frequency upon the dose at the irradiation *in vitro*. *Genetika* 4:130–137, 1968.
158. SHARPE HBA: Pitfalls in the use of chromosome aberration analysis for biological radiation dosimetry. *Brit J Radiol* 42:943–944, 1969.
159. SHARPE HBA, DOLPHIN GW, DAWSON KB, FIELD EO: Methods for computing lymphocyte kinetics in man by analysis of chromosomal aberrations sustained

during extracorporeal irradiation of the blood. *Cell Tissue Kinet* 1:263–271, 1968.
160. SHARPE HBA, SCOTT D, DOLPHIN GW: Chromosome aberrations induced in human lymphocytes by X-irradiation *in vitro*: the effect of culture techniques and blood donors on aberration yield. *Mutat Res* 7:453–461, 1969.
161. STADLER LJ: The experimental modification of heredity in crop plants. I. Induced chromosomal irregularities. *Sci Agric* 11:557–572, 1931.
162. TAYLOR JH, WOODS PS, HUGHES WL: Organization and duplication of chromosomes as revealed by autoradiographic studies using tritium-labelled thymidine. *Proc Nat Acad Sci (US)* 43:122–128, 1957.
163. THODAY JM, READ J: Effect of oxygen on the frequency of chromosome aberrations produced by X-rays. *Nature* 160:608, 1947.
164. TROSKO JE, CHU EHY, CARRIER WL: The induction of thymine dimers in ultraviolet-irradiated mammalian cells. *Radiat Res* 24:667–672, 1965.
165. TSUBOI A, TERASIMA T: Rejoining of single breaks of DNA induced by X-rays in mammalian cells: effects of metabolic inhibitors. *Mol Gen Genet* 108: 117–128, 1970.
166. UN Report. Radiation-induced chromosome aberrations in human cells. United Nations Scientific Committee on the Effects of Atomic Radiation, New York, General Assembly 26th Session, Suppl 13(A/7613) 98–155, 1969.
167. VISFELDT J: Radiation-induced chromosome aberrations in human cells. Doctoral thesis published as a Riso Report 117, Riso, Danish Atomic Energy Commission, 1966.
168. WATSON GE, GILLIES NE: The oxygen enhancement ratio for X-ray-induced chromosomal aberrations in cultured human lymphocytes. *Int J Rad Biol* 17: 279–283, 1970.
169. WENNSTROM J: Effect of ionizing radiations on the chromosomes in meiotic and mitotic cells. *Commentat Biol Soc Sci Fenn Helsinki* 45:5–60, 1971.
170. WINKELSTEIN A, CRADDOCK CG, MARTIN DC, LIBBY RI, NORMAN A, SASAKI MS: Sr^{90}-Y^{90} extracorporeal irradiation in goats and man. *Radiat Res* 31: 215–229, 1957.
171. WOLFF S: Interpretation of induced chromosome breakage and rejoining. *Radiat Res* Suppl 1:453–462, 1959.
172. WOLFF S: Radiation studies on the nature of chromosome breakage. *Amer Naturalist* 94:85–93, 1960.
173. WOLFF S: Radiation genetics. *In* Mechanisms in Radiobiology (Errera M, Forssberg A, eds), vol 1, New York, Academic Press, 419–475, 1961.
174. WOLFF S: Are sister chromatid exchanges sister strand crossovers or radiation-induced exchanges? *Mutat Res* 1:337–343, 1964.
175. WOLFF S, LUIPPOLD HE: Metabolism and chromosome-break rejoining. *Science* 122:231–232, 1955.
176. WOLFF S, LUIPPOLD HE: Chromosome splitting as revealed by combined X-ray and labelling experiments. *Exp Cell Res* 34:548–556, 1963.
177. WOLFF S, ATWOOD KC, RANDOLPH ML, LUIPPOLD HE: Factors limiting the number of radiation-induced chromosome exchanges. I. Distance: evidence from noninteraction of X-ray and neutron-induced breaks. *J Biophys Biochem Cytol* 4:365–372, 1958.

MITOTIC ABNORMALITIES AND CANCER

TARVO OKSALA
Department of Genetics, University of Turku, Turku, Finland

and

EEVA THERMAN
Department of Medical Genetics, University of Wisconsin, Madison, Wisconsin

---

Introduction 240
Modifications in the Mitotic Cycle and Endopolyploidy 241
    Endoreduplication 241
    Endomitosis 244
    C-Mitosis 244
    Restitution 248
Origin and Structure of Multipolar Divisions 248
An Increased Metaphase/Prophase Ratio, Multipolar Mitoses, and Cancer . . . 250
    Human Tumors 250
    Primary Tumors Induced in Mice 253
    Transplanted Animal Tumors 253
A Low Metaphase/Prophase Ratio and Cancer 254
Other Mitotic Irregularities and Cancer 256
Correlations Between Mitotic Aberrations and Cancer 257
Significance of Mitotic Aberrations in Cancer 258
Use of Mitotic Aberrations as a Tool in Cancer Diagnosis 260
Acknowledgements 260
Literature Cited 261

INTRODUCTION

Since the 1950s the chromosome constitution of an ever-increasing number of primary and transplantable cancers has been analyzed. Apart from certain tumors whose chromosome constitution does not seem to deviate from that of the normal tissue of its origin, most cancers have abnormal chromosome complements that show a wide range of variation, both between tumors and within the same tumor. Such cancers owe their growth mainly to one or more stemlines with a characteristic chromosome constitution, which, however, may change, especially in response to new environments.

Attempts to find a common chromosomal denominator for malignant neoplasias of a certain type have in general failed. The one exception is chronic myeloid leukemia, which is causally connected with a specific chromosome aberration. Similar claims for other types of cancer still seem to require confirmation.

The changes in chromosome number and structure in cancer cells have been dealt with in numerous reviews, also in the present volume. Much less attention has been paid to the processes leading to such aberrations. This is especially true of the last 15 years, during which treatments with colchicine and hypotonic solutions have become standard methods to study mammalian chromosomes. Unfortunately, slides prepared by means of such techniques yield very little information about the course of mitosis.

Although some of the mitotic aberrations in cancer were described as early as the end of the last century, most of the papers published before 1950 have now only historical interest (for earlier literature see, e.g., refs. [55, 15]). We have tried to review all the papers since that time that add to our knowledge of the mitotic processes in malignant cells. Most of such studies are based on fixed material. For instance, the determination of the relative duration of the mitotic phases—especially the metaphase/prophase ratio (M/P ratio), often referred to in the present paper—has been based on counts of these stages. The number of living cells that have been followed through mitosis is relatively small (for literature see, e.g., refs. [54, 32, 44]). Measurements of the stages in such cells have been of little help, probably because accurate delimitation of the phases in vivo, especially of the prophase, is almost impossible.

In the present article the mitotic aberrations that are the cause of the variation in the chromosome numbers in cancer are reviewed against the background of similar processes in nonmalignant cells. No single aberration has been found to be specific for cancer. However, in malignant tumors, abnormalities are not only greatly increased in frequency, but seem to occur in characteristic combinations.

MODIFICATIONS OF THE MITOTIC CYCLE AND ENDOPOLYPLOIDY

An interphase and the subsequent mitosis constitute a mitotic cycle. The essential feature of a normal cycle is regular alternation between chromosome reproduction and segregation of sister chromatids. This segregation is accomplished by means of the spindle that two centrosomes (centrioles) form between them (Fig. 1). A normal mitotic cycle is characteristic of embryonic and meristematic tissues as well as of stemlines in malignant tumors. However, chromosome reproduction and segregation may in different ways be out of step with each other [39, 41].

For instance, prophase may be shortened relative to the metaphase. In a cell population of this type, the metaphase/prophase ratio (M/P ratio) is increased relative to a control population. Such a situation has been observed in the last premeiotic mitoses of the dragonfly spermatogonia in which the prophase was clearly shorter than in earlier divisions [38]. In malignant growth this trend has been found to be widespread and is correlated with the formation of multipolar divisions (Fig. 1), discussed later in this paper.

An opposite phenomenon is the unusually long duration of the prophase. If a whole cell population is thus affected, the M/P ratio is smaller than in a control population. Such a tendency has been observed, for instance, in the young classes of secondary spermatogonia in the dragonflies [38] and in certain malignant tumors, to be discussed later.

The relationship of the spindle and the chromosomes may be disturbed to the extent that the spindle is aborted at some stage, or altogether absent, which results in the duplication of the chromosome complement —in other words, polyploidy. This type of polyploidy is known as endopolyploidy* and may be caused by different processes [50, 2, 8, 61].

Endoreduplication

Endoreduplication is the most common cause of endopolyploidy. It consists of—at least two—chromosome replications without an intervening mitosis (Fig. 1). This results each time in a doubling of the chromosome complement and in an increase in the volume of the nucleus which, if the process continues, may reach gigantic size (Fig. 2). Chromosome repro-

---

* The present authors do not wish to enter into a discussion on the terminology of the various processes leading to endopolyploidy (see ref. [3]). The terms are used here in the sense of Levan and Hauschka [19], since these have been established in the literature.

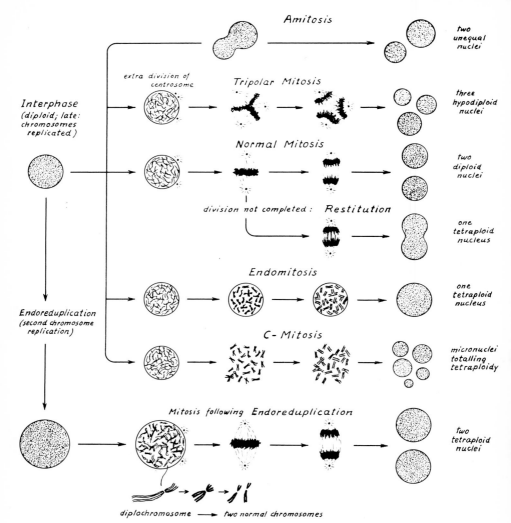

Figure 1. Normal mitosis and its modifications. *Normal mitosis*: chromosomes condense in prophase; after the nuclear membrane dissolves, two centrosomes form the spindle that separates the daughter chromosomes into two diploid nuclei; after an incomplete anaphase, a tetraploid *restitution* nucleus may be formed. *Endomitosis*: after normal prophase, chromosomes divide within the nuclear membrane and form a tetraploid nucleus. *C-mitosis*: after prophase, the nuclear membrane dissolves, but no spindle is formed; chromosomes scatter in the cell and form micronuclei with reduced chromosome numbers. *Endoreduplication*: chromosomes replicate a second time, which results in a tetraploid nucleus (with a diploid number of diplochromosomes); if mitosis follows, diplochromosomes separate into normal chromosomes that divide normally. *Tripolar mitosis:* three centrosomes form a spindle with three poles, which results in three nuclei with reduced chromosome numbers. *Amitosis* (pathological): nucleus is fragmented into two parts with unequal chromosome constitutions. (For the diagram we are grateful to Dr. Klaus Patau.)

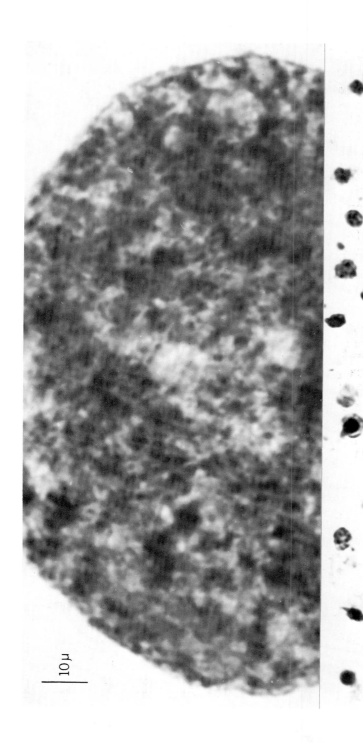

Figure 2. Section of a highly polyploid nucleus from a human squamous cell carcinoma of the cervix, compared with normal stroma nuclei (same magnification).

duction is thus all that is left of the mitotic cycle. If, as often happens, mitosis is restored after two or more chromosome replications, a tetraploid or more highly polyploid chromosome constitution is revealed (Fig. 1). In the first metaphase after two chromosome reproductions, two sister chromosomes lie side by side or form a diplochromosome in which an undivided centromere holds together four chromatids [50, 8, 61]. Even larger bundles of chromatids have been observed [22], especially in animal (Fig. 3) and plant tumors [19, 51]. Subsequent mitoses in such polyploid cells are normal.

In certain plants, of which the best known is the spinach [1, 24, 2], endoreduplication and normal polarized mitoses alternate, which leads to a stepwise increase in polyploidy. Endoreduplication is widespread in the differentiated tissues of both plants and animals [50, 2, 8], seeming in many cases to be correlated with the differentiation process itself [37]. It is common also in many malignant tumors, as will be discussed below.

Endomitosis

Endomitosis was the first of the processes that lead to endopolyploidy to be described and analyzed [2, 7, 8, 19]. Endomitosis takes place—often repeatedly—within the intact nucleus (Figs. 1 and 3); in the interphase, the chromosomes obviously replicate normally; the prophase also appears normal; in endometaphase, the two daughter chromatids are seen as separate bodies side by side—often more condensed than in normal metaphase; in endoanaphase, they separate; in endotelophase, despiralization takes place; and a nucleus with double the number of chromosomes is formed. The nuclear membrane does not dissolve; no spindle is formed, and endomitosis can thus be regarded as a highly abortive mitosis. In its characteristic form it is found in the differentiating glandular tissues of insects [7, 8] and in tapetum cells of plants [2, 3], but it occurs also in some malignant tumors [19, 40, 59; Fig. 3].

C-Mitosis

C-mitosis has obtained its name from the fact that it resembles a mitosis arrested with colchicine (Figs. 1 and 4f). It is more like normal mitosis than endomitosis, but it too is abortive, since no nuclear division takes place. A normal interphase precedes C-mitosis, and a normal prophase is followed by the disappearance of the nuclear membrane. However, no spindle is formed, and the overcontracted chromosomes lie scattered in the cell. Usually the chromosomes fall into daughter chromatids, which

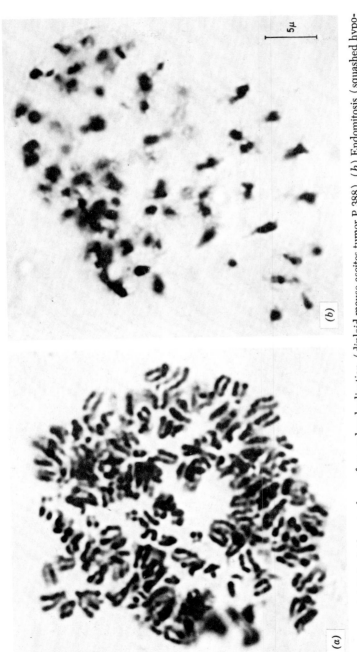

Figure 3. (a) Polyploid metaphase after endoreduplication (diploid mouse ascites tumor P 388). (b) Endomitosis (squashed hypotetraploid Ehrlich mouse ascites tumor; in the upper part the nuclear membrane is visible; the chromosomes divide within it).

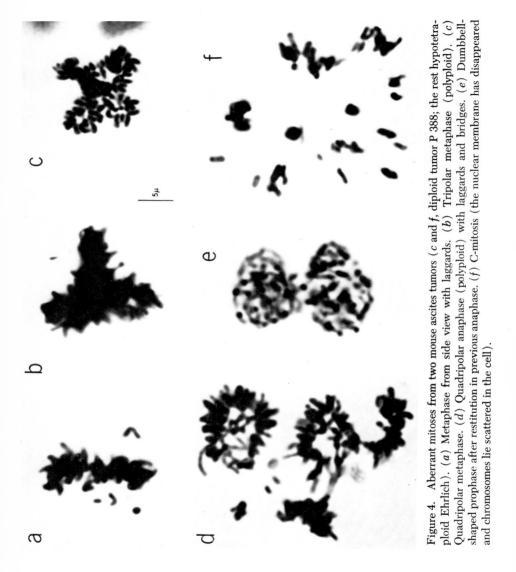

Figure 4. Aberrant mitoses from two mouse ascites tumors (c and f, diploid tumor P 388; the rest hypotetraploid Ehrlich). (a) Metaphase from side view with laggards. (b) Tripolar metaphase (polyploid). (c) Quadripolar metaphase. (d) Quadripolar anaphase (polyploid) with laggards and bridges. (e) Dumbbell-shaped prophase after restitution in previous anaphase. (f) C-mitosis (the nuclear membrane has disappeared and chromosomes lie scattered in the cell).

may be included in one nucleus with double the normal number of chromosomes, or in several micronuclei with variable numbers of chromosomes. C-mitosis is found most frequently in malignant tissues [60, 40]. In nonmalignant tissues its occurrence is rare and sporadic, although in three abnormal axolotl embryos, C-mitosis was reported to be the prevailing mode and to lead to polyploid nuclei [5].

Restitution

In typical restitution* the mitotic apparatus functions normally up to a certain stage. However, in metaphase or anaphase the chromosomes are united into one nucleus, with a resultant one-step increase in ploidy (Fig. 1). This process is clearly pathological and usually signals cell degeneration. Accordingly, it is common in malignant tissues (Fig. 4e).

ORIGIN AND STRUCTURE OF MULTIPOLAR DIVISIONS

The structure of a multipolar metaphase depends on a number of factors [53]. The centrosomes (or centrioles) seem to repel each other in that they take positions as far as possible from each other on the surface of the prophase nucleus. Hence, the number of the centrosomes and the shape of the nucleus mainly determine the structure of the metaphase configuration (Fig. 4b–4d). Each centrosome may form spindles in any direction. If two centrosomes are too far apart, a spindle will fail to form between them. In very large cells the centrosomes seem to lose touch altogether, and the chromosomes collect in unorganized groups.

Although multipolar divisions are practically absent in normal human [58, 55, 15] or mouse tissues [46], as well as in nonmalignant tissues in general [53], they seem to occur in mouse tissue culture [47]. The upset of timing affected both the chromosomes and the spindle apparatus in tissue cultures of the mouse and of two other rodents [47]. About half of the metaphases that displayed diplochromosomes were also tripolar.

Multipolar divisions can be induced by various treatments. Thus X-rays, especially at higher doses, were found to increase the number of multipolar mitoses [23]. Very low concentrations of colchicine and certain other substances had the same effect [17]. Multipolar divisions were found also in cells recovering from a stronger dose of colchicine [43, 18], and their frequency was increased by vinblastine, an alkaloid that has similar

---

* The term restitution has often been used in a wider sense to include all processes through which the chromosomes in a cell end up in one nucleus instead of two.

effects to those induced by colchicine [29]. Such effects of colchicine-type substances were found to be caused by their action on the centriole [49]. The presence of these drugs prevented the normal separation of centrioles during prophase. If this situation lasted long enough, the procentrioles had time to mature, and when the drug was removed, a multipolar spindle was formed [49].

Multipolar divisions probably arise mainly through additional divisions of the centrosomes and not through cell and nuclear fusions, which also are known to take place in cancer [12, 26, 35, 45, 62]. This is supported by the following observations [53, 54]. Tripolar, and not quadripolar divisions, were the first to appear and also the most frequent. In addition, other mitoses with an uneven number of poles were not uncommon. Tripolar spindles were found, not only in polyploid but also in small near-diploid cells, and their appearance and frequency were correlated with the M/P ratio. The frequencies of divisions with an increasing number of poles seemed to fit a decreasing geometric series.

There is, however, no doubt that cell and nuclear fusions are an additional source of multipolar divisions in cancer; daughter nuclei and cells after a multipolar division seem especially prone to fuse (e.g., refs. [12, 26, 32, 34, 35]). Fusions occur also between cells of two different mouse ascites tumors grown in the same host [45]. Mouse tumor cells have also been found to fuse with host cells [62]. Fusion in vitro by means of Sendai virus of cells from widely different organisms is now a well-established procedure. Multipolar divisions with high numbers of poles are also formed when nuclei fuse or divide synchronously in syncytia of human cells induced by measles virus [11]. The structure of multipolar divisions in such syncytia does not in principle differ from those found in cancer cells. Whether spontaneous fusions occur in primary cancer tissue, and with what frequency, is unknown.

The hypothesis has been repeatedly put forward that somatic segregation after cell fusion might come about through a multipolar division in which whole chromosome sets would regularly segregate [36, 42]. No real evidence for this hypothesis has been presented, and the whole idea appears improbable, since there is no evidence of any regulating mechanism in a tripolar or quadripolar division that could accomplish an orderly segregation (see ref. [11]). The mechanism of the preferential elimination of human chromosomes from, for instance, mouse/man hybrid cells is not known. This and possibly a more or less complete somatic segregation could come about through selection acting on cells with different chromosome constitutions, resulting from various mitotic irregularities such as loss of chromosomes in abnormal anaphases. One mitotic aberration is of special interest in this connection. In normal, but especially in abnormal or treated mammalian cells, an extended or interphase-like chromosome

(Fig. 5) has been observed occasionally in a metaphase plate (e.g., refs. [60, 63, 9, 31, 52]). Such a chromosome is obviously eliminated in the next anaphase, and this could be a mechanism for the loss of specific chromosomes.

AN INCREASED METAPHASE/PROPHASE RATIO, MULTIPOLAR MITOSES, AND CANCER

Human Tumors

Although mitotic abnormalities in cancer cells, especially multipolar mitoses, had been described by the end of the last century, the question of the order in which they appear and how they are related to each other was raised only recently. In 1950 Timonen and Therman [58] compared mitoses in normal endometrium and malignant tumors of the human female genital tract. In the normal tissues, the frequencies of metaphase and prophase were about equal. The first and most constant aberration found in practically all the malignant tissues was an increase in the number of metaphases relative to prophases. Of the total of 200 prophases, metaphases, and anaphases counted in each case, anaphases were not significantly different between normal and malignant tissues. This indicated that the relative average duration of the prophases and metaphases was changed in cancer cells. Many of these tumors were also characterized by the occurrence of multipolar divisions, which were absent in normal tissues. Other mitotic aberrations in cancer cells occurred more sporadically.

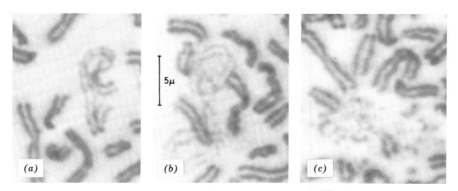

Figure 5. Part of three metaphase plates from hypotetraploid Ehrlich mouse ascites tumor treated with 1-methyl-2-benzylhydrazine. (a) Less extended, (b) more extended, (c) interphase-like chromosome.

These observations were confirmed in 10 samples of normal endometrium and 30 cases of cancer of the female genital tract [55]. In nonmalignant cells the M/P ratio was generally 1 or less. It was increased in all malignant tissues, and when it reached a certain level, the first multipolar divisions appeared (Fig. 6). This study was expanded to include 100 samples each of normal and malignant tissues [54]. The first tripolar divisions appeared when the M/P ratio reached 4 to 6, and their frequency was positively correlated with a rising ratio. The frequencies of higher multipolar mitoses fitted a decreasing geometric series (see also refs. [53, 25, 15]): $a, ak^{-1} ak^{-2}, \ldots, ak^{-n}$, in which a is the frequency of tripolar divisions, $n+3$ the number of poles, and $k<1$ a constant, characteristic for each tumor.

The correlation of M/P and the frequency of multipolar divisions was further demonstrated in tissue cultures of HeLa cells [13]. On the second day of the culture the M/P was 3.86 and about 40% of the divisions were multipolar, whereas on the eighth day the ratio was 1.64 and the percentage of multipolar divisions was down to 7.4. During the same time, the mitotic index in the periphery of the culture went down from almost 12% to about 1% (in other areas of the culture the change was similar, although somewhat less drastic).

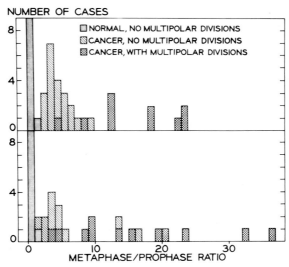

Figure 6. M/P ratio and the occurrence of multipolar divisions. Top: 10 cases of normal human endometrium and 30 cases of cancer of the female genital tract (data from ref. [55]). Bottom: 10 cases of fetal Fallopian tubes and 23 cases of cancer of the same organ (data from ref. [15]). The two samples are not significantly different either in regard to M/P values or to the occurrence of multipolar divisions.

These findings were borne out by a comparison of 10 human fetal Fallopian tubes and 23 cases of cancer of the same organ [15]. In the nonmalignant tissues the M/P varied between 0.57 and 1.07, in the malignant neoplasias between 1.7 and 36.6 (Fig. 6). Multipolar divisions occurred in more than half the cancer cases, whereas none was found in the fetal tissues. In general, a positive correlation between the cytological abnormalities and the malignancy of the tumor was obvious.

The M/P ratio seems usually to be more or less constant within a given tumor, but exceptions occur. Timonen and Lehto [57] provided data on 15 samples from five epidermoid carcinomas of the cervix. In four, there was no indication of heterogeneity, but in one there was quite significant variation from sample to sample (Table 1). Multanen [33] found that in laryngeal carcinoma, too, an increased M/P was positively correlated with malignancy as judged by histological findings, the presence of metastases, the recurrence rate, and a survival of less than 5 years.

Two other studies were concerned with primary human cancers. Iwanaga [14] cytologically analyzed 30 samples of fetal tissue (epithelium of stomach, small intestine) and some 50 malignant tumors, mainly of the digestive tract. The M/P ratio in the normal fetal tissues was less than 0.5 (this ratio seems in general to be lower in fetal than in adult tissues, see ref. [15]) and above 1.0 in the malignant tissues. Multipolar divisions and multinuclear cells—unfortunately, the author has pooled these— were frequent in all malignant samples, but absent in the normal tissues. In another study, comparing 10 cases of nonmalignant with 10 cases of malignant cervical tissue [28], the M/P ratio was also found to be higher in all the malignant samples. However, the claim was made that not only the malignant, but also all the nonmalignant, tissues showed multipolar divisions; this seems inexplicable, since no multipolar divisions have been

Table 1. Tests of Homogeneity within Cancers on the M/P Ratio in 15 Samples from 5 Cancers*

Cancer	N†	d.f.	χ^2	P
1	4×100	3	16.970	$P = 0.00072$
2	4×100	3	0.962	
3	3×100	2	0.241	$\chi^2_7 = 5.313$
4	2×100	1	1.170	$P = 0.62$
5	2×100	1	2.940	
	15×100	10	22.283	$P = 0.014$

* Data from Timonen and Lehto [57].
† Numbers of mitoses (prophase and metaphase only).

found in any untreated normal tissue biopsies by other investigators [58, 55, 14, 46, 15].

Two studies have disagreed with the findings of an increased and variable M/P ratio in malignant tissues as compared with a lower and relatively constant ratio in normal cells. However, in the first of the two papers [6] tissues from different animals were compared and no "early" prophases were included in the M/P determination. In the second study [4], part of the work was done on autopsy material, which is completely unusable for such an analysis. Neither paper contained illustrations to show the authors' concepts of the mitotic stages.

Primary Tumors Induced in Mice

An interesting experimental study on the development of mitotic aberrations during the change of a tissue from normal through dysplasia to cancer was done by Scarpelli and von Haam [46]. Mice were painted intravaginally with 1% 3,4-benzpyrene in acetone to induce cancer; inflammation was induced by using 4% croton oil in acetone. The mean values of M/P were 0.96 in the control, 1.42 in the inflamed tissue, 1.77 in dysplasia, 7.3 in carcinoma in situ, and 11.2 in invasive carcinoma (Fig. 7). No multipolar divisions were observed in the first three tissues, but in carcinoma in situ 29 out of 1000 mitoses were multipolar, and in the invasive carcinoma the corresponding number was 52/1000. Other mitotic aberrations were also greatly increased in carcinoma in situ and in invasive carcinoma.

Transplanted Animal Tumors

Transplantable cancers, particularly in ascites form, offer many advantages for the investigation of tumor development. A study of mitotic aberrations occurring during the development of the near-tetraploid Ehrlich ascites tumor in the mouse was made by Oksala [40]. The relative frequencies of the mitotic stages (excluding telophase) were compared on the third and seventh day after inoculation, and the frequencies of certain mitotic aberrations were determined.

During these four days the M/P ratio increased from 1.14 to 1.78. The combined frequencies of metaphases and anaphases rose 15%, whereas that of prophases decreased correspondingly by 22%. In principle this could be interpreted as depending either on a prolongation of the metaphase and anaphase stages or on a shortening of the prophase. Oksala has accepted the latter explanation as the more probable, since the

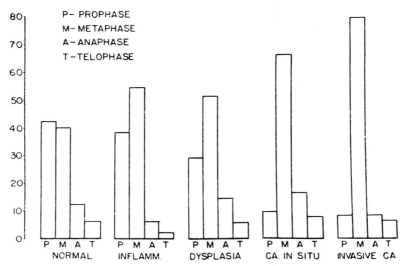

Figure 7. Percentages of mitotic stages in normal tissue and four types of cervical lesions in the mouse. Based on 1000 cells in mitosis (Chart from ref. [46]). (For the permission to publish this diagram the authors are grateful to Dr. D. G. Scarpelli, The University of Kansas Medical Center, and to *Cancer Research*.)

spindle could hardly be slowed down to such extent in a relatively young tumor. It is not impossible that the shortening of the prophase might be connected with the relatively fast growth rate of the Ehrlich tumor.

Of the various mitotic aberrations, multipolar mitoses were the most consistent; they were found in all the ascites samples, whereas endomitoses were practically absent. Whether the shortened prophase and the occurrence of multipolar divisions are correlated with a high mitotic index (in other words, a shortened interphase)—a situation found in tissue culture of HeLa cells [13]—is still an open question.

Maruyama [30] also determined the M/P ratio in the Ehrlich tumor and in four lines of rat tumors. Interestingly enough, he found no multipolar divisions in the Ehrlich cancer that had the lowest M/P ratio, but they did occur in the rat tumors, and their frequency showed a positive correlation with increasing M/P values.

A LOW METAPHASE/PROPHASE RATIO AND CANCER

Oksala first showed that not all malignant tumors are characterized by an increase in the M/P ratio [40]. The M/P ratio in the hyperdiploid Landschütz ascites tumor of the mouse decreased from 0.91 on the third day to 0.55 on the seventh in contrast to the ratio in the near-tetraploid Ehrlich ascites tumor. On the third day the ratio was thus nearly equal to that of the Ehrlich tumor (1.14). The day 3 ascites samples were not any more representative of the pure stemlines, since divergence in the development of the two tumors had already started. It can therefore be assumed that the M/P ratio in the stemline of each tumor—which consists of normally dividing cells and is mainly responsible for the growth of the tumor—was about 1.0. In the day 7 Landschütz sample, the combined frequency of metaphases and anaphases had decreased by 26% and the proportion of prophases had correspondingly increased 30%. The most probable explanation for this change is a prolongation of the prophase, since it can hardly be assumed that the metaphases and anaphases could be accelerated to such an extent when the tumor grows older (for detailed analysis of the duration of the mitotic stages, see [40]). The Landschütz tumor also appears to grow more slowly than the Ehrlich tumor, since under the same conditions the host mice with this tumor lived 20-21 days as compared to 14-15 days of the Ehrlich hosts.

Also, in contrast to the Ehrlich tumor, typical endomitoses constituted some 2% of all the scored stages on the third day, and 4.7% on the seventh day. Their frequency showed a positive correlation with that of the prophases. Another feature that distinguished this cancer from the Ehrlich tumor was that multipolar mitoses were practically absent.

Oksala predicted that a Landschütz-type human cancer might be found, if looked for, and indeed among some 300 tumors of the female genital tract one such neoplasm was discovered [59]. In this highly anaplastic cervical carcinoma, the M/P ratio was only 1.52. Nuclear growth seemed to take place mainly through true endomitosis, and multipolar divisions were absent. It is possible that among cancers with a relatively low M/P ratio, this type of tumor might not be too rare. However, they are more difficult to recognize than the opposite kind, since endomitotic stages are considerably more sensitive to destruction by unsuitable cytological technique than the multipolar divisions (see ref. [57]) and are also otherwise more difficult to detect.

Levan and Hauschka [19] described endomitosis and/or endoreduplication in some 15 transplantable mouse ascites tumors. They assumed, with some reservation, that the frequency of polyploid cells (relative to the stemline) reflected the rate of occurrence of these two processes. This does not seem justified to us, since it is unclear whether the polyploidization had taken place recently (as in one clone) or in the remote past. Unfortunately, the authors did not record other aspects of mitosis. We would have predicted that the tumors with a relatively high frequency of endomitoses would also have been characterized by a low M/P ratio and by the absence of multipolar divisions.

In most malignant tumors some metaphases display the characteristics of C-mitosis. However, their frequency varies in such an unpredictable fashion that this aberration is probably best interpreted as a secondary phenomenon, caused possibly by anoxia and necrosis.

OTHER MITOTIC IRREGULARITIES AND CANCER

In addition to the aberrations in mitosis discussed above, a number of others are also found more or less frequently in cancer (e.g., ref. [27]).

An almost total absence of "early" prophases has been reported in human cancers, for instance. All recognizable prophases in these cells seem to correspond in appearance to "late" prophases in normal cells [55, 15].

Lagging chromosomes [58] occur frequently and may be connected with an abnormal functioning of the spindle. These lie, individually or in groups, outside the metaphase plate (Fig. 4a), sometimes at the poles; laggards are also often seen between anaphase groups (Fig. 4d). Most lagging chromosomes form micronuclei and are probably always lost as a consequence.

Formation of ring-shaped metaphase plates, which give rise to doughnut-shaped interphase nuclei [58, 40], is another result of a spindle anomaly in cancer cells.

An increased "stickiness" of the chromosomes is common in cancer cells [53, 58, 27] and may lead to the formation of numerous bridges in anaphase (Fig. 4d), which in turn may result in a dumbbell-shaped restitution nucleus (Figs. 1 and 4e). As an example of the frequency of such aberrations, it may be mentioned that more than half of the anaphases in a tumor may show laggards and bridges [58, 55, 15]. The chromosomes in cancer cells often revert to interphase directly from metaphase or anaphase; they form one nucleus, which thus contains double the number of chromosomes.

Amitosis (Fig. 1), which formerly was regularly invoked in the interpretation of various cytological phenomena in cancer cells, has been shown to be practically nonexistent. One case in which this term might apply legitimately is mouse lymphosarcoma 6C3HED [20], in which some of the telophase nuclei break up into micronuclei, which subsequently are synchronous in their mitoses.

Individual chromosomes which seem to remain in interphase, while the rest of the chromosomes in the tumor cell are in metaphase, has been mentioned above.

"*Partial endoreduplication,*" which is a reverse timing aberration to the interphase-like chromosomes in metaphase, implies the double reduplication of a chromosome segment while the rest of the chromosome complement has undergone only one replication. This is perhaps of more theoretical than practical importance, although if such "partial endoreduplication" affected a whole chromosome, it would lead to trisomy (e.g., ref. [16]).

"*Heterochromatization,*" in some degree, is a common finding in tumor cells. In most malignant cells, in contrast to normal cells, the interphase nuclei show an increase in number and size of heterochromatic bodies. Indeed, exfoliative cytology as a method of diagnosis is in part based on this characteristic. Levan and Hauschka [19] mentioned that many of the mouse ascites tumors studied by them showed characteristic heterochromatin patterns. Heterochromatization of a sizable chromosome segment, as found in a mouse mammary carcinoma [21], has not yet been explained. This segment appeared negatively heteropycnotic in a way never seen in normal mouse chromosomes. The behavior of the affected chromosome was also otherwise highly erratic. It would be very interesting to know what different banding techniques would reveal about such a region. The physiological implications of this increased and differential heterochromatization in malignant cells is as yet completely unknown.

In addition to these recognizable and explicable mitotic irregularities, there is a confusing picture presented by an anaplastic cancer. Even apart from the presence of clearly degenerating cells, these tumors may display nuclei of bizarre shapes and sizes—such as doughnut, multilobed, or giant—as well as micronuclei, or several nuclei per cell. These dying cells and aberrant nuclei are an end result of mitotic irregularities, but lack of oxygen and other metabolic factors may also cause, or contribute to, cell death.

CORRELATIONS BETWEEN MITOTIC ABERRATIONS AND CANCER

Malignant tumors seem to fall into two main classes in regard to mitotic aberrations [40]. In the Ehrlich type, which is represented by the vast majority of cancers, the M/P ratio is increased. At a certain point in the development of such tumors, multipolar divisions appear, and the frequency of such divisions is positively correlated with the malignancy of the tumor, as measured by various criteria. In the much rarer Landschütz type of tumor, the M/P ratio is low or even decreased during tumor development. Instead of multipolar divisions, endomitotic stages are found. Some of the other mitotic irregularities may be correlated with these basic ones, but most of them tend to appear sporadically.

Obviously an increase in the M/P ratio reflects an increase in the duration of the metaphase or a decrease in that of the prophase, or possibly both. In his study of two mouse ascites tumors, Oksala [40], who compared the relative frequencies of prophases, metaphases, and anaphases, came to the conclusion that the increase of the M/P ratio in the Ehrlich tumor is caused mainly by a shortening of the prophase during tumor development and that the decrease of this ratio in the Landschütz tumor results from a lengthening of the prophase. As mentioned earlier, studies on living cells have so far not been able to solve the problem of which stage is shortened and which lengthened. However, accurate estimates of the duration of the mitotic phases might possibly be obtained by means of autoradiography [48].

Positive correlation of increased M/P values with the occurrence of multipolar divisions is beyond doubt. The correlation of a decreased M/P with the occurrence of endomitotic stages seems probable. The possibility that an increased M/P ratio and multipolar divisions on the one hand and a decreased M/P ratio and endomitosis on the other hand would be independently caused by a fundamental change in the cancer cell cannot definitely be excluded at this point. However, a couple of preliminary studies indicate that the change in the relative duration of the mitotic phases is the primary phenomenon, which in turn gives rise to either multipolar divisions or endomitoses. Lettré and Lettré [17] in a short note reported that colchicine and certain other substances, at concentrations sufficiently low so as not to destroy the spindle mechanism but only to slow it down, caused an increase in the relative frequency of metaphases and a great increase in the frequency of multipolar divisions in normal and malignant mammalian cells. An opposite effect was achieved in onion roots by treatment with 3'-deoxyadenosine, which caused the accumulation of prophases [10]. Such cells did not continue into metaphase but went through endomitosis inside the nuclear membrane.

SIGNIFICANCE OF MITOTIC ABERRATIONS IN CANCER

An interesting question is whether a change in the M/P ratio in incipient cancers appears before any changes occur in chromosome structure and/

or number. If this turns out to be the general rule, it would militate against the suggestion that chromosome mutations are generally a cause of cancer. This question could be studied by combining chromosome analysis (preferably by means of fluorescence microscopy or other banding techniques) with a study of mitotic changes in induced tumors [46], or perhaps in a variety of human cervical tissues, from precancerous lesions through carcinoma in situ to invasive carcinoma.

Although the relationship of mitotic aberrations to the origin of cancer remains unclear for the present, their role in the progression of a malignant tumor is obvious. It has long been known that primary cancers may, stepwise, become more malignant. Numerous articles, also in the present volume, deal with changes in the chromosome constitution of the stemline during the evolution of a tumor. Table 2 summarizes the various processes that add to or delete from the chromosome complement of a cell. These aberrations create a great variety of new cell types, and subsequent selection promotes those which divide most actively under the particular prevailing conditions.

In summary, it can be stated that no completely new cytological phenomena have been observed in cancer cells. All the mitotic aberrations described in malignant cells are also found in nonmalignant cells. Some, like endoreduplication, are widespread; others occur sporadically; and still others are extremely rare. Thus even highly anaplastic tumors are characterized on the one hand by greatly increased frequencies of ordinarily rare abnormalities, and on the other hand by specific combinations of aberrations.

Table 2. POSSIBLE MEANS OF CHANGING THE CHROMOSOME NUMBER IN CANCER CELLS

Loss of Chromosome or Chromosome Segment	Gain of Chromosome or Chromosome Segment
Lagging or nondisjunction of chromosomes	Nondisjunction of chromosomes
Failure of metaphase alignment	Failure of metaphase alignment
Multipolar divisions	Unbalanced segregation in chromatid translocations
Extended or interphase-like chromosomes in metaphase	Endoreduplication
Break of chromosome and loss of fragment	Endomitosis
Unbalanced segregation in chromatid translocations	C-mitosis
	Restitution from metaphase or anaphase
	Failure of cell division after mitosis
	Cell fusion

USE OF MITOTIC ABERRATIONS AS A TOOL IN CANCER DIAGNOSIS

Scarpelli and von Haam [46] demonstrated in induced mouse tumors that when the tissue changed from dysplasia to carcinoma in situ, the M/P ratio jumped from 1.77 to 7.3. Human neoplasias have generally not displayed such a clearcut gap between the values in malignant and nonmalignant tissues. In spite of this, the M/P ratio has proved valuable in the diagnosis of neoplasia of the female genital tract [56]. Especially in conditions considered borderline in respect to malignancy—such as leukoplakia, carcinoma in situ, adenomatous hyperplasia, and adenoma malignum—the M/P ratio has turned out to be a good indicator of the subsequent course of the disease. (The borderline value of the M/P ratio between malignant and nonmalignant tissue was about 1.5.) Indeed, it was through this method that the conclusion was reached that carcinoma in situ represents the initial stage of cervical cancer.

An example of how the M/P ratio could have overcome the difficulties encountered in diagnosis is a case of adenoma malignum [56]. A biopsy specimen had been examined by Finnish, Swedish, and American pathologists, and the diagnoses were contradictory. The patient preferred to believe her condition to be nonmalignant and refused treatment. An operation was performed later, but the carcinoma by then was inoperable, and the patient died after a few months. A determination of the M/P ratio was made afterward from the original slides and turned out to be 2.4. The tumor had been malignant from the start.

As stressed above, multipolar divisions are nearly absent in untreated nonmalignant tissue. The occurrence of tripolar mitoses, not to mention divisions with more poles, is, therefore, a convincing sign of malignancy.

As should be clear from the present article, the cytological picture in cancer is often very confusing, displaying a combination of mitotic aberrations and the resulting abnormal nuclei, and the degeneration and death of cells as a result of anoxia and necrosis. Successful differentiation between relevant and irrelevant phenomena presupposes suitable cytological techniques (see ref. [57]), such as Feulgen squashes in the case of solid tumors, and above all a clear understanding of normal and abnormal mitotic processes. If careful training in this critical area of cytology is not made available to our diagnosticians in pathology, the considerable advances made in the research laboratories cannot be effectively applied where they are most meaningful—in diagnosing borderline cases and saving lives.

Acknowledgements.–The work has been supported by a National Institutes of Health grant GM-15422, a grant from the Brittingham Trust, and American Cancer Society Institutional grant IN-35. This is paper No. 1580 from the Genetics Laboratory.

LITERATURE CITED

1. BERGER CA: Reinvestigation of polysomaty in Spinacia. *Bot Gaz* 102:759–769, 1941.
2. D'AMATO F: Polyploidy in the differentiation and function of tissues and cells in plants. A critical examination of the literature. *Caryologia* 4:311–358, 1952.
3. D'AMATO F: A brief discussion on "endomitosis." *Caryologia* 6:341–344, 1954.
4. DAVID H: Der Prophasenindex menschlicher Tumoren. *Arch Geschwulstforsch* 13:258–263, 1958.
5. FANKHAUSER G, HUMPHREY RR: The rare occurrence of mitosis without spindle apparatus ("colchicine mitosis") producing endopolyploidy in embryos of the axolotl. *Proc Nat Acad Sci (US)* 38:1073–1082, 1952.
6. FARDON JC, PRINCE JE: A comparison of the ratios of metaphase to prophase in normal and neoplastic tissues. *Cancer Res* 12:793–795, 1952.
7. GEITLER L: Die Entstehung der polyploiden Somakerne der Heteropteren durch Chromosomenteilung ohne Kernteilung. *Chromosoma (Berlin)* 1:1–22, 1939.
8. GEITLER L: Endomitose und endomitotische Polyploidisierung. *Protoplasmatologia* 6C:1–89, 1953.
9. GERMAN J: Chromosomal breakage syndromes. *Birth Defects: Original Article Ser* 5(5):117–131, 1969.
10. GIMÉNEZ-MARTÍN G, GONZÁLES-FERNÁNDEZ A, DE LA TORRE C, FERNÁNDEZ-GÓMEZ ME: Partial initiation of endomitosis by 3'-deoxyadenosine. *Chromosoma (Berlin)* 33:361–371, 1971.
11. HENEEN WK, NICHOLS WW, LEVAN A, NORRBY E: Polykaryocytosis and mitosis in a human cell line after treatment with measles virus. *Hereditas* 64:53–84, 1970.
12. HIRONO I: Some observations on the mitosis of living malignant tumor cells. *Acta Pathol Jap* 1:40–47, 1951.
13. HSU TC: Cytological studies on HeLa, a strain of human cervical carcinoma. I. Observations on mitosis and chromosomes. *Tex Rep Biol Med* 12:833–846, 1954.
14. IWANAGA K: Morphological comparison of mitosis in human fetal cell and human cancer cell (in Japanese). *J Nagasaki Med Soc* 30:1078–1095, 1955.
15. LEHTO L: Cytology of the human Fallopian tube. *Acta Obstetr Gynecol Scand* 42, Suppl 4:1–95, 1963.
16. LEJEUNE J, BERGER R, RETHORÉ M-O: Sur l'endoréduplication sélective de certains segments du génome. *CR Acad Sci Paris* 263:1880–1882, 1966.
17. LETTRÉ H, LETTRÉ R: Einige Beobachtungen über die experimentelle Erzeugung multipolarer Mitosen. *Z Krebsforsch* 60:1–8, 1954.
18. LEVAN A: Colchicine-induced C-mitosis in two mouse ascites tumours. *Hereditas* 40:1–64, 1954.
19. LEVAN A, HAUSCHKA TS: Endomitotic reduplication mechanisms in ascites tumors of the mouse. *J Nat Cancer Inst* 14:1–43, 1953.
20. LEVAN A, HAUSCHKA TS: Nuclear fragmentation—a normal feature of the mitotic cycle of lymphosarcoma cells. *Hereditas* 39:137–148, 1953.
21. LEVAN A, HSU TC: The chromosomes of two cell strains from mammary carcinomas of the mouse. *Hereditas* 46:231–240, 1960.

22. LEVAN A, HSU TC: Repeated endoreduplication in a mouse cell. *Hereditas* 47: 69–71, 1961.
23. LEVIS AG: Effetti dei raggi X sulla mitosi di cellule di mammiferi coltivate in vitro. *Caryologia* 15:59–87, 1962.
24. LORZ A: Cytological investigations on five chenopodiaceous genera with special emphasis on chromosome morphology and somatic doubling in Spinacia. *Cytologia* 8:241–276, 1937.
25. MAKINO S, KANÔ K: Cytological observations on cancer. II. Daily observations on the mitotic frequency and the variation of the chromosome number in tumor cells of the Yoshida sarcoma through a transplant generation. *J Fac Sci Hokkaido Univ* 10:225–242, 1951.
26. MAKINO S, NAKAHARA H: Cytological studies of tumors, X. Further observations on the living tumor cells with a new hanging-drop method. *Cytologia* 18:128–132, 1953.
27. MAKINO S, YOSIDA TH: Cytological studies on cancer. I. Morphological and statistical observations on the abnormal mitosis in tumor cells of the Yoshida sarcoma through a transplant generation. *J Fac Sci Hokkaido Univ* 10:209–224, 1951.
28. MANNA GK: The relative frequencies of different mitotic stages with some of their abnormalities in non-neoplastic and neoplastic human cervix uteri. *Proc Zool Soc* 15:1–10, 1962.
29. MARTIN GM, SPRAGUE CA: Vinblastine induces multipolar mitoses in tetraploid human cells. *Exp Cell Res* 63:466–467, 1970.
30. MARUYAMA Y: Cytological studies of neoplasm. III. Prophase index and its relation to mitotic abnormalities in experimental ascites tumors. *Med J Shinshu Univ* 4:311–319, 1959.
31. MOORHEAD PS: Virus effects on host chromosomes. In Genetic Concepts and Neoplasia. Baltimore, Williams & Wilkins Company, 281–306, 1970.
32. MOORHEAD PS, HSU TC: Cytologic studies of HeLa, a strain of human cervical carcinoma. III. Durations and characteristics of the mitotic phases. *J Nat Cancer Inst* 16:1047–1066, 1956.
33. MULTANEN I: Histo-cytological malignancy and clinical picture in epidermoid cancer of the larynx. *Acta Oto-Laryngol* Suppl 135:1–63, 1958.
34. NAKAHARA H: Cytological studies of tumors. XI. Observations of the multipolar division in tumor cells of ascites tumor of rats by phase microscopy. *J Fac Sci Hokkaido Univ* 11:473–480, 1953.
35. OFTEBRO R, WOLF I: Mitosis of bi- and multinucleate HeLa cells. *Exp Cell Res* 48:39–52, 1967.
36. OHNO S: Cytologic and genetic evidence of somatic segregation in mammals, birds and fishes. *In Vitro* 2:46–60, 1966.
37. OKSALA T: Über Tetraploidie der Binde- und Fettgewebe bei den Odonaten. *Hereditas* 25:132–144, 1939.
38. OKSALA T: Zytologische Studien an Odonaten. II. Die Entstehung der meiotischen Präkozität. *Ann Acad Sci Fenn* IV, A, 5:1–33, 1944.
39. OKSALA T: Timing relationships in mitosis and meiosis. *Caryologia* 6, Suppl: 272–281, 1954.

40. OKSALA T: The mitotic mechanism of two mouse ascites tumours. *Hereditas* 42:161–188, 1956.
41. PATAU K, DAS NK: The relation of DNA synthesis and mitosis in tobacco pith tissue cultured in vitro. *Chromosoma (Berlin)* 11:553–572, 1961.
42. PERA F, SCHWARZACHER HG: Die Verteilung der Chromosomen auf die Tochterzellkerne multipolarer Mitosen in euploiden Gewebekulturen von Microtus agrestis. *Chromosoma (Berlin)* 26:337–354, 1969.
43. PETERS JJ: A cytological study of mitosis in the cornea of Triturus viridescens during recovery after colchicine treatment. *J Exp Zool* 103:33–60, 1946.
44. RICHART RM, LERCH V, BARRON BA: A time-lapse cinematographic study in vitro of mitosis in normal human cervical epithelium, dysplasia, and carcinoma in situ. *J Nat Cancer Inst* 39:571–577, 1967.
45. RUTISHAUSER A, HAEMMERLI G, STRÄULI P: Cytogenetik transplantabler tierischer und menschlicher Tumoren. *Neujahrsblatt Naturforsch Ges Zürich* 107:1–85, 1963.
46. SCARPELLI DG, VON HAAM E: A study of mitosis in cervical epithelium during experimental inflammation and carcinogenesis. *Cancer Res* 17:880–884, 1957.
47. SCHMID W: Multipolar spindles after endoreduplication. *Exp Cell Res* 42:201–204, 1966.
48. SPARVOLI E, GAY H, KAUFMANN BP: Duration of the mitotic cycle in Haplopappus gracilis. *Caryologia* 19:65–71, 1966.
49. STUBBLEFIELD E: Centriole replication in a mammalian cell. *In* The Proliferation and Spread of Neoplastic Cells. Baltimore, Williams & Wilkins Company, 175–189, 1968.
50. THERMAN E: The effect of indole-3-acetic acid on resting plant nuclei. I. Allium cepa. *Ann Acad Sci Fenn* A, IV, 16:1–40, 1951.
51. THERMAN E: Dedifferentiation and differentiation of cells in crown gall of Vicia faba. *Caryologia* 8:325–348, 1956.
52. THERMAN E: Chromosome breakage by 1-methyl-2-benzylhydrazine in mouse cancer cells. *Cancer Res* 32:1133–1136, 1972.
53. THERMAN E, TIMONEN S: Multipolar spindles in human cancer cells. *Hereditas* 36:393–405, 1950.
54. THERMAN E, TIMONEN S: The prophase index and the occurrence of multipolar divisions in human cancer cells. *Hereditas* 40:313–324, 1954.
55. TIMONEN S: Mitosis in normal endometrium and genital cancer. *Acta Obstetr Gynecol Scand* 31, Suppl 2:1–88, 1950.
56. TIMONEN S: Prophase index in the diagnosis of gynecological cancer. *Ann Chir Gynaecol Fenn* 44:222–233, 1955.
57. TIMONEN S, LEHTO L: Method of determining the ratio of metaphases to prophases in normal and malignant cells. *Ann Chir Gynaecol Fenn* 44:99–107, 1955.
58. TIMONEN S, THERMAN E: The changes in the mitotic mechanism of human cancer cells. *Cancer Res* 10:431–439, 1950.
59. TIMONEN S, THERMAN E: Endomitotic nuclear growth in a human cervical carcinoma. *Ann Chir Gynaecol Fenn* 45:237–244, 1956.

60. TJIO JH, LEVAN A: Chromosome analysis of three hyperdiploid ascites tumours of the mouse. *K Fysiogr Sällsk Handl (Lund)* NF, 65:1–38, 1954.
61. TSCHERMAK-WOESS E: Karyologische Pflanzenanatomie. *Protoplasma* 46:798–834, 1956.
62. WIENER F, FENYÖ EM, KLEIN G, HARRIS H: Fusion of tumour cells with host cells. *Nature New Biol* 238:155–159, 1972.
63. ZUR HAUSEN H: Chromosomal changes of similar nature in seven established cell lines derived from the peripheral blood of patients with leukemia. *J Nat Cancer Inst* 38:683–696, 1967.

CANCER AS A CLONE

CHROMOSOME CHANGES AND THE CLONAL EVOLUTION OF CANCER

PETER C. NOWELL

Department of Pathology, School of Medicine, University of Pennsylvania,
Philadelphia, Pennsylvania

Introduction 268
Temporal Correlation between Malignancy and Appearance of Clones with
    Abnormal Chromosome Complements 268
    Solid Tumors 269
    Leukemias 270
Relationship between Chromosome Changes and Metabolic Alterations in
    Malignancy 274
Nature and Stability of Chromosome Changes 274
    Chromosome Changes in Natural Progression of Neoplasia 274
        Animal Tumors 276
        Human Tumors 279
    Effect of Therapy on Karyotype 279
    Reversibility of Chromosome Changes 280
Significance of Chromosome Changes to the Concept of Clonal Evolution
    in Neoplasia 280
    The Unicellular Origin of Cancer 280
    Tumor Progression as a Microevolutionary Phenomenon 281
    Tumor Progression Is Irreversible 282
    The End Stage: Individual Therapy for Each Tumor? 283
Summary 283
Acknowledgements 284
Literature Cited 284

INTRODUCTION

As information continues to accumulate on the chromosome changes in cancer, and new "banding" techniques are applied, a number of generalizations that were first stated several years ago apparently remain valid: (1) Most established neoplasms do demonstrate chromosome changes, although it is equally clear that visible cytogenetic alteration is not *required* for the neoplastic state. (2) Specific types of tumors generally do not have characteristic chromosome abnormalities, although a degree of nonrandomness with respect to cytogenetic change does exist in some human and animal neoplasms. (3) Chromosome alterations that are observed in individual tumors usually indicate a clonal pattern of growth. The neoplastic cells may all have the same chromosome change or may show related changes, suggesting, in both instances, origin of the tumor from a single aberrant cell. Even in tumors having considerable variation in chromosome number among the neoplastic cells (the case with most human solid malignancies) the underlying clonal nature of the neoplastic growth is frequently indicated by the presence of characteristic abnormal marker chromosomes.

It is the purpose of this brief chapter to focus on tumor chromosomes as related to the clonal aspect of neoplastic growth, with particular emphasis on the time when cytogenetically identifiable stemlines first appear during tumor development and on the nature and stability of the chromosome alterations as tumors progress. From this point of view I will consider the significance of stemline karyotype changes, both to the natural history of tumors and to our thinking concerning therapeutic approaches. I will draw on personal investigations with experimental and human tumors, as well as on cytogenetic observations made by others concerning the clonal derivation of tumors. However, it is not planned to review the subject exhaustively, nor is any claim made for the originality of the theoretical concepts advanced.

TEMPORAL CORRELATION BETWEEN MALIGNANCY AND APPEARANCE OF CLONES WITH ABNORMAL CHROMOSOME COMPLEMENTS

Most tumors, by the time they reach macroscopic size and demonstrate malignant characteristics, have cytogenetic alterations in the neoplastic cells, and these changes generally indicate a clonal growth pattern. Attempts have been made to determine for both solid tumors and leukemias the stage in the development of a neoplasm when these stemline cytogenetic patterns commonly first appear. Although malignancy is more easily defined for solid tumors in classical terms of invasion and metastasis, more data have been obtained on neoplastic and preneoplastic states in the hematopoietic system because many proliferating cells are readily accessible for chromosome study. Some of the observations on both solid and hemic neoplasms will be reviewed here.

Solid Tumors

It has been difficult to obtain adequate numbers of dividing cells for cytogenetic investigations from normal or "premalignant" solid tissues by direct techniques. Culture methods provide sufficient material but are always subject to questions concerning sampling and selection, particularly in attempts to identify cell clones existing in vivo.

In general, stemlines with chromosome abnormalities appear to be extremely rare in non-neoplastic solid tissues. In the few instances where they have been described, specific etiologic mechanisms such as a genetic disorder (e.g., xeroderma pigmentosum) or ionizing radiation have been incriminated [1]. As such non-neoplastic clones are more common in the hematopoietic system, they will be discussed in more detail in the next section.

Premalignant and *preinvasive* lesions of human solid tissues have been most extensively investigated in the uterine cervix. The data indicate that considerable karyotypic variation from cell to cell is usually present, both in dysplastic lesions and carcinoma in situ, and occasionally evidence for definite clones, as indicated by marker chromosomes, can also be obtained [2, 3]. Preinvasive human and experimental tumors of the bladder, bowel, and skin have also revealed chromosome changes indicating clonal growth in some instances, as have a number of benign human meningiomas [3–5]. Only in the very earliest stages of tumor induction have stemline patterns seemed consistently absent. In experimental studies by Stich [6] on the rodent liver and thymus, cells with a spectrum of random chromosome changes were observed during the pre-neoplastic period shortly after exposure to chemical carcinogens; cytogenetic patterns indicating definite cell clones were demonstrable only at later stages of tumor development.

Considerable cytogenetic information is available on those solid tumors that are frankly *malignant* and on chromosome preparations that have been obtained not only from the primary tumors themselves, but also from malignant effusions and metastases. Cytogenetic data on human solid malignancies indicate that an underlying stemline pattern of chromosome change is discernible in nearly every case [3, 7, 8], although the presence of a wide variety of random alterations will frequently blur this impression when one looks at rapidly growing tumors or malignant effusions. The presence of one or more marker chromosomes common to most of the cells in a given tumor usually strongly suggests a clonal growth pattern, and much of the observed cytogenetic variability is apparently related to secondary stemlines derived from the initial neoplastic clone. The chromosome picture is generally similar in experimental solid tumors, whether the malignancies are primary or transplanted. Under certain experimental conditions, it has been possible to obtain some tumors in which the individual cells show almost no karyotypic variation around the single predominant abnormal stemline [8–10].

Thus, chromosome changes indicative of clonal patterns of tumor growth have been demonstrated throughout solid tumor development from the earliest preinvasive lesions to the end stages of frank malignancy; however, definite cytogenetically defined stemlines have generally been less common in early lesions. It has also been shown that neoplasms can exist, in all stages of development, without any demonstrable chromosome abnormalities. Absence of all abnormality is least common in the frankly malignant tumors, particularly in human cancers; however, under certain experimental conditions of induction or transplantation, definitely invasive and metastasizing neoplasms with an apparently normal diploid karyotype can be produced. This is true, for instance, of several primary rodent tumors induced by RNA viruses, and also of certain chemically induced transplantable rat hepatomas selected for their slow growth rate [8–11]. In all tumor studies, a constant potential source of error is in distinguishing between non-neoplastic cells and neoplastic cells with a normal karyotype. With most transplantable animal tumors this can be overcome by passage into the opposite sex and use of sex chromosome differences as markers. However, if these diploid neoplasms can be followed long enough, chromosome abnormalities usually do appear, and these cytogenetic changes indicate a clonal pattern of tumor growth as in other malignancies (see Table 2).

Leukemias

Considerable data have been compiled on the time relationships in the hematopoietic system between appearance of chromosome abnormalities and the evolution of a neoplastic state. As already noted, ready access to many dividing cells has made it relatively easy to obtain cytogenetic information on hemic cells. This advantage, however, has been partially outweighed by the greater difficulty in defining "malignancy," and even "neoplasia," since cells metastasize physiologically in the hematopoietic system and excessive proliferation of primitive elements is not restricted to neoplastic conditions.

In both man and animals it has been possible in the hematopoietic tissues, much more readily than in other tissues, to demonstrate clones of cells with karyotypic changes in conditions that are clearly *not* neoplastic. In humans, such findings have generally been associated with previous exposure to ionizing radiation or with genetic disorder [1, 12]. Ford and others [13, 14] have shown that in mice and rats exposed to sublethal doses of ionizing radiation and then allowed to recover, large clones of cells with radiation-induced chromosome abnormalities frequently populate their hematopoietic tissues and appear to function normally. In fact, such radiation-induced chromosome markers have been used by many investigators as a means of identifying proliferating and differentiating lymphoid and hematopoietic cells at various sites in the body of the original host or transplant recipients [13, 14]. Apparently these large non-neoplastic clones are able to evolve in the sublethally irradiated rodent because repopulation of the entire hematopoietic system can occur from very few hemic stem cells. Radiation-altered liver cells, on the other hand, do not demonstrate similar clonal proliferation unless they have acquired a significant selective (neoplastic) growth advantage over adjacent liver cells [15].

It has been pointed out that the chromosome changes observed in these functionally normal hemic clones are of the "balanced" type. In such clones, all centromeres are preserved, and the chromosome changes presumably involve reciprocal translocations with conservation of essentially all genetic material [13, 16]. Conversely, in the *neoplastic* clones discussed throughout this chapter, the chromosome rearrangements involve both structurally abnormal chromosomes and changes in chromosome number that do represent significant gain or loss of genetic material from the cell. It may be that such major genetic change is incompatible with successful proliferation and function of somatic cells unless it confers a significant growth advantage (e.g., initiates neoplasia) or occurs in an already neoplastic cell [13, 16].

As expected, then, *non-neoplastic* aneuploid clones appear to be quite rare in the hematopoietic tissues of both man and animals, although there is one major exception: the 45,X chromosome complement. The 45,X clones, apparently spontaneously developed, have been observed among both peripheral blood lymphocytes and marrow cells of a number of elderly humans. In many somatic tissues, the loss of the second X from a female cell or of the Y from a male cell seems to confer no significant selective disadvantage. Such clones have, in some cases, been related to significant marrow dysfunction, including perhaps neoplasia [3]; in other instances there has been no indication of associated hemic disorder [17], but actually evidence for normal immunological function of 45,X cells [18]. To date, marrow clones with other types of chromosome change have not been detected in normal humans without a history of irradiation or of blood disorder; but the number of marrow studies on such persons has been relatively small. (It is assumed, and has usually been demonstrated, that the 45,X clones and other variants discussed here are somatic aberrations arising in the hematopoietic tissues and are not examples of constitutional mosaicism of gametic or zygotic origin.)

In the vaguely defined *premalignant* group of hemic disorders, marrow clones with chromosome abnormalities are not rare. These clones are usually somewhat easier to identify than those associated with early solid neoplasms, because the range of karyotypic variation around the predominant stemline is generally much less among the hemic cells than in pre-invasive lesions of the cervix and of other solid tissues. However, the stages in neoplastic development are difficult to compare in the two circumstances; what hematopoietic disorder does one consider comparable to cervical dysplasia or carcinoma in situ?

In any event, a number of studies, including some of our own, have indicated that abnormal stemlines can frequently be found in the bone marrow of individuals suffering from blood disorders which clinical experience indicates are clearly premalignant but which cannot yet be considered frank leukemia [3, 19]. In Table 1 are some recently published

Table 1. Relationship Between Marrow Chromosome Findings and Clinical Course in "Preleukemia"*

Chromosome Findings	Original Diagnosis:	Total Patients	Dead with Leukemia	
			1–3 months	3+ months
Abnormal marrow karyotype	Myeloproliferative Syndrome:			
	Polycythemia Vera	4†	0	0
	Other	7	4	1
	Pancytopenia	5	4	0
	Total			9/16
Normal marrow karyotype	Myeloproliferative Syndrome:			
	Polycythemia Vera	5	0	0
	Other	10	0	2
	Pancytopenia	11	1	1
	Miscellaneous	9	0	0
	Total			4/35

* For further details concerning these data, refer to ref. [19].
† Includes 3 patients treated with ^{32}P.

observations on a series of 51 patients considered "preleukemic," primarily on the basis of a myeloproliferative disorder or idiopathic pancytopenia [19]. Only 4 of 35 patients in this series who had a normal marrow karyotype (all of whom had been followed for at least a year) developed leukemia, whereas 9 of the 16 patients with a cytogenetically abnormal marrow clone proceeded promptly to a frank clinical leukemia. It appears that those in the latter group had, in fact, at the time of study, a leukemic clone proliferating in the marrow which was not yet sufficiently large to permit the clinical diagnosis. However, it is also clear that a few preleukemic patients with grossly abnormal clones may remain hematologically static for years, without ever progressing to a clearly defined leukemia [19]. This may be particularly true of individuals with polycythemia vera, whether they are exposed to therapeutic radiation or not [3, 20; P. C. Nowell, unpublished].

In the circumstances where there is a definite *malignant* process (leukemia) of one type or another, there are extensive data on the nature of the cytogenetic changes observed [3, 8]. When present, chromosome abnormalities in the acute leukemias clearly indicate a clonal pattern of tumor growth and usually show much less karyotypic variation around the basic stemline than in solid neoplasms. The same is probably true of chronic lymphocytic leukemia, although dividing neoplastic cells for study have been difficult to obtain in this disorder. The solid lymphomas also show stemline patterns, perhaps with fewer cells showing random chromosome aberrations than are seen in solid tumors of other organs.

Chronic granulocytic leukemia is well recognized as a special situation. This is the one neoplasm consistently characterized by the same chromosome abnormality in nearly every case. The stemline bearing the Philadelphia chromosome appears to be present from the very earliest stage at which the disease can be recognized clinically or hematologically [3, 8].

In all of the leukemias, even more commonly than in solid tumors, frank malignancy can develop without any visible chromosome change. This is indicated in Table I and is particularly true of the acute leukemias, where more than half of the cases may have an apparently normal diploid karyotype [3].

Thus, both in the solid neoplasms and in the leukemias, cytogenetic changes suggesting a clonal pattern of growth can be observed at all stages of the neoplastic process. They are probably more common in the later stages, or at least more readily identified, but the difficulty in clearly defining what constitutes a neoplasm, particularly in the hematopoietic system, makes even this simple statement difficult to substantiate. Therefore, some attempts have been made to relate chromosome changes to specific aspects of malignancy other than the classical morphological criteria, and these will be mentioned before considering in more detail the nature and possible significance of cytogenetic abnormalities as related to tumor progression.

RELATIONSHIP BETWEEN CHROMOSOME CHANGES AND METABOLIC ALTERATIONS IN MALIGNANCY

Attempts to correlate specific biochemical alterations with karyotype have generally been unsuccessful in malignant cells. Enzyme patterns have, for instance, been extensively studied in the series of transplantable Morris hepatomas; but to date no particular metabolic alteration has been associated with any of the various cytogenetic changes described in these neoplasms [10]. Similar results have been obtained with primary hepatomas in the rat [21]. The Philadelphia chromosome and reduced leukocyte alkaline phosphatase are consistent findings in the neoplastic granulocytes of chronic granulocytic leukemia, but even this well-established association is not present in some patients [3, 8]. Similarly, it has not been possible to correlate the nature of specific immunological products of neoplastic cells to characteristic alterations in karyotype. No consistent relationship has been demonstrated between specific immunoglobulins produced in various monoclonal gammopathies and particular chromosome patterns, either in our own laboratory (unpublished), or by others [3, 22]. Attempts to relate the degree of chromosome abnormality to the various biological characteristics grouped together under the general heading of "tumor progression" have been somewhat more successful. This phenomenon will be considered in the next section.

NATURE AND STABILITY OF CHROMOSOME CHANGES

Chromosome Changes in Natural Progression of Neoplasia

The discussion thus far has at least implied that the cytogenetically abnormal cell clones observed late in the course of neoplastic disease tend to deviate further from normal than those found in earlier lesions. This has been the general conclusion of most studies carried out since modern techniques for chromosome study have become available. Quite early, this led Levan, Makino, and others [23–25] to postulate a concept of tumor progression correlated with sequential alterations in chromosome morphology. Figure 1, from a 1961 paper by Hauschka [25], illustrates this formulation. The legend explains variations postulated to occur in tumor phenotype; many of these would be expected to have associated changes in karyotype.

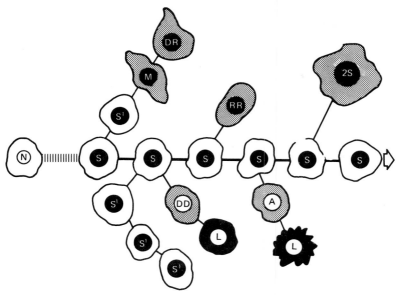

Figure 1. Diagram illustrating tumor progression through stemline variability. The nuclear symbols represent the following cell types: N, nonmalignant cell; S, main stemline; S^1, secondary stemlines; 2S, polyploid stemcell variant; M, metastatic cell; DR, drug-resistant mutant; RR, cell with increased radiation resistance; DD, drug-dependent mutant; A, antigenic mutant; L, mutations lethal for individual tumor cells (reproduced from Hauschka, 1961 [25]).

Refinements and modifications have been added subsequently, but the basic hypothesis still appears valid today. As tumors progress, losing their specialized functions and improving their capacity for growth and invasion, their chromosome patterns tend to diverge farther and farther from normal. Where it has been possible to follow this process sequentially in a given tumor, later cytogenetic changes frequently are clearly superimposed on preceding ones.

At present, as has already been discussed, no particular chromosome or type of rearrangement (other than the Ph^1) appears to be uniquely involved in this process of tumor progression. Both structural abnormalities and changes in chromosome number are observed. As the newer banding techniques are more generally applied, the importance of a number of other apparently nonrandom aberrations described in various human and animal tumors may be better understood.

The general trend in chromosome number during tumor progression is usually toward an increase, and frequently the first demonstrable alteration may be the gain of a single chromosome [11, 16]. Ohno has discussed the deleterious effects of monosomy and the potential advantages to the tumor cell of chromosome duplication, particularly with respect to both lethal and useful recessive mutations [16]. These concepts may help to explain why tumors with a hypodiploid DNA content are uncommon, and why many mammalian solid malignancies may show change from a hyperdiploid karyotype to one that is near-triploid or near-tetraploid in later progression [11, 16].

It is not known whether the same mechanism(s) that operate to initiate and propel the biological progression of neoplastic cells are also important in the production of sequential chromosome changes, but at least the two phenomena clearly proceed in parallel. It has been suggested that an early step in carcinogenesis might be the activation of a genetic locus for nondisjunction or other mitotic irregularity [8]. One or more such active loci could be responsible for repeated genetic changes during tumor development, which would explain both the frequency of new chromosome rearrangements, with opportunity for selection of more variant stemlines, and the progressive functional deviation of the neoplastic cells.

Animal Tumors. Another approach to the relationship between tumor progression and chromosome change has been to consider possible mechanisms of action of different carcinogens, particularly in animal tumors, with the thought that different classes of agents might have variable effects on both the initial nature and subsequent stability of the alterations in karyotype. Highly cytotoxic and mutagenic chemical carcinogens (and perhaps ionizing radiation as well) appear to produce a spectrum of chromosome patterns in the damaged tissue, permitting selection of one, or a few, stemlines of cells that best fit the particular local and systemic conditions necessary for their neoplastic growth [6]. Once established, these stemlines could remain relatively stable, particularly in experimental systems in which tumors are purposefully selected for slow growth rate and differentiated characteristics (e.g., certain of the Morris hepatomas). However, even a very low rate of continuing genetic change would ultimately lead to a more deviant karyotype and associated tumor progression, if the usual selective growth pressures were allowed to operate.

A somewhat different sequence of events can be visualized for animal tumors initiated by oncogenic RNA viruses. As already noted, these agents may cause little initial chromosome damage but can produce definite malignancies that reach macroscopic size without any visible chromosome alterations, as in the Rous sarcoma systems [8, 9, 11]. However, both RNA and DNA oncogenic viruses may persist in the tumor cell, in one guise or another, and retain their capacity to interact with the genome, thus providing a possible mechanism for subsequent chromosome rearrangements during later stages of tumor development [1, 26]. Continuing mutagenic action by the initiating agent might also be expected in neoplasms induced by long-lived radioisotopes deposited in the tissues.

Whatever the operative mechanisms, it does appear generally true that the more extensively a malignancy deviates from normal in its various biological and metabolic characteristics, the less stable its chromosome pattern, and the more likely one is to observe random chromosome changes in every cell generation and the continuing evolution of more aberrant clones. This would explain why solid cancers (which also generally require greater functional deviation from normal than do the leukemias in order to express their malignant capacities) tend to show less stable and more deviant karyotypic patterns.

A few specific examples will illustrate these generalizations concerning the nature and relative stability of chromosome changes in animal neoplasms at different stages of development. Both Mark and Mitelman have made serial studies of a number of primary sarcomas induced in rodents by the RNA-containing Rous virus [9, 11]. Initially, many of these sarcomas had a normal karyotype, but they developed individual and diverging stemlines roughly paralleling tumor progression. We have been able to follow the evolution of chromosome changes through several transplant generations in a number of rat hepatomas chemically induced by Morris [10]. These data are summarized in Table 2. Three tumors, two diploid and one aneuploid, maintained their characteristic karyotypes over the course of several transplant generations (7 to 30 months). Eight other tumors, five diploid and two hyperdiploid when first examined, progressed over the same general time span to increasingly deviant karyotypes, generally with higher chromosome numbers. In most instances, karyotype changes were accompanied by a significant increase in tumor growth rate, despite a conscious effort in this series to retard tumor progression by selection for the slowest-growing tumors in each generation. This approach probably accounts for the relatively high degree of both phenotypic and karyotypic stability in these solid tumors. This view is supported by the chromosome findings in Tumor $9618A_2$, an exceptional subline that was isolated because of a sudden change to an unusually rapid growth rate; the subline demonstrated marked karyotypic and histological deviation in its second transplant generation as compared to the normal diploid karyotype of the well-differentiated, slow-growing parental strain.

A final example of stability of chromosome change in an animal tumor deserves special mention. Canine venereal sarcoma is a malignancy of dogs occurring worldwide. Chromosome studies of this tumor from widely separated geographical locations have revealed highly aneuploid, but very

Table 2. Stability and Progression of Chromosome Changes in Several Transplantable Rat Hepatomas (Morris)

Tumor	Transplant Generations Studied	Time span* (months)	Chromosome Number†		Karyotype	
			1st Study	2nd Study	1st Study	2nd Study
9098	5 and 8	7	42	42	Normal diploid	Normal diploid
7800	22 and 26	10	42	42	Normal diploid	Normal diploid
9618A	2 and 7	45	42	42	Normal diploid	Pseudodiploid
9618A₂‡	2	‡		41		Hypodiploid with rearrangements
9108	7 and 9	9	42	42, 43	Normal diploid	Hyperdiploid clone
9618B	2 and 3	13	42	42, 43	Normal diploid	Hyperdiploid clone
66	2 and 4	12	42	42, 43	Normal diploid	Hyperdiploid clone with rearrangements
9121	7 and 26	28	42	43	Normal diploid	Hyperdiploid with rearrangements
16	2 and 6	12	42	84	Normal diploid	Tetraploid
6	2 and 11	30	44	44	Hyperdiploid	Hyperdiploid (no change)
39A	2 and 10	30	46, 47	47	Hyperdiploid (related clones)	Hyperdiploid (same 47 clone)
38B	2 and 23	39	43	48	Hyperdiploid	Hyperdiploid with rearrangements

† Chromosome number of predominant clone(s).
‡ Rapidly growing subline of 9618A, which was analyzed only once.
* Time between first and second study for each tumor line.

similar, karyotypes [16, 27; also, Makino, this volume]. The cytogenetic data suggest that this is a spontaneously transplantable malignancy requiring no human intervention and that many generations of spontaneous passage have resulted in a tumor that is currently quite stable karyotypically.

Human Tumors. Sequential studies of chromosomes in human solid cancers are scarce because repeated biopsies have rarely been obtained on the same tumor. Marker chromosomes have provided evidence for the same stemline in metastases and malignant effusions as in the primary tumor, but considerable cytogenetic variability and instability has been the rule in these neoplasms, which have usually been studied late in their natural course [1, 3].

In contrast, the abnormal chromosome patterns observed in the human leukemias have usually appeared to be quite stable. Several acute leukemias have been reported in which the same abnormal karyotype was present both at the time of original diagnosis and in a subsequent exacerbation following a period of remission [3, 8]. The most extensive data, however, are in chronic granulocytic leukemia, where the Ph^1 characteristically persists throughout the course of the disease. However, in the terminal stage, when the neoplastic cells suddenly demonstrate more malignant characteristics, an additional chromosome change is often present. The new neoplastic clone, containing both the Ph^1 and an additional alteration that varies from case to case, partially or completely replaces the more benign stemline (Ph^1 only) from which it derived [3, 8, 16].

Effect of Therapy on Karyotype

Few data are available on the effect of therapy on the nature or stability of chromosome changes in human malignancies. In one study of a solid tumor and in several cases of acute leukemia, the same stemline was observed both before and after a remission induced by therapy [3]. In this connection, it is important to recognize that the absence of demonstrable changes during a remission in leukemia is not considered to indicate a reversion of the neoplastic cells to a normal karyotype, but simply the reduction of the neoplastic clone to a level that makes it undetectable in the hematopoietic tissues by standard cytogenetic procedures.

If rapid clinical progression of a neoplastic process (e.g., the terminal phase of chronic granulocytic leukemia) is the result of further genetic change in the tumor cells, as the additional chromosome alterations would suggest, several forms of therapy may pose a significant hazard. Ionizing radiation and many chemotherapeutic agents are mutagenic, and it is certainly possible that progression from a relatively benign process such as chronic leukemia or polycythemia vera to a more malignant one could result from additional genetic changes induced by these agents and indicated in some cases by further alterations in karyotype.

Reversibility of Chromosome Changes

The considerations in the previous section raise the question of whether the chromosome alterations characterizing cell clones in neoplasia might ever be reversible. The necessity for distinguishing between a cytogenetic reversion within individual cells and a decrease in clone size within a total cell population that temporarily obscures the presence of an abnormal clone, has already been noted. There have been occasional reports suggesting that clones of transformed cells in tissue culture, including some with demonstrable chromosome abnormalities, could revert to normal function, and even to a normal chromosome complement, under certain conditions; but it is difficult to rule out problems of sampling and cell selection [28]. The possibility that such a phenomenon might occur in vivo seems even more unlikely. Reverse mutation is a very rare event and, more important, the tumor clones under consideration have, by definition, been selected for their neoplastic characteristics. As will be discussed further below, even with changing conditions in vivo, continued selection for increasingly malignant variants would be expected, rather than selection for cells that had reverted to a normal karyotype.

SIGNIFICANCE OF CHROMOSOME CHANGES TO THE CONCEPT OF CLONAL EVOLUTION IN NEOPLASIA

The various observations that have been discussed concerning patterns of chromosome change related to tumor cell clones may provide some useful insights into the fundamental nature of neoplasia and the problems associated with its treatment. The findings suggest at least four general conclusions concerning the neoplastic state: (1) cancer is usually unicellular in origin, (2) tumor progression is a microevolutionary process, (3) this process is irreversible and is individual to each tumor, and (4) approaches to therapy must therefore take both individuality and irreversibility into account.

The Unicellular Origin of Cancer

The fact that most individual neoplasms consist of cells with a common chromosome abnormality or a limited number of stemlines, frequently with related cytogenetic changes, clearly suggests that each tumor has been derived from a single abnormal cell. This conclusion does not rule out the possibility that application of carcinogen to an organ or to an area of skin can produce many potentially neoplastic cells; it simply indicates that as a tumor evolves from such an area, the progeny of one or at most a very few cells ultimately overgrow the site and account for the macroscopic neoplasm. It further indicates that most tumors do not have the capacity to induce a neoplastic change in adjacent normal cells, unless one makes the unlikely assumption that such induction also involves production of an identical chromosome change.

This concept of a unicellular origin for most neoplasms has been given increasing attention in recent years and is supported not only by cytogenetic findings such as those cited here, but also by phenotypic evidence, enzyme data [29], and observations on the myeloma proteins in particular [30]. It should be emphasized, however, that this formulation does not require that the gross chromosome changes observed in the fully developed tumor were also present in the original neoplastic cell. As has already been discussed, many neoplasms are initiated without visible karyotype alterations; the chromosome changes indicating clonal growth which are observed later simply support the concept of an *evolving* population, stretching backward toward a single cell of origin.

Tumor Progression as a Microevolutionary Phenomenon

The cytogenetic data also fit the second conclusion, that the natural history of tumor progression is a microevolutionary process with the continuing production of variant cells. Among these variants is an occasional mutant with a growth advantage over the parental line, permitting it to overgrow not only the normal cells in the area but the parental tumor line as well. Gross chromosome changes may appear early or late in this process, but once present they may demonstrate the sequential nature of genetic alterations taking place. Prior to the appearance of karyotype changes, evolutionary stages in the tumor cell population may be difficult to distinguish from one another and, in some cases, from adjacent normal cells.

Evidence supporting this concept is primarily indirect; in most instances only "snapshots" of the postulated evolutionary process at various stages have been available, revealing cell clones with chromosome alterations of increasing degrees of deviation from normal. Occasionally, it has been possible to follow the course of an individual tumor and to note the development of sequential cytogenetic changes associated with the progression of the tumor to more malignant biological and metabolic characteristics. Data on the Rous sarcoma systems, on the Morris hepatomas in rats, and on chronic granulocytic leukemia in man have already been cited.

It can be argued that this demonstrated evolution of chromosome changes in various neoplasms is unrelated to their malignant progression, even though the two phenomena are proceeding in parallel. This argument that visible genetic alterations in tumors as observed at the chromosome level are simply irrelevant "noise" in the system cannot be dismissed out of hand, since it has not been possible to relate specific cytogenetic changes to particular characteristics of malignancy, metabolic or otherwise. However, it seems unnecessarily tortuous reasoning to dismiss these major genetic rearrangements as unrelated to the acquisition of heritable selective growth advantage which permits the tumor cells to evolve in a clonal fashion.

Tumor Progression Is Irreversible

Stemline chromosome alterations speak strongly to the irreversibility of the neoplastic state as it is usually observed clinically. If the clonal pattern we find in most fully developed tumors is the end stage of a multistep evolutionary process, in which mutants have been repeatedly selected for their increasingly efficient capacity to function as malignant cells, the possibility of restoring this population to normal seems extremely remote. Even if by some trick of genetic engineering one could restore a significant number of the tumor cells to a normal karyotype, any remaining aberrant cells, retaining their selective advantages, would once again overgrow the normal elements and reestablish the malignancy.

Rather than suggest restoration of a normal karyotype, one can consider the alternative possibility that proper manipulation of the internal milieu might induce even highly aneuploid neoplastic cells to differentiate and function normally, since presumably most if not all of the normal genetic information remains within the aneuploid genome. However, this seems a much more probable occurrence in tumors that retain a diploid chromosome complement (e.g., many human acute leukemias and certain experimental neoplasms induced by RNA viruses) than in aneuploid tumors with major cytogenetic rearrangements. In the few circumstances in which tumor cell populations have apparently changed to a more differentiated state with some loss of neoplastic characteristics (e.g., human neuroblastoma, mouse teratoma, plant tumors), the cells involved have generally been diploid, or the chromosome complement has not been defined [28]. Perhaps the best-documented exception involves the transplantation of aneuploid nuclei from a frog carcinoma into enucleated frog ova [31]. In this very special environment, some degree of development toward a recognizable tadpole was occasionally observed. Such reports have suggested to some workers the possibility that many aneuploid malignancies might be stimulated to revert to normal differentiated function [28]. However, most evidence suggests little likelihood that in vivo conditions can, in fact, be provided to a growing tumor which would cause an aneuploid cell population, clonally selected for neoplastic growth characteristics, to undergo such a developmental course.

The End Stage: Individual Therapy for Each Tumor?

All of the foregoing considerations suggest that in considering the therapy of human cancer one should recognize that most tumors as seen clinically are in the later stages of an extensive, highly individual evolutionary process. Not only is the population of cells with which the physician must deal clonal in nature, but also the genetic complement of the predominant clone differs from case to case. The few diploid neoplasms, such as some acute leukemias, may be exceptional and at least theoretically amenable to a common therapeutic agent or even to functional reversal. The majority of tumors, however, are aneuploid, and the physician must think in terms of eradication rather than reversal. Furthermore, the genetic individuality of each neoplasm, and probable associated metabolic individuality, make eradication of a large group of tumors unlikely by any single agent. Perhaps immunotherapy, specific for each patient, will have to be much more actively considered in conjunction with the more general modalities of surgery, radiation, and chemotherapy. Working against the therapist, however, will be the very processes that have produced the aneuploid clones of cells that are observed. Karyotypic instability and selective growth pressures will continue to operate in nearly every tumor, tending to generate a resistant clone for every therapeutic agent devised.

SUMMARY

(1) Most tumors, by the time they reach macroscopic size, have chromosome abnormalities that indicate a clonal pattern of neoplastic growth. A few neoplasms are exceptional in that karyotype alterations may not be observed.

(2) The time of appearance of cell clones that are identifiable cytogenetically may be individually variable, but they have been observed at all stages of tumor development.

(3) Chromosome abnormalities usually progress in parallel with other neoplastic characteristics, evolving toward more deviant karyotypes as the tumors become more malignant.

(4) These observations suggest (a) that most neoplasms are unicellular in origin; (b) that they develop through a microevolutionary process, with sequential selection of more neoplastic variants and more aneuploid karyotypes; and (c) that the fully developed human malignancy is therefore usually seen in an irreversible and individualized stage that may require individually designed therapy for systemic eradication.

Acknowledgements.–Original studies supported by U.S. Public Health Service Grant CA-10320.

LITERATURE CITED

1. GERMAN J: Genes which increase chromosomal instability in somatic cells and predispose to cancer. *In* Progress in Medical Genetics (Steinberg AG, Bearn AG, eds) vol 8, New York, Grune & Stratton, 61–101, 1972.
2. AUERSPERG N, COREY MJ, WORTH A: Chromosomes in preinvasive lesions of the human uterine cervix. *Cancer Res* 27:1394–1401, 1967.
3. SANDBERG AA, HOSSFELD DK: Chromosomal abnormalities in human neoplasia. *Ann Rev Med* 21:379–408, 1970.
4. MARK J: Chromosomal patterns in human meningiomas. *Eur J Cancer* 6:489–498, 1970.
5. MCMICHAEL H, WAGNER JE, NOWELL PC, HUNGERFORD DA: Chromosome studies of virus-induced rabbit papillomas and derived primary carcinomas. *J Nat Cancer Inst* 31:1197–1215, 1963.
6. STICH HF: Chromosomes of tumor cells. I. Murine leukemias induced by one or two injections of 7, 12-dimethylbenz(a)anthracene. *J Nat Cancer Inst* 25:649–661, 1960.
7. MAKINO S, SASAKI MS, TONOMURA A: Cytological studies of tumors. XI. Chromosome studies in fifty-two human tumors. *J Nat Cancer Inst* 32:741–777, 1964.
8. NOWELL P: Chromosome changes in primary tumors. *In* Progress in Experimental Tumor Research (Homberger F, ed), vol 7, New York, Basel and Karger, 83–103, 1965.
9. MARK J: Rous sarcomas in mice: the chromosomal progression in primary tumours. *Eur J Cancer* 5:307–315, 1969.
10. NOWELL P, MORRIS HP, POTTER VR: Chromosomes of "minimal deviation" hepatomas and some other transplantable rat tumors. *Cancer Res* 27:1565–1579, 1967.
11. MITELMAN F: The chromosomes of fifty primary Rous rat sarcomas. *Hereditas* 69:155–186, 1971.
12. NOWELL P: Biological significance of induced human chromosome aberrations. *Fed Proc* 28:1797–1803, 1969.
13. FORD CE: The use of chromosome markers. *In* Tissue Grafting and Radiation (Micklem HS, Loutit JF, eds), New York, Academic Press, 197–208, 1966.

14. NOWELL P, HIRSCH B, FOX D, WILSON DB: Evidence for the existence of multipotential lympho-hematopoietic stem cells in the adult rat. *J Cell Physiol* 75:151–158, 1970.
15. NOWELL P, CRAIG D, MATTHEWS F, COLE L: Chromosome abnormalities in liver and marrow of mice irradiated with fast neutrons, gamma rays, and X-rays: effect of dose rate. *Radiat Res* 24:108–118, 1965.
16. OHNO S: Genetic implication of karyological instability of malignant somatic cells. *Physiol Rev* 51:496–526, 1971.
17. O'RIORDAN ML, BERRY E, TOUGH I: Chromosome studies on bone marrow from a male control population. *Brit J Haematol* 19:83–90, 1970.
18. NOWELL P: Unstable chromosome changes in tuberculin-stimulated leukocyte cultures from irradiated patients. Evidence for immunologically committed, long-lived lymphocytes in human blood. *Blood* 26:798–804, 1965.
19. NOWELL P: Marrow chromosome studies in "preleukemia," further correlation with clinical course. *Cancer* 28:513–518, 1971.
20. MILLARD RE, LAWLER SD, KAY H, CAMERON C: Further observations of patients with a chromosomal abnormality associated with polycythaemia vera. *Brit J Haematol* 14:363–374, 1968.
21. HORI SH, SASAKI M: Glucose 6-phosphate dehydrogenase isoenzyme patterns and chromosomes in primary liver tumors of the rat. *Cancer Res* 29:880–891, 1969.
22. HOUSTON EW, RITZMAN SE, LEVIN WC: Chromosomal aberrations common to three types of monoclonal gammopathies. *Blood* 29:214–232, 1967.
23. LEVAN A: Chromosomes in cancer tissue. *Ann NY Acad Sci* 63:774–789, 1956.
24. MAKINO S: The concept of stemline cells as progenitors of a neoplastic population. *In* Proceedings of the International Genetic Symposia, Tokyo and Kyoto, 1956, 177–181, 1957.
25. HAUSCHKA TS: The chromosomes in ontogeny and oncogeny. *Cancer Res* 21:957–974, 1961.
26. MOORHEAD PS: Virus effects on host chromosomes. *In* Genetic Concepts and Neoplasia, Baltimore, Williams & Wilkins Company, 282–306, 1970.
27. WEBER W, NOWELL P, HARE D: Chromosome studies of a transplanted and a primary canine venereal sarcoma. *J Nat Cancer Inst* 35:537–547, 1965.
28. BRAUN AC: The Cancer Problem. New York, Columbia University Press, 1969.
29. LINDER D, GARTLER SM: Glucose-6-phosphate dehydrogenase mosaicism: utilization as a cell marker in the study of leiomyomas. *Science* 150:67–69, 1965.
30. MILSTEIN CB, FRANGIONI B, PINK J: Studies on the variability of immunoglobulin sequence. *Cold Spring Harbor Symp Quant Biol* 32:31–36, 1967.
31. DIBERARDINO M, KING T: Transplantation of nuclei from the frog renal adenocarcinoma. I. Chromosomal and histologic analysis of tumor nuclear-transplant embryos. *Develop Biol* 11:217–242, 1965.

CLONAL EVOLUTION IN THE MYELOID LEUKEMIAS

JEAN DE GROUCHY and CATHERINE TURLEAU

Clinique de Génétique Médicale and Unité de Recherche INSERM,
Hôpital des Enfants Malades, Paris, France

Introduction 287

Models for Clonal Evolution in Leukemia 289
   Model 1. Duplication of the Ph1 Chromosome 289
   Model 2. Acquisition of a Supernumerary Chromosome 293
   Model 3. Acquisition and Duplication of Supernumerary Chromosomes . . 293
   Model 4. Loss of Chromosomes 293
   Model 5. Loss and Acquisition of Chromosomes 295
   Model 6. Structural Rearrangements 295
      The Ph1 Marker 295
      The i(17q) Marker 295
   Model 7. Polyploidization 300
   Frequency with which the Various Models Are Observed 300

Possible Mechanisms and Significance of Clonal Evolution 301

"Laws" Governing Clonal Evolution 302

The General Concept of the Clonal Evolution of Malignancy 303

Literature Cited 305

INTRODUCTION

Clonal evolution is frequently observed in neoplastic processes. It is recognized most easily in leukemias and more particularly during blastic transformation of chronic myelogenous leukemia (CML). That it also occurs in other forms of neoplasia is likely. Yet it may be difficult to recognize as such due to the comparatively late stage of evolution at which observations are made, as for instance in solid tumors.

Clonal evolution is defined as rearrangements of the karyotype that occur step by step in an apparently orderly fashion. The process is comparable to the phylogenic evolution of a species. In both instances, under favorable conditions, all the intermediate steps can be observed (i.e., the fossils in the case of a phylum), while under other conditions only the final step can be identified. In the case of a clonal evolution in which many cell populations are present, one type may predominate, corresponding to the "stem-line" defined by Makino [70].

For some time it has been known that changes in chromosome numbers take place in leukemic cells. At first this seemed to be a disorderly process, until in 1963 Lejeune and his colleagues [62] demonstrated, in a case of leukoblastosis in an individual with trisomy 21, a systematic evolution of the karyotype by successive acquisitions and duplications of supernumerary chromosomes. The patient was a newborn girl exhibiting immediately after birth an erythroblastic crisis, followed by a leukoblastic crisis that regressed spontaneously. A picture of acute leukoblastosis then developed progressively, and the baby died at 2½ years of typical leukemia. A skin biopsy showed trisomy for chromosome No. 21. Three months before death, and prior to any treatment, two consecutive karyotypic analyses of blood cells were performed. They revealed numerical rearrangements compatible with the following sequence of events: acquisition of a group G chromosome, duplication of the extra G, acquisition of a D, duplication of the extra D, etc. The complete sequence is shown in Table 1. Progressive acquisition of abnormalities and duplication of the supernumerary chromosomes were the two outstanding features of this particular chromo-

Table 1. Clonal Evolution in a Case of Leukoblastosis in a Trisomic 21 Patient*

Number of Chromosomes	Karyotype
47	tri 21
48	tri 21 + G
49	tri 21 + 2G
50	tri 21 + 2G + 1D
51	tri 21 + 2G + 2D
52	tri 21 + 2G + 2D + 1C
53	tri 21 + 2G + 2D + 2C
54	tri 21 + 2G + 2D + 2C + 1F
55	tri 21 + 2G + 2D + 2C + 2F
62	tri 25 + 2G + 2D + 2C + 2F + ⋯

* First example of clonal evolution reported by Lejeune et al. [62].

some evolution. Other comparable types of evolution have since occasionally been reported in trisomy 21 [9, 11, 85].

This first observation as well as further observations by Ford and Clarke [22] have established the concept of clonal evolution in human leukemias. This is now well documented by a rather impressive number of reports. In a majority of cases, sequential changes precede and herald the onset of blastic transformation in chronic myeloid leukemia.

MODELS FOR CLONAL EVOLUTION IN LEUKEMIA

As first suggested by Grouchy and Nava [35, 40], clonal evolution appears to proceed according to certain patterns or models involving numerical changes, structural rearrangements, or a combination of both. However, as more cases have been analyzed, it has become apparent that several of these particular evolutionary models can be associated with specific patterns of progressive clinical symptoms. Future studies must therefore attempt to characterize completely such models and to discover still unrevealed clinical features associated with each.

This chapter is therefore aimed toward individually analyzing known models for such evolution in leukemia and describing particular features of the clinical picture for each.

Model 1. Duplication of the Ph^1 Chromosome

The Philadelphia chromosome, or Ph^1, is discussed elsewhere in this book ([72], see also model 6, here). It was described for the first time in 1960 by Nowell and Hungerford [76]. Ph^1 is a marker chromosome, resulting from a structural rearrangement, that is, partial deletion of the long arm of the group G chromosome, No. 22, as shown by Caspersson et al. [7], using the fluorescing technique. The deletion is a postzygotic event, as demonstrated by the existence of monozygotic twins, only one of whom is leukemic [13, 29, 51, 59]. The Ph^1 has also been shown to be present before any clinical symptoms of the disease [54, 77]. Being a postzygotic event, Ph^1 occurs in a single tissue, probably in a single cell, and can become established only if it occurs in hematopoietic cells [84]; a skin biopsy obtained from a leukemic patient, for instance, does not reveal the presence of the marker. Its presence in all three types of marrow cells—erythrocytic, granulocytic, and megakaryocytic—is now well documented by a series of clinical and experimental observations [25, 105]. Conversely, this fact has been considered by several authors as proof of a single-cell origin for these three cell types [90, 98–100, 105, 106]. Others, however, have severely questioned the validity of this concept [92]. The specificity of the Ph^1 chromosome seems well documented. With few exceptions, most patients with CML show the Ph^1, and it is present only in CML or in closely related myeloproliferative syndromes [44, 47, 48]. However, apparently true cases of CML are known to exist without the Ph^1. Their relative frequency varies considerably from one laboratory to another—about 26% in a recent report on 107 patients by Sandberg et al. [90]. In our own experience, the prevalence of CML patients without a Ph^1 does not exceed 2 to 3%. The reasons for this discrepancy are not clear. It could be due, in part, to different diagnostic criteria [32]. Whatever their true frequency, cases of CML without the Ph^1 do raise difficult questions. The answers to these questions could, in the future, prove important to the understanding of the etiology of the disease. Lastly, the presence of the Ph^1 has an immense diagnostic as well as prognostic value. It is always present in bone marrow. It may disappear in blood during remission, where its reappearance foreshadows a new relapse.

Whether one considers the Ph^1 chromosome itself as the first step of clonal evolution is a purely academic matter. In any case, it is perhaps the simplest model, and the first to be discussed here will be the duplication of the Ph^1. Duplication of Ph^1 was first directly related to the onset of blastic transformation [20, 56, 88, 89], but further observations have shown that Ph^1 may be duplicated during remission long before transformation. In fact cytogenetic and clinical observations from the literature and from our own laboratory are heterogeneous. In reviewing the literature, Nava et al. [73] pointed out that these investigations could be grouped as follows:

(1) In the first group, duplication of the Ph^1 is the only chromosome change. It may occur during blastic transformation or be present during remission as long as one or two years before blastic crisis [12, 19, 45, 46, 55, 75, 94].

(2) In the second group, duplication and even triplication of the Ph^1 is only one of several chromosome changes. Clones of 48 to 52 chromosomes can be observed. Duplication of the Ph^1 may be the first event, but usually its chronology in the clonal evolution is not known [4, 21, 52, 73, 75, 82, 86, 92, 96, 97]. Figure 1 shows a karyotype of a patient with duplication of the Ph^1 and acquisition of one group F and one group C chromosome.

(3) Lastly, duplication of the Ph^1 may be isolated (as in group 1) or not (as in group 2). In this group, the clinical picture is remarkable: it is characterized by considerable adenopathies, blastic infiltration of the organs, or reticulosarcomatous tumors. These tumors may precede or even take the place of the blastic crisis, and death usually occurs most

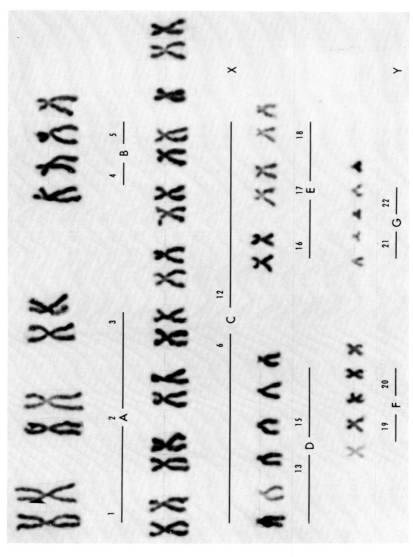

Figure 1. Duplication of Ph¹ and acquisition of a chromosome of group C and one of group F in patient with CML.

rapidly after their appearance. They have been considered as representing "extra-medullar blastic crisis" rather than true tumors [8, 14, 30, 58, 68, 73, 75]. Figure 2 shows a typical example in which duplication of the Ph¹ is a part of a relatively complex clonal evolution in a patient having both CML and a reticulosarcomatous mass, probably of blastic origin [73]. Such observations suggest an association between a particular model

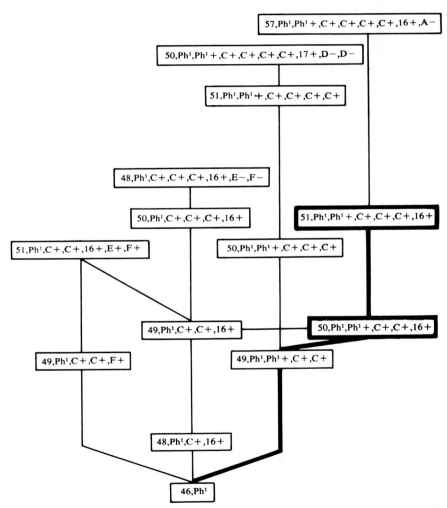

Figure 2. Duplication of Ph¹ as part of complex clonal evolution in patient with CML and a reticulosarcomatous mass [72].

of clonal evolution and the capacity of blastic cells to undergo intra- or extramedullary invasion.

Model 2. Acquisition of a Supernumerary Chromosome

Acquisition of an extra chromosome, most often belonging to group C, is a simple model of a clonal evolution. It is a relatively frequent mode of evolution in CML, but the abnormality can also be observed in acute leukemia [10, 49, 50, 53, 56, 57, 61, 75, 87, 91, 104, 107]. However, it has not so far been possible to associate this model with a particular type of clinical evolution (i.e., the change of the clinical disease with time).

Model 3. Acquisition and Duplication of Supernumerary Chromosomes

The above-mentioned clonal evolution described by Lejeune et al. [62] clearly obeyed model 3. Acquisition of successive supernumerary chromosomes, followed by their duplication, represents one of the most important models of chromosomal evolution. A typical example is provided by a patient studied in our laboratory during the onset of blastic transformation of her CML [31]. Chromosome analysis revealed hyperploidy between 47 and 74 chromosomes, with a mode of 49. Clonal evolution was demonstrated by the existence of all intermediary cellular types between a clone with 47 and clone with 52 chromosomes, including two Ph¹s, one extra G, two extra Fs, and two extra 17-18s. Figure 3 shows that different pathways are possible to reach this last stage, depending on the sequence of acquisition and duplication of extra chromosomes. All of these intermediate steps have indeed been found to be present.

This model of clonal evolution has been observed in various types of leukemias. It does not appear to be very specific and is not associated with any particular clinical picture [31, 64, 75].

Model 4. Loss of Chromosomes

Loss of the Y chromosome is another instance of the association of a particular type of chromosomal evolution with an uncommon clinical evolution. Remarkable examples of this model have been observed in a group of male patients with CML. The karyotypes showed loss of the Y chromosome in the clone with the Ph¹ chromosome, prolonged clinical evolution and reduced fertility [1, 15, 17, 33, 60, 80, 93, 95]. Whether it was always the Y which was lost was not certain in all patients, however,

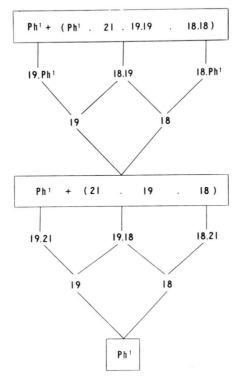

Figure 3. Clonal evolution showing acquisition and duplication of supernumerary chromosomes [31].

since it could be questioned in some cells whether the lost chromosome was not a group G. With this one reservation in mind, it can be stated that loss of a Y represented a feature common to this group of male patients, all of whom were also experiencing a prolonged clinical evolution. Indeed, in several cases the evolutionary period was as long as 8 or 10 years, with hardly any treatment. Family studies show that most of these patients failed to reproduce, and azoospermia could be demonstrated in one patient at the time of chromosomal study [33]. No obvious explanation can be found to account for this observation.

Hypodiploidy as a result of chromosome loss has been observed in other patients. The affected chromosome may belong to group G, as in some instances of acute leukemia [88], or to group C or D [75]. Clinical evolution does not seem modified in these instances.

Model 5. Loss and Acquisition of Chromosomes

Loss of chromosomes associated with progressive acquisition of supernumerary chromosomes is a frequent model of evolution in leukemias. The chromosomes most often involved belong to groups G, C, E, and D. The evolutionary patterns show various degrees of complexity depending on the number of pathways [2, 23, 56, 76, 90].

Model 6. Structural Rearrangements

In nonleukemic hemopathies, marker chromosomes have also been observed. Grouchy et al. [37] reported partial deletion of one arm of an F chromosome in acquired idiopathic sideroblastic anemia. This same marker was also found by Kay et al. [53] in polycythemia vera and was secondarily shown to result from partial deletion of the long arm of a chromosome No. 20 [83] (Fig. 4). The long marker chromosome, the size of a group A chromosome, described in Waldenström's macroglobulinemia [5, 6], is another example. It is remarkable that these diseases, although they cannot be considered malignant per se, do evolve toward malignancy with a high frequency. (This point will be discussed again.)

The Ph^1 marker. This is the most remarkable instance of a structural rearrangement in the course of human leukemias and, in fact, of all human neoplasias. (A detailed description of this chromosomal aberration, and its correlation with the clinical course of CML, is given elsewhere in this volume [72]; see also model 1, here.) Other marker chromosomes had been observed first in solid tumors, and then, after the discovery of the Ph^1 chromosome in myeloid leukemias, markers were also occasionally reported in other leukemias. Although such markers may appear to be isolated chromosome abnormalities, they are more frequently associated with numerical changes (i.e., losses or acquisitions of chromosomes).

The i(17q) Marker. The most outstanding of these markers is the isochromosome for the long arm of chromosome No. 17: i(17q). It occurs in practically half of the cases of CML in acute phase with clonal evolution. Pedersen [79] first reported three patients with CML showing loss of a 17-18 and acquisition of a C. Engel [18] and Stich et al. [96] drew attention to the marker chromosome and suggested that it was an isochromosome for the long arm of the missing 17-18. Grouchy et al. [41] then described 11 similar cases and confirmed the existence of the i(17q).

Eighteen patients with an i(17q) were investigated in our laboratory [103]. Additional patients have been reported in the literature. Figure 5

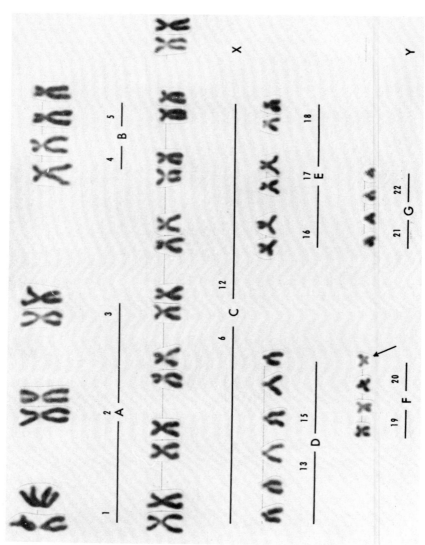

Figure 4. Karyotype showing deletion of the long arm of a No. 20 chromosome in a case of polycythemia vera.

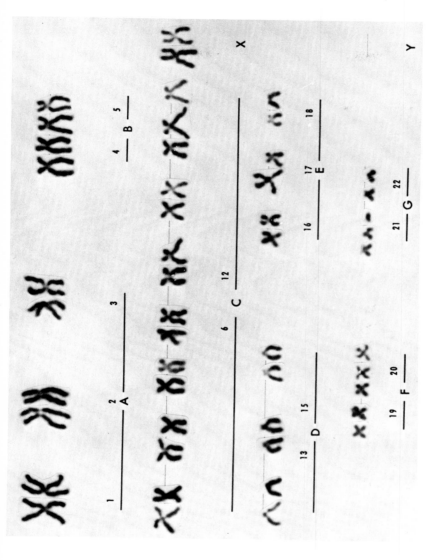

Figure 5. Isochromosome for long arm of chromosome No. 17, i(17q), and acquisition of a group C, a group F, and a group G chromosome.

297

shows a typical karyotype with the Ph[1] chromosome and the i(17q) marker chromosome replacing a 17. The hypothesis that the marker chromosome is indeed an i(17q) was confirmed statistically by showing with chromosome measurements that: (1) the marker was mediocentric, (2) the missing 17-18 was a 17 and not an 18, and (3) the arm length of the marker equaled 17q rather than 18q [41]. By using the recent banding techniques, it was confirmed that the marker was indeed an i(17q) [69].

Figure 6 is a composite representation of the clonal evolution obtained by adding all available karyotypes from all patients with the i(17q) marker [41]. It demonstrates that the appearance of the i(17q) is only the first step in the main path of chromosomal evolution, which then proceeds by successive acquisition of group C chromosomes and duplication of the Ph[1]. Some patients show only the i(17q) with no other abnormality. Others show part or all of the progressive changes in the main pathway [103].

It can also be seen from Fig. 6 that certain karyotypes do not appear at first examination to be compatible with the model. These are either cells having lost a No. 17 chromosome without having acquired the marker chromosome, or cells having acquired the marker without having lost the 17. These cells are rare, approximately 15% of all cells. They can be explained, however. Those cells having the marker and the 17 may have undergone a duplication of the remaining 17—in other words a "tetrasomization" of the long arm of the 17—whereas those having lost the 17 without having acquired the marker may be complementary to the cells of the major clone.

Figure 6 also indicates that while the Ph[1] and the i(17q) markers may represent the major theme of the evolution, there are cells present that are witness to new evolutionary attempts. These show, in particular, duplication of the Ph[1] and acquisition of extra C chromosomes.

From a purely clinical standpoint, patients with the i(17q) marker do not depart markedly from average. Yet bone-marrow preparations reveal a remarkable feature (unpublished observations by C. Turleau, J. de Grouchy, and J. Fréteaux), namely, the presence of two populations of blasts: (1) myelobasts with more or less abundant granulations and a nucleus often dystrophic and (2) blasts of variable size, with a high nucleocytoplasmic ratio, a basophilic cytoplasm often vacuolated, no granulations, and a nucleus frequently carrying large nucleoli. The ratio of these seemingly undifferentiated blasts is roughly proportional to the frequency of cells carrying the i(17q). It is thus possible that there is a direct relationship between the presence of the marker and the existence of a double population of blasts. This particular model would therefore represent a further example of an association between a unique type of clonal evolution and a unique cytological feature [103].

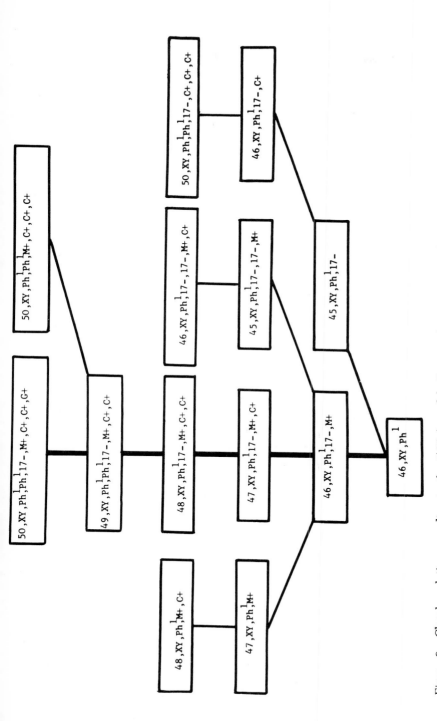

Figure 6. Clonal evolution according to the i(17q) model. Compound representation obtained from several different patients, as described in the text.

As mentioned earlier, the i(17q) is not the only marker chromosome observed in leukemias. In fact, any type of rearrangement can be identified (e.g., deletions, ring formations, translocations, pericentric inversions) [75]. In many instances, the real nature of the rearrangement is not clear, and it is hoped that the new techniques (fluorescence, denaturation, etc.) will help in the future to decipher the true constitution of these markers and hence allow us better to understand their significance.

Model 7. Polyploidization

Polyploidization of neoplastic cells is a striking phenomenon observed in solid tumors and in acute leukemia. "Replacement" of pseudodiploid cell lines by cells that tend toward polyploidization is frequently seen in the advanced stages of malignant invasion [101]. Often, two cell lines—one diploid or pseudodiploid and one poly- or parapolyploid—are present simultaneously with the same marker chromosomes or numerical changes. One such case studied by Nava et al. [74] led to the conclusion that polyploidization is most probably due to independent duplication of haploid sets of chromosomes.

Frequency with which the Various Models Are Observed

A series of 210 CML patients with the Ph^1 chromosome has been studied in our laboratories. The frequency of additional abnormalities (60/210) was higher during the acute phase than during the chronic phase of the disease (81 vs. 15%) (refer to Table 2 and ref. 102).

Table 2. Distribution by Leukemic Phase of 210 CML Patients with a Ph^1 Chromosome According to Presence or Absence of Additional Abnormalities*

Abnormality	Number of Patients (%)
Chronic phase (80%)	
Ph^1 alone	142 (85)
Additional abnormalities†	26 (15)
Total	168 (100)
Acute phase (20%)	
Ph^1 alone	8 (10)
Additional abnormalities†	34 (81)
Total	42 (100)

* Grouchy and Turleau (1973), unpublished results.
† See Table 3 for detailed data.

Table 3 shows the distribution of these 60 cases among the seven models of clonal evolution we have specified in the previous section. It can be seen that model 6 was most often represented; the i(17q) marker chromosome was observed in almost half of all patients during the acute phase. Next in importance during the acute phase were model 3 (acquisition of several extra chromosomes) and model 2 (acquisition of a group C chromosome).

POSSIBLE MECHANISMS AND SIGNIFICANCE OF CLONAL EVOLUTION

Acquisition and duplication of normal or of marker chromosomes are essential to the progression of clonal evolutions. When an extra chromosome is present, this chromosome itself tends to duplicate. In other words, the cell tends to become tetrasomic for the chromosome involved.

These processes might result from two mechanisms: non-disjunction or selective endoreduplication. The observation of complementary cells (i.e., cells missing the homologous chromosome) would be an argument in

Table 3. DISTRIBUTION BY LEUKEMIC PHASE OF 60 CML PATIENTS WITH A PH^1 CHROMOSOME ACCORDING TO SEVEN DEFINED CLONAL EVOLUTIONARY MODELS*

	Model	Chronic Phase	Acute Phase	Total
1.	Duplication of the Ph^1			
	Isolated	2		2
	With other abnormalities		(16)	(16)
	Extramedullary crisis		2	2
2.	Acquisition of a single supernumerary			
	Group C	5	5	10
	Group G	2	1	3
3.	Acquisition of several supernumeraries	0	7	7
4.	Loss of chromosomes			
	G-Y	4		4
	17-18	1		1
5.	Loss and acquisition	1	2	3
6.	Structural rearrangement			
	i(17q)	2	16	18
	Other	8	1	9
7.	Polyploidization	1		1
	Total	26	34	60

* Karyotypic details of those 60 (of 210) patients listed in Table 2 as having "additional abnormalities."

favor of non-disjunction. Such cells have apparently never been found, perhaps because they would be nonviable. Selective endoreduplication, defined by Lejeune [66] as the result of double replication of part of the genome (a whole chromosome or part of a chromosome), may explain the acquisition of supernumerary chromosomes in the absence of complementary cells. Indeed several observations show the presence in normal or malignant cells of one extra chromosome (or of part of a chromosome arm) closely lined up against its homologue [38]. Selective endoreduplication could, therefore, be an important mechanism underlying clonal evolution.

Loss of chromosomes can also be explained by non-disjunction. But here again complementary cells are not found. This would be in favor of another possible mechanism, namely, chromosome lag at anaphase, which results in chromosome loss without production of complementary cells.

Whatever the true significance of clonal evolution—a quest by the cell for a new equilibrium, a process responsible for acute transformation of leukemia, an imminent process from which the cell cannot escape—it does not represent a haphazard event. It is hardly conceivable that haphazard mitotic accidents would repeatedly produce the same chromosome rearrangements and that these rearrangements would furthermore be consistent with an apparently systematic evolutionary process, as shown by the numerous well-documented reports now available. It must also be emphasized that in many instances the abnormal clone represents all or almost all of the cell population. This observation is again in favor of a unique, single-cell origin and not of repeated accidents. Numerous observations of clonal evolution before any treatment exclude the hypothesis that they are a consequence of therapy, although therapeutic agents most probably do act differentially upon clones differing in karyotype.

The existence of different models of clonal evolution is not clearly understood. However, the fact that in some instances different patterns of evolution appear to be associated with specific clinical patterns of evolution can no longer be disputed. Future studies must therefore attempt to discover the still unrevealed, specific clinical features associated with certain models of clonal evolution.

"LAWS" GOVERNING CLONAL EVOLUTION

As suggested by Lejeune [65], clonal evolution must obey certain laws, since it has been demonstrated to be systematic. A first law is the existence of a *"variant commun"* of which the Ph^1 chromosome is the most remarkable example. A *variant commun* is conceived of as a rearrangement (numerical or structural), specific for each type of neoplasia, and constantly present during the course of evolution. Theoretically, its occurrence depends on many factors: the degree of fragility of specific chromosomes, the more or less predictable evolutionary pathways, the different selective pressures imposed by different tissues of the organism. Yet intensive clonal evolution may obscure this *variant commun* by secondary fortuitous rearrangements.

The second law proposed by Lejeune is the control of "forbidden combinations" by metabolic imperatives. For instance, in the following reaction

$$A \rightarrow B \rightarrow C$$
$$\downarrow$$
$$D$$

A is a substance present in limited amounts, C is essential to the life of the cell, and D is a second metabolic pathway for B. If there is a quantitative relation between enzymatic activities and the number of genes which govern their synthesis (and hence of the number of chromosomes that carry these genes), then one can speculate that an excess activity of B→D (due to the duplication of the corresponding chromosome) will be incompatible with the survival of the cell if the rate of production of A is not also increased. Thus the duplication of the chromosome carrying the "B→D" gene may occur only after the duplication of the chromosome controlling the production of A or the rate of its transformation into B. The demonstration that an enzyme, L-asparaginase, has a therapeutic effect in certain types of neoplasia is compatible with the hypothesis of metabolic control in clonal evolution [78].

Another imperative controlling "forbidden combinations" could be an immunological one, which would prevent survival of a clone producing antigenic qualities liable to be recognized and rejected by the host. There are undoubtedly other mechanisms conceivable that would regulate chromosomal evolution in neoplasias.

THE GENERAL CONCEPT OF THE CLONAL EVOLUTION OF MALIGNANCY

We have suggested in the past that between normality and death due to neoplasia there exists a "continuum" in which it is extremely difficult to tell precisely when malignancy really begins [34]. As recently pointed out by Baikie [3] and Pedersen [81], the chronic phase of CML may be considered not as a leukemic condition but rather as a preleukemic state having a very high probability of terminating in an acute myeloid leukemia, namely, the blastic phase.

According to the chromosomal theory of carcinogenesis [40, 65, 67, 71], chromosomal aberrations are the common pathway through which carcinogenic factors induce malignancy. These factors are external and internal. External factors are well documented; they include oncogenic viruses, carcinogenic chemicals, and ionizing radiation, all of which are well known to produce chromosome damage.

Internal factors are far more complex. One is chromosome interaction, defined to mean that the presence of a given chromosome abnormality increases the risk of production of another chromosome abnormality [18, 63]. Also, it is clear that there is a strong association between the presence of congenital chromosome aberrations and the development of malignancy [3, 40, 63]. The prevalence of leukemia in trisomy 21 is increased by a factor of 20 when compared with normals [9, 11]. Likewise, the risk of malignant transformation of the gonads is extremely high in certain cases of intersexuality associated with specific abnormal karyotypes [24]. These observations tend indeed to show how difficult it may be to pinpoint the precise onset of malignancy.

An impressive group of internal factors includes genetic diseases such as Fanconi's anemia and Bloom's syndrome, which are associated with chromosome damage as well as with high risk of malignancy [26-28, 36, 42]. Other probable conditions in this group are scleroderma [16] and incontinentia pigmenti [43]. Still other gene mutations with no obvious phenotypic effect may well be responsible for chromosome breakage and thus would explain a high prevalence of carcinogenesis in certain families [108].

Other internal factors that are believed to produce chromosome changes and to increase the risk of carcinogenesis are aging and intensive cellular proliferation.

Finally, it is likely, although impossible to demonstrate (if only on account of the permanent background of irradiation), that chromosome damage may occur spontaneously. A permanent background of chromosome rearrangements has been shown in normal somatic cells, where the rate of occurrence of marker chromosomes would be higher than 1 in 10,000 mitoses [39].

Chromosome rearrangements are thus produced with a frequency and a degree of specificity that is dependent on the etiologic factor and probably also on the type of tissue involved. The resulting abnormal cells will most often be eliminated, except for rare occasions when the rearrangement is "favorable." Then a new clone of cells is born which is not necessarily malignant *stricto sensu* immediately, as in the case of sideroblastic anemia, polycythemia vera, Waldenström's macroglobulinemia, and—why not?—CML.

The newly arisen clone will establish itself and, according to selective pressures and to specific laws, the karyotype will evolve in a very subtle fashion until malignancy (blastic transformation) is achieved. The clone will then grow dramatically until it overgrows the host.

The comparison with the evolution of species holds true. The minute life appeared in the depth of an ocean in the form of a blue alga, then evolution was inevitable. Through the combined action of selective pressures and newly arising chromosome and gene mutations, it has never ceased to produce new species, many of which have overgrown and disappeared. Both types of evolution—of species on the surface of Earth, of clones within an organism—both are inescapable, one as the other.

LITERATURE CITED

1. Atkin NB, Taylor MC: 45 chromosomes in chronic myeloid leukaemia. *Cytogenetics* 1:97–103, 1962.
2. Atkins L, Goulian M: Multiple clones with increase in number of chromosomes in the G group in a case of myelomonocytic leukemia. *Cytogenetics* 4:321–328, 1965.
3. Baikie AG: Chronic granulocytic leukaemia: the metamorphosis of a conditioned neoplasm to an autonomous one. *In* Proceedings of the Fourth Congress of the Asian Pacific Society of Hematologists, 197–205, 1969.
4. Bauke J, Neubauer A, Schoffling K: Occurrence of cell line with 55 chromosomes and two Philadelphia chromosomes at a stage preceding the terminal blastic phase in a case of chronic myeloid leukemia. *Deut Med Wochenschr* 92:301–305, 1967.
5. Bottura C, Ferrari I, Veiga AA: Chromosome abnormalities in Waldenström's macroglobulinaemia. *Lancet* 1:1170, 1961.
6. Broustet A, Hartmann L, Moulinier J, Staeffen J, Moretti G: Étude cytogénétique de dix-huit cas de maladie de Waldenström. *Nouv Rev Fr Hémat* 7:809–826, 1967.
7. Caspersson T, Gahrton G, Lindsten J, Zech L: Identification of the Philadelphia chromosome as a number 22 by quinacrine mustard fluorescence analysis. *Exp Cell Res* 63:238–240, 1970.
8. Castro-Sierra E, Wolf U, Gorman LZ, Merker H, Obrecht P: Clinical and cytogenetic findings in the terminal phase of chronic myelogenous leukaemia. *Humangenetik* 4:62–73, 1967.
9. Cawein M, Lappat EJ, Rackley JW: Down's syndrome and chronic myelogenous leukemia. *Arch Int Med* 116:505, 1965.
10. Chapelle A de la, Wennstrom J, Wasastjerna C: Apparent C trisomy in bone marrow cells. Report of 2 cases. *Scand J Haematol* 7:112–122, 1970.
11. Demayo AP, Kiossoglou KA, Erlandson ME, Notterman RF, German J:

A marrow chromosomal abnormality preceding clinical leukemia in Down's syndrome. *Blood* 29:233–241, 1967.

12. DOUGAN L, WOODLIFF HJ: Presence of two Ph^1 chromosomes in cells from a patient with chronic granulocytic leukaemia. *Nature* 205:405–406, 1955.

13. DOUGAN L, SCOTT ID, WOODLIFF HJ: A pair of twins, one of whom has chronic granulocytic leukaemia. *J Med Genet* 3:217–219, 1966.

14. DUVALL CP, CARBONE PP, BELL WR, WHANG J, TJIO JH, PERRY P: Chronic myelocytic leukemia with two Philadelphia chromosomes and prominent peripheral lymphadenopathy. *Blood* 29:654–666, 1967.

15. ELVES MW, ISRAELS MCG: Cytogenetic studies in unusual forms of chronic myeloid leukaemia. *Acta Haemat* 38:129–141, 1967.

16. EMERIT I, HOUSSET E, GROUCHY J DE, CAMUS JP: Chromosomal breakage in diffuse scleroderma. A study of 27 patients. *Eur J Clin Biol Res* 16:684–694, 1971.

17. ENGEL E, JENKINS DE, TIPTON RE, MCGEE BJ, ENGEL DE MONTMOLLIN M: Ph^1 positive chronic myelogenous leukemia, with absence of another G chromosome in a male. *New Eng J Med* 273:738–742, 1965.

18. ENGEL E, MCGEE BJ, HARTMANN RC, ENGEL DE MONTMOLLIN M: Two leukemic peripheral blood stemlines during acute transformation of chronic myelogenous leukemia in a D/D translocation carrier. *Cytogenetics* 4:157–170, 1965.

19. ENGEL E, MCKEE CL: Double Ph^1 chromosomes in leukaemia. *Lancet* 2:337, 1966.

20. ERKMAN B, CROOKSTON JH, CONEN PE: Double Ph^1 chromosome in chronic granulocytic leukemia. *Cancer* 20:1963–1975, 1967.

21. FITZGERALD PH: Complex pattern of chromosome abnormalities in the acute phase of chronic granulocytic leukaemia. *J Med Genet* 3:258–264, 1966.

22. FORD CE, CLARKE M: Cytogenetic evidence of clonal proliferation in primary reticular neoplasms. *Can Cancer Res Conf Proc* 5:129–146, 1963.

23. FORTEZA-BOVER G, BAGUENA-CANDELLA R: Anomalias cromosomicas en la leucemia mieloide cronica. *Rev Inf Med Terap* 7:426, 1962.

24. FRASIER SD, BASHORE RA, MOSIER HD: Gonadoblastoma associated with pure gonadal dysgenesis in monozygous twins. *J Pediatr* 64:740–745, 1964.

25. FREI E, TJIO JH, WHANG J, CARBONE PP: Studies on the Philadelphia chromosome in patients with chronic myelogenous leukemia. *Ann NY Acad Sci* 113:1073, 1964.

26. GERMAN J, ARCHIBALD R, BLOOM D: Chromosomal breakage in a rare and probably genetically determined syndrome of man. *Science* 148:506–507, 1965.

27. GERMAN J, CRIPPA LP: Chromosomal breakage in diploid cell line from Bloom's syndrome and Fanconi's anemia. *Ann Genet* 9:143–154, 1966.

28. GERMAN J: Genes which increase chromosomal instability in somatic cells and predispose to cancer. *In* Progress in Medical Genetics (Steinberg AG, Bearn AG, eds), vol 8, New York, Grune & Stratton, 61–101, 1972.

29. GOH KO, SWISHER SN: Identical twins and chronic myelocytic leukemia. Chromosomal studies of a patient with chronic myelocytic leukemia and his normal identical twin. *Arch Int Med* 115:475–478, 1965.

30. Grouchy J de, Nava C de, Bilski-Pasquier G: Duplication d'un Ph[1] et suggestion d'une évolution clonale dans une leucémie myéloïde chronique en transformation aiguë. *Nouv Rev Fr Hématol* 5:69–78, 1965.
31. Grouchy J de, Nava C de, Bilski-Pasquier G: Analyse chromosomique d'une évolution clonale dans une leucémie myéloïde. *Nouv Rev Fr Hémat* 5:565–590, 1965.
32. Grouchy J de, Nava C de, Bilski-Pasquier G, Piguet H, Bousser J: Études chromosomiques dans des leucémies myéloïdes subaiguës. *In* Comptes Rendus du Dixième Congrès de la Société Européenne Hématologique, 1965.
33. Grouchy J de, Nava C de, Bilski-Pasquier G: Chromosome Ph[1] et perte de petits acrocentriques dans une leucémie myéloïde chronique à évolution prolongée chez un homme. *Ann Génét* 9:73–77, 1966.
34. Grouchy J de: Aberrations chromosomiques et processus malins. La notion d'un continuum entre l'état normal et la mort. *Ann Génét* 9:55–57, 1966.
35. Grouchy J de, Nava C de, Cantu JM, Bilski-Pasquier G, Bousser J: Models for clonal evolutions. A study of chronic myelogenous leukemia. *Am J Hum Genet* 18:485–503, 1966.
36. Grouchy J de: Genetic diseases, chromosome rearrangements, and malignancy. *Ann Int Med* 65:603–607, 1966.
37. Grouchy J de, Nava C de, Zittoun R, Bousser J: Analyses chromosomiques dans l'anémie sidéroblastique idiopathique acquise. Une étude de six cas. *Nouv Rev Fr Hémat* 6:367–387, 1966.
38. Grouchy J de, Nava C de, Bilski-Pasquier G, Zittoun R, Bernadou A: Endoreduplication sélective d'un chromosome surnuméraire dans un cas de myélome multiple (maladie de Kahler). *Ann Génét* 10:43–45, 1967.
39. Grouchy J de: Le fonds permanent de remaniements chromosomiques. Fréquence de survenue de chromosomes marqueurs dans les cellules normales. *Ann Génét* 10:46–48, 1967.
40. Grouchy J de, Nava C de: A chromosomal theory of carcinogenesis. *Ann Int Med* 69:381–391, 1968.
41. Grouchy J de, Nava C de, Feingold J, Bilski-Pasquier G, Bousser J: Onze observations d'un modèle précis d'évolution clonale au cours de la leucémie myéloïde chronique. *Eur J Cancer* 4:481–492, 1968.
42. Grouchy J de, Nava C de, Marchand JC, Feingold J, Turleau C: Études cytogénétique et biochimique de huit cas d'anémie de Fanconi. *Ann Génét* 15:29–40, 1972.
43. Grouchy J de, Bonnette J, Brussieu J, Roidot M, Begin P: Cassures chromosomiques dans l'*incontinentia pigmenti*. Étude d'une famille. *Ann Génét* 15:61–65, 1972.
44. Gruenwald H, Kiossoglou KA, Mitus WJ, Dameshek W: Philadelphia chromosome in eosinophilic leukemia. *Am J Med* 39:1003–1010, 1965.
45. Hammouda F, Quaglino D, Hayhoe FGJ: Blastic crisis in chronic granulocytic leukaemia. Cytochemical, cytogenetic and autoradiographic studies in 4 cases. *Brit Med J* 1:1275–1281, 1964.
46. Hampel KE: Diplo-Ph[1]-chromosom bei des myeloischen leukamie. *Klin Wochenschr* 42:522, 1964.

47. HEATH CW, MOLONEY WC: The Philadelphia chromosome in an unusual case of myeloproliferative disease. *Blood* 26:471–478, 1965.
48. HOSSFELD DK, HAN T, HOLDSWORTH RN, SANDBERG AA: Chromosomes and causation of human cancer and leukemia. VII. The significance of the Ph^1 in conditions other than CML. *Cancer* 27:186–192, 1971.
49. HUNGERFORD DA: Chromosome studies in human leukaemia. 1. Acute leukaemia in children. *J Nat Cancer Inst* 27:983–1011, 1961.
50. HUNGERFORD DA, NOWELL PC: Chromosome studies in human leukaemia. III. Acute granulocytic leukaemia. *J Nat Cancer Inst* 29:545–566, 1962.
51. JACOBS EM, LUCE JK, CAILLEAU R: Chromosome abnormalities in human cancer. Report of a patient with chronic myelocytic leukemia and his non-leukemic monozygotic twin. *Cancer* 19:869, 1966.
52. KAMADA N, UCHINO H: Double Ph^1 chromosomes in leukaemia. *Lancet* 1:1107, 1967.
53. KAY HEM, LAWLER SD, MILLARD RE: The chromosomes in polycythaemia vera. *Brit J Hematol* 12:507–528, 1966.
54. KEMP NH, STAFFORD JL, TANNER R: Chromosome studies during early and terminal chronic myeloid leukaemia. *Brit Med J* 1:1010–1014, 1964.
55. KENIS Y, KOULISHER I: Étude clinique et cytogénétique de 21 patients atteints de leucémie myéloïde chronique. *Eur J Cancer* 3:83–93, 1967.
56. KIOSSOGLOU KA, MITUS WJ, DAMESHEK W: Chromosomal aberration in acute leukemia. *Blood* 26:610–640, 1965.
57. KIOSSOGLOU KA, MITUS WJ, DAMESHEK W: Cytogenetic studies in the chronic myeloproliferative syndrome. *Blood* 28:241–257, 1966.
58. KNOSPE WH, KLATT RW, BERGIN JW, JACOBSON CB, CONRAD M: Cytogenetic changes in chronic granulocytic leukaemia during blastic crisis: two Ph^1-chromosomes and hyperdiploidy. *Am J Med Sci* 254:816–823, 1967.
59. KOSENOW W, PFEIFFER RA: Chronisch-myeloisch Leukömie bei enieigen Zwillingen. *Deut Med Wochenschr* 22:1170–1176, 1969.
60. LAWLER SD, GALTON DAG: Chromosome changes in the terminal stages of chronic granulocytic leukemia. *Acta Med Scand Suppl* 445:312–318, 1966.
61. LAWLER SD, KAY H, BIRBECK M: Marrow dysplasia with C trisomy and anomalies of the granulocyte nuclei. *J Clin Pathol* 19–21, 1966.
62. LEJEUNE J, BERGER R, HAINES M, LAFOURCADE J, VIALATTE J, SATGE P, TURPIN R: Constitution d'un clone à 54 chromosomes au cours d'une leucoblastose chez une enfant mongolienne. *CR Acad Sci Paris* 256:1195–1197, 1963.
63. LEJEUNE J: Autosomal disorders. *Pediatrics* 32:326–337, 1963.
64. LEJEUNE J, BERGER R, CAILIE B, TURPIN R: Évolution chromosomique d'une leucémie myéloïde chronique. *Ann Génét* 8:44–49, 1965.
65. LEJEUNE J: Leucémies et cancers. *In Les Chromosomes Humains* (Turpin R, Lejeune J, eds), Paris, Gauthier Villars, 1965.
66. LEJEUNE J, BERGER R, RETHORÉ M-O: Sur l'endoréduplication sélective de certains segments du gênome. *CR Acad Sci Paris* 263:1880–1882, 1966.
67. LEVAN A: Chromosome abnormalities and carcinogenesis. *In*: Handbook of Molecular Cytology. (Lima-de-Faria A, ed), Amsterdam and London, North Holland Publishing Company, 717–731, 1969.

68. LIBRE EP, MCFARLAND W: Chronic myelogenous leukemia. Possible association with reticulum cell sarcoma. *Arch Int Med* 119:626, 1967.
69. LOBB DS, REEVES BR, LAWLER SD: Identification of isochromosome 17 in myeloid leukemia. *Lancet* 1:849, 1972.
70. MAKINO S: The chromosome cytology of the ascites tumors of rats, with special reference to the concept of the stem-line cell. *Int Rev Cytol* 6:26–84, 1957.
71. MOURIQUAND C: Chromosomes, leucémies et cancers. *Grenoble Med Chir*, 6: 31–40, 1968.
72. MULDAL S, LAJTHA LG: This volume, pp. 451–480.
73. NAVA C DE, GROUCHY J DE, THOYER C, BOUSSER J, BILSKI-PASQUIER G, FRETEAUX J: Évolutions clonales et duplication du Ph[1]. Présentation de deux cas. *Ann Génét* 12:83–93, 1969.
74. NAVA C DE, GROUCHY J DE, THOYER C, TURLEAU C, SIGUIER F: Polyploidisations et évolutions clonales. *Ann Génét* 12:237–241, 1969.
75. NAVA C DE: Les anomalies chromosomiques au cours des hémopathies malignes et nonmalignes. Une étude de 171 cas. *Monogr Ann Génét* 1:89, 1969.
76. NOWELL PC, HUNGERFORD DA: Chromosome studies on normal and leukemic human leukocytes. *J Nat Cancer Inst* 25:85–109, 1960.
77. NOWELL PC: Marrow chromosome studies in "preleukemia." Further correlation with clinical course. *Cancer* 28:513–518, 1971.
78. OLD LJ, BOYSE EA, CAMPBELL HA, BRODEY RS, FIDLER J, TELLER JD: Treatment of lymphosarcoma in the dog with L-asparaginase. *Lancet* 1:447–448, 1967.
79. PEDERSEN B: Two cases of chronic myeloid leukaemia with presumably identical 47-chromosome cell-lines in the blood. *Acta Pathol Microbiol Scand* 61: 497–502, 1964.
80. PEDERSEN B: Males with XO Ph[1]-positive cells: a cytogenetic and clinical subgroup of chronic myelogenous leukemia. *Acta Pathol Microbiol Scand* 72: 360–366, 1968.
81. PEDERSEN B: Karyotype evolution in human leukaemia. The relation between karyotypes, cellular phenotypes and clinical progression. *In* Proceedings of the Fourth International Congress on Human Genetics, Excerpta Medica, 166–175, 1972.
82. PEGORARO L, PILERI A, ROVERA G, GAVOSTO F: Presence of three Philadelphia-chromosomes during blastic transformation in a case of chronic myeloid leukemia. *Tumori* 53:315–321, 1967.
83. REEVES BR, LOBB D, LAWLER SD: Identity of the abnormal F-group chromosome associated with polycythaemia vera. *Humangenetik* 14:159–161, 1972.
84. REISMAN LE, MITANI M, ZUELZER WW: Chromosome studies in leukemia. 1. Evidence for the origin of leukemic stem-lines from aneuploid mutants. *New Engl J Med* 270:591–597, 1964.
85. RETHORÉ M-O, PRIEUR AM, GRISCELLI C, MOZZICONACCI P, LEJEUNE J: Évolution clonale au cours d'une leucémie aiguë myéloblastique chez un enfant trisomique 21. *Ann Génét* 14:193–198, 1971.
86. RIGO SJ, STANNARD M, DOWLING DC: Chronic myeloid leukemia associated with multiple chromosome abnormalities. *Med J Aust* 2:70–72, 1966.

87. Rowley JD, Blaisdell RK, Jacobson LO: Chromosome studies in preleukemia. 1. Aneuploidy of group C chromosomes in three patients. *Blood* 27:782–799, 1966.
88. Ruffié J, Lejeune J: Deux cas de leucose aiguë myéloblastique avec cellules sanguines normales et cellules haplo (21 ou 22). *Rev Fr Etud Clin Biol* 7: 644–647, 1962.
89. Ruffié J, Ducos J, Bierme R, Colombies P, Salles-Mourlan AM: Multiple chromosomal abnormalities in an acute exacerbation of myeloid leukaemia. *Lancet* 1:609–610, 1965.
90. Sandberg AA, Ishihara TC, Crosswhite LH, Hauschka TS: Comparison of chromosome constitution in chronic myeloid leukemia and other proliferative disorders. *Blood* 20:393–423, 1962.
91. Sandberg AA, Ishihara TC, Crosswhite LH: Group C trisomy in myeloid metaplasia with possible leukemia. *Blood* 24:716–725, 1964.
92. Sandberg AA, Hossfeld DK, Ezdinli EZ, Crosswhite LH: Chromosomes and causation of human cancer and leukemia. VI. Blastic phase, cellular origin, and the Ph^1 in CML. *Cancer* 27:176–185, 1971.
93. Serra A, Sargentini S, Patrono C, Ferrara A, Laghi V: 45, X, G−, Ph^1+ cell line in bone marrow and blood of a CML affected male. *Ann Génét* 13: 239–243, 1970.
94. Smalley RV: Double Ph^1 chromosomes in leukaemia. *Lancet* 2:591, 1966.
95. Speed DE, Lawler SD: Chronic granulocytic leukaemia: the chromosomes and the disease. *Lancet* 1:403–407.
96. Stich W, Back F, Dormer P, Tsirimbas A: Dopple Philadelphia Chromosom und Isochromsom 17 in der terminalen Phase der chronischen myeloischen Leukamie. *Klin Wochenschr* 44:334–337, 1966.
97. Streiff F, Peters A, Gilgenkrantz S: Double Ph^1 chromosomes in leukaemia. *Lancet* 2:1193–1194, 1966.
98. Tjio JH, Carbone PP, Whang J, Frei E: The Philadelphia chromosome and chronic myelogenous leukemia. *J Nat Cancer Inst* 36:567–584, 1966.
99. Tough IM, Jacobs PA, Court Brown WM, Baikie AG, Williamson ERD: Cytogenetic studies on bone marrow in chronic myeloid leukaemia. *Lancet* 1:844–846, 1963.
100. Trujillo JM, Ohno S: Chromosomal alteration of erythropoietic cells in chronic myeloid leukemia. *Acta Haemat* 29:311–316, 1963.
101. Trujillo JM, Cork A, Drewinko B, Hart JS, Freireich EJ, Case report: tetraploid leukemia. *Blood* 38:623–637, 1971.
102. Turleau C, Trébuchet C, Frétault J, Finaz C, Nava C de, Grouchy J de: Étude cytogénétique de 200 cas de leucémie myéloïde chronique (LMC) avec Ph^1. *In* Proceedings of the Fourth International Congress on Human Genetics, Excerpta Medica, 1971.
103. Turleau C: Évolutions clonales avec isochromosomes 17 dans les syndromes myéloprolifératifs. Une étude de 20 cas. Doctoral thesis, Paris, in preparation.
104. Weinstein AW, Weinstein ED: A chromosomal abnormality in acute myeloblastic leukemia. *New Eng J Med* 268:253–255, 1963.
105. Whang-Peng J, Frei E, Tjio JH, Carbone PP, Brecher G: The distribution

of the Philadelphia chromosome in patients with myelogenous leukemia. *Blood* 22:664–673, 1963.

106. WHANG-PENG J, HENDERSON ES, KNUTSEN T, FREIREICH EJ, GART JJ: Cytogenetic studies in acute myelocytic leukemia with special emphasis on the occurrence of Ph[1] chromosome. *Blood* 36:448–457, 1970.

107. WINKELSTEIN A, SPARKES RS, CRADDOCK CG: Trisomy of a group C in a myeloproliferative disorder. Report of a case. *Blood* 27:722–733, 1966.

108. ZUELZER WW, THOMPSON RI, MASTRANGELO R: Evidence for a genetic factor related to leukemogenesis and congenital anomalies: chromosomal aberrations in pedigree of an infant with partial D trisomy and leukemia. *J Pediat* 72:367–376, 1968.

UTILIZATION OF MOSAIC SYSTEMS IN THE STUDY OF THE ORIGIN AND PROGRESSION OF TUMORS

STANLEY M. GARTLER

Departments of Medicine and Genetics, University of Washington, Seattle, Washington

Introduction 314

Methods and Markers 315

Kinetics of Tumor Initiation 317
    Single-phenotype Tumors 317
        Possible Explanations other than Single-cell Origin 318
        Single-cell or Unicentric Origin 318
    Double-phenotype Tumors 320

Tumor Progression and the Problem of Initial Cell Populations 322
    Cervical Cancer 322
    Chronic Myelogenous Leukemia 323

Relationship between Carcinogenic Factors and Initial Cell Populations . . . 324
    Virus-associated Tumors 324
    Radiation-related Tumors 325

Metastatic Tumors 325

Recurrent Tumors 326

Tumor Cell Autonomy 327

Recent Studies of Special Interest 329
    Human Ovarian Teratoma: Meiotic Origin 329
    Is All Clonal Growth Neoplastic? 330

Summary 331

Acknowledgements 331

Literature Cited 331

INTRODUCTION

The question of whether a tumor is unicellular or multicellular in origin is one of long standing in cancer research. Early ideas favored a unicentric (single-cell) origin, possibly because of the prevailing notion of somatic mutation as a causal factor. Subsequently, a multicentric or "field" theory gained prominence [23]. Experimental evidence was minimal, and these different views were based primarily on gross surgical and pathological observations of different cancers at late stages of growth and development. To answer this question and others related to tumor cell kinetics, one must either be able to diagnose the initial stages of tumorigenesis or to utilize independent markers to infer something about the kinetics of early tumor development. Unfortunately, early diagnosis is usually not possible. Perhaps the most useful tool for such studies have been mosaic cell populations, which are the subject of this review.

A mosaic individual is one consisting of two or more cell types that differ in genetic expression. The cell types may be derived from one or more zygotes; in the latter case, the term chimera is preferred by some workers. To be maximally useful to the investigator, the mosaicism should occur early in development, and the different cell types should be distinguishable by a sensitive and reliable assay. Mosaic populations provide a built-in set of autonomous cell markers that may be used to investigate a variety of developmental phenomena, both normal (e.g., tissue morphogenesis, cell lineage relationships) and abnormal (single- vs. multiple-tumor origin; metastasis or independent origin).

The earliest use of mosaics for developmental studies was made by Sturtevant [48] in 1929 to analyze cell lineage relationships between different morphological structures in *Drosophila*. Since that time, and especially in recent years, work in this area has expanded considerably, to cover a number of different organisms and a variety of mosaic systems [20, 38, 39].

Three different mosaic systems have been used for the study of mammalian tumors. The first of these, hemopoietic chimerism, can be induced in mice by injecting donor bone marrow into lethally irradiated recipients. This technique has been useful in answering developmental questions related to the hemopoietic system and has also been important recently in furthering our understanding of leukemia recurrence in man [12, 53].

A second technique involves the experimental production of generalized mosaicism in laboratory mice. Preimplantation embryos of two or more different genotypes may be surgically manipulated in a variety of ways so as to form a single embryo of mixed origin [33, 49]. This is then introduced into the uterus of a suitable female, where it develops into an animal containing cells of all the parental types. These experimentally structured chimeric mice have been utilized for a wide range of developmental problems and recently for the study of tumor formation and tumor cell autonomy [34, 35].

The third source of mosaicism, and the most useful for tumor studies in man, has been X chromosome inactivation [9, 22, 31]. In cells that contain two X chromosomes, one of the two Xs becomes genetically inactive early in development. When inactivation of either the maternally or the paternally derived X has occurred, it is stable to cell division, so that a cell with an active maternal X, for instance, will give rise to a clone of cells all of which have an active maternal X. Thus, in a person heterozygous for a pair of X-linked alleles, a natural system of mosaicism is made available for developmental investigations. The X-linked glucose-6-phosphate dehydrogenase (G6PD) variants, polymorphic in African and Mediterranean populations, are excellent markers for this purpose. The first use of mosaic systems for the study of tumor formation was made with G6PD heterozygotes in our laboratory [16, 26].

METHODS AND MARKERS

The G6PD mosaic system is the main one we shall be considering here. There are four relatively frequent alleles and over 20 rare ones known at this locus [47]. The important characteristics of the four common alleles are given in Table 1. Due to X-chromosome inactivation, a G6PD heterozygote ($Gd^A Gd^B$, e.g.) will have two kinds of somatic cells—those producing product A and those producing product B. The initial event of inactivation is random and occurs early in development, probably around the time of blastocyst formation. Thus somatic cells expressing A or B should be present in approximately equal proportions prior to the time of any known tumor formation. It follows that a tumor arising from a

Table 1. Properties of Four Common G6PD Alleles

G6PD Allele[*]	Relative G6PD Activity		Electrophoretic Pattern
	Red Cells	Nucleated Cells	
Gd B	1.00	1.00	Slow
Gd B-	0.05	0.15	Slow
Gd A	1.00	1.00	Fast
Gd A-	0.15	0.85	Fast

[*] See ref. [15].

single cell in such a patient will be either A or B. A tumor arising from multiple cells could be composed of both A and B cells and therefore would have the electrophoretic bands of the products determined by both alleles. Multiple tumors arising by cellular metastasis from a primary tumor of single-cell origin would be expected to have one and the same G6PD form.

There are several possible pitfalls in determining whether a given tumor is clonal (of single-cell origin) or not. The first of these concerns the patch size in the mosaic hosts. With few exceptions adult G6PD heterozygotes are mosaic in all somatic tissues [28]; in a small percentage of cases the blood may have only a single G6PD type, while other tissues in the same individual will be mosaic [15]. This is apparently due to selective overgrowth of one cell type by the other; the selective difference in such cases is probably not due to the different G6PD types but to genes linked to the G6PD locus [15, 41]. A single-phenotype tumor arising in a tissue consisting of one and the same cell phenotype has no significance, and it is important to recognize such instances. In mosaic tissues, except for epithelial, the two-cell types are well mixed; that is, there is no tendency for cells of like type to be contiguous (to form patches or to be variegated), as would be expected if coherent clonal growth were characteristic of tissue growth. Furthermore, multiple small samples from the same individual tend to exhibit the same degree of mosaicism. All this indicates that the structure of the mosaicism in non-epithelial tissues is decidedly mixed and that the "patch" size (number of adjacent cells of like type) is very small. The structure of mosaicism in chimeric mice, in contrast to the evenly mixed structure described above, normally exhibits significant variegation (i.e., the patch size is considerably larger [51, 52]). The reasons for this difference in the fine mosaic structure of these two systems are not understood, but the difference must be kept in mind in the interpretation of single-phenotype tumors. If the patch size is very large (variegation marked), then a single-phenotype tumor could develop from multiple cells of the same type merely by chance. Thus in chimeric mice, a larger number of samples of single-phenotype tumors would be required to draw a firm conclusion about single-cell origin. In the case of X-chromosome inactivation, the normal epithelial tissue of the mosaics exhibits some variegation; for example, epidermal biopsies the size of normal hair follicles may consist of cells of only one type [18, 19].

Second, in the case of a double-phenotype tumor, it is important to check for contamination of the tumor with normal tissue before concluding multicellular origin. The G6PD electrophoretic variants are the most useful in mosaic studies, and it is important to keep in mind that a minor component of less than 5% of the tumor tissue is usually not detectable. Therefore, a double-phenotype tumor should be considered significant only if the minor mosaic component is greater than the degree of normal tissue admixture. For example, a tumor including 5% of a minor mosaic component and 10% of normal tissue is not necessarily of multicellular origin; the double phenotype is more probably due to the normal tissue components.

A third theoretical source of error would be depression of the inactive X; in that case, a tumor could be of single-cell origin but due to derepression of the inactive X, both alleles might be expressed and a double-phenotype tumor would then result. In the case of G6PD mosaics, this can be checked quite simply; G6PD is a dimer, and when both alleles are active, three products are formed: A, B, and the hybrid product, AB. In the case of monomeric enzymes, the tumor would have to be cloned in a cell culture system to answer such an objection.

The tumor we analyze is usually far removed in time and growth from its inception. Since tumors are not subject to the same growth restrictions as normal cells, cellular evolution can occur during their growth. Thus cell variants may arise and replace the original population. This is especially important for tumors of multicellular origin, in which case, through selective overgrowth, a single-phenotype tumor could result. This process will be considered in more detail in the following section.

KINETICS OF TUMOR INITIATION

Single-phenotype Tumors

The first tumor to be studied using mosaics has been the leiomyoma of the uterus. Leiomyomas are relatively frequent, benign neoplasms, homogeneous in structure and easily separable from adjoining normal tissues.

This work was begun in our laboratory with Dr. David Linder and has since been extended by Linder and others [26, 27, 29, 50]. Over 200 non-necrotic tumors from 25 individuals heterozygous at the G6PD locus have now been examined by electrophoresis. All of these were "single-phenotype" tumors (i.e., composed entirely of cells all of which were either A or B type); furthermore, both A and B tumors were found in individual patients. Only one tumor, a necrotic one, showed a major band and a trace of minor G6PD component, suggesting non-tumor tissue contamination. Normal myometrial specimens as small as 1 mm^3 and immediately adjacent to the leiomyomas were found to contain both A and B bands, whereas tumors measuring up to several centimeters in diameter were consistently single-phenotype throughout.

Possible Explanations for Single-Phenotype Tumors, other than Single-cell Origin. What mechanisms, other than single-cell origin, might account for a single-phenotype tumor? First, since the tumor cells have a normal female karyotype with a single sex chromatin body, the single G6PD phenotypes could not be ascribed to chromosome loss.

Second, it might be argued that a single-phenotype tumor could start from multiple cells simply by chance sampling of all cells of the same G6PD type. The chance of such an event would depend on the fine structure of the tissue mosaicism; the greater the variegation, or larger the patch size, the greater the chance that a number of adjacent cells will be of the same mosaic type. We have carried out an analysis of the data gathered on leiomyomas and can statistically exclude the possibility that the more than 200 tumors studied could all be single-phenotype tumors simply due to chance sampling of two or more adjacent cells of like type [26].

A third possible explanation for such tumors could be that through selective overgrowth, one cell type has predominated in a population originally mixed. Although it is true that it would take only a slight growth rate advantage for one cell type in a tumor of several centimeters in diameter to account for 95% of the tumor (a minor component of 5% or less would not be detectable), two arguments can be made against this possibility. Microdissection of a number of tumors, especially the core, has failed to reveal any trace of a double component; if the tumor had started from a mixture of cell types, one would expect to find significant levels of both cell types in the early part of the tumor. Also, if selection were the reason for the single-phenotype leiomyomas, one would expect that all of the tumors in a patient would show the same single phenotype, since selection would be acting on X chromosomes, which carry the only known genetic differences between the cell types.

Finally, a more complex explanation for single-phenotype tumors could be that repeated selection has occurred during the growth of the tumor—that is, that there has been a growth cycle, followed by extensive cell death, and then reseeding from a small group of cells to start the growth cycle again. This would be analogous to genetic drift and, given a large heterogeneous starting cell population, could account for a single-phenotype tumor [4]. However, the histology and growth patterns indicate that such cycling growth and death is not characteristic of leiomyomas.

Single-cell or Unicentric Origin. The leiomyoma data effectively demonstrate a single-cell origin. The presence of both A and B tumors in single patients indicates an independent origin also for each of multiple tumors.

It has been argued on the basis of statistical analysis that cancer must be a multi-hit process [1]. The data for benign leiomyomas do not fit this idea. The usual somatic mutation rates have been estimated to be in the neighborhood of 1×10^{-6}. The target tissue for leiomyomas, the myometrium, consists of approximately 1×10^{10} cells. If two mutations per cell were required for expression of leiomyoma, this tumor would be rare, and multiple independent tumors would be out of the question—and yet they are common.

There are no known factors causal to leiomyomas, but it is conceivable that they arise as result of somatic mutation disturbing the cell's ability to respond to tissue growth regulation. It is not at all unlikely that the first step in neoplastic progression—loss of response to tissue growth restrictions, for example—could be effected by a single mutational step; this stage would be exemplified by benign neoplasms, such as leiomyomas. In contrast, malignant neoplasms, which have not only lost the capacity to respond to growth regulation but are also invasive or metastatic or both, might require multiple mutational steps.

The results of most tumor studies with mosaic systems are compatible with a single-cell origin (e.g., chronic myelogenous leukemia, the Burkitt lymphoma, and warts—Table 2). However, it has not been possible in any

Table 2. Tumors of Single-Cell (Unicellular) Origin as Determined by Studies with G6PD Heterozygotes

Tumor	References
Leiomyoma	27, 26, 50
Chronic myelogenous leukemia	10
Burkitt lymphoma	11, 14
Lipoma	16
Nodular goiter	29
Myeloma, plasmacytoma	29, 32
Warts	37
Metastatic adenosarcoma, lung	29
Lymphosarcoma	3, 14
Paroxysmal nocturnal hemoglobinuria	42
Ovarian teratoma	29, 30
Melanoma	14
Neuroblastoma	14
Cervical dysplasia	43, 46
Carcinoma (palate)	14

of these cases to exclude thoroughly the alternative explanations for single-cell origin referred to in the leiomyoma discussion. It is possible, especially for malignant tumors, that clonal selection to a malignant state from a first-stage benign tumor could have occurred. However, whether the first stage was multiple or single is not always clear. We will consider this question under tumor progression, and certain of the other single phenotype tumors will be considered in more detail in other parts of this paper.

Double-phenotype Tumors

The hypothesis that sequential mutational events are required for cancer predicts that tumors would always be of single-cell origin, since the probability that two or more mutational events could occur in several adjacent cells is so unlikely that it can be discounted. However, as we shall see, there are now several well-documented examples of tumors of multicellular origin.

The first demonstration of a tumor of multicellular origin, using the mosaic system, was hereditary trichoepithelioma. Susceptibility to development of this neoplasm is transmitted as a simple dominant. A family segregating for this tumor and the Mediterranean variant of G6PD deficiency (quantitative variant) was investigated, and it was shown that the mosaic composition of the tumors was the same as that of adjoining normal tissue [17], indicating that the tumors consisted of both kinds of cells. Since this tumor is heterogeneous with respect to tissue composition, the significance of the mixed cell population could be questioned.

More recently, it has been possible to study another hereditary tumor, multiple neurofibromatosis, in G6PD electrophoretic variant heterozygotes [13]. This dominantly inherited tumor is histologically homogeneous. The data from this study demonstrated two types of cells in every tumor biopsy studied, and the mosaic composition of the tumors was again essentially the same as the adjacent normal tissue. Thus both hereditary tumors studied seemed to be of multicellular origin; in fact, since the mosaic composition of the tumors was like that of normal tissue, the starting cell population must have been many cells. It seems possible that, since in hereditary tumors an initial susceptibility exists in every cell of the subject, the initiating tumor event would have a large target. Presumably the event in these instances is not mutational (i.e., a rare event); otherwise, as Knudson [24, 25] has suggested, we would expect a single-cell target. Another possibility is that once a single cell is transformed, it then becomes capable of recruitment, thus forming a double-phenotype tumor.

A test of this notion of recruitment could be made by comparison of tumor phenotypes for hereditary and nonhereditary cancer of the same type—for example, retinoblastoma. The hereditary form of this tumor is bilateral and multiple, while the nonhereditary tumor is unilateral and single and has a negative family history. Knudson's interpretation [24] is that two mutations are necessary for the expression of the retinoblastoma cell; in the hereditary form, one of the mutations is already present in every cell and, therefore, enough second mutations can occur to form multiple tumors. In the nonhereditary form, both mutations must occur in a somatic cell in order to transform it; the probability of this occurrence is so low that, at most, only single tumors would be expected. If these two forms could be studied in G6PD heterozygotes, one would expect from Knudson's hypothesis that both would show single phenotypes whereas, if the second step is not mutational, a double phenotype would be predicted for the hereditary form and a single for the nonhereditary.

Besides the two hereditary tumors mentioned above, there are several other types of neoplasms that have been reported to have double phenotypes, and therefore a multicellular origin (Table 3). Except for the hereditary tumors, there is no apparent common pattern among them.

When known carcinogenic agents have been used, and when host susceptibility is not critical, it might be expected that the titer of the carcinogen would determine whether the tumor was of single-cell or multicellular origin. Unfortunately, there is little evidence available at this time that bears on this point; the one pertinent case will be discussed in a later section. Several other instances of double-phenotype tumors will also be considered in later sections.

Table 3. Tumors of Multicellular (Multicentric) Origin as Determined by Studies with G6PD Heterozygotes

Tumor	References
Trichoepithelioma	17
Neurofibromatosis	13
Carcinoma, breast	32
Carcinoma, colon	3
Chronic lymphocytic leukemia	32

TUMOR PROGRESSION AND THE PROBLEM OF INITIAL CELL POPULATIONS

Cervical Cancer

As mentioned earlier, a widely held notion regarding cancer development is that multiple steps are involved in the progression from a normal tissue to frank malignancy. A series of mutational events (i.e., a mutational "threshold") may be required before any abnormal expression or, alternatively, each step may alter the tissue in a recognizably progressive manner. For example, a first event may permit a cell to escape general tissue growth regulation (e.g., inability to enter G_o) with a resulting hyperplasia, but a second step might be required for invasiveness or metastasis to develop. This second event could involve a change in cell surface antigens that permitted growth in an otherwise restricted tissue.

Cancer of the cervix, one of the most widely studied cancers, might be considered prototypic of such a progression. Three stages are widely recognized: cervical dysplasia, carcinoma in situ, and invasive carcinoma. Are these different stages related ontogenetically, and if so, what are the cellular dynamics at each stage? Two independent studies of cervical dysplasia in G6PD heterozygotes have yielded results indicating a clonal origin for this abnormal growth pattern [43, 46]. If the invasive stage of the disease progresses ontogenetically from cervical dysplasia, then invasive carcinoma in G6PD heterozygotes should show single-phenotype malignancies of the same G6PD phenotypes as the cervical dysplasia from which they supposedly develop. As yet, a study of the three stages in the same heterozygous patient has not been possible. The two studies on invasive cervical carcinoma in heterozygotes are conflicting; one study [43] reported only single-phenotype tumors, while another report [46] showed some apparently double-phenotype tumors, as well as single ones. The amount of contaminating stromal cells in invasive carcinoma samples may be considerable and, as has already been mentioned, unless this is carefully controlled, a double-phenotype tumor may only indicate normal tissue contamination. True double-phenotype cervical cancers would indicate either that the disease is not a progressive one in the sense discussed above, or that at the invasive steps the disease becomes "infectious"—that is, that cells not involved in the early dysplasia stage become transformed, a notion suggested earlier. This important point remains to be clarified by further work.

At one time the prevailing concept regarding the development of cervical cancer was the so-called field theory [23]. This hypothesis implied that after the initial change to cervical dysplasia, carcinoma develops from many cells in the cervical dysplasia field rather than by rare transformation of single cells (i.e., a non-genetic explanation). Since cervical dysplasia is clonal in origin, the available mosaic systems do not permit resolution between single-cell and multicellular origin of invasive carcinoma (i.e., a single-phenotype carcinoma would be expected in either case). Hypothetically, a mosaic-generating system would be needed to resolve this question; that is, following clonal selection from a mosaic system (A from A/B), a new mosaicism is induced in the selected single-phenotype region (A to A/C). It would then be possible to analyze the next step in neoplastic development in terms of single or multicellular events. Such a system is not available today, but it is conceivable that it could be developed with transplantation techniques in experimental animals.

Chronic Myelogenous Leukemia

Chronic myelogenous leukemia is characterized cytogenetically by a specific chromosomal deletion called the Ph^1 chromosome (deletion of the distal portion of chromosome No. 22) [6, 40]. Chromosomal markers, like the Ph^1 chromosome, have been cited as a basis for considering various tumors to be monoclonal. If irradiation, for example, were the causative agent, clonal derivation would seem reasonable, since it would be almost impossible to expect that the Ph^1 chromosome could be produced simultaneously in several cells. However, irradiation is not the only agent that produces chromosome changes, and it is conceivable that integration of a viral nucleic acid at a specific chromosomal site may be the initial event in this neoplasm. Under such conditions, numerous transformed cells with the same chromosomal aberration could result. As indicated more than once in this review, to study tumor cell kinetics effectively, a mosaicism must exist prior to the tumor. Thus when three patients with chronic myelogenous leukemia who were also G6PD heterozygotes were shown to express single G6PD phenotypes in their erythrocytes and granulocytes, the problem appeared solved [10]. Unicentric origin was indicated and, furthermore, the presence of erythrocytes and granulocytes having the same phenotype supported the idea of a common stem cell for these members of the hemopoietic series.

It may be premature to conclude that chronic myelogenous leukemia is of single-cell origin. It is possible that the single-cell phase could apply only to a later stage of the disease. The cases we were able to study were in late stages of the disease, when the bone marrow was essentially 100% Ph^1-positive. It is now known that years before any clinical or overt hematological signs are evident, Ph^1-positive cells can be present in the bone marrow, though at levels well below 100% [5]. The single G6PD phenotype found in the tumor cells of heterozygotes in late stages of the disease does not prove that this was also true at the beginning or early stages. The first Ph^1-positive cells may not be malignant, but rather may represent a first step in tumor progression. Analysis of early cases will be most difficult, as they are rarely found; furthermore, separation of the Ph^1-positive from the Ph^1-negative cells would be necessary for an evaluation, and this is not possible at present. As should be appreciated, the problem of tumor progression will not be easy to solve. In theory, early Ph^1-positive cases in G6PD heterozygotes offer an approach to sequential mosaicism; and it would, therefore, be worthwhile to attempt to search for such cases for intensive study.

RELATIONSHIP BETWEEN CARCINOGENIC FACTORS AND INITIAL TUMOR CELL POPULATION

Virus-associated Tumors

Two naturally occurring neoplasms associated with virus have now been investigated in G6PD heterozygotes: the Burkitt lymphoma, which may be caused by EB virus, and the common wart (verucca vulgaris), which is known to be virus-induced. Twenty-four lymphoid tumors in 12 patients with the Burkitt lymphoma all showed single G6PD phenotypes [11, 14]; similarly, six warts from six patients all showed single G6PD phenotypes [37]. These data are compatible with a single-cell origin of these tumors; assuming a viral etiology, the results suggest that under natural conditions the viral titers are such that the target is a single cell. Another implication is that the neoplasm or transformed cell does not produce significant amounts of infectious virus, but that the tumor spreads primarily by proliferation of the cell initially transformed.

A recent study in marmosets by Chu and Rabson [7] on induction of malignant lymphoma with a herpesvirus appears to support this interpretation. Most marmosets are natural hemopoietic chimeras as a result of bone-marrow exchange between litter mates early in development. A lymphoma cell line was established from a mesenteric lymph node of an XX/XY marmoset by herpesvirus infection, herpesvirus saimiri. The lymphoma cell line had both XX and XY cells in approximately a 50–50 ratio and appeared to be a stable mosaic. This demonstrates multicellular origin of the transformed cell line and also indicates that the mosaic system can remain stable under malignant conditions (i.e., no selection occurs). It is possible that the difference in phenotypes between similar natural tumors and this induced one is due to the differences in infectious virus titer. That is, in the induced tumor the virus titer was so high that many cells were transformed and consequently a tumor of multicellular origin resulted, whereas in natural viral tumors the virus titer was very low, and therefore only single-cell transformation occurred. An alternative explanation is that the cell transformed by herpesvirus saimiri becomes infectious and transforms (or recruits) neighboring cells, thus forming a double-phenotype tumor. By varying the titer of infectious virus in the marmoset system, these alternative explanations of the double-phenotype lymphoma could be resolved.

Radiation-related Tumors

There is considerable evidence indicating a causal connection between radiation exposure and subsequent development of cancer. The mutations and chromosomal aberrations known to be induced by irradiation have often been cited as possible steps in the pathway between irradiation and cancer. As yet, controlled experiments using irradiation-induced tumors have not been done in appropriately marked animals. A recent statistical analysis of radiation-induced mammary tumors in mice suggests a multicellular involvement in tumor initiation. The essence of this work is that at low irradiation dosages a much higher incidence of tumors was observed than was expected [45]. These results were interpreted as suggesting that the irradiation target was larger than one cell. An experimental model satisfying these data would be the induction by irradiation of an infectious factor (a virus?) that could then be shown to serve as the active carcinogen. A single irradiated cell could then transform adjacent cells, thus giving rise to an apparently large target size. Tumor induction by radiation has not yet been carried out in mosaic animals, but chimeric mice would be extremely useful in such experiments.

METASTATIC TUMORS

Mosaics can be used effectively for studying the spread of tumors. For example, in the case of leiomyomas, the presence of both A and B tumors in the same patient demonstrated that these multiple tumors arose independently and not by metastasis. The most thoroughly studied tumor involving apparent metastasis is the Burkitt lymphoma [14]. This viral neoplasm has now been studied in 12 subjects who were also heterozygous for G6PD. In all cases, individual tumors expressed only a single

G6PD phenotype, indicating a single-cell origin. Furthermore, multiple lymphomas in single patients were always of the same single phenotype, suggesting cellular metastasis from a primary tumor.

One of the most interesting tumors studied from this point of view was a primary carcinoma of the colon and its various metastases examined by Buetler et al. [3]. The primary tumor exhibited a double-G6PD phenotype, while the various metastatic nodules exhibited single-G6PD phenotypes, either A or B. The interpretation was made that the single-phenotype nodules were the result of metastasis of single cells from the primary double-phenotype tumor. It is interesting that although the primary tumor was apparently of multicellular origin, the metastatic sites were seeded by single cells. This might possibly have been due to a tumor structure that allowed only single cells to break away from the primary; however, another possibility is that tumor progression to a metastatic state is such a rare event that only single cells would be expected to be capable of metastasis.

RECURRENT TUMORS

One of the more important applications of mosaic systems to tumor studies is to the question of tumor recurrence. When a neoplasm recurs after therapy or spontaneous remission in a mosaic heterozygote, it is possible to determine whether the recurrence derives from residual transformed cells or is a new neoplasm. In the case of a single-phenotype tumor, recurrence of tumor with the same phenotype suggests that the original tumor was not entirely removed or did not regress completely; on the other hand, appearance of a new phenotype suggests that an infectious process could be involved in tumor recurrence. Recurrence from an originally double-phenotype tumor would, of course, be subject to various interpretations. For example, recurrences of different single-phenotype tumors suggests reseeding from residual cells of the primary. Thus far, very few cases of tumor recurrence in mosaics have been studied. Fialkow [14] has examined four recurrent tumors in as many G6PD-heterozygous persons with the Burkitt lymphoma; three tumor recurrences were of the same single-phenotype as the original neoplasm, but one was different. Recurrences of the same phenotypes might be expected, but a recurrent neoplasm with a new mosaic cell type is difficult to explain. As indicated above, such an event could suggest an infectious factor; that is, the neoplastic cells might release an infectious factor and transform other cells. However, if this were a significant process in the development of the Burkitt lymphoma, double-phenotype tumors would be expected, and the multiple tumors would not all be of the same phenotype. It seems more likely that the recurrent new-phenotype tumors in the Burkitt lymphoma may represent a reinitiation of the entire disease process.

In acute lymphoblastic leukemia, recurrence has consistently taken place following destruction of the patient's marrow and transplantation of normal bone marrow [12, 14]. The question arises as to whether such recurrence is due to repopulation by residual leukemia cells or to transformed bone-marrow cells from the donor. Where partial destruction of host bone marrow resulted from cyclophosphamide therapy, a method that permits a chimeric state to be maintained in the bone marrow, the recurrence was shown by cytogenetic markers to be due to residual leukemia cells of the host [14]. However, in the case of irradiation where host bone marrow destruction was essentially complete, when recurrence occurred, the leukemia cells were found to be of donor origin [12]. It has been suggested from the results of the latter case that the donor cells are transformed by host infectious factor. An alternative explanation is that the donor cells contain rare variants that are permitted to expand in the host leukemic environment. On the assumption of a high-titer infectious factor, transformed donor cells would have been expected in the chimeric cyclophosphamide cases, as well as in the X-irradiation transplants. On the other hand, rare donor-cell variants in the cyclophosphamide cases might not have been able to compete with the remaining host cell population.

TUMOR CELL AUTONOMY

What is the significance of the cell genotype in tumor development? More specifically, in the case of hereditary tumors, does the genotype exert a *direct* effect (i.e., is a neoplastic growth autonomous, in the sense that it is determined completely by cell genotype) or an *indirect* effect (i.e., do genetically susceptible cells merely create a tumorigenic environment)?

This question can be examined using chimeric mice produced experimentally by aggregation of preimplantation embryos from genetically susceptible and nonsusceptible strains. An independently segregating enzyme difference is also incorporated into the chimeric animals as a marker, and by the electrophoretic type found in tumor tissue, the tumor genotype can be determined (susceptible or mixed). As mentioned earlier, mosaicism in chimeric mice exhibits relatively great variegation; consequently, it is important to note the degree of mosaicism in the normal tissue surrounding the tumor. As emphasized earlier, a single-phenotype tumor developing in nonmosaic normal tissue of the same phenotype can contribute little to the understanding of tumor development.

Condamine et al. [8] studied chimeric mice from hepatoma-susceptible and nonsusceptible strains. The majority of tumors (9/12) were composed entirely of susceptible-strain cells, and the authors concluded that tumor development was essentially cell-autonomous. However, the nontumorous liver tissue was predominantly of the susceptible-cell type (more than 90%) and the tumor distribution (9/12 susceptible, 2/12 mixed, and 1/12 nonsusceptible) is compatible with a random development of tumors, that is, unrelated to cell genotype. Thus, the results of this study could be interpreted as suggesting that the tumor-susceptible genotype leads to a tumor-inducing environment, rather than that the genotype predisposes certain cells to neoplastic transformation (i.e., that the cells are autonomous).

The C3H mouse strain is susceptible to mammary tumors, as well as to hepatomas, whereas the C57 strain is resistant to both [34, 35]. Chimeric mice derived from these strains have been used in studies of cell autonomy in mammary tumor development. The results were similar to those found in the hepatoma work involving these same strains. The majority of tumors were from C3H cells, although some appeared mixed, and a minority were C57. The apparently mixed tumors were quite interesting in that transplantation studies with them revealed that the C57 cell population always grew out as normal cells. Only the C3H cells in these tumors gave rise to tumors upon transplantation, which suggested that the C57 cells were merely entrapped. This is a somewhat sophisticated and complex means of detecting normal tissue admixture in tumors. It is likely that histological analysis of tumor sections also would have revealed a significant proportion of normal cells, and histological examination should be a routine procedure for any mosaic or mixed tumor.

As in the hepatoma study, the majority of normal cells in the susceptible tissue were of C3H origin. This makes interpretation in terms of cell autonomy difficult. Mintz [36] points out that normal C3H cells have a growth-rate advantage over C57 in both liver and mammary gland; in a sense this could represent a premalignant condition in the C3H tissue and, if so, it would be most interesting. Under such conditions it is surprising that any tumors would develop from the nonsusceptible (C57) cell type, but apparently they do, which suggests that autonomy of tumor development may not be a simple matter.

In another study of tumor cell autonomy, these same workers have analyzed lung tumors in chimeric mice derived from BALB/c (susceptible) and C57 B1/6 (resistant) parents [36]. In these animals the normal tissue mosaicism showed roughly equal numbers of both cell types, and yet the tumors were largely, and possibly exclusively, of susceptible-cell type. This is compatible with a cell autonomy interpretation of lung-tumor development in BALB/c cells.

The chimeric mouse system offers theoretically some very exciting approaches to the study of tumor cell kinetics and the dynamics of cancer development, since such mice may be derived from multiple aggregation of pre-implantation embryos to yield triple or even quadruple mosaics.

RECENT STUDIES OF SPECIAL INTEREST

Meiotic Origin: Human Ovarian Teratomas

One of the most interesting recent studies of tumor origin using genetic markers was done by Linder [29, 30] on human ovarian teratomas. One theory suggested to explain the appearance of these tumors has been that they arise from germ cells that have undergone meiosis. If this is true, then all loci for which the patient is heterozygous will be subject to detectable gene loss due to segregation. Over 20 tumors were studied from patients heterozygous at one or more of one X-linked locus (G6PD) and at three autosomal loci (PGM_1, PGM_3, 6PGD). Examples of gene loss were found for all the loci examined. Since these are unlinked genes, it is probable that the causal factor is affecting the entire genome, and segregation in meiosis best fits this requirement. Each tumor is uniform in cell phenotype for the loci examined, which indicates their origin from single transformed cells. Since the teratomas were all diploid and always XX, they must have arisen either from a first meiotic product or from fusion of two second meiotic products. Either stage of meiosis (considering crossing-over) would permit either retention of both alleles or loss of one of them at a heterozygous locus.

Another interesting feature of these tumors is that they show X-chromosome inactivation (i.e., single sex chromatin body), even though the germ cells from which they are derived do not [21]. This result implies that neither fertilization nor the paternal genome is necessary for X-chromosome inactivation.

In the same study, Linder and Power showed that mucinous cystadenomas, which are often associated with teratomas, are not metastases of the teratoma. Enzyme markers showed that in the same ovary a teratoma exhibited so-called gene loss, while the associated mucinous cystadenoma did not.

Is All Clonal Growth Neoplastic?

As I have pointed out before, clonal growth is not characteristic of normal development. Does it follow that all clonal growth is abnormal or neoplastic? The available evidence would appear to support such a deduction, and the last two studies have a special bearing on this interpretation.

The first study is that of Oni et al. [42], who utilized G6PD mosaicism in an investigation of paroxysmal nocturnal hemoglobinuria. The red cells of such patients are heterogeneous with respect to sensitivity to acid hydrolysis. One explanation for this could be that the acid-sensitive cells are derived from a single-stem cell that arose by somatic mutation. If this were true, then all of the acid-sensitive cells in a G6PD heterozygote should be of the same G6PD phenotype, which was the case, even though whole-blood assays showed the presence of both G6PD types. The fact that a significant fraction of the red cell population was acid-sensitive suggests either that the original mutant stem cell had a growth-rate advantage over normal cells or that it was not subject to normal growth-regulating signals of the hemopoietic system. In this sense, paroxysmal nocturnal hemoglobinuria resembles a neoplasm.

The final study is that of Benditt and Benditt [2], who examined atherosclerotic plaques obtained at autopsy from G6PD heterozygotes. The normal arterial wall exhibited a fine mixture of A and B cells. The plaques were distinctly different, with some apparently pure A and pure B plaques detected in each subject. A number of the plaques showed both A and B cells, but the authors argue that these lesions may represent a mixture of plaques and/or normal tissue contamination. The authors feel that their results are compatible with a clonal origin of the atheromas and suggest that consideration should be given to tumorigenic-like factors in the development of atherosclerotic plaques. The classical idea of plaque origin involves wound repair following mechanical or chemical injury, and this aspect of atherosclerosis has been reviewed recently by Ross and Glomset [44]. The study of wound repair with mosaics or other good marker systems is yet to be carried out, and nothing can be stated definitively as to whether wound repair is of single-cell or multicellular origin. However, it seems unlikely, if not impossible, that a macroscopic wound repair would be monoclonal. The nature of repair, for chemical type injury or irritation and microscopic sized wounds, remains to be determined, and it is this type of injury that is pertinent to atherosclerosis. Furthermore, if the lesion has a cyclic pattern (i.e., injury–proliferation–regression–injury, etc.), then the repair could be multicellular in origin but would appear as a single phenotype upon analysis; indeed such a lesion has been suggested as a possibility in atherosclerosis [44]. This concept of single phenotypes arising as a result of repeated multicellular samplings was discussed in the section entitled "Possible Explanations for Single-phenotype Tumors, other than Single-cell Origin." The findings in the atherosclerotic plaque study are very stimulating, but more important, they point out the need for mosaic-type marker analyses of wound repair.

SUMMARY

The utilization of mosaic systems to study various aspects of tumor initiation, growth, and development has been the subject of this review. Tumors mainly of single-cell origin, but some of multicellular origin as well, have been described. Of special interest are two hereditary tumors shown to be of multicellular origin. Whether this is a general pattern for genetic tumors remains to be seen, particularly with respect to the malignant genetic tumors. Several spontaneous tumors of presumed viral origin have proved to be of single-cell origin; however, a marmoset lymphoma induced in the laboratory by high titers of virus was shown to be multicellular in origin. Thus in the case of exogenous agents it may be simply the quantity of the carcinogen that determines uni- or multicentric origin.

Other questions relating to tumor cell dynamics that have been investigated with mosaic system are metastasis and recurrence, both extensively studied in the Burkitt lymphoma system. Tumor cell autonomy has been explored using chimeric mice, and it is expected that much more tumor work with these interesting animals will be carried out in the future.

A variety of other studies using mosaic systems are reviewed, including some concerned with tumor progression, the origin of ovarian teratomas, and the suggestion that paroxysmal nocturnal hemoglobinuria and atherosclerotic plaques may result from a neoplastic process. It is concluded that with the increasing potential of mosaic systems, considerably more progress will be made in understanding tumor cell dynamics and that a number of conditions now poorly understood will prove to have originated in a manner similar to that by which tumors originate.

Acknowledgements.–This work was supported in part by U.S. Public Health Service Research Grant GM 15253. The author is a Research Career Awardee of the National Institutes of Health.

LITERATURE CITED

1. Ashley DJB: The two "hit" and multiple "hit" theories of carcinogenesis. *Brit J Cancer* 23:313–328, 1969.

2. BENDITT EP, BENDITT JM: Evidence for a monoclonal origin of human atherosclerotic plaques. *Proc Nat Acad Sci (US)* 70:1753–1756, 1973.
3. BEUTLER E, COLLINS Z, IRWIN LE: Value of genetic variants of glucose-6-phosphate dehydrogenase in tracing the origin of malignant tumors. *New Eng J Med* 276:389–391, 1967.
4. BUHLER WL: Single cell against multicell hypothesis of tumor formation. In Proceedings of the Fifth Berkeley Symposium on Mathematical Statistics and Probabilities, 635–637, 1967.
5. CANELLOS GP, WHANG-PENG J: Philadelphia chromosome positive preleukaemic state. *Lancet* 2:1227–1228, 1972.
6. CASPERSSON T, GAHRTON G, LINDSTEN J, ZECH L: Identification of the Philadelphia chromosome as a number 22 by quinacrine mustard fluorescence analysis. *Exp Cell Res* 63:238–240, 1970.
7. CHU EW, RABSON AS: Chimerism in lymphoid cell culture line derived from lymph node of marmoset infected with Herpesvirus saimiri. *J Nat Cancer Inst* 48:771–773, 1972.
8. CONDAMINE H, CUSTER RP, MINTZ B: Pure strain and genetically mosaic liver tumors histochemically identified with the β-glucuronidase marker in allophenic mice. *Proc Nat Acad Sci (US)* 68:2032–2036, 1971.
9. DAVIDSON RG, NITOWSKY HM, CHILDS B: Demonstration of two populations of cells in the human female heterozygous for glucose-6-phosphate dehydrogenase variants. *Proc Nat Acad Sci (US)* 50:481–485, 1963.
10. FIALKOW PJ, GARTLER SM, YOSHIDA A: Clonal origin of chronic myelocytic leukemia in man. *Proc Nat Acad Sci (US)* 58:1468–1471, 1967.
11. FIALKOW PJ, KLEIN G, GARTLER SM, CLIFFORD P: Clonal origin for individual Burkitt tumors. *Lancet* 1:384–386, 1970.
12. FIALKOW PJ, THOMAS ED, BRYANT JI, NEIMAN PE: Leukaemic transformation of engrafted human marrow cells in vivo. *Lancet* 1:251–255, 1971.
13. FIALKOW PJ, SAGEBIEL RW, GARTLER SM, RIMOIN DL: Multiple cell origin of hereditary neurofibromas. *New Engl J Med* 284:298–300, 1971.
14. FIALKOW PJ: Use of genetic markers to study cellular origin and development of tumors in human females. *Adv Cancer Res* 15:191–223, 1972.
15. GANDINI E, GARTLER SM, ANGIONI G, ARGIOLAS N, DELL'ACQUA G: Developmental implications of multiple tissue studies in glucose-6-phosphate dehydrogenase deficient heterozygotes. *Proc Nat Acad Sci (US)* 61:945–948, 1968.
16. GARTLER SM, LINDER D: Selection in mammalian mosaic cell populations. *Cold Spring Harbor Symp Quant Biol* 29:253–260, 1964.
17. GARTLER SM, ZIPRKOWSKY L, KRAKOWSKI A, EZRA R, SZEINBERG A, ADAM A: Glucose-6-phosphate dehydrogenase mosaicism as a tracer in the study of hereditary multiple trichoepithelioma. *Am J Hum Genet* 18:282–287, 1966.
18. GARTLER SM, GANDINI E, ANGIONI G, ARGIOLAS N: Glucose-6-phosphate dehydrogenase mosaicism: utilization as a tracer in the study of the development of hair root cells. *Ann Hum Genet* 33:171–176, 1969.
19. GARTLER SM, GANDINI E, HUTCHISON HT, CAMPBELL B, ZECCHI G: Utilization of a sex-linked marker to study hair follicle variegation and development in man. *Ann Hum Genet* 35:1–7, 1971.

20. GARTLER SM, NESBITT MN: Sex chromosome markers as indicators in embryonic development. *Adv Biosci* 6:225–254, 1970.
21. GARTLER SM, LISKAY RM, CAMPBELL BK, SPARKES R, GANT N: Evidence for two functional X chromosomes in human oocytes. *Cell Differ* 1:215–218, 1972.
22. GARTLER SM, CHEN S, FIALKOW PJ, GIBLETT ER, SINGH S: X chromosome inactivation in cells from an individual heterozygous for two X-linked genes. *Nature NB* 236:149–150, 1972.
23. JOHNSON LD, EASTERDAY CL, GORE H, HERTIG AT: The histogenesis of carcinoma *in situ* of the uterine cervix. *Cancer* 17:213–229, 1964.
24. KNUDSON AG: Mutation and cancer: statistical study of retinoblastoma. *Proc Nat Acad Sci (US)* 68:820–823, 1971.
25. KNUDSON AG, STRONG LC: Mutation and cancer: neuroblastoma and pheochromocytoma. *Am J Hum Genet* 24:514–532, 1972.
26. LINDER D, GARTLER SM: Glucose-6-phosphate dehydrogenase mosaicism: utilization as a cell marker in the study of leiomyomas. *Science* 150:67–69, 1965.
27. LINDER D, GARTLER SM: Problem of single cell versus multicell origin of a tumor. *In* Proceedings of the Fifth Berkeley Symposium on Mathematical Statistics and Probabilities, 625–633, 1965.
28. LINDER D, GARTLER SM: Distribution of glucose-6-phosphate dehydrogenase electrophoretic variants in different tissues of heterozygotes. *Am J Hum Genet* 17:212–220, 1965.
29. LINDER D: Gene loss in human teratomas. *Proc Nat Acad Sci (US)* 63:699–704, 1969.
30. LINDER D, POWER J: Further evidence for post-meiotic origin of teratomas in the human female. *Ann Hum Genet* 34:21–30, 1970.
31. LYON MF: Chromosomal and subchromosomal inactivation. *Ann Rev Genet* 2:31–52, 1968.
32. MCCURDY PR: *In* Hereditary Disorders of Erythrocyte Metabolism (Beutler E, ed), New York, Grune & Stratton, 121, 1968.
33. MINTZ B: Formation of genotypically mosaic mouse embryos (abstract). *Am Zool* 2:432, 1962.
34. MINTZ B, SLEMMER G: Gene control of neoplasia. Genotypic mosaicism in normal and preneoplastic mammary glands of allophenic mice. *J Nat Cancer Inst* 43:87–95, 1969.
35. MINTZ B: Neoplasia and gene activity in allophenic mice. In Twenty-third Annual Symposium on Fundamental Cancer Research, M. D. Anderson Hospital, University of Texas, 476–517, 1970.
36. MINTZ B, CUSTER RP, DONNELLY AJ: Genetic diseases and developmental defects analyzed in allophenic mice. *Int Rev Exp Pathol* 10:143–179, 1971.
37. MURRAY RF, HOBBS J, PAYNE B: Possible clonal origin of common warts (Verruca vulgaris). *Nature* 232:51, 1971.
38. NANCE WE: Genetic tests with a sex-linked marker: glucose-6-phosphate dehydrogenase. *Cold Spring Harbor Symp Quant Biol* 29:415–425, 1964.
39. NESBITT MN, GARTLER SM: The application of genetic mosaicism to developmental problems. *Ann Rev Genet* 5:143–162, 1971.

40. Nowell PC, Hungerford DA: A minute chromosome in human chronic granulocytic leukemia. *Science* 132:1497, 1960.
41. Nyhan WL, Bakay B, Connor JD, Marks JF, Keele DK: Hemizygous expression of glucose-6-phosphate dehydrogenase in erythrocytes for the Lesch-Nyhan syndrome. *Proc Nat Acad Sci (US)* 65:214–218, 1970.
42. Oni SB, Osunkoya BO, Luzzato L: Paroxysmal nocturnal hemoglobinuria: evidence for monoclonal origin of abnormal red cells. *Blood* 36:145–152, 1970.
43. Park I, Jones HW: Glucose-6-phosphate dehydrogenase and the histogenesis of epidermoid carcinoma of the cervix. *Am J Obst Gynecol* 102:106–109, 1968.
44. Ross R, Glomset JA: Atherosclerosis and the arterial smooth muscle cell. *Science* 180:1332–1339, 1973.
45. Rossi HH, Kellerer AM: Radiation carcinogenesis at low doses. *Science* 175:200–202, 1972.
46. Smith JW, Townsend DE, Sparkes RS: Glucose-6-phosphate dehydrogenase polymorphism: a valuable tool to study tumor origin. *Clin Genet* 2:160–162, 1971.
47. Standardization of procedures for the study of glucose-6-phosphate dehydrogenase. *WHO Tech Rep Ser* 366, 1967.
48. Sturtevant AH: The claret mutant type of Drosophila simulans, a study of chromosome elimination and cell-lineage. *Z Wiss Zool* 135:323, 1929.
49. Tarkowski AK: Mouse chimaeras developed from fused eggs. *Nature* 190:857–860, 1961.
50. Townsend DE, Sparkes RS, Baluda MC, McClelland G: Unicellular histogenesis of uterine leiomyomas as determined by eletrophoresis of glucose-6-phosphate dehydrogenase. *Am J Obstetr Gynecol* 107:1168–1173, 1970.
51. Wegmann TG, Gilman JG: Chimerism for three genetic systems in tetraparental mice. *Develop Biol* 21:281–291, 1970.
52. Wegmann TG: Enzyme patterns in tetraparental mouse liver. *Nature* 225:462–463, 1970.
53. Wu AM, Till JE, Siminovitch L, McCulloch EA: Cytological evidence for a relationship between normal hematopoietic colony-forming cells and cells of the lymphoid system. *J Exp Med* 127:455–464, 1967.

CYTOGENETICS OF CANINE VENEREAL TUMORS: WORLDWIDE DISTRIBUTION AND A COMMON KARYOTYPE

S. MAKINO

Chromosome Research Unit, Faculty of Science, Hokkaido University,
Sapporo, Japan, 060

Introduction 336
Early Studies of Canine Venereal Tumors: Worldwide Distribution 336
Nature of the Disease 337
Cytogenetic Aspects: A Common Karyotype 343
    Studies in Japan 343
    Studies in the United States 356
    Studies in France 356
    Studies in Jamaica 359
    Cellular DNA Content 359
Discussion 361
    Cytogenetics and Cancer 361
    The Stemline Concept 361
    Tumor by Cellular "Contagion" 362
    Mechanism of Karyotypic Change 363
    Hypotheses to Explain the Universal Karyotype 364
    Is This Tumor Virus-induced ? 366
    Addendum in Proof 367
Acknowledgements 367
Literature Cited 367

INTRODUCTION

Comparative studies in vertebrates have been valuable in establishing biological principles important for the understanding of human diseases, including tumor formation. About 100 years ago in France, the study of tumors in horses, mules, donkeys, cattle, dogs, cats and swine [75] was begun and, as result of the technical limitation of the times, gross and histological aspects were emphasized. Now, however, emphasis should also be placed on cytogenetic studies of such material, in view of the fact that cancer's origin from preexisting normal cells, as well as its progression and regression, necessarily involves cell division. Interest has been heightened by the finding of many types of chromosome abnormalities in almost all cases studied. Extensive cytogenetic studies of transplantable tumors in rodents have shown that specific cells which are the main contributors to the growth and development of the tumor in a new host are of a "stemline" lineage having a characteristic (abnormal) chromosome complement [40]. The cytogenetics of human tumors has been shown to be similar in many ways to that of animal tumors [25, 33, 42, 44, 47].

The venereal tumor in dogs is an unusual one in that it apparently is transmitted from dog to dog by direct contact and cell transfer. In comparative studies, the important question regularly arises whether a given disorder in the animal is comparable to that in man. Although presently without known counterpart in man, this unusual canine neoplasm appears to deserve careful evaluation; its cytogenetics is the subject of this paper.

EARLY STUDIES OF CANINE VENEREAL TUMOR: WORLDWIDE DISTRIBUTION

The first successful experimental transmission of this tumor was accomplished by Nowinsky [59] in 1876. His systematic charting of the growth of transplanted tumors attracted the interest of investigators in cancer etiology, and this basic work in oncology led to the expansion of cancer research at that time. However, it was Sticker [76, 77] who, at the beginning of this century, studied and described this tumor in detail. It has been known by a variety of names: Sticker's tumor; venereal granuloma; transmissible venereal sarcoma; transmissible lymphosarcoma; histiocytoma; round-cell sarcoma, and so on [74].

Sticker [77] described a tumor of this kind in Germany, and cases had also been reported from England by Smith and Washbourn [73] and Dunstan [11]. More recently, Kaalund-Jorgensen and Thomsen [26] described eight cases among 1400 dogs in Denmark. During the first decade of this century, these venereal tumors were known as a local disease in native dogs of British New Guinea before the arrival of Europeans [69].

In 1906 Beebe and Ewing [5] first mentioned occurrence of the tumor in the United States. Stubbs and Furth [78] in 1934 reported five cases among 30,000 dogs in Philadelphia and stated that the tumor seemed more common than in the two previous decades. Karlson and Mann [27] have since indicated that the disease is common in Texas but is rare in Iowa, Ohio, Kansas, Michigan, Minnesota, New York, and Pennsylvania; according to Bloom et al. [6], it has shown a tendency to decrease in most regions of the United States, except for Texas. However, the tumor still seems to be endemic in some parts of the world (e.g., in Puerto Rico [65], in southern France [54], and in Illinois [31]), and at the present time distribution appears to be worldwide, through Europe and South America, toward Asia.

In Japan, dogs affected by this tumor are seen rather frequently. The cases reported to date seem to cover almost all parts of the country and tend to increase especially in city zones. A considerable number of cases have been reported from the city of Tokyo [15, 22, 23, 49, 70, 88, 89, 99, etc.]. Shirasu [70] reviewed the epidemiologic and clinical features of the disease of dogs in Japan. Basing his statements on data collected during some 10 years in Japan, Saeki [66] noted an outstandingly high incidence in occurrence of this tumor, in striking contrast to other kinds of tumors in animals. Recently Koike and Honjo [29] reported its incidence as 2.74% around Sapporo and 1.98% around Gifu. Watanabe et al. [91-95] have studied several cases in Nagasaki from pathological and cytological standpoints.

THE NATURE OF THE DISEASE

In early reports the nature of this disease was considered to be inflammatory [54], but more recently authors have agreed that it appears to fulfill the histologic criteria of a malignancy [5, 6, 12, 23, 27, 74, 77, 78]. Smith and Jones [74] have mentioned that the venereal tumor of dogs occupies a unique position among neoplastic conditions, and they have summarized many of the interesting features characterizing the tumor. Tumor cells are found freely suspended in the tissue fluids from the tumors. Natural transmission occurs through coitus, and experimental transmission can be accomplished simply by transplantation of fresh tumor cells. Tumors are usually restricted to the external genitalia in both sexes. Metastasis seldom occurs, although rarely metastatic nodes may appear in the skin of the face and shoulders, for instance, and in the visceral organs. Long survival of tumor-bearing animals is common, and, once having had the disease, animals seem to gain immunity to some extent. The sensitivity of the tumor to radiation is high.

Pathologically, the tumor has been categorized as a round-cell sarcoma by many authors [22, 23, 49, 73, 74, 77, 89, 92, 94, 99]. Histologically, it consists of closely arranged, large, irregularly shaped cells which are spherical or polyhedral in general appearance and bear some resemblance to lymphoid cells. They are closely packed and their boundaries are indistinct; supported by vascular connective-tissue trabeculae of varying sizes, they sometimes display pseudoalveolar arrangements (Figs. 1 and 2). Smith and Jones [74] have reported that both in total volume and in size of the nucleus, these cells are larger than lymphocytes, approximating more nearly the dimensions of the maternal lymphoblasts of lymph nodes. The cytoplasm is finely granular, and occasionally, though not in most instances, cytoplasmic staining by hematoxylin and eosin may be poor. The nuclei are large and round or oval in appearance, are vesicular with minute and clumps of chromatin granules, and usually have one prominent nucleolus, eccentrically located. The numerous mitotic figures constitute a striking feature of the tumor, and several mitotic phases are often present in a single field.

In the female, tumors of the vaginal wall are seen as granular, reddish nodules or submucosal masses, sometimes attaining the size of a hen's egg and infiltrating the external labia and adjacent parts (Figs. 3 and 4). The tumor generally increases in size following parturition and frequently tends to bleed, ulcerate, and undergo necrosis. In the male, the clinical picture varies. In the early stages, single or multiple small, reddish nodules are seen on the affected site of the penis, and these grow to more than 5 cm in diameter, occasionally infiltrating the scrotum. Metastasis of the tumor does occur, but rarely, to regional lymph nodes or to some visceral organs [94]. As the growth enlarges, hemorrhage frequently occurs (Fig. 1). This is often associated with superficial infection, and the affected sites produce a bloody, serous, or purulent discharge.

Inflammatory cells are often observed in the growing tumor tissue, particularly in the hollows of the ulcerated surface of the tumor, where the fluid stagnates and there are many isolated cells with delicate fibers [92]. One of the striking features of this tumor is that the serous, hemorrhagic, and purulent fluid contains a number of the large, round tumor cells interspersed with leucocytes, erythrocytes, micrococci, and bacilli. Fluid sometimes drips from the tumor, and it has been found to contain a number of the freely suspended tumor cells in process of

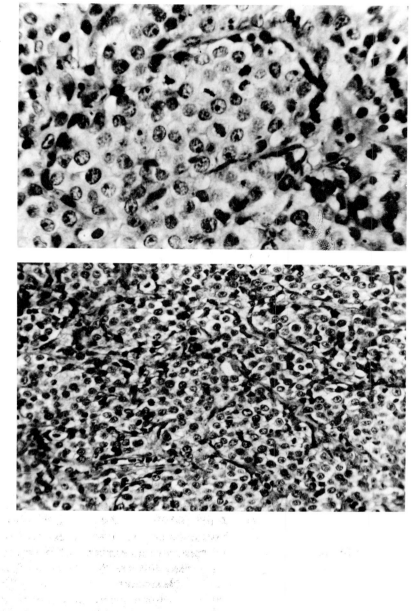

Figures 1 and 2. Histological sections of venereal tumor, at low and high powers, respectively. Courtesy of Dr. F. Watanabe.

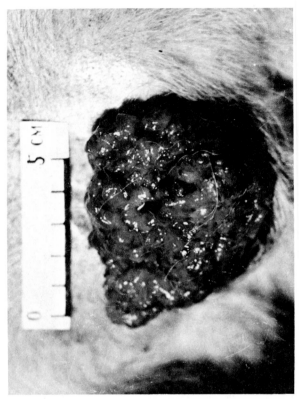

Figure 3. Venereal tumor growing around the external genitalia of a female dog.

division (Figs. 5 and 6) [93, 94]. Ulceration of the larger and more superficial masses is frequently followed by eventual disappearance of the growth; many tumors show a tendency to break down and regress after some months of progression [12].

Many attempts have been made to transplant the tumor to other mammals as well as to dogs. Sticker [77] and Wade [90] successfully transplanted the tumor into the subcutaneous tissue and the peritoneal cavity of foxes. Stubbs and Furth [78] reported a temporary propagation of the tumor in the subcutaneous tissue of mice exposed to 400 r of X-irradiation. Shirasu [70] successfully transplanted the tumor into the cheek pouch of the hamster, with or without cortisone pretreatments. Using dogs, De Monbreun and Goodpasture [10] transplanted viable cell sus-

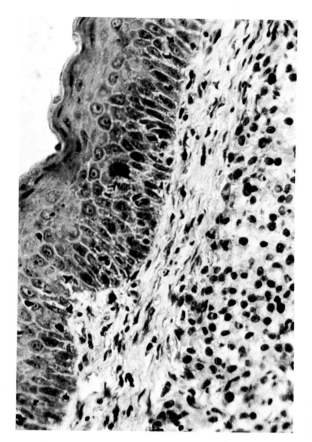

Figure 4. Section of a tumor growing near the external portion of the labia.

pensions, either by scarification of skin or genital mucosa or by subcutaneous implantation.

Karlson and Mann [27] reported the experimental passage of the tumor through 40 generations of dogs over a period of about 7 years. Watanabe and his associates [91, 94] have suggested the possibility that this tumor is capable of both natural and artificial transfer through contact with the hemorrhagic fluid stagnated from the tumor surface. It now seems apparent that transfer of the serous effusion containing the suspended, proliferating tumor cells from an affected dog to a healthy dog, through the injured mucous membrane of the external genitalia by means of direct contact on coition, does result in malignant tumor growth.

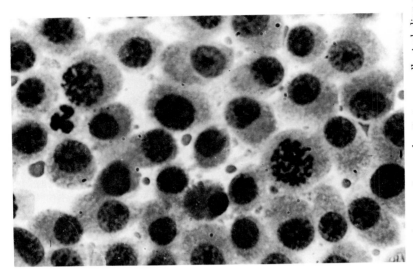

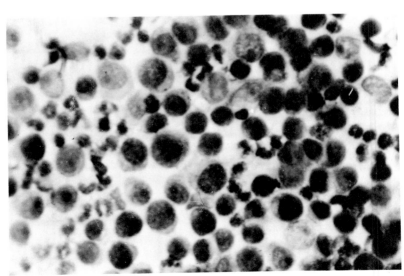

Figures 5 and 6. Smears of tissue fluid stagnating on the tumor surface: pyknotic and nonpyknotic tumor cells, including some in mitosis; lymphocytes; granulocytes; other tissue cells, at low and high powers, respectively. Courtesy of Dr. F. Watanabe.

So far, transmission by subcellular, filterable material has failed or results have been inconclusive [4].

CYTOGENETIC ASPECTS: A COMMON KARYOTYPE

Since 1956, the author, in collaboration with his colleagues, Watanabe, Takayama, Sofuni, and Ojima, has carried out a survey of the chromosomes of the dog venereal tumors found in several greatly separated localities in Japan; special regard was given to the morphology and behavior of the chromosomes in relation to the transmission, distribution, and growth of the tumors [41, 46, 71, 80–81, 84, 94, 95]. Recently three important papers have been published on the studies of a similar scheme in the venereal tumors by Weber, Nowell, and Hare [96] in Philadelphia, by Barski and Cornefert-Jensen [4] in Paris, and by Thorburn et al. [85] in Jamaica. Results were identical or similar to those obtained by the Japanese authors mentioned above. The cytogenetic features of the venereal tumors of the dog are reviewed here with particular attention to the karyotypic characteristics, as they can be related to the pathology of the disease.

Studies in Japan

Although there were a few fragmented reports on the chromosomes of the dog venereal tumors [81, 91, 94], Takayama and Makino [81] seem to have been the first to analyze the chromosomes in detail by use of the hypotonic squash technique. It was indeed through their studies that the remarkable similarities were established among the karyotypes of tumors obtained from different parts of Japan. Additional and supplementary studies followed, with confirmatory results by Makino, Sofuni, and Takayama [46], Sofuni and Makino [71], Takayama and Ojima [84], Watanabe and Matsunaga [95], and Koike [30]. A total of 24 primary tumors supplied samples for chromosomes studies in both sexes of dogs from Hyogo Prefecture, Nagasaki Prefecture, Osaka, and two cities in Hokkaido, Sapporo and Otaru (Table 1).

In most cases, the tumors were protruding from the labia in females (Fig. 3) or were infiltrating the root of the penis in males. Particularly in breeding season, the tumors were purulent, with hemorrhagic fluid stagnated in shallow depressions on their surfaces. In addition to the above primary tumors, one case of a transplanted tumor was also studied cytogenetically [30, 61]. To droplets of tissue fluid on clean slides was added a nearly equal amount of distilled water or of a hypotonic solution of any formula. After 15 to 20 minutes, preparations were stained with acetic dahlia (0.75 g of dahlia crystal in 30% acetic acid) for 5 to 10 minutes and squashed under the coverslips. Alternatively, the tissue fluid was fixed with acetic methanol (3:1) after the hypotonic treatment on a slide, air-dried, and stained with dahlia solution [84]. Biopsy pieces were minced in Hank's solution, and the suspension was transferred to 10 ml of balanced salt solution. After treatment with hypotonic solution,

Table 1. Chromosome-Number Distributions in 24 Cases of Primary Venereal Tumors of Dogs from Different Locations in Japan*

Case Designation	Locality	Ref.	≤ 52	52	53	54	55	56	57	58	59
Hy-I	Sasayama	81								4	42
Hy-II	Sasayama	81	1					1		2	40
Hy-III	Sasayama	81							3	13	39
Sa	Sapporo	81									§
Ot	Otaru	81									§
Hy-IV	Hyogo	83									30
Os-I	Osaka	83								32	
Sa-I	Sapporo	46	1	1	1	2	2	6	7	7	20
Sa-II	Sapporo	71	2	2	1	1	2	4	1	2	17
Sa-III	Sapporo	71			1			10	1	1	53
Sa-IV	Sapporo	71						5	2	4	34
Sa-V	Sapporo	71	1		1			1	6	3	27
Sa-VI	Sapporo†	71	4					3	3	2	44
Sa-VI	Sapporo‡	71									
Sa-VII	Sapporo	71	1					1	1	8	11
Sa-VIII	Sapporo	71	1					1	1	8	11
Sa-IX	Sapporo	71					1	1	3	2	35
Sa-X	Sapporo	71					2	5		13	40
Os-s	Senriyama	84					1	1	3	5	10
Hy-n	Nishinomiya	84							2	1	41
Na-I	Nagasaki	95					2	1	7	6	19
Na-II	Nagasaki	95					2	1	6	4	19
Na-III	Nagasaki	95					2	2	7	4	21
Sa-XI	Sapporo	29	1		3		1	3	4	3	3	18
Sa-XII	Sapporo	29							2	3	2	13

such cell suspensions were centrifuged, air-dried, and stained with Giemsa [30, 40, 61, 95]. These analyses were made from tumor fluids, or from biopsies, using hypotonic treatment and either squash or air-drying techniques [30, 40, 61, 64, 84, 95]. We used monolayer cultures of lung and kidney tissues from unaffected male and female puppies for normal karyotype controls [44, 101].

By use of the older testis-section method, Minouchi [51], Makino [39], and some others had already established the diploid chromosome number to be 78 in both sexes of the dog. Confirmative and supplementary evidence has been provided by recent investigators based on current technical methods involving tissue culture and colchicine and hypotonic pretreatments [2, 19, 63, 81, 97]. It was shown that the 78 chromosomes in the normal somatic complement comprised 76 acrocentric autosomes and a pair of metacentric sex elements, XX in the female and XY in the male (Figs. 7–9).

Chromosome Number. The microscopic appearance of the tumor fluids has already been described (Figs. 5 and 6). The mitotic rate has been reported by two groups

60	61	62	63	64	65	72–116†,‡	117	118	119	121–137	Number of Cells Observed
2	1					1		1			50
5						1		1			50
38	6	1									100
											30
											32
		1									48
1	1	1									34
	1	2									69
4	1					1				1	52
4	1										44
2	3		1			15†			9		86
						6‡	8		3	4	71
						1					23
						1					23
1	2										45
9	2					2		1	1		75
											20
2											49
3	13	8	9		5						73
4	18	10	12	6	7						89
5	17	6	10	5	5						84
1	2	4	2	1							42
											20

* All animals were female except Hy-I and Na-II.

† In Sa-VI there were: 3 cells with 100 chromosomes, 3 with 114, 2 with 113, and 1 cell each with 98, 101, 107, 108, 112, 115, or 116.

‡ In the second study of Sa-VI there were: 2 cells with 113 chromosomes and 1 cell each with 74, 107, 108, or 114.

§ In a few cells only.

of investigators. Watanabe and Azuma [94] found it to be 1.55 to 1.85%, with 60% of the cells in metaphase; Takayama [80] found considerable variation among samples and reported that the rate varied from 0.65 to 2.35% with an average of 1.35%. (These mitotic rates were similar to those for the Yoshida and MTK sarcomas, two ascitic tumors of rats [18, 87].) Abnormal mitoses, such as multipolar division, bridge formation and irregular scattering of chromosomes, and chromosome stickiness were rather rare. Watanabe and Azuma [94] found 18 such abnormal cells in 1030 dividing cells observed (1.75%). It is evident, therefore, that the majority of tumor cells in the process of division show a regular mitotic behavior.

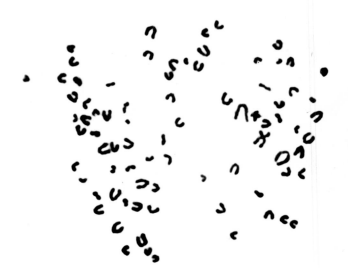

Figure 8. Metaphase of a normal male dog.

Figure 7. Metaphase of a normal female dog.

346

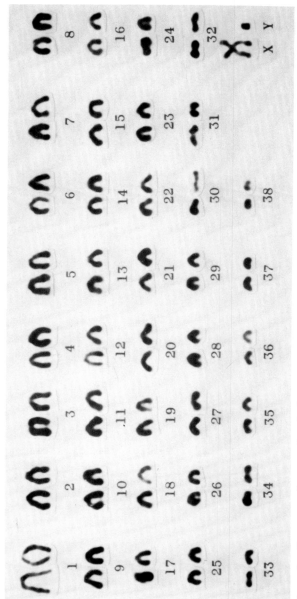

Figure 9. Karyotype of a normal male dog.

Descriptions of the cytogenetic findings in 24 tumors from six different and greatly separated locations of Japan follow. These tumors are given the following designations: Hy-n (from Nishinomiya City, Hyogo Prefecture), Hy (from Sasayama City, Hyogo Prefecture), Na (from suburbs of Nagasaki City, Nagasaki Prefecture), Os (from vicinities of Osaka City), Ot (from Otaru City, Hokkaido), and Sa (from suburbs of Sapporo City, Hokkaido). Results are summarized in Table 1. The data indicate that, although the number of cells which permitted reliable counting of chromosomes was dissimilar in different cases studied, and there was a variation or a wide scattering in number of the chromosomes in all of the cases, the modal number was 59 in the vast majority of these cases (Figs. 10 through 14). There was frequent variation between 53 and 62, but cases in which chromosome examinations were available in many reliable metaphases usually had a mode of 59 chromosomes: 89% in Hy-n, 84% in Hy-I, 80% in Hy-II, 78% in Sa-IX, 77% in Sa-III, 65% in Sa-IV, 61% in Sa-V, and 53% in Sa-X.

Three cases among these 24 seemed exceptional: Hy-III (with 59 and 60 chromosomes), Sa-VIII (with 58 and 59), and Os-I (with 58). It cannot be stated with certainty whether the chromosome counts in these cases were decisive or not. Although we are aware of cases of murine and human tumors in which the existence of sublines with different preponderant modal numbers have been found in addition to the main line, it seems probable that the results in those instances were due to poor preparations or to inadequate numbers of dividing cells (e.g., ref. [83]). This conclusion is supported by three studies by American, French and Jamaican investigators, who have shown the modal chromosome number of the canine venereal tumor cells to be invariably 59. It is most interesting that findings on these tumors should agree so closely, though different tumors were obtained from different localities and at different times. This seems to indicate that they undergo in all cases a similar striking alteration from their original or normal tissue. It appears that the tumor cells with the modal chromosome number 59 are none other than the stemline cells of this venereal tumor, and that, as such, they are analogous to such cells in mouse and rat tumors (i.e., the essential contributors to tumor formation).

Some interesting sequential changes were observed in the stemline cells derived from a specimen Sa-VI. In the first sample, the stemline cells had the 59 chromosomes (s-range) characteristic of this tumor in 73% of the cells, whereas the second sample, taken two months later, had chromosome numbers in 2s-range in over 75% of observed cells (Table 1). The 2s-number fluctuated from 107 to 123, with a mode of 117 in 38%. In the third sample, taken after another two months, the 2s-cells were increased considerably to 91% (Fig. 15). Thus Sa-VI showed the stemline cells with 59 chromosomes (s-range) in the early growth period of the tumor, while in the later stage the modal cells were replaced by the 2s-chromosomes which then formed the stem lineage, probably as a result of selective competition under new physiological conditions. The 2s-cells probably originated through the process of endoreduplication from the s-cells. This same Sa-VI tumor also showed a metastatic growth in ventral lymph nodes [71].

Chromosome Morphology. The morphology of the chromosomes of the tumor cells was analyzed in many cases on the basis of two to six reliable metaphases of stem cells in each case. Every cell with the modal number 59 of well-delineated chromosomes had 17 metacentric or submetacentric and 42 acrocentric chromosomes (Figs. 16 to 18). Takayama and Makino [81] at first reported the number of meta- and submetacentric elements as 16, and that of acrocentrics as 43 in some Hy cases,

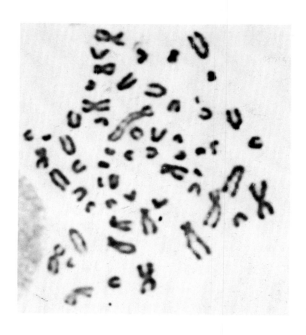

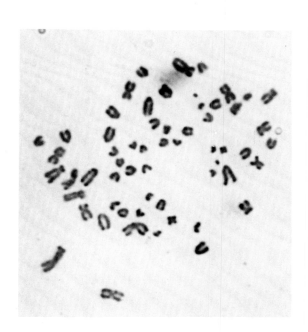

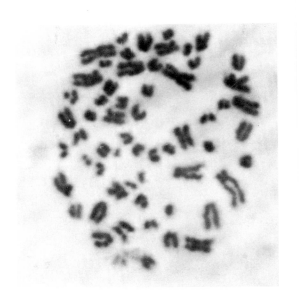

CYTOGENETICS OF CANINE VENEREAL TUMORS 351

Figures 10 through 14. Metaphases of stemline cells from different primary venereal tumors, 59 chromosomes in each: 10 from Sa-V, 11 from Sa-X, 12 from Hy-n (Takayama and Ojima, 1967 [84]), 13 from Na-II, and 14 from Na-III. Courtesy of Dr. F. Watanabe. Acetic dahlia stain. X3500 (approx).

but detailed examinations with improved techniques (of both Sa and Hy cases) have confirmed that the number of metacentrics and submetacentrics is 17, and that of the acrocentrics, 42.

Detailed identification and comparison of individual chromosomes, made of cells studied from different tumor animals coming from different localities, revealed that the chromosomes of the stem cells exhibited no identifiable morphological difference among the cells from different tumor animals.

This similarity among tumors was further supported by detailed analysis of the biarmed chromosomes (meta- and submetacentrics). The first and second chromosomes, by length, were easily distinguishable from the others by their large size and submedian–median structure. The third to seventh elements showed very gradual size reduction, being nearly metacentric in appearance. The chromosomes ranging in length from the eighth to the fifteenth rank were rather difficult to differentiate because of their similar size and shape. The tenth- to seventeenth-rank chromosomes, which were metacentric, were identifiable by their relative smallness. The remaining 42 chromosomes were acrocentric and of decreasing lengths. The rate of size reduction appeared to be somewhat greater than that of the normal complement of chromosomes for the species, though actual measurements were not carried out. It is apparent that in this karyotype, although the total number of chromosomes decreased, there was an increase in the number of meta- or submetacentric (biarmed) chromo-

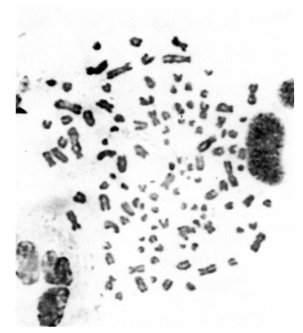

Figure 15. From Sa-VI, showing 118 chromosomes (third sampling).

somes (to be discussed later). The findings suggest that considerable structural deviation from the normal chromosome complement has taken place—a situation quite comparable to that found in some murine and human tumors—that is, that each tumor is characterized by its own particular karyotype, distinguishable from other tumors as well as from the normal somatic tissue [32, 40].

Takayama and Ojima [84] and Watanabe and Matsunaga ([95] and personal communication) recently reported the presence of a marker chromosome in canine venereal tumors (Hyogo Prefecture and Nagasaki areas). In direct squashes from tumor tissues with a modal number of 59 and 16 to 17 biarmed chromosomes, they found one of the larger submetacentric or metacentric chromosomes to display negative heteropyknosis either of its long arm or throughout most of its length. This marker seems to be similar to that found by Weber et al. [96] in both a transplanted and a primary venereal tumor of the dog in the Philadelphia area. It seems probable that our failure to find such a marker in the earlier studies [41, 46, 71, 81] is attributable to the use of less suitable technical procedures or possibly to inattention. In our more recent studies, using improved techniques, we also have been successful in demonstrating this marker chromosome in Japanese specimens. (Refer to Addendum; [61]).

In addition to the primary tumors just described, we had an opportunity to observe the chromosomes of an experimentally transplanted venereal tumor. In 1967 a venereal tumor was removed from the external genitalia of a female dog found in Sapporo City, and it has since been maintained by serial subcutaneous transfers in the abdominal regions of dogs by Koike [30] in the School of Veterinary Medicine,

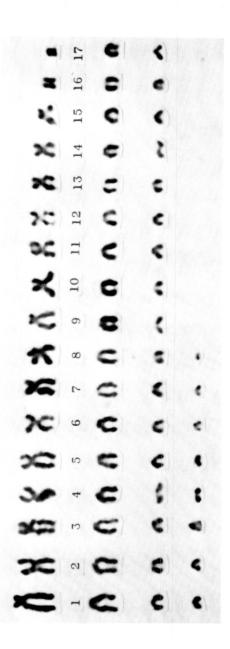

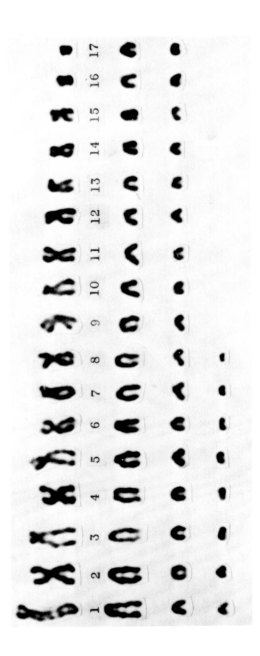

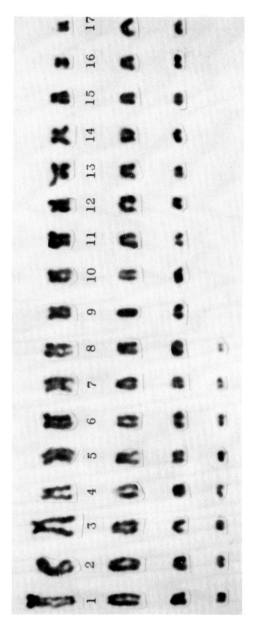

Figures 16 through 18. Stemline karyotypes of primary venereal tumors, each with 17 metacentric and 42 acrocentric chromosomes (16 from Sa-VIII, 17 from Sa-X, 18 from Sa-IV).

355

Hokkaido University. The chromosomes were analyzed in samples from the seventeenth to twentieth passages. In 42 metaphases from the eighteenth passage, for example, most showed the chromosome numbers ranging from 54 to 61, around a mode of 59 [30]. Well-spread cells contained 59 chromosomes (Figs. 19 and 20). The chromosomes in these cells were strikingly similar to those described for cells from primary tumors with respect to both biarmed and acrocentric chromosomes. One large submetacentric chromosome with negative heteropyknosis of both chromatids of the long arm was present. Histological pictures were quite the same between the original and the transplanted tumors ([30]; Figs. 21 and 22).

Studies in the United States

In 1965 Weber et al. [96] studied two venereal sarcomas obtained from Pennsylvania dogs. One tumor had been removed from the external genitalia of an East Stroudsburg male and carried in adult dogs by serial subcutaneous transplantation; it was studied in its fourth and fifteenth passages. The other tumor was studied directly from a Philadelphia male, using biopsies of masses in three areas: the penis and the subcutaneous tissues of the flank and the scapula—the last two areas apparently representing metastases. For all tissues studied from each tumor, whether primary or transplanted, the findings were similar, with 59 chromosomes in most cells.

Karyotype analysis of representative cells from these tumors revealed that there were 15 metacentric or submetacentric chromosomes and 44 acrocentric ones. Among the 15 biarmed chromosomes, one large submetacentric chromosome was remarkable by displaying negative heteropyknosis of both chromatids of the long arm. Although the appearance of this marker chromosome varied slightly from cell to cell, it could be identified in all well-spread metaphases. Among the other 14 metacentric and submetacentric elements, only 6 could be readily arranged in homologous pairs. In the remaining 44 (acrocentric) chromosomes, the location of the centromere in several smaller chromosomes was indistinct. Leukocyte cultures from the dog with the spontaneous venereal tumor revealed only cells with the karyotype of a normal male.

Thus although these two tumors had arisen in locales 100 miles apart, their chromosome complements were strikingly similar in both number and morphology. The tumor cells had a chromosome mode of 59, 15 meta- or submetacentrics, and 44 acrocentrics. One large submetacentric was distinguishable from the other chromosomes by a negatively heteropyknotic long arm. It is thus apparent that the venereal tumors of dogs in Japan and in Pennsylvania have nearly identical karyotypes, with one slight difference: the karyotype of the Japanese tumor cells included 17 metacentrics and 42 acrocentrics, whereas those in Pennsylvania included 15 metacentrics and 44 acrocentrics.

Studies in France

Just after the report by Weber et al. [96], we learned that Barski and Cornefert-Jensen [4] had also studied the chromosomes of venereal tumors of the dog obtained from three different locations of France: two primary tumors from the Paris area (one of them also transplanted) and another carried in transplant for three years in southern France. All primary tumors as well as the transplanted ones showed proliferating, well-vascularized, malignant growth that histologically fitted the classical description of the Sticker venereal sarcoma [77, 92].

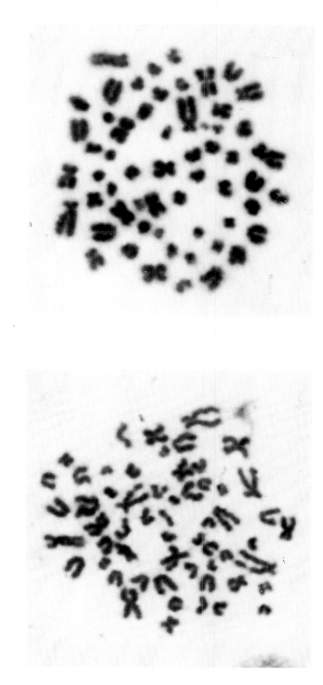

Figures 19 and 20. Metaphase chromosomes of a transplanted venereal tumor at the eighteenth passage (Fig. 19) and the twentieth passage (Fig. 20). Courtesy of Dr. T. Koike and Mr. M. Oshimura.

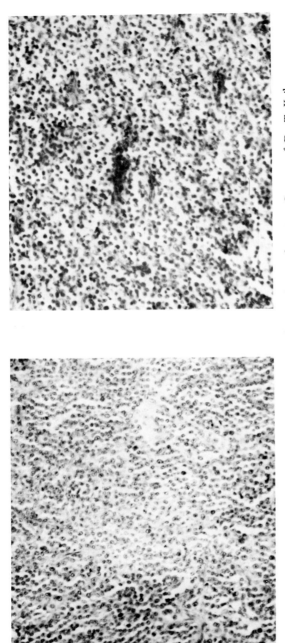

Figures 21 and 22. Histological sections of a primary tumor and its ninth transplantation. Courtesy of Dr. T. Koike.

Detailed descriptions were made on the chromosomes in one of the primary tumors found in Villejuif (south of Paris). In 55 metaphases studied by aceto-orcein staining, the frequency distribution of chromosome number ranged between 49 and 60, with a mode at 57-58. The cells contained 11 to 23 biarmed chromosomes, with a sharp mode of 17. Many biarmed chromosomes were indistinguishable from corresponding ones reported by the Japanese authors. The general pattern of the karyotype, and particularly a negatively heteropyknotic marker chromosome, were reminiscent of the karyotype described by Weber et al. [96] for the American specimens. A transplant of the above tumor had a similar karyotype.

Another case of the primary tumor from north Paris showed between 50 and 61 chromosomes per cell, 58-59 being most frequent. The metacentric chromosomes showed a striking similarity to those reported for Japanese specimens. A transplanted tumor which had originated in a Toulouse dog was found to have from 59 to 61 chromosomes per cell.

The authors concluded that in all cases studied, the karyotype of the tumors strikingly differed from that of the normal dog; the modal chromosome numbers were in each case close to 59, with 13 to 17 unusual metacentric chromosomes, many of which were markers similar or identical to those found previously in venereal sarcomas of Japanese and American dogs.

Studies in Jamaica

Thorburn et al. [85] studied the chromosomes of venereal tumors of Jamaican dogs. It was shown that eight dogs had a modal number of 59 in cells derived from the exudate as well as from the tissue of the tumors. There was one exceptional case that showed a tetraploid modal number of around 118. The karyotype, varying very slightly among different animals, showed 15 to 17 to be metacentric or submetacentric chromosomes and the remainder to be acrocentrics. There were two consistent marker chromosomes, one a long submetacentric and the other a very long acrocentric that showed a tendency to display a secondary constriction halfway along its long arm. There are also two less consistent markers, a long metacentric and a long submetacentric, sometimes showing negative heteropyknosis. Thus the chromosomal findings are, in essential points, similar to those reported from Japan, the United States, and France.

Cellular DNA Content

As an essential ingredient of the chromosomes, DNA plays a significant role in the process of cell division. Furthermore, cytophotometric data have indicated that the normal amount of DNA per nucleus is proportional to the chromosome number [37, 79]. Some variation in the amount of DNA occurs with metabolic activity of the cell [36], and it would be reasonable to expect that this would also occur as a result of certain physiological or pathological disturbances. Certainly DNA is known to govern the growth and differentiation of a cell. The DNA of the cell may retain its chromosomal architectural integrity, but if it be shifted, it results in an alteration in its expression on cellular function. Thus it is important to inquire into the state of the DNA in tumor cells, since carcinogenesis by viruses, chemicals, and radiation can be considered to be dependent, ultimately, on modification of the DNA structure and its function in the chromosomes [67].

Individual tumor cells from three dogs with venereal tumors (Ot, Sa, and one of Hy) were assayed for DNA content by use of Feulgen microspectrophotometry, according to the procedure of Pollister and Ris [62]. As a control, liver cells from a normal puppy were examined with the same technical process and the relationship of the chromosomal constitutions to the DNA content was analyzed [60, 82].

The results of DNA measurements are shown with histograms (Fig. 23). It can be seen that individual cells from the three tumors had nearly the same mean DNA content (i.e., from 3.4 to 3.5 arbitrary units, with a mean around 3.5). The mean DNA content in the control liver cells was 2.8. It can be concluded that the DNA content is higher in tumor than in normal cells.

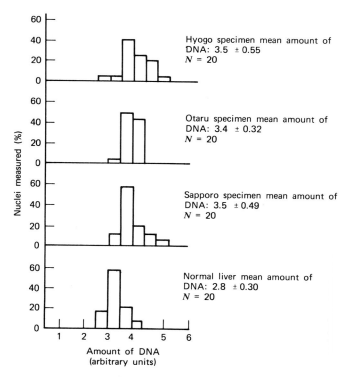

Figure 23. Histogram showing DNA contents in three venereal tumors and in normal liver tissue. The mean value for the tumor cells was 3.5±0.32; for normal cells, 2.8±0.30.

DISCUSSION

The literature in oncology includes a number of articles about chromosomes in cancer tissue. In the earlier literature, cytogenetic data emerged as by-products of pathological studies, and evidence was confused; high mitotic rates, remarkable mitotic irregularities, and striking abnormalities in chromosome number were reported as universal qualities of malignant tissues. Opinions concerning the significance of the chromosome abnormalities were diverse. Two major concepts were held: one attributed to the cytogenetic changes a predominant role in the initiation and course of cancer, while the other considered the chromosomal alterations to be secondary events. The two concepts have led to an animated controversy [13, 32, 38, 40, 67, 68].

Cytogenetics and Cancer

Obviously, knowledge of cancer cytogenetics has been much less satisfactory than that of other fields of cancer research. Indeed, the technical difficulties in handling the tumor tissue in the early work blocked inquiry into details of the essential features of the chromosome changes in relation to malignancy. However, as early as 1914, Boveri [7] had suggested some important etiological connections between chromosomal abnormalities and malignant changes in terms of the somatic mutation theory. (A summary of Boveri's contributions appears elsewhere in this volume [97].) The essential points in his theory are that mitotic abnormalities result in different chromosomes in individual cells and that those changes are reflected in the production of cells with new properties, one of which may be malignancy. The discovery of ascites tumors of rats [100] and the conversion of induced or spontaneous solid neoplasms into an ascitic phase [28] have greatly facilitated detailed analyses of the chromosomes of tumor cells; their use constitutes a new approach to studying the etiology of cancer and to understanding the mechanisms of malignant growth and the genetic pattern of the neoplasm. Karyotypic features of tumors have gained significance and have aroused increasing interest in medical fields because of the correlations between chromosome irregularities and the general clinical and pathological properties of some tumors.

The Stemline Concept

Current data collected on the chromosomes of transplantable murine tumors have suggested the stemline concept of tumor growth—that is, the appearance of a predominant cell population with a characteristic modal number of chromosomes and a particular pattern of chromosomal aberrations that are perpetuated on transplantation. Thus a specific stemline is thought to characterize each tumor and at the same time to distinguish it from other tumors and from the normal somatic tissues of the host. Such stemlines have been found in both animal and human tumors. Some chromosomes in the stemline usually differ in morphology from the normal, an observation that suggests a correlation between chromosome mutation and the neoplastic condition [*16, 24, 33, 38, 40, 43, 44, 47, 67, 68, 86*]. The stemline of each tumor is maintained because the chromosomes, although rearranged, are stable and are passed successfully through the usual stages of cell division. The behavior of different types of tumors can be explained by the existence of stemline cells, these being the cells that keep the tumor going. Available data indicate that the constancy of the stemline karyotype, though it is rather pronounced, is not always permanent, and that under certain circumstances or during serial transfers it has undergone numerical and structural changes [*16, 32, 43, 48*]. This shift is accompanied by a change also in the behavior and properties of the tumor.

In spontaneous and experimentally induced animal tumors, as well as in primary human tumors, the karyotypes of the cells usually show a kaleidoscopic picture, with each tumor having a chromosome constitution different from each other tumor. To our knowledge, furthermore, there has been observed in animal and human tumors little definite correlation between the tumor karyotype and the pathology [*14, 34, 35, 44, 47, 67, 68*]. An exception among human tumors, of course, is chronic myelogenous leukemia, in which the specific chromosome aberration, referred to as the Ph[1] chromosome, is almost always present [3, 58]. The exception among animal tumors is the venereal tumor of dogs, which we have just described, in which there is a striking similarity in the stemline chromosome picture among tumors obtained from widely separated geographic areas; the same, or very similar, stemline characteristics have been reported in specimens from Japan, the United States, France and Jamaica, as has already been mentioned. In both modal number and structural aberrations, the tumors are in all cases very much the same. The clinical features of the tumor also appear to be the same throughout the world.

Tumor by Cellular "Contagion"

It is well known that the tumor cells of the dog venereal tumors are readily transmissible and can be repeatedly transplanted to homologous, genetically unrelated dogs; the tumor cells maintain their specific and stable chromosome constitution through several in vivo transfers [4, 30, 61, 96]. As has already been said, in the serous fluid stagnated on the tumor surfaces there are numerous actively dividing tumor cells having the characteristic chromosomes. Under these conditions, it is quite reasonable to expect that this tumor would be transmitted in natural condition by "cellular contagion," as has been demonstrated by Watanabe and his associates [92-94].

There is another animal tumor that appears to be transmissible by natural contagion of tumor cells. A reticulum cell sarcoma of the Syrian hamster, described by Brindley and Banfield [8], seems to be similar in several aspects to the canine venereal tumor. It arose spontaneously in a Syrian hamster and was maintained by serial subcutaneous transfers. When tumor-free animals were caged with tumor-bearing animals, the majority of non-tumor animals developed tumors similar in histology to the original tumor. Even feeding of tumor tissue resulted in tumor development in some animals. The karyotype of each transferred tumor appeared identical to that of the parental tumor. The findings support the view of direct implantation of tumor cells as the mechanism of contagion [9]. Also in this connection, Manolov has discussed the possibility of such a mechanism for transmission of the Burkitt lymphoma (to be published).

Mechanism of Karyotypic Change

The mechanism accounting for the deviation of the karyotype of the venereal tumor from the normal karyotype of the dog is not yet known. Evidence was presented that the total number of chromosomes decreased inversely with the increase in number of metacentric (biarmed) chromosomes in the tumor. Such a relationship could be taken as evidence favoring the origin of the biarmed chromosomes by a centric fusion-fission mechanism. Barski and Cornefert-Jensen [4] explained the origin of the tumor karyotype in terms of Robertsonian centric fusion, arguing that the fundamental chromosome number of the tumor cells, as estimated from the numbers of chromosome arms, closely approximated the normal number of the dog [74, 76].

On the other hand, information is available to indicate that the majority of murine and human tumors are heteroploid. Chromosome changes in cellular transformation have been demonstrated in neoplastic mouse cells by Hsu and Klatt [20] to follow a sequence: diploidy to tetraploidy to heteroploidy. Another possible sequence, diploidy to subdiploidy to heteroploidy, has been proposed by Hsu et al. [21]. In the author's view, tumor cells have undergone profound alteration in the structure of their chromosomes.

It is probable that the situation is the same for canine venereal tumors. Both the numerical and the morphological deviations of the tumor chromosomes may be regarded as a reflection of structural alterations of the chromosomes. Most likely, the biarmed chromosomes of this tumor, whether they are of centric fusion origin or not, represent structural rearrangements of the chromosome material and have involved translocation, deletion, and inversion (particularly pericentric) (see Addendum; [61]).

In connection with the above discussion, the cytogenetic resemblance of chromosome changes as seen in certain canine sarcomas to those in the dog venereal tumor is of interest. Working on the chromosomes of a case of the canine fibrosarcoma, Mori in association with Sonoda et al. [52, 72] found in the lymph-node cultures a hypodiploid chromosome number with two distinct modes of 54 and 56. The karyotypes of the modal cells differed grossly from normal, and the presence of 14 to 23 biarmed elements in addition to acrocentric ones was noted. Thus the stemline karyotype of this disease generally resembles that of the venereal tumor, especially in the number and morphology of the biarmed chromosomes. Miles et al [50], working on the chromosomes of an in vitro cell strain that had been derived from a biopsy from a lymphosarcoma in a dog, found a stable karyotype consisting of 43-44 chromosomes, 26-27 of which were biarmed. It is interesting to see that the centric fusion (or any other mechanism that results in multiple biarmed chromosomes) occurs within a certain limited range of change, as is the general tendency of chromosome rearrangements in the cellular transformation of certain dog tumors. Miles et al. [50] have stated that if centric fusion is a frequent event in canine tumor cells or canine tissue culture, then the finding of the same total number of chromosomes and the same numbers of metacentrics in different tumors does not necessarily indicate that the tumors have a common cell lineage; rather, this similarity may indicate that karyotypes may have evolved by a similar mechanism. This is an important sequence of cellular transformation in carcinogenesis, but further confirmatory data are needed from dogs and other animals. Levan [34] has pointed out that nonrandomness of the chromosome changes has occurred in advanced human cancer stemlines.

Hypotheses to Explain the Universal Karyotype of Canine Tumors

The following possibilities seem to be worth considering to explain the apparent universality of the karyotype of the canine venereal tumors

[*41*, *81*]: (1) All the canine venereal tumors studied so far may have been derived ultimately from a single tumor in a single dog. (2) A certain definite constitution might be indispensable to continued active proliferation. (3) The chromosomal similarity may be merely superficial. Each of these will be briefly discussed below.

(1) Though the venereal tumors of the dog are transmissible from dog to dog by copulation, it is very surprising to see that the tumors found in geographically and diametrically distant locations such as Japan, the United States, France, and Jamaica—many thousands of miles apart—have a similar stemline karyotype. It is worth mentioning, however, that a long distance forms no serious bar to the transmission of this tumor by copulation because of the considerable mobility of a dog. Furthermore, the tumor-bearing dogs are usually able to live for several years and act as normal dogs do and may therefore have an effective opportunity for dispersion of this disease. In addition, the stability of the tumor karyotype, as well as of the histology, has been rigidly maintained in artificially transplanted tumors.

(2) On the other hand, one may reasonably assume that a certain definite chromosome constitution may be indispensable for the proliferation of the tumor. We are aware of several examples indicating that the chromosomes of human tissue cells of different origins growing in vitro tend to shift after many culture generations in certain, rather rigid patterns [32]. Levan [32] has shown that the chromosome number of most of the mouse tumors studied so far tends to gather in a hypotetraploid range. In the mouse, tumor changes of this type seem to promise high viability. Here, the following speculation might be made: at the start of the tumor development, a venereal tumor in a given locale might have had a stemline with a karyotype different from that of a tumor appearing elsewhere. In the course of development of each, however, several similar changes might have taken place in the chromosomes, and the tumor cells that arrive at the combination most successful for active reproductive ability within a certain limited change might have formed the stemline. The stemlines with similar karyotypes, as seen in the venereal tumors of the dog, might have been established by a sequence of such changes [*41*]. This view seems to warrant some attention.

(3) Consideration should also be given to the possibility that the apparent similarities in the karyotypes of the tumors are only superficial. It is most probable that the morphological likeness of the chromosomes does not necessarily guarantee their structural identity; witness the fact that structural and mutational changes in a chromosome can occur without concomitant change in the length or arm ratio of the chromosome. Apparently there may be many structural rearrangements in tumor chromosomes that may be closely correlated with the differences in neoplastic

properties characteristic of each tumor. Therefore, although the chromosomes of the tumors are morphologically similar to each other, their genetic constitution may not be identical. There are several animal tumors that show in their chromosomes no observable morphological deviation from normal tissue cells but differ from one another in neoplastic properties [17, 40]. It should be added, however, that the constancy of the stemline is not always permanent, and the stemline chromosomes tend to undergo transitions [16, 32, 43]: evidence has been presented in murine tumors that during serial transfers transitions have taken place in the stemline chromosomes as well as in the type and property of disease. The application of the new banding techniques (using quinacrine or Giemsa staining) to the chromosomes of the canine venereal tumor promises to explain the origin and nature of the biarmed marker chromosomes. The investigations along this line are currently being done in both normal and tumor cells by our associates. (Note: see Addendum; [61]).

Is the Venereal Tumor Virus-induced?

Another problem possibly to be considered here is that the canine venereal tumor might be a virus-induced neoplasm. Many of the results reported in the literature suggest that this is *not* the case: (1) the tumor could not be produced using a cell-free filtrate of a tumor extract [10, 53, 78], (2) viruslike particles were not demonstrated in the tumor, and (3) the transmission of the tumor was accomplished by use of viable tumor cells only; the tumor was not transmitted with frozen, heated, glycerin-treated, or desiccated cells [10, 78]. Recently Ajello [1] reported the presence of inclusion bodies in the cells of this tumor, but it cannot be concluded that they are virus particles.

Investigations have provided evidence that some viruses are capable of inducing numerical and structural chromosome abnormalities [55, 56]. The chromosome abnormalities induced by virus injection include at least three types of chromosome changes: single-chromosome breaks, chromosome pulverization, and alterations of mitotic mechanisms. No oncogenic virus has been shown capable of producing extensive cytogenetic alterations and of repeatedly producing cells with new and similar karyotypes, which could represent both the gain of metacentric chromosomes with the elimination of acrocentric ones, such as has occurred in the canine venereal tumor. Varied inconstant chromosome changes have usually been reported in virus-induced or virus-associated tumors.

The reasons that the karyotype of the canine venereal tumor is apparently the same in dogs all over the world and that this seemingly unique karyotype remains the same when the tumor is transplanted from dog to dog experimentally, remains unexplained. However, the findings so far support the view that the veneral tumors in diverse geographic locales have a common origin in a single dog (i.e., that the canine venereal tumor is naturally transplanted). Whether a virus was originally involved in the appearance of this tumor is unknown [41, 92, 96].

Application of the recently introduced Q- or G-banding techniques very soon should throw light on the origin and nature of the marker chromosomes and thus provide an explanation for the chromosome constancy and for the stability of this tumor.

Addendum in Proof

Oshimura et al. [61] studied the chromosomes of two primary and one transplanted venereal tumors of the dog obtained in Sapporo. The stemline of each contained from 57 to 59 chromosomes, which included 16-17 metacentrics and submetacentrics and 40 to 42 acrocentrics. The next to the largest submetacentric chromosome was remarkable in appearance because of negative heteropyknosis in its long arm. Comparative analyses of normal with the malignant cells, using quinacrine fluorescence and modified Giemsa-banding methods, suggested that extensive structural reorganization had taken place in the latter. The banding patterns in several of the larger biarmed and acrocentric chromosomes of the tumor cells appeared identical in both primary and transplanted tumors, with a few exceptions. Studies using ^{3}H-uridine autoradiography in combination with banding techniques indicated that the negatively-heteropyknotic segment of the second largest submetacentric element corresponds to the heterochromatin associated with the nucleolus organizer of interphase cells.

Acknowledgements.–It is my pleasant duty to express cordial thanks to my many colleagues with whom I have had pleasure of working on cytogenetic studies, particularly to Dr. Fumitomo Watanabe, Nagasaki University; Dr. Susumu Takayama and Dr. Yoshio Ojima, Kwansei Gakuin University; Dr. Toshio Sofuni, ABCC; and Dr. Toshio Koike, Hokkaido University, for their cooperation in placing important data at my disposal. I am also grateful to Dr. James German, The New York Blood Center, and to Dr. Motomichi Sasaki, Hokkaido University, for their encouragement and assistance in many ways with invaluable advice. This work was supported in part by grant No. 90375 for the study of cancer from the Ministry of Education, Japan.

LITERATURE CITED

1. AJELLO P: The presence of inclusion bodies in the cells of the Sticker tumor. *Boll Soc Ital Biol Sper* 37:247–248, 1961.

2. Awa A, Sasaki M, Takayama S: An *in vitro* study of the somatic chromosomes in several mammals. *Jap J Zool* 12:257–265, 1959.
3. Baikie AG, Court Brown WM, Buckton KE, Harnden DG, Jacobs PA, Touch IM: A possible specific chromosomal abnormality in human chronic myeloid leukemia. *Nature* 188:1165–1166, 1960.
4. Barski G, Cornefert-Jensen F: Cytogenetic study of Sticker venereal sarcoma in European dogs. *J Nat Cancer Inst* 37:787–797, 1966.
5. Beebe SP, Ewing J: A study of the so-called infectious lymphosarcoma of the dog. *J Med Res* 15:209–227, 1906.
6. Bloom F, Paff GH, Noback CR: The transmissible venereal tumor of the dog. Studies indicating the tumor cells are mature end cells of reticulo-endothelial origin. *Am J Pathol* 27:119, 1951.
7. Boveri T: Zur Frage der Entstehung maligner Tumoren. Jena, Gustav Fischer, 1914.
8. Brindley DC, Banfield WG: A contagious tumor of the hamster. *J Nat Cancer Inst* 26:949–957, 1961.
9. Cooper HL, Mackey CM, Banfield WG: Chromosome studies of a contagious reticulum cell sarcoma of the Syrian hamster. *J Nat Cancer Inst* 33:691–706, 1964.
10. DeMonbreun WA, Goodpasture EW: An experimental investigation concerning the nature of contagious lymphosarcoma of dogs. *Am J Cancer* 21:295–321, 1934.
11. Dustan J: Ineffective venereal tumors in dogs. *J Comp Path* 17:358–359, 1904.
12. Feldman WH: Neoplasms of Domesticated Animals. Philadelphia, WB Saunders Company, 1932.
13. Hamerton JL: Sex chromatin and human chromosomes. *Int Rev Cytol* 12:168, 1961.
14. Hansen-Melander E, Kullander S, Melander Y: Chromosome analysis of a human ovarian cystocarcinoma in the ascites form. *J Nat Cancer Inst* 16:1067–1081, 1956.
15. Hataya M, Usui K, Ichigi H, Fujita T: Effects of X-ray irradiation on the transmissible venereal tumors of the dog. *Gann* 49:307–318, 1958.
16. Hauschka TS: Cell population studies on mouse ascites tumors. *Trans NY Acad Sci* 16:64–73, 1953.
17. Hauschka TS, Levan A: Inverse relationship between chromosome ploidy and host specificity of sixteen transplantable tumors. *Exp Cell Res* 4:457–467, 1953.
18. Hirono I: Cytological studies on the effect of colchicine upon Yoshida sarcoma cells. *Nagoya J Med Sci* 17:59–66, 1954.
19. Hsu TC, Pomerat CM: Mammalian chromosomes *in vitro*. II. *J Hered* 43:167–172, 1953.
20. Hsu TC, Klatt O: Mammalian chromosomes *in vitro*. X. Heteroploid transformation in neoplastic cells. *J Nat Cancer Inst* 22:313–339, 1959.
21. Hsu TC, Billen D, Levan A: Mammalian chromosomes *in vitro*. XV. Patterns of transformation. *J Nat Cancer Inst* 27:515–541, 1961.
22. Ichikawa K: Tumors in laboratory animals. *Gann* 11:213, 1916.

23. IMAMAKI K: Ueber vergleihende Pathologie der Hundegeschwülste. 3. Ein Beitrag zur Kenntnis der transplantablen Rundzellensarkoma bei Hunden. Gann 26:29–33, 1932.
24. ISHIHARA T, MOORE GE, SANDBERG AA: Chromosome constitution of cells in effusions of cancer patients. *J Nat Cancer Inst* 27:893–933, 1961.
25. ISING U, LEVAN A: The chromosomes of two highly malignant human tumors. *Acta Pathol Microbiol Scand* 40:13–24, 1959.
26. KAALUND-JORGENSEN J, THOMSON AS: Das übertragbare venerische Sarkoma bei Hunden. *Z Krebsforsch* 45:385–398, 1937.
27. KARLSON AG, MANN FC: The transmissible venereal tumor of dogs; observations of 40 generations' experimental transfers. *Ann NY Acad Sci* Art 6, 54: 1197–1213, 1952.
28. KLEIN G: Immediate transformation of solid into ascites tumors. Studies on a mammary carcinoma of an inbred mouse strain. *Exp Cell Res* 8:213–225, 1955.
29. KOIKE T, HONJO H: Some statistical studies of the transmissible venereal tumors of the dog (in Japanese). *Jui-Chikusan Shimpo* 558:11–16, 1972.
30. KOIKE T, MATSUBARA K, KARASAWA T: Notes on the chromosomes of the primary and transplanted venereal sarcomas of the dog. *Chromosome Inf Serv* (CIS) 14:33–34, 1973.
31. LACROIX JV, RISER WH: Transmissible lymphosarcoma of the dog. *N Am Vet* 28:451–453, 1947.
32. LEVAN A: Chromosomes in cancer tissue. *Ann NY Acad Sci* 63(5):774–783, 1956.
33. LEVAN A: Chromosome studies on some human tumors and tissues of normal origin grown *in vivo* and *in vitro* at the Sloan-Kettering Institute. *Cancer* 9:648–663, 1956.
34. LEVAN A: Non-random representation of chromosome types in human tumor stemlines. *Hereditas* 55:28–38, 1966.
35. LEVAN A: Chromosome abnormalities and carcinogenesis. *In* Handbook of Molecular Cytology (Lima-de-Faria A, ed), 1969.
36. LEUCHTENBERGER C, SCHRADER F: Variation in the amount of DNA in cells of the same tissue and its correlation with secretory function. *Proc Nat Acad Sci (US)* 38:99–105, 1952.
37. LEUCHTENBERGER C, LEUCHTENBERGER R, DAVIS AM: A microspectrophotometric study of the DNA content in cells of normal and malignant human tissues. *Am J Pathol* 30:65–85, 1954.
38. MAKINO S: A cytological study of the Yoshida sarcoma, an ascites tumor of white rats. *Chromosoma* (Berlin) 4:649–674, 1952.
39. MAKINO S: A contribution to the study of chromosomes in some Asiatic mammals. *Cytologia* 16:288–301, 1952.
40. MAKINO S: The chromosome cytology of the ascites tumors of rats, with special reference to the concept of the stemline cell. *Int Rev Cytol* 6:25–84, 1957.
41. MAKINO S: Some epidemiologic aspects of venereal tumors of dogs as revealed by chromosome and DNA studies. *Ann NY Acad Sci* Art 3, 108:1106–1122, 1963.

42. MAKINO S, SASAKI MS, FUKUSCHIMA T: Cytological studies of tumors. XLI. Chromosomal instability in human chorionic lesions. *Okajima's Folia Anat Jap* 40:439–465, 1956.
43. MAKINO S, SASAKI M: Cytological studies of tumors. XXI. A comparative ideogram study of the Yoshida sarcoma and its subline derivatives. *J Nat Cancer Inst* 20:465–487, 1958.
44. MAKINO S, ISHIHARA H, TONOMURA A: Cytological studies of tumors. XXVII. The chromosomes of thirty human tumors. *Z Krebsforsch* 63:184–208, 1959.
45. MAKINO S, SASAKI M: A study of somatic chromosomes in a Japanese population. *Am J Hum Genet* 13:47–63, 1961.
46. MAKINO S, SOFUNI T, TAKAYAMA S: Cytological studies of tumors. XXXIX. A further study of the chromosomes in venereal tumors of the dog. *Nucleus* 5:115–122, 1962.
47. MAKINO S, SASAKI MS, TONOMURA A: Cytological studies of tumors. XL. Chromosome studies in fifty-two human tumors. *J Nat Cancer Inst* 32:741–777, 1964.
48. MATANO K, MAKINO S: Cytological studies of tumors. XXVII. Alterations of cell population in the MTK sarcoma-II after some experimental procedures. *Tex Rep Biol Med* 19:613–624, 1961.
49. MATSUI Y: Transplantable sarcoma-like neoplasm of dogs. *Gann* 4:123, 1909.
50. MILES CP, MOLDVANNU G, MILLER DG, MOORE A: Chromosome analysis of canine lymphosarcoma: two cases involving probable centric fusion. *Am J Vet Res* 31:783–790, 1970.
51. MINOUCHI O: The spermatogenesis of the dog, with special reference to meiosis. *Jap J Zool* 1:255–268, 1928.
52. MORI M: Chromosomes of a canine fibrosarcoma. *Chromosome Inf Serv* (CIS) 10:32, 1969.
53. MOULTON JE: Tumors in domestic animals. Berkeley and Los Angeles, University of California Press, Chapter 8, 164–168, 1961.
54. NANTA A, MARQUES LASSERRE BAZEX, BRU, PUGET: Les tumeurs veneriennes du chien. 'Candre' nosologique. *Rev Vet Toulouse* 101:298–310, 1950.
55. NICHOLS WW: Studies on the role of viruses in somatic mutation. *Hereditas (Lund)* 55:1–27, 1966.
56. NICHOLS WW: Interactions between viruses and chromosomes. *In* Handbook of Molecular Cytology (Lima-de-Faria A, ed), 732–750, 1969.
57. NOWELL PC, HUNGERFORD DA: A minute chromosome in human granulocytic leukemia. *Science* 132:1497, 1960.
58. NOWELL PC, FERRY S, HUNGERFORD DA: Chromosomes of primary granulocytic leukemia in the rat. *J Nat Cancer Inst* 30:687–703, 1963.
59. NOWINSKY MA: Zur Frage über die Impfung der Krebsigen Geschwulste. *Zbl Med Wiss* 14:790–791, 1876.
60. OJIMA Y, INUI N, MAKINO S: Cytochemical studies on tumor cells. V. Measurement of DNA by Feulgen-microspectrophotometry in some human uterine tumors. *Gann* 51:271–276, 1960.
61. OSHIMURA M, SASAKI M, MAKINO S: Chromosomal banding patterns in primary

and transplanted venereal tumors of the dog. *J Nat Cancer Inst* 51:1197–1204, 1973).
62. POLLISTER AW, RIS H: Nucleoprotein determination in cytological preparations. *Cold Spring Harbor Symp Quant Biol* 12:147–157 (1947).
63. REITER MB, GILMORE VH, JONES TC: Karyotype of the dog (*Canis familiaris*). Mamm Chromosome Newsl 12:170, 1963.
64. ROTHFELS KH, SIMINOVITSCH L: An air-drying technique for flattening chromosomes in mammalian cells grown *in vitro*. *Stain Technol* 33:73–77, 1958.
65. RUST JH: Transmissible lymphosarcoma in the dog. *J Am Vet Med Assoc* 14:10–14, 1949.
66. SAEKI Y: Cancer of domestic animals in Japan (in Japanese). *Jui-Chikusan Shimpo* 350:13–20, 1963.
67. SANDBERG AA: The chromosomes and causation of human cancer and leukemia. *Cancer Res* 26:2064–2081, 1966.
68. SANDBERG AA, YAMADA K: Chromosomes and cancer. *Chem Abst*, 15:58–74, 1965.
69. SELIGMANN CG: On the occurrence of new growths among the natives of British New Guinea. *In* Third Scientific Report Invert. Imp. Cancer Research Fund. 26–40, 1908.
70. SHIRASU Y: Polyp in dogs (in Japanese). *Jap J Vet Sci* 11:245–252, 1958.
71. SOFUNI T, MAKINO S: A supplementary study of the chromosomes of venereal tumors of the dog. *Gann* 54:149–154; 1963.
72. SONODA M, NIIYAMA M, MORI M: A case of canine fibrosarcoma with abnormal chromosomes. *Jap J Vet Res* 18:145–151, 1970.
73. SMITH G, WASHBOURN JW: Infective sarcoma in dogs. *Brit Med J* 2:1807–1810, 1898.
74. SMITH HA, JONES C: Veterinary pathology. Philadelphia, Lea and Febiger, 1966.
75. STEWART HL: Perspectives in comparative oncology. *Nat Cancer Inst Monogr* 32:1–16, 1969.
76. STICKER A: Transplantables Lymphosarcoma des Hundes. Ein Beitrag zur Lehre der Krebsübertragbarkeit. *Z Krebsforsch* 1:413, 1904.
77. STICKER A: Transplantables rundzellen Sarkom des Hundes. Ein Beitrag zur Lehre der Krebsübertragbarkeit. *Z Krebsforsch* 4:227, 1906.
78. STUBB EL, FURTH J: Experimental studies on venereal sarcoma of dog. *Am J Pathol* 10:275–286, 1934.
79. SWIFT HH: The desoxyribose nucleic acid content of animal nuclei. *Physiol Zool* 23:169–198, 1950.
80. TAKAYAMA S: Existence of a stem cell lineage in an infectious venereal tumor of the dog. *Jap J Genet* 33:56–64, 1958.
81. TAKAYAMA S, MAKINO S: Cytological studies of tumors. XXXV. A study of chromosomes in venereal tumors of the dog. *Z Krebsforsch* 64:253–261, 1961.
82. TAKAYAMA S, OJIMO Y: A study of the desoxyribonucleic acid content in some dog venereal tumors. *Jap J Genet* 36:206–209, 1961.
83. TAKAYAMA S, OJIMO Y: An additional note on the chromosomes of the venereal tumor of the dog. *Kwansei Gakuin Univ Ronko*, 9:99–107, 1962.

84. TAKAYAMA S, OJIMA Y: Consistency of karyotypes in canine venereal sarcoma cells. *Kwansei Gakuin Univ Ann St* 16:115–124, 1967.
85. THORBURN MJ, GWYNN RJR, RAGBEER MS, LEE BI: Pathological and cytogenetic observations on the naturally occurring canine venereal tumour in Jamaica (Sticker's tumour). *Brit J Cancer* 22:720–727, 1968.
86. TJIO JH, LEVAN A: Comparative idiogram analysis of the rat and Yoshida rat ascites sarcoma. *Hereditas (Lund)* 42:218–234, 1956.
87. TONOMURA A: Cytological studies of tumors. XVI. Cytological difference of MTK-sarcoma II and Takeda sarcoma, with preliminary experiments of double inoculation with the two tumors. *J Fac Sci Hokkaido Univ Ser VI Zool* 12:158–168, 1954.
88. TUTIE Y, HATAYA M: Experimentelle Studien über die biologische Behandlung der übertragbaren Geschwulst beim Hunde. *Jap J Vet Sci* 3:631–649, 1941.
89. UEDA A: An experimental study of infectious round cell sarcoma of the dog (in Japanese). *Obihiro Chikusan Gakujutsuho* 1:18–26, 1951.
90. WADE H: An experimental investigation of infective sarcoma of the dog, with a consideration of its relationship to cancer. *J Pathol Bacteriol* 12:384, 1908.
91. WATANABE F: Observation of the venereal tumor of dogs through fluid infection. In Proceedings of the International Genetics Symposium, 1956, 202–205, 1957.
92. WATANABE F, URANO K: Study of the clinically so-called infectious venereal sarcoma of dogs. II. Histological observations of venereal tumors of dogs. *Gann* 45:331–333, 1954.
93. WATANABE F, HOSAKA T, AZUMA M: Study of the clinically so-called infectious venereal sarcoma of dogs. III. Chromosomes of free sarcoma cells in the secretion from the venereal sarcoma of dogs, with particular regard to aqueous infection by transmission of free tumor cells. *Gann* 46:404–406, 1955.
94. WATANABE F, AZUMA M: Cytological confirmation of fluid infection in the venereal tumor of dogs. *Gann* 47:23–35, 1956.
95. WATANABE F, MATSUNAGA T: Chromosome studies of the infectious venereal tumor of the dog. In Proceedings of the Twelfth International Congress on Genetics 1:203, 1968.
96. WEBER WT, NOWELL PC, HARE CD: Chromosome studies of a transplanted and a primary canine venereal sarcoma. *J Nat Cancer Inst* 35:537–547, 1965.
97. WOLF U: Theodor Boveri and his book On the Problem of Malignant Tumors. This volume, pp 1–20, 1974.
98. WURSTER DA, BENIRSCHKE K: Comparative cytogenetic studies in the order Carnivora. *Chromosoma (Berlin)* 24:336–382, 1968.
99. YAMAGIWA K: Transmissible venereal tumor of a dog. *Gann* 3:475–477, 1908.
100. YOSHIDA T: The Yoshida sarcoma, an ascites tumor. *Gann* 40:1–21, 1949.
101. YOUNGNER JS: Monolayer tissue cultures. I, *Proc Soc Exp Biol Med (NY)* 85:202–205, 1954.

CYTOGENETICS OF CERTAIN SPECIFIC CANCERS

CHROMOSOMES IN HUMAN MALIGNANT TUMORS: A REVIEW AND ASSESSMENT

N. B. ATKIN

Department of Cancer Research, Mount Vernon Hospital, Northwood, Middlesex, England

Introduction	376
Study of the Chromosomes of Human Tumors: Some Problems in Methodology	377
Mechanical Injury and Chromosome Numbers	377
Choice of Tumor and Method of Study	378
Cytogenetic Assessment of Tumors	379
Karyotyping of Tumor Cells	379
Chromosome Complements of Human Malignant Tumors	380
Search for Common Features in a Given Tumor and Its Metastases	380
Common Features in Chromosome Patterns of Tumors in General (Irrespective of Site)	381
Chromosomes of Tumors at Specific Sites	384
Stomach	384
Colon, Cecum, Appendix, and Rectum	385
Bronchus	387
Breast	390
Ovary	390
Testis	394
Corpus Uteri	394
Bladder and Urethra	395
Malignant Melanoma	398
Lymphomas and Related Conditions	399
Miscellaneous Tumors	399
Discussion of Genesis and Significance of Chromosome Changes in Human Cancer	401
Relationship of the Chromosome Changes to the Site and Histologic Type of the Tumor	404
Karyotypic Changes in Relation to Chromosome Number	404
Chromosomes in Nonmalignant and Premalignant Conditions	406

A Possible Pathway Leading to Aneuploidy 406
Diagnostic Value of Cytogenetic Studies 407
Chromosome Changes and Malignant Transformation 409

Acknowledgements 411

Literature Cited 411

INTRODUCTION

The last decade has seen a steady accumulation of data on the chromosomes of human tumors, but still there are many gaps in our knowledge, in part because the types of tumors that have received most study are naturally those that are most readily available and produce the best preparations.

While revealing great variation from tumor to tumor, the data have failed to disclose any simple change, common to all cancers, which might be comparable to that giving rise to the Philadelphia chromosome in chronic myeloid leukemia; either there is no such change, or it is at present undetectable. However, it must be emphasized that the improved resolution that can be achieved with the use of the new chromosome-banding techniques has as yet made little impact on the field of cancer cytogenetics.

Perhaps we should look for common changes, specific to a given tumor site or histological type, which may be more readily discernible in the premalignant than in the invasive phase of tumor development. An example of such a change may be the loss of a No. 22 chromosome in meningiomas, which are benign tumors with some malignant potential [135, 224, 225].

The presence in some tumors of cells with normal, or apparently normal, karyotypes requires special consideration. Many tumors, albeit largely experimental tumors of animals, have been found to have normal chromosome complements in the early stages, and this has colored much of the thinking concerning chromosomes and cancer.

Often, there are questions of interpretation that may be difficult to answer. For example, are the cells with normal karyotypes stromal or inflammatory cells or are they tumor cells? Are the karyotypes really normal or might they have undergone some small but significant change?

The objectives in this chapter will be to review this substantial body of data, to discuss the possible significance of chromosome changes in neoplasia, and to attempt to discover and to state whatever generalizations there may be to explain the evolution of the aneuploid karyotypes found in malignant cells.

THE STUDY OF THE CHROMOSOMES OF HUMAN TUMORS: SOME PROBLEMS IN METHODOLOGY

Mechanical Injury and Chromosome Numbers

Since they have a bearing on the interpretation of the data, it is necessary first to refer to some technical considerations. Visualization of metaphase chromosome groups in solid tumors requires that the material be "processed." This usually involves pretreatment of the minced tissue with hypotonic solutions, perhaps exposure to Colcemid, flattening and spreading of the chromosomes on slides by squashing or air-drying, and application of staining procedures such as the new banding techniques. To ensure satisfactory spreading it may be necessary to employ fairly vigorous treatment; for example, the material may need to be resuspended several times in fresh fixative. If the treatment is too vigorous, many disrupted metaphases may result. This is most evident when comparing samples of the same material given different treatments; isolated chromosomes or small groups of chromosomes may be found in some preparations but not others. Unfortunately, it is impossible to be absolutely sure whether a given metaphase is in fact complete. The presence of many broken metaphases will result in distortion of the distribution curve of chromosome numbers, although the position of the mode or modes may not be appreciably altered. Ford's comments in 1957 [70] on the supposed variation in chromosome number in normal somatic tissues, which diminishes as techniques improve, are apposite to the present problem. There may be a bias favoring inclusion of broken metaphases, since they tend to exhibit better spreading and chromosome morphology, as is often evident from the superior appearance of small groups of obviously detached chromosomes. Problems of technique and analysis of data have also been discussed by Berger [35, 36].

Data concerning DNA content of interphase tumor cells suggest that the variation in chromosome number about the mode is on the average only a little greater than in normal tissues [6]; certainly, interphase DNA values corresponding to, say, 35 chromosomes or less are very seldom encountered in tumors (or normal somatic tissues, for that matter).

Choice of Tumor and Method of Study May Affect the Quality of the Data

A similar technical problem concerns the selection of tumors for chromosome study. It is unfortunately true at present that it is usually impossible to obtain successful chromosome preparations from every tumor studied. A major difficulty may be a paucity of dividing cells. Various subtle features of the tumor, including the type and degree of differentiation, may influence the spreading and morphology of metaphase chromosomes. In all, therefore, there may be a bias in favor of selecting the more malignant tumors with frequent mitoses.

By employing culture techniques, it may be possible to obtain larger numbers of divisions than could be obtained from the uncultured tumor. However, long-term cultures may be suspect, since the cells studied may either have originated from nonmalignant cells also present in the tumor, or, if malignant, they may have undergone some change in their karyotypes in vitro. In short-term cultures, the possibility of such changes having occurred during culture is obviously smaller [108]. In my own laboratory, we find that it is usually advantageous with tumors such as lymphomas and malignant testicular tumors, which we have found amenable to short-term culture techniques, to make preparations from both uncultured material and material cultured for varying periods up to 96 hours; one or the other may prove to be more productive, although the results are generally comparable.

Because of possible hazards, there have been few attempts to give Colcemid systemically to the patient before removal of a tumor in order to increase the number of metaphases. However, this method has proved to be useful in a recent study of bladder tumors [193], although it was restricted to patients over 50 in view of possible genetic damage to the germ cells and to those undergoing transurethral resection in case the Colcemid interfered with the healing of open operative procedures in the cystectomy patients.

An interesting observation is that, while to some extent dependent on the techniques, the morphology of the chromosomes of tumor cells tends on the whole to be less good than those of normal cells. This may be noticeable when comparing the chromosomes of (aneuploid) tumor cells and (diploid) normal or presumed normal cells coexisting in the same preparation, whether these are from solid tumors (see Fig. 4) or malignant effusions. A similar difference in chromosome morphology was noted between aneuploid and diploid cells from leukemia patients [20, 155]. The poor morphology comprises an indistinct outline, an ill-defined primary constriction, and sometimes an overcontraction of the chromosome, rendering it more difficult to classify. Varying degrees of spreading of chromosomes within a metaphase group, such as greater spreading of those lying peripherally, can present a problem (Fig. 4b); however, the varying intensity of staining of the chromosomes that results can often be compensated for by the experienced observer.

Cytogenetic Assessment of Tumors

Besides an assessment of the karyotypes of the tumor cells, their common features as well as their degree of variation, a complete cytogenetic assessment would take account of other features dependent on their chromosome complements. These include qualities of the interphase tumor cells such as X-chromatin [4] and fluorescent Y-chromatin [12], as well as nuclear protrusions due to large abnormal chromosomes [13]. Cancer cytogenetics, as a branch of pathology, has now been extended to include observations on histological sections; besides the above-mentioned features of interphase cells, the presence of protruding chromosome arms in metaphases, anaphases, and telophases may be recognizable in sections [18, 43]. The histologic terms "small-celled" and "large-celled" as applied to tumors may commonly, though by no means invariably, relate to the number of chromosomes in the tumor cells. Of course, the presence of giant cells in tumors is a direct result of polyploidization, whether by fusion of cells or some other mechanism.

Estimation of the modal DNA content of interphase tumor cells is also a useful technique and one that is not subject to some of the disadvantages attendant on chromosome studies such as infrequency or breakage of metaphases. The modal chromosome number can be estimated to an accuracy of about 10%, bearing in mind that there may be a slight discrepancy between the actual number of chromosomes and that estimated from the DNA value (the latter being on average too high, by about 4%, in one study [22]).

Karyotyping of Tumor Cells

In most published studies, karyotypes of tumor metaphases have been arranged according to the classification recommended by the Denver and London Conferences [164]. Where feasible, each chromosome is assigned to one of the lettered groups A to G, those in group A usually being further classified as A1, A2, and A3, and those in group E as E16 and E17/18. Within group G, Y chromosomes may be distinguished. Chromosomes that do not fit in, that is, do not resemble in morphology some normal chromosome, are then characterized as "markers."

It is debatable whether this is the best method for tumor cells that may have undergone many structural as well as numerical changes; but to anyone familiar with this system, it may seem to be the most satisfactory (or satisfying) arrangement available. Certainly, by any other arrangement so far devised, it would be less easy to recognize a karyotype that has undergone only minor changes, such as the gain or loss of a chromosome or two. It should go without saying that although a chromosome has been assigned to a normal group or pair number, it may in fact have undergone some unrecognized structural change and indeed may be quite foreign to the group or pair in its origin. In what follows, therefore, designations such as "A3 chromosome" or "in group B" should generally be taken to refer only to chromosomes of the given morphologic type. This rather unsatisfactory situation should, of course, change with the employment of the banding techniques that permit a more accurate identification of each chromosome [181].

CHROMOSOME COMPLEMENTS OF HUMAN MALIGNANT TUMORS

Before considering in detail the chromosomes of tumors at each of the common sites, some observations will be made on the karyotypic features of human cancer in general. In the detailed account of individual sites that follows, any exceptions or modifications to the general picture will be mentioned, together with any information that is available on premalignant lesions and benign tumors.

Search for Common Features Within a Particular Tumor and Its Metastases

In many studies on tumors it is found that the number of karyotypes of good quality that are available for study is strictly limited, although there may be other karyotypes which are usable, though less good. Often, therefore, it is only possible to describe some limited features of the karyotypes, which are then taken, with greater or lesser assurance, to be representative of the tumor as a whole. Nevertheless, in some studies there have been large numbers of cells with chromosome complements which were countable and reasonable numbers of analyzable metaphases, in spite of the inevitable presence of broken metaphases already mentioned. As a generality, it is found that in any given sample of tumor tissue the chromosome numbers of metaphases are grouped more or less closely around a mode, with a secondary mode of metaphases with doubled chromosome complements. On karyotype analysis, similarities are found which are common to most if not all metaphases within the sample. One or two diploid cells may be present, and these can usually be presumed to be nonmalignant stromal cells; however, as will be discussed later, substantial numbers of diploid metaphases have been found in tumors at some sites such as the bladder and corpus uteri.

The most noticeable feature of the metaphases in a tumor may be one or more marker chromosomes, especially where these are larger than the largest normal chromosome. Identical markers are often found in virtually every metaphase. On detailed examination, it may be apparent that the number of chromosomes of each type (recognizable as, or indistinguishable from, A1, group B, E16 chromosomes, and so on) tends to be the same in every metaphase and twice this number in metaphases that have doubled their complement. Although the similarities are usually close enough to indicate a clonal relationship, minor variations or at least uncertainties are common. Where a reasonable number of metaphases (say 10 or more) with the modal number are available, most may appear to have the same karyotype; on the other hand, it may seem that every one is different! Perhaps it is significant that the differences or uncertainties tend to involve chromosomes with near-median centromeres rather than those which are subtelocentric or acrocentric. Differential contraction of the two arms may present a greater problem in the former. However, in spite of the uncertainties, it is highly probable that the *degree* of variation within a tumor—such as minor variations in chromosome number, variations in karyotype among metaphases with the same chromosome number, or the degree of polyploidization—is itself a variable characteristic that may have a significant relationship to the behavior pattern of the tumor. Data obtained in this laboratory on carcinoma of the large bowel suggest that the greater the variation in the karyotypes (see the tumor of the corpus uteri in Fig. 5), the worse the prognosis.

Common Features in Chromosome Patterns of Tumors in General

The following considerations apply to tumors whose malignancy is not in doubt, but not to those tumors at sites such as the corpus uteri and bladder that are more or less well differentiated and show little or no evidence of local invasion and (at the time of study) none of metastasis. Nearly every tumor can be characterized by its modal chromosome number which, however, may fall anywhere within a wide range. Nevertheless at most sites the modal chromosome numbers tend to fall into one or another of two more or less well-defined groups, one situated around the diploid level and the other in the triploid-tetraploid range. This tendency to fall into two groups is reflected in the distributions of modal DNA values of tumors at various common sites (Fig. 1). Within the diploid range, chromosome numbers may vary from a little below 40 to 50 or so; tumors with modal chromosome numbers of 37 or 38 are quite common at some sites such as the ovary and breast, but modes below this are exceptional, although a malignant melanoma with a modal number of 29 or 30 has been reported [34]. At the upper end of the scale, malignant cells in an effusion from a patient with carcinoma of the breast had a mode of 133 chromosomes [95] and a primary seminoma had a mode of 156 chromosomes [137]; DNA data suggest that some carcinomas of the cervix have similarly high modes [8].

Certain common features of the karyotypes should be noted which are to some extent independent of the modal chromosome number. First is the presence of one or more marker chromosomes: markers are very commonly, and perhaps universally, present in tumors with hypodiploid chromosome numbers, while they are frequently but less regularly present in other tumors. From data obtained in this laboratory, for instance, markers were present in each of 11 hypodiploid ovarian carcinomas and in 18 of 21 with modal numbers of 46 or more; they were present in each of 12 hypodiploid cervical carcinomas but in only 8 of 18 tumors at this site with modes of 46 or more.

The second notable feature is a tendency toward certain changes in the relative numbers of chromosomes that can be fitted into each of the normal groups or subgroups. In particular, there are fewer than the expected numbers of chromosomes in groups B, D, and G; that is, the numbers of chromosomes in these groups are fewer in proportion to the total number of chromosomes present than in cells with normal karyotypes. Conversely, there are often more A3, group C, E16, and group F chromosomes than expected; the tendency, on the whole, appears to be toward an increase in chromosomes with medially placed centromeres as compared to those with distally placed centromeres [7]. Although commonly some group B, D, and G chromosomes have "disappeared," it is not known whether this represents actual loss of the chromosomes or transfer of their material to new chromosomes following structural rearrangements. It should be stressed that each individual tumor varies with respect to the direction and extent of any change in the number of chromosomes of a particular type; the tendency is usually clear, how-

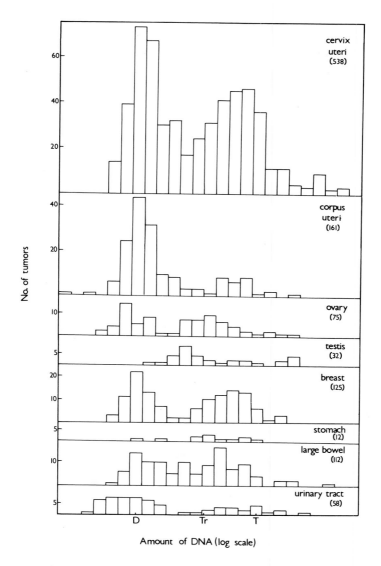

Figure 1. Modal DNA values of 1113 malignant tumors. The figures in parentheses indicate the number of tumors at each site. Urinary tract tumors were mainly from the bladder. D, Tr, and T: diploid, triploid, and tetraploid levels, respectively (cf. Table 1). Included here are data from the tumors listed in Table 1, as well as from 46 tumors of the large bowel, 26 of the breast, and 36 of the corpus uteri which were not listed in Table 1.

ever, when the average extent to which chromosomes of each type is under- or overrepresented in a series of tumors is calculated. Thus the numbers of group B, D, and G chromosomes in series of 12 carcinomas of the cervix, 11 of the large bowel, and 20 of the ovary were on the average reduced to 62 to 97% of the expected number [7]; see also Levan [119] and Steenis [201]. (The postulate that an increase in the proportion of E16 chromosomes is of special significance in human cancer was based on a semiautomatic system of chromosome analysis [147], but the method used has since been criticized [30].)

The bimodality of the distributions of modal chromosome numbers (see Table 1) and DNA values (Fig. 1) of series of tumors suggests that chromosome doubling has occurred at some stage in the evolution of a proportion of the tumors. However, as we discuss later, other interpretations are possible, as suggested by the finding that tumors in the higher range are commonly near-triploid rather than near-tetraploid.

In the ensuing detailed consideration of the cytogenetic data on tumors at individual sites, it is hoped that the coverage of published studies has been reasonably complete. Some of the older papers refer to chromosome numbers rather than morphology. Although no detailed account will be given of the numerous papers on DNA content and sex chromatin, a few individual studies will be mentioned when it is felt that they have contributed to an understanding of the tumors at a particular site. Leukemias and tumors of the cervix uteri and of the nervous system have not been included, since they form the subject of other chapters in this book [150, 195 and Mark, this volume].

THE CHROMOSOMES OF TUMORS AT SPECIFIC SITES

Stomach

There have been a number of studies of malignant effusions but only a few of primary stomach tumors. Among the former, modes of about 49-51 [23, 56, 128] and 60-65 [128, 177, 187, 206] appear to be common. However, modes as low as 40 and as high as 108 have been reported, the latter from a solid primary tumor [126]. A sessile adenomatous polyp showing malignant transformation had a mode of 58 chromosomes: out of 21 counts, eight were of 58 and four were of 46, one of the latter on analysis showing a diploid karyotype [138].

Several of the studies on effusions include repeated observations which sometimes showed small changes; thus Šlot [187] found that the modal number in one case had changed from 60 to 64 over a period of two months. Ising and Levan observed two ring chromosomes in a malignant effusion that had a modal number of 80 [98, 118].

There have been several other studies on gastric carcinoma [34, 93-96, 143, 194].

Colon, Cecum, Appendix, and Rectum

In contrast to the stomach, there have been numerous studies of tumors of the large intestine; data have been published on modal chromosome numbers, with or without details of the karyotypes, of at least 50 malignant tumors mostly obtained on the primary tumor [61, 95, 96, 112, 123, 124, 128, 138, 176, 222, 223]. In my laboratory, the chromosomes of 119 carcinomas have been studied by Miss M. C. Baker (Table 1), including samples from two or more regions of most of the tumors. The distribution of modal chromosome numbers in these cases agrees quite well with that of modal DNA values (Fig. 1). There are accumulations of tumors near the diploid level (from about 40 to 50 chromosomes) and again near the triploid level (60 to 75 chromosomes).

Based on the fairly extensive data on carcinoma of the large intestine, some generalizations may be made that are probably equally applicable to tumors at other sites: (1) Generally, all the malignant cells are aneuploid. (2) Their chromosome patterns indicate that they belong to a clone. (3) Their karyotypes tend to show the change in distribution that is characteristic of the aneuploidy of malignant cells: there are relatively few chromosomes in groups B, D, and G. (4) One or more easily recognizable marker chromosomes are frequently but not invariably present. (5) With a few exceptions, the chromosome patterns of the cells in samples from different regions of the primary tumor, or in the primary as compared with a metastasis, are indistinguishable, or at least are sufficiently similar to indicate that each belongs to the same clone. Sometimes two regions differ in that one has a mode that is about twice that of the other, the similar chromosome pattern indicating that the higher has been derived from the lower by chromosome doubling. The occasional exceptions, in which the chromosomes of two regions are dissimilar, may represent the confluence of two tumors. One example has previously been briefly reported [7]: two different regions of a carcinoma of the cecum varied histologically and also had quite different karyotypes with modes of 41-42 and 68 chromosomes, respectively. It appears that *different* tumors from the same patient are as likely to have different chromosome patterns as if they came from different patients; in a patient from Miss Baker's series, a carcinoma of the cecum and, 14

TABLE 1. MODAL CHROMOSOME NUMBERS OF 119 CARCINOMAS OF THE COLON, RECTUM, AND CECUM, 77 OF THE BREAST, AND 58 OF THE CORPUS UTERI*†

Site of Carcinoma	Characteristics	37-39	40-42	43-45	46-48	49-51	52-55	56-59	60-63	64-68	69-73	74-78	79-84	85-90	91-96	97-103	104-110	111-118	119-127	128-136	Total
Colon, rectum, and cecum	Lymph nodes not involved	—	2 (1)	5 (1)	15 (4)	6	6	2	8	4	5	8	2	1	—	—	—	—	—	1	65
	Lymph nodes involved	4 (4)	2 (2)	7 (2)	7 (1)	3 (1)	3	2	2	5	6	8	3	2	—	—	—	—	—	—	54
	Total	4 (4)	4 (3)	12 (3)	22 (5)	9 (1)	9	4	10	9	11	16	5	3	—	—	—	—	—	1	119
Breast	Lymph nodes not involved	5	5 (1)	3	6	1	2	1	3	6	5	4	4	2	—	—	—	—	—	—	47
	Lymph nodes involved	1	4 (2)	5 (1)	1	—	—	1	5 (1)	1	3	3	4	1	—	—	1	—	—	—	30
	Total	6 (1)	9 (3)	8 (1)	7	1	2	2	8 (1)	7	8	7	8	3	—	—	1	—	—	—	77
Corpus uteri	Well differentiated	—	—	—	7 (1)	1	1	—	—	1	—	—	—	—	—	—	—	—	—	—	10
	Moderately well differentiated	—	1	4	19 (1)	5	1	1	—	1	3	1	—	—	1	—	1	—	—	—	32
	Poorly differentiated	—	1 (1)	—	8 (1)	1	1	1	—	—	—	1	—	—	—	—	1	—	—	—	16
	Total	—	2 (1)	4	34 (3)	7	3	1	—	2	3	1	—	—	1	—	1	—	—	—	58

* The chromosome number classes correspond approximately to the DNA classes in Fig. 1. Figures in parentheses indicate the numbers of tumors having a prominent secondary mode at double the level of the primary mode. A carcinoma of the cecum with lymph node involvement which had modes of about 42 and 68 chromosomes in different regions (see text) has been regarded as two tumors. Also included in Fig. 1 are 46 tumors of the large bowel, 28 of the breast, and 36 of the corpus uteri.

† Data of Miss M. C. Baker.

months later, two carcinomas of the sigmoid colon, each had its distinctive chromosome pattern.

Malignant ascites from a patient with a mucocele of the appendix and pseudomyxoma peritonei was studied on four occasions over a period of three weeks. Modes of 54 or 55 chromosomes were found, with 3 marker chromosomes, but there was a considerable amount of minor variation among the karyotypes [*174*].

Carcinomas of the large bowel may often commence in polyps: the smallest carcinomas are seen as malignant transformations in polyps rather than as arising *de novo* from the mucosa [*200*]. Studies on polyps, therefore, have a direct relevance to carcinoma at this site [*7, 28, 61, 123, 124, 138, 143*]. Our own and the published data on polyps can be summarized as follows: (1) Cells with normal karyotypes may be present (these may or may not be epithelial cells). (2) Cells with abnormal karyotypes may be found alone, or may be present together with diploid cells. (3) The chromosome abnormalities commonly, though not invariably, involve simple changes such as the addition of one or two group C chromosomes, and their pattern may indicate or suggest a clonal relationship. The chromosome numbers of cells with abnormal karyotypes are often near-diploid (e.g., 47 or 48 chromosomes), but a polyp with a mode of about 80 chromosomes has been described [*7, 28*]. (4) Especially where there is histological evidence of early malignant transformation, more complex changes, including structural alterations, may be found. Examples of hyperdiploid karyotypes from a carcinoma and polyp of the colon are shown in Figs. 2 and 3, respectively.

Bronchus

Although it is one of the commonest of malignant tumors, there have been only a few studies of carcinoma of the bronchus [*31, 58, 63, 64, 78, 95, 98, 108, 118, 143, 198*]. Nearly all the published modal chromosome numbers, whether from the primary tumor or a malignant effusion, are in the region of 60 or above, but the true distribution for the different histological types is as yet unknown. Greisen [*83*] has shown from DNA measurements that all the main histological types of bronchial carcinoma may be near or below diploid.

Falor [*63*] has studied a carcinoid adenoma that yielded three karyotypes with 45 chromosomes without any consistent pattern (no markers were seen), and a cylindromatous adenoma showing local invasion that had a mode of 46 chromosomes, including markers.

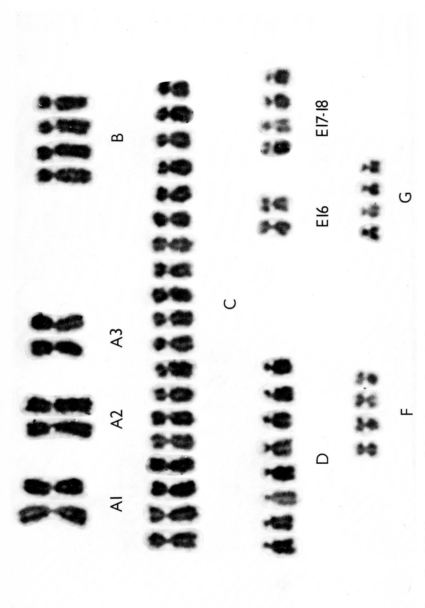

Figure 2. Colloid carcinoma of the colon, moderately differentiated, from a woman aged 75; 51 chromosomes. Three extra group C and two extra group D chromosomes are present, but no markers.

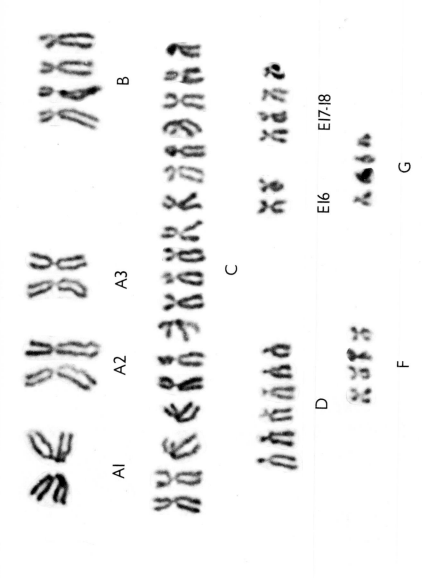

Figure 3. Hyperplastic villous polyp of the colon from a woman aged 71; 48 chromosomes. Two extra group C chromosomes are present.

Breast

Although the total number of tumors studied is not large, there have been a number of studies of primary and secondary carcinoma of the breast [*31, 43, 46, 75, 78, 84, 86, 87, 94–96, 103, 109, 112, 126, 143, 165, 197, 198, 205, 209*]. Like the DNA data (Fig. 1), Miss Baker's data on 77 primary tumors show (Table 1) that they fall into two groups. In the near-diploid group, hypodiploid tumors preponderate, as they do in the ovary (q.v.). The modes of most tumors in the higher group fall within the range of 60 to 85 chromosomes. Among the published reports is one of a fairly well-differentiated, infiltrating duct-carcinoma in a woman aged 77. The tumor, which had remained stationary over a 6-month period, had a mode of 46 chromosomes; 11 metaphases which were karyotyped showed a diploid complement [*205*]. A metastatic breast carcinoma [*143*] yielded only diploid karyotypes, but areas of lymphocytes were present, and the identity of the cells in the chromosome preparations was uncertain. Most carcinomas of the breast, however, are aneuploid, with marker chromosomes.

A giant fibroadenoma in a patient aged 19 yielded nine cells with 46 and eight with 47 chromosomes; analysis of three of the former showed diploid karyotypes, while four with 47 chromosomes had an extra group F chromosome [*159*]. In a lobular carcinoma in situ, one of two karyotyped cells with 47 chromosomes also had an extra group F chromosome [*205*]; in the same study, another lobular carcinoma in situ and four benign cystic hyperplasias yielded only a few diploid karyotypes.

Ovary

The chromosomes of carcinoma of the ovary have received fairly extensive study [*9, 13, 15, 31, 34, 41, 44, 47, 53, 55, 60, 66, 71–73, 81, 86, 87, 90, 94–96, 99, 104, 112, 114, 128, 130, 140, 142, 158, 172, 185, 186, 188, 198, 204, 207, 223*]. Most of the earlier publications relate to cells in malignant effusions.

A study of chromosome numbers and DNA values (see also Fig. 1) showed a tendency resembling that of carcinoma of the breast for the tumors to fall into two groups, the chromosome counts indicating that many of the tumors in the lower (near-diploid) group were *hypo*diploid [*9*]. Marker chromosomes, frequently as large as group A chromosomes or larger, are present in most but not all ovarian carcinomas, according to both the published reports and our own unpublished findings.

Among our cases, five had bilateral ovarian tumors, and in all these a similar aneuploid chromosome pattern was seen in each ovary. In one case, however, there were also a number of diploid metaphases in one of the ovarian tumors: eight of the 16 metaphases analyzed were diploid. Since few mitoses were seen among the stromal cells in histological sections and squashes of this tumor, it seems that some, at least, of the diploid metaphases were tumor cells. Possibly, the primary tumor was in this ovary; if so, it was apparently only the aneuploid cells (with a mode of 38 chromosomes) that had metastasized, because apart from occasional diploid cells that were probably not tumor cells, these constituted the cell population in mitosis in the other ovary, in an omental metastasis, and in malignant ascitic fluid—all of which had a similar hypodiploid pattern. It was noteworthy that the diploid cells generally had clear-cut chromosomes of good morphology, while those of the aneuploid cells tended to be more contracted with the poorer morphology more characteristic of malignant cells (Fig. 4).

A possible example of a metastasizing tumor with a diploid karyotype is provided by a folliculoma; the primary tumor, an omental metastasis, and malignant ascitic fluid were studied, and all had apparently normal karyotypes [54].

On the basis of DNA measurements, Bader and colleagues [25] found that benign, borderline, and Grade I malignant ovarian tumors consistently had modes in the diploid region. However, in a study from my laboratory the modal DNA content of a mucinous cystadenoma was compatible with a near-triploid chromosome complement; six other ovarian cystadenomas had DNA modes at the diploid level [17]. There have been few chromosome studies of benign ovarian tumors. Material from a papilliferous cystadenoma cultured for 12 days yielded cells with 47 and 48 chromosomes which had one and two extra group C chromosomes, respectively [74]; however, it has been suggested from the published photomicrograph that this tumor was in fact a well-differentiated mucinous cystadenocarcinoma [169]. In another report [32], three aneuploid metaphases with markers and chromosome numbers of 60, 150, and 300 were found in uncultured material from a mucinous cystadenoma; it was suggested that the aneuploidy implied that areas of early malignant transformation were present.

Only female diploid karyotypes were found in six benign ovarian teratomas [168]. A further benign teratoma in culture yielded diploid cells and cells with 47 chromosomes including an extra member in group C [180]. A low grade malignant teratoma in a patient aged 14, examined after culture for 29 and 169 days, showed on both occasions a mode of 47 chromosomes with an extra chromosome resembling an A3 [2]. A malignant teratoma in a child aged 9 had a mode of 46 chromosomes; nine cells were analyzed, all of which showed a slight departure from

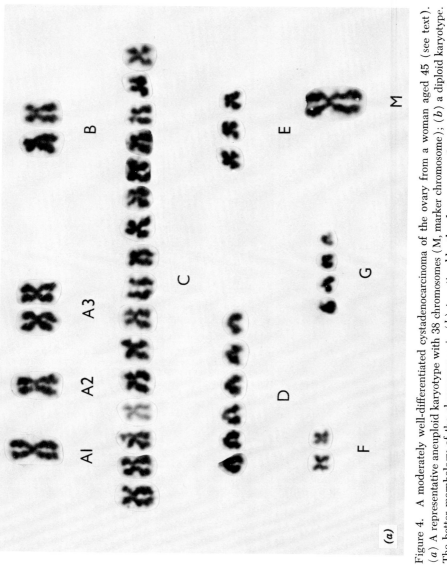

Figure 4. A moderately well-differentiated cystadenocarcinoma of the ovary from a woman aged 45 (see text). (a) A representative aneuploid karyotype with 38 chromosomes (M, marker chromosome); (b) a diploid karyotype. The better morphology of the chromosomes in (b) is noticeable; the chromosomes that appear paler were situated

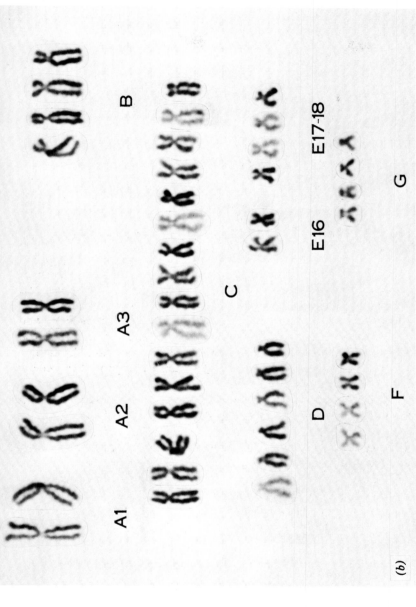

(b) at the edge of the group and were more stretched. In several diploid metaphases from this tumor, a difference was seen between the A2 chromosomes, one (shown at left) having a constriction at the centromeric region.

a diploid karyotype: an A3 chromosome was replaced by a metacentric chromosome of the size of the group C chromosomes [204].

Testis

In contrast to the ovary, a notable feature of the some 70 seminomas and malignant teratomas of the testis that have been studied is the absence of tumors with chromosome numbers at or below the diploid level [12, 15, 38, 65, 76, 116, 117, 136, 137, 142, 166, 170]; this is reflected in the DNA data shown in Fig. 1. In contrast, one benign teratoma had a modal DNA value in the diploid region (unpublished data). Lelikova and colleagues noted a nonrandom distribution of chromosomes in both malignant teratomas [116] and seminomas [117], particularly a deficiency of group B and excess of group C chromosomes. Berger and Martineau [38] found an excess of group F and a deficiency of E17-18 chromosomes in seminomas. Comparing the ratios of various pairs of groups, they found that in tumors combining the features of both seminomas and teratomas, the ratio of A3 to group B chromosomes was increased, while the ratios B/F and E17-18/F were decreased. The ratio A3/B was also increased in pure teratomas, and the ratios B/F and E17-18/F were decreased in pure seminomas. The pure tumors could also be distinguished by other features of the various ratios, and the tumors combining the features of seminomas and teratomas appeared to occupy an intermediate position in this respect.

Nearly all malignant testicular tumors have at least one large or medium-size marker chromosome (see Fig. 6); however, we have studied a seminoma without any obvious markers (unpublished data). It is well known that testicular teratomas are frequently X-chromatin positive. The reason for this is unknown, but it is clear that studies of the sex chromosomes are of special interest in this type of tumor. Quinacrine fluorescence observations on metaphases and interphases have confirmed, or at least suggested, the presence of Y chromosomes in both seminomas and malignant teratomas, including among the latter some that were X-chromatin positive [12].

Corpus Uteri

Malignant tumors of the corpus uteri show a tendency to fall into two groups having near-diploid and high modes, respectively (Table 1 and Fig. 1). This pattern bears some relation to histological differentiation: while most undifferentiated tumors have high modes, the more well-differentiated tumors are usually near-diploid (Fig. 5) [8]. Among the latter, modes of 46 to 49 chromosomes are common. Some of these tumors show small deviations such as a single group C trisomy or at most small numbers of additional and perhaps missing chromosomes; these minimal deviations may be associated with the presence of slight, if any, myometrial invasion [26]. In contrast, tumors that are of high malignancy, as judged by their histological picture and capacity to invade and metastasize, generally show evidence of structural as well as more extensive numerical changes. A correlation between degree of malignancy and extent of the chromosome changes is generally confirmed (where sufficient details are given) by published studies [3, 47, 66, 102, 106, 139, 143, 161, 194, 199, 208, 210–212].

However, Tseng and Jones [208] studied a series of seven tumors which included not only a well-differentiated carcinoma but also a more malignant tumor, a grade 3 adenoacanthoma, in both of which over half the cells karyotyped were diploid. They discussed the origin of these diploid cells and concluded that in the latter tumor, at least, they were probably tumor cells. One highly malignant tumor studied in this laboratory also yielded predominantly diploid metaphases [26].

Bladder and Urethra

In a study of 64 transitional-cell carcinomas of the bladder, Spooner and Cooper [193] found that whereas most of the undifferentiated tumors had modes in the triploid or tetraploid range, all the 34 well-differentiated and moderately well-differentiated tumors fell into the diploid range, including 10 in which at least 50% of the metaphases had a normal karyotype; 5 of these 10 in fact contained almost solely normal karyotypes. Histological examination showed that the cells with normal karyotypes were dividing tumor cells, rather than stromal or inflammatory cells. The remaining tumors had aneuploid karyotypes, frequently showing consistencies within each tumor. Marker chromosomes were present in some tumors of all histological grades. The poorly and moderately well-differentiated tumors especially were characterized by a deficiency of group B chromosomes; on the other hand, the well-differentiated tumors tended to show a gain of group D chromosomes.

A number of the patients received Colcemid injections before removal of the tumor in order to increase the number of mitoses available for study; the rate of entry of cells into mitosis was then estimated by

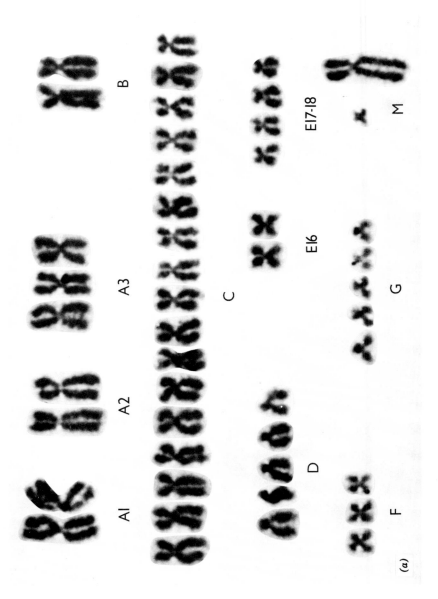

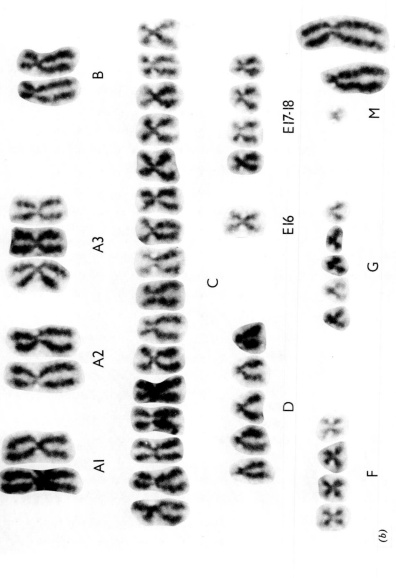

Figure 5. A mixed mesodermal tumor of the corpus uteri with an adenoacanthomatous structure from a 76-year-old woman. Two metaphases with 47 chromosomes (the modal number). There were minor differences among the metaphases with 47 chromosomes from this tumor (M, marker chromosomes).

dividing the number of mitoses per 1000 cells by the time in hours between the Colcemid injection and the tumor resection. An interesting finding was that mitotic activity was correlated with ploidy level rather than with differentiation; near-diploid tumors of all three grades had low rates of proliferation.

Two-thirds of the well-differentiated and moderately well-differentiated (and only one of the poorly differentiated) tumors studied by Spooner and Cooper [193] showed no signs of invasion; these results confirm earlier reports [62, 110, 182] that noninvasive bladder tumors are generally near-diploid.

Duruman [59] studied two bladder "papillomas" that yielded only diploid karyotypes, two carcinomas in situ that had aneuploid cells with marker chromosomes, and 19 carcinomas that were generally aneuploid. Rigby and Franks [171] succeeded in maintaining a cell line derived from a well-differentiated bladder carcinoma for over 80 transfer generations. The line was believed to be of tumor cell origin and had a chromosome number of 48 with an extra group C and group D chromosome. A few other authors have studied tumors of the bladder [198, 223] and the urethra [126].

Malignant Melanoma

Chromosome studies on some 38 cases of malignant melanoma have been reported [10, 31, 34, 39, 78, 80, 96, 105, 142, 167, 198, 214, 219, 220]. Modal chromosome numbers ranged widely but were quite frequently in the triploid region. A summary of representative karyotypes from 22 tumors that have been reported in detail [10, 34, 39, 96, 105, 142, 214] shows an average chromosome number of 65. The average number of chromosomes of each type, expressed as a percentage of the expected number, was as follows:

A1	A2	A3	B	C	D	E	F	G
88%	101%	105%	74%	101%	90%	108%	98%	90%

A reduction in the proportion of chromosomes in groups D, G, and, especially, B is to be noted. In addition, the karyotypes included on average two marker chromosomes.

We have observed a large marker with a pronounced secondary constriction in its long arm in two tumors, including one published case [10], and large markers with constrictions have been described or illustrated in at least three other tumors [96, 167, 214].

Lymphomas and Related Conditions

This short review will be confined to those reticuloses that manifest themselves as lymph node or other "tumors," including Hodgkin's disease whose cytogenetic features certainly favor its classification as a neoplastic condition. (The Burkitt lymphoma is dealt with elsewhere in this volume; see ref. [40].)

Chromosome studies have been reported in lymphosarcoma [1, 27, 45, 48, 67–69, 145, 175, 191, 203, 218], reticulum cell sarcoma [15, 45, 48, 67, 77, 80, 100, 125, 126, 145, 146, 157, 190, 191], follicular lymphoma [191], mycosis fungoides [156], Hodgkin's disease [27, 45, 48, 67, 68, 145, 146, 179, 184, 191, 196, 216], and miscellaneous or unclassified reticuloses [29, 67, 68, 80]. In general, where sufficient numbers of cells have been analyzed, aneuploidy with evidence of clone formation has been found; however, there may also be substantial numbers of diploid, perhaps normal, dividing cells in both direct and cultured preparations. Apart from Hodgkin's disease, the modal chromosome numbers of the aneuploid cells tend to be in the diploid region.

In Hodgkin's disease, dividing cells are often scanty, and aneuploid cells have not always been found, but where present, they may be hyperdiploid or, more frequently, hypertriploid or hypotetraploid. The study in my laboratory of one case (nodular sclerotic type, female aged 19) showed related cell lines in a supraclavicular node and, three months later with no intervening treatment, in para-aortic and splenic hilar nodes.

Spiers and Baikie [192] have suggested that abnormalities of the E17/18 chromosomes play a special role in the formation of some neoplasms of the reticuloendothelial system.

Miscellaneous Tumors

In *myeloma (plasmacytoma)*, aneuploid cells with chromosome numbers in the diploid range, with or without marker chromosomes, have usually been found, most of the studies having been made on bone marrow [42, 45, 57, 89, 183]. In two cases, however, chromosome numbers in the range of 60 to 85 were found [121].

Hydatidiform moles are usually euploid; it is interesting that they may show triploidy or diploid–triploid mosaicism, which may also be the chromosome constitution of the associated fetus [19, 127, 129, 160]; indeed, the placentae of triploid conceptions frequently undergo partial or complete hydatidiform change. In *chorioadenoma destruens*, however, increasing numbers of aneuploid cells have been found [*129*], while *choriocarcinomas* present the typical chromosome abnormalities of malignant tumors [*101, 129*].

In reports on two carcinomas of the *Fallopian tube* [*52, 213*], the modal chromosome numbers were 74 and 42, respectively; both tumors had marker chromosomes.

Chromosome counts from an area of Bowen's disease of the *vagina* showed a mode of 56 to 59 [*126*], and a mode of 75 was found in carcinoma in situ adjacent to an early vaginal carcinoma [*194*].

There have been a few reports on carcinoma of the *vulva* [*105, 139, 162*]. A significant degree of aneuploidy was found in carcinoma in situ but not in condyloma acuminata or Paget's disease [*105, 221*].

A well-differentiated squamous carcinoma of the *esophagus* proved to be hypodiploid (42-43 chromosomes) with a long acrocentric marker [*123*], while malignant ascites from another epidermoid esophageal carcinoma had a mode of about 50, including markers [*112*]. Among other squamous cell carcinomas, chromosome numbers, in some cases with details of karyotypes, have been reported for carcinoma of the *larynx* [*125, 223*], *maxilla* [*125, 126*], and *skin* [*126*].

A malignant ascites from a carcinoma of the *liver* had a modal chromosome number of 75 without obvious markers [*128*].

An *adrenocortical* carcinoma in a female infant aged one year showed 47 chromosomes, the karyotype being made up of a diploid set with an additional, marker chromosome [*34*].

A follicular carcinoma of the *thyroid* cultured for 3 to 8 days yielded counts which were mainly in the range of 55 to 58, no recognizable markers being seen [*189*]. In the same study, and also in cultures, a microfollicular thyroid adenoma showing nuclear atypia from a patient with a history of X-ray therapy for acne gave counts of 46; analysis, however, showed a variety of chromosome abnormalities in these cells. A secondary carcinoma of the thyroid in culture was reported to show only diploid karyotypes [*144*]. Aneuploidy was found in secondary ascites from a carcinoma of the *pancreas* [*34*].

Cox studied seven *nephroblastomas* in patients aged up to 13 years [*49*]. In one, the mode was in the range of 55 to 58 chromosomes, but in the others it was at or near 46. In fact, three patients, including the two youngest (aged 11 and 18 months), yielded diploid karyotypes only. Two additional tumors had some hyperdiploid cells with extra group C or A1 chromosomes, while the tumor in the oldest patient had a consistent karyotype with 46 chromosomes in which an A1 chromosome was missing, and in which an extra group C-like chromosome might have been derived from the A1. In another report, a *mesonephroma* (site not stated) had a minor mode at 92, and a large marker was illustrated [202].

A series of seven *embryonic sarcomas* studied in direct preparations as well as in cultured material from patients aged up to 7 years yielded 69 predominantly aneuploid karyotypes, 5 of the tumors having large markers [153]. An embryonic sarcoma of the urogenital sinus (male, aged 2½ yrs) and a highly malignant embryonic tumor of the mediastinum (female, aged 8) were aneuploid, with modal numbers of about 58 and 48, respectively [34].

A metastatic embryonic *rhabdomyosarcoma* (male, aged 16) had a mode of about 60 chromosomes [82]. All metaphases had a number of double-minute chromatin bodies, but no other markers were seen. Two successive recurrences following radiotherapy of a rhabdomyosarcoma in a girl aged 3 years showed modes of about 70 and 67-68, respectively; among the markers was one in the second specimen which had not been present in the first [215]. A similar tumor (female, aged 3 months) showed only diploid karyotypes, while an undifferentiated sarcoma (female, aged 10 years) was aneuploid with 47 chromosomes including a large marker [50].

In a series of tumors of *bone* [97], six sarcomas had modes of 58 or above in direct preparations, while eight benign giant-cell tumors appeared to be diploid both in direct preparations and in long-term cultures. A periosteal sarcoma had a mode of 42 chromosomes, including markers [34].

Cultured for eight days, a benign *sacrococcygeal teratoma* from a 2-day-old male infant yielded a few diploid karyotypes, as well as one metaphase with 47 and one with 48 chromosomes, the latter having two additional members in group E [111].

In three studies which included a total of five *mesotheliomas*, all the tumors appeared to be aneuploid [24, 112, 197].

A *thymoma* (female, aged 7 years), probably benign, yielded only diploid karyotypes [50].

DISCUSSION OF THE GENESIS AND SIGNIFICANCE OF PARTICULAR CHROMOSOME CHANGES IN HUMAN CANCER

We have seen that the great majority of human cancers exhibit gross chromosome changes, and that the pattern within each tumor indicates a clonal relationship. The various common features of the abnormal karyotypes and their mode of genesis is now considered in more detail.

The *gain* or *loss* of individual chromosomes may be a consequence of non-disjunction, and the formation of new chromosomes, a consequence of breakage followed by reunion. Breakage may involve chromosomes or chromatids; the structural changes that follow may be intra- or interchromosomal. An alternative hypothesis suggests the selective endoreduplication of whole chromosomes or segments of chromosomes [115].

In appearance, *marker chromosomes* may be 1.5 to 2 times larger than the largest or somewhat smaller than the smallest of the normal chromosomes. Naturally, only those abnormal chromosomes which deviate significantly from normal are recognizable as markers, using conventional cytogenetic staining methods. The differences may involve length, centromere position, and other features such as secondary constrictions (as in a melanoma [10] and seminoma [Fig. 6]). Ring chromosomes are encountered from time to time. We have seen them in tumors of the common sites, including a rectal polyp that showed areas of atypia in the histologic sections [28]; in a carcinoma of the cervix, a ring was consistently present in a minor hypodiploid line but was only present (in duplicate) in a small proportion of the hypotetraploid cells which constituted the major line [5]. Dicentric markers may be difficult to distinguish from chromosomes with prominent secondary constrictions; one was seen in 25% of the metaphases in a cervical carcinoma and may have accounted for the anaphase and telophase bridges that were seen in this tumor [21]. It has been suggested that dicentric chromosomes with short intercentric distances, which are difficult to recognize, may occur in tumors as a result of translocations, for example, between two group G chromosomes [151]. Metacentric chromosomes may be isochromosomes; an isochromosome for the long arm of chromosome E17, for instance, may regularly appear in the course of chronic myeloid leukemia [88], and abnormal medium-sized metacentric chromosomes with even and equal fluorescence in both arms have been observed in lymphomas after staining with quinacrine [67, 69].

Multiple "double-minute" chromatin bodies were first described in neurogenic tumors of young patients [51]. They are rare in tumors of adults, although they have been described in a carcinoma of the ovary [158] and we have observed them in two carcinomas of the rectum.

High chromosome numbers may be achieved by a doubling of the complete set. Tumors with modal numbers in the triploid or tetraploid region often have markers that are *not* present in duplicate; this, together with the presence of an uneven number of chromosomes in one or more of the normal groups, indicates that changes must have occurred subsequent to any doubling [7]. The commonly observed near-triploid numbers could be achieved by a series of non-disjunctions, or by a combination of chromosomal loss through non-disjunction and a complete dou-

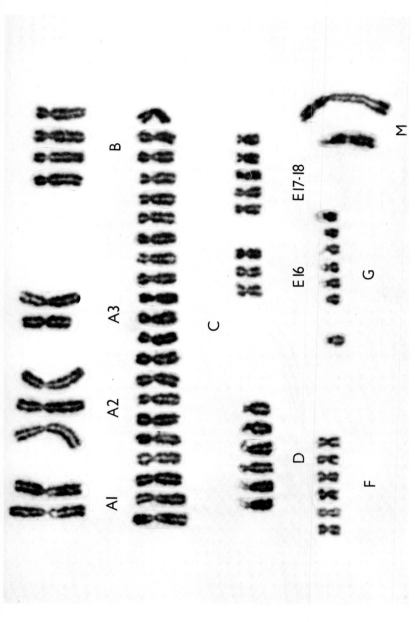

Figure 6. Seminoma with 61 chromosomes. The shorter of the two marker chromosomes (M) has a secondary constriction in its long arm and satellites on its short arm; this chromosome showed satellite association with other acrocentric chromosomes. Same as Case 17 in a previous publication [15].

bling. Other mechanisms might, however, account for near-triploid complements; in cultured marrow from a patient with chronic myeloid leukemia and localized lymphadenopathy, de Nava and colleagues [152] noted among the pseudodiploid and related pseudotetraploid leukemia cells a near-triploid cell that could have been derived from a pseudodiploid cell by duplication of a single (abnormal) haploid set. (It is of interest that sporadic triploid cells have been observed in blood and fibroblast cultures derived from patients with congenital abnormalities and in fibroblasts derived from placenta; their origin is speculative, however [163].)

On the basis of studies in the mouse, it has been suggested that fusion of tumor cells with host cells (with a consequent increase in the chromosome number) is a mechanism whereby the degree of malignancy of a tumor might be increased [217].

Relationship of the Chromosome Changes to the Site and Histologic Type of the Tumor

Well-differentiated tumors at some sites have been found to show minimal chromosome changes, or even cells with normal karyotypes; these will be discussed later. Reference has already been made to possible differences between seminomas and malignant teratomas of the testis with respect to the proportions of chromosomes in the various normal groups. Also, it has already been noted that there may be differences in the proportions of tumors at the various ploidy levels according to the site (for instance, the frequency of hypodiploid tumors in the breast and ovary and their absence in the testis). Large markers have frequently been observed in tumors at sites such as the breast, ovary, and testis; but despite the fact that markers of similar appearance are present in different tumors, there is no real evidence for the existence of markers that are specific to tumors at any given site or group of sites. The presence of large markers may be related to the chromosome number (see below) rather than to the site. Perhaps detailed comparisons among tumors at various sites will reveal subtle differences in the karyotypes similar to those described for seminomas and teratomas.

Karyotypic Changes in Relation to Chromosome Number

In *hypodiploid* tumors the constancy of the various features characteristic of malignant cell karyotypes is impressive: the presence of marker chromosomes, often as large as or larger than group A chromosomes, and the reduction in the numbers of chromosomes in groups B, D, and G (e.g., to two or three in each). Such karyotypes apparently do not occur in preinvasive lesions, and they may, therefore, be indicative of malignancy [16]. DNA estimations show that hypodiploid tumors, like tumors in general, tend to have slightly more DNA than the amount expected on the basis of their chromosome number [22]. They may, therefore, show no net loss of chromosomal material as compared with diploid cells, or only a small loss which is not in proportion to the reduction in chromosome number. The presence of large markers may have some relation to this conservation of material. An interesting observation relating to the hypodiploid tumors of the ovary and other sites that we have studied is that while there is some spread of chromosome numbers immediately around the mode, this very rarely extends above 46.

Pseudodiploid tumors show the same characteristic features of their karyotypes, although perhaps less constantly.

Hyperdiploid tumors (in the range of 47 to about 57 chromosomes) sometimes show only minimal changes, as will be discussed later. There may only be additional chromosomes, or gains combined with losses, and marker chromosomes are quite frequently absent (Fig. 2). Frequently, though not so consistently, there are relatively few group B, D, and G chromosomes (the D's may be present in excess). The distribution of chromosome numbers often has a positive tail, analysis of the hypermodal cells showing a fairly consistent pattern of additional, with perhaps some lost, chromosomes. In a carcinoma of the rectum studied in this laboratory (Miss M. C. Baker), closely related karyotypes with 50 to 55 chromosomes were found, but whereas the mode in one region of the tumor was at 51, in another region it was at 54 or 55 chromosomes.

Tumors in the *triploid* and *tetraploid* ranges frequently show the same changes in the proportions of the normal chromosomes as hypodiploid tumors, and markers are usually, although by no means constantly, present (less easily recognizable abnormal chromosomes, especially in the middle size range, may of course be present). In some tumors, particularly those with modes near or above the tetraploid level, a few cells with related karyotypes but only half the number of chromosomes may be found; such near- or hyper-diploid cells may predominate in other regions of the tumor. As already mentioned, where markers are present in near-diploid cells, these markers are generally present in duplicate in near-tetraploid cells from the same tumor. However, many tumors with high modes (particularly where the mode is in the hypertriploid or hypotetraploid region) are lacking a minor component with half the number of chromosomes; in these tumors the markers are usually *not* present in duplicate.

Chromosomes in Nonmalignant and Premalignant Conditions

A study of hyperplasia of the endometrium has shown only normal karyotypes [26]. The same is true of hyperplastic lymphadenopathies studied in this laboratory (Miss M. C. Baker), apart from one enlarged lymph node, which was the possible seat of a viral infection and yielded metaphases with one to three extra minute chromosomes and an occasional additional group D chromosome [7]. There is little information on benign tumors, in which mitoses are generally infrequent; however, culture techniques have led to interesting findings in meningiomas [135]. We have found minimal aneuploidy, usually in the form of C-trisomy, in cultured thyroid tissue from patients with Hashimoto's disease [7, 14].

The major part of our information on premalignant conditions comes from the cervix uteri [195] and the large bowel, although reference has already been made to data from a few other sites. In both the cervix uteri and the large bowel there may be some correlation between the degree of aneuploidy and the extent of the histologic changes. Thus, in mild dysplasia of the cervix, and in large-bowel polyps without atypia, there may be only scanty normal karyotypes, or mixtures of normal karyotypes and aneuploidy, usually in the diploid region. In the more severe forms (e.g., carcinoma in situ of the cervix, and polyps with areas of atypia or possible early malignant change), aneuploidy is more pronounced, often showing evidence of clone formation, and sometimes including structural chromosome changes. In the cervix, but not in the large bowel, the clones often have modal chromosome numbers in the triploid or tetraploid region.

A Possible Pathway Leading to Aneuploidy

The presence of trisomies and several minor changes in large-bowel polyps suggests that they represent the first stage of an evolutionary pathway. According to this model, such cells have a selective advantage, as indeed their presence in polyps and early carcinomas of the endometrium suggests and, once an actively growing clone is established, the way is open for further changes, which may include structural chromosome rearrangements. A liability to these secondary changes might have a similar basis to the liability to leukemia in Down's syndrome and might be related to the effect of altered enzyme activity in the trisomic cells on the production of further chromosome changes. A similar hypothesis was put forward by Sasaki and colleagues [178] to account for their interesting finding that leukocytes from patients with congenital abnormalities in which there was trisomy for the whole or a part of a chromosome showed an increased radiosensitivity in terms of the production of exchange aberrations.

Diagnostic and Prognostic Value of Cytogenetic Studies

Where pathological and clinical criteria are equivocal, chromosome studies may be of diagnostic value, since the presence of a clone of cells with an extensively altered karyotype might in the present state of our knowledge provide a strong indication that a malignant condition is present. Clones with less extensive changes could be interpreted more cautiously, although they might suggest that a premalignant lesion, at least, is present. These would include clones in which the only abnormality is a single trisomy, or in which the karyotype is pseudodiploid with evidence for the presence of one or more balanced translocations. (It is of interest vis à vis the unbalanced karyotypes of malignant tumors that, in aneuploid clones of apparently benign cells that arise as a late result of radiotherapy, balanced pseudodiploid karyotypes are usually present, i.e., there has been a rearrangement of the chromosomal material without any loss or duplication.)

For tumors at a number of sites, DNA estimations may be of prognostic significance; these would include the cervix and corpus uteri [8], the breast [11], and the ovary [9]; tumors with near-diploid modes have a better prognosis, except for squamous cell carcinomas of the cervix among which hypotetraploid tumors are more favorable. A potential clinical application of DNA estimations is suggested by a recent study in which 12 benign pheochromocytomas had near-diploid DNA modes, while 3 of similar histological appearance, but which had metastasized, were hyperdiploid or near-triploid [122].

Sex chromatin may indicate a chromosome change, when it presents an anomalous pattern (e.g., triple sex chromatin [4]). The presence of sex chromatin in a malignant melanoma in a male patient whose metastases regressed [10] and in a Burkitt lymphoma in a male child in which most of the tumor cells had a female diploid karyotype [134], raises questions concerning its origin and possible relation to the unusual clinical course of some tumors of these types.

Protruding chromosome arms in dividing cells in histologic sections may indicate a large marker chromosome [18, 43] and thus may have diagnostic or prognostic significance. Figure 7 shows an example in a giant cell tumor of tendon sheath which recurred after operation.

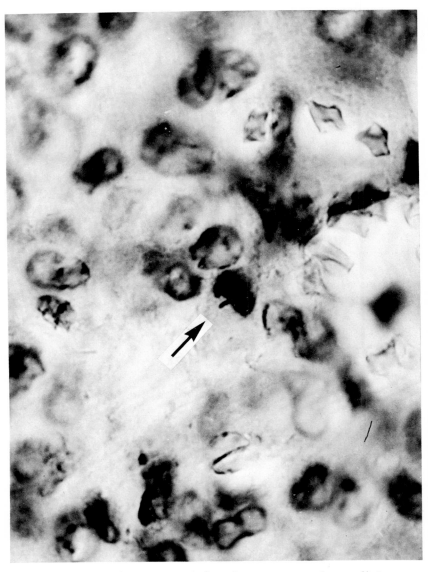

Figure 7. Metaphase showing a protruding chromosome arm (arrowed) in a recurrent giant cell tumor of tendon sheath of a toe (girl aged 8). Such protrusions, indicative of the presence in the tumor's chromosome complement of a large abnormal chromosome, were also seen in the primary tumor. (Hematoxylin and eosin stain, X 2100.)

Chromosome Changes and Malignant Transformation

It is well known that carcinogenic agents—whether viruses, ionizing radiations, or chemical substances—generally cause chromosome breakage; certain inborn anomalies in which there is an excessive liability to chromosome breakage carry an increased risk of malignant disease [79, 91]. There is, therefore, a clear link between chromosome breakage and cancer; however, a key question concerns the sequence of events that occurs between the initial chromosome damage and the acquisition of malignant properties by the cell or its descendants. In man, at least, it seems that this is usually a lengthy process of sequential changes (clonal evolution).

A further question is whether (and in what way) the aneuploidy of malignant cells directly determines their abnormal behavior. A number of authors have recently discussed this and similar problems with particular reference to human cancer [33, 37, 85, 107, 109, 113, 120, 141, 154, 165, 173]. While aneuploidy can safely be regarded as a characteristic of most forms of cancer in man, and while degrees of aneuploidy bear some relation to the degrees or stages of malignancy, it is at present impossible to answer in detail the question as to how an unbalanced chromosome set is related to the properties of malignancy or premalignancy.

In many experimental animal tumors, the presence of normal chromosome complements in the early stages of the development of the tumor has suggested to several authors that chromosome changes are only concerned with the "progression" of tumors to higher degrees of malignancy. In man, there indeed appears to be a "progression" which is related to chromosome change, but this progression involves the earlier stages (i.e., from normality to the beginning of invasion). Once established, human cancers are usually already aneuploid, with highly unbalanced karyotypes, and serial studies indicate that they tend to retain the same chromosome complements or to undergo only small changes—somewhat comparable, in fact, to the differences that may be found when different regions of a tumor are compared. Moreover, progression in terms of changes in the histological picture in the direction of greater anaplasia, increased growth rate, and greater liability to metastasize is the exception rather than the rule; recurrences are usually similar to the primary tumor in these respects.

Some recent findings in rats and Chinese hamsters on experimental tumors which are initially diploid may be of great significance: there is apparently a regular and predetermined sequence of chromosome changes in both species which, however, varies (i.e., different chromosomes are involved) according to whether a virus or a chemical carcinogen is the inducing agent [148, 149]. This raises the exciting possibility that different pathways of chromosomal evolution in human tumors may be related to different causative agents.

The normal, or apparently normal, karyotypes found in some human tumors raise several questions. There is the obvious possibility of some small change, corresponding in extent to the change resulting in the formation of the Philadelphia chromosome in chronic myeloid leukemia, that has not yet been detected. On the other hand, it is feasible that at times quite gross changes may have occurred that have not altered the appearance of the karyotype (although they might now be revealed by banding techniques). The presence of diploid cells and cells in which the only abnormality is a single trisomy in some early, usually well-differentiated tumors of the corpus uteri and bladder raises the possibility that these cells, though tumor cells, are not themselves malignant. The malignancy of these tumors on histological and clinical grounds is in fact often in doubt and, even when it is not in doubt, the situation might be that some of the cells—namely, those with diploid complements or only minor changes—constitute the remnants of precursors of the aneuploid malignant cell line which are still present in the tumor (e.g., the ovarian carcinoma discussed earlier).

The essential feature of the aneuploidy of malignant cells might be that it constitutes the attainment of an unbalanced combination of chromosomes and parts of chromosomes that ensures the ability to grow and divide and at the same time a release from normal controls; the ability of cells or groups of cells to differentiate may, however, be retained. [It has been suggested that malignant transformation is determined by the balance between chromosomes carrying factors for the expression and others carrying factors for the suppression of malignancy [92]. The frequency of structural alterations among the changes occurring in the karyotypes of cancer cells suggests that the balance necessary for malignancy may require chromosome breakage and reunion, rather than (or in addition to) the gain or loss of whole chromosomes. Perhaps in the course of human (and animal) evolution, selective forces have operated to ensure that dangerous combinations of genes may result only from unlikely events—namely, particular chromosome rearrangements as opposed to the gain or loss of chromosomes.

Fluorescent banding techniques have already proved to be useful in the study of abnormal chromosomes in lymphomas [67–69, 131–133] and meningiomas [225]. When the new techniques for chromosome identification have been applied to many more tumors, including those which have so far received little or no study (basal cell and prostatic carcinomas, e.g.), it is to be hoped that questions on the relationship between aneuploidy and cancer in its many forms, including their developmental stages, can be answered in more detail.

Acknowledgements.—The cooperation of the clinical and pathological staff of Mount Vernon Hospital is gratefully acknowledged. Thanks are due to my colleague Miss M. C. Baker for her help throughout this work; to Miss S. E. Gaze for reading and discussing the manuscript and for help in the preparation of the karyotypes; and to Miss M. Sears, Miss L. Killingback, and Mr. R. Muller for technical assistance. I also thank Mr. D. Astwood for photographic work and Mrs. C. T. Elledge and Mrs. B. J. Langdon for secretarial services. This work was supported by a grant from the Cancer Research Campaign.

LITERATURE CITED

1. ADAM M, THORBURN MJ, GIBBS WH, BROOKS SEH, HANCHARD B: Clonal evolution in two patients with autoimmune disease and lymphoreticular neoplasia. *Brit J Cancer* 24:266–276, 1970.
2. ARIAS-BERNAL L, JONES HW JR: Chromosomes of a malignant ovarian teratoma. *Am J Obstet Gynec* 100:785–789, 1968.
3. ATKIN NB: A single heteropycnotic chromosome in a human tumour. *Exp Cell Res* 20:214–215, 1960.
4. ATKIN NB: Triple sex chromatin, and other sex chromatin anomalies, in tumours of females. *Brit J Cancer* 21:40–47, 1967.
5. ATKIN NB: A carcinoma of the cervix uteri with hypodiploid and hypotetraploid stem-lines. *Eur J Cancer* 3:289–291, 1967.
6. ATKIN NB: Perimodal variation of DNA values of normal and malignant cells. *Acta Cytol* 13:270–273, 1969.
7. ATKIN NB: Cytogenetic studies on human tumors and premalignant lesions: the emergence of aneuploid cell lines and their relationship to the process of malignant transformation in man. *In* Genetic Concepts and Neoplasia (Cumley RW, ed), Proceedings of the Twenty-third Annual Symposium on Fundamental Cancer Research, 1969, University of Texas, M. D. Anderson Hospital and Tumor Institute, Baltimore, Williams & Wilkins Company, 36–56, 1970.
8. ATKIN NB: Cytogenetic factors influencing the prognosis of uterine carcinoma. *In* Modern Radiotherapy; Gynaecological Cancer, 2nd ed (Deeley TJ, ed), London, Butterworth & Company, 138–154, 1971.
9. ATKIN NB: Modal DNA value and chromosome number in ovarian neoplasia. A clinical and histopathologic assessment. *Cancer* 27:1064–1073, 1971.
10. ATKIN NB: Sex chromatin positive metastatic melanoma in a male with a favourable prognosis. *Brit J Cancer* 25:487–492, 1971.
11. ATKIN NB: Modal deoxyribonucleic acid value and survival in carcinoma of the breast. *Brit Med J* 1:271–272, 1972.
12. ATKIN NB: Y bodies and similar fluorescent chromocentres in human tumours including teratomas. *Brit J Cancer* 27:183–189, 1973.

13. ATKIN NB, BAKER MC: A nuclear protrusion in a human tumor associated with an abnormal chromosome. *Acta Cytol* 8:431–433, 1964.
14. ATKIN NB, BAKER MC: Chromosome abnormalities, neoplasia, and autoimmune disease. *Lancet* 1:820–821, 1965.
15. ATKIN NB, BAKER MC: Chromosome abnormalities as primary events in human malignant disease: evidence from marker chromosomes. *J Nat Cancer Inst* 36:539–557, 1966.
16. ATKIN NB, BAKER MC: Possible differences between the karyotypes of pre-invasive lesions and malignant tumours. *Brit J Cancer* 23:329–336, 1969.
17. ATKIN NB, BAKER MC: Chromosome and D.N.A. abnormalities in ovarian cyst-adenomas. *Lancet* 1:470, 1970.
18. ATKIN NB, BRANDÃO HJS: Evidence for the presence of a large marker chromosome in histological sections of a carcinoma *in situ* of the cervix uteri. *J Obstet Gynaecol Brit Commonw* 75:211–214, 1968.
19. ATKIN NB, KLINGER HP: The superfemale mole. *Lancet* 2:727–728, 1961.
20. ATKIN NB, TAYLOR MC: A case of chronic myeloid leukaemia with a 45-chromosome cell-line in the blood. *Cytogenetics* 1:97–103, 1962.
21. ATKIN NB, BAKER MC, WILSON S: Stem-line karyotypes of 4 carcinomas of the cervix uteri. *Am J Obstet Gynecol* 99:506–514, 1967.
22. ATKIN NB, MATTINSON G, BAKER MC: A comparison of the DNA content and chromosome number of fifty human tumours. *Brit J Cancer* 20:87–101, 1966.
23. AWANO I, TUDA F: The chromosomes of stomach cancers and myelogenous leucemias in comparison with normal human complex. *Jap J Genet* 34:220–225, 1959.
24. AYRAUD N, KERMAREC J: Étude cytogénique de huit tumeurs d'origine mesotheliale. *Bull Cancer* 55:91–110, 1968.
25. BADER S, TAYLOR HC JR, ENGLE ET: Deoxyribonucleic acid (DNA) content of human ovarian tumors in relation to histological grading. *Lab Invest* 9:443–459, 1960.
26. BAKER MC: A chromosome study of seven near-diploid carcinomas of the corpus uteri. *Brit J Cancer* 22:683–695, 1968.
27. BAKER MC, ATKIN NB: Chromosomes in short-term cultures of lymphoid tissue from patients with reticulosis. *Brit Med J* 1:770–771, 1965.
28. BAKER MC, ATKIN NB: Chromosome abnormalities in polyps and carcinomas of the large bowel. *Proc Roy Soc Med* 63:9–10, 1970.
29. BAUKE J, SCHÖFFLING K: Polyploidy in human malignancy. Hypopentaploid chromosome pattern in malignant reticulosis with secondary sideroachrestic anemia. *Cancer* 22:686–694, 1968.
30. BENDER MA, KASTENBAUM MA, LEVER CS: Chromosome 16: a specific chromosomal pathway for the origin of human malignancy? *Brit J Cancer* 26:34–42, 1972.
31. BENEDICT WF, BROWN CD, PORTER IH: Long acrocentric marker chromosomes in malignant effusions and solid tumors. *NY J Med* 71:952–955, 1971.
32. BENEDICT WF, ROSEN WC, BROWN CD, PORTER IH: Chromosomal aberrations in an ovarian cystadenoma. *Lancet* 2:640, 1969.

33. BERGER R: Leucémies, cancers et anomalies chromosomiques. *Gaz Med France* 75:961–976, 1968.
34. BERGER R: Sur la méthodologie de l'analyse des chromosomes des tumeurs. Thesis (*Sciences Naturelles*), Paris, 1–180, I–XXXII, 1968.
35. BERGER R: Chromosomes et tumeurs. I. Techniques de préparation. *Pathol Biol* 17:1051–1058, 1969.
36. BERGER R: Chromosomes et tumeurs humaines. II. Méthodes d'analyse des anomalies des chromosomes dans les tumeurs. *Pathol Biol* 17:1059–1068, 1969.
37. BERGER R: Chromosomes et tumeurs humaines. *Pathol Biol* 17:1133–1151, 1969.
38. BERGER R, MARTINEAU M: Analyse caryotypique de tumeurs testiculaires. *Eur J Cancer* 6:61–66, 1970.
39. BERGER R, LEJEUNE J, LACOUR J: Évolution chromosomique d'un mélanome malin. *Rev Eur Etud Clin Biol* 16:476–481, 1971.
40. BLOOM A: This volume, pp. 565–599.
41. BORCH S VON DER, STANLEY MA, KIRKLAND JA: Chromosome studies on direct and cultured preparations from malignant tissues. *Pathology* 1:243–250, 1969.
42. BOTTURA C: Chromosome abnormalities in multiple myeloma. *Acta Haematol* 30:274–279, 1963.
43. BRANDÃO HJS, ATKIN NB: Protruding chromosome arms in histological sections of tumours with large marker chromosomes. *Brit J Cancer* 22:184–191, 1968.
44. BRIGATO G, FRANCO G, CIPANI F, ZANOIO L: Studio cromosomico sulle cellule del liquido ascitico nei tumori ovarici maligni. *Attual Ostetr Ginecol* 11:666–675, 1965.
45. CASTOLDI GL: I cromosomi nelle reticulo-linfopatie. *Haematologica* 55:185–224, 1970.
46. CASTOLDI GL, SCAPOLI GL, SPANEDDA R: Sull'evoluzione clonale del cariotipo delle cellule neoplastiche in versamento pleurico da adenocarcinoma mammario. *Tumori* 11:3–22, 1968.
47. CECCO L DE, RUGIATI S, SBERNINI R: Aspetti citogenetici dei carcinomi uterini. *Quad Clin Ostetr Ginecol* 21:562–573, 1966.
48. COUTINHO V, BOTTURA C, FALCAO RP: Cytogenetic studies in malignant lymphomas: a study of 28 cases. *Brit J Cancer* 25:789–801, 1971.
49. COX D: Chromosome constitution of nephroblastomas. *Cancer* 19:1217–1224, 1966.
50. COX D: Chromosome studies in 12 solid tumours from children. *Brit J Cancer* 22:402–414, 1968.
51. COX D, YUNCKEN C, SPRIGGS AI: Minute chromatin bodies in malignant tumours of childhood. *Lancet* 2:55–58, 1965.
52. CURCIO S: Studio citogenetico di un carcinoma primitivo della salpinge. *Arch Ostetr Ginecol* 71:450–456, 1966.
53. CURCIO S: Analisi cromosomica di un caso di ca. ovarico. Aspetti della replicazione del DNA. *Arch Ostetr Ginecol* 71:436–449, 1966.
54. CURCIO S: Analisi cromosomica di un folliculoma. Estratto dagli Atti del III

Congresso della Società Italiana di Citologia Clinica e Sociale. (Castellammare di Stabia 14–16 Aprile 1966) *Arch Obstetr Ginec* Suppl 71: 139–143, 1966.
55. CURCIO S, SARTORI R: Citogenetica delle neoplasie ginecologiche. *Arch Ostetr Ginecol* 71:423–435, 1966.
56. DALLA PICCOLA B, TATARANNI G: Modello di evoluzione clonale del cariotipo in versamento pleurico neoplastico. *Arch Ital Patol Clin Tumori* 12:3–21, 1969.
57. DAS KC, AIKAT BK: Chromosomal abnormalities in multiple myeloma. *Blood* 30:738–748, 1967.
58. DAVIDSON E, BULKIN W: Long marker chromosome in bronchogenic carcinoma. *Lancet* 2:227, 1966.
59. DURUMAN NA: Chromosome abnormalities in bladder tumors. *Hacettepe Bull Med Surg* 2:15–23, 1969.
60. EICKE J, EMMINGER A, STRAUSS C, MOHR U, WRBA H: Cytogenetisch-karyologische Studien an klinisch behandelten gynäkologischen Tumoren. *Z Krebsforsch* 67:205–212, 1965.
61. ENTERLINE HT, ARVAN DA: Chromosome constitution of adenoma and adenocarcinoma of the colon. *Cancer* 20:1746–1759, 1967.
62. FALOR WH: Chromosomes in noninvasive papillary carcinoma of the bladder. *JAMA* 216:791–794, 1971.
63. FALOR WH: Chromosomes in bronchial adenomas and in bronchogenic carcinomas. *Am Rev Resp Dis* 104:198–205, 1971.
64. FALOR WH, GORDON M, KACZALA OA: Chromosomes in bronchoscopic biopsies from patients with bronchial adenoma, bronchogenic carcinoma, and from heavy smokers. *Cancer* 24:198–209, 1969.
65. FISCHER P, GOLUB E: Similar marker chromosomes in testicular tumours. *Lancet* 1:216, 1967.
66. FISCHER P, GOLUB E, HOLZNER JH: Cromosomenzahl und DNS-Wert bei malignen Tumoren des weiblichen Genitaltraktes. *Z Krebsforsch* 58:200–208, 1966.
67. FLEISCHMANN T, HÅKANSSON CH, LEVAN A: Fluorescent marker chromosomes in malignant lymphomas. *Hereditas* 69:311–314, 1971.
68. FLEISCHMANN T, GUSTAFSSON T, HÅKANSSON CH, LEVAN A: The fluorescent pattern of normal chromosomes in biopsies of malignant lymphomas, and its computer display. *Hereditas* 70:75–88, 1972.
69. FLEISCHMANN T, HÅKANSSON CH, LEVAN A, MULLER T: Multiple chromosome aberrations in a lymphosarcomatous tumor. *Hereditas* 70:243–258, 1972.
70. FORD CE: Discussion of paper by BM Slizynski. *In* Symposium on Nuclear Sex (Robertson Smith D, Davidson WM, eds), London, W. Heinemann. 10–11, 1958.
71. FORTEZA BOVER G, BÁGUENA CANDELA R, TORTAJADA MARTINEZ M: Citogenética de las ascitis carcinomatosas en ginecologia. *Rev Esp Obstetr Ginecol* 23:301–316, 1964.
72. FRACCARO M, MANNINI A, TIEPOLO L, ZARA C: High frequency of spontaneous recurrent chromosome breakage in an untreated human tumour. *Mutat Res* 2:559–561, 1965.

73. Fraccaro M, Tiepolo L, Gerli M, Zara C: Analysis of karyotype changes in ovarian malignancies. *Panminerva Med* 8:1–19, 1966.
74. Fraccaro M, Mannini A, Tiepolo L, Gerli M, Zara C: Karyotypic clonal evolution in a cystic adenoma of the ovary. *Lancet* 1:613–614, 1968.
75. Fritz-Niggli H: Die Chromosomen in menschlichen Mamma-Karzinom. *Acta Un Int Cancer* 12:623–636, 1956.
76. Galton M, Benirschke K, Baker MC, Atkin NB: Chromosomes of testicular teratomas. *Cytogenetics* 5:261–275, 1966.
77. Ganner E: Monoklonales abnormes Karyogramm. 45,XX,2D-,(17-18)−,2C+, Gp+ in den Tumorzellen eines Rethothelsarkoms. *Blut* 19:416–419, 1969.
78. Genes IS: Chromosome constitution of cells of ascitic and pleural fluids in cancer (in Russian). *Vop Onkol* 16:39–45, 1970.
79. German JL: This volume, pp. 601–617.
80. Goh K-O: Large abnormal acrocentric chromosome associated with human malignancies. *Arch Int Med* 122:241–248, 1968.
81. Goodlin RC: Karyotype analysis of gynecologic malignant tumors. *Am J Obstetr Gynecol* 84:493–500, 1962.
82. Granberg I, Mark J: The chromosomal aberration of double-minutes in a human embryonic rhabdomyosarcoma. *Acta Cytol* 15:42–45, 1971.
83. Greisen O: The bronchial epithelium. Nucleic acid content in morphologically normal, metaplastic and neoplastic bronchial mucosae. A microspectrophotometric study of biopsy material. *Acta Otolaryng Suppl* 276:1–110, 1971.
84. Gropp H, Pera F, Lohmann H, Wolf U: Untersuchungen über die Anzahl der X-Chromosomen beim Mammacarcinom. *Z Krebsforsch* 69:326–334, 1967.
85. Grouchy J de, Nava C de: A chromosomal theory of carcinogenesis. *Ann Int Med* 69:381–391, 1968.
86. Grouchy J de, Vallée G, Lamy M: Analyse chromosomique directe de deux tumeurs malignes. *CR Acad Sci Paris* 256:2046–2048, 1963.
87. Grouchy J de, Vallée G, Nava C de, Lamy M: Analyse chromosomique de cellules cancéreuses et de cellules médullaires et sanguines irradiées *in vitro*. *Ann Genet* 6:9–20, 1963.
88. Grouchy J de, Nava C de, Feingold J, Bilski-Pasquier G, Bousser J: Onze observations d'un modèle précis d'évolution caryotypique au cours de la leucémie myéloïde chronique. *Eur J Cancer* 4:481–492, 1968.
89. Guillan BA, Ranjini R, Zelman S, Hocker EV, Smalley RL: Multiple myeloma with hypogammaglobulinemia. Electron microscopic and chromosome studies. *Cancer* 25:1187–1192, 1970.
90. Hansen-Melander E, Kullander S, Melander Y: Chromosome analysis of a human ovarian cystocarcinoma in the ascites form. *J Nat Cancer Inst* 16:1067–1081, 1956.
91. Harnden DG: This volume, pp. 619–636.
92. Hitotsumachi S, Rabinowitz Z, Sachs L: Chromosomal control of reversion in transformed cells. *Nature* 231:511–514, 1971.
93. Ishihara T: Cytological studies of tumors. XXXI. A chromosome study in a human gastric carcinoma. *Gann* 50:403–408, 1959.

94. ISHIHARA T, SANDBERG AA: Chromosome constitution of diploid and pseudodiploid cells in effusions of cancer patients. *Cancer* 16:885–895, 1963.
95. ISHIHARA T, KIKUCHI Y, SANDBERG AA: Chromosomes of twenty cancer effusions: correlation of karyotypic, clinical, and pathologic aspects. *J Nat Cancer Inst* 30:1303–1361, 1963.
96. ISHIHARA T, MOORE GE, SANDBERG AA: Chromosome constitution of cells in effusions of cancer patients. *J Nat Cancer Inst* 27:893–933, 1961.
97. ISHII S: Chromosome studies on human bone tumors in *in vivo* and *in vitro*. *Gann* 56:251–260, 1965.
98. ISING U, LEVAN A: The chromosomes of two highly malignant human tumours. *Acta Pathol Microbiol Scand* 40:13–24, 1957.
99. JACKSON JF: Chromosome analysis of cells in effusions from cancer patients. *Cancer* 20:537–540, 1967.
100. KAJII T, NEU RL, GARDNER LI: Chromosome abnormalities in lymph node cells from patient with familial lymphoma. Loss of no. 3 chromosome and presence of large submetacentric chromosome in reticulum cell sarcoma tissue. *Cancer* 22:218–224, 1968.
101. KATAYAMA KP: Chromosomes and pelvic cancer. *Clin Obstetr Gynecol* 12:435–458, 1969.
102. KATAYAMA KP, JONES HW JR: Chromosomes of atypical (adenomatous) hyperplasia and carcinoma of the endometrium. *Am J Obstetr Gynecol* 97:978–983, 1967.
103. KATAYAMA KP, MASUKAWA T: Ring chromosomes in a breast cancer. *Acta Cytol* 12:159–161, 1968.
104. KATAYAMA KP, TOEWS HA: Chromosomes of metastatic ovarian carcinoma treated with a progestogen and alkylating agents. *Am J Obstetr Gynecol* 104:997–1003, 1969.
105. KATAYAMA KP, WOODRUFF JD, JONES HW JR, PRESTON E: Chromosomes of condyloma acuminatum, Paget's disease, *in situ* carcinoma, invasive squamous cell carcinoma and malignant melanoma of the human vulva. *Obstetr Gynecol* 39:346–356, 1972.
106. KIM SW, KANG YS, KIM SR: A chromosome study on uterine carcinoma. *J Kor Cancer Res Assoc* 2:25–28, 1967.
107. KOLLER PC: Chromosomes in neoplasia. *In* Cellular Control Mechanisms and Cancer (Emmelotand P, Mülhock O, eds), Amsterdam, Elsevier Publishing Company, 174–189, 1964.
108. KOTLER S, LUBS HA: Comparison of direct and short-term tissue culture technics in determining solid tumor karyotypes. *Cancer Res* 27:1861–1866, 1967.
109. KROMPOTIC E, ZELLNER JM: Medical cytogenetics. Part XV. Chromosomes in malignancies. *Chic Med Sch Quart* 27:171–185, 1968.
110. LAMB D: Correlation of chromosome counts with histological appearances and prognosis in transitional-cell carcinoma of bladder. *Brit Med J* 1:273–277, 1967.
111. LAURENT M, ROUSSEAU M-F, NEZELOF C: Étude caryotypique d'un tératome sacro-coccygien. *Ann Anat Pathol* 13:413–422, 1968.
112. LEGRAND E: Étude cytogénétique de neuf épanchements néoplasiques (Con-

sidérations sur l'aneuploidie et les chromosomes anormaux des cellules cancéreuses), *Imprimerie Baillet, Bordeaux,* 176 pp, 1968.

113. LEJEUNE J: Aberrations chromosomiques et cancer. UICC Monograph series, vol 9. *Proceedings of the Ninth International Cancer Congress,* Tokyo October 1966 (Harris RJC, ed), Berlin, Springer-Verlag, 71–85, 1967.

114. LEJEUNE J, BERGER R: Sur une méthode de recherche d'un variant commun des tumeurs de l'ovaire. *CR Acad Sci Paris* 262:1885–1887, 1966.

115. LEJEUNE J, BERGER R, RETHORÉ M-O: Sur l'endoréduplication selective de certains segments du génome. *CR Acad Sci Paris* 263:1880–1882, 1966.

116. LELIKOVA GP, LASKINA AV, ZAKHAROV AF, POGOSYANTS EE: Cytogenetic study of teratoid testicular tumors in man (in Russian). *Vop Onkol* 16:32–38, 1970.

117. LELIKOVA GP, LASKINA AV, ZAKHAROV AF, POGOSYANTS EE: Cytogenetic study of human seminomas (in Russian). *Vop Onkol* 17:20–28, 1971.

118. LEVAN A: Self-perpetuating ring chromosomes in two human tumours. *Hereditas* 42:366–372, 1956.

119. LEVAN A: Non-random representation of chromosome types in human tumor stemlines. *Hereditas* 55:28–38, 1966.

120. LEVAN A: Some current problems of cancer cytogenetics. *Hereditas* 57:343–355, 1967.

121. LEWIS FJW, MACTAGGART M, CROW RS, WILLS MR: Chromosomal abnormalities in multiple myeloma. *Lancet* 1:1183–1184, 1963.

122. LEWIS PD: A cytophotometric study of benign and malignant phaeochromocytomas. *Virchows Arch Abt B Zellpathol* 9:371–376, 1971.

123. LUBS HA, CLARK R: The chromosome complement of human solid tumors. I. Gastrointestinal tumors and technic. *New Engl J Med* 268:907–911, 1963.

124. LUBS HA, KOTLER S: The prognostic significance of chromosome abnormalities in colon tumors. *Ann Int Med* 67:328–336, 1967.

125. MAEDA M, TABATA T, SAITO H: Chromosomes of human tumors. VI. A comparative chromosome survey of the primary and its secondary tumors. *Wakayama Med Rep* 9:225–234, 1965.

126. MAKINO S, ISHIHARA T, TONOMURA A: Cytological studies of tumors. XXVII. The chromosomes of thirty human tumors. *Z Krebsforsch* 63:184–208, 1959.

127. MAKINO S, SASAKI MS, FUKUSCHIMA T: Triploid chromosome constitution in human chorionic lesions. *Lancet* 2:1273–1275, 1964.

128. MAKINO S, SASAKI MS, TONOMURA A: Cytological studies of tumors. XL. Chromosome studies in fifty-two human tumors. *J Nat Cancer Inst* 32:741–777, 1964.

129. MAKINO S, SASAKI MS, FUKUSCHIMA T: Cytological studies of tumors. XLI, Chromosomal instability in human chorionic lesions. *Okajima Folia Anat Jap* 40:439–465, 1965.

130. MANNINI A, FRACCARO M, GERLI M, TIEPOLO L, ZARA C: Studi citogenetici in una serie di carcinomi ovarici prima del trattamento. Estratto degli Atti del III Congresso della Società Italiana di Citologia Clinica e Sociale (Castellammare di Stabia 14–16 aprile 1966). *Arch Obstetr Ginecol* Suppl 71:1–15, 1966.

131. MANOLOV G, MANOLOVA Y: Marker band in one chromosome 14 from Burkitt lymphomas. *Nature* 237:33–34, 1972.
132. MANOLOV G, MANOLOVA Y, LEVAN A, KLEIN G: Fluorescent pattern of apparently normal chromosomes in Burkitt lymphomas. *Hereditas* 68:160–163, 1971.
133. MANOLOV G, MANOLOVA Y, LEVAN A, KLEIN G: Experiments with fluorescent chromosome staining in Burkitt tumors. *Hereditas* 68:235–244, 1971.
134. MANOLOV G, LEVAN A, NADKARNI JS, NADKARNI J, CLIFFORD P: Burkitt's lymphoma with female karyotype in an African male child. *Hereditas* 66:79–100, 1970.
135. MARK J: This volume, pp. 481–495, 497–517.
136. MARTINEAU M: A similar marker chromosome in testicular tumours. *Lancet* 1:839–842, 1966.
137. MARTINEAU M: Chromosomes in human testicular tumours. *J Pathol* 99:271–282, 1969.
138. MESSINETTI S, ZELLI GP, MARCELLINO LR, ALCINI E: Benign and malignant epithelial tumors of the gastroenteric tract. Chromosome analysis in study and diagnosis. *Cancer* 21:1000–1010, 1968.
139. MESSINETTI S, ZELLI GP, MARCELLINO LR, MOSCARINI M: L'analisi cromosomica nello studio di alcune lesioni proliferative e neoplastiche dell'apparato genitale femminile. *Progr Med* 26:382–401, 1970.
140. MEUGÉ C: Étude cytogénétique de trois épanchements néoplasiques chez des malades atteintes de tumeurs ovariennes. *Imprimerie Baillet, Bordeaux*, 127 pp, 1967.
141. MILES CP: Chromosomal alterations in cancer. *Med Clin N Am* 50:875–885, 1966.
142. MILES CP: Chromosome analysis of solid tumors. I. Twenty-eight nonepithelial tumors. *Cancer* 20:1253–1273, 1967.
143. MILES CP: Chromosome analysis of solid tumors. II. Twenty-six epithelial tumors. *Cancer* 20:1274–1287, 1967.
144. MILES CP, GALLAGHER RE: Chromosomes of a metastatic human cancer. *Lancet* 2:1145–1146, 1961.
145. MILES CP, GELLER W, O'NEILL F: Chromosomes in Hodgkin's disease and other malignant lymphomas. *Cancer* 19:1103–1116, 1966.
146. MILLARD RE: Chromosome abnormalities in the malignant lymphomas. *Eur J Cancer* 4:97–105, 1968.
147. MINKLER JL, GOFMAN JW, TANDY RK: A specific common chromosomal pathway for the origin of human malignancy. II. *Brit J Cancer* 24:726–740, 1970.
148. MITELMAN F: Predetermined sequential chromosome changes in serial transplantation of Rous rat sarcomas. *Acta Pathol Microbiol Scand* 80:313–328, 1972.
149. MITELMAN F, MARK J, LEVAN G, LEVAN A: Tumor etiology and chromosome pattern. *Science* 176:1340–1341, 1972.
150. MULDAL S, LAJTHA LG: This volume, pp. 451–480.
151. MULDAL S, ELEJALDE R, HARVEY PW: Specific chromosome anomaly asso-

ciated with autonomous and cancerous development in man. *Nature* 229: 48–49, 1971.
152. Nava C de, Grouchy J de, Thoyer C, Turleau C, Siguier F: Polyploidisation et évolutions clonales. *Ann Genet* 12:237–241, 1969.
153. Nezelof C, Laurent M, Rousseau M-F, Ayraud N, Urano Y: Le sarcome embryonnaire. Étude caryotypique de 7 observations. *Bull Cancer* 54:423–446, 1967.
154. Nowell PC: Chromosome changes in primary tumors. *Progr Exp Tumor Res* 7:83–103, 1965.
155. Nowell PC, Hungerford DA: Chromosome studies in human leukemia. II. Chronic granulocytic leukemia. *J Nat Cancer Inst* 27:1013–1035, 1961.
156. Obara Y, Makino S, Mikuni C: Cytologic studies of tumors. LI. Notes on some abnormal chromosome features in a case of mycosis fungoides. *Proc Jap Acad* 46:561–566, 1970.
157. Obara Y, Sasaki M, Makino S, Mikuni C: Cytologic studies of tumors. L. Clonal proliferation of four stemlines in three hematopoietic tissues of a patient with reticulosarcoma. *Blood* 37:87–95, 1971.
158. Olinici CD: Double-minute chromatin bodies in a case of ovarian ascitic carcinoma. *Brit J Cancer* 25:350–353, 1971.
159. Olinici CD, Simu G: Histologic and cytogenetic observations on a case of giant fibroadenoma. *Neoplasma* 17:663–665, 1970.
160. Paterson WG, Hobson BM, Smart GE, Bain AD: Two cases of hydatidiform degeneration of the placenta with foetal abnormality and triploid chromosome constitution. *J Obstetr Gynecol Brit Commonw* 78:136–142, 1971.
161. Paulete-Vanrell J, Camacho de Osorio O: Estudio cromosomico de un carcinoma de endometrio. *Actas Ginecotocolog* 20:298–324, 1966.
162. Paulete-Vanrell J, Laguardia AM, Camacho de Osorio O: Determinacion del numero de cromosomas de un carcinoma vulvar humano tratado con "Trenimon." *Actas Ginecotocolog* 18:98–112, 1964.
163. Pawlowitzki IH, Cenani A: Sporadic triploid cells in human blood and fibroblast cultures. *Humangenetik* 5:65–69, 1967.
164. Penrose LS: Chicago conference: standardization in human cytogenetics. *Birth Defects: Original Article Ser* 2:1–21, 1966.
165. Porter IH, Benedict WF, Brown CD, Paul B: Recent advances in molecular pathology: a review. Some aspects of chromosome changes in cancer. *Exp Mol Pathol* 11:340–367, 1969.
166. Quiroz-Gutiérrez A, Alfaro-Kofman S, Marquez-Monter H: Chromosome markers in a seminoma. *Lancet* 2:306, 1967.
167. Quiroz-Gutierrez A, Islas GM, Robles INH: Observaciones sobre la accion de la metilhidracina en algunos melanomas. II. Estudios cromosomicos. *Rev Med Hosp Gen* 31:645–651, 1968.
168. Rashad MN, Fathalla MF, Kerr MG: Sex chromatin and chromosome analysis in ovarian teratomas. *Am J Obstetr Gynecol* 96:461–465, 1966.
169. Richart RM, Ludwig AS Jr: Alterations in chromosomes and DNA content in gynecologic neoplasms. *Am J Obstetr Gynecol* 104:463–471, 1969.

170. Rigby CC: Chromosome studies in ten testicular tumours. Brit J Cancer 22: 480–485, 1968.
171. Rigby CC, Franks LM: A human tissue culture cell line from a transitional cell tumour of the urinary bladder: growth, chromosome pattern and ultrastructure. Brit J Cancer 24:746–754, 1970.
172. Rugiati S, Ragni N, Cecco L de: Analisi citogenetica di un gruppo di tumori maligni dell'ovario. Quad Clin Ostetr Ginecol 21:617–627, 1966.
173. Sandberg AA, Hossfeld DK: Chromosomal abnormalities in human neoplasia. Ann Rev Med 21:379–408, 1970.
174. Sandberg AA, Yamada K: Chromosomes and causation of human cancer and leukemia. I. Karyotypic diversity in a single cancer. Cancer 19:1869–1878, 1966.
175. Sandberg AA, Ishihara T, Kikuchi Y, Crosswhite LH: Chromosomes of lymphosarcoma and cancer cells in bone marrow. Cancer 17:738–746, 1964.
176. Sandberg AA, Ishihara T, Moore GE, Pickren JW: Unusually high polyploidy in a human cancer. Cancer 16:1246–1254, 1963.
177. Sasaki MS: Cytological effect of chemicals on tumors. XII. A chromosome study in a human gastric tumor following radioactive colloid gold (Au^{198}) treatment. J Fac Sci Hokkaido Univ Ser VI, Zool 14:566–575, 1961.
178. Sasaki MS, Tonomura A, Matsubara S: Chromosome constitution and its bearing on the chromosomal radiosensitivity in man. Mutat Res 10:617–633, 1970.
179. Seif GSF, Spriggs AI: Chromosome changes in Hodgkin's disease. J Nat Cancer Inst 39:557–570, 1967.
180. Serr DM, Padeh B, Mashiach S, Shaki R: Chromosomal studies in tumors of embryonic origin. Obstetr Gynecol 33:324–332, 1969.
181. Shaw MW, Chen TR: This volume, pp. 135–150.
182. Shigematsu S: Significance of the chromosome in vesical cancer. In International society of Urology, Thirteenth Congress, London 1964, vol. 2, London, E & S Livingstone, 111–121, 1965.
183. Siebner H, Aly FW, Braun HJ: Chromosomenbefund bei yD-Plasmocytom. Klin Wochenschr 47:884–885, 1969.
184. Sinks LF, Clein GP: The cytogenetics and cell metabolism of circulating Reed-Sternberg cells. Brit J Haematol 12:447–453, 1966.
185. Siracký J: Ploidy studies in ovarian cancer during cystostatic treatment. Neoplasma 16:427–433, 1969.
186. Šlot E: A karyologic study of the cancer of the ovary and the cancer cells in the ascitic effusions. Neoplasma 14:3–10, 1967.
187. Šlot E: Spontaneous changes of the stemline cells in human carcinomas. Neoplasma 14:629–639, 1967.
188. Šlot E: Spontaneous structural aberrations of chromosomes in human tumour cells of the effusions. Neoplasma 17:189–195, 1970.
189. Socolow EL, Engel E, Mantooth L, Stanbury JB: Chromosomes of human thyroid tumors. Cytogenetics 3:394–413, 1964.
190. Spiers ASD, Baikie AG: Reticulum cell sarcoma: demonstration of chromo-

somal changes analogous to those in SV40-transformed cells. *Brit J Cancer* 21:679–683, 1967.
191. SPIERS ASD, BAIKIE AG: Cytogenetic studies in the malignant lymphomas and related neoplasms. Results in twenty-seven cases. *Cancer* 22:193–217, 1968.
192. SPIERS ASD, BAIKIE AG: A special role of the group 17,18 chromosomes in reticuloendothelial neoplasia. *Brit J Cancer* 24:77–91, 1970.
193. SPOONER ME, COOPER EH: Chromosome constitution of transitional cell carcinoma of the urinary bladder. *Cancer* 29:1401–1412, 1972.
194. SPRIGGS AI: Karyotype changes in human tumour cells. *Brit J Radiol* 37:210–212, 1964.
195. SPRIGGS AI: This volume, pp. 423–450.
196. SPRIGGS AI, BODDINGTON MM: Chromosomes of Sternberg-Reed cells. *Lancet* 2:153, 1962.
197. SPRIGGS AI, BODDINGTON MM: Karyotype analysis in the diagnosis of malignancy. *In* The Cytology of Effusions in the Pleural, Pericardial and Peritoneal Cavities and of Cerebrospinal Fluid, 2nd ed, London, William Heinemann, 40–41, 1968.
198. SPRIGGS AI, BODDINGTON MM, CLARKE CM: Chromosomes of human cancer cells. *Brit Med J* 2:1431–1435, 1962.
199. STANLEY MA, KIRKLAND JA: Cytogenetic studies of endometrial carcinoma. *Am J Obstetr Gynecol* 102:1070–1079, 1968.
200. STEARNS MW: National conference on carcinoma of the colon and rectum: questions and answers. *CA* 21:233–234, 1971.
201. STEENIS H VAN: Chromosomes and cancer. *Nature* 209:819–821, 1966.
202. STRAUB DG, LUCAS LA, MCMAHON NJ, PELLETT OL, TEPLITZ RL: Apparent reversal of X-condensation mechanism in tumors of the female. *Cancer Res* 29:1233–1243, 1969.
203. TJIO JH, MARSH JC, WHANG J, FREI E III: Abnormal karyotype findings in bone marrow and lymph node aspirates of a patient with malignant lymphoma. *Blood* 22:178–190, 1963.
204. TOEWS HA, KATAYAMA KP, JONES HW JR: Chromosomes of normal and neoplastic ovarian tissue. *Obstetr Gynecol* 32:465–476, 1968.
205. TOEWS HA, KATAYAMA KP, MASUKAWA T, LEWISON EF: Chromosomes of benign and malignant lesions of the breast. *Cancer* 22:1296–1307, 1968.
206. TONOMURA A: The cytological effect of chemicals on tumors. VIII. Observations on chromosomes in a gastric carcinoma treated with Carzinophilin. *Gann* 51:47–53, 1960.
207. TORTORA M: Chromosome analysis of four ovarian tumors. *Acta Cytol* 11:225–228, 1967.
208. TSENG P-Y, JONES HW JR: Chromosome constitution of carcinoma of the endometrium. *Obstetr Gynecol* 33:741–752, 1969.
209. WAKABAYASHI M, ISHIHARA T: Cytological studies of tumors. XXVI. Chromosome analysis of a human mammary carcinoma. *Cytologia* 23:341–348, 1958.
210. WAKONIG-VAARTAJA R: A human tumour with identifiable cells as evidence for the mutation theory. *Brit J Cancer* 16:616–618, 1962.

211. WAKONIG-VAARTAJA R: Chromosomes in gynaecological tumours. *Aust N Z J Obstetr Gynaecol* 3:170–177, 1963.
212. WAKONIG-VAARTAJA R, HUGHES DT: Chromosome studies in 36 gynaecological tumours: of the cervix, corpus uteri, ovary, vagina and vulva. *Eur J Cancer* 3:263–277, 1967.
213. WEISE W, BÜTTNER HH: Chromosome analysis of a primary carcinoma of the Fallopian tube. *Humangenetik* 15:196–197, 1972.
214. WHANG-PENG J, CHRETIEN P, KNUTSEN T: Polyploidy in malignant melanoma. *Cancer* 25:1216–1223, 1970.
215. WHITE L, COX D: Chromosome changes in a rhabdomyosarcoma during recurrence and in cell culture. *Brit J Cancer* 21:684–693, 1967.
216. WHITELAW DM: Chromosome complement of lymph node cells in Hodgkin's disease. *Can Med Assoc J* 101:74–81, 1969.
217. WIENER F, FENYÖ EM, KLEIN G: Fusion of tumour cells with host cells. *Nature New Biol* 238:155–159, 1972.
218. WISNIEWSKI L, KORSAK E: Cytogenetic analysis in two cases of lymphoma. Comparison between lymphosarcoma and reticulosarcoma. *Cancer* 25:1081–1086, 1970.
219. WITKOWSKI R: Chromosomenbefunde bei Melanomzellen. *Derm Wochenschr* 156:345–347, 1970.
220. WITKOWSKI R, ZABEL R: Chromosomenuntersuchungen an Metastasen eines malignen Melanoms beim Menschen. *Derm Wochenschr* 158:255–260, 1972.
221. WOODRUFF JD, DAVIS HJ, JONES HW JR, RECIO RG, SALIMI R, PARK I-J: Correlated investigative technics of multiple anaplasias in the lower genital canal. *Obstetr Gynecol* 33:609–616, 1969.
222. YAMADA K, SANDBERG AA: Preliminary notes on the chromosomes of eleven primary tumors of colon. *Proc Jap Acad* 42:168–172, 1966.
223. YAMADA K, TAKAGI N, SANDBERG AA: Chromosomes and causation of human cancer and leukemia. II. Karotypes of human solid tumors. *Cancer* 19:1879–1890, 1966.
224. ZANG KD, SINGER H: Chromosomal constitution of meningiomas. *Nature* 216:84–85, 1967.
225. ZANKL H, ZANG KD: Cytological and cytogenetical studies on brain tumors. IV. Identification of the missing G chromosome in human meningiomas as no. 22 by fluorescence technique. *Humangenetik* 14:167–169, 1972.

CYTOGENETICS OF CANCER AND PRECANCEROUS STATES OF THE CERVIX UTERI

A. I. SPRIGGS

Laboratory of Clinical Cytology, Churchill Hospital, Oxford, England

---

Introduction 423
Pathology of Cervical Tumors 424
Early Cytogenetic Studies 428
Recent Cytogenetic Studies 429
    Methods 429
    Invasive Carcinoma 431
    Microcarcinoma (Microinvasive Carcinoma) 434
    Carcinoma In Situ and Dysplasia 437
    Mild Dysplasia 440
Origin and Significance of Chromosome Changes 442
    Polyploidy 442
    Aneuploidy 444
    Breakage 444
Summary 445
Literature Cited 446

INTRODUCTION

In the study of naturally occurring human cancer, the cervix uteri has a special importance for several reasons. In the first place, it is the only site in the body where supposedly precancerous epithelial changes are being discovered daily, as a result of a deliberate policy of "screening" the female population by cytological techniques. Secondly, the cervix is an organ that can be inspected and subjected to biopsy; indeed, the whole of the area where cancer most commonly develops can be removed, leaving the patient still capable of conceiving and bearing children. Thirdly, although normal squamous epithelium is unfavorable material for direct cytogenetic study, the abnormal epithelia of so-called carcinoma in situ and of invasive carcinoma sometimes furnish preparations of a quality good enough for direct karyotype analysis.

For these reasons, the cervix offers a unique opportunity to investigate cytogenetic changes in the developmental stages of a common spontaneous cancer. It can be argued that if chromosome changes are seldom or inconstantly present in precancerous stages, then they can be dismissed as late or inessential features of the established cancer, perhaps due to adverse conditions in tumor tissue. If, on the other hand, chromosome abnormalities are regularly found in lesions that precede cancer by many years, it can reasonably be held that they play an important part, perhaps even a necessary part, in carcinogenesis at this site; the type and sequence of such abnormalities then become of the greatest interest for our understanding of the neoplastic process.

PATHOLOGY OF CERVICAL TUMORS

Much the commonest histologic type of malignant cervical tumor is squamous carcinoma. In its fully developed form it does not differ in any important respect from squamous carcinoma of other mucosal surfaces. The lesions believed to be precancerous are characterized by a failure in normal differentiation so that mitotic activity occurs at levels of the epithelium well above the basal layer and sometimes near the surface. The nuclei tend to be enlarged, pleomorphic, and hyperchromatic. It is these nuclear changes which permit detection by cervical smear (Figs. 1 to 3).

International and national committees on nomenclature [19, 30] have recommended classifying these epithelial, noninvasive lesions into two categories. Where there is no differentiation of the surface layers, the lesion is called carcinoma in situ.* Where there is surface maturation, the lesion is called dysplasia [60]. This classification is unsatisfactory, partly because it is not accepted by some of the principal authorities on cytodiagnosis [43], but also because there is a spectrum of appearances, among which those with a few layers of flattened or even keratinized cells are not known to have a different biological potential from those

---

* This term is too well entrenched to change, but it must be noted that carcinoma in situ is not cancer in the sense used here, but rather a condition believed to be precancerous. In some gynecologic literature it is classified as squamous carcinoma, stage 0.

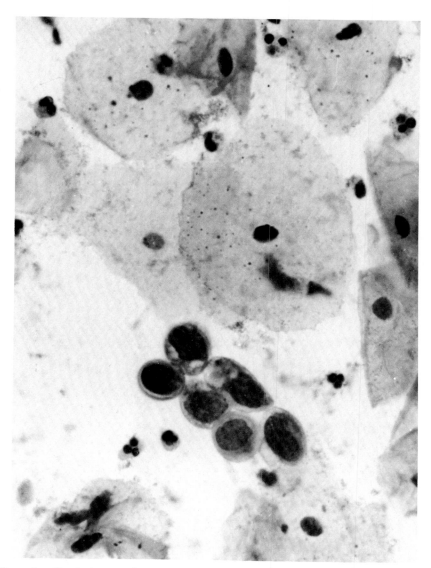

Figure 1. Cervical smear from a case of microcarcinoma (Case 20 of Spriggs et al. [71]). The group of six darkly staining cells have the cytological characters of neoplasia, chief of which is increased nuclear size without a correspondingly large cytoplasm. Normal squamous epithelial cells and polymorphonuclear leucocytes are also shown. (Papanicolaou, X 800.)

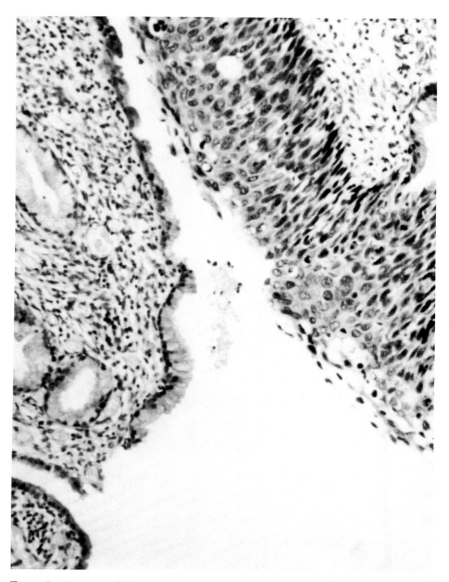

Figure 2. Sections of the posterior lip of the cervix uteri from the same case as Fig. 1, showing an endocervical gland opening lined on one side by simple columnar epithelium and on the other by severe dysplasia, approaching to carcinoma in situ at right. Elsewhere in the cervix there was microcarcinoma. (Hematoxylin and eosin stain, X 200.)

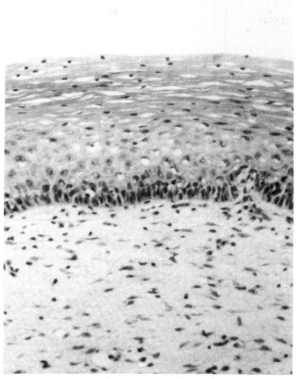

Figure 3. For comparison, an adjacent area of normal squamous epithelium from the same section shown in Fig. 2. (Hematoxylin and eosin, X 200.)

without. Moreover, the superficial layers can be lost during handling of the biopsy. Most pathologists use the terms "dysplasia" and "carcinoma in situ," but in various different senses of their own choosing. The literature has to be read with these ambiguities in mind. Some papers are written by pathologists who are aware of the problem but do not trouble to give their own definitions, while others are by authors who may not even be aware of how little uniformity there is in interpretation among diagnostic centers.

In spite of the difficulties in investigation imposed by the obligation to treat patients, there seems little reason to doubt that most cancers of the cervix are preceded by carcinoma in situ or dysplasia. The evidence for believing that these lesions are really the early stages of the cancerous process can be found summarized elsewhere [27, 68, 69]. But these conditions need not necessarily progress to cancer. Certainly they can persist unchanged for many years, and it is thought that they may often regress spontaneously, particularly the mild dysplasias [56, 61].

EARLY CYTOGENETIC STUDIES

Studies on the cervix uteri have played an important part in the advances in knowledge described in other chapters of this volume. Well before the technical breakthrough in 1956, when it became possible to analyze the human karyotype, studies had been made on the chromosomes of the cervix uteri. It was known that abnormal divisions were frequent in cervical carcinoma. Koller in 1947 [42] and Timonen soon after [76] had described lagging chromosomes, stickiness, hollow spindles, colchicine effect, multipolar divisions, and structural changes. The striking "3-group metaphases" of carcinoma in situ were described and depicted by Parmentier and Dustin in 1951 [54].

Malignant tumors often consist of cells with enlarged and hyperchromatic nuclei, which has strongly suggested that they might have an increased DNA content, not simply on account of a high proportion being in S or G_2 phases, but owing to a departure from the normal $2n$ chromosome number. Ultraviolet spectrophotometry performed on cervical smears by Mellors et al. [49] revealed that "positive" smears were indeed associated with increased nucleic acid levels. At that time, misleading chromosome counts done without hypotonic pretreatment indicated that cells from the *normal* cervix (and endometrium), as well as from tumors of these organs, had widely variable chromosome numbers. On the contrary, microspectrophotometric data on Feulgen-stained tissues tended to support the view that cells from normal epithelium have a constant modal DNA content, about 10% above that of the lymphocyte, whereas those from malignant tumors, including cervical carcinoma, have a wider scatter and abnormal modal peaks [12, 44]. The constancy of the diploid number in normal tissues was also inferred from chromosome counts of cells in culture, including cells from the cervix uteri [77].

Just before the introduction of hypotonic techniques, 29 carcinomas of the human cervix were studied cytogenetically by Manna [47, 48]. Allowing for the technical difficulties, Manna's findings were remarkably similar to those found later by other workers. In normal cervix as well as endometrium, there was a peak of chromosome numbers at the normal level (thought at that time to be 48). In addition, there was considerable scatter in values to the left, but not to the right. This scatter to the left has been found by subsequent observers in the case of normal endometrium and bone marrow and is believed to be an artifact caused by cell breakage [21], though Manna thought it to be genuine. He found cervical carcinoma, on the other hand, to show a scatter of numbers to the right as well as to the left of a modal value, and there was variation from case to case. These curves agreed well with the stemline concept of tumor growth, which had recently gained acceptance.

The introduction of hypotonic pretreatment made it possible to determine the normal human chromosome number and therefore to recognize deviations from it. Nevertheless, it was and still is technically difficult to make reliable direct preparations from tissues. The first series of substantial numbers of solid tumors was reported by Makino and his colleagues in 1959 [46]. These studies included five uterine carcinomas, but whether endometrial or cervical was not stated. Studies by Tonomura [78], also from the Zoological Institute at Hokkaido University, dealt with six cervical carcinomas and clearly established that like many other types of carcinoma they consist of aneuploid populations of cells, each with unique abnormalities including marker chromosomes, and each with variation of counts around one, or occasionally two, stemline modes.

Makino's paper of 1959 also inaugurated the cytogenetic study of epithelial precancerous states. Leuchtenberger et al. [44] had already shown that nuclei from a senile keratosis of the skin had a distribution of DNA values resembling that of carcinoma. Makino and his colleagues now reported a case described as Bowen's disease of the vagina, in which chromosome counts ranged from 54 to 116, clustering at the 56 to 59 level.

It was now possible to attack the controversial problem of carcinoma in situ of the cervix uteri. Connie Clarke, in C. E. Ford's laboratory at Harwell, performed counts and limited analyses on Feulgen squashes from six cases [70]. Aneuploid cells were present, sometimes along with cells containing 46 or 92 chromosomes. Furthermore, in two cases in which microcarcinoma was subsequently found in another part of the cervix, there was either a distinct abnormal stemline number or a marker chromosome common to a number of different cells. These were the hallmarks of a neoplastic clone, and they indicated that clonal proliferation had occurred in lesions that had been presumed on other grounds to be the early stages of cervical cancer.

RECENT CYTOGENETIC STUDIES

Methods

Direct Preparations. Most of the subsequent work on the cytogenetics of carcinoma of the cervix and its precursors has been done using direct preparations, intended to display the chromosomes of cells already in division when the biopsy was taken. Preparations have been made by the usual techniques for solid tissues. Preliminary incubation for one to several hours in the presence of colchicine was sometimes used, a step intended simply to improve the presentation of cells already dividing by damaging the spindle fibers and contracting the chromosome arms. In most cases the final results leave much to be desired, and new technical methods will do more to increase knowledge in the future than the accumulation of more and more data using the present methods. The new quinacrine- and Giemsa-banding techniques cannot at present be usefully applied to these direct biopsy preparations, because they also require that the chromosomes be clearly presented, without overlapping or distortion.

Tissue Culture. Some of these technical difficulties have been overcome by the use of tissue culture. However, there is certainly a differential in vitro growth of some cells in preference to others, and the cell line that is finally selected may not necessarily be derived from the neoplastic cells at all. Richart [63] has found cultures from carcinoma in situ to consist of cells with normal karyotypes, which conflicts with the evidence of DNA distributions and chromosome counts made directly. It has been suggested [18] that the "epithelioid" cells developing in culture are not in fact of epithelial derivation, but this question remains to be settled [62].

DNA Estimations. The results of microdensitometry of Feulgen-stained nuclei will be mentioned later in this chapter. This method of DNA assay is not accurate enough to show small deviations but is of great value in detecting gross abnormalities, particularly abnormal modal peaks, and it confirms the abnormally wide scatter of values with abnormal modes found in carcinoma in situ and dysplasia of the cervix [3, 22, 67, 74, 87].

Enzyme Determinations. The question of the clonal derivation of preinvasive and invasive cervical lesions has also been studied in a few cases by identification of the isoenzymes for X-linked glucose-6-phosphate dehydrogenase (G6PD) in patients who happen to be heterozygous for these. The principle and methodology of this test, as well as its limitations, are discussed in this volume by Gartler [28].

Park and Jones [53] reported that carcinomas of the cervix have a single electrophoretic band, representative of a single enzyme type, except in material containing substantial amounts of stroma; they interpreted this as supporting single-cell origin. In a series of patients reported by Smith et al. [66], six cases of dysplasia and carcinoma in situ showed single bands, indicating single-cell origin, but three out of seven invasive carcinomas showed double bands. It is too early to interpret these findings with confidence.

Invasive Carcinoma

As in most other kinds of malignant tumor, carcinomas of the cervix consist of populations of cells with abnormal karyotypes, each tumor usually showing a single cytogenetically unique stemline. Chromosome counts above and below the modal number are regularly found, and this almost certainly represents genuine karyotypic variation rather than artifact. The scatter of values is least, however, when preparations are of high quality [10].

Chromosome Numbers. Table 1 lists the modal chromosome numbers

Table 1. Invasive Carcinoma:
Biopsy Samples Classified by Modal Chromosome Number

Number of Chromosomes							Samples with Insufficient Cells or No Mode	Totals*		Ref. for Series
40–49	50–59	60–69	70–79	80–89	90–99	100+		Entries	Cases	
	4	1					1	6	6	78
			1					1	1	29
3	2							5	5	8
3	2	2	1					8	8	11
13	3	3	4	5	1	2	8	39	39	39, 81, 83, 85
3	3	1						7	7	25
	1							1	1	26
				1				1	1	50
			1					1	1	1
3		1						4	4	10
7	1	3						11	11	9
5	1		1	1				8	8	17
2	3	4	2	1				12	12	57
8	3	2	6	2			2	23	23	36
		1						1	1	86
47	23	13	18	11	3	2	11	128	128	Totals

* Each entry represents a separate sample. When two samples were taken from the same cervix, and the counts were concordant, they were entered as one; this was done because cells with concordant numbers are assumed to come from the same clone.

from published cases, wherever these have been given separately. Each entry represents a whole specimen, and when the same case clearly appears in several publications it is entered only once. (The totals are shown graphically in Fig. 4.)

Modal numbers below 50 are the commonest, and hypodiploids are particularly frequent. (Probably cells with low chromosome numbers are still viable on account of translocations that prevent monosomy or nullisomy for essential genes.) Modal numbers from 50 to 79 are also frequent, but higher ones become progressively rarer. This pattern of modal-number distribution differs somewhat from the curves usually shown. Most authors have presented distribution graphs in which all available

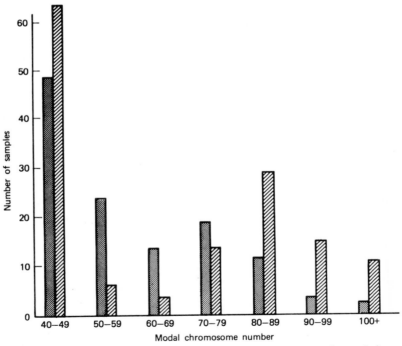

Figure 4. Histograms showing the distribution of modal counts obtained from the literature (see Tables 1 and 3); dotted columns, invasive carcinoma; hatched columns, carcinoma in situ and dysplasia, excluding those entered as "mild dysplasia" in Table 4. Each entry represents a separate sample. When two samples were taken from the same cervix, they are both entered if their modal counts differ, but only one is entered if the counts were concordant. (This is done because concordant counts are assumed to come from the same clone.)

counts from many cases are pooled [31, 36, 40, 41, 82]. The resultant curve embodies two different distributions. First, there is the distribution of chromosome counts within one lesion, usually showing a marked modal peak but sometimes with secondary peaks. Second, there is a wide variation in the position of the modal peak in different cases. Moreover, no correction is made for the fact that each case contributes a different number of counts, some many and others few, so that the final distribution curve has little real meaning.

The distribution obtained by Atkin from DNA measurements, where each case is assigned a modal value, is shown in his Fig. 1 in this volume [6]. The highest peak of DNA values is just above diploid, and there is a second lower one just below the tetraploid level. DNA measurements have the advantage that every case can be analyzed, whereas chromosome counts are frequently unsuccessful. The chromosomes of tumor cells have slightly more DNA per chromosome than do normal cells [11], which may in part explain the difference in the positions of the upper peaks in the histograms of DNA content and of chromosome numbers. The higher chromosome counts may perhaps have been underestimated, because the chance of performing accurate counts diminishes as the chromosome number rises.

Atkin and Richards [13] have demonstrated that the prognosis differs in the two broad groups, cases with higher and those with lower DNA content. The latter have a somewhat lower probability of survival after treatment.

Abnormalities of the Karyotype. Several authors [8, 9, 17, 31, 82] have provided karyotype data in the form of tables dividing the chromosomes into Patau groups [55]. Results are consistent, in that the numbers of groups B, D, and G chromosomes (averaged from a number of cells from each tumor) are nearly always reduced, relative to the other groups, whereas group C chromosomes are increased in number. This may be general among human carcinomas [9]. It should be emphasized that these are net losses and gains and are by no means shown in every cell. It has not yet been proved (e.g., by banding patterns) that Robertsonian ("centric fusion") translocations explain the abnormal karyotypes of cervical cancer, but this would not be surprising, since reductions in groups D and G seem to be a common phenomenon in neoplasia [45, 51, 75].

Compared with photographs of real cells and of chromosome arrays, entries in tables are a poor source of information and embody errors of interpretation which the reader has no means of judging. The literature of the world contains about 30 illustrations showing the chromosomes of primary cervical carcinoma cells [1, 4, 5, 8, 10, 17, 31, 38, 65, 79, 86]. In addition to abnormalities in number, there is frequently evidence of structural rearrangement in the form of marker chromosomes, and these are different from case to case. The commonest markers are long metacentrics and long acrocentrics; but rings, dicentrics, small metacentrics, and minute chromosomes have also been recorded. No specific feature has been identified.

Evidence for Clonal Evolution. The interpretation of the findings so far described is, of course, that the whole-cell population of the malignant epithelium in a particular cervical tumor is of clonal derivation and is usually descended from a single altered cell.* In a few cases there is evidence of two separate stemlines, but this is usually explained by polyploidization and the two lines are related [1]. Some equivocal or contrary evidence from studies on isoenzymes of G6PD [53, 65] is not conclusive enough to counterbalance the extremely convincing cytogenetic evidence, and each tumor can be assumed to be a more or less autonomous cell population that grows by proliferation of its own cells and not by recruiting further epithelial cells of its host.

"Microcarcinoma" (Microinvasive Carcinoma)

The histological appearance of a microinvasive carcinoma suggests incipient or perhaps abortive invasion, or else shows equivocal evidence of invasive growth. Since this state of affairs can only be identified after removal of the tissue for examination, its natural course is not known. After removal of the lesion, the prognosis is uniformly good, and some authorities therefore prefer to classify it with carcinoma in situ rather than as stage I carcinoma.

The microinvasive points are isolated, or occur at intervals in an epithelium that otherwise would be classified as showing carcinoma in situ or dysplasia, so that a piece taken for cytogenetic study does not necessarily represent the same lesion as is studied histologically from an adjacent piece. The cytogenetic evidence therefore concerns abnormal epithelium taken from cervices that harbor microcarcinoma in some part.

Chromosome Numbers. The distribution of modal numbers, as extracted from the literature, is given in Table 2. The number of cases is

---

* In this chapter, the word "clone" is used to mean a population of cells all of which are descended from the same (abnormal) ancestral somatic cell. All of the cells of a clone do *not* necessarily have identical karyotypes; the word is here used to include variants from the main stemline.

Table 2. "Microcarcinoma" (Microinvasive Carcinoma): Biopsy Samples Classified by Modal Chromosome Number

Number of Chromosomes							Samples with Insufficient Cells or No Mode	Totals[*]		Ref. for Series
40–49	50–59	60–69	70–79	80–89	90–99	100+		Entries	Cases	
1	1							2	2	70
1								1	1	41
1								1	1	17
1		1	1					3	2	36
3		1					2	6	5	31
3			2		1		3	9	9	71
1								1	1	72
11	1	1	1	3	1		5	23	21	Totals

[*] See footnote, Table 1. (When counts were discordant, they were recorded as separate entries.)

small, but there appears to be a different chromosome distribution in microcarcinoma from that found in invasive carcinoma. The most frequent higher modes are in the 80 to 89 region; also (not visible in the table), the 40 to 49 group almost all had 46 chromosomes or above. In these respects, the picture is like that of carcinoma in situ, to be described later.

Abnormalities of the Karyotype. In the first Oxford series [70], two of the samples were taken from cervices subsequently found to contain microcarcinoma. In one case, chromosome numbers were mostly close to diploid or tetraploid, but four of the cells with 46 chromosomes showed a long marker. In the second case, there was a distinct clone having a modal number of 54, and some scatter of counts above and below, as well as a few cells with twice this stemline number. Six of the cells with counts of 52 to 54 showed a long acrocentric marker. Nine additional cases were reported from the same center [71], seven of which showed marker chromosomes of various types. In one case (ref. [71], case 26) without a marker, chromosome numbers clustered between 75 and 88 in samples taken from both the anterior and posterior lips of the cervix. In another (ref. [71], case 20), only one photographic karyotype was made from each of the two samples, anterior and posterior, but both showed a long acrocentric chromosome similar in both cells (Fig. 5). In a third (ref. [71], case 27), anterior and posterior lip samples showed a similar ring marker chromosome. Another with a ring marker in two separate biopsies has subsequently been reported [72]. In these cases with ring markers,

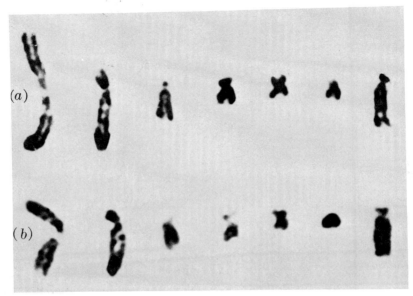

Figure 5. Parts of the karyotypes of two cells from the case shown in Figs. 1 to 3. In each, the chromosomes shown are A1, B, D, E, F, G, and a marker chromosome placed last. (a) From posterior lip of cervix (severe dysplasia): chromosome count, 46. (b) From anterior lip of cervix (carcinoma in situ, near to a microinvasive focus): chromosome count, 44.

The presence of a similar marker chromosome in both areas strongly suggests that the same clone occupies both cervical lips.

the microinvasive focus was in each case at a distance from the area sampled, and it seems certain that there were extensive areas of clonal origin that in most parts of the cervix only showed the histological picture of carcinoma in situ or dysplasia. Further evidence of clonal proliferation is given by the finding of a similar pseudodiploid karyotype in most of the cells from a single sample involved in microcarcinoma [73]. Supporting data from concordant chromosome numbers were provided by Jones et al. [36]. An area of marked dysplasia in the posterior lip and an area in the anterior lip showing microcarcinoma both had the same abnormal modal peak at 63–64; however, their other case showed discordant modes in two samples (46 and 87). Discordant chromosome counts were also found by Granberg [31] in one of five cases in which two samples were examined.

Carcinoma in situ adjacent to occult invasive carcinoma may also be mentioned here. In one case [20], in which the vagina was involved rather than the cervix, three samples showed concordant counts around

70 to 77. Another [7] showed a near-diploid clone containing in its complement a long subacrocentric marker chromosome.

It has not been proved that the noninvasive and microinvasive or invasive areas are derived from the same altered cell; but since the noninvasive portions have all the cytogenetic hallmarks of neoplasia, and since in some cases cells of undoubtedly common ancestry are distributed widely over the cervical epithelium, there is a very strong likelihood that microcarcinoma does develop as a subline from carcinoma in situ or dysplasia and that it is a developmental stage of cervical cancer.

Carcinoma in Situ and Dysplasia

Because of the confusion of nomenclature used in different centers, it seems best to combine the conditions of carcinoma in situ and dysplasia, only setting apart the cases of "dysplasia" which the various authors make clear are of the mildest type.

Chromosome Numbers. All observers have found abnormalities of the chromosomes in most cases of carcinoma in situ and dysplasia. There is a scatter in chromosome numbers in each case, with or without a recognizable modal peak either at 46 or in some abnormal position. Even when the modal count is 46, aneuploid cells have nearly always been present, too.

Table 3 shows the distribution of the main modal regions or clusters, as derived from inspection of the data in the literature. This is more arbitrary than the distribution shown in Tables 1 and 2, because in some cases there was no clearly defined mode and in others there were two or more.* The totals from Tables 1 and 2 are presented graphically in Fig. 4. The patterns of invasive carcinoma and of the putative precancerous states are distinctly different. Among invasive carcinomas, as we have seen, there are many with hypodiploid modes; the upper peak is in the 70 to 79 region, but modes in the 50s are even more common, while modal numbers above 80 are relatively rare. In contrast, microcarcinoma, carcinoma in situ, and dysplasia often have diploid and rarely hypodiploid modes (not shown separately in the tables); those in the 50 to 69 region are rare, and there are many above 80. Modal values

---

* We have not used the rigid definition of a "mode" given by Kirkland and Stanley [40] (i.e., that 50% of the counts must be within 10 adjacent numbers). This would result in the rejection of a clone of many cells with identical karyotypes, provided they were outnumbered by others with scattered chromosome numbers. Since there is no null hypothesis, nor any expected distribution from which one might look for deviations, there seems to be no justification for a statistical test for the significance of clusters.

Table 3. Carcinoma In Situ and Dysplasia: Biopsy Samples Classified by Modal Chromosome Number

Number of Chromosomes							Samples with Insufficient Cells or No Mode	Totals*		Ref. for Series
40–49	50–59	60–69	70–79	80–89	90–99	100+		Entries	Cases	
1					2		1	4	4	70
5		1	2					8	8	20
4	1		6	2		1	13	27	27	39, 84, 85
			2	1		1	1	5	5	41
2							3	5	5	15
5		3		1	2		1	12	10	16
2		1	1					4	4	26
14	2	1	1	9	4	4	4	39	39	36
6	1		2			1	3	13	13	24
3		2	4	2	1	1	3	16	15	71
20	2		1	4	3		5	35	34	31
62	6	3	13	26	14	10	34	168	164	Totals

* See footnote, Table 1. (When counts were discordant, they were recorded as separate entries.)

determined by Feulgen microspectrophotometry also have indicated a predominance of stemlines in the diploid and hypotetraploid areas [3]. The bimodal distribution is also evident in pooled results of chromosome counts from multiple cases, which, as pointed out above, provide a blend rather than a distillate of information [31, 36, 40, 82].

The difference between the distribution of modal numbers of chromosomes in precancerous and invasive disease is most easily explained by postulating that invasion tends to be associated with a loss of chromosomes, the numbers in general being shifted downward. In Table 3 it will be seen that the Granberg series [31] differs from the others in that there are modal numbers of exactly 46 in the majority of cases of carcinoma in situ (21 out of 32). Others have found this much more rarely—for instance, 2 out of 23 cases of carcinoma in situ and dysplasia reported by Spriggs et al. [71]. Granberg found some normal euploid cells in 16 cases and in 2 (Nos. 4 and 11/1) all of the cells analyzed were consistent with normal or broken normal cells. These findings, so much at variance with those of other workers, might perhaps be explained by the inclusion of many cases which, in other hands, would not have furnished any countable metaphases. (The author states that in the latter part of the series a karyotype analysis was possible in "nearly all.")

Abnormalities of the Karyotype. Because of the technical difficulties in obtaining good preparations, so that reports differ even in the frequencies of chromosome numbers, it is even harder to make reliable generalizations about the more detailed abnormalities of the karyotype in carcinoma in situ and dysplasia. Photographs of karyotypes have been published from at least 32 cases [16, 20, 23, 24, 32, 34, 35, 38, 41, 58, 59, 68, 71, 84, 85, 88]. It can be assumed that these particular photographs have been selected either because they were the best ones available, or because they make a particular point (e.g., marker chromosomes). They clearly demonstrate the need for a new technical breakthrough to permit chromosome preparations of much better quality to be made from solid tissues.

Patau group distributions can be made on the best cells, but tables of results have to be interpreted with caution.* In these precancerous lesions, as in the case of invasive carcinoma, chromosomes are sometimes found to be missing from groups B and D [31, 85], although Atkin and Baker [9] in a summary of data from the literature found only a moderate deficit in group B. Most of the published karyotypes have been clearly abnormal but contained no distinct marker chromosomes. This is in contrast to the findings in invasive carcinoma, in which marker chromosomes are very common. Nevertheless, various markers have been reported from carcinoma in situ and dysplasia, either sporadic or occurring in a number of related cells. The type resembling a No. 1 or No. 2 chromosome with a pericentric inversion, giving a chromosome with a high arm ratio, has even been given a special name, Api [16]. Long acrocentrics, very small metacentrics, minutes, and rings have all been seen [71].

Granberg [31] has emphasized the frequency of stretched areas in the long arm of chromosome No. 2, but in the absence of control material from tissues that are equally hard to process, it is difficult to assess its significance. There is certainly no justification for proposing any specificity of a marker chromosome in this disease, and the frequency with which aneuploid cells occur without obvious markers does not support the idea that chromosome breakage and rearrangement are of primary importance in producing the abnormal karyotypes.

Evidence for Clonal Evolution. Whenever there is a marked cluster of chromosome numbers at an abnormal level, there is a likelihood that the cells concerned are related by descent from a common ancestral cell with that chromosome number. In the attempt to discover whether

---

* It is a salutary exercise to rearrange "blind" any of the published karyotype arrays or cutouts from original photographs. As there is no expected idiogram against which to compare an aneuploid cell, there is usually room for a number of different versions.

extensive areas are occupied by the same clone, multiple samples have been studied from the same cervix, and information from markers is sometimes forthcoming. In one case (ref. [71], case 13), samples from the anterior and posterior cervical lips both showed chromosome numbers clustered in the region 76 to 83; both contained cells with a similar long acrocentric marker. On the contrary, in one of the cases from Adelaide (ref. [84], case H 210), two of three different samples showed clearly different modes, one being 47 and the other about 96; a long marker was present in the first of these only. More frequently, the evidence for clonal evolution comes from the comparison of counts in two or more areas, without the help of markers. In the series of Auersperg et al. [16], multiple biopsies were taken from 10 cervices carrying moderate or severe dysplasia or carcinoma in situ. Some of the discordant counts are explained by inclusion of areas of normal epithelium; allowing for this, in two cases the counts agreed and in four they did not. In the series from Baltimore of Jones et al. [36], two biopsies were taken from each of three cases, and in two of these cases, modes from the separate areas were concordant (diploid and near diploid modes); in the third, they were discordant. Spriggs et al. [71] found concordant counts in four out of six cases, including the one mentioned above. Using DNA estimations on Feulgen-stained nuclei, Brandão [22] found concordant abnormal peaks in two samples from one case, and Atkin [3], using scrapings from a relatively wide area rather than multiple small pieces, found what appeared to be a stemline, usually with an abnormal DNA value, in all of 18 cases.

Evidently, the extensive proliferation of a clone is only demonstrable with confidence in a small proportion of cases, but it can be suspected in many more, probably the majority. At the same time, there is sometimes more than one mode, as well as many counts outside the modal region, and in some cases there are not distinct modes at all. Where photographic karyotyping has been done, it is frequently impossible to recognize any real relationship between different cells from the same lesion. Even allowing for technical imperfections, it seems almost certain that some lesions contain many competing cell lines that do not necessarily all have a common origin in a single, chromosomally altered cell.

Mild Dysplasia

Because of the lack of uniformity of nomenclature, there is conflicting information about the findings in cervical "dysplasia." However, in some series there is separate mention of lesions that may be regarded as "minimum-deviation" changes. These have been referred to as dysplasia [39], minimal dysplasia [16], and mild dysplasia [36, 71].

Except in the series of Spriggs et al. [71], all of seven recorded cases had counts at or near diploid, though some had abnormal karyotypes. Kirkland and Stanley [40] presented pooled chromosome distributions for nine cases; there was a high peak at 46, with scatter above and below, but no tetraploids. However, in the four cases of Spriggs et al. [71], with nothing more than mild dysplasia in the whole-cone biopsy, chromosome numbers in three were mostly or all in the tetraploid region (Table 4).

In most mild dysplasias, the mitotic rate is too low to furnish cytogenetic data, and therefore, those reported cannot be considered definitive. Mild dysplasia is detected by the finding of "superficial dyskaryosis" in cervical smears—that is, enlarged and hyperchromatic nuclei in otherwise well-differentiated squamous cells. The enlargement and hyperchromasia are not due to DNA synthesis preparatory to cell division, because the superficial cells in mild dysplasia do not synthesize DNA or divide, and it seems most likely, therefore, that these nuclei are polyploid. DNA estimations would help to settle this point. In Figs. 3 and 4 of Wilbanks et al. [87], histograms from two cases had a wide distribution of values, mainly between diploid and tetraploid but extending above; higher multiples were shown in the case illustrated in their Fig. 5. In case 2 of Brandão [22], a very mild dysplasia, there were many nuclei in the tetraploid class. Recently, Wagner et al. [80] have reported DNA estimations for 14 cases of dysplasia, 8 of which were classified as mild, as well as for 3 cases of carcinoma in situ. In mild and moderate dysplasia, the modes were located at the diploid level and its multiples, and although the full data were not given, it was concluded that "the atypical cells which characterize mild dysplasia generally reflect polyploidization of the epithelium."

Table 4. Mild Dysplasia:
Biopsy Samples Classified by Modal Chromosome Number

Number of Chromosomes							Samples with Insufficient Cells or No Mode	Totals*		Ref. for Series
40–49	50–59	60–69	70–79	80–89	90–99	100+		Entries	Cases	
4							2	6	6	39
2								2	2	16
2								2	2	36
					3		1	4	4	71
8					3		3	14	14	Totals

* See footnote, Table 1.

The finding of normal chromosome counts on the one hand and polyploid DNA levels on the other could be explained by supposing that normal mitoses occur in the basal layer, while endomitosis produces polyploid end cells at a more superficial level.

It is particularly tantalizing that the information on mild dysplasia is so incomplete. It seems almost certain that many or most of these lesions regress spontaneously [56], and the relationship of chromosome changes to reversibility is of the utmost interest.

ORIGIN AND SIGNIFICANCE OF CHROMOSOME CHANGES IN THE CERVIX UTERI

Polyploidy

So far as is known, cervical epithelium, in its normal state and in conditions of inflammation and regeneration, consists of diploid cells. Direct evidence of this from chromosome counts is minimal [13, 34], but the DNA data are consistent with this supposition [3] and it will be assumed as a starting point. As was mentioned above, the mildest and presumably earliest cytological change believed to bear any relationship to cancer is characterized by nuclear enlargement and hyperchromasia in otherwise well-differentiated superficial cells. Similar appearances can be produced experimentally with podophyllin, which induces polyploidy [37], and these lesions are certainly reversible.

Cases of more severe dysplasia and carcinoma in situ with modal counts near 92 are common, and those with hypotetraploid counts are most probably derived from tetraploids by chromosome loss.

Tetraploidy may be produced by endomitosis or endoreduplication [52] or by cell fusion. Apparent endomitoses are common in carcinoma [14], carcinoma in situ, and dysplasia (Fig. 6), and such chromatin condensation may be seen simultaneously in many adjacent cells. In some of these cells Atkin [2] has shown the DNA content to be typical of resting nuclei rather than of nuclei that have completed DNA synthesis; he therefore proposes that the condensed appearance results from a prolonged telophase pattern (rather than from endomitosis). Perhaps both conditions occur. Endoreduplication has sometimes been demonstrated by the finding of diplochromosomes [38, 71]. In case 3 of Spriggs et al. [71] about 25% of the cells in mitosis showed diplochromosomes. This was the same case from which Fig. 6 is taken.

Cell fusion as a cause of polyploidy remains an unproved possibility in cervical lesions. The current hypothesis that infection by herpes virus,

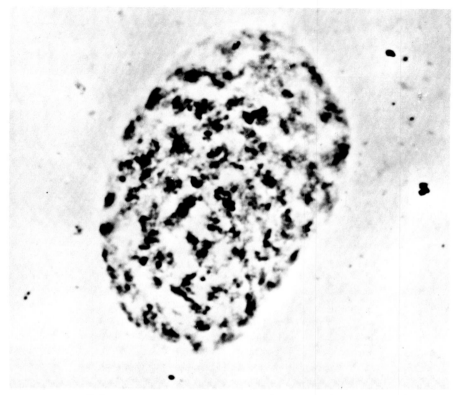

Figure 6. Cell from cervical scraping in a case of mild dysplasia (case 3 of Spriggs et al. [71]). The appearance of condensed chromosomes within an intact nuclear membrane has been attributed to endomitosis, and in this case about 25% of cells in mitosis showed evidence of endoreduplication (diplochromosomes). (Lacto-acetic orcein spread, phase contrast, X 2200.)

which is also known to be an agent for cell fusion in vitro [64], may be related to subsequent carcinoma in situ and carcinoma of the cervix, makes the suggestion an attractive one.

Polyploidy is, then, a highly probable explanation for the morphological findings of an increased nuclear size and density in some of the mildest dysplasias, and it would serve very well to explain the later development of hypotetraploid clones by chromosome loss; but direct evidence for it from chromosome counts is very slender and the question awaits better evidence.

Aneuploidy

If cells are found having one or a few chromosomes more than the normal 46, this is a significant finding. Loss of chromosomes is much less easy to prove, because with present techniques loss of chromosomes during preparation [21] is common. Under these conditions, the best evidence for genuine aneuploidy is clustering of chromosome numbers, with a mode at a level other than 46 or its multiples.

All studies on direct preparations from cervical lesions have shown aneuploidy in most cases of carcinoma in situ. Most authors have failed to find more than a few cells with normal karyotypes, but, as we have seen, Granberg [31] differs in that many of her cases of carcinoma in situ (though not of microinvasive or invasive carcinoma) had modal numbers of 46, with normal karyotypes, associated with relatively few cells having numbers above and below.

The mechanisms of chromosome gain and loss are not known specifically for this material, but it is common to see abnormal mitoses in sections and they can sometimes be observed in smears. Figure 7 shows a sheet of cells in a cervical smear from carcinoma in situ with several synchronously dividing cells at metaphase. In each cell there are chromosomes that have failed to reach the metaphase plate and have been left behind on either side. These so-called 3-group metaphases are particularly common in carcinoma in situ but occur in other tumors and also in cells that have been experimentally poisoned [41, 54]. It has been shown that the misplaced chromosomes are pairs of sister chromatids, still together [33], and this can be seen also in the preparation, of which Fig. 7 is the photograph. It is not known whether particular chromosomes are preferentially involved.

This and other types of abnormal division could easily result in aneuploidy, but it must be admitted that the divisions seen in the milder dysplasias usually appear regular, and "3-group metaphases" are more often a feature of classical carcinoma in situ and microcarcinoma than of the lesser abnormalities.

Chromosome Breakage

None of the observers who have studied direct preparations of the human cervix has noted any particular tendency to chromosome breakage. Exaggerated secondary constrictions producing unusually long No. 2 chromosomes have been noted [31], but chromosome fragments and figures resulting from chromatid exchanges are seldom mentioned. Markers are of course a sure sign that chromosome breakage and recombination have

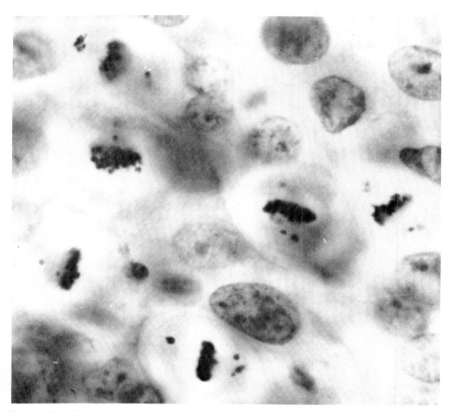

Figure 7. Sheet of cells in a cervical smear from a case of carcinoma in situ. A number of cells are in synchronous metaphase, and all show chromosomes remaining near the poles ("3-group metaphase"). This is a common feature of carcinoma in situ of the cervix. (Papanicolaou, X 1500.)

taken place, and there are many reports of these; but they are distinctly more common in microcarcinoma and invasive carcinoma [31, 71]. Visible chromosome breaks and recombination figures are therefore not a regular feature of the putative precancerous states of the cervix, even if they are actually present in an occult form.

SUMMARY

Because of some conflicting findings, it is only possible to make a tentative summary. *Normal* and *inflammatory* conditions of the cervix are not known to be associated with any deviation from euploidy. The mildest changes regarded as possibly on the road to neoplasia (*mild dysplasia*) have in one series shown chromosome numbers near to tetraploid (some of which could represent normal tetraploid cells), but most other series have had numbers near to, or exactly at, the diploid number of 46. In *severe dysplasia* and *carcinoma in situ*, some or all of the cells are aneuploid. There is a wide scatter in chromosome numbers within one sample, but in many cases clustering of numbers in an abnormal region suggests the formation of an aneuploid clone. Karyotypic evidence of clonal growth has also been obtained in a few cases. According to some authors, cells with normal karyotypes are commonly present along with the abnormal ones. In *microcarcinoma* (microinvasive carcinoma) there is good evidence from karyotype analysis and from the presence of markers that all or a substantial part of the lesion consists of a neoplastic clone. This is also the typical finding for established *invasive carcinoma* of the cervix.

LITERATURE CITED

1. ATKIN NB: A carcinoma of the cervix uteri with hypodiploid and hypotetraploid stem-lines. *Eur J Cancer* 3:289–291, 1967.
2. ATKIN NB: A high incidence of cells with a condensed (telophase) chromatin pattern in human tumors and carcinoma-in-situ. *Acta Cytol* 11:81–85, 1967.
3. ATKIN NB: The use of microspectrophotometry. *Obstetr Gynecol Surv* 24:794–804, 1969.
4. ATKIN NB: Cytogenetic studies on human tumors and premalignant lesions: the emergence of aneuploid cell lines and their relationship to the process of malignant transformation in man. In Genetic Concepts and Neoplasia, Twenty-third Annual Symposium on Fundamental Cancer Research 1969, M. D. Anderson Hospital, University of Texas. Baltimore, Williams & Wilkins Company, 36–56, 1970.
5. ATKIN NB: Cytogenetic factors influencing the prognosis of uterine carcinoma. In Modern Radiotherapy: Gynaecological Cancer (Deeley TJ, ed), London, Butterworth & Company, 138–154, 1971.
6. ATKIN NB: Chromosomes in human malignant tumors: a review and assessment. This volume, pp 375–422, 1974.
7. ATKIN NB, BAKER MC: Chromosomes in carcinoma of the cervix. *Brit Med J* 1:522–523, 1965.
8. ATKIN NB, BAKER MC: Chromosome abnormalities as primary events in human malignant disease: evidence from marker chromosomes. *J Nat Cancer Inst* 36:539–557, 1966.
9. ATKIN NB, BAKER MC: Possible differences between the karyotypes of preinvasive lesions and malignant tumours. *Brit J Cancer* 23:329–336, 1969.
10. ATKIN NB, BAKER MC, WILSON S: Stem-line karyotypes of 4 carcinomas of the cervix uteri. *Am J Obstetr Gynecol* 99:506–514, 1967.

11. ATKIN NB, MATTINSON G, BAKER MC: A comparison of the DNA content and chromosome number of fifty human tumours. *Brit J Cancer* 20:87–101, 1966.
12. ATKIN NB, RICHARDS BM: Deoxyribonucleic acid in human tumours as measured by microspectrophotometry of Feulgen stain. *Brit J Cancer* 10:769–786, 1956.
13. ATKIN NB, RICHARDS BM: Clinical significance of ploidy in carcinoma of cervix: its relation to prognosis. *Brit Med J* 2:1445–1446, 1962.
14. ATKIN NB, Ross AJ: Polyploidy in human tumours. *Nature* 187:579–581, 1960.
15. AUERSPERG N, COREY MJ, AUSTIN G: Chromosomes in cervical lesions. *Lancet* 1:604–605, 1966.
16. AUERSPERG N, COREY MJ, WORTH A: Chromosomes in preinvasive lesions of the human uterine cervix. *Cancer Res* 27:1394–1401, 1967.
17. AUERSPERG N, WAKONIG-VAARTAJA T: Chromosome changes in invasive carcinomas of the uterine cervix. *Acta Cytol* 14:495–501, 1970.
18. AUERSPERG N, WORTH A: Growth patterns *in vitro* of invasive squamous carcinomas of the cervix—a correlation of cultural, histologic, cytogenetic and clinical features. *Int J Cancer* 1:219–238, 1966.
19. BETTINGER HF, REAGAN JW (Co-chairmen): Proceedings of the International Committee on Histological Terminology for Lesions of the Uterine Cervix. *In* Proceedings of the First International Congress on Exfoliative Cytology, Vienna, 1961. Philadelphia, Lippincott, 283–297, 1962.
20. BODDINGTON MM, SPRIGGS AI, WOLFENDALE MR: Cytogenetic abnormalities in carcinoma-in-situ and dysplasias of the uterine cervix. *Brit Med J* 1:154–158, 1965.
21. BOWEY CE, SPRIGGS AI: Lost chromosomes in endometrial cells. *J Med Genet* 5:58–59, 1968.
22. BRANDÃO HJS: DNA content in epithelial cells of dysplasias of the uterine cervix. Histologic and microspectrophotometric observations. *Acta Cytol* 13:232–237, 1969.
23. DE CECCO L, RUGIATI S, SBERNINI R: Aspetti citogenetici dei carcinomi uterini. *Quad clin Ostetr Ginecol* 21:562–573, 1966.
24. DEHNHARD F, BREINL H, KNÖRR-GÄRTNER H: Chromosomenbefunde bei Krebsvorstadien der Cervix uteri (Zusammenfassende Darstellung erster Ergebnisse). *Minerva Ginecol* 23:78–81, 1971.
25. FISCHER P, GOLOB E, HOLZNER JH: Chromosomenzahl und DNS-Wert bei malignen Tumoren des weiblichen Genitaltraktes. *Z Krebsforsch* 68:200-208, 1966.
26. FISCHER P, GOLOB E, HOLZNER JH: Zytogenetische Untersuchungen am Portioepithel bei positivem und bei zweifelhaftem Abstrichbefund. *Krebsarzt* 22:289–294, 1967.
27. FOULDS L: Neoplastic Development, Vol 2, London, Academic Press, in press.
28. GARTLER SM: Utilization of mosaic systems in the study of the origin and progression of tumors. This volume, pp. 313–334, 1974.
29. GOODLIN RC: Karyotype analysis of gynecological malignant tumors. *Am J Obstetr Gynecol* 84:493–500, 1962.
30. GOVAN ADT, HAINES RM, LANGLEY FA, TAYLOR CW, WOODCOCK AS: The

histology and cytology of changes in the epithelium of the cervix uteri. *J Clin Pathol* 22:383–395, 1969.
31. GRANBERG I: Chromosomes in preinvasive, microinvasive and invasive cervical carcinoma. *Hereditas* 68:165–218, 1971.
32. GRANBERG I, TRANEUS A, SILFVERSWÄRD C: Chromosome pattern in a patient with cervical carcinoma in situ and atypical hyperplasia of the endometrium. *Acta Obstetr Gynecol Scand* 51:47–53, 1972.
33. HENEEN WK, NICHOLS WW, NORRBY E: Polykaryocytosis and mitosis in a human cell line after treatment with measles virus. *Hereditas* 64:53–84, 1970.
34. JONES HW, DAVIS HJ, FROST JK, PARK IJ, SALIMI R, TSENG PY, WOODRUFF JD: The value of the assay of chromosomes in the diagnosis of cervical neoplasia. *Am J Obstetr Gynecol* 102:624–640, 1968.
35. JONES HW, KATAYAMA KP, STAFL A, DAVIS HJ: Chromosomes of cervical atypia, carcinoma *in situ* and epidermoid carcinoma of the cervix. *Obstetr Gynecol* 30:790–805, 1967.
36. JONES HW, WOODRUFF JD, DAVIS HJ, KATAYAMA KP, SALIMI R, PARK IJ, TSENG PY, PRESTON E: The evolution of chromosomal aneuploidy in cervical atypia, carcinoma *in situ* and invasive carcinoma of the uterine cervix. *Johns Hopkins Med J* 127:125–135, 1970.
37. KAMINETZKY HA, JAGIELLO GM: Differential chromosomal effects of carcinogenic and noncarcinogenic substances. *Am J Obstetr Gynecol* 98:349–355, 1967.
38. KATAYAMA KP: Chromosomes and pelvic cancer. *Clin Obstetr Gynecol* 12:435–458, 1969.
39. KIRKLAND JA: Mitotic and chromosomal abnormalities in carcinoma *in situ* of the uterine cervix. *Acta Cytol* 10:80–86, 1966.
40. KIRKLAND JA, STANLEY MA: The cytogenetics of carcinoma of the cervix. *Aust N Z J Obstetr Gynecol* 7:189–193, 1967
41. KIRKLAND JA, STANLEY MA, CELLIER KM: Comparative study of histologic and chromosomal abnormalities in cervical neoplasia. *Cancer* 20:1934–1952, 1967.
42. KOLLER PC: Abnormal mitosis in tumours. *Brit J Cancer* 1:38–47, 1947.
43. Koss L: Diagnostic Cytology and its Histopathologic Bases, 2nd ed, New York, Pitman, 1968.
44. LEUCHTENBERGER C, LEUCHTENBERGER R, DAVIS AM: A microspectrophotometric study of the deoxyribose nucleic acid (DNA) content in cells of normal and malignant tissues. *Am J Pathol* 30:65–85, 1954.
45. LEVAN A: Non-random representation of chromosome types in human tumor stem lines. *Hereditas* 55:28–38, 1966.
46. MAKINO S, ISHIHARA T, TONOMURA A: Cytologic studies of tumors. XXVII. The chromosomes of thirty human tumors. *Z Krebsforsch* 63:184–208, 1959.
47. MANNA GK: A study on chromosomes of the human nonneoplastic and neoplastic uterine tissues. *In* Proceedings of the International Genetics Symposia, 1956. Cytologia Suppl, 182–187, 1957.
48. MANNA GK: A study on the chromosome number of human neoplastic uterine cervix tissue. *In* Proceedings of the Zoological Society of Calcutta, Mookerjee Memorial Volume, 95–112, 1957.

49. MELLORS RC, KEANE JF, PAPANICOLAOU GN: Nucleic acid content of the squamous cancer cell. *Science* 116:265–269, 1952.
50. MILES CP: Chromosome analysis of solid tumors II. Twenty-six epithelial tumors. *Cancer* 20:1274–1287, 1967.
51. MULDAL S, ELEJALDE R, HARVEY PW: Specific chromosome anomaly associated with autonomous and cancerous development in man. *Nature* 229:48–49, 1971.
52. OKSALA T, THERMAN E: Mitotic abnormalities and cancer. This volume, pp. 239–263, 1974.
53. PARK IJ, JONES HW: Glucose-6-phosphate dehydrogenase and the histogenesis of epidermoid carcinoma of the cervix. *Am J Obstetr Gynecol* 102:106–109, 1968.
54. PARMENTIER R, DUSTIN P: Réproduction expérimentale d'une anomalie particulière de la métaphase des cellules malignes ("métaphase à trois groupes"). *Caryologia* 4:98–109, 1951.
55. PATAU K: The identification of individual chromosomes, especially in man. *Am J Hum Genet* 12:250–276 (1960).
56. PATTEN SF: Dysplasia of the uterine cervix. In New Concepts in Gynecological Oncology (Lewis GC, Wentz WB, Jaffé RM, eds), Philadelphia, Davis, 33–44, 1966.
57. PAULETE-VANRELL J: DNA content and chromosome number in twenty-five human carcinomas. *Oncology* 24:48–57, 1970.
58. PESCETTO G: Considerazioni e rilievi sugli aspetti citogenetici delle neoplasie ginecologiche. *Rass Clin Sci* 43:329–337, 1967.
59. PESCETTO G: Problemi prognostici nelle displasie cervicali uterine. In Scritti mediche in onore del Prof. G. Dellepiane, 817–822, 1968.
60. REAGAN JW, PATTEN SF: Dysplasia: a basic reaction to injury in the uterine cervix. *Ann NY Acad Sci* 97:662–682, 1962.
61. RICHART RM: Natural history of cervical intraepithelial neoplasia. *Clin Obstetr Gynecol* 10:748–784, 1967.
62. RICHART RM, LUDWIG AS: Alterations in chromosomes and DNA content in gynecologic neoplasms. *Am J Obstetr Gynecol* 104:463–471, 1969.
63. RICHART RM, WILBANKS GD: The chromosomes of human intraepithelial neoplasia: Report of 14 cases of cervical intraepithelial neoplasia and review. *Cancer Res* 26:60–74, 1966.
64. ROIZMAN B: Polykaryocytosis. *Cold Spring Harbor Symp Quant Biol* 27:327–342 (1962).
65. SALIMI R, JONES HW: Chromosomes of adenocarcinoma of the cervix uteri with a ring and a minute marker chromosome. *J Surg Oncol* 2:17–22, 1970.
66. SMITH JW, TOWNSEND DE, SPARKES RS: Genetic variants of glucose-6-phosphate dehydrogenase in the study of carcinoma of the cervix. *Cancer* 28:529–532, 1971.
67. SANDRITTER W, FISCHER R: Der DNS-Gehalt des normalen Plattenepithels, des Carcinoma in situ und des invasiven Carcinoms der Portio. In Proceedings of the First International Congress on Exfoliative Cytology, Vienna. Philadelphia, J. B. Lippincott, 189–194, 1962.

68. SPRIGGS AI: Twenty-five years of cervical cytology. *In* Trends in Clinical Pathology. Essays in Honour of Gordon Signy. London, BMA, 252–268, 1969.
69. SPRIGGS AI: Population screening by the cervical smear. *Nature* 238:135–137, 1972.
70. SPRIGGS AI, BODDINGTON MM, CLARKE CM: Carcinoma-in-situ of the cervix uteri. *Lancet* 1:1383–1384, 1962.
71. SPRIGGS AI, BOWEY CE, COWDELL RH: Chromosomes of precancerous lesions of the cervix uteri. *Cancer* 27:1239–1254, 1971.
72. SPRIGGS AI, COWDELL RH: Ring chromosomes in carcinoma *in situ* of the cervix as evidence of clonal growth. *J Obstetr Gynaecol Brit Commonw* 79:833–840, 1972.
73. STANLEY MA, KIRKLAND JA: Chromosome analysis in a progressive lesion of the cervix. *Acta Cytol* 13:76–80, 1969.
74. STEELE HD, MANOCHA SL, STICH HF: Deoxyribonucleic acid content of epidermal in-situ carcinomas. *Brit Med J* 2:1314–1315, 1963.
75. VAN STEENIS H: Chromosomes and cancer. *Nature* 209:819–821, 1966.
76. TIMONEN S: Mitosis in normal endometrium and genital cancer. *Acta Obstetr Gynaecol Scand* 31: Suppl 2, 1950.
77. TJIO JH, PUCK TT: Genetics of somatic mammalian cells. II. Chromosomal constitution of cells in tissue culture. *J Exp Med* 108:259–268, 1958.
78. TONOMURA A: A chromosome survey in six cases of human uterine cervix carcinomas. *Jap J Genet* 34:401–406, 1959.
79. TORTORA M: Chromosome studies in gynecologic cancer. *Arch Ostetr Ginecol* 68:437–444, 1963.
80. WAGNER D, SPRENGER E, BLANK MH: DNA-content of dysplastic cells of the uterine cervix. *Acta Cytol* 16:517–522, 1972.
81. WAKONIG-VAARTAJA R: Chromosomes in gynaecological malignant tumours. *Aust NZ J Obstetr Gynaecol* 3:170–177, 1963.
82. WAKONIG-VAARTAJA R, HUGHES DT: Chromosomal anomalies in dysplasia, carcinoma-in-situ, and carcinoma of cervix uteri. *Lancet* 2:756–759, 1965.
83. WAKONIG-VAARTAJA R, HUGHES DT: Chromosome studies in 36 gynaecological tumours: of the cervix, corpus uteri, ovary, vagina and vulva. *Eur J Cancer* 3:263–277, 1967.
84. WAKONIG-VAARTAJA R, KIRKLAND JA: A correlated chromosomal and histopathologic study of pre-invasive lesions of the cervix. *Cancer* 18:1101–1112, 1965.
85. WAKONIG-VAARTAJA T: Chromosomes in precancerous lesions and in carcinoma of the uterine cervix. *Bull NY Acad Med* 45:22–38, 1969.
86. WAKONIG-VAARTAJA T, AUERSPERG N: Cytogenetics of gynecologic neoplasms. *Clin Obstetr Gynecol* 13:813–830, 1970.
87. WILBANKS GD, RICHART RM, TERNER JY: DNA content of cervical intraepithelial neoplasia studied by two-wavelength Feulgen cytophotometry. *Am J Obstetr Gynecol* 98:792–799, 1967.
88. WOODRUFF JD, DAVIS HJ, JONES HJ, RECIO RG, SALIMI R, PARK IJ: Correlated investigative technics of multiple anaplasias in the lower genital canal. *Obstetr Gynecol* 33:609–616, 1969.

CHROMOSOMES AND LEUKEMIA

S. MULDAL AND L. G. LAJTHA

Christie Hospital and Holt Radium Institute, Manchester, England

---

Introduction	451
Proliferative Diseases of the Hematopoietic System	452
Myeloproliferative Diseases	452
Dysproteinemias	453
Lymphoproliferative Diseases	454
General Cytogenetic Aspects	454
Cytogenetics of Myeloproliferative Diseases	456
Chronic Myeloid Leukemias	456
The Ph[1] Chromosome	456
Variant Karyotypes	459
Two Ph[1] Chromosomes and Blastic Crisis	460
Origin of Ph[1] Chromosome	460
Properties of Ph[1]-positive Cells	461
Ph[1], LAP, and Mongolism	461
Conditions Associated with Chronic Myeloid Leukemia	462
Acute Myeloid Leukemia	462
Fanconi's Anemia and Bloom's Syndrome	464
Cytogenetics of Lymphoproliferative Diseases	464
Chronic Lymphatic Leukemia	464
Lymphomas and Paraproteinemias	465
Acute Lymphoblastic Leukemia	467
Concluding Remarks	468
Literature Cited	470

INTRODUCTION

The object of this chapter is to give a somewhat selective review of the cytogenetic work carried out on myeloproliferative disorders and related neoplasms during the last 15 years. This is a somewhat inopportune moment to try to assess the general significance of chromosome change in leukemia, because new and more accurate information has just begun to accumulate through use of the new banding techniques. We therefore emphasize those areas where the most information is already available and sufficiently substantiated. The choice of particular references, out of nearly a thousand on the subject, is not meant to reflect on the quality of those quoted or left out but was the result of space limitations. Certain clinical aspects have been mentioned for the benefit of the nonmedical reader, emphasizing overlapping features and the range of disorders under consideration.

PROLIFERATIVE DISEASES OF THE HEMATOPOIETIC SYSTEM

Differential diagnosis of the proliferative diseases of the hematopoietic system is extremely difficult in many cases. While any cell in the system may participate in progressive proliferation, constitutional factors are likely to modify the clinical expression in the individual, and the numerous transition forms of these diseases often makes the hematological identification uncertain. The concept of malignancy, which may at times be difficult to define in solid tumors, is considerably more vague in the hematopoietic system, in which case the malignant cell populations usually are systemically disseminated. For this reason, chromosome studies are of special importance here in clarifying diagnosis, particularly when the definitive new banding techniques for chromosome identification are used as an adjunct to the hematological findings.

Myeloproliferative Diseases

The myeloproliferative family of diseases is characterized by sustained, progressive proliferation of cells of the myeloid, erythroid, and megakary-

---

Abbreviations used frequently in text:

ALL, acute lymphoblastic leukemia
AML, acute myeloid leukemia
ASG, acetic–saline Giemsa
CLL, chronic lymphocytic leukemia
CML, chronic myeloid leukemia
EL, erythroleukemia
ISA, idiopathic sideroeytic anemia
LAP, leukocyte alkaline phosphatase
PHA, phytohemagglutinin
PV, polycythemia vera
QM, quinacrine mustard

Chromosome terminology.—Denver groups: A(1-3), B(4-5), C(X,6-12), D(13-15), E(16-18), F(19-20), G(Y,21-22); short arm = p; long arm = q; isochromosome = i; 20q− = long arm deletion of chromosome No. 20.

ocytoid series, or any combination of the three. Those generally accepted to be neoplastic are the acute and chronic myeloid (or granulocytic) leukemias (AML, CML), acute erythremic myelosis, and erythroleukemia (EL); those not widely accepted as neoplasias are polycythemia vera (PV), idiopathic thrombocythemia (IT), and myelosclerosis (with myeloid metaplasia, MMP).

The Myeloid Leukemias. Both the acute (AML) and chronic (CML) forms of this type of leukemia are relatively well defined; the subacute variety is not, although it resembles the acute form most. Sarcomatous, solid tumor growths may occur in either the acute or chronic form of the disease; those associated with acute myeloid leukemia have a green color and are termed "chloroma." When granulocytic differentiation is not evident, it is difficult to distinguish the acute form of this leukemia from other leukemias, although the absence of granulocytes is usually diagnostic of the acute form, as opposed to the chronic form of myeloid leukemia.

Acute erythremic myelosis is a rapidly progressive, irreversible, systemic proliferation of erythroblasts and pro-erythroblasts, characterized by a preponderance of these cell types in the circulation.

True *erythroleukemia* (Di Guglielmo's) (EL) is characterized by progressive proliferation of immature and atypical cells of both the erythroid and myeloid series. Solid growths composed of reticulum cells, erythroblasts, and myeloblasts have been reported and are probably analogous to the myeloid sarcoma of acute myeloid leukemia [158]. An acute myelosis may occur rarely, in which case atypical megakaryocytes are present, as well as erythroblasts and myeloid cells.

Many of the conditions not accepted as neoplasms are "preleukemic states," and the following transitions are recognized (after Rappaport [158]):

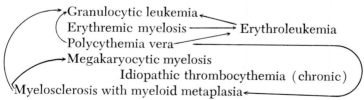

Dysproteinemias

In the proliferative diseases with dysproteinemias, the proliferating cells correspond to the producers of immunoglobins IgG, IgA, and IgM. Three conditions are clearly recognized: *multiple myeloma, Waldenström's macroglobulinemia,* and *heavy chain disease.* In Waldström's macroglobulinemia, lymphocytes are the proliferating cells; heavy chain disease is a malignant myeloma.

Lymphoproliferative Diseases

The lymphoproliferative family of diseases consists of aleukemic solid-growing types and freely disseminating types. The former includes lymphomas and lymphosarcomas (e.g., *Hodgkin's disease, reticulosis*), and the latter includes the lymphatic leukemias.

Acute lymphoblastic leukemia (ALL) is characterized by proliferating lymphoblasts. When the blast cells are relatively unrecognizable, the term "stem-cell leukemia" is applied. *Chronic lymphocytic leukemia (CLL)* is characterized by increased numbers of small or middle-sized lymphocytes in the peripheral blood.

GENERAL CYTOGENETIC ASPECTS

The development of autonomy implies the capacity of indefinite propagation, which is not a general property of normal mammalian cells, except for germ cells. Cell lines with acquired autonomy (whether neoplastic or not) are usually characterized by chromosomal instability, which is expressed as nullo-, mono-, or polysomy for various chromosomes or chromosome parts. Even when the commonest cell in a line—the modal or stem cell—appears normal, the number of cells that deviate from the mode in chromosome number is greater than in normal tissue. It can be demonstrated that most solid tumors have a monoclonal origin [59]. A malignant cell population is characterized by a more or less distinct modal number, which can be diploid, pseudodiploid, or aneuploid, and a nonmodal range of aneuploid cells. Nonmodal aneuploidy may be narrow or widely spread. This "aneuploid spread" can lead to changes in the modal chromosome constitution by allowing selection for new modes with better survival potentials; that is, karyotypic evolution can occur. The rate of clonal evolution must depend upon the properties of the population and the specificity of the host's responses. The relative lack of "exposed" cell surfaces in the solid tumor may allow large numbers of tumor cells to escape from such host responses. Even cells that may not contribute much to the growth rate of the tumor can nevertheless add to the karyotypic heterogeneity observed.

The general cytogenetic picture in leukemias and related diseases differs from that of the solid tumor. While in the solid tumors aneuploidy often appears at a significant level as soon as the growth is detectable [*81*], in leukemia—particularly in the chronic types—a period of diploidy precedes the aneuploidy that is often observed in the final phase of the disease. The apparently low tolerance for aneuploidy in these "free" cells will, therefore, lead to a selection for much more subtle deviations than for gross aneuploid spread. The significance of this feature of leukemia may be deduced from the contrasting condition in which tumefaction occurs (as in lymphomas, lymphosarcomas, giant follicular lymphomas, reticular cell sarcomas, and leukosarcomas), where aneuploid spread may be just as wide as in other solid tumors [*65*].

Leukemia may therefore be looked upon as an extension of an early stage of neoplastic transformation, a stage that is very short and elusive in solid tumors but is one of the most interesting stages to the student of chromosome changes in neoplasms. This early stage is attenuated in the leukemias, which implies that both apparently unchanged as well as changed karyotypes can occur at what is seemingly the same phase of neoplastic development. When karyotypic changes occur at an early stage, there may be no aneuploid spread, and the aneuploid mode (or the pseudodiploid mode with some structural rearrangement in the karyotype) shows great stability over a considerable length of time. For this reason, progression and clonal evolution are very slow and seem to gather momentum only just before death—that is, if the leukemia has run its full course without intervention from some lethal complication.

The extent to which genes involved in the same metabolic processes are spread throughout the chromosome set determines the stability of the diploid (and haploid) karyotype. From the study of spontaneous abortions, it is clear that a large number of abnormal karyotypes (excepting X and autosomal nullosomy) are intrinsically viable. The effect of monosomy or partial monosomy is the immediate uncovering of recessive genes for hemizygous expression. Furthermore, new mutations in the monosomic region will be expressed immediately, thus enhancing the "effective" mutation rate. Trisomy or partial trisomy will reduce the effective mutation rate, and the immediate expression of the trisomy will be dosage effects. Generally speaking, trisomy is tolerated much better than monosomy.

Leukemic cells show chromosomes with "fuzzy" edges and do not yield preparations of the same quality as do PHA-stimulated lymphocytes. The new banding techniques, nevertheless, have greatly aided the recognition of individual chromosomes in leukemic cells. Most of the observations to be quoted in the paper, however, have not been made using these modern techniques. The true identity and structure of extra chromosomes associated with certain disorders has not, therefore, necessarily been ascertained, even after repeated observations. Nor does the presence of an extra chromosome imply that simple nondisjunction had occurred; there is reason to believe that most extra or marker chromosomes are derived from chromatid-exchange configurations which most often separate the two chromosomes involved in reciprocal translocation into different daughter cells. Application of the banding techniques is already yielding information to this effect. In three reinvestigated cases where 17qi chromosomes could be recognized by banding techniques, the karyotypes had previously been reported to have an extra group C chromosome [130]. Similarly, the common 20q− chromosome must have masqueraded in an F−, G+ cell, and cells with the 20pi must have been assumed to have intact group F chromosomes. Deletions of the long arm of chromosome No. 17 or No. 18 can give pseudo-Ph^1 chromosomes [79, 80, 199]. This finding suggests that the past reports on the nature of chromosome abnormalities in leukemia must be viewed with a certain degree of uncertainty.

CYTOGENETICS OF MYELOPROLIFERATIVE DISEASES

Pitfalls must be avoided in studying the cytogenetics of myeloproliferative disease; some of the difficulties will be mentioned first. The peripheral blood picture is often not sufficiently informative in leukemia; therefore, it is necessary to study marrow samples directly, or short-term marrow cultures, and these can be supplemented with peripheral blood cultures. Furthermore, chromosome analysis should be carried out before the patient is treated; the effects of the therapy may be studied thereafter using serial samples. A number of the chromosome studies reported are of uncertain value because these conditions were not fulfilled. Many cases have been inadequately studied, mainly due to insufficiency of mitotic cells or inadequate quality of the chromosome preparations; nevertheless, such cases have been reported, perhaps in order to swell the numbers. In cases in which no abnormalities were found, many fewer cells have usually been scored than when abnormalities were present. There is a tendency, also, to report only abnormal cases and not to report apparently normal findings. In addition, the hematological data may be insufficient for classification of the disease.

Chronic Myeloid Leukemia (CML)

The Ph^1 Chromosome. The breakthrough in the techniques of blood culture led to the discovery that CML is characterized by a deletion in one of the smallest chromosomes [148, 149], which loses half its chromatin content [9, 150], producing the "Philadelphia" or Ph^1 chromosome. The chromosome at the site of the Giemsa-positive band, a site presumably with telomeric properties. The morphology of the Ph^1 chromosome is extremely consistent in our material, although variations are claimed to occur [7, 191]. It may represent a spontaneously occurring mutation. The increased incidence of CML after radiotherapy for ankylosing spondylitis [73], or its occurrence within a short time (3 years) of exposure to the atomic bomb explosions of 1945 [95], does not prove that the Ph^1 chromosome is induced by radiation (although Ph^1-like chromosomes have been seen in patients exposed to radiation who have not, to date, developed CML [80]). There is also no evidence that Ph^1-negative cases of CML are induced by radiation. It is more likely that the greater stress on normal cell proliferation under these conditions increases the probability of establishing a Ph^1-positive clone. Similar stresses on the bone-marrow population by constitutional tendencies to chromosome fracture are also found in Fanconi's anemia, Bloom's syndrome, and ataxia telangiectasia [82, 93], disorders in which an increased incidence of leukemia is also found [177].

Properties of Ph^1-positive Cells. The short arm of the Ph^1 chromosome is intact [157], although it is subject to coiling variations probably associated with a shortened long arm. A normal frequency of association with other acrocentric chromosomes is found, although the overall association of acrocentrics in Ph^1 positive cells has been reported to be lower than in normal lymphocytes. The Ph^1-positive cells should actually be compared with normal myeloid cells; the difference may indicate a shorter interphase time. The degree of segmentation in the leukemic polymorphonuclear leukocytes is lower than normal [169]. When compared with normal lymphocytes, CML cells have a slower recovery rate after Colcemid [142], with abnormal centriole structure in the cells; presumably their microtubules have a different binding capacity for Colcemid. Leukemic polymorphs have reduced motility [15]. Ph^1-positive cells will form colonies in agar and will differentiate into apparently mature metamyelocytes, polymorphs, and macrophages [144].

The Ph^1 Chromosome, LAP, and Mongolism. It is now of historic interest only that the elevated level of leukocyte alkaline phosphatase (LAP) in mongolism and the lowered LAP levels of CML were thought to indicate that a LAP gene was located in the deleted part of the affected group G chromosome [3]. Although granulocytes in CML usually have reduced LAP levels, there are also cases with normal LAP levels; reduced LAP levels also occur in Ph^1-negative cases [56]. The level of LAP often returns to normal during remission, although all cells examined in the bone marrow may remain Ph^1-positive. The concept of the LAP gene assignment to a group G chromosome was, therefore, controversial [83] even before it was known that different group G chromosomes were involved in CML (No. 22) and in mongolism (No. 21).

Mongolism has an increased incidence of leukemia [97], mainly AML and ALL, and such individuals will express the usual chromosome changes expected of each [98, 105, 111, 190, 194]. The trisomy 21 persists in the leukemic cells with or without other superimposed changes [22]. Ph^1-positive CML appears not to have been described in mongolism [170], which is not surprising, however, if the incidences of the two conditions are considered. The susceptibility to leukemia in mongolism may be related to a nonleukemic, leukemoid reaction that frequently occurs in these patients [165, 170]. Stresses on the hemopoietic and lymphatic systems are undoubtedly caused by the frequent and chronic infections from which these patients tend to suffer; the situation may be similar to the increased incidence of leukemia in germ-free animals returned to a normal environment [144].

Conditions Associated with Chronic Myeloid Leukemia (CML)

CML is sometimes accompanied by a very striking eosinophilia. This has been investigated sufficiently well to suggest that there appears to be a frequency of Ph^1-positive [108, 120] and Ph^1-negative cases similar to that in the standard type of CML [51]. In view of the finding that the Ph^1 chromosome occurs in all three bone-marrow series [159] and that the existence of CML may be considered to indicate that the leukemic change had occurred in a common ancestral cell, one might therefore expect that the Ph^1 chromosome would also be found in erythroleukemia and thrombocythemia. However, such findings are relatively rare [72, 106, 168], and the diagnosis is not always sufficiently reliable [32].

EL (Di Guglielmo's) in acute form has been reported to be Ph^1-positive nine times [115], but various features suggest that these actually represented a subacute form [170]. The rare chronic form of EL has been reported as Ph^1-positive [25, 32, 99]. Thrombocythemia has been reported Ph^1-positive on three occasions, negative on one [201]. Myelosclerosis with myeloid metaplasia [119, 121] has been found Ph^1-positive on many occasions, and Ph^1-negative cases have also been reported [125].

Conditions described as atypical panmyelosis, agnogenic myeloid metaplasia [69], myeloid metaplasia without myelofibrosis, and the like [71, 92, 112, 114, 116, 151] have been reported as Ph^1-positive or Ph^1-negative. The Ph^1 chromosome has also been found in PV, although this did not always proceed to develop into CML [170]. Other chromosome complements [5, 140] such as $-Y$ (45,X in a male) observed in CML, have been described in PV [109], EL [91] and aleukemic leukemia [29, 70, 166]. PV has also had the 20q− marker in about 20% of cases [126, 145, 160]; this marker is more frequent in idiopathic siderocytic anemia (ISA) [43] and can occur in "refractory anemia" [5]. Group C trisomy, which is often associated with clonal evolution in the blastic crisis of CML, is even more common in myelosclerosis with myeloid metaplasia.

The fact that in some cases of CML some normal cells persist together with Ph^1-positive cells indicates that cells with the Ph^1 mutation do not always overrun the whole bone-marrow progeny [171]. It would be interesting to determine whether the normal cells in such cases belong to a particular cellular series—for example, the erythroid or the myeloid—or whether the Ph^1-negative cells occur in all three series [129]. It is also possible, however, that in such cases a Ph^1 chromosome having duplicated its deleted long arm in a subclone makes the karyotype appear normal. This dichotomy is more marked in Ph^1-positive cases other than CML, but it should be remembered that even in a "100%" Ph^1-positive marrow, a small number of cells (less than 10%) cannot with certainty be declared Ph^1-positive [168].

Although the Ph^1 chromosome is by far most typical for CML, the fact that it can be found, although not commonly, in conditions other than CML points to important affinities between the myeloproliferative disorders and suggests that the genetic constitution of the patient probably modifies the expression of the disease [28].

Acute Myeloid Leukemia (AML)

A number of the above-mentioned conditions develop into AML [23, 55, 153]. It is, therefore, not surprising that some AML cases may show a Ph^1 chromosome [6, 23, 113, 140, 198] and sometimes even two [115]. About half of the AML cases, however, show apparently normal diploid myeloblasts [12, 114, 162, 172] with a more or less marked aneuploid spread [174]. In the other half of the cases, aneuploidy in mode may be hypodiploidy or hyperdiploidy [11, 12]. Hypodiploidy is more marked in AML than in CML. Compared with acute lymphoblastic leukemia (ALL; see below), the abnormal stemlines in AML have a hypodiploidy involving 25% of the cases, which is not seen in ALL [175]; but in ALL the hyperdiploidy may involve much higher chromosome numbers. Aneuploid modes in AML are associated with a variable degree of nonmodal spread, which here as in other conditions does not involve the chromosomes randomly. Clonal evolution is usually difficult to demonstrate, although it can occur [68]. The aneuploid modes appear heterogeneous, and no consistent changes in chromosomes within the Denver groups seem to have been found [19, 27, 60, 132]. The changes in the karyotypes are similar to those seen in the more advanced clones of CML in blastic crisis. Erythroleukemia [45, 195], both the erythremic myelosis and Di Guglielmo's type [46], do not differ greatly from each other cytogenetically. A number of refractory anemias have been reported with chromosome changes similar to those in AML, but, as mentioned above, these may already have been in transition into acute leukemia [44]. In general, group C trisomy appears to be the most frequent abnormality in AML [101, 183]. A group of hypodiploid cases consists of older males with 45,X modal cells. Even though the original clone may be maintained, there is often progressive increase in aneuploidy during the course of the disease. In contrast to the situation in ALL, good remission in these older studies was not common in such cases, and aneuploidy was not clearly seen to disappear from the bone marrow when a relative clinical remission was achieved, except in a few reported cases where the leukemic cells may have had an origin different from that of the normal diploid cells observed [170]. As previously stated, differential diagnosis of acute leukemias is by no means easy or always possible, and AML must therefore represent a motley group of terminal conditions.

Fanconi's Anemia and Bloom's Syndrome

Both Fanconi's anemia and Bloom's syndrome are characterized by abnormally high frequencies of chromosome breaks and gaps in cultured lymphocytes, to a lesser extent in fibroblasts, and much less frequently in direct marrow preparations [176]. Both syndromes are associated with a high incidence of acute leukemia [74]. Although such chromosome breakability is likely to vary from time to time and is not likely to involve the chromosomes randomly [75], no regularity has been observed so far; the significant feature may be an occasional pancytopenia, which permits mutant clones to be established more easily.

CYTOGENETICS OF LYMPHOPROLIFERATIVE DISEASES

Chronic Lymphatic Leukemia (CLL)

Very little is known about the chromosome constitution of the affected cells in CLL, mainly because of the technical difficulties involved in its study. Few cells are found in division in the patient, and the lymphocytes often fail to grow in culture [124]. Due to the presence of normal cells in the bone marrow, it is difficult to know whether one is observing leukemic cells in division or not. CLL cells show a characteristic, reduced response to PHA [124].

A CLL-specific chromosome abnormality was claimed when the "Christchurch" (Ch^1) chromosome, a Gp− chromosome, was found in a family in which two carriers of Ch^1 had developed CLL [84]. The association between CLL and this segregating chromosome was claimed to be fortuitous, and the same or similar chromosomes segregating in other families was not associated with CLL [1]. Some years later, CLL developed in a third Ch^1-carrier in the original family, and this report claimed that other Ch^1-carrying family members without CLL showed increased aneuploidy [63]. Another report claims to find a reduction in the size of group G chromosome in males following irradiation [80].

CLL patients have a three- to fourfold chance of developing secondary cancers and an eight- to tenfold chance of developing cancer of the skin. Some authors compare CLL with immunodeficiency syndromes, such as Bruton's agammaglobulinemia, the Swiss form of agammaglobulinemia, and the Wiscott-Aldrich syndrome. CLL patients show a decreased incidence of allergic symptoms and reduced IgE levels [141] (patients with lymphomas lose previous symptoms of allergy just prior to the presumed onset of growth of their lymphomas). The findings that allergic symptoms are decreased throughout the lives of patients with CLL might be explained by the presence of some hereditary immunodefect. This is only expressed late in life, as hypogammaglobulinemia develops and an immunologically unreactive type of lymphocyte proliferates [154]. If this is the case, then it would be surprising if CLL would have a single chromosome abnormality as its cause [61].

The group A-sized marker chromosome(s) that apparently occurs as a supernumerary chromosome in about 80% of the Waldenström macroglobulinemias [14, 20, 26, 57] is also recorded in CLL. A number of other structural and numerical abnormalities that have been claimed indicate the heterogeneity of CLL, or a lack of causal relationship between chromosome abnormalities and CLL. A number of normal chromosome constitutions have been recorded [13, 61, 90, 152, 154] but, as pointed out above, the proof that it is the CLL cells that had been analyzed is frequently lacking.

Lymphomas and Paraproteinemias

While the connection between CLL and immunological disorders may be arguable, the association between lymphomas and paraproteinemias is not [13, 38, 50, 100, 107, 198]. The cytogenetic findings here are not very different from those obtained in other solid tumors [65]; the chromosome numbers are in the hyperdiploid range predominantly. In Hodgkin's disease, where the Sternberg-Reed cells are postulated to be of basic importance, bone-marrow aspirations rarely contain these cells, and normal diploid cells and megakaryocytes with normal chromosome morphology are usually found [13]. In lymph nodes, and sometimes in peripheral blood, cells may be found with a hypotetraploid mode [181, 186]. Since the Sternberg-Reed cells are polyploid (judging from their morphology), such hypotetraploid metaphases have been assumed to represent Sternberg-Reed cells, implying that the disease is neoplastic rather than immunological. The majority of cells, however, are diploid and apparently without chromosome abnormalities.

Marker chromosomes found in other lymphomas [184], observed more than once, are a group of D-sized metacentric isochromosomes, of uncertain origin even after fluorescence studies [64, 65], and the group A-sized markers: a long metacentric isochromosome, a submetacentric chromosome, and an acrocentric chromosome [13, 17, 20, 65, 173, 202]. Identifiable group C variation has been reported to involve monosomy 6, and monosomy 9, and trisomy 7 [65]. The group D-sized, unidentifiable isochromosome appears to be associated with monosomy 17 or 18. A short- or long-arm deletion of a No. 17 or No. 18 (the "Melbourne chromosome") has been found in some cases of Hodgkin's disease [179], in follicular lymphomas, and in reticulum cell sarcoma [185]. Chromosome studies of various types of lymphomas also show normal diploid metaphases, presumably representing nonmalignant cells.

The Burkitt lymphoma is usually accompanied by a diploid chromosome constitution [103, 117], and when aneuploidy occurs, no great numerical changes take place. Several marker chromosomes have been reported; for example, a long acrocentric marker replacing chromosome No. 2, a long submetacentric, replacing chromosome No. 1 (the "Madison chromosome")—similar to the markers in malignant lymphoma [134, 135] —and a chromosome No. 9 with a marked secondary constriction [103].

The supernumerary group A-sized marker(s) in Waldenström's macroglobulinemia is present in about 80% of the IgM and in about 30% of the IgA and IgG gammopathies. These markers may be seen in PHA-stimulated lymphocytes and in the bone marrow in a variable percentage of cells (from 5 to 50%) and might therefore also be overlooked. Such markers have not been seen in fibroblasts or in cells from an unaffected monozygotic twin of an affected person. Such large chromosomes (when observed in the complements of a person born with abnormal chromosome constitution) have not been group A chromosome-derived, but have been instead giant-sized, duplicated X chromosomes, associated with Turner's syndrome; in such cases they are tolerated because they are inactivated. However, both gammaglobulin IgM [163] and factor VIII levels in serum [136] are (fortuitously?) proportional to the number of X chromosomes (XX<XXX<XXXX; XY<XXY).

Other abnormalities have been occasionally recorded in gammopathies, the frequency of abnormalities being higher in IgA and IgG gammopathies than in IgM.

Acute Lymphoblastic Leukemia (ALL)

As with mongolism, a high maternal age is associated with ALL, which is the leukemia typical of childhood. The possibility of a prenatal predisposition must therefore be kept in mind [131]. Recently, it has been shown that fetal lymphocytes have ready access to the maternal lymphatic system and may reside there for a considerable length of time after parturition [178]; so, presumably, maternal lymphocytes can also enter the fetus before immunocompetence has been developed. An acquired immunological tolerance to these foreign cells in the fetus might impair the individual's ability to detect abnormal cells in later life.

About half the studied cases of ALL appear to have a normal karyotype [35, 172], and the rest, which are pseudodiploid and aneuploid, include hyperdiploids with chromosome numbers up to 90 [122, 162, 172]. This extreme hyperdiploidy is only found in a minor proportion of cases (6%). Nevertheless, this apparent tolerance to circulating cells with high chromosome numbers may indicate the involvement of cells that themselves represent part of the immunological control mechanism [133]. When ALL is compared with AML, in which 25% of the abnormal modes are hypodiploid and where, megakaryocytes apart, modal numbers rarely reach 50 chromosomes, this becomes evident.

A varied spectrum of chromosome abnormalities is present, and no particular Denver group is exceptionally frequently involved in the aneuploidy [96]. Gains in groups D and E and losses in group G may be somewhat commoner than for other groups. Claims that deletions in group D and E chromosomes are associated with decreased response to therapy [89], and that gains in these groups improved prognosis, require confirmation. There appear to be indications that when aneuploidy is observed in the bone marrow of ALL in full clinical remission, this may be a sign of an early relapse [172]. Past reports suggest that while in ALL in full clinical remission, the aneuploidy often disappears [161, 172], in AML, even in remission, some instability in the marrow remains. The recent use of the more intensive chemotherapy of AML seems to have changed this picture; immunotherapy, in particular, can also cause the aneuploidy in AML to disappear for a time.

CONCLUDING REMARKS

Comments will be restricted to the disorders best studied. The unique position of chronic myeloid leukemia as the only neoplastic disease with a specific chromosome abnormality may be explained by its long induction period compared with those of solid tumors (which usually show a phase of nonproliferation) and by being detectable relatively early. The free exposure of the entire cell surfaces of leukocytes allows only subtle changes to escape detection by the immunological system and the proliferation-controlling factor in the hemopoietic system [144, 146, 180], both of which are highly specific. CML is *"carcinogenesis in slow motion."* The predisposing Ph^1 mutation is apparently not detected, but not until temporary episodes of defective proliferative control occur will the number of Ph^1-positive cells be boosted sufficiently to give the picture of leukemia. Constitutional factors and environmental challenges may modify the clinical events so that these episodes do not always synchronize well with the intrinsic malignant evolution [66].

One particular event, however, is well synchronized with the clinical picture, that is, *the appearance of two Ph^1 chromosomes and the onset of blastic crisis.* As a "dominant mutation," the Ph^1 chromosome is likely to show a dosage effect when duplicated, and the progressive increase in the frequency of the two-Ph^1-chromosome cell proves its selective advantage. As to the origin of the two-Ph^1-chromosome line: if we assume that the Ph^1 mutation is a simple terminal deletion, surviving only if the break occurs in the distal edge of the Giemsa-positive band in the long arm of chromosome No. 22, the heterochromatin in this band may provide a *telomere.* Such a telomere may, however, be prone to repeated loss of its terminal material during successive mitotic cycles. When this material loss leads to terminal fusion of sister chromatids, subsequent nondisjunction may take place. (Such terminal attachments have actually been observed.) From these configurations some chromosomes may presumably recover and exist in the cell as two Ph^1 chromosomes, sometimes of unequal size, and a new and more malignant cell line [49, 67, 164] may emerge. The visible G band could contain a large amount of surplus DNA (moderately repetitious?), so that, even if short sequences of DNA were lost in most mitoses, the chromosome could remain viable for many more divisions than are required to produce leukemia. The same mechanism might, incidentally, also lead to a chromosome $p^*q-:-q$ (where one p and the centromere* is lost)—that is, an almost normal-appearing No. 22 chromosome with a possible value of two Ph^1 chromosomes (Ph^1-negative CML?). The appearance of this type of "telomere" may explain the extraordinary instability of some marker chromosomes in neoplasms (e.g., progressive shortening of the Y chromosome in lymphoma [65]).

What can at the moment be said about the nature of the chromatin lost from the marker chromosomes most frequently observed in the myeloid leukemias? In all instances studied, active genetic material has been lost. The chromatin missing from the Ph^1 chromosome is neither late-labeling nor positively staining with acetic–saline Giemsa (ASG) and quinacrine mustard (QM). It is, therefore, presumably genetically active euchromatin. The 20q–, found mostly in PV and ISA, has also lost chromatin that is presumed to be active; the preserved short arm is the one heavily stained with the ASG technique. Both 17p– and 17q– involve the loss of some presumptively active chromatin. Dq–, if this is 15q–, again involves loss of distal, active chromatin, while the proximal (late-labeling, Giemsa- and fluorescent-positive) part of the arm remains. These are found as partial monosomies in preleukemic and leukemic cells. (In the complements of individuals with congenital abnormalities, these chromosomes are not involved, either as partial monosomies or as partial trisomies. The imbalances compatible with life affect the more heavily staining, QM staining, and late-labeling regions, viability declining in the order of the absolute amount of euchromatin [e.g., 21, 18, and 13] in the extra chromosome. In partial congenital monosomy, the severity of defects and viability is also related to the amount of euchromatin lost.)

In both partial monosomy and partial trisomy in leukemic cells, there is a relative increase in the total amount of inactive chromatin, which might serve as a selection gradient in karyotypic progression. This progressive loss of euchromatin could "simplify" the metabolism of the cells but could also make them less specific and responsive to controlling factors. The loss of euchromatin *and* the loss of the heterochromatin-rich Y chromosome would set the cell back on the selection gradient and would explain the better prognosis when the leukemic clone is 45X, –Y, and Ph^1-positive [123].

It is unlikely that all the genes missing (or present) on a deleted chromosome (such as Ph^1) can be related to neoplastic development. Therefore, individual gene mutations may also provide a pathway towards malignancy [30], and the more genes involved, the longer is the time required. Since the mutations seem to destroy the normal function of these genes, a localized, potential structural instability might often be the consequence. If this results in chromosome breaks, a whole family of aberrations might be generated, characteristic for the particular tumor: chromosome No. 22 loss in meningiomas [137, 203]; gain of group C chromosomes and loss from groups B, D, and G in testicular tumors [139]; and loss of group D chromosomes in gliomas [138] and carcinoma of the cervix [81]. The potential instability is modified by extrinsic factors, so that aneuploidy ("cytological malignancy") may not always appear at the same point in time.

As has been said in the introduction, the present is an inopportune time for evaluating the significance of chromosome abnormalities in neoplasms. The new techniques have not been available long enough to permit accumulation of an adequate amount of significant material for generalization, although the quality of the new information is incomparably better than that of the old. In fact, the banding methods are still being improved, and, therefore, it might be premature in a review of this kind to present "typical" illustrations for leukemia, particularly in the case of paraproteinemias. However, it should not be thought that the banding techniques can solve every problem. The specificity of the patterns is limited. One can expect that identification of missing or extra chromosomes will be possible, but the identification and structural definition of abnormal chromosomes demands minimum-change cells, just as with the old techniques. However exciting the new techniques may be, the importance cannot be overemphasized of concentrating studies on early, untreated material, of making serial studies of the material, and of reporting only on material adequately studied with high-quality preparations and supplemented with full information on the patient's constitutional karyotype.

LITERATURE CITED

1. ABBOTT CR: The Christchurch chromosome. *Lancet* 1:1155–1156, 1966.
2. ADAMS A, FITZGERALD PH, GUNZ FW: A new chromosome abnormality in chronic granulocytic leukaemia. *Brit Med J* 2:1474–1476, 1961.
3. ALTER A, DOBKIN G, POURFAR M, ROSNER F, LEE SL: Genes on the Mongol chromosome. *Lancet* 1:506, 1963.
4. ATKIN NB, TAYLOR MC: A case of chronic ML with 45 chromosome cell line in blood. *Cytogenetics* 1:97–103, 1962.
5. ATKINS L, GOULIAN M: Multiple clones with increase in numbers of chromosomes in G-group in a case of myelomonocytic leukaemia. *Cytogenetics* 4:321–328, 1965.
6. BAGUENA-CANDELA R, FORTEZA BOVER G: Estudio citogenetico de una leucosis aguda del tipo mioloblastico-promielocitico con cromosoma filadelfia (Ph^1). *Med Espan* 298:1–8, 1964.
7. BAIKIE AG: Chromosomes and leukaemia. *Acta Haematol* 36:157–173, 1966.

8. BAIKIE AG: Cytogenetic studies in leukaemia. *In* Proceedings of the Eighth Congress of the European Society of Haematologists 108, 1962.
9. BAIKIE AG, COURT BROWN WM, BUCKTON KE, HARNDEN DG, JACOBS PA, TOUGH IM: A possible specific chromosome abnormality in human chronic myeloid leukaemia. *Nature* 188:1165–1166, 1960.
10. BAIKIE AG, GARSON OM, SPIERS ASD, FERGUSON J: Cytogenetic studies in familial leukaemia. *Aust Ann Med* 18:7–8, 1969.
11. BAIKIE AG, COURT BROWN MW, JACOBS P, MILNE JS: Chromosome studies in human leukaemia. *Lancet* 2:425–426, 1960.
12. BAIKIE AG, JACOBS PA, McBRIDE JA, TOUGH IM: Cytogenetic studies in acute leukaemia. *Brit Med J* I:1564–1567, 1961.
13. BAKER MC, ATKIN NB: Chromosomes in short term tissue culture of lymphoid tissue from five patients with reticulosis. *Brit Med J* I:770–771, 1965.
14. BANERJEE AR: Chromosome abnormalities in Waldenström's macroglobulinaemia. *Hum Chromosome Newsl* 14:2–3, 1964.
15. BANERJEE KB, HANSJOERG S, HOLLAND JF: Comparative studies on localised leucocyte mobility in patients with CML. *Cancer* 29:437–441, 1972.
16. RAUKE J, NEUBAUER A, SCHOFFLING K: Uber das Auftreten einer Stammlinie mit 55 Chromosomen und Diplio-Philadelphia-Chromosom vor der terminalen Blastenkrise einer chronischen myeloischen Leukämie. *Deut Med Wochenschr* 92:301–305, 1967.
17. BAUKE J, SCHÖFFLING K: Polyploidy in human malignancy. Hypopentaploid chromosome pattern in malignant reticulosis with secondary sideroachrestic anaemia. *Cancer* 22:686–694, 1968.
18. BAUKE J: Chronic myeloid leukaemia: chromosome studies in a patient and his non-leukaemic identical twin. *Cancer* 24:643–648, 1969.
19. BECAK W, BECAK ML, SARAIVA LG: Chromosoma acrocentrico gigante et un caso di sindrome de Di Guglielmo. *Sangre* 12:65–70, 1967.
20. BENIRSCHKE K, BROWNHILL L, EBAUGH FG: Chromosome abnormality in Waldenström's macroglobulinaemia. *Lancet* 1:594–595, 1962.
21. BERGER R: Chromosomes et leucémies humaines, la notion d'evolution clonale. *Ann Genet* 8:70–82, 1965.
22. BLATTNER RJ: Chromosomes in chronic myeloid leukaemia and in acute leukaemia associated with mongolism. Comments on current literature. *J Pediatr* 59:145–146, 1961.
23. BLATTNER RJ: The Philadelphia chromosome. *Lancet* 1:433–434, 1961.
24. BLOOM GE, GERALD PS, DIAMOND LK: CML with Ph[1] chromosome in childhood leukaemia. *Pediatrics* 38:295–299, 1966.
25. BORGERS WH, WALD N, KIM J: Non-specificity of chromosome abnormalities in human leukaemias. *Clin Res* 10:211–213, 1962.
26. BOTTURA C, FERRARI I, VEIGA AA: Chromosome abnormalities in Waldenström's macroglobulinaemia. *Lancet* 1:1170, 1961.
27. BOUSSER J, TANZER J: Syndrome de Klinefelter et leucémie aiguë, à propos d'un cas. *Nouv Rev Fr Hematol* 3:194–197, 1963.

28. Bowen P, Lee CSN: Ph^1 chromosome in a diagnosis of chronic myeloid leukemia; a case with features simulating myelofibrosis. *Bull Johns Hopkins Hosp* 113:1–12, 1963.
29. Boyd E, Pinkerton PH, Hutchison HE: Chromosomes in aleukaemic leukaemia. *Lancet* 2:444–445, 1965.
30. Burch PRJ: A biological principle and its converse: some implications for carcinogenesis. *Nature* 195:241–245, 1962.
31. Carbone PP, Tjio JH, Whang J, Block JB, Kremer WB, Frei E III: The effect of treatment on patients with chronic myeloid leukaemia (hematological and cytogenetic studies). *Ann Int Med* 59:622–628, 1963.
32. Castoldi G, Yam LT, Mitus WJ, Crosby WH: Chromosomal studies in erythroleukaemia and chronic erythremic myelosis. *Blood* 31:202–215, 1968.
33. Castro-Sierra E, Gorman LZ, Merker H, Obrecht P, Wolf V: Clinical and cytogenetic findings in the terminal phase of chronic myeloid leukaemia. *Humangenetik* 4:62–73, 1967.
34. Clein GP, Flemans RJ: Involvement of the erythroid series in blastic crisis of CML. Further evidence of the presence of the Ph^1 chromosome in erythroblasts. *Brit J Haematol* 12:754–758, 1966.
35. Conen PE: Chromosome studies in leukemia. *Can Med Assoc J* 96:1599–1605, 1967.
36. Court Brown WM, Tough IM: Cytogenetic studies in CML. *In* Advances in Cancer Research (Haddow, Weinhouse, eds), vol 7, New York, Academic Press, 351–381, 1963.
37. Crawfurd M, Pegrum G: Chronic myeloid leukaemia. *Lancet* 1:1044, 1964.
38. Das KC, Aikat BK: Chromosome abnormalities in multiple myeloma. *Blood* 30:738–747, 1967.
39. Grouchy J de, Nava C de, Bilski-Pasquier G: Analyse chromosomique d'une evolution clonale dans une leucémie myéloïde. *Nouv Rev Fr Hematol* 5:565–590, 1965.
40. Grouchy J de, Nava C de, Bilski-Pasqiuer G, Bousser J: Ph^1 and loss of a G chromosome after prolonged evolution in a male. *Ann Genet* 9:73–77, 1966.
41. Grouchy J de, Nava C de, Cantu JM, Bilski-Pasquier G, Bousser J: Models for clonal evolution: a study of chronic myelogenous leukemia. *Am J Hum Genet* 18:485–503, 1966.
42. Grouchy J de, Nava C de, Feingold J, Bilski-Pasquier G, Bousser J: Onze observations d'un modèle précis d'evolution caryotypique au cours de la leucémie myéloïde chronique. *Eur J Cancer* 4:481–492, 1968.
43. Grouchy J de, Nava C de, Zittoun R, Bousser J: Analyses chromosomiques dans l'anémie sideroblastique idiopathique acquise. *Nouv Rev Fr Hematol* 6:367–388, 1966.
44. DeMayo AP, Kiossoglou KA, Erlandson ME, Notterman RF, German J: A marrow chromosome abnormality preceding clinical leukemia in Down's syndrome. *Blood* 29:233–241, 1967.
45. Diekmann L, Rickers H, Pfeiffer RA, Baumer A: Morphologische, immunozytologische und zytogenetische Untersuchungen bei einem Kind mit Erytroleukämie. *Blut* 18:321–333, 1969.

46. GRADO F DI, MENDES FT, SCHROEDER F: Ring-chromosome in a case of DiGuglielmo's syndrome. *Lancet* 2:1243–1244, 1964.
47. DOUGAN L, ONESTI P, WOODLIFF HJ: Cytogenetic studies in chronic myeloid leukaemia. *Aust Ann Med* 16:52–61, 1967.
48. DOUGAN L, WOODLIFF HJ: Presence of 2 Ph^1 chromosomes in cells from a patient with CML. *Nature* 205:405–406, 1965.
49. DUVALL CP, CARBONE PP, BELL WR, WHANG J, TJIO JH, PERRY S: Chronic myeloid leukaemia with 2 Ph^1 chromosomes and prominent lymphadenopathy. *Blood* 29:652–665, 1967.
50. ELVES MW, ISRAELS MCG: Chromosomes and serum proteins. A linked abnormality. *Brit Med J* II:1024–1026, 1963.
51. ELVES MW, ISRAELS MCG: Cytogenetic studies in unusual forms of chronic myeloid leukaemia. *Acta Haematol* 38:129–141, 1967.
52. ENGEL E, MCGEE BJ, HARTMANN RC, ENGEL DE MONTMOLLIN M: Two leukaemic peripheral blood stemlines during acute transformation of CML in a D/D-translocation carrier. *Cytogenetics* 4:157–170, 1965.
53. ENGEL E, MCGEE BJ: Double Ph^1 chromosomes in leukaemia. *Lancet* 2:337, 1966.
54. ERKMAN B, CROOKSTON JH, CONEN PE: Double Ph^1 chromosomes in chronic granulocytic leukemia. *Cancer* 20:1963–1975, 1967.
55. ERKMAN B, HAZLETT B, CROOKSTON JH, CONEN PE: Hypodiploid chromosome pattern in acute leukemia following polycythaemia vera. *Cancer* 20:1318–1325, 1967.
56. EZDINI EZ, SOKAL JE, CROSSWHITE LH, SANDBERG AA: Philadelphia chromosome positive and negative CML. *Ann Int Med* 72:175–182, 1970.
57. FERGUSON J, MACKAY IR: Macroglobulinaemia with chromosome abnormalities. *Aust Ann Med* 12:197–201, 1963.
58. FIALKOW PJ, GARTLER SM, YOSHIDA A: Clonal origin of CML in man. *Proc Nat Acad Sci (US)* 58:1468–1471, 1967.
59. FIALKOW PJ, KLEIN G, GARTLER SM, CLIFFORD P: Clonal origin for individual Burkitt tumours. *Lancet* 1:384–386, 1970.
60. FITZGERALD PH, ADAMS A, GUNZ FW: Chromosome studies in adult acute leukaemia. *J Nat Cancer Inst* 32:395–417, 1964.
61. FITZGERALD PH, ADAMS A: Chromosome studies in CCL and lymphosarcomas. *J Nat Cancer Inst* 34:827–939, 1965.
62. FITZGERALD PH: A complex pattern of chromosome abnormalities in the acute phase of CML. *J Med Genet* 3:258–264, 1966.
63. FITZGERALD PH, HOMER JW: Third case of CCL in a carrier of the inherited Ch^1 chromosome. *Brit Med J* 3:752, 1969.
64. FLEISCHMANN T, HAKANSSON CH, LEVAN A: Fluorescent marker chromosome in malignant lymphomas. *Hereditas* 69:311–314, 1971.
65. FLEISCHMANN T, HAKANSSON CH, LEVAN A, MOLLER T: Multiple chromosome aberrations in a tumor (malign. lymphoma). *Hereditas* 70:243–258, 1972.
66. FLEISCHMANN EV, VOLKOVA MA: Cytogenetic studies of cell dynamics in CML. *Probl Gematol* 15:33–36, 1970.

67. FLEISCHMANN EV, VOLKOVA MA, DUBROVSHAYA VS, MARGULIS MI, KRAVCHENKO GP: Concerning sarcomatosis of the lymphnodes in chronic granulocytic leukaemia. *Probl Gematol* 16:30–36, 1971.
68. FORD CE, CLARKE CM: Cytogenetic evidence of clonal proliferation in primary reticular neoplasms. *Can Cancer Res Conf Proc* 5:129–146, 1963.
69. FORRESTER RH, LOURO JM: Ph[1] chromosome abnormality in agnogenic myeloid metaplasia. *Ann Int Med* 64:622–627, 1966.
70. FREIREICH EJ, WHANG J, TJIO JH, LEVIN RH, BRITTIN GM, FREI E III: Refractory anaemia, granulocytic hyperplasia of bone marrow, and a missing chromosome in marrow cells. A new syndrome? *Clin Res* 12:284–289, 1964.
71. FREY I, SIEBNER H: Osteomyelofibrose mit Philadelphia-(Ph[1])-Chromosom. *Med Welt* 42:2275–2279, 1968.
72. FRICK PG: Primary thrombocythaemia, clinical, haematological and chromosomal studies. *Helv Med Acta* 35:20–29, 1969.
73. GAVOSTO F, PILERI A, PEGORARO L: X-rays and Ph[1] chromosome. *Lancet* 1:1336–1337, 1965.
74. GERMAN J: Bloom's syndrome. I. Genetical and clinical observations in the first 27 patients. *Am J Hum Genet* 21:196–227, 1969.
75. GMYREK D, WITKOWSKI R, SYLIM-RAPOPORT I, JACOBASCH G: Chromosomenaberrationen und Stoffwechselstörungen der Blutzellen bei Fanconi-Anämie vor und nach Ubergang in Leukose am Beispiel einer Patientin. *Deut Med Wochenschr* 92:1701–1707, 1967.
76. GOH KO, SWISHER SN: Specificity of Ph[1] chromosome: cytogenetic studies on cases of CML and myeloid metaplasia. *Ann Int Med* 61:609–624, 1964.
77. GOH KO, SWISHER SN, TROUP SB: Submetacentric chromosome in chronic myeloid leukaemia. *Arch Int Med* 114:439–443, 1964.
78. GOH KO, SWISHER SN: Identical twins and chronic myelocytic leukaemia. *Arch Int Med* 115:475–478, 1965.
79. GOH KO: Total-body irradiation and human chromosomes: cytogenetic study of the peripheral blood and bone marrow leucocytes seven years after total-body irradiation. *Radiat Res* 35:155–160, 1968.
80. GOH KO, RIDGE O: Smaller G(Gp−) and t(Gp−; Dp+) chromosomes. *Am J Dis Child* 115:732–738, 1968.
81. GRANBERG I: Chromosomes in preinvasive, microinvasive and invasive cervical carcinoma. *Hereditas* 68:165–218, 1971.
82. GROPP A, FLATZ G: Chromosome breakage and blastic transformation of lymphocytes in ataxia telangiectasia. *Humangenetik* 5:77–79, 1967.
83. GROPP A, FISCHER R, NIEDERHALT G, KLEASE MP, HENSEN S: Philadelphiachromosom und alkalische Leukocyten-phosphatase bei chronische Myelose. *Klin Wochenschr* 46:177–187, 1968.
84. GUNZ FW, FITZGERALD PH, ADAMS A: An abnormal chromosome in chronic lymphatic leukaemia. *Brit Med J* II:1097, 1962.
85. HAMMOUDA F, QUAGLINO D, HAYHOE FGJ: Blastic crisis and CML. Cytochemical, cytogenetic and autoradiographic studies of 4 cases. *Brit Med J* I:1275–1281, 1964.

86. HAMPEL KE: Diplo-Ph¹-Chromosom bei der myeloischen Leukämie. *Klin Wochenschr* 42:522, 1964.
87. HAMPEL KE, PALME G: Chromosome studies in acute and chronic myeloid leukaemia. *Neoplasma* 11:113–121, 1964.
88. HARDISTY RM, SPEED DE, TILL M: Granulocytic leukaemia in childhood. *Brit J Haematol* 10:551–566, 1964.
89. HART JS, TRUJILLO JM, FREIREICH EJ, GEORGE SL, FREI E III: Cytogenetic studies and their clinical correlates in adults with leukaemia. *Ann Int Med* 75:353–360, 1971.
90. HAYHOE FGJ, HAMMOUDA F: Blood disorders (general review). *In* Current Research in Leukaemia, Cambridge, Cambridge University Press, 55–76, 1965.
91. HEATH CW, BENNETT JM, WHANG-PENG J, BERRY EW, WIERNICK PH: Cytogenetic findings in erythroleukaemia. *Blood* 33:453–467, 1969.
92. HEATH CW, MOLONEY WC: The Philadelphia chromosome in an unusual case of myeloproliferative disease. *Blood* 26:471–478, 1965.
93. HECHT F, KOLER RD, RIGAS DA, DAHNKE GS, CASE MP, TISDALE V, MILLER RW: Leukaemia and lymphocytes in ataxia telangiectasia. *Lancet* 2:1193, 1966.
94. HEMEL J, HIRSCHHORN K: The origin of some narrow fibroblasts. *Blood* 38: 26–29, 1971.
95. HEYSELL R, BRILL AM, WOODBURY LA, NISHIMURA ET, GHOSE J, HOSHINO T, YAMASAKI M: Leukaemia in Hiroshima atomic bomb survivors. *Blood* 15:313–331, 1960.
96. HILTON HB, LEWIS IC, TROWELL HR: C-group trisomy in identical twins with acute leukaemia. *Blood* 35:222–226, 1970.
97. HOLLAND SN, DOLL R, CARTER CO: The mortality from leukaemia and other causes among patients with Down's syndrome and among their parents. *Brit J Cancer* 16:177–186, 1962.
98. HONDA F, PUNNETT HH, CHARNEY E, MILLER G, THIEDE HA: Serial cytogenetic and haematologic studies on a mongol with trisomy-21 and acute congenital leukemia. *J Pediat* 65:880–887, 1964.
99. HOSSFELD DK, HAN T, HOLDSWORTH RN, SANDBERG AA: Chromosomes and causation of human cancer and leukaemia. VII. The significance of the Ph¹ chromosome in conditions other than CML. *Cancer* 27:186–192, 1971.
100. HOUSTON EW, TITZMANN SE, LEVIN WC: Chromosome aberrations common to three types of monoclonal gammopathies. *Blood* 29:214–230, 1967.
101. HUNGERFORD DA, NOWELL PC: Chromosome studies in human leukaemia. III. Acute myeloid leukaemia. *J Nat Cancer Inst* 29:545–553, 1962.
102. HURDLE AD, GARSON OM, BUIST DGP: Clinical and cytogenetic studies in chronic myelomonocytic leukaemia. *Brit J Haematol* 22:773–782, 1972.
103. JACOBS PA, TOUGH IM, WRIGHT DA: Cytogenetic studies in Burkitt's lymphoma. *Lancet* 2:1144–1146, 1963.
104. JACKSON JF, HIGGINS LC JR: Group C monosomy in myelofibrosis with myeloid metaplasia. *Arch Int Med* 119:403–406, 1967.
105. JOHNSTON AW: The chromosomes in a child with leukaemia. *New Engl J Med* 264:391, 1961.

106. JENSEN MK: Chromosome studies in potential leukaemic myeloid disorders. *Acta Med Scand* 183:535–542, 1968.
107. KANSOW U, LANGE B, NIEDERHALT G, GROPP A: Chromosomenuntersuchungen bei Paraproteinämien. *Klin Wochenschr* 45:1076–1084, 1967.
108. KAUER GL JR, ENGLE RL JR: Eosinophilic leukaemia (CML with Ph[1] chromosome). *Lancet* 2:1340, 1964.
109. KAY HEM, LAWLOR SD, MILLARD RE: The chromosomes in polycythaemia vera. *Brit J Haematol* 12:507–527, 1966.
110. KEMP NH, STAFFORD JL, TANNER R: Chromosome studies in early and terminal chronic myeloid leukaemia. *Brit Med J* I:1010–1014, 1964.
111. KEMP NH, STAFFORD JL, TANNER RK: Acute leukaemia and Klinefelter syndrome. *Lancet* 2:434–435, 1961.
112. KENIS Y, KOULISCHER L: Étude clinique et cytogénétique de 21 patients atteints de leucémie myéloïde chronique. *Eur J Cancer* 3:83–93, 1967.
113. KHAN MH, MARTIN H: Myeloblastenleukämie mit Philadelphia chromosom. *Klin Wochenschr* 45:821–824, 1967.
114. KIOSSOGLOU KA, MITUS WJ, DAMESHEK W: Chromosome abnormalities in acute leukaemia. *Blood* 26:610–641, 1965.
115. KIOSSOGLOU KA, MITUS WJ, DAMESHEK W: Two Ph[1] chromosomes in AML. A study of 2 cases. *Lancet* 2:665–668, 1965.
116. KIOSSOGLOU KA, MITUS WJ, DAMESHEK W: Cytogenetic studies in chronic myeloproliferative syndromes. *Blood* 28:241–252, 1966.
117. KOHN G, MELLMANN WJ, MOORHEAD PS, LOFTUS J, HENLE G: Involvement of C group chromosomes in five Burkitt lymphoma cell lines. *J Nat Cancer Inst* 38:209–222, 1967.
118. KOSENOW W, PFEIFFER RA: Chronisch-myeloische Leukämie bei eineiigen Zwillingen. *Deut Med Wochenschr* 22:1170–1176, 1969.
119. KOULISCHER L, FRÜHLING J, HENRY J: Observations cytogénétique dans la maladie de Vaquez. *Eur J Cancer* 3:193–201, 1967.
120. KRAUSS S, SOKAL JE, SANDBERG AA: Comparison of Ph[1] positive and negative chronic myeloid leukaemia. *Ann Int Med* 61:625–635, 1964.
121. KRAUS S: CML with features simulating myelofibrosis with myeloid metaplasia. *Cancer* 19:1321–1332, 1966.
122. LAMPERT F: Cellurärer DNS-Gehalt und Chromosomenzahl bei der akuten Leukämie im Kindesalter und ihre Bedeutung für Chemotherapie und Prognose. *Klin Wochenschr* 45:763–768, 1967.
123. LAWLER SD, GALTON DAG: Chromosome changes in terminal stages of chronic myeloid leukaemia. *Acta Med Scand Suppl* 445:312–318, 1966.
124. LAWLER SD, PENTYCROSS CR, REEVES BR: Chromosomes and transformation of lymphocytes in lymphoproliferative disorders. *Brit Med J* 4:213–219, 1968.
125. LAWLER SD, KAY HEM, BIRBECK MSC: Marrow dysplasia with C-trisomy and anomalies of the granulocytic nuclei. *J Clin Pathol* 19:214–219, 1966.
126. LAWLER SD, MILLARD RE, KAY HEM: Further cytogenetic investigations in polycythaemia vera. *Eur J Cancer* 6:223–233, 1970.
127. LEJEUNE J, CAILLE B, TURPIN R: Chromosomal evolution of a chronic myeloid leukaemia (in French). *Ann Genet* 8:44–49, 1965.

128. Levan A, Nichols WW, Norden A: A case of CML with two leukaemic stemlines in the blood. *Hereditas* 49:434–441, 1963.
129. Levan A, Nichols WW, Hall B, Löw B, Nilsson SB, Norden A: Mixture of Rb positive and negative erythrocytes and chromosome abnormalities in a case of polycythaemia. *Hereditas* 52:89–105, 1964.
130. Lobb DS, Reeves BR, Lawler SD: Identification of isochromosome 17 in myeloid leukemia (Giemsa). *Lancet* 1:849–850, 1972.
131. MacMahon B, Levy M: Prenatal origin of childhood leukaemia. Evidence from twins. *New Engl J Med* 270:1082–1088, 1964.
132. Namunes P, Lapidus PH, Abbot JA, Roath S: Acute leukaemia and Klinefelter syndrome. *Lancet* 2:26, 1961.
133. Mancinelli S, Durant JR, Hammack WJ: Cytogenetic abnormalities in a plasmacytoma. *Blood* 33:225–233, 1969.
134. Manolov G, Manolova Y, Levan A, Klein G: Experiments with fluorescent chromosome staining in Burkitt tumours. *Hereditas* 68:235–244, 1971.
135. Manolov G, Manolova Y: A marker band in one chromosome 14 in Burkitt lymphomas. *Hereditas* 69:300, 1971.
136. Mantle DJ, Pye CP, Hardisty RM, Vessey MP: Plasma factor VIII concentrations in XXX women. *Lancet* 4:58–59, 1971.
137. Mark J: Chromosomal patterns in human meningiomas. *Eur J Cancer* 6:489–498, 1970.
138. Mark J: Chromosomal characteristics of neurogenic tumours in adults. *Hereditas* 68:61–100, 1971.
139. Martineau M: Chromosome changes in human testicular tumours. *J Pathol* 99:271–282, 1969.
140. Mastroangelo R, Zuelzer WW, Thompson RI: The significance of the Ph[1] chromosome in acute myeloblastic leukaemia. Serial studies in a critical case. *Pediatrics* 40:834–841, 1967.
141. McCormick DP, Amman AJ, Kimishige I, Miller DG, Hong R: A study of allergy in patients with malign lymphoma and CLL. *Cancer* 27:93–99, 1971.
142. McGill M, Brinkley BR: Mitosis in human leukaemia lymphocytes during Colcemid inhibition and recovery. *Cancer Res* 32:746–755, 1972.
143. Merker H, Schneider G, Burmeister P, Wolf U: Chromosomentranslokation bei chronisch-myeloproliferativem Syndrom. *Klin Wochenschr* 46:593–600, 1968.
144. Metcalf D: The nature of leukaemia. Neoplasm or disorder of haemopoietic regulation? *Med J Aust* 2:739–746, 1971.
145. Millard RE, Lawler SD, Kay HEM, Cameron CB: Further observations on patients with a chromosome abnormality associated with polycythaemia vera. *Brit J Haematol* 14:363–374, 1968.
146. Morley AA, Baikie AG, Galton DAG: Cyclic leucocytosis as evidence for retention of normal homeostatic control in CML. *Lancet* 2:1320–1323, 1967.
147. Muldal S, Taylor JJ, Asquith P: Non-random karyotype progression in chronic myeloid leukaemia. *Int J Radiat Biol* 12:219–226, 1967.
148. Nowell PC, Hungerford DA: Chromosome studies in normal and leukaemic human leucocytes. *J Nat Cancer Inst* 25:85–109, 1960.

149. NOWELL PC, HUNGERFORD DA: A minute chromosome in human CML. *Science* 132:1497, 1960.
150. NOWELL PC, HUNGERFORD DA: Chromosome studies in human leukaemia. II. Chronic granulocytic leukaemia. *J Nat Cancer Inst* 27:1013–1035, 1961.
151. NOWELL PC, HUNGERFORD DA: Chromosome studies in human leukaemias. IV. Myeloproliferative syndrome and other atypical myeloid disorders. *J Nat Cancer Inst* 29:911–931, 1962.
152. NOWELL PC, HUNGERFORD DA: Chromosome changes in human leukaemia and a tentative assessment of their significance. *Ann NY Acad Sci* 113:654–662, 1964.
153. NOWELL PC: Marrow chromosome studies in "preleukaemia." *Cancer* 28:513–518, 1971.
154. OPPENHEIM JJ, WHANG J, FREI E III: Immunologic and cytogenetic studies of chronic lymphocytic leukaemia cells. *Blood* 26:121–132, 1965.
155. PEDERSEN B: Cytogenetic evolution in CML. Doctoral thesis, University of Köbenhavn, 1–131, 1969.
156. PEGORARO L, PILERI A, ROVERA G, GAVOSTO F: Trisomia del chromosoma Filadelfia nella crisi blastica di un caso d'leucemia mioloidé cronica. *Tumori* 53:315–321, 1967.
157. PRIETO F, EGOZCUE J, FORTEZA G, MARCO F: Identification of the Philadelphia (Ph[1]) chromosome. *Blood* 35:23–26, 1970.
158. RAPPAPORT H: Tumours of the haemopoietic system. Washington, D.C., Armed Forces Institute of Pathology, 1966.
159. RASTRICK JM, FITZGERALD PH, GUNZ FW: Direct evidence for presence of Ph[1] chromosome in erythroid cells. *Brit Med J* 1:96–98, 1968.
160. REEVES BR, LOBB DS, LAWLER SD: Identity of the abnormal F-group chromosome associated with polycythaemia vera. *Humangenetik* 14:159–161, 1972.
161. REISMAN LE, ZUELZER WW: Chromosome studies in leukaemia: recurrence of similar aneuploid stem lines during successive relapses (Society for Pediatric Research, Thirty-fourth Annual Meeting, Washington, DC). *Abstr J Pediatr* 65:117, 1964.
162. REISMAN LE, MITANI M, ZUELZER WW: Chromosome studies in leukaemia. Evidence for origin of leukaemic stemlines from aneuploid mutants. *New Engl J Med* 270:591–597, 1964.
163. RHODES K, MARKHAM RL, MAXWELL PM, MONK-JONES ME: Immunoglobins and the X-chromosome. *Brit Med J* 1:439–441, 1969.
164. ROLEVIC Z, MARKOVIC V, JANKIC M, KALICANIN P, BOSKOVIC D, RUDIVIC R: Double Ph[1] chromosome in CML. *Lijec vjesn* 92:1415–1423, 1970.
165. ROSS JD, MOLONEY WC, DESFORGES JF: Ineffective regulation of granulopoiesis masquerading as congenital leukaemia in mongol child. *J Pediat* 63:1–10, 1963.
166. ROWLEY JD: Loss of Y chromosome in myelodysplasia. A report of 3 cases studied with quinacrine mustard. *Brit J Haematol* 21:717–728, 1971.
167. ROWLEY JD: A new consistent chromosomal abnormality in chronic myelogenous leukaemia identified by quinacrine fluorescence and Giemsa staining. *Nature* 243:290–293, 1973.

168. Rowley JD, Blaisdell RK, Jacobson LO: Chromosome studies in preleukaemic aneuploidy in C-chromosomes in 3 patients. *Blood* 27:782–799, 1966.
169. Ryabov SI, Katseoman AE: A change in the number of "drumsticks" in mature neutrophils in women with chronic myelosis. *Probl Gematol* 9:15–18, 1964.
170. Sandberg AA, Hossfeld DK: Chromosome abnormalities in human neoplasma. *Ann Rev Med* 379–400, 1970.
171. Sandberg AA, Hossfeld DK, Ezdinli EZ, Crosswhite LH: Chromosomes and causation of human cancer and leukaemia. VI. Blastic phase, cellular origin and Ph^1 chromosome in CML. *Cancer* 27:176–185, 1971.
172. Sandberg AA, Ishihara T, Kikuchi Y, Crosswhite LH: Chromosomal differences among acute leukaemias. *Ann NY Acad Sci* 113:663–676, 1964.
173. Sandberg AA, Ishihara T, Kikuchi Y, Crosswhite LH: Chromosomes of lymphosarcomas and cancer cells in bone marrow. *Cancer* 17:738–746, 1964.
174. Sandberg AA, Ishihara T, Miwa T, Hauschka TS: The in vivo chromosome constitution of marrow from 34 human leukaemias and 60 non-leukaemic controls. *Cancer Res* 21:678–689, 1961.
175. Sandberg AA, Takagi N, Sofuni T, Crosswhite LH: Chromosomes and causation of human cancer and leukaemia. V. Karyotypic aspects of acute leukaemia. *Cancer* 22:1268–1282, 1968.
176. Schmid W: Chromosome breakage in familial panmyelopathy (typus Fanconi). *Schweiz Med Wochenschr* 97:1057–1059, 1967.
177. Schroeder TM, Kurth R: Spontaneous chromosome breakage and high incidence of leukaemia in inherited disease. *Blood* 37:96–109, 1971.
178. Schroeder J, de la Chapelle A: Fetal lymphocytes in maternal blood. *Blood* 39:153–162, 1972.
179. Seif GSF, Spriggs AI: Chromosome changes in Hodgkin's disease. *J Nat Cancer Inst* 39:557–570, 1967.
180. Shadduck RK, Winkelstein A, Nagabhushanam GN: Cyclic leukaemic cell production in CML. *Cancer* 29:399–401, 1972.
181. Sinks LF, Clein GP: The cytogenetics and cell metabolism of circulating Reed-Sternberg cells. *Brit J Haematol* 12:447–453, 1966.
182. Smalley RV: Double Ph^1 chromosome in leukaemia. *Lancet* 2:591, 1966.
183. Smalley RV, Bouroncle BS: Hyperploidy in a patient with Di Guglielmo's syndrome. *Arch Int Med* 120:599–601, 1967.
184. Spiers ASD, Baikie AG: Cytogenetic studies in malignant lymphomas and related neoplasms. *Cancer* 22:193–217, 1968.
185. Spiers ASD, Baikie AG: A specific role of group 17-18 chromosomes in reticulo-endothelial neoplasias. *Brit J Cancer* 24:77–91, 1970.
186. Spriggs AI, Boddington MM: Chromosomes of Sternberg-Reed cells (Hodgkin's). *Lancet* 2:153, 1962.
187. Stich W, Back F, Dörmer P, Tsirimbas A: Double Ph^1 and isochromosome 17 in terminal phase of CML. *Klin Wochenschr* 44:334–337, 1966.
188. Tassoni EM, Durant JR, Becker S, Kravitz B: Cytogenetic studies in multiple myeloma. 14 cases. *Cancer Res* 27:806–810, 1967.

189. TJIO JH, CARBONE PP, WHANG J, FREI E III: The Ph^1 chromosome and chronic myeloid leukaemia. *J Nat Cancer Inst* 36:567–584, 1966.
190. TOUGH IM, COURT BROWN WM, BAIKIE AC, BUCKTON KE, HARNDEN DG, JACOBS PA, KING MJ, MCBRIDE JA: Cytogenetic studies in chronic myeloid leukaemia and acute leukaemia associated with mongolism. *Lancet* 1:411–417, 1961.
191. TOUGH IM, COURT BROWN WM, BAIKIE AG, BUCKTON KE, HARNDEN DG, JACOBS PA, WILLIAMS JA: Chronic myeloid leukaemia: cytogenetic studies before and after splenic irradiation. *Lancet* 2:115–120, 1962.
192. TOUGH IM, JACOBS PA, COURT BROWN WM, BAIKIE AG, WILLIAMSON ERD: Cytogenetic studies of bone marrow in chronic myeloid leukaemia. *Lancet* 1:844–846, 1963.
193. TRUJILLO JM, OHNO S: Chromosomal alteration of erythropoietic cells in chronic myeloid leukaemia. *Acta Haematol* 29:311–316, 1963.
194. WARKANEY J, SCHUBERT WK, THOMPSON JN: Chromosomal anomalies in mongolism associated with leukaemia. *New Engl J Med* 268:1–4, 1963.
195. WEATHERALL DJ, WALKER S: Changes in chromosomal and haematological patterns in a patient with erythroleukaemia. *J Med Genet* 2:212–219, 1965.
196. WHANG J, FREI E III, TJIO JH, CARBONE PP, BRECHNER G: The distribution of Ph^1 chromosome in patients with chronic myeloid leukaemia. *Blood* 22:664–673, 1963.
197. WHANG-PENG J, CANELLOS GP, CARBONE PP, TJIO JH: Clinical implications of cytogenetic variants in chronic myelocytic leukaemia. *Blood* 32:755–756, 1968.
198. WHANG-PENG J, HENDERSON ES, KNUTSEN T, FREIREICH E, GART JJ: Cytogenetic studies in AML with specific emphasis on the occurrence of the Ph^1 chromosome. *Blood* 36:448–456, 1970.
199. WOLF U, MERKER H, BOCKELMANN W: Chromosomenuntersuchungen bei chronischen myeloischer Leukämie. *Klin Wochenschr* 44:12–19, 1966.
200. WOODLIFF HJ, DOUGAN L, ONESTI P: Cytogenetic studies in twins, one with CML. *Nature* 211:533, 1966.
201. WOODLIFF HJ, ONESTI P, DOUGAN L: Karyotypes in thrombocythaemia. *Lancet* 1:114–115, 1967.
202. YAM LT, CASTOLDI GL, GARVEY MG, MITUS WJ: Functional, cytogenetic, and cytochemical study of the leukaemic reticulum cell. *Blood* 32:90–101, 1968.
203. ZANG KD, SINGER H: Chromosomal constitution of meningiomas. *Nature* 216:84–85, 1967.

CHROMOSOME PATTERNS IN BENIGN AND MALIGNANT TUMORS IN THE HUMAN NERVOUS SYSTEM

JOACHIM MARK

Department of Pathology, Central Hospital, Skövde, Sweden, and
Department of Neurosurgery, Lund's Hospital, Lund, Sweden

Introduction 481

Nomenclature 482

Tumors Arising in the Nervous System 482
    Astrocytic Malignant Gliomas 484
    Oligodendrogliomas 488
    Ependymomas 489
    Medulloblastomas and Neuroblastomas 490
    Retinoblastomas and Optical Gliomas 491
    Neurinomas 491
    Pituitary Adenomas 492

Tumors Arising Outside the Nervous System 492

Concluding Remarks 494

Acknowledgements 494

Literature Cited 494

INTRODUCTION

The cytogenetic features of about 235 human tumors in the nervous system have been explored, mainly during the past five years. Meningiomas account for almost half of these, and the chromosome changes observed are dealt with in a separate chapter in this volume [12]; these tumors have been given special attention, partly because of the extensive information accumulated about them, but particularly because of their unique cytogenetic features. The other tumors in the nervous system to be summarized in this chapter include several different malignant neurogenic tumors in adults and children, a variety of metastatic neoplasms, and two types of benign tumors: the neurinomas and the pituitary adenomas. The purpose is to survey the essential chromosomal findings in each of these tumor types and also, in some cases and in a limited way, to discuss the possible relation between the chromosomal observations and the clinical and pathological manifestations of the tumor.

NOMENCLATURE

Before discussion of the chromosomal characteristics of the different tumor types, it is necessary to clarify the meaning of some special symbols and terms to be used, and also to give a brief description of the principles of the statistical treatment applied to the chromosomal observations. Here, as in the chapter on meningiomas, S is a symbol for the stemline of a given cell population; similarly s (s_1, s_2, etc.) is a symbol for the sidelines, that is, other karyotypes found in a frequency of 10% or more in the tumor cell population. The "chromosomal representation" in a certain tumor or tumor stemline is a term for the pattern of involvement of 10 classes of normal chromosome types—A1, A2, A3, B, C, D, E(16), E(17-18), F, and G—and the possible occurrence of different abnormal chromosome types (i.e., markers). To obtain a measure of the character of this chromosomal representation in the various tumor types, I have applied the statistical principles outlined earlier in detail by Levan [2]. The chromosome numbers expected in the 10 classes were calculated for each tumor stemline number, assuming the same proportions between the classes as in normal male and female cells; the differences between observed and expected numbers were then determined for each class. The average differences for the various tumor types are shown in Table 1; the figures for standard error, and the significance of the deviations from zero for the astrocytic gliomas have been omitted. As a consequence of the method, markers were classified separately; since the expectation for this class is always zero, only positive values are to be found in the markers class in Table 1.

TUMORS ARISING IN THE NERVOUS SYSTEM

The tumors to be discussed here originated in the nervous system, though not always from nervous tissue.

Table 1. The Mean Chromosomal Deviation from the "Normal" in Nine Different Human Tumor Types in the Nervous System*

Tumor Type	Number of Tumors Studied	Specific Chromosome Groups (and Numbers)										Marker Chromosomes†
		1	2	3	A	B	C	D	16	E 17-18	F	G	

Tumor Type	Number of Tumors Studied	1	2	3	A	B	C	D	16	E 17-18	F	G	Marker Chromosomes†
Tumors arising in the nervous system													
Astrocytic gliomas	50	−0.3	0.0	+0.2		−0.1	−0.6	−1.0	0.0	−0.2	+0.2	−0.1	+2.0
Oligodendrogliomas	5	−0.5	+0.1	−0.1		−0.2	+0.5	−0.7	+0.1	0.0	−0.4	−0.2	+1.4
Ependymomas	8	+0.4	+0.4	−0.2		+0.1	−0.2	−0.4	+0.4	+0.1	+0.1	−0.3	+0.6
Medulloblastomas‡	10	+0.1	+0.2	0.0		+0.5	+0.2	−0.3	−0.1	−0.5	−0.2	−0.3	+0.4
Neuroblastomas‡	14	−0.1	−0.2	0.0		−0.4	+0.2	−0.7	0.0	0.0	0.0	−0.1	+1.4
Neurinomas	7	0.0	0.0	0.0		−0.1	−0.2	0.0	0.0	−0.3	0.0	−0.3	+0.9
Meningiomas§	21	0.0	0.0	+0.1		+0.1	+0.1	−0.2	0.0	0.0	+0.2	−0.9	+0.8
Pituitary adenomas	7	−0.2	−0.4	+0.2		0.0	+0.5	−0.2	0.0	−0.2	+0.4	−0.2	+0.1
Tumors arising outside the nervous system													
Metastases‖	20	−0.5	0.0	+0.7		−0.7	+0.5	−2.2	−0.1	−0.5	−0.4	−2.2	+5.4

* The units of expression represent the mean differences between observed and expected (tumors/"normal") numbers of chromosomes per group, and of markers, for each type of tumor. Calculations were made according to Levan [2]. This type of analysis has been termed the "chromosomal representation" of a tumor type; see also text, Introduction.
† The double-minute marker chromosomes are excluded from the calculations.
‡ Data for primary and secondary medulloblastomas and neuroblastomas have been pooled; only cases with complete karyotype analyses have been included.
§ Data derived from Mark [6].
‖ Eleven metastases from lung carcinomas, two from renal carcinomas, two from breast carcinomas, one each from a pancreatic and a colonic carcinoma, and three from malignant melanomas of the skin.

Astrocytic Malignant Gliomas

The astrocytic malignant gliomas, the predominant type of gliomas in adults, are tumors usually located in the cerebral hemispheres. The soft texture of these neoplasms, and especially the frequently high mitotic rate, make them very favorable material for chromosome studies. Excepting my own cases (50 cerebral and one cerebellar), however, no more than 13 glioblastomas have been reported in the literature (for references to all reported cases, see refs. [5, 7]). Two of those 13 occurred in children. Both had hypertriploid–hypotetraploid modal populations, but no karyotypic data were given; such information was also missing in the reports of about half of the remaining 11 cases found in adults. The available data describing numerical and structural changes in the chromosomes, however, agreed roughly with those from my own material. The distribution of chromosome numbers in my study of 50 tumors (the cerebellar glioblastoma excluded) is illustrated in Figs. 1a and 2. About three-quarters

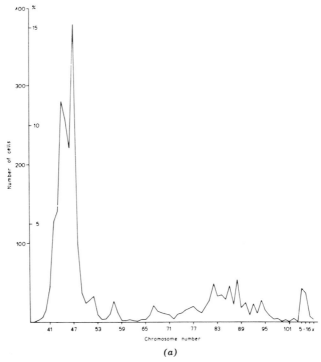

Figure 1a. Distribution of chromosome numbers in 50 astrocytic gliomas [7].

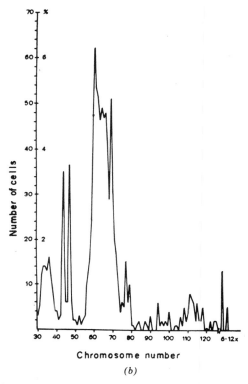

Figure 1b. Distribution of chromosome numbers in 20 metastases to the brain from tumors originating elsewhere [10].

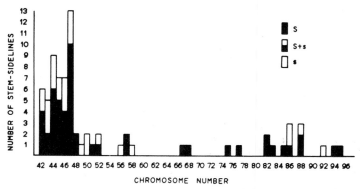

Figure 2. Histogram showing distribution of stemline numbers (S, blackened areas); sideline numbers (s, open areas); and stemline and sideline numbers (S+s) in the 50 astrocytic gliomas [8].

of the gliomas had diploid or near-diploid S-numbers, and those with near-tetraploid S-numbers were more than twice as common as those with a near-triploid S. The distribution of the stemline-sideline numbers shows that the karyotypic evolution in these gliomas usually does not stop at the diploid number (as is often seen in several types of neurogenic tumors in children, such as the medulloblastomas and neuroblastomas, below) but proceeds with about equal frequency in both hyperdiploid and hypodiploid directions. The sharp peak at a hyperdiploid mode of $S=47$ (Fig. 1a) indicates that loss of chromosomes is a greater threat to viability than gain of chromosomes; loss more often upsets the genetic balance, necessitating compensatory changes and hence the greater diversity of the hypodiploid S-numbers [7]. The chromosome pattern in the polyploid region shows that the doubling products of S-s-cells and related variant cells are very liable to chromosome losses. The "flattened" modes in the triploid-to-tetraploid range also demonstrate the increased tolerance to numerical deviations once a cell becomes polyploid. The small population of cells with high-polyploid chromosome numbers ($5n$ to $16n$) is probably composed of the giant cells frequently found during histologic examination of these particular tumors.

The karyotype analyses of the 50 cerebral astrocytic gliomas revealed no case with a normal diploid S; on the contrary, most tumors showed extensive deviations in the normal chromosome groups (A-C), and markers were found in almost three-quarters of the cases. In the S-karyotypes, the chromosome groups affected most often were C and D; groups G and A (No. 1) were also involved relatively often but considerably less often than groups C and D. The nature of this involvement of various chromosome groups was elucidated further by a statistical analysis of the relative chromosomal representation in all 50 stemline karyotypes (Table 1) as well as in the different S-groups (e.g., the hypodiploid group and the pseudodiploid groups—not shown in the table). The negative values in Table 1 indicate fewer chromosomes than expected. According to the results of this analysis, the karyotypic evolution in the astrocytic gliomas was characterized by a proportional loss of chromosomes (particularly in groups D and C and less frequently of chromosome No. 1), as well as by a tendency toward a gain of No. 3 chromosomes.

When tumors with and without markers were compared, excluding the double-minute markers (Table 1), the loss of chromosomes in groups D and C, and the loss of No. 1 chromosomes could be partly ascribed to their common participation in the formation of markers. Among the many different types of marker chromosomes observed in the gliomas, the "double-minutes" deserve special attention (Fig. 3). These often varied extensively in number from cell to cell and were usually extremely small

Figure 3. About 20 "double-minutes" (arrows to some of them) in a hyperdiploid cell from an astrocytic glioma (X 1250).

rod- or dot-shaped, paired structures; their "chromatids" exhibited a pronounced tendency to become widely separated, and whether or not they have a centromere is unknown. The mechanism(s) responsible for their origin and continued replication in the tumor cell population remains unexplained. This rare type of marker was found in most or all cells of four gliomas and also occasionally in nine additional gliomas. Thus double-minute markers are by no means restricted only to some tumor types found in children, as was apparently suspected earlier [1]. It may be, instead, that this aberration is particularly prone to appear in tumors derived from neurogenic tissues, regardless of the age of the patient.

Analysis of the relation between the cytogenetic changes and the clinical and pathological parameters of these tumors revealed several interesting trends. The age of the patient appeared to be a factor; the results suggested that gliomas with a pseudodiploid S (i.e., a stemline with 46 chromosomes but with an abnormal karyotype) were most likely to appear early, whereas those with hyperdiploid and tetraploid stemlines tended to develop in older patients. Thus the earlier appearance of gliomas of certain S-groups seemed to suggest a correlation between degree of malignancy and specific karyotypic features. The *sex* of the patient appeared to have an effect both on the frequency of different S-groups (hypodiploid, pseudodiploid, etc.), and on the karyotypic pattern in certain S-groups (triploid and tetraploid). When the *site* of the glioma and also the *duration* of symptoms (prior to operation) were considered, there was an influence of sex, too. Thus there was a correlation between the male sex and three tumor qualities: development of hyperdiploid stemlines, a parietal location, and a long period of symptoms prior to operation. This is of interest in view of recent observations by Westermark et al. [15]; their results indicated that cell lines could be established from malignant gliomas much more frequently from parietally or temporally located tumors in males. An effort was also made, in my own study, to find a relation between the histopathological characteristics of the gliomas and the chromosomal features; though nonrandom trends were observed, the patterns were too vague to permit definite conclusions.

The one cerebellar glioblastoma mentioned in the beginning of the present section was the only tumor of this type studied. It will suffice here to say that the cytogenetic results were similar to those in the cerebral cases in many respects.

Unfortunately, nothing is known at present about the cytogenetics of either the cerebral or the cerebellar astrocytomas. Direct preparations from these more mature tumors have been repeatedly unsuccessful due to the low mitotic rate. Many astrocytomas, especially cerebellar ones, however, have been studied from primary cultures. The chromosomal picture has been normal in these cases (Mark, unpublished results), but the results must be regarded as highly preliminary because it was impossible to exclude stromal overgrowth.

Oligodendrogliomas

Oligodendrogliomas are generally found in adults, but in a much lower frequency than the astrocytic malignant gliomas. This is reflected in the number of cases studied cytogenetically. Thus only one case could be found in the literature ([3]; in the present survey this tumor is given the arbitrary number, 04), except for the three cases (01, 02, and 03) I have reported previously [7, 8]. Out of these four tumors, two were well-differentiated oligodendrogliomas (02 and 04) with special chromosomal characteristics: one of them (02) had a hyperdiploid stemline, $S=47$, and an S-karyotype that differed from the normal only by a minute marker; the other (04) had a 92-chromosome stemline and an S-karyotype of a doubled normal cell. This comparatively slight departure from the normal in the mature forms of the oligodendrogliomas (which are usually more benign than the astrocytic type) is a noticeable feature. Recent findings in a fifth oligodendroglioma (05) of mature type, studied by the Giemsa-banding technique (Mark, unpublished), added further support to the above results. This hypodiploid tumor from a male had an S-karyotype that differed from the normal only by the loss of the Y chromosome (Fig. 4); there were also many cells with 46-chromosomes and completely normal karyotypes (46,XY).

The other two tumors (01 and 03) of the present group were immature

Comparatively few tumors metastatic to the brain have been studied cytogenetically. Only three cases could be found in the literature (for reference, see ref. [10]). Because karyotypic data were either missing or incomplete in these cases, the survey below will refer almost exclusively to my own observations in a series of 20 different metastatic tumors found in the brain: 11 metastases from lung carcinomas, 2 from renal carcinomas, 2 from breast carcinomas, 1 each from a pancreatic and a colonic carcinoma, and 3 from malignant melanomas of the skin.

The numerical changes in the chromosomes of metastatic tumors (Fig. 1b) were completely different from those found in the tumor groups already discussed (see, e.g., Fig. 1a). Thus in about three-quarters of the metastases, the stemlines had a modal number in the triploid region—specifically, in the hypotriploid zone. Other cases with pentaploid or extreme hypodiploid modes (S-numbers around 35) illustrate further the advanced character (i.e., the late stage) of the karyotypic evolution of most of the metastatic neoplasms studied. The karyotype analyses in the majority of these tumors revealed a pronounced variability among the cells studied in a particular neoplasm, and in several cases it was difficult to find even two cells in one tumor with the same karyotype. A substantial part of the variation could be attributed to the markers, which were found in all but one tumor stemline. Many tumors contained one or several marker types with a notable tendency toward variation in size and morphology, indicating that these particular markers were unstable and were often involved in further structural rearrangements. Also contributing to the variation were the losses or gains of one or several marker types from the stemline karyotype. This numerical variation in the marker set was not unexpected in view of the high numbers of markers found in many of the metastases (about half of the tumors had seven or more markers in their S) and the common occurrence of mitotic irregularities.

The changes in chromosome numbers were accompanied by changes in the chromosomal representation in the metastases; all cells analyzed in the modal region of each tumor were included in the calculations. Most of the values were on an average higher than in the other tumor groups (Table 1). The most striking features were the high proportional losses in groups D and G and the high positive values for markers. Actually, a separate analysis demonstrated a correlation between the frequency of markers and the extent of the decrease of the acrocentrics (i.e., the D and G chromosomes).

The brain metastases displayed more extreme numerical and structural changes in the chromosomes than did cancerous effusions (for reference, see ref. [10]). It is possible that there are neoplastic properties related to a capacity to metastasize to the CNS. In three patients, two different brain metastases from the same primary tumor were examined; cytogenetic analyses indicated that different subclones from the same primary tumor had a similar invasive capacity.

CONCLUDING REMARKS

In the present survey (and in the chapter dealing with the meningiomas in this volume, [12]), I have reviewed the chromosomal findings in all human tumor types in the nervous system that have been sufficiently well analyzed to permit valid conclusions. A large amount of data forms the basis for the review. Because of this, and the limited space available here, it has been necessary to exclude many details and also to omit all tables of cytogenetic "raw data"; these can be found in the references provided.

There is no counterpart among studies of the tumors of other organs or organ systems for the comprehensive cytogenetic studies that have been made of the neoplasms in the nervous system. The observations that I have summarized here and in my paper on meningiomas, therefore, constitute a promising area for more detailed studies using the banding techniques recently introduced (e.g., see Shaw and Chen in this volume [14]). Thus far, these new techniques have been applied mainly to the meningiomas, in which their use has rapidly clarified the specificity of the group G anomaly. Those results suggest that other specific chromosomal aberrations may be found and characterized in other tumor types in the nervous system. It can also be expected that the new techniques will contribute to a better understanding of other special cytogenetical problems, as for instance the nature of the double-minute aberration.

Acknowledgements.–The present investigation was supported by grants from the Swedish Cancer Society and by grants from John and Augusta Persson's Foundation for Medical Research.

LITERATURE CITED

1. Cox D, Yunken C, Spriggs AI: Minute chromatin bodies in malignant tumours of childhood. *Lancet* 2:55–58, 1965.
2. Levan A: Non-random representation of chromosome types in human tumor stemlines. *Hereditas* 55:28–38, 1966.
3. Lubs HA, Salmon JH: The chromosome complement of human solid tumours. II. Karyotypes of glial tumours. *J Neurosurg* 22:160–168, 1965.
4. Mark J: Chromosomal analysis of a human retinoblastoma. *Acta Ophthalmol* 48:124–135, 1970.

5. MARK J: Chromosomal characteristics of neurogenic tumours in children. *Acta Cytol* 14:510–518, 1970.
6. MARK J: Chromosomal patterns in human meningiomas. *Eur J Cancer* 6:489–498, 1970.
7. MARK J: Chromosomal characteristics of neurogenic tumours in adults. *Hereditas* 68:61–100, 1971.
8. MARK J: The chromosomes in two oligodendrogliomas and three ependymomas in adults. *Hereditas* 69:145–149, 1971.
9. MARK J: Chromosomal characteristics of human pituitary adenomas. *Acta Neuropathol* 19:99–109, 1971.
10. MARK J: Chromosomal characteristics of secondary human brain tumours. *Eur J Cancer* 8:399–407, 1972.
11. MARK J: The chromosomal findings in seven human neurinomas and one neurosarcoma. *Acta Pathol Microbiol Scand (A)* 80:61–70, 1972.
12. MARK J: The human meningioma—a benign tumor type with specific chromosome characteristics. This volume, pp. 497–517, 1974.
13. SANDBERG AA, SAKURAI M, HOLDSWORTH RN: Chromosomes and causation of human cancer and leukemia. V. Dms chromosomes in a human neuroblastoma. *Cancer* 29:1671–79, 1972.
14. SHAW MW, CHEN TR: The application of banding techniques to tumor chromosomes. This volume, pp. 135–150, 1974.
15. WESTERMARK B, PONTEN J, HUGOSSON R: Determinants for the establishment of permanent tissue culture lines from human gliomas. *Acta Pathol Microbiol Scand (A)*, 81:791–805, 1973.

THE HUMAN MENINGIOMA: A BENIGN TUMOR WITH SPECIFIC CHROMOSOME CHARACTERISTICS

JOACHIM MARK

Department of Pathology, Central Hospital, Skövde, Sweden,
and Department of Neurosurgery, Lund's Hospital, Lund, Sweden

Introduction 498
Recent Studies on 23 Meningiomas Previously Unreported 498
    Tumor Sources and Methods 498
    Chromosome Numbers 499
    Descriptions of Stemline Karyotypes 503
        46-Chromosome Stemlines 503
        45-Chromosome Stemlines 503
        44- and 43-Chromosome Stemlines 504
        Near-triploid Stemline 504
        Chromosome Changes During Growth In Vitro 504
        Multiple Tumors 506
        Different Areas of One Tumor 506
Survey of All 105 Meningiomas Studied to Date 507
    Tumor Sources 507
    Chromosome Numbers 507
    Descriptions of Stemline Karyotypes 508
    Marker Chromosomes 513
        Incidence of Marker Chromosomes 513
        Ring and Dicentric Chromosomes 513
        Markers in Group G 514
Summary 516
Acknowledgements 516
Literature Cited 516

INTRODUCTION

During the last few years it has become increasingly clear that chromosomal abnormalities are common in several types of benign human tumors [9]. Cytogenetic studies have been particularly rewarding in the meningiomas, in which evolution of the karyotype usually is characterized by the loss of one or several chromosomes of group G [6, 17, 18]. Using the recently introduced fluorescent method for chromosome identification, it has been possible in several stemlines to determine exactly which of the group G chromosomes has been lost (tumor numbers underscored in the last column of Table 2). The results from different studies have agreed well, showing that pair No. 22 is selectively affected [10, 11, 21]. In the light of these recent findings, and also because a considerable amount of data has been accumulated about the meningiomas, now is an appropriate time to review the chromosomal characteristics of this tumor. The review presented here is based on the studies of meningiomas already reported (both those analyzed with conventional methods and those with fluorescence techniques) and, in addition, studies of 23 tumors not reported earlier. The 23 new cases—all of them studied with conventional methods —will be described first, and then a survey of all the cases will be made.

RECENT STUDIES ON 23 MENINGIOMAS PREVIOUSLY UNREPORTED

Brief descriptions of the patients and their tumors and of the methods used will first be given. Chromosomal observations will then be presented from primary cultures (i.e., not subcultured cells) of the 23 new meningiomas, grouped according to their stemline numbers. Finally, summaries will be given of some specific problems, namely, chromosomes in the serial in vitro passages of seven tumors, chromosomes in the multiple tumors of two patients, and chromosomes in different areas of the same tumor in four meningiomas.

Tumor Sources and Methods

The clinical source and the pathological features of the 23 new meningiomas are given in Table 1. All but one of the patients (M43) were over 50 years of age at the time of tumor removal, and, as is characteristic for these tumors, females outnumbered males. Each of the 23 meningiomas was located intracranially. Tumors M33 and M34 were different meningiomas removed from the same patient, and M37, M38, and M39 were different tumors all removed from one patient. The meningiomas were classified according to their histology, as in previous reports (see [6]). Two belonged to the rare myxomatous type (M30 and M43). No sign of malignancy was found in any tumor.

Cultured cells were used for the chromosome studies. Thus pieces of tumor tissue were explanted in vitro using milk-dilution bottles and Parker's medium 199, supplemented with 10% unfiltered, inactivated calf serum [6]. One or several chromosome preparations were made from the primary cultures of all meningiomas (growth period in vitro 5 to 13 days). Seven of the tumors (M29, 31, 35, 36, 37, 40, and 48) were also studied in preparations from in vitro passages (usually performed once a week; split ratio, 1:2). The number of subcultures varied between 3 and 15 (Table 1, third column).

The chromosomes were studied in squash preparations, stained with orcein, and 40 or more cells were examined in each preparation. Karyotype analyses were made from photographs, using 10 or more cells per preparation. S is used as a symbol for the stemline of a certain population of tumor cells (i.e., the most frequent karyotype of the population). Similarly, s (s_1, s_2, etc.) is used for the sideline (i.e., other karyotypes found in a frequency of 10% or more in the tumor cell population). Cells with other karyotypes than those of S-s-cells are termed variant cells. The abbreviations used for different marker types (both in text and Tables 1 and 2) are defined below:

M-type: biarmed marker with arms of equal size, arm ratio 1.00
m-type: biarmed marker, arm ratio between 1.00 and 1.67
sm-type: biarmed marker, arm ratio between 1.67 and 3.00
st-type: biarmed marker, arm ratio between 3.00 and 7.00
t-type: biarmed marker, arm ratio greater than 7.00

As shown in the fourth column of Table 2, it was possible only in some cases to transcribe the above-mentioned marker types in more specific terms, using the nomenclature of the "Chicago Conference" [2].

Chromosome Numbers

Table 1 shows the number of chromosomes per cell and the karyotypic features of the stemlines and the sidelines of the 23 meningiomas. Except for the hypotriploid tumor (M50), all meningiomas had diploid or hypodiploid S-numbers. This was the case also for the sidelines found in about two-thirds of the tumors. The spread was mainly hypomodal, and

Table 1. DETAILED CHARACTERIZATION OF 23 MENINGIOMAS NOT PREVIOUSLY REPORTED

Tumor No.	Histologic Type	Days in Vitro	Sex, Age of Patient	Number of Cells with Numbers of Chromosome:													
				39	40	41	42	43	44	45	46	47	48–56	57	58	59	60	±4–16n[†]
M28	Transitional	8	F, 66	2	3	5	7	30	4	1	—	—	—	—	—	—	—	2
M29	Transitional	7	F, 66	2	2	6	22	24	34	4	—	—	—	—	—	—	—	6
		22		—	1	—	2	4	21	—	13	—	—	—	—	—	—	7
		65		—	—	—	—	—	—	7	34	—	—	—	—	—	—	—
M30	Myxomatous	12	M, 61	—	—	1	—	3	21	17	9	—	—	—	—	—	—	2
M31	Transitional	7	F, 62	2	4	3	3	6	9	54	—	—	1	—	—	—	—	3
		23		—	—	—	—	—	6	42	—	—	—	—	—	—	—	—
M32	Transitional	10	M, 64	—	—	—	1	1	4	28	11	1	—	—	—	—	—	8
M33[§]	Syncytial	8	M, 63	—	—	13	1	—	4	32	—	—	—	—	—	—	—	2
M34[§]	Syncytial	7	M, 63	—	—	—	1	3	8	33	4	—	—	—	—	—	—	1
M35	Transitional	10	M, 51	2	4	2	2	4	4	52	28	—	—	—	—	—	—	2
		57		—	—	—	—	—	3	28	17	—	—	—	—	—	—	2
		89		—	—	—	1	—	—	3	46	—	—	—	—	—	—	—
M36	Angioblastic	9	M, 61	1	1	3	2	—	—	30	6	—	—	—	—	—	—	1
		53		—	—	—	—	—	—	35	10	—	1	—	—	—	—	2
M37[||]	Transitional	6	F, 69	—	1	1	1	4	7	36	—	—	—	—	—	—	—	—
		56		—	—	—	—	—	1	—	3	46	—	—	—	—	—	—
M38[||]	Syncytial	8	F, 69	—	—	—	1	1	4	40	2	—	—	—	—	—	—	—
M39[||]	Transitional	8	F, 69	—	—	—	—	—	3	10	34	3	—	—	—	—	—	—
M40	Transitional	8	F, 51	—	—	—	—	4	6	10	72	2	—	—	—	—	—	—
		38		—	—	—	—	2	2	3	40	2	—	—	1	—	—	—
M41	Transitional	10	M, 51	—	—	—	—	1	1	4	26	11	1	—	—	—	—	6
M42	Transitional	12	F, 72	—	—	—	—	—	—	3	46	1	—	—	—	—	—	—
M43	Myxomatous	7	F, 39	—	—	—	—	—	—	1	10	38	—	—	—	—	—	1
M44	Transitional	5	F, 70	—	—	—	—	—	1	1	7	41	—	—	—	—	—	—
M45	Angioblastic	9	F, 53	—	—	—	—	—	1	—	4	40	1	—	—	—	—	2
M46	Transitional	8	M, 60	—	—	—	—	—	—	1	1	46	—	—	—	—	—	2
M47	Transitional	9	F, 58	—	—	—	—	—	—	5	7	40	—	—	—	—	—	2
M48	Transitional	12	F, 66	4	6	—	—	—	6	16	66	—	—	—	—	—	—	2
		58		—	1	3	1	—	5	3	28	—	—	—	—	—	—	—
		118		—	—	—	—	—	—	1	37	1	—	—	—	—	—	1
M49	Transitional	7	M, 56	—	—	—	—	—	1	2	47	—	—	—	—	—	—	—
M50	Transitional	13	M, 51	—	—	—	1	—	—	1	9	—	8	3	24		1	1

*M = Specimens from J. Mark laboratory, Department of Pathology, Central Hospital, Skövde, Sweden.

[†] $4n$ (tetraploid) to $16n$ range.

‡Definitions of abbrevations used appear in "Tumor Sources and Methods" section; nomenclature of the "Chicago Conference": number of chromosomes per cell, followed by missing (−) or extra (+) chromosomes.

§Different tumors from the same patient.

||Different tumors from the same patient.

Tumor No.	Total Cells	Karyotype of Stemline	Karyotype of Sideline(s)
M28	54	43, −C, −D, −G	
M29	100	44, −C, −G	
	48	44, −C, −G	46
	41	46	
M30	53	44, −A1, −A3, −C, +t	s_1 = 45, −G; s_2 = 46
M31	85	45, −D, −G, +ring chr	s_1 = 45, −G; s_2 = 44, −D, −G
	48	45, −G	44, −C, −G
M32	54	45, −C, −G, +m	46, −G, +m
M33§	52	45, −G	41, −A1, −2C, −D, −G
M34§	50	45, −G	44, −E17−18, −G
M35	100	45, −G	46
	50	45, −G	46
	50	46	
M36	44	45, −G	46
	48	45, −G	46
M37‖	50	45, −G	45, −C, −G, + dic
	50	46	
M38‖	48	45, −G	
M39‖	50	45, −G	44, −C, −G
M40	94	45, −G	
	50	45, −G	
M41	50	45, −G	46
M42	50	45, −G	
M43	50	46	45, −G
M44	50	46	45, −G
M45	48	46	
M46	50	46	
M47	54	46	s_1 = 45, −G; s_2 = 44, −C, −2G, + dic
M48	100	46	45, −C
	41	46	
	40	46	
M49	50	46, −G, +sm	46
M50	49	58, +2A3, +2B, +4C, +D, +F, +G, +t	46

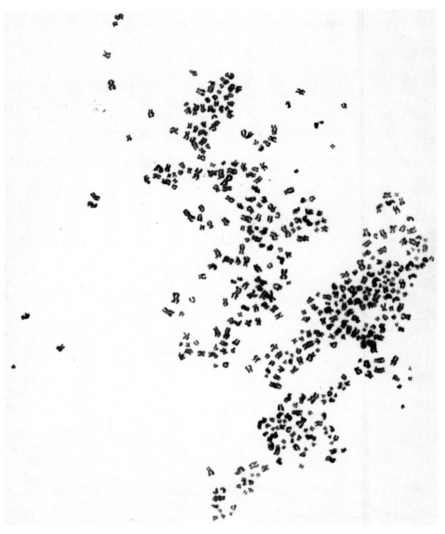

Figure 1. High-polyploid metaphase in M29 with about 330 chromosomes ($\times 780$).

hyperdiploid counts were rare or absent (except in the hypotriploid case). A few polyploid cells occurred in many of the tumors. These were usually in the double-stemline range, but in two meningiomas, M29 and M32, several metaphases had a higher ploidy. One such metaphase with about 330 chromosomes, from M29, is depicted in Fig. 1.

Descriptions of Stemline Karyotypes

46-Chromosome Stemlines. This group consisted of six tumors with normal diploid S (M43-48) and one with a pseudodiploid S (M49). Among those tumors with normal diploid stemlines, five sidelines were found, three showing a monosomy in group G and one a monosomy in group C as the only deviation from normal; the fifth sideline, s_2 of M47, contained a dicentric chromosome (Fig. 2e), apparently formed by the missing group C chromosome and one of the two lost group G chromosomes. The pseudodiploid S of M49 differed from the normal only by the substitution of one group G autosome and by one small marker of sm-type, apparently a group G chromosome with a deletion of approximately the distal third of its long arm (Fig. 2f); a considerable number of the 46 chromosome cells were normal, however, constituting a sideline carrying four G autosomes of normal size.

45-Chromosome Stemlines. This class consisted of the 12 tumors, M31-42. No less than 10 of them showed the karyotypic abnormality characteristic of meningiomas, namely, the loss of one group G chromosome. The majority of them, however, also contained remnants of the original, normal diploid S, and in three tumors (M35, M36, and M41) these cells were frequent enough to constitute a sideline. Two other sidelines differed from their stemline by the loss of one more chromosome, a group C in M39, and one E17 or 18 in M34; a third sideline with 45 chromosomes (M37) showed a dicentric obviously formed by the missing group C and G chromosomes (also, refer to the s of M47, Table 1); a fourth sideline with 41 chromosomes (M33) displayed more complex changes but shared monosomy in group G with the stemline.

Two 45-chromosome meningiomas (M31 and M32) contained structural abnormalities. Of them, M31 had lost one group D and one group G chromosome and gained one ring chromosome (Fig. 2b). From two sideline karyotypes of this tumor it appeared that the formation of the ring had been superimposed on the original stemline with monosomy G, and that the ring was often involved in non-disjunctional variation. It seems probable that the lost group D chromosome had participated in the formation of the monocentric ring, perhaps together with other unknown chromosomes.

The stemline of M32 had lost one group C and one group G chromosome and gained one minute marker of the m-type (Fig. 2c). Since satellite structures were often discernible on the shorter arm of this m-marker, it seems probable that it corresponded to the lost group G autosome, from which most of the long arm had been deleted. A 46-chromosome sideline was found, in which the m-marker had also been substituted for one group G chromosome, but without the loss of a group C chromosome.

44- and 43-Chromosome Stemlines. Only three tumors (M28, M29, and M30) belonged to this class. The $S=44$ in M30 had lost two group A chromosomes and one group C chromosome and gained one marker of the t-type, somewhat longer than the longest of the group D chromosomes (Fig. 2a). The stemline as well as the 46-chromosome sideline (s_2) which had a normal karyotype showed no deviation in group G, but the 45-chromosome sideline (s_1) had lost one member of group G. Accordingly, the stemline at the time of investigation was probably not derived from the s_1 with monosomy G but rather represented an independently evolved line. The remaining two tumors of this class (M28 and M29) had closely related karyotypes. Both stemlines had lost one group C and one group G chromosome and, in addition, M28 had lost one group D chromosome.

Near-triploid Stemline. The only tumor in this class (M50) had gained from one to four chromosomes in six different groups, as well as one marker of the t-type (Fig. 2g). The origin of this t-marker could not be traced. Whereas the few 46-chromosome cells analyzed had normal karyotypes, the single 45-chromosome cell had lost one group G chromosome. The findings in M50 were reminiscent of those in tumor M12, previously reported, whose stemline was in the hyperdiploid–hypotriploid range [6]. As was actually shown in M12, an original loss of a group G chromosome could easily be concealed by changes superimposed during the progressive alteration in the karyotype, such as the development in both M12 and M50 of hyperdiploid–hypotriploid stemlines.

Chromosome Changes during Growth In Vitro. Seven meningiomas were studied once or twice during serial in vitro passage. Column four of Table 1 shows the duration of the in vitro growth. One of the tumors (M48) had a normal, diploid S in primary culture; the S-karyotype of apparently normal chromosomes in the stemline pattern will be dealt with in the present section, and then, in the following section, structurally changed chromosomes will be discussed.

In the remaining 104 tumors (karyotype No. 42 excluded) studied, there were no less than 44 different stemline karyotypes. These are listed in Table 2 in order of increasing S-numbers. Two karyotypes were outstanding by their frequency: No. 28, characterized by monosomy in group G, was found in 31 tumors and No. 29, with a normal karyotype, in 23 tumors. One karyotype (No. 30), characterized only by a deletion in a G chromosome, was found in five tumors; one karyotype (No. 21), monosomic for both a group G and a group D chromosome, was found in three tumors; two karyotypes (Nos. 11 and 13), both monosomic for a group G chromosome in addition to other changes, were found in two tumors each; all of the other 38 karyotypes were found in only one tumor each.

In spite of this variability, however, the meningiomas do have important karyotypic features which they share in common and which distinguish them from other types of human tumors. The data in Table 2 indicate that the normal chromosomes vary in the meningioma stemlines in a nonrandom way, the deviations affecting mainly chromosomes of groups G, C, D, and, to a lesser degree, chromosomes Nos. E17-18 and A1.

The actual involvement of the various identifiable chromosome pairs and groups is summarized in Table 3. The percentages of involvement are given both for all of the 104 tumors (line 1) and for a specific arbitrarily selected subgroup of 43 tumors (line 2). The subgroup consisted of all tumors remaining after the exclusion of: normal diploid tumors (karyotype No. 29 of Table 2), diploid tumors in which the only aberration was that one or two group G chromosomes had undergone change (karyotypes Nos. 30, 31, and 32), and 45-chromosome tumors in which the only aberration was monosomy G (karyotype No. 28). In this subgroup, the focus should be expected to have shifted toward the karyotypic evolution subsequent to the usual primary step, loss or change of a member of group G [6, 11].

The distribution of aberrations among the chromosome types in these 104 tumors revealed the predominant involvement to be in groups G, C, and D (in that order). The pattern of aberrations was similar in the subgroup of 43 heteroploid tumors. However, in the heteroploid tumors, group C approached group G in importance, and more than 50% of the stemlines deviated in group D; also, aberrations were increased in chromosomes Nos. 1 and 17-18. In recent studies of meningiomas using fluorescence analysis, in which all individual pairs were identified, Mark et al. [11] reported similar results. In that analysis, in addition to the

Table 2. Karyotypes of All 105 Meningiomas Studied to Date

Karyotype No.*	Number of Chromosomes	Chromosome Number or Group										Markers†	Number of Stemlines	Tumor Identification†
		A1	A2	A3	B	C	D	E16	E17-18	F	G			
1	38	−1				−5	−3	−1	−1§		−1	+st, +t$_1$, +t$_2$	1	III
2	40		−1§		−1	−1	−1	−1	−1§		−2	+t(=Bp−)	1	T689
3	40					−2	−2				−2		1	T796
4	40		−1§			−3	−1				−1		1	T659
5	41					−1	−1		−1§	−2	−1	+m(=Gq−?)	1	II
6	41					−1	−2			−1	−1		1	T759
7	41	−1				−5	−1				−3	+M(=Gq−?), +M$_2$, +M$_3$, +sm$_1$, +t	1	M1
8	42	−1					−1				−1	+ring chr	1	M14
9	42				−1		−2		−1		−1		1	T584
10	43					−1	−1		−1§		−1	+m(=Gq−)	1	T585
11	43		−1§			−1	−1		−1§		−1	+m−sm(=Ap−)	2	T765, 768
12	43	−1					−3					+m−sm(=Alp−), +sm−st(=Cqs)	1	T669
13	43					−1	−1				−1		2	M28, T919
14	43		−1§			−1	−1				−1	+st$_1$, +st$_2$	1	T1274
15	43	−1				−1				−1	−1	+t, +ring chr	1	M15
16	43	−1				−3					−1		1	M20
17	44								−1§		−1		1	T1215
18	44	−1					−1						1	T606
19	44										−1		1	T465
20	44	−1									−1	+t	1	M30
21	44			−1									3	M16, 29, T616
22	45					−1	−1				−1	+ring chr	1	M31
23	45					−1	−1						1	M6
24	45								−1		−1	+t	1	M7

25	45			−1§	+1	−1		−1		−2	+m$_2$, +st, +t$_1$(=Gp−q−?), +t$_2$	1	M8
26	45					−1				−1	+m(=Gq−)	1	M32
27	45					−1						1	M17
28	45									−1		31	M2-5, 21-24, 33-42, T543, 562, 582, 612, 630, 658, 674, 691, 693, 716, 1314, 1326, 1356
29	46											23	M18, 27, 43-48, T349, 496, 503, 559, 600, 621, 623, 653, 680, 682, 687, 776, 830, 837, 863
30	46									−1	+M(=G22pi)	1	M25
31	46									−1	+m−sm(=Gq−)	5	M26, 49, T1059, 1082, I
32	46									−2	+m$_1$(=Gp+q−), +m$_2$(=Gq−)	1	T1046
33	46				−1					−1	+sm−st(=Cp−), +m(=Gp+)	1	T870
34	46				+1							1	T579
35	46				+1	−1		+1				1	M9
36	46				−1					−1	+m	1	M10
37	46				−1					−2	+sm, +2t(=2Gp−q−)	1	M11
38	47										+t(=Gp−q−?)	1	M13
39	47				+1					−1	+t(=G22p−q−)	1	M19
40	47									+1	?	1	T907
41	47				+1							1	T1012
42‖	50–54	?	?	?	?	?	+2(+3)?	?	?			1	T781
43	53		+1		+2	+2		+1§				1	T976
44	57			+2	+4	+2		+1	+1		+t	1	M12
45	58		+2	+2	+4	+1		+1	+1		+t	1	M50

* Arbitrary numbering, according to increasing number of chromosomes in karyotype.
† Definitions of abbreviations used appear in "Tumor Sources and Methods" section.
‡ I–III = Porter and coworkers [1, 14]; T = Zang and coworkers [17–21]; M = Mark and coworkers [5–7, 10, 11]. Underlined numbers = tumors studied by the fluorescence technique.
§ The chromosomes affected in this group were not identified specifically.
‖ Complete karyotypic data were not available for tumor T781.

Table 3. Percentage of Involvement of Various Chromosomes in Stemlines of All Meningiomas Studied to Date (104 Cases*) and of a Heteroploid Subgroup

Meningiomas	Chromosome Number or Group									
	A1	A2	A3	B	C	D	E16	E17-18	F	G
1. All cases	9	3	5	5	30	23	3	10	7	69
2. Heteroploid subgroup†	23	6	13	12	72	54	6	24	16	79

* Karyotype No. 42 excluded (data incomplete).
† 43 cases; see text for definition of this subgroup.

selective involvement of pair 22, involvement of chromosomes Nos. 1, 8, 12, 17, 7, 6, and 11 was also often found. The only discrepancy was in group D, which showed an infrequent involvement in the fluorescence studies; work is in progress to elucidate this discrepancy.

In chromosome groups with frequent involvement, both losses and gains of chromosomes have taken place; losses predominate, however, in keeping with the observation that the majority of heteroploid stemlines are hypodiploid. It is a reasonable assumption that both gains and losses are the result of non-disjunction and that high incidences of losses and gains in specific chromosomes indicate that these chromosomes are more liable to non-disjunction than others. For unknown reasons, cells with gains seem usually to be inferior in viability to those with losses in meningiomas.

The preferential pattern, discussed above, was also apparent from the karyotypic deviations recorded in the few meningiomas having an abnormal stemline without involvement of Group G, namely, the karyotype Nos. 12, 20, 23, 27, 36, 38, 41, 43, and 44 (Table 2). Except for karyotype No. 38, in which the only change was a gain of one marker of t-type (which was possibly derived from a group G chromosome), all the other eight karyotypes had undergone deviations in group C, group D, or both. Furthermore, the groups C and D were involved in all cases in which the only change was a gain or loss of one normal chromosome. Thus it seems that in rare cases the karyotypic evolution of meningiomas may commence directly with those steps which usually follow the initial loss or change of a group G chromosome in the majority of the tumors. This conclusion is supported by the common occurrence (both in the tumors I studied and those studied in Germany) of variant cells with monosomy in group C, as well as by the occurrence of sidelines with the loss of one C chromosome as the only change (e.g., tumor M48 of Table 1).

In my own cases, variant cells (refer to "Tumor Sources and Methods"

6. MARK J: Chromosomal patterns in human meningiomas. *Eur J Cancer* 6:489–498, 1970.
7. MARK J: Chromosomal aberrations and their relation to malignancy in meningiomas: a meningioma with ring chromosomes. *Acta Pathol Microbiol Scand* 79:193–200, 1971.
8. MARK J: Chromosomal characteristics of neurogenic tumours in adults. *Hereditas* 68:61–100, 1971.
9. MARK J: The chromosomal findings in seven human neurinomas and one neurosarcoma. *Acta Pathol Microbiol Scand* 80:61–70, 1972.
10. MARK J, LEVAN G, MITELMAN F: Identification by fluorescence of the G chromosome lost in human meningiomas. *Hereditas* 71:163–168, 1972.
11. MARK J, MITELMAN F, LEVAN G: On the specificity of the G abnormality in human meningiomas studied by the fluorescent technique. *Acta Pathol Microbiol Scand* 80:812–820, 1972.
12. MITELMAN F: The chromosomes of fifty primary Rous rat sarcomas. *Hereditas* 69:155–186, 1971.
13. MITELMAN F, MARK J, LEVAN G, LEVAN A: Tumor etiology and chromosome pattern. *Science* 176:1340–1341, 1972.
14. PORTER IH, BENEDICT WF, BROWN CD, PAUL B: Recent advances in molecular pathology (a review); some aspects of chromosome changes in cancer. *Exp Mol Pathol* 11:340–367, 1969.
15. SAKSELA E, MOORHEAD PS: Aneuploidy in the degenerative phase of serial cultivation of human cell strains. *Proc Nat Acad Sci (US)* 50:390–395, 1963.
16. SANDBERG AA, HOSSFELD DK: Chromosomal abnormalities in human neoplasia. *Ann Rev Med* 21:379–408, 1970.
17. SINGER H, ZANG KD: Cytologische und cytogenetische Untersuchungen an Hirntumoren. I. Die Chromosomenpathologie des menschlichen Meningeoms. *Humangenetik* 9:172–184, 1970.
18. ZANG KD, SINGER H: Chromosomal constitution of meningiomas. *Nature* 216:84–85, 1967.
19. ZANKL H, SINGER H, ZANG KD: Cytological and cytogenetical studies on brain tumors. II. Hyperdiploidy, a rare event in human primary meningiomas. *Humangenetik* 11:253–257, 1971.
20. ZANKL H, ZANG KD: Cytological and cytogenetical studies on brain tumours. III. Ph[1]-like chromosomes in human meningiomas. *Humangenetik* 12:42–49, 1971.
21. ZANKL H, ZANG KD: Cytological and cytogenetical studies on brain tumours. IV. Identification of the missing G chromosome in human meningiomas as no. 22 by fluorescence technique. *Humangenetik* 14:167–169, 1972.

SPECIAL APPROACHES

CELL HYBRIDIZATION IN THE STUDY OF THE MALIGNANT PROCESS, INCLUDING CYTOGENETIC ASPECTS

ORLANDO J. MILLER

Department of Human Genetics and Development and Department of Obstetrics and Gynecology, College of Physicians and Surgeons, Columbia University, New York, New York

---

Cell Hybridization 522
   Spontaneous Cell Hybridization 522
   Conditional Lethals and Selective Systems 523
   Virus-induced Cell Fusion 524
   Chromosome Loss from Hybrid Cells 525
   Further Developments 525

Clues to Etiology of Cancer 525
   Cell Hybridization and Viral Oncogenesis 526
      Theories of Viral Carcinogenesis 526
      Viral Transformation of Cells 527
      Rescue of Oncogenic Viruses by Cell Fusion 527
      Detection of Unknown Tumor Viruses 531
   Cell Hybridization and Chemical Carcinogenesis 532
   Effects of Cell Hybridization on Malignancy 533
      Early Evidence that Malignancy May Be a Dominant Character . . . 533
      Later Evidence that Malignancy Is a Graded Character 534
      Evidence for Suppression of Malignancy 534
   Chromosome Loss and Malignancy 535
      Studies in the Mouse 535
      Exceptions to the Chromosome Loss Model 540
      Extension to Man 541
   Effects of Endogenous Factors 542
      Defective DNA Repair 542
      Immunological Factors 543
      The Cell Membrane 546
   Cell Hybridization in Initiation of Malignancy 550

Clues to Tumor Progression 550

Possible Involvement of Cell Hybridization 550
Tumor Cell Differentiation 552

Clues to Tumor Therapy 553
Tumor Immunity 553
Drug Resistance 554

Conclusion 554

Acknowledgements 555

Literature Cited 555

CELL HYBRIDIZATION

Fusion of animal cells in culture has been known to occur for many years (see, e.g., the review by Lewis, ref. [90]). Fusion of two or more nuclei in multinucleated cells has been directly observed and recorded by time lapse photography [102]. When two nuclei in a common cytoplasm enter mitosis at the same time, the chromosomes generally become aligned on a single equatorial plate and are thus distributed to a single pair of daughter cells with, approximately, identical hybrid chromosome constitutions [105]. Rarely a multipolar mitosis occurs [158], but the infrequency may indicate that the centrioles of one parental cell are dominant [9].

Spontaneous Cell Hybridization

The spontaneous formation of viable somatic hybrids in which the single nucleus contains genetic material contributed by two different parental cells was first demonstrated by Barski and his associates [12, 13] in mixed cultures of N1 (high cancer) and N2 (low cancer) cells. After several months large metaphase spreads were observed whose chromosome constitution was close to that predicted if one added the two parental sets together. This work did not receive general acceptance at first, because the evidence that the large cells were true hybrids and not simply polyploid cells rested mainly on cytological observations. Metaphase spreads of these large "M" cells contained fairly distinctive marker chromosomes, some characteristic of N1 type cells and others of N2 type cells. In mixed cultures in which M cells were relatively uncommon, they might have occurred simply as the result of overlapping of two independent metaphase spreads. Furthermore, heteroploid mouse cell lines are notoriously variable, despite their tendency to retain a generally fairly characteristic karyotype over long periods of time [11, 34, 69, 103]. New markers do occur in cultured cells, and doubts about the hybrid nature of the large M cells could not be wholly dispelled. In retrospect, the supporting evidence of Barski and his associates seems quite good. They isolated the presumptive hybrid cells, using either in vitro cloning or in vivo selection of cells able to grow progressively (i.e., malignant cells), starting with an inoculum containing mainly M and N2 cells. They showed that the cells had a mixed phenotype as well as karyotype [10, 13]. Confirmatory investigations were carried out by Ephrussi and Sorieul [46]. Ephrussi was among the first to recognize the tremendous potential of cell hybrids for genetic analysis, and with his associates he has pioneered in the exploitation of this method (see review, ref. [44]). However, their first study, which was again based solely on chromosome analysis, was insufficient to convince everyone that they were in fact looking at hybrid cells.

More convincing was the demonstration that the presumptive hybrid cells carried on their surfaces the *H-2* histocompatibility antigens of both parental types [53, 126] and that both parental forms of the enzyme β-glucuronidase (differing in heat stability) were present in hybrid cells that arose in mixed cultures of NCTC 2455 (N2) mouse cells of C3H/He origin, and Py-198-1, polyoma virus-transformed cells of noninbred Swiss mouse origin [51]. Hybrids also occurred spontaneously when normal diploid mouse cells were mixed with malignant N1 cells of the same species [115]. Low temperature was found to favor growth of some hybrid cells in comparison to their parent cells.

Conditional Lethals and Selective Systems

Littlefield [91] introduced the use of conditional lethals to eliminate unfused parental cells. His studies were crucial in establishing the reality of spontaneous cell hybridization. He introduced two biochemical selection systems which are still the ones most widely used to obtain hybrid lines free of parental cell contamination. He produced enzyme-deficient cell lines by exposing mouse L cells to 5-bromodeoxyuridine (BrdU) or 8-azaguanine (Aza). In normal cells each of these agents is incorporated into the nucleic acids of the cells through the mediation of an enzyme of a nucleic acid salvage pathway: BrdU by thymidine kinase (TK) and Aza by inosinic acid pyrophosphorylase (IMP, also called hypoxanthineguanine phosphoribosyl transferase, HGPRT). The incorporation of either chemical agent led to cell death, but mutant cells appeared which were able to survive in the presence of the drug because they lacked the enzyme (TK or IMP) necessary for its incorporation. Thus A9 cells were IMP deficient (IMP⁻) and B82 cells were TK deficient (TK⁻). The A9 and B82 cells, however, with a block in the salvage pathway for RNA or DNA synthesis, respectively, are dependent on *de novo* synthesis of RNA and DNA and are killed when this pathway is blocked by aminopterine, even when excess hypoxanthine and thymidine are present in the HAT medium to support the salvage pathway. The use of HAT medium, containing hypoxanthine, aminopterine and thymidine, was devised by Szybalska and Szybalski [129]. Littlefield used it to eliminate IMP⁻ and TK⁻ parental cells from a mixed culture, leaving only TK⁺/IMP⁺ hybrid cells.

Kusano et al. [87] used a third enzyme-deficient cell line (APRT⁻, adenine phosphoribosyl transferase), which they developed by growing cells in the presence of 2,6-diaminopurine or 2-fluoroadenine. The APRT⁻ cells cannot survive if alanosine is added to the medium, and hybrids can thus be easily selected. Kao et al. [75] have used nutritional mutants (auxotrophs) as tools in selecting somatic hybrid cells.

Virus-induced Cell Fusion

A major advance in the experimental utilization of cell fusion was the discovery that animal cells from *different species* would undergo cytoplasmic fusion in the presence of the Sendai strain of parainfluenza virus to produce viable heterokaryons in which nuclei of both species resided in a common cytoplasm [62]. Cell fusion by Sendai virus has been extensively explored and developed by Okada and his associates from 1957 on [104]. Virus-induced cell fusion was, in fact, a fairly well-known phenomenon, and many viruses, as well as other agents, were known to induce cytoplasmic fusion [110]. However, the generality of this phenomenon, and its revolutionary importance as a tool for genetic analysis and studies of evolution, only became clear as a result of the work of Harris and his associates [62, 63]. They pointed out many of the implications of their findings (e.g., the generality of the synthetic processes and the regulatory signals present in higher vertebrates). They also noted the tendency of the allogeneic nuclei in heterokaryons to become synchronized in terms of DNA synthesis and mitosis, and the rare occurrence of mitoses potentially capable of producing a hybrid cell line.

Spurred on by the work of Harris and his associates, Weiss and Ephrussi [147] showed that interspecific mouse/rat hybrids could be produced by growing two cell lines in mixed culture, and that such cultures were capable of long-term growth. Since then, more hybrids have been produced by the mixed culture technique, and still others by the Sendai virus technique. In general, a higher proportion of cells form viable hybrids after treatment with Sendai virus than spontaneously [28]. However, by mixing a relatively small number of cells (e.g., 10^2 to 10^4) of one type with a large number (e.g., 10^6) of another type, an "effective mating rate" of up to one in 100 cells (1%) of the rarer class has been achieved [32, 148], which is even higher than the hybridization rate of one in 500 to 1000 cells achieved by Weiss [145] with the Sendai virus technique.

Chromosome Loss from Hybrid Cells

The next major advance in cell hybridization was based on the discovery [149] that human chromosomes were rapidly lost from man/mouse hybrid cells during continued in vitro cultivation. The tendency of cell hybrids to lose chromosomes had been noted in other systems (e.g., see Marin and Littlefield, ref. [93]), and had been found to occur to a marked extent even in intraspecific mouse/mouse hybrids [43], but the selective loss of chromosomes derived from one of the two parents was something new and intriguing.

Further Developments

Several reviews dealing with the application of cell hybridization to cancer research are available [9, 44, 59, 60, 141]. In addition, a short review has recently appeared listing about 150 hybrid cell lines (but none of the multinucleate, heterokaryotic systems) used in experiments published up to 1972 [115]. Poste [108] has prepared a comprehensive review on cell hybridization and its applications, with emphasis on the mechanisms of virus-induced cell fusion. A better understanding of this subject has led to the development of a chemical means of inducing fusion, using lysolecithin. Viable hybrid lines can be produced in this way [30], perhaps reducing the risks of chromosome breakage, which is known to occur in virus-induced hybrids [96].

CLUES TO CANCER ETIOLOGY

Dulbecco [40] has discussed two contrasting theories in viral carcinogenesis. One idea is that the malignant cell has gained additional genetic material. The other is that a genetic regulatory element has been inactivated or eliminated from the cell. The provirus theory of carcinogenesis (Temin, [130]) fits the first model. The viral oncogene hypothesis (Huebner and Todaro [70], and Todaro and Huebner [134]) fits the second. Clearly, one or both of these ideas about carcinogenesis could be correct with respect to a given type of tumor, whether or not it has a viral etiology. However, the models have been developed so far in the field of tumor virology that it seems advisable to start at this point, and then to consider the role of other exogenous or endogenous factors in malignancy.

Cell Hybridization and Viral Oncogenesis

The effects of virus infection on chromosomes [58] and a possible means by which the viral DNA might be integrated into the host-cell genome [152] are discussed elsewhere in this volume as they relate to cancer. In this paper, I will emphasize the usefulness of cell hybridization in analyzing some of the problems associated with cancer.

Theories of Viral Carcinogenesis. We examine here two broad theories of viral carcinogenesis.

(1) The provirus hypothesis. Temin [130] showed by molecular hybridization that cells infected with Rous sarcoma virus (RSV) have new DNA not found in uninfected cells and that this new DNA is complementary to RNA isolated from purified RSV. He concluded that the virus acts as a carcinogen by adding new genetic information to the cell. The discovery of an RNA-dependent DNA polymerase, or reverse transcriptase, vindicated Temin's idea that replication of C-type RNA viruses requires the formation of a DNA template, or provirus (Baltimore, ref. [8]; Temin and Mizutani, ref. [131]). Several reactions take place: formation of a DNA transcript on an RNA template, replication of a complementary DNA strand to produce a double-stranded DNA provirus, and repeated transcription of this provirus to give many copies of viral RNA. These copies are translated, giving rise to virus-specific antigens, reverse transcriptase, and other enzymes, and transforming proteins that may render the cell malignant. Clearly, in this system, new genetic information has been added to the cell, and this is in some way responsible for its malignancy.

(2) The virogene–oncogene hypothesis. Huebner and Todaro [70, 134] proposed an alternative hypothesis, one they felt would account for the carcinogenic effect of such diverse nonviral agents as chemicals, radiation, and the aging process itself, while still recognizing the growing amount of evidence that malignant cells of many different types con-

tain C-type RNA virus particles. Their idea is that in vertebrates the genetic information necessary for specifying all C-type viral functions resides in the genome and can be transmitted vertically, from parent to progeny, generation after generation. However, these genes are usually not expressed because genetic regulators in other parts of the genome keep them repressed. Nevertheless, the repression can be overcome and oncogenic virus released, as already shown in mice [112] and in birds. All kinds of agents could thus lead to cancer by causing mutation or loss of a regulatory gene and activation of the viral oncogene already present in the genome. In this model, cancer is the result of loss of a genetic regulatory function rather than a gain of genetic information.

Viral Transformation of Cells. Cells in which the complete lytic cycle of virus multiplication can occur are called permissive or susceptible. Cells that allow adsorption and penetration of virus but not the complete lytic cycle are called nonpermissive. Soon after virus infection of nonpermissive cells, they may become transformed, losing contact inhibition of motility and growth. However, cell transformation is not synonymous with malignancy. For example, Kit et al. [78] infected mouse kidney cells with simian vacuolating virus 40 (SV40) and produced a transformed line. The cells, which contained the SV40 T antigen, were inoculated into BALB/c mice at the twenty-sixth and seventy-first passages. At passage 26 the cells did not produce tumors, but at passage 71 they did. Barski et al. [11] described similar observations in spontaneously transformed mouse lines.

Todaro and his associates developed a plaque assay for measuring the transformability of human diploid cells by SV40. They found a marked increase in the number of transformed colonies when the cells came from 21-trisomic individuals [135] or those with Fanconi's anemia [132]—conditions in which there is a marked increase in the incidence of malignancies, especially of leukemia. They concluded that a parallel exists between in vivo susceptibility to neoplasia and in vitro susceptibility to a tumor virus. This has been confirmed in the case of xeroderma pigmentosum [136], a disease in which there is defective DNA repair of ultraviolet-induced damage and a markedly increased incidence of skin cancers. It is thus apparent that host factors play an important role not only in susceptibility to malignancy but also in susceptibility to virus-induced transformation.

Rescue of Oncogenic Viruses by Cell Fusion. Cell fusion has been useful in showing that the complete viral genome persists in the cells of virus-induced tumors. Malignant transformation of cells by a virus is not generally associated with the production of infectious virus or cell death. Nevertheless, the viral genome persists in the transformed cells, as shown by the presence of viral antigens and by the occasional production of infectious virus spontaneously, or after treatment of the cells with 5-iododeoxyuridine [55], 5-bromodeoxyuridine [49], mitomycin C [151], or X-ray [49], or as a result of cell fusion. The reappearance of virus has been called activation, reactivation, induction, "rescue," or detection of transforming virus and, as this proliferation of terms suggests, its mechanism is unclear [141].

(1) DNA viruses. Gerber [52] found that cells derived from an SV40-induced hamster ependymoma (EPA) did not usually produce infectious virus but that virus could be recovered if the cells were mixed with permissive cells in the presence of Sendai virus. Convincing evidence that SV40 virus production occurs in heterokaryons produced by fusion of transformed cells with permissive cells has been provided by work in several laboratories. Koprowski et al. [86] used an SV40-transformed skin line, W98VaE, and African green monkey kidney cells, AGMK, which are permissive for SV40. Co-cultivation of the two cell lines did not lead to any viral release. However, when heterokaryons were produced by fusing cells of these two types with inactivated Sendai virus, infectious SV40 was produced. Watkins and Dulbecco [144] and Kit et al. [79] obtained similar results. They noted that virus production requires cytoplasmic but not nuclear fusion and that both types of nuclei showed cytological signs of viral multiplication. They concluded that the ability to produce virus is dominant over nonproduction and suggested that the permissive cell contributes a factor that is lacking in the transformed cell. This was confirmed by the appearance of viral replication, first in the TSV-5 transformed hamster cell nuclei and secondarily, after release of infectious particles into the cytoplasm, in the CV-1 African green monkey kidney indicator nuclei [151].

After fusion of transformed cells with permissive cells, only a small proportion, usually about 5%, of heterokaryons produce virus [144]. This proportion can be greatly increased by treating the virus-transformed cells with 5-iododeoxyuridine or 8-azaguanine for several hours before fusion [140]. Such antimetabolites have been used to obtain virus release from transformed cells even without virus-induced fusion. Treatments that damage DNA may tend to activate the viral genome [49], which is integrated into host DNA and covalently bound [113, 138]. Watkins [142] suggested that virus integration is maintained by a protein repressor, analogous to the situation in lysogeny of *Escherichia coli* K12 with phage. He also suggested that integrated virus becomes detached from the host DNA in a minority of SV40 virus-transformed, nonpermissive cells but that replication cannot occur because of the nonpermissiveness of the cell. In such cases, fusion with a permissive cell provides the missing metabolites essential for a lytic cycle of virus replication.

Cell hybridization has been used to show that in SV40-transformed cells the virus is integrated into the chromosomes. Weiss et al. [150] fused cells of an 8-azaguanine-resistant subline of an SV40-transformed human cell line (WI-18-VA-2) with cells of the BrdU-resistant, 3T3 mouse line. When examined after about 20 generations, the hybrid cells contained the nuclear T antigen characteristic of SV40-transformed cells, although they were estimated to contain only 5 to 10 human chromosomes. After further growth, a number of T antigen negative (T^-) sublines were obtained, and these were estimated to have only 0 to 3 human chromosomes. Weiss [145] extended these experiments in an attempt to find out whether the loss of the T antigen was correlated with the loss of a specific human chromosome. While she was able to confirm that T^- cells in general had fewer chromosomes than T^+ cells, the method of study, which was not based on a chromosome banding technique, was inadequate for a critical test of the hypothesis. On the other hand, the studies do support other lines of evidence that the SV40 genome is integrated into the chromosomal DNA of transformed human cells.

Kit and Brown [77] showed that if defective virus is used to transform cells, infectious virus could not be obtained by fusing these transformed cells with susceptible CV-1 African green monkey cells. When they transformed human cells with UV-irradiated SV40, no infectious virus was released by fusion with CV-1 cells. Similarly, only 4 of 12 fusions between CV-1 cells and human cells transformed at a low multiplicity of virus led to the release of active virus. Presumably, defective SV40 genomes were integrated in the lines yielding little virus.

Knowles et al. [85] studied two SV40-transformed cell lines that did not yield infectious virus after fusion with susceptible African green monkey kidney (AGMK) cells. When cells from their two lines (hamster F5-1 and human W98Vac) were fused with each other, especially if permissive AGMK cells were present, too, infectious virus was detected. They suggested that recombination between viral genomes of the two types, each presumably defective in a different function, may have led to the production of a normal genome, or else that complementation occurred when both classes of nuclei were present in a common cytoplasm.

Wild-type strains of the small oncogenic polyomavirus or of adenovirus have not been detected in transformed cells by fusion with permissive cells [25, 142]. However, a temperature-sensitive mutant of polyoma, ts-a, has been detected by fusion of ts-a transformed Syrian hamster kidney (BHK) cells with the permissive 3T3 mouse cells [127].

Basilico et al. [14] used cell fusion between BHK and 3T3 cells to demonstrate that the permissiveness of the hybrids depended on the maintenance of a diploid or near-diploid complement of mouse chromosomes. Such hybrids supported viral replication. This finding supports the idea that nonpermissive cells lack something essential for the full lytic cycle of virus replication and that its presence in permissive cells requires the presence of most of the genome.

Glaser and Rapp [55] fused cells of a Burkitt lymphoblastoid cell line P3J-HR-1 with the HeLa-cell derivative, D98. The hybrids thus produced did not contain detectable Epstein-Barr virus (EBV) markers; but after treatment with 5-iododeoxyuridine, the hybrid cells produced both EBV-specific antigens and virus particles. This study indicates that EBV genomes can be present for a long time without phenotypic expression and that synthesis of the virus can then be induced in heterokaryons.

(2) RNA tumor viruses. The malignant potential of the Rous sarcoma virus (RSV), which is a C-type RNA virus [18], has been established in several mammalian as well as avian species. RSV-transformed mammalian cells are generally nonpermissive, but the virus can frequently be recovered by co-cultivation of nonpermissive transformed cells with permissive chicken cells, as was first demonstrated by Svoboda and Klement in 1963 [128]. This effect, later attributed to cell fusion, can be accomplished either spontaneously or by using Sendai virus [27, 121, 137].

Using both immunofluorescent and autoradiographic techniques, Machala et al. [92] showed that in lines from which virus could be released by fusion virtually every RSV-transformed hamster cell produced infectious virus upon fusion with a permissive chicken cell. This is in marked contrast to the very small proportion of SV40-transformed cells in which virus can be detected by fusion, and probably indicates that the RSV genome is integrated into the cell in a different way than is SV40 and can be more easily released. The viral genome appears to be present in the form of a DNA provirus [8, 131], as might be expected of C-type RNA viruses that replicate by way of a DNA template. Thus Hill and Hillova [66] found infectious DNA in transformed, nonpermissive hamster cells; with this they infected permissive chicken cells and produced RSV identical to that used in transforming the cells. Since there is infectious DNA in RSV-transformed, nonpermissive cells, the nonpermissiveness of these cells is presumably due to a transcriptional block in the synthesis of viral RNA or a translational block in the synthesis of viral proteins. This is supported by the finding, using metabolic inhibitors, that RNA and protein synthesis are required for virus production following fusion of transformed cells with permissive cells, while DNA synthesis is not [137].

Hybrid cells have been used to investigate the genetic regulation of C-type virus production. Fenyö and her associates [47] fused A9 mouse fibroblastic cells with mouse cells of an immunosensitive Moloney lymphoma (YAC) and with cells of an immunoresistant subline (YACIR). The virus assay was based on the induction of membrane immunofluorescence in negative indicator cells (JLS-V9) by culture supernatants. Both YAC and A9 cells released infectious virus, based on the appearance of Moloney virus-induced cellular antigen on the surface of the indicator cells, but YACIR cells did not. The YAC-A9 hybrid cells released infectious virus in amounts comparable to the parent cells; but no infectious virus could be detected in the medium of the YACIR-A9 cells, even though these cells had just as much of the specific, virus-induced, cell–surface antigen as the YAC-A9 or YAC cells. Thus the production of the specific surface antigen and the production of infectious virus were not invariably coupled. Furthermore, since the A9 cells released the C-type virus particles that are detected by the assay system used, while the YACIR-A9 cells did not, the mechanisms leading to suppression of the release of such infectious particles in the YACIR cells appeared to be dominant in the YACIR-A9 hybrid, and to act at some point in the viral cycle subsequent to the formation of the virus-specific cell surface antigen.

Cell hybridization has been used also to analyze the nature of the nonpermissiveness of human cells for mouse leukemia virus (MLV) [132]. Fusion of human cells with permissive mouse cells produced a nonpermissive hybrid. However, five hybrid clones derived from a human KL × mouse 3T3-4E (TK$^-$) cross were permissive for MLV synthesis. All five clones had lost many of their human chromosomes, and it seems likely that the nonpermissiveness of human cells for MLV may be due to a function specified by genes on one or more chromosomes, that is dominant in the hybrid.

Cell fusion can clarify some of the other complexities in the area of viral oncology. Sarma et al. [114] observed that most strains of avian and murine sarcoma virus (MSV) are defective (i.e., unable to produce their viral envelope without help from a closely related, nondefective, helper leukemia virus). They obtained a rescue of the defective MSV by cocultivation of hamster tumor cells infected with defective MSV with cat cells infected with a helper feline leukemia virus (FeLV), and by inoculation of the tumor cells into FeLV-infected newborn cats. Cell fusion was presumably involved here, though not demonstrated.

Detection of Unknown Tumor Viruses. Cell fusion of malignant and normal cells may be useful for detection of a hitherto undetected tumor virus integrated in some manner into the genome of a malignant cell [86]. A possible example, involving bovine lymphosarcoma, has been reported

[29]. Watkins [141] discussed the possible application of cell fusion to the search for human tumor viruses and pointed out a number of serious pitfalls in this system. The major problem is finding a cell type that is permissive for the putative tumor virus. If the etiologic agent is a DNA virus similar to SV40, and if man is the natural host, then human cells might all be permissive for the virus. If so, then any human line transformed by this virus would, by analogy with permissive SV40-transformed cells, be likely to contain defective virus that could only be detected by triple fusion experiments of the type described earlier. If human cells are not permissive, then the natural host would have to be sought among other species (e.g., a domestic animal). On the other hand, if the virus in question is a C-type virus, for which the Rous sarcoma virus could serve as a model, similar problems arise. A more general difficulty is that of determining from which parental cell any virus detected after cell fusion has come.

Cell Hybridization and Chemical Carcinogenesis

There is growing evidence that chemical carcinogens can lead to cell transformation quite similar to that induced by oncogenic viruses. For example, the polycyclic hydrocarbons benzo(α)pyrene and 3-methylcholanthrene can transform hamster cells in vitro, and the transforming activity can be obtained free of the usual toxic effect of these compounds if the induction of aryl hydrocarbon hydroxylase, which converts these agents to more toxic derivatives, is blocked [39]. The similarity in the effects of oncogenic viruses and polycyclic hydrocarbons, both of which can produce chromosome breaks, cellular transformation, and malignancy, led Allison and Mallucci [3, 4] to suggest that the lysosomes are the primary target of carcinogens and that the activation and release of lysosomal enzymes are responsible for the progressive changes just described. They found that hydrocarbon carcinogens were concentrated in lysosomes and that freshly virus-infected monkey, chick, and human cells showed activation of lysosomal enzymes. If their hypothesis is correct, the importance of evaluating such factors as chromosome loss and defective DNA mechanisms becomes obvious.

Cell hybridization has thus far contributed very little to the understanding of chemical carcinogenesis. However, the studies of Harris, Klein, and their associates [61, 82, 156], to be described in the next section, clearly showed that the malignancy of a methylcholanthrene-induced ascites sarcoma, MSWBS, is a reversible characteristic, since it was suppressed by cell hybridization and reappeared after (presumably) specific chromosome loss.

One can envision the use of cell hybridization to answer many of the unsolved problems regarding chemical carcinogenesis—for example, in defining more exactly the series of progressive changes cells undergo from the time of exposure to the chemical to the appearance of frank cancer. Cell hybridization may also permit tests of specific hypotheses of chemical carcinogenesis, and thus speed the development of a fuller understanding of this important area.

Effects of Cell Hybridization on Malignancy

Early Evidence that Malignancy May Be a Dominant Character. In 1961 Barski and his associates [13], using cultured cells in vitro, discovered spontaneous formation of hybrid cells and reported their use for the genetic analysis of malignancy. They mixed cells of the highly malignant (high tumor) mouse line NCTC 2472 (or N1) with cells of the very much less malignant (low tumor) mouse line NCTC 2555 (or N2). When these two types of cells were grown together for several months, a new M type of cell appeared which was larger in size and shared karyological, morphological, and behavioral characteristics with both N1 and N2 cells, indicating its hybrid nature. The tumor-forming ability of four clones of these hybrid M cells in each case resembled that of the more malignant of the parents.

Barski and Cornefert [10] extended these studies by making further clones of the hybrid cells. Fourteen of 15 clones were highly malignant, producing tumors as quickly and with as small a cell inoculum as with N1 (i.e., about 10^4 cells). They repeated the process, mixing N1 and N2 cells, and again observed the appearance of large M cells with many characteristics of each parental cell type and an almost additive chromosome constitution. After 78 days in mixed culture, the mixed-cell population was cloned. Five pure cell lines were established. All were of the hybrid M cell type, and each was just as malignant as the N1 parent.

Scaletta and Ephrussi [115] showed that hybrid cells also arose spontaneously when the highly malignant N1 cells were grown in mixed culture with normal CBA mouse fibroblasts carrying the T6 chromosome marker. These hybrids, which contained either one or two sets of N1 chromosomes to one set of CBA, were also malignant, producing tumors in at least two thirds of animals receiving 10^6 cells. Studies such as these led to the conclusion that malignancy is a dominant trait in hybrid cells.

The view was further supported by the observation that the spontaneous hybrids that arose in mixed cultures of polyoma-transformed malignant mouse cells and nonmalignant mouse cells were highly malignant [36]. These hybrids also showed another characteristic of polyoma virus-transformed cells, production of a T antigen [35], which also behaved as a dominant trait. However, this would be expected, since its appearance was due to the addition of new genetic information, the viral genome, to the cell. It came as no surprise, then, when another virus-induced characteristic of one parental cell type, malignancy, was also expressed in the hybrid.

Later Evidence that Malignancy Is a Graded Character. The idea that malignancy usually acts as a dominant trait gained such widespread acceptance that exceptions to this generalization tended to be overlooked. Although most of the early studies involving hybrid cells used systems in which malignancy could be treated as an all-or-none affair, of course, malignancy is not an all-or-none trait, but rather a graded characteristic, with all grades from precancerous to highly malignant (i.e., from benign cell hyperplasia to tumors capable of metastasis).

In 1967 Defendi et al. [36] reported a malignant hybrid derived from two almost nonmalignant parents, and the same year Silagi [122] derived relatively nonmalignant hybrid clones from spontaneous fusion of mouse melanoma cells and mouse A9 cells, an 8-azaguanine-resistant, inosinic acid pyrophosphorylase-deficient line. In the latter case, tumors appeared in 12/18 mice inoculated with the parental melanoma cells, in 2/20 mice inoculated with the A9 cells, and in 6/10, 2/8, 4/9, 0/11, 1/10, and 5/10 mice inoculated with six different melanoma/A9 hybrid clones.

Murayama and Okada [99], on the other hand, found a different situation in one type of hybrid. They fused the Lettré subline of Ehrlich ascites tumor cells (ETC) with a nonmalignant, 8-azaguanine-resistant subline of the L cells (LAG). The number of cells required to produce tumors in 50% of the hosts was only 10^2 for the malignant ETC parental cells but was 10^6 for the hybrid cells. Thus, in terms of the 50% tumor dose, the hybrid cells are only about 10^{-4} as malignant as the malignant parent. This intermediate tumor-forming capacity of the hybrid cells was a stable trait during alternative in vivo and in vitro passages, and Murayama and Okada concluded that all the hybrid cells had an intermediate tumor-forming capacity. In their system, malignancy acted as a graded character, neither dominant nor recessive, but showing additive genetic influences.

Evidence for Suppression of Malignancy. Harris and his associates [59–61], in work to be described, also obtained evidence that malignancy is not simply a dominant trait. In contrast to Murayama and Okada's results, however, their findings are better explained by the idea that the greatly increased number of hybrid cells one must inoculate to obtain tumors is due to the complete suppression of malignancy in most of the hybrid cells but that instability of the chromosome complement generates a small population of cells that can be just as malignant as the more malignant parental cell type. Whether the malignancy of the hybrid cells is at some intermediate level or at a high or low level, the role of specific genetic factors may be clarified by correlating changes in the genetic makeup of hybrid and parental lines with a shift in the degree of malignancy. This may lead to an understanding of the regulatory functions whose loss can lead to the development of cancer. Ephrussi et al. [45] also observed the suppression of malignancy by somatic cell hybridization. They produced hybrids between A9 cells and cells of a malignant teratoma arising in a strain 129 mouse. Injection of more than 10^6 cells into each of nine C3H × 129 F1 mice produced no tumors.

If the neoplastic transformation of cells by a virus is due to the permanent gain of viral genetic material, then fusion with a normal cell should produce a hybrid that is also malignant. This is generally not the case, despite early reports to the contrary [36]. SEWA is a highly malignant polyoma virus-induced sarcoma that arose in a mouse of the A.SW strain. It bears the specific polyoma transplantation antigen, PTA. Hybrids of SEWA and A9 cells are far less malignant than the SEWA parent [59, 61]; so, too, are the majority of hybrids between SEWA and diploid mouse CBA/T6T6 cells that have retained approximately a full diploid complement of CBA chromosomes. Comparable results have been observed in hybrids of the Moloney virus-induced lymphoma, YACIR, with another highly malignant cell line, MSWBS, a methylcholanthrene-induced sarcoma; that is, the malignant phenotype has been suppressed by fusion of one malignant tumor with another [59]. In all these cases, the virus-specific antigens were expressed in the hybrid, indicating the continued presence of the viral genome despite the lack of expression of the malignant characteristics. Thus, the addition of the viral genetic material to the cell is not enough to assure its malignancy. It has been suggested that oncogenic viruses may not play a *direct* role in the production of malignancy but only an *indirect* one by producing a heritable perturbation in cell regulation [40], perhaps by increasing instability in the chromosome complement [60].

Chromosome Loss and Malignancy

Studies in the Mouse. Barski and Cornefert [10] carried out chromosome analyses on many of their hybrid lines. The results indicated that the large M cells were indeed hybrids. They contained marker chromosomes characteristic of each parent and a modal number of chromosomes (varying from 100 to 115 in different lines) only slightly less than expected from the fusion of an average N1 (modal number 55) with an average N2 (modal number 62) cell. The method of selection of the hybrid cells used for extensive chromosome study involved inoculation of mixed cultures into animals. Thus, only hybrid cells capable of progressive growth in vivo were selected. These were cloned after two months, or 12 passages, and the clones were analyzed. Their modal chromosome numbers varied from 100 to 116. It was not possible, of course, to determine the parental origin of more than a handful of the chromosomes: the single, long telocentric marker in N1 and the 9 to 15 metacentric markers in N2. Nor was it possible, given the in vivo method of selection of the hybrids, to compare the chromosomal makeup of the hybrid cells that arose in vitro with that of the malignant cells derived from them.

Silagi [122] did make such a comparison between a melanoma/A9 hybrid clone and cells of the tumor that arose following the injection of the hybrid cells into histocompatible mice. She found that the tumor cells had, on the average, one chromosome less than the hybrid cells from which they arose, but the methods then available were not adequate to identify this chromosome.

The importance of chromosome loss for the expression of malignancy in hybrid cells produced by fusion of highly malignant cells with nonmalignant cells has been stressed by Harris and his associates, who have demonstrated convincingly that malignancy can behave as a recessive trait in hybrid cells. Initially we studied hybrids obtained by fusing A9 cells with the Klein near-tetraploid variant of Ehrlich ascites tumor cells [61]. The Ehrlich tumor was derived from a mammary carcinoma many years ago. It was highly malignant, and injection of even a few Ehrlich cells will produce a tumor in any strain of mouse. The line used had a modal number of about 75 chromosomes. The A9 cell line [91] is an 8-azaguanine-resistant derivative of the mouse L cell line with a modal number of about 57 chromosomes and a limited ability for progressive growth in vivo; only 12% of X-irradiated newborn syngeneic mice developed tumors after inoculation of from 5×10^4 to 10^6 A9 cells [82].

Harris and his associates found that Ehrlich/A9 hybrid cells were no more malignant than cells of the A9 parent; that is, the high degree of malignancy of the Ehrlich ascites tumor parent was *suppressed* in the hybrid. Occasionally, however, a tumor did arise in an animal inoculated with 10^5 to 10^6 hybrid cells, and such tumors were subsequently highly malignant. Chromosome analysis of five Ehrlich/A9 hybrid cell tumors showed them to have a much reduced modal chromosome number, 81 to 89, in comparison to the parental hybrid cell population, 128. This suggested that the hybrid cell population is generally nonmalignant but that malignant variants are generated as a result of chromosome loss. Harris, Klein, and their associates showed that it was not simply an overall reduction in chromosome number that is important, because prolonged growth of Ehrlich/A9 cells in vitro led to a comparable reduction in chromosome number but no increase in malignancy. They concluded, therefore, that the generation of malignant variants in the nonmalignant hybrid cell population requires the loss of *specific* chromosome(s) and not just an overall reduction in chromosome number [82].

When Ehrlich tumor cells were fused with freshly cultured, diploid, embryonic fibroblasts from CBA mice homozygous for the T6 translocation marker, the hybrid cells usually produced tumors if 10^6 cells were inoculated [20, 26]. Here too, however, chromosome loss may have been responsible for the malignancy of those hybrids, for they showed marked instability of the chromosome complement. By the time chromosome analysis could be carried out, substantial chromosome losses had already occurred in all the hybrids, and further chromosome loss was frequently noted on comparing the tumors with the hybrid cells injected. Selection of the cells in vivo is clearly involved in the production of tumors, even when some reduction in chromosome numbers had already occurred in vitro.

When freshly isolated, diploid embryonic fibroblasts (CBA/T6T6) were fused with cells of the polyoma-induced SEWA tumor line, most of the hybrids showed the same substantial loss of chromosomes by the time enough cells had been generated to permit chromosome analysis. Twelve clonal SEWA/CBA T6 hybrid lines were analyzed for both transplantability and for distribution of chromosome numbers. Three of five clones with chromosome numbers approximating the sum of the two parental cell types showed reduced take incidences (i.e., incidence of tumor appearance; transplantability) as compared with the 90 to 100% takes following inoculation of the same numbers of cells (1.4×10^4 to 3.6×10^6) from each of seven clones showing loss of chromosomes or from the hybrid mass culture from which they were derived. The 40 or 50 tumors produced by injection of SEWA/CBA T6 hybrid cells were dominated by cells containing reduced numbers of chromosomes, and no tumor was composed of cells whose chromosome complements corresponded to the sum of the two parental chromosome sets [155].

Further evidence of the importance of chromosome loss was gained by a study of hybrids produced by fusing normal diploid ACA strain fibroblasts with TA3 tumor cells, which are virtually euploid and diploid. One of these hybrids was found to contain considerably more chromosomes (117) than the sum of the two parental sets (80); it was apparently derived by fusion of two ACA cells and one TA3 cell. Two clones derived from this hybrid contained similar high numbers of chromosomes, modes 112 and 114, and all three hybrid lines showed a reduced take incidence after inoculation of 1.2×10^4 to 3.6×10^6 cells. The tumors that arose had fewer chromosomes, with modes usually in the range of 80 to 90 and none greater than 103 [155].

From such studies, Harris and his associates concluded that the malignancy of various tumor cells can be suppressed by diploid fibroblasts, as well as by heteroploid derivatives of L cells, but the rapid chromosome losses sustained by hybrids between tumor cells and diploid fibroblasts leads to a greater rate of production of malignant segregants.

The findings of Belehradek and Barski [15] also indicate that there must be chromosome loss before malignancy can be expressed in hybrid cells resulting from fusion of malignant and nonmalignant cells. They fused cells of an 8-azaguanine-resistant, malignant subline of the C3H-derived NCTC 2472 line (N1/Aza) with cells of a nonmalignant line derived from BALB/c embryos (EBA-N). The hybrid cells (HyEN) had a highly variable number of chromosomes with a modal number of 116, whereas the N1/Aza parent had a mode of 47 and the EBA-N parent 74 to 75. The hybrid line was cloned in its twelfth passage, and the malignancy of six clones checked. One was not malignant and still had about the same number of chromosomes, 112 to 118. Two of the malignant clones had greatly reduced modal numbers, 98 and 79, while three did not. However, the tumors that arose from two of these three showed marked reductions in chromosome number, and so did the tumor that arose from one of the malignant lines with an already reduced chromosome number.

The major supporting evidence for the role of chromosome loss in the return of malignancy has come from studies of hybrids between A9 cells and cells of various other, highly malignant, mouse tumor lines (viz., SEWA, MSWBS, YAC, and YACIR [61, 82]). The SEWA tumor is a polyoma virus-induced ascites sarcoma that arose in an A.SW mouse. It carries the polyoma-specific transplantation antigen in addition to the H-2^s antigen complex and grows only in A.SW strain mice; it has a modal chromosome number of 43. The MSWBS tumor is a methylcholanthrene-induced ascites sarcoma whose growth is also restricted to A.SW mice; it has a modal number of only 28 or 29 chromosomes. The YAC tumor is a Moloney virus-induced ascites lymphoma derived from an A/Sn mouse; it has an approximately diploid chromosome complement. The immunoresistant YACIR tumor was derived from YAC by serial passage in strain A mice immunized against the surface antigen associated with Moloney virus. YACIR cells have a much lower concentration of this surface antigen and are resistant to the cytotoxic action of anti-Moloney antisera; their modal chromosome number is about 41.

The hybrids between A9 cells and cells of the highly malignant mouse tumor cells just described were hardly any more malignant than A9 cells, showing take incidences of only 5 to 40% with inocula of 3×10^4 to 4×10^6 cells, whereas the malignant parental cell lines themselves produced 100% take incidence in syngeneic mice with inocula of even 100 cells [61, 82]. In terms of the 50% tumor dose, the hybrid cells are thus only about 10^{-5} as malignant as the parental tumor cells. Chromosome studies on the hybrid cells and on the tumors derived from them indicate that the tumors arose by selective overgrowth of cells from which certain specific, but as yet unidentified, chromosomes were eliminated rather than by progressive growth of the hybrid cell population as a whole. That is, the A9 cells appear to have contributed a factor to the hybrid cells that suppressed their malignancy, and the loss of certain chromosomes appears to have abolished this suppression.

Further support for this idea came from analysis of hybrids between some of the same highly malignant cell lines (Ehrlich, SEWA, MSWBS, and YACIR) and a malignant derivative of A9 called A9HT. In every case, the hybrids were highly malignant. The number of chromosomes in the tumors was not different from that seen in the hybrids produced in vitro, indicating that virtually all the hybrid cells were malignant [156]. The A9HT cells appeared to have lost a factor, present in A9 cells, that suppressed the ability of the cells to grow progressively in vivo; this is presumably the same factor whose loss from Ehrlich/A9 and similar hybrids is associated with a return of malignancy.

If the ability of the A9 cell to suppress malignancy is due to a factor specified by a gene carried by a specific chromosome whose loss endows the cells with a more malignant phenotype, then karyotypic comparisons of A9 and A9HT, using the much more accurate chromosome-banding techniques now available, might provide evidence of this in spite of the heterogeneity of these L-cell derivatives. The first stage of this attempt to identify specific chromosome differences between A9 and A9HT lines was recently reported [2]. Using fluorescence microscopy of quinacrine-stained chromosomes, we showed that virtually every chromosome in these lines has a characteristic banding pattern. The origin of at least two-thirds of the chromosomes could be specified, and the remainder of the chromosomes could be identified and assigned marker numbers so that the mean number of copies per cell of every chromosome in each cell line could be compared. Since only a single A9 and a single A9HT cell line were included in this initial study, it was not possible to distinguish important from fortuitous differences in the chromosome makeup. Examination of additional A9 and A9HT lines might reveal whether there are any significant chromosome differences between highly malignant and much less malignant cell populations. Unpublished data on two more A9 lines and six A9HT lines derived from them show that there may be one consistent difference: loss of a chromosome No. 12, so that all A9HT lines have a single copy of this chromosome instead of the two usually found in the A9 lines (Miller et al., in preparation, 1973).

The hypothesis that "the fundamental lesions responsible for malignancy are indeed some form of genetic or epigenetic loss" [60] has a number of testable consequences:

(1) Hybrids between a malignant and a nonmalignant cell type should be nonmalignant unless chromosome loss or gene mutation occurs; in heteroploid lines, the former is much more frequent and the latter can perhaps be disregarded. It is interesting in this regard that the original malignant, hybrid M cells studied by Barski and Cornefert [10] had 2 to 17 chromosomes fewer than the sum of the modal number of chromosomes in the N1 and N2 parent lines. It is unnecessary, therefore, to accept this as a case in which malignancy acted as a dominant trait, as first suggested. In most of the subsequent studies, the hybrids have remained nonmalignant until chromosome loss has occurred.

(2) Hybrids between two malignant cell types can be malignant, assuming the two types of cells have undergone the same genetic or epigenetic functional loss. This appears to be the case in hybrids between A9HT and Ehrlich, SEWA, MSWBS, or YACIR cells [156] and in the MBA/YACIR hybrids [60].

(3) Hybrids between two malignant cell types can be nonmalignant if the two types of cells have undergone different functional losses. One example is known. MSWBS/YACIR hybrids have a chromosome constitution approximating the sum of the two parental chromosome sets [156]; but they gave substantially lower take incidences than the highly malignant parent cells, with some of the hybrid clones being no more malignant than the comparable hybrids between either parent tumor cell and the A9 cells. The different behavior of MBA and MSWBS when hybridized with YACIR cells suggests that the basic lesion responsible for malignancy may not be identical in the two sarcomas, even though both were induced by methylcholanthrene [156].

Exceptions to the Chromosome Loss Model. Murayama-Okabayashi et al. [100] used a sequential double-hybridization technique to prepare hybrids with varying numbers of Ehrlich (E) and L cells. The hybrids showed a regular downward progression in 50% tumor-forming dose (TFD_{50}), with 2E/1L, E/L, and 1E/2L each showing almost a step-wise, tenfold difference in TFD_{50} from each other. They concluded that the effect of L cells on the tumor-forming capacity of the hybrids is not fully described by calling it "suppression of the malignancy of Ehrlich cells by L cells" but that their effect must be understood as the sum of several unspecified factors. They also produced hybrids between Ehrlich cells and diploid embryonic mouse fibroblasts. These hybrids were not as malignant as the Ehrlich cells but were more malignant than the E/L hybrids, providing further support for the conclusion that malignancy can be a graded character. However, the chromosome studies they reported indicate that chromosome loss is a general characteristic of the malignant hybrids. Their results, by revealing a greater degree of complexity in the system than most studies have shown, enrich the hypothesis that malignancy is the result of specific genetic or epigenetic losses and lend further support to the idea that cell hybridization is a useful tool in discovering the nature of such loss.

Most of the studies on malignancy of cell hybrids have been carried out in the mouse. Berebbi and Barski [16] have extended this work to Chinese hamsters. They produced hybrids of the nearly diploid (modal chromosome number, 21) and malignant DC3F line and the isologous nonmalignant variant DC3F/AD/Aza (modal chromosome number, 22), which is resistant to both actinomycin D and 8-azaguanine. The hybrid cells had 42 to 44 chromosomes, including markers from the DC3F/AD/Aza parent, but were as malignant as the DC3F parent. The tumors produced by the hybrid cell had a modal number of 44 chromosomes and thus no apparent loss of chromosomes. This system may prove to be an exception to the generalization that malignancy is the result of a genetic loss. On the other hand, chromosome-banding studies have revealed a large amount of variation in Chinese hamster lines [76]. Such an analysis in the present case might reveal a specific chromosome loss despite the unchanged total number of chromosomes.

Extension to Man. Yoshida [159] used the technique of cell hybridization to study the relationship of tumorigenicity to karyotypic profiles in human tumors, using man/mouse hybrids. He fused cells of the human diploid line, B46M, derived from a Burkitt lymphoma, with cells of the mouse aneuploid line, LM(TK)⁻ Cl-1-D. B46M formed tumors on heterotransplantation into 8/10 cortisone-treated newborn mice, while Cl-1-D cells did not. Two clones of the hybrid cells, Cl-7 and Cl-12, were transplantable at the tenth and twelfth passages, with somewhat reduced take incidences (6/18 and 5/20). Each clone had retained about the same 15 to 20 human chromosomes, whose identity was checked by quinacrine fluorescence; they included chromosomes Nos. 3, 6–8, 10, 12–14, and 16–22. At the thirty-first and thirty-second passages, the tumor incidence was further reduced (1/15 in Cl-7 and 1/13 in Cl-12), and the tumor nodules that did appear regressed within two weeks. Only a few human chromosomes remained at this stage in the development of the hybrid lines.

It is clear that the Cl-1-D cells did not suppress the malignancy of the human cells with which they were fused. The loss of malignancy was associated with the loss of human chromosomes. However, further studies are needed to clarify this rather complex but exciting situation. Can a single chromosome from a malignant human cell endow a nonmalignant mouse cell with the ability to grow progressively in a suitable host? This approach offers a means of correlating chromosomes with cancer in a most interesting way, one that holds promise of leading to a better understanding of the molecular mechanisms underlying malignancy. At a rather superficial level, one might suggest that more rapid in vivo growth is possible if the defect in the salvage-pathway for DNA synthesis in the LM(TK⁻) Cl-1-D cells is corrected [*111*] by having human chromosome No. 17 present [*96*]. However, Wiener et al. [*154*] described highly malignant B82HT cells that were deficient in the same enzyme, thymidine kinase.

Effects of Endogenous Factors

Defective DNA Repair. Xeroderma pigmentosum (XP) is an autosomal recessive disorder in man characterized by increased sensitivity to ultraviolet light. Exposure to sunlight leads to severe damage to the skin, which usually progresses to basal or squamous cell carcinomas or other malignant tumors. After exposure of cultured XP fibroblasts to UV light, unscheduled DNA synthesis or repair replication occurs at a much reduced rate or not at all, apparently because XP cells are unable to excise UV-induced thymidine dimers effectively [*117*].

The cause of the oncogenicity of UV light in patients with xeroderma pigmentosum is unclear, although the inability to excise pyrimidine dimers from the DNA of exposed cells is probably involved in some way. Viruses may play a role. Cultured cells from one XP individual showed a thirteenfold increase in the number of transformed clones induced by SV40 virus in comparison to cultured cells from his heterozygous mother [*136*].

Cell hybridization has provided some information about the defect in XP cells. When XP cells were growing close to normal cells in mixed cultures, they showed no repair replication after UV-exposure, nor did they when treated with extracts of normal cells. In heterokaryons resulting from fusion of XP cells with normal cells induced by Sendai virus, however, XP nuclei showed repair replication, even when RNA synthesis was blocked by Daunomycin [*54*]. Clearly, the repair enzyme must have diffused from the normal nucleus to the XP nucleus. Hybridization of XP cells with golden hamster cells produced hybrids with the same resistance to UV-irradiation as the hamster cells, indicating that the hamster enzymes can correct the specific defect in the XP cells [56].

Clinically, two forms of xeroderma pigmentosum have been distinguished: the classical form, and the more severe de Sanctis-Cacchione variant. Cell fusion has been used to show that the genetic defect is different in the two forms. That is, functional complementation, with normal repair replication, was observed in heterokaryons containing nuclei of any of three classical XP cases and nuclei of a de Sanctis-Cacchione variant [38].

Immunological Factors. Five immunological factors should be mentioned here.

(1) Antigenic expression in hybrids. Various studies have shown that hybrid cells usually carry the surface antigens of both parents. Spencer et al. [126] examined N1/N2 (=M) and N2/Py198 (a polyoma-transformed, Swiss mouse cell line) hybrids and found co-dominance of several H-2 antigenic components and of two other parental antigens. Similar results were reported by Gershon and Sachs [53] in a hybrid between L cells (of C3H origin) and a Swiss mouse tumor cell line, by Scaletta and Ephrussi [115] in hybrids between diploid CBA and N1 mouse cells, and by Silagi [122] in hybrids between A9 cells (L cell origin) and C57BL melanoma cells. Klein, Harris, and their associates have carried out extensive experiments indicating that the mechanisms regulating the expression of histocompatibility and virus-induced antigens are not closely linked to those regulating the ability of cells to grow progressively in syngeneic hosts [57, 61, 81–83].

(2) Antigen suppression in hybrids. Histocompatibility antigens are suppressed, or masked, in Ehrlich ascites tumor cells. When these cells are fused with A9 cells, which carry $H-2^k$ and virus-induced antigens, both classes of antigens are suppressed (i.e., they can be detected with difficulty if at all [61, 81]). The $H-2^k$ antigen complex is strongly expressed in some of the malignant segregants that sometimes arise when the generally nonmalignant, hybrid cells are inoculated into X-irradiated newborn syngeneic mice. Since these segregants always have a reduced number of chromosomes, it seems likely that the mechanism of antigen suppression in Ehrlich cells and Ehrlich hybrids is the presence of a genetic regulatory locus. Consequently, one would expect suppression to be a dominant trait in Ehrlich hybrids, and it is.

Suppression of antigens has also been observed in other tumor cells. For example, the immunoresistant YACIR subline of the murine leukemia virus (MLV)-induced YAC lymphoma shows very weak expression of the MLV antigen. When YACIR cells are hybridized with A9 cells, full expression of the MLV antigen is reestablished [47]. Interestingly, the H-2 antigens are not suppressed in YACIR cells, suggesting the presence of a specific defect in the mechanism for expression of viral-specific surface antigen. Because of this defect, which is corrected when YACIR cells are fused with A9 cells, antigen suppression acts as a recessive trait.

Antigen suppression also appears as a recessive trait in the TA3/Ha ascites sarcoma sublime of strain A (H-2^a) mammary carcinoma. A related tumor subline, TA3/St, has a high concentration of H-2^a antigens and is transplantable only to a syngeneic strain. The TA3/Ha cells, on the other hand, have a markedly reduced amount of H-2^a antigens and are transplantable across histocompatibility barriers. Fusion with normal cells corrected the defect in antigen expression characteristic of TA3/Ha cells. Nine hybrid clones produced by fusion of the immunoresistant TA3/Ha cells with diploid ACA (H-2^f) fibroblasts showed full expression of H-2^a antigens, and eight of the nine showed full expression of the H-2^f antigens [83].

Cell hybridization has thus made it possible to demonstrate the existence of two different kinds of regulation of antigenic expression. One might ask whether it might also be of use in determining whether either of these mechanisms is involved in the regulation of malignancy.

(3) Histocompatibility antigens and malignancy. Cell hybridization has been instrumental in showing that known histocompatibility factors are not responsible for the suppression of malignancy when malignant cells are fused with nonmalignant or less malignant cells, when their ability for progressive growth is tested in X-irradiated syngeneic newborn hosts. For example, Ehrlich/A9 hybrid cells have a very low take incidence despite the suppression of H-2^k antigens in the hybrid cells [61, 81]. Furthermore, the H-2^k antigens are very strongly expressed in some of the malignant segregants with reduced chromosome numbers that arose from these hybrids, providing further support for the conclusion that the ability of the hybrid cells to grow progressively in X-irradiated newborn syngeneic mice is not correlated with the suppression of histocompatibility antigens.

(4) Virus-induced antigens and malignancy. When tumors are induced by virus, the tumor cells express new, virus-induced, tumor-specific transplantation antigen(s) (the TSTA), which are presumably specified directly by the viral genome, or are directly controlled by the small number of viral proteins codable by the very small viral genome. Production of these antigens can lead to immunological reaction and rejection of tumor cells in autologous, syngeneic, and allogeneic hosts, unless the hosts are immunologically incompetent, X-irradiated newborns [124]. For example, tumors induced in different species by polyoma virus carry such identical, or at least highly cross-reactive antigen(s), the polyoma virus-induced

transplantation antigen (PTA). PTA is very stable, in that no PTA-negative variants have been obtained by serial propagation of polyoma-induced tumor cells in preimmunized, syngeneic mice for more than 40 passages [123]. Some surface antigens are much less stable under the same kind of selection, raising the question of whether the membrane changes associated with the appearance of PTA are essential for the malignancy of polyoma-induced tumors.

This question has been answered by cell hybridization. Klein and Harris [84] studied a series of hybrids prepared by fusing cells of the polyoma-induced SEWA ascites sarcoma with freshly isolated diploid mouse CBA fibroblasts. The parental SEWA tumor gave a 100% take incidence when no more than 100 cells were inoculated into syngeneic mice. Three of twelve independently derived SEWA/CBA hybrid clones gave a low take incidence, even when more than 10^6 hybrid cells were inoculated into newborn, X-irradiated syngeneic A.SW × CBA F1 mice. The inoculated mice which failed to develop tumors were allowed to grow to adulthood, and 20 of them were challenged with 10^5 SEWA cells. Ten of the 20 failed to develop tumors, presumably because the hybrid cells had been immunogenic in the syngeneic mice due to the presence of the polyoma-induced transplantation antigen (PTA).

The persistence of PTA in hybrid cells that have a markedly reduced degree of malignancy indicates that the presence of this antigen and of at least part of the viral genome are not sufficient to render a cell malignant. The hybrid cell, as discussed above, regains its ability to grow progressively in vivo only after the loss of the genetic determinants contributed by the normal cell which enable the hybrid cell to respond to growth regulating forces [61, 82, 155].

A9 cells carry two presumptive virus-induced antigens: the L cell antigen, which is assumed to be associated with type-C virus particles, and a surface FMR antigen that can be induced by Friend, Moloney, or Rauscher leukemia viruses. The expression of these antigens is suppressed when A9 cells are hybridized with Ehrlich cells [61, 81]. The hybrid cells have a low degree of malignancy despite the suppression of these antigens and of the H-2^k surface antigens. Furthermore, the virus-induced antigens are expressed independently of each other or of H-2 antigens in the tumors with lower chromosome numbers that arose from these hybrids [57] Clearly, there is no close correspondence between the expression (or suppression) of any of these antigens and the malignancy of the cell.

(5) Weak antigens and malignancy. The studies described thus far clearly indicate that the mechanisms determining the ability of the hybrid cells to grow progressively in vivo (i.e., their malignancy) are independent of those that determine the expression of either histocompatibility or

virus-induced antigens, which are carried by different chromosomes [81]. That is, known immunological factors cannot account for the suppression of malignancy in hybrid cells or for the differences in tumorigenicity between one hybrid cell population and another when tested in X-irradiated, newborn syngeneic hosts [60].

There is no evidence that other, weaker antigens play a role, either. On the other hand, it is interesting that cell hybridization is such a powerful tool that it can be used to show that some tumor cells (e.g., Ehrlich) carry weak antigens of unknown origin [26, 143]. When hybrid cells were injected into hosts across a strong histocompatibility barrier, the hosts mounted an immunological reaction against the hybrid cells. This reaction was directed not only against the strong histocompatibility or species antigens but even against weak antigens of unknown origin in the unfused tumor cells. This was shown by challenging the immune animals with Ehrlich tumor cells themselves and observing their failure to grow progressively.

The Cell Membrane. Several lines of evidence suggest that changes at the cell surface are implicated in malignant transformation. Transformed cells have an altered ultrastructure of the cell periphery, with a decreased number of microfilaments and microvilli, and the presence of bulging pseudopodia [95]. In normal cells the α-filaments are intimately associated with regions of cell-to-cell attachment and may play a role in their development. The loss of these microfilaments may be associated with the lack of normal intercellular connections between transformed cells, described by Azarnia and Lowenstein [6].

Transformed cells are capable of taking up nutrients from the medium at considerably greater rates than their normal counterparts [73], and they no longer show the same density-dependent growth regulation, or contact inhibition of growth, exhibited by normal cells. Various changes in the cell membrane have been demonstrated in transformed cells. For example, virus-transformed mouse cell lines have a marked decrease in the synthesis of tri- and tetrahexosyl gangliosides, probably due to a reduction in the amount of a sialyl transferase [31]. They also show an increase in the synthesis of fucose-containing glycoproteins in the cell membrane that is mediated by an increase in the amount of a different and relatively specific sialyl transferase [139]. Transformed cells are more readily agglutinated by several plant lectins (for an excellent review on lectins, see Sharon and Lis [118]) and show a different arrangement of surface receptors [21, 101]. Transformed cells also tend to have a lower level of cyclic adenosine 3′, 5′-monophosphate (cAMP), possibly due to factors affecting cell membrane-bound adenyl cyclase. Agents that block some of these membrane receptor sites or raise the level of cAMP can produce reversion to density-dependent cell growth. Cell hybridization has been used to clarify several facets of these problems.

(1) Contact inhibition and transformation. Normal cells show a density-dependent growth regulation in which cell mobility and cell division are inhibited when cultured cells have grown to confluence or a certain density. This is called contact inhibition of mobility or growth. After exposure to an oncogenic virus, a carcinogenic hydrocarbon, or a variety of other agents, cells may become transformed and, like cancer cells, no longer exhibit contact inhibition. They frequently exhibit marked instability of their karyotypes, with a hypotetraploid modal number but with large numerical and structural variation.

Reversal of cell transformation can occur, either spontaneously or as a result of a variety of treatments, including hybridization with another type of cell. Weiss et al. [148] mixed BALB/c 3T3 mouse cells, which are extremely sensitive to contact inhibition and show saturation densities of only 4×10^4 cells/cm^2, with LM(TK$^-$) clone 1D cells, which show minimal contact inhibition and achieve saturation densities of 70×10^4 cells/cm^2. Hybrids arose spontaneously, and all the hybrid clones showed contact inhibition, with saturation densities of 4 to 8×10^4 cells/cm^2. The normal trait, contact inhibition, thus appeared to be dominant. On continued growth, variants appeared that were no longer contact inhibited, presumably because of the loss of a 3T3 chromosome carrying a regulatory locus. Chromosome control of the reversible changes from normal to transformed and back has been suggested by Hitotsumachi et al. [67, 68] on the basis of their findings in hamster cells transformed with either polyoma virus or dimethylnitrosamine.

A different model of the control of contact inhibition is needed to account for the recent findings of Levisohn and Thompson [89]. They produced hybrids between a rat hepatoma tissue culture line, HTC, and a mouse L cell derivative, B82(TK$^-$). Neither of the parental lines showed contact inhibition of cell division, but the hybrid HL cells did. The restoration of regulated growth appears to be due to some sort of complementation, with each parent supplying something the other lacks. The malignancy of the hybrid cells was apparently not evaluated; but in view of the correlation between loss of contact inhibition of growth and malignant potential [107] and the findings of Harris and his associates, it would not be surprising if the malignancy of the HTC cells had been suppressed in the hybrids. On the other hand, Wyke [157] found that polyoma-transformed BHK-21 cells could regain contact inhibition while remaining malignant.

(2) Surface receptors. A different approach has been to investigate the effects of lectins, which show highly specific binding at specific membrane sites [72, 118]. For example, the jackbean lectin, concanavalin-A, agglutinates many types of transformed and malignant cells, but binding is specifically blocked by α-methyl-mannopyranoside.

Inbar et al. [71] reported that concanavalin-A inhibited ascites tumor development, an effect presumably mediated by its agglutinating effect. De Micco and Berebbi [37] compared the agglutinating activity of concanavalin-A on cells of the highly tumorigenic DC3F Chinese hamster line, its nontumorigenic derivative DC3F/ADX/Aza, and a series of hybrids between the two. In general, the take incidence on transplantation closely paralleled the agglutinability of the cells. Clonal lines of the original hybrid showed a wide range of agglutinability, which was correlated inversely with the presence of a marker chromosome not seen in either parental line. Every cell in clone 2, for example, had this marker. Cells of clone 2 were the poorest agglutinators in the whole series, and they failed to produce any tumors. This marker chromosome may carry a regulatory locus that influences both a glycoprotein on the cell surface and the malignancy of the cells.

The evidence of a relationship between a cell membrane glycoprotein and malignancy was strengthened by Burger and Noonan [24]. They found that concanavalin-A, in the monovalent form produced by tryptic digestion of the crude extract, bound to polyoma-transformed 3T3 mouse cells, which are not contact inhibited, and restored their growth pattern to that of normal cells.

Wheat germ agglutinin (WGA) is another plant lectin that agglutinates cells that are no longer contact inhibited or normal cells that have been treated with trypsin [23]. Growth regulation thus appears to depend upon the ability of cells to synthesize a trypsin-sensitive group that masks the WGA receptor site [106].

The change in the cell surface in transformed cells appears to be under the control of a viral gene. Evidence for this comes from experiments involving a temperature-sensitive mutant of polyoma virus, ts-3 [41]. Ordinarily, infection of BALB/3T3 cells by polyoma virus leads to enhanced agglutination of the cells by wheat germ agglutinin or concanavalin-A. When 3T3 mouse or BHK hamster cells were infected by ts-3, the same cell surface alterations occurred at the permissive temperature but not at the nonpermissive temperature. The increased agglutinability required cellular DNA synthesis but not viral DNA synthesis, indicating probable involvement of a host gene as well as a viral gene.

(3) Cyclic AMP. Cyclic AMP (cAMP) is involved in growth regulation

and, more specifically, in contact inhibition of cell division. Cyclic AMP produced marked growth inhibition of four transformed lines (L cells, HeLa, HEp-2, and Fl amnion) but only a minimal effect on the human diploid WI-38 strain [65]. Cyclic AMP levels are low in transformed cells [119]. When growth of untransformed 3T3 mouse cells is stimulated by protease, the cAMP level falls [22]. Agents that increase cAMP (e.g., prostaglandin E_1, isoproterenol, and adrenalin), suppress growth [119], Dibutyryl cyclic AMP plus theophylline, an inhibitor of the phosphodiesterase which hydrolyzes cyclic nucleotides, restores spontaneously- or virally transformed 3T3 mouse cells to contact-inhibited growth [120]. More important, cAMP can suppress the tumorigenicity of arbovirus-transformed hamster cells [109].

Cyclic AMP is the common intracellular second messenger of a number of hormones and biogenic amines; that is, it is the final common pathway by which a variety of agents trigger differentiated functions. Its effects on the cell membrane may go far toward explaining the well-known antagonism between growth and differentiation. Ballard and Tomkins [7] showed, by measuring cell adhesion, electrophoretic mobility, and antigenic properties, that glucocorticoids not only induce higher levels of tyrosine aminotransferase in hepatoma tissue culture (HTC) cells but also induce a cell surface adhesive factor. This reduces the net negative change in the cells and enables them to attach to glass instead of growing in suspension.

(4) Intercellular connections. The involvement of the cell membrane in the regulation of malignancy is suggested by recent cell hybridization studies of Azarnia and Loewenstein [6]. They had shown earlier that three strains of rat cancer cells do not make junctions of a kind that connects the cell interiors to each other, whereas normal cells do [5, 19]. Hybrids between one of these cancer cell strains and each of three different nonmalignant cell types were produced by the Sendai virus technique. The modal number of chromosomes in the hybrids corresponded approximately to the sum of the modal numbers of the parental cell types. All the hybrids behaved like the nonmalignant parent; that is, they did not produce tumors when 10^6 cells were injected into rats of the strain from which these lines were derived, they showed contact inhibition with saturation densities of 10^4 cells/cm^2 or less (in contrast to the saturation density of 10^6 cells/cm^2 of the parental cancer cell), and they showed cell coupling as measured electrically and by diffusion to adjacent cells of a fluorescent tracer injected into one cell [6]. The authors concluded that defective junctional coupling may be an etiological factor in certain kinds of cancer and suggested that this type of intercellular junction may be a pathway of growth-controlling molecules.

Cell Hybridization in Initiation of Malignancy

Cell hybridization may be involved in the initiation of the malignant process. The first evidence that cell hybridization can occur in vivo was obtained by Lengerova and Zeleny [88]. They inoculated, into lethally irradiated mice, a mixture of two kinds of embryonic liver cells, one homozygous for the T6 translocation-marker chromosome, the other lacking it. Subsequently, they found about 2% polyploid cells and about 3% (100 in 3000 metaphases) heterozygous $T6^+/T6^-$ cells among those that repopulated the bone marrow and spleen of the recipient mice. The polyploid cells presumably arose by in vivo somatic cell hybridization, and the T6 heterozygotes by diploidization of the tetraploid hybrid cells. There is independent evidence that tetraploid cells can undergo a regular tetrapolar mitosis to become diploid again, with segregation of genetic markers [94].

Direct evidence that cell fusion can initiate the changes that produce a malignant cell has come from studies in Syrian hamsters. Elkort et al. [42] treated cells of a hamster cell line with β-propiolactone-inactivated Sendai virus, observed cell fusion, and inoculated the fusion mixture into the cheek pouches of eight young hamsters, injecting each pouch with 10^3 to 10^6 viable cells. Tumors occurred in all 16 pouches but in only one of 6 pouches inoculated with cells of the same line not treated with Sendai virus. No virus control was included in the experiment, and one cannot rule out in vivo fusion as the tumorigenic event. In either case, however, cell fusion may be involved in the resultant tumorigenesis.

In view of the large number of viruses and other agents that can induce cell fusion [108, 110], the proposed mechanism must be regarded as an important working hypothesis concerning a potential cause of cancer. In the natural history of most tumors, fusion can only occur between cells of the host. The hybrid cells will have an increased number of chromosomes, usually being tetraploid. Tetraploid cells tend to lose chromosomes, rarely by a regular tetrapolar mitosis to become diploid, or more commonly by irregular mitoses to become aneuploid. The parallel with the findings in transformed cells and some tumors is suggestive, to say the least.

CLUES TO TUMOR PROGRESSION

Possible Involvement of Cell Hybridization

Cell fusion may be involved at a later stage in the natural history of tumors, too (i.e., in their progression). Agnish and Federoff [1] suggested, on the basis of chromosome studies on cell populations isolated from the near-tetraploid (Klein) variant of Ehrlich ascites tumor cells, that the major cell type, with a mode of 75 chromosomes, was accompanied by a population of larger cells with a stable mode of 110 chromosomes. Since they also found host lymphocytes in the ascites fluid, they suggested that the large cells had risen by in vivo fusion of host lymphocytes with the Ehrlich cells.

Recent observations of Rothschild and Black [111] could be explained by in vivo cell hybridization, although their evidence was insufficient to prove it. They developed a thymidine kinase deficient (TK^-) subline of transplantable, SV40-transformed hamster cells by growing them in the presence of BUdR. These TK^- cells formed tumors poorly and only after a prolonged latent period. Six tumors thus produced contained TK^+ as well as TK^- cells, and the authors concluded that nucleic-acid–salvage-pathway enzymes may play a rate-limiting role in tumorigenesis. Chromosome studies on four of these tumors showed them to have a wide range of chromosome numbers, 48 to 148, with modes of over 50, over 80, and two over 110 chromosomes, far more than found in the TK^- parental line. Rothschild and Black suggested cell hybridization as a possible mode of origin of these malignant TK^+ cells, although they decided that this was probably not the right explanation. In view of the new finding of Wiener and his associates, this explanation should probably be reconsidered.

Wiener et al. [153] transplanted small numbers of A9HT or B82HT tumor cells into CBA mice homozygous for the T6 marker. When tumors appeared, they were explanted and grown in vitro in HAT medium. The malignant A9HT and B82HT cells were killed by the aminopterin in this medium, because they were deficient in one of the salvage-pathway enzymes, inosinic acid pyrophosphorylase or thymidine kinase, respectively. However, in each experiment, a few cells survived and grew into HAT-resistant clones. The cells in these clones were found to contain an increased number of chromosomes, including the T6 marker, unequivocally indicating hybridization with host cells. Since the cells had been removed from the host and grown in culture, it is theoretically possible that cell fusion did not occur in vivo at all, but only later, under the peculiar conditions of in vitro growth. The same objection does not appear to be valid for the evidence of in vivo as well as in vitro spontaneous hybridization of mouse sarcoma-180 and L5178Y lymphoblasts obtained by Janzen et al. [74]. They injected sarcoma-180 and L-5178Y lymphoblasts simultaneously into the thigh muscle of 36 C3H mice and studied the resultant tumors. Judged by karyological evidence, 25 of 800 (3.1%) of the metaphases from the tumors were hybrid cells, while spontaneous hybridization occurred in 9 of 200 (4.5%) of cells grown in vitro. This study was not definitive because of the very small number of chromosome markers available in the two lines used and the resultant weakness of the karyological evidence of hybridization.

In spite of the difficulties in designing unambiguous tests, experimental methods are clearly at hand for investigating the role of cell hybridization in tumor initiation and progression. The results appear to be most promising.

Tumor Cell Differentiation

Cell hybridization has been used to analyze the regulation of phenotypic expression and the problem of cell differentiation (i.e., the occurrence of specialized phenotypes in restricted cell lineages). Many of these studies have involved tumor cell lines, and they have provided a great deal of information about the regulation of the state of differentiation of tumor cells.

Finch and Ephrussi [48] produced hybrids between multipotential mouse teratoma cells and clone 1D (TK-) cells or cells of an IMP- subline of N1 cells. Seven hybrid clones were studied. All formed undifferentiated tumors, like those produced by one parent, whereas the other parent formed tumors containing neural tissue, cartilage, bone, and mesenchyme. In other words, the one parent's potential for tumor cell differentiation was suppressed in the hybrids.

The expression of the differentiated functions of hepatoma cells [146] and of neuroblastoma cells [98] was suppressed in hybrids produced by fusion with undifferentiated and heteroploid mouse 3T3 or L cells. One hepatic cell function, tyrosine aminotransferase inducibility, reappeared in one hybrid line in which 30 to 40% of the chromosomes originally present had been lost. The malignancy of these hybrids was not evaluated.

Many studies have been carried out on melanoma cells. When such cells were fused with fibroblastic cells, dihydroxyphenylalanine oxidase disappeared and melanin production ceased [33]. The loss of this function was not closely correlated with the malignant potential of the hybrid cells; they varied tremendously in this trait, from highly malignant to nonmalignant [122].

Fougère et al. [50] showed that the production of pigment in melanoma × fibroblast hybrids is dependent on gene dosage; that is, pigment synthesis occurred if two melanoma cells hybridized with a single fibroblastic cell. Again, no information is available concerning the malignancy of these hybrids.

The importance of chromosome segregation in determining the phenotype of hybrid cells has been suggested by Wiener et al. [153, 154], who fused cells of the YACIR mouse lymphoma line with A9 fibroblastic cells. The hybrids had a markedly reduced take incidence when compared with the lymphoma parent. Ten of 16 tumors that arose were characterized. Three had the histological appearance of a sarcoma, like the rare tumors produced by the A9 parent; six had an intermediate morphology with A9-like and YACIR-like cells intermingled; and one had the YACIR-like morphology of a lymphoid tumor. Analysis of the chromosome constitutions of the hybrid tumor cells did not provide decisve information about the chromosomal basis of either the fibroblastic or the lymphocytic form of differentiation, because the parental origin of the chromosomes, especially the acrocentrics, could not be clearly distinguished. All the hybrid tumors had a reduced number of chromosomes.

The retention of the fibroblastic morphology despite extensive loss of chromosomes, and the continued presence of many biarmed chromosomes, presumably most of A9 origin, suggested that the A9 chromosomes may have been preferentially retained in most cases. The tumor with a lymphocytic morphology had undergone drastic chromosome loss, including most of the biarmed chromosomes. More YACIR chromosomes must therefore have been retained, and this combination may have permitted the expression of the YACIR phenotype. The mixed tumors are most simply explained in terms of different types of chromosome segregation in various cells within the same tumor.

It is interesting that the determinants of malignancy are not linked to the determinants of morphological differentiation. Thus the malignancy of the YACIR cells was expressed in hybrids with either a lymphocytic or a fibroblastic morphology [153].

CLUES TO TUMOR THERAPY

Tumor Immunity

Watkins and Chen [143] used cell hybridization as a means of getting mice to mount an immunological reaction to Ehrlich ascites tumor (EAT) cells. Ordinarily, these cells are so highly malignant that the LD_{50} due to progressively growing tumors is about 10 cells. Ehrlich cells were hybridized with polyoma-transformed hamster fibroblasts (PYY). Mice were inoculated with these PYY/EAT hybrid cells. After some initial growth, the allografts were rejected. When challenged with EAT cells, the LD_{50} of the immunized mice was found to be about 10^5 cells. Immunization had thus increased the LD_{50} by a factor of about 10^4. Chen and Watkins [26] also succeeded in immunizing Swiss mice against Ehrlich tumor cells by injecting CBA mouse × Ehrlich hybrid cells. It is thus unnecessary to immunize across a species barrier to bring about the desired result.

This method of tumor immunization, using hybrids between a tumor with weak transplantation antigens and a cell line with transplantation antigens that are very strong in the tumor host animal, may be useful in controlling certain kinds of tumors, and further exploration of this system is indicated.

Drug Resistance

Cell hybridization has been used as a means of evaluating the phenotypic expression of resistance to such agents as vinblastin sulfate and cytosine arabinoside. Chinese hamster cells that were resistant to vinblastin conferred this resistance on hybrids with nonresistant cells, while the resistance to cytosine arabinoside was totally suppressed in comparable hybrids [64]. Actinomycin D resistance to malignant Chinese hamster DC3F/ADIV cells also acted as a dominant trait when the cells were hybridized with amethopterin-resistant DC3F/A3 cells of low tumor-forming capacity but the same malignant DC3F parentage [125]. The amethopterin resistance and associated high folate reductase activity of the hybrid cells were intermediate between the very high values of the DC3F/A3 parent and the low values of the DC3F/ADIV parent. Of five hybrid lines, two were malignant, one had an intermediate level of malignancy, and two were nonmalignant, indicating a rather complex relationship among several variables operating in this system.

CONCLUSION

Cell hybridization has played an increasingly important role in cancer research as means have been developed to utilize the tremendous potential the technique offers for genetic analysis. Studies on cell hybrids have already clarified the central role of chromosomal factors in malignancy, whether spontaneous or induced by a virus or a polycyclic hydrocarbon. And while additional genetic information (e.g., a viral genome) may trigger a chain of events leading to cancer, the evidence, much of it gained from cell hybridization, favors the idea that a specific genetic or epigenetic functional loss is the key event in the onset of cancer, no matter what the underlying etiologic factor.

It should be emphasized that the utilization of cell hybridization is still in its infancy. The number of somatic cell genetic markers (i.e., those with recognizable expression at the cellular level) is still very limited, and satisfactory methods for obtaining temperature-sensitive mutants or other conditional lethals are just being developed. The biochemistry of cells in culture is a rapidly advancing field, but even such basic considerations as the nutritional requirement of specialized cell types are very poorly understood.

Cell hybridization, which permits complementation analysis in mammalian cell systems, should play a central role in the solution of some of these problems, especially those involving more complex systems, such as the cell membrane. In fact, at the present time, it is difficult to envision any method that has so much to offer toward increasing our understanding of the interplay of exogenous and endogenous factors in the etiology and progression of neoplastic disease.

Acknowledgements.–The work reported here was supported in part by Public Health Service Research Grants Nos. CA 12504 from the National Cancer Institute and GM 18153 from the National Institute of General Medical Sciences. The author is a Career Scientist of the Health Research Council of the City of New York.

LITERATURE CITED

1. AGNISH ND, FEDEROFF S: Tumor cell population of the Ehrlich ascites tumors. *Can J Genet Cytol* 10:723–746, 1968.
2. ALLDERDICE PW, MILLER OJ, MILLER DA, WARBURTON D, PEARSON PL, KLEIN G, HARRIS H: Chromosome analysis of two related heteroploid mouse cell lines by quinacrine fluorescence. *J Cell Sci* 12:263–274, 1973.
3. ALLISON AC, MALLUCCI L: Uptake of hydrocarbon carcinogens by lysosomes. *Nature* 203:1024–1027, 1964.
4. ALLISON AC, MALLUCCI L: Histochemical studies of lysosomes and lysosomal enzymes in virus-infected cell cultures. *J Exp Med* 121:463–476, 1965.
5. AZARNIA R, LOEWENSTEIN WR: Intercellular communications and tissue growth. V. A cancer cell strain that fails to make permeable membrane junctions with normal cells. *J Membrane Biol* 6:368–385, 1971.
6. AZARNIA R, LOEWENSTEIN WR: Parallel correction of cancerous growth and of a genetic defect of cell-to-cell communication. *Nature* 241:455–457, 1973.
7. BALLARD PL, TOMKINS GM: Hormone induced modification of the cell surface. *Nature* 224:344–345, 1969.
8. BALTIMORE D: RNA-dependent DNA polymerase in virions of RNA tumour viruses. *Nature* 226:1209–1211, 1970.
9. BARSKI G: Cell association and somatic cell hybridization. *Int Rev Exp Pathol* 9:151–190, 1970.
10. BARSKI G, CORNEFERT FR: Characteristics of "hybrid"-type clonal cell lines obtained from mixed cultures *in vitro*. *J Nat Cancer Inst* 28:801–821, 1962.
11. BARSKI G, BILLARDON C, JULLIEN PM, CARSWELL E: Evolution *in vitro* et cancerisation des cellules pulmonaires de souris. *Int J Cancer* 1:541–556, 1966.
12. BARSKI G, SORIEUL S, CORNEFERT FR: Production dans des cultures *in vitro*

de deux souches cellulaires en association de cellules de caractere "hybride." *CR Acad Sci Paris* 251:1825–1827, 1960.
13. BARSKI G, SORIEUL S, CORNEFERT FR: "Hybrid" type cells in combined cultures of two different mammalian cell strains. *J. Nat Cancer Inst* 26:1269–1291, 1961.
14. BASILICO C, MATSUYA Y, GREEN H: The interaction of polyoma virus with mouse-hamster somatic hybrid cells. *Virology* 41:295–305, 1970.
15. BELEHRADEK J, BARSKI G: Karyological patterns and expression of malignancy in some homologous mouse somatic hybrid cells. *Int J Cancer* 8:1–9, 1971.
16. BEREBBI M, BARSKI G: Hybridization entre deux lignes cellulaires de hamster chinois à pouvoir tumorigène différent. *CR Acad Sci Paris* 272:351–354, 1971.
17. BEREBBI M, MEYER G: Transformation cellulaire maligne et modifications caryologiques: apport de l'hybridation somatique de lignes de hamster chinois. *Int J Cancer* 10:418–435, 1972.
18. BERNHARD W: The detection and study of tumour viruses with the electron microscope. *Cancer Res* 20:712–727, 1960.
19. BOREK C, HIGASHINO S, LOEWENSTEIN WR: Intercellular communication and tissue growth. IV. Conduction of membrane junctions of normal and cancerous cells in culture. *J Membrane Biol* 1:274–293, 1969.
20. BREGULA U, KLEIN G, HARRIS H: The analysis of malignancy by cell fusion II. Hybrids between Ehrlich cells and normal diploid cells. *J Cell Sci* 8:673–680, 1971.
21. BURGER MM: A difference in the architecture of the surface membrane of normal and virally transformed cells. *Proc Nat Acad Sci (US)* 62:994–1001, 1969.
22. BURGER MM, BOMBIK BM, BRECKENRIDGE B McL, SHEPPARD JR: Growth control and cyclic alterations of cyclic AMP in the cell cycle. *Nature New Biol* 239:161–163, 1972.
23. BURGER MM, GOLDBERG AR: Identification of a tumor-specific determinant on neoplastic cell surfaces. *Proc Nat Acad Sci (US)* 57:359–366, 1967.
24. BURGER MM, NOONAN KD: Restoration of normal growth by covering of agglutinin sites on tumour cell surface. *Nature* 228:512–515, 1970.
25. BURNS WH, BLACK PH: Induction experiments with adenovirus and polyoma virus transformed lines. *Int J Cancer* 4:204–211, 1969.
26. CHEN L, WATKINS JF: Evidence against the presence of H2 histo-compatibility antigens in Ehrlich ascites tumour cells. *Nature* 225:734–735, 1970.
27. COFFIN JM: Rescue of Rous sarcoma virus from Rous sarcoma virus-transformed mammalian cells. *J Virol* 10:153–156, 1972.
28. COON HC, WEISS MC: A quantitative comparison of formation of spontaneous and virus-produced viable hybrids. *Proc Nat Acad Sci (US)* 62:852–859, 1969.
29. CORNEFERT-JENSEN FR, HARE WCD, STOCK ND: Studies on bovine lymphosarcoma: formation of syncytia and detection of virus particles in mixed-cell cultures. *Int J Cancer* 4:507–519, 1969.
30. CROCE CM, SAWICKI W, KRITCHEVSKY D, KOPROWSKI H: Induction of homokaryocyte, heterokaryocyte and hybrid formation by lysolecithin. *Exp Cell Res* 67:427–435, 1971.
31. CUMAR FA, BRADY RO, KOLODNY EH, MCFARLAND VW, MORA PT: Enzymatic

block in the synthesis of gangliosides in DNA virus transformed tumorigenic mouse cell lines. *Proc Nat Acad Sci (US)* 67:757–764, 1970.
32. DAVIDSON R, EPHRUSSI B: Factors influencing the "effective mating rate" of mammalian cells. *Exp Cell Res* 61:222–226, 1970.
33. DAVIDSON RL, EPHRUSSI B, YAMAMOTO K: Regulation of melanin synthesis in mammalian cells, as studied by somatic hybridization. *Proc Nat Acad Sci (US)* 56:1437–1440, 1966.
34. DE BRUYN W, HANSEN-MELANDER E: Chromosome studies in the MB mouse lymphosarcoma. *J Nat Cancer Inst* 28:1333–1354, 1962.
35. DEFENDI V, EPHRUSSI B, KOPROWSKI H: Expression of polyoma-induced cellular antigens in hybrid cells. *Nature* 203:495–496, 1964.
36. DEFENDI V, EPHRUSSI B, KOPROWSKI H, YOSHIDA MC: Properties of hybrids between polyoma-transformed and normal mouse cells. *Proc Nat Acad Sci (US)* 57:299–305, 1967.
37. DE MICCO P, BEREBBI M: Tumorigenicity and agglutination by concanavalin A of Chinese hamster cells and their hybrids. *Int J Cancer* 10:249–253, 1972.
38. DE WEERD-KASTELEIN EA, KEIJZER W, BOOTSMA D: Genetic heterogeneity of xeroderma pigmentosum demonstrated by somatic cell hybridization. *Nature New Biol* 238:80–83, 1972.
39. DI PAOLO JA, DONOVAN PJ, NELSON RL: Transformation of hamster cells *in vitro* by polycyclic hydrocarbons without cytotoxicity. *Proc Nat Acad Sci (US)* 68:2958–2961, 1971.
40. DULBECCO R: Viral carcinogenesis. *Cancer Res* 21:975–980, 1961.
41. ECKHART W, DULBECCO R, BURGER MM: Temperature-dependent surface changes in cells infected or transformed by a thermosensitive mutant of polyoma virus. *Proc Nat Acad Sci (US)* 68:283–286, 1971.
42. ELKORT RJ, HANDLER AH, KIBRICK S, KLEINMAN L: The relationship of somatic cell hybridization to carcinogenesis. *Eur J Cancer* 8:259–261, 1972.
43. ENGEL E, MCGEE BJ, HARRIS H: Recombination and segregation in somatic cell hybrids. *Nature* 223:152–155, 1969.
44. EPHRUSSI B: Hybridization of Somatic Cells. Princeton NJ, Princeton University Press, 1972.
45. EPHRUSSI B, DAVIDSON RL, WEISS MC: Malignancy of somatic cell hybrids. *Nature* 224:1314–1315, 1969.
46. EPHRUSSI B, SORIEUL S: Nouvelles observations sur l'hybridation *in vitro* de cellules de souris. *CR Acad Sci Paris* 254:181–182, 1962.
47. FENYÖ EM, GRUNDNER G, KLEIN G, KLEIN E, HARRIS H: Surface antigens and release of virus in hybrid cells produced by the fusion of A9 fibroblasts with Moloney lymphoma cells. *Exp Cell Res* 68:323–331, 1971.
48. FINCH BW, EPHRUSSI B: Retention of multiple developmental potentialities by cells of a mouse testicular teratocarcinoma during prolonged culture in vitro, and their extinction upon hybridization with cells of permanent lines. *Proc Nat Acad Sci (US)* 57:615–621, 1967.
49. FOGEL M: Induction of virus synthesis in polyoma-transformed cells by DNA antimetabolites and by irradiation after pretreatment with 5-bromodeoxyuridine. *Virology* 49:12–22, 1972.
50. FOUGÈRE C, RUIZ F, EPHRUSSI B: Gene dosage dependence of pigment

synthesis in melanoma X fibroblast hybrids. *Proc Nat Acad Sci (US)* 69:330–334, 1972.
51. GANSHOW R: Glucuronidase gene expression in somatic hybrids. *Science* 153:84–85, 1966.
52. GERBER P: Studies on the transfer of subviral infectivity from SV40-induced hamster tumor cells to indicator cells. *Virology* 28:501–509, 1966.
53. GERSHON D, SACHS L: Properties of a somatic hybrid between mouse cells with different genotypes. *Nature* 198:912–913, 1963.
54. GIANNELLI F, CROLL PM, LEWIN SA: DNA repair synthesis in human heterokaryons formed by normal and UV sensitive fibroblasts. *Exp Cell Res* 78:175–185, 1973.
55. GLASER R, RAPP F: Rescue of Epstein-Barr virus from somatic cell hybrids of Burkitt lymphoblastoid cells. *J Virol* 10:288–296, 1972.
56. GOLDSTEIN S, LIN CC: Survival and DNA repair from somatic cell hybrids after ultraviolet irradiation. *Nature New Biol* 239:142–145, 1972.
57. GRUNDNER G, FENYÖ EM, KLEIN G, KLEIN E, BREGULA U, HARRIS H: Surface antigen expression in malignant sublines derived from hybrid cells of low malignancy. *Exp Cell Res* 68:315–322, 1971.
58. HARNDEN DG: Viruses, chromosomes, and tumors: the interaction between viruses and chromosomes. This volume, pp. 151–190, 1974.
59. HARRIS H: Cell Fusion. Oxford, Clarendon Press, 1970.
60. HARRIS H: The Croonian lecture, 1971. Cell fusion and the analysis of malignancy. *Proc Roy Soc London B* 179:1–20, 1971.
61. HARRIS H, MILLER OJ, KLEIN G, WORST P, TACHIBANA T: Suppression of malignancy by cell fusion. *Nature* 223:363–368, 1969.
62. HARRIS H, WATKINS JF: Hybrid cells derived from mouse and man: artificial heterokaryons of mammalian cells from different species. *Nature* 205:640–646, 1965.
63. HARRIS H, WATKINS JF, FORD CE, SCHOEFL GI: Artificial heterokaryons of animal cells from different species. *J Cell Sci* 1:1–30, 1966.
64. HARRIS M: Phenotypic expression of drug resistance in hybrid cells. *J Nat Cancer Inst* 50:423–429, 1973.
65. HEIDRICK ML, RYAN WL: Cyclic nucleotides on cell growth *in vitro*. *Cancer Res* 30:376–378, 1970.
66. HILL M, HILLOVA J: Recovery of the temperature sensitive mutant of Rous sarcoma virus from chicken cells exposed to DNA extracted from hamster cells transformed by the mutant. *Virology* 49:309–313, 1972.
67. HITOTSUMACHI S, RABINOWITZ Z, SACHS L: Chromosomal control of chemical carcinogenesis. *Int J Cancer* 9:305–315, 1972.
68. HITOTSUMACHI S, RABINOWITZ Z, SACHS L: Chromosomal control of reversion in transformed cells. *Nature* 231:511–514, 1972.
69. HSU TC: Chromosomal evolution in cell populations. *Int Rev Cytol* 12:69–161, 1961.
70. HUEBNER RJ, TODARO GJ: Oncogenes of RNA tumor viruses as determinants of cancer. *Proc Nat Acad Sci (US)* 64:1087–1094, 1969.
71. INBAR M, BEN-BASSAT H, SACHS L: Inhibition of ascites tumor development by concanavalin-A. *Int J Cancer* 9:143–149, 1972.

72. INBAR M, SACHS L: Interaction of the carbohydrate-binding protein concanavalin A with normal and transformed cells. *Proc Nat Acad Sci (US)* 63:1418–1425, 1969.
73. ISSELBACHER KJ: Increased uptake of amino acids and 2-deoxy-D-glucose by virus-transformed cells in culture. *Proc Nat Acad Sci (US)* 69:585–589, 1972.
74. JANZEN HW, MILLMAN PA, THURSTON OG: Hybrid cells in solid tumors. *Cancer* 27:455–459, 1971.
75. KAO FT, CHASIN L, PUCK TT: Genetics of somatic mammalian cells, X. Complementation analysis of glycine-requiring mutants. *Proc Nat Acad Sci (US)* 64:1284–1291, 1969.
76. KATO H, YOSIDA TH: Differential responses of several aneusomic cell clones to ultraviolet irradiation. *Exp Cell Res* 74:15–20, 1972.
77. KIT S, BROWN M: Rescue of simian virus 40 from cell lines transformed at high and at low input multiplicities by unirradiated or ultraviolet-irradiated virus. *J Virol* 4:226–230, 1969.
78. KIT S, KURIMURA T, DUBBS DR: Transplantable mouse tumor line induced by infection of SV40-transformed mouse kidney cells. *Int J Cancer* 4:384–392, 1969.
79. KIT S, KURIMURA T, SALVI ML, DUBBS DR: Activation of infectious SV40 DNA synthesis in transformed cells. *Proc Nat Acad Sci (US)* 60:1239–1246, 1968.
80. KIT S, KURIMURA T, TORRES RA DE, DUBBS DR: Simian virus 40 deoxyribonucleic acid replication. I. Effect of cyloheximide on the replication of SV40 deoxyribonucleic acid in monkey kidney cells and in heterokaryons of SV40-transformed and susceptible cells. *J Virol* 3:25–32, 1969.
81. KLEIN G, GARS U, HARRIS H: Isoantigen expression in hybrid mouse cells. *Exp Cell Res* 62:149–160, 1970.
82. KLEIN G, BREGULA U, WIENER F, HARRIS H: The analysis of malignancy by cell fusion. I. Hybrids between tumour cells and L cell derivatives. *J Cell Sci* 8:659–672, 1971.
83. KLEIN G, FRIBERG S, HARRIS H: Two kinds of antigen suppression in tumor cells revealed by cell fusion. *J Exp Med* 135:839–849, 1972.
84. KLEIN G, HARRIS H: Expression of polyoma-induced transplantation antigen in hybrid cell lines. *Nature New Biol* 237:163–164, 1972.
85. KNOWLES BB, JENSEN FC, STEPLEWSKI Z, KOPROWSKI H: Rescue of infectious SV40 after fusion between different SV40-transformed cells. *Proc Nat Acad Sci (US)* 61:42–45, 1968.
86. KOPROWSKI H, JENSEN FC, STEPLEWSKI Z: Activation of production of infectious tumor virus SV40 in heterokaryon cultures. *Proc Nat Acad Sci (US)* 58:127–133, 1967.
87. KUSANO T, LONG C, GREEN H: A new reduced human-mouse somatic cell hybrid containing the human gene for adenine phosphoribosyltransferase. *Proc Nat Acad Sci (US)* 68:82–86, 1971.
88. LENGEROVA A, ZELENY V: *In* Symposium on the Mutational Process. Genetic Variations in Somatic Cells. Prague, Academia, 79–83, 1965.
89. LEVISOHN SR, THOMPSON EB: Contact inhibition and gene expression in HTC/L cell hybrid lines. *J Cell Physiol* 81:225–232, 1973.
90. LEWIS WH: The formation of giant cells in tissue cultures and their similarity to those in tuberculous lesions. *Am Rev Tuberc* 15:616–628, 1927.

91. LITTLEFIELD JW: Selection of hybrids from mating of fibroblasts *in vitro* and their presumed recombinants. *Science* 145:709–710, 1964.
92. MACHALA O, DONNER L, SVOBODA J: A full expression of the genome of Rous sarcoma virus in heterokaryons formed after fusion of virogenic mammalian cells and chicken fibroblasts. *J Gen Virol* 8:219–229, 1970.
93. MARIN G, LITTLEFIELD JW: Selection of morphologically normal cell lines from polyoma-transformed BHK 21/13 hamster fibroblasts. *J Virol* 2:69–77, 1968.
94. MARTIN GM, SPRAGUE CA: Parasexual cycle in cultivated tumor somatic cells. *Science* 166:761–763, 1969.
95. MCNUTT NS, CULP LA, BLACK PH: Contact-inhibited revertant cell lines isolated from SV40-transformed cells, IV. Microfilament distribution and cell shape in untransformed, transformed and revertant Balb/3T3 cells. *J Cell Biol* 56:412–428, 1973.
96. MILLER OJ, ALLDERDICE PW, MILLER DA, BREG WR, MIGEON BR: Human thymidine kinase gene locus: assignment to chromosome 17 in a hybrid of man and mouse. *Science* 173:244–245, 1971.
97. MILLER OJ, COOK PR, MEERA KHAN P, SHIN S, SINISCALCO M: Mitotic separation of two human X-linked genes in man–mouse somatic cell hybrids. *Proc Nat Acad Sci (US)* 68:116–120, 1971.
98. MINNA J, NELSON P, PEACOCK J, GLAZER D, NIRENBERG M: Genes for neuronal properties expressed in neuroblastoma X L cell hybrids. *Proc Nat Acad Sci (US)* 68:234–239, 1971.
99. MURAYAMA F, OKADA Y: Appearance, characteristics and malignancy of somatic hybrid cells between L and Ehrlich ascites tumor cells formed by artificial fusion with UV-HVJ. *Biken J* 13:11–23, 1970.
100. MURAYAMA-OKABAYOSHI F, OKADA Y, TACHIBANA T: A series of hybrid cells containing different ratios of parental chromosomes formed by two steps of artificial fusion. *Proc Nat Acad Sci (US)* 68:38–42, 1971.
101. NICOLSON G: Difference in topology of normal and tumor cell membranes shown by different surface distributions of ferritin-conjugated concanavalin-A. *Nature New Biol* 233:244–246, 1971.
102. OFTEBRO R: Further studies on mitosis in bi- and multinucleate HeLa cells. *Scand J Clin Lab Invest* 22:Suppl 106, 79–96, 1968.
103. OHNO S: Genetic implication of karyological instability of malignant somatic cells. *Physiol Rev* 51:496–526, 1971.
104. OKADA Y: Analysis of giant polynuclear cell formation caused by HVJ virus from Ehrlich's ascites tumor cells. I. Microscopic observations of giant polynuclear cell formation. *Exp Cell Res* 26:98–107, 1962.
105. PERA F, SCHWARZACHER HG: Formation and division of binucleated cells in kidney cell cultures of Microtus agrestis. *Humangenetik* 6:158–162, 1968.
106. POLLACK RE: Cellular and viral contributions to maintenance of the SV40-transformed state. *In Vitro* 6:58–65, 1970.
107. POLLACK R, GREEN H, TODARO GJ: Growth control in cultured cells: selection of sublines with increased sensitivity to contact inhibition and decreased tumor-producing ability. *Proc Nat Acad Sci (US)* 60:126–133, 1968.
108. POSTE G: Mechanisms of virus-induced cell fusion. *Int Rev Cytol* 33:157–252, 1972.

109. REDDI PK, CONSTANTINIDES SM: Partial suppression of tumor production by dibutyryl cyclic AMP and theophylline. *Nature* 238:286–287, 1972.
110. ROIZMAN B: Polykaryocytosis. *Cold Spring Harbor Symp Quant Biol* 27:327–340, 1962.
111. ROTHSCHILD H, BLACK PH: Effect of loss of thymidine kinase activity on the tumorigenicity of clones of SV40-transformed hamster cells. *Proc Nat Acad Sci (US)* 67:1042–1049, 1970.
112. ROWE WP, HARTLEY JW, LANDER MR, PUGH WE, TEICH N: Non-infectious AKR mouse embryo cell-line in which each cell has the capacity to be activated to produce infectious murine leukemia virus. *Virology* 46:866–876, 1971.
113. SAMBROOK J, WESTPHAL H, SRINIVASAN PR, DULBECCO R: The integrated state of viral DNA in SV40 transformed cells. *Proc Nat Acad Sci (US)* 60:1288–1295, 1968.
114. SARMA PS, LOG T, HUEBNER RJ: Trans-species rescue of defective genomes of murine sarcoma virus from hamster tumor cells with helper feline leukemia virus. *Proc Nat Acad Sci (US)* 65:81–87, 1970.
115. SCALETTA LJ, EPHRUSSI B: Hybridization of normal and neoplastic cells *in vitro*. *Nature* 205:1169–1171, 1965.
116. SELL EK, KROOTH RS: Tabulation of somatic cell hybrids formed between lines of cultured cells. *J Cell Physiol* 80:453–462, 1972.
117. SETLOW RB, REGAN JD, GERMAN J, CARRIER WL: Evidence that xeroderma pigmentosum cells do not perform the first step in the repair of ultraviolet damage to their DNA. *Proc Nat Acad Sci (US)* 64:1035–1041, 1969.
118. SHARON N, LIS H: Lectins: cell-agglutinating and sugar-specific proteins. *Science* 177:949–959, 1972.
119. SHEPPARD JR: Difference in the cyclic adenosine 3′, 5′-monophosphate levels in normal and transformed cells. *Nature New Biol* 236:14–16, 1972.
120. SHEPPARD JR: Restoration of contact-inhibited growth to transformed cells by dibutyryl adenosine 3′, 5′-cyclic monophosphate. *Proc Nat Acad Sci (US)* 68:1316–1320, 1971.
121. SHEVLIAGHYN VJ, BIRYULINA TI, TIKHONOVA ZN, KARAZAS NV: Activation of Rous virus in the transplated golden hamster tumour with the aid of artificial heterokaryon formation. *Int J Cancer* 4:42–46, 1969.
122. SILAGI S: Hybridization of a malignant melanoma line with L cells *in vitro*. *Cancer Res* 27:1953–1960, 1967.
123. SJÖGREN HO: Studies on specific transplantation resistance to polyoma-virus-induced tumors, IV. Stability of the polyoma cell antigen. *J Nat Cancer Inst* 32:661–666, 1964.
124. SJÖGREN HO: Immunology of virus-induced tumors. In Modern Trends in Medical Virology (Heath RB, Waterson AP, eds) London, Butterworths & Company, 207–222, 1962.
125. SOBEL JS, ALBRECHT AM, RIERM H, BIEDLER JL: Hybridization of actinomycin D- and amethopterin-resistant Chinese hamster cells *in vitro*. *Cancer Res* 31:297–301, 1971.
126. SPENCER RA, HAUSCHKA TS, AMOS DB, EPHRUSSI B: Co-dominance of isoantigens in somatic hybrids of murine cells grown *in vitro*. *J Nat Cancer Inst* 33:893–903, 1964.

127. SUMMERS DF, VOGT M: In The Biology of Oncogenic Viruses (SILVESTRI LG, ed) Amsterdam, North Holland Publishing Company, 306, 1970.
128. SVOBODA J, KLEMMENT V: Formation of delayed tumors in hamsters inoculated with Rous virus after birth and finding of infectious Rous virus in induced tumor P1. *Folio Biol. (Praha)* 9:403–411, 1963.
129. SZYBALSKA EH, SZYBALSKI W: Genetics of human cell lines. IV. DNA-mediated heritable transformation of a biochemical trait. *Proc Nat Acad Sci (US)* 48:2026–2034, 1962.
130. TEMIN HM: Nature of the provirus of Rous sarcoma. *Nat Cancer Inst Monogr* 17:557–570, 1964.
131. TEMIN HM, MIZUTANI S: RNA-dependent DNA polymerase in virions of Rous sarcoma viruses. *Nature* 226:1211–1213, 1970.
132. TENNANT RW, RICHTER CB: Murine leukemia virus: restriction in fused permissive and nonpermissive cells. *Science* 178:516–518, 1972.
133. TODARO GJ, GREEN H, SWIFT MR: Susceptibility of human diploid fibroblast strains to transformation by SV40 virus. *Science* 153:1252–1254, 1966.
134. TODARO GJ, HUEBNER RJ: The viral oncogene hypothesis: new evidence. *Proc Nat Acad Sci (US)* 69:1009–1015, 1972.
135. TODARO GJ, MARTIN GM: Increased susceptibility of Down's syndrome fibroblasts to transformation by SV40. *Proc Soc Exp Biol Med* 124:1232–1236, 1967.
136. VELDHUISEN G, POUWELS PH: Transformation of xeroderma-pigmentosum cells by SV40. *Lancet* 2:529–530, 1970.
137. VIGIER P: Persistence of Rous sarcoma virus in transformed nonpermissive cells: characteristics of virus induction following Sendai virus-mediated fusion with permissive cells. *Int J Cancer* 11:473–483, 1973.
138. WALL R, DARNELL JE: Presence of cell and virus specific sequences in the same molecules of nuclear RNA from virus transformed cells. *Nature New Biol* 232:73–76, 1972.
139. WARREN L, FUHRER JP, BUCK CA: Surface glycoproteins of normal and transformed cells: a difference determined by sialic acid and a growth-dependent sialyl transferase. *Proc Nat Acad Sci (US)* 69:1838–1842, 1972.
140. WATKINS JF: The effects of some metabolic inhibitors on the ability of SV40 virus in transformed cells to be detected by cell fusion. *J Cell Sci* 6:721–737, 1970.
141. WATKINS JF: Cell fusion in the study of tumor cells. *Int Rev Exp Pathol* 10:115–141, 1971.
142. WATKINS JF: Fusion of cells for virus studies and production of cell hybrids. *Methods Virol* 5:1–32, 1971.
143. WATKINS JF, CHEN L: Immunization of mice against Ehrlich ascites tumour using a hamster/Ehrlich ascites tumour hybrid cell line. *Nature* 223:1018–1022, 1969.
144. WATKINS JF, DULBECCO R: Production of SV40 virus in heterokaryons of transformed and susceptible cells. *Proc Nat Acad Sci (US)* 58:1396–1403, 1967.
145. WEISS MC: Further studies on loss of T-antigen from somatic hybrids between mouse cells and SV40-transformed human cells. *Proc Nat Acad Sci (US)* 66:79–86, 1970.

146. WEISS MC, CHAPLAIN M: Expression of differentiated functions in hepatoma cell hybrids: reappearance of tyrosine aminotransferase inducibility after the loss of chromosomes. *Proc Nat Acad Sci (US)* 68:3026–3030, 1971.
147. WEISS MC, EPHRUSSI B: Studies of interspecific (rat × mouse) somatic hybrids. I. Isolation, growth, and evolution of the karyotype. *Genetics* 54:1095–1108, 1966.
148. WEISS MC, EPHRUSSI B, SCALETTA LJ: Loss of T-antigen from somatic hybrids between mouse cells and SV40-transformed human cells. *Proc Nat Acad Sci (US)* 59:1132–1135, 1968.
149. WEISS MC, GREEN H: Human–mouse hybrid cell lines containing partial complements of human chromosomes and functioning human genes. *Proc Nat Acad Sci (US)* 58:1104–1111, 1967.
150. WEISS MC, TODARO GJ, GREEN H: Properties of a hybrid between lines sensitive and insensitive to contact inhibition of cell division. *J Cell Physiol* 71:105–107, 1968.
151. WEVER GH, KIT S, DUBBS DR: Initial site of synthesis of virus during rescue of simian virus 40 from heterokaryons of simian virus 40-transformed and susceptible cells. *J Virol* 5:578–585, 1970.
152. WHITEHOUSE HLK: Chromosome integration of viral DNA: the open-replicon hypothesis of carcinogenesis. This volume, pp 41–76, 1974.
153. WIENER F, COCHRAN A, KLEIN G, HARRIS H: Genetic determinants of morphological differentiation in hybrid tumors. *J Nat Cancer Inst* 48:465–486, 1972.
154. WIENER F, FENYÖ EM, KLEIN G, HARRIS H: Fusion of tumour cells with host cells. *Nature New Biol* 238:155–159, 1972.
155. WIENER F, KLEIN G, HARRIS H: The analysis of malignancy by cell fusion. III. Hybrids between diploid fibroblasts and other tumour cells. *J Cell Sci* 8:681–692, 1971.
156. WIENER F, KLEIN G, HARRIS H: The analysis of malignancy by cell fusion. IV. Hybrids between tumour cells and a malignant L cell derivative. *J Cell Sci* 12:253–261, 1973.
157. WYKE J: Phenotypic variation and its control in polyoma-transformed BHK 21 cells. *Exp Cell Res* 66:209–223, 1971.
158. YAMANAKA T, OKADA Y: Cultivation of fused cells resulting from treatment of cells with HVJ. II. Division of binucleated cells resulting from fusion of two KB cells by HVJ. *Exp Cell Res* 49:461–469, 1968.
159. YOSHIDA MC: A preliminary note on the chromosomes and tumorigenicity in human-mouse somatic cell hybrids. *Jap J Genet* 47:211–213, 1972.

EPSTEIN-BARR VIRUS INFECTION OF LYMPHOID CELLS AND THE CYTOGENETICS OF ESTABLISHED HUMAN LYMPHOCYTE CELL LINES

ARTHUR D. BLOOM, JEANNE A. McNEILL, AND FRANK T. NAKAMURA

Department of Human Genetics, University of Michigan Medical School, Ann Arbor, Michigan

Introduction	565
Establishment of Lymphocyte Lines and the Presence of the EB Virus	568
Lymphomatous Disease and the EB Virus	572
Oncogenicity of Lymphocyte Cell Lines	579
Cytogenetics of Lymphocyte Cell Lines from Normal Donors	580
Cytogenetics of Lymphocyte Cell Lines from Abnormal Donors	583
Abnormal Chromosome Constitution	583
Infectious Mononucleosis	586
Burkitt Lymphoma	587
Malignant Lymphoma other than Burkitt	590
Leukemia	590
Summary	592
Acknowledgements	593
Literature Cited	593

INTRODUCTION

The presence of the Epstein-Barr (EB) virus in virtually all established* lymphocyte lines has now been demonstrated by immunofluorescent tech-

---

* The term "established" is used here to describe cell lines which may be serially passaged, for a more or less indefinite period of time in the case of lymphocyte lines.

niques, by DNA–DNA, and by DNA–RNA hybridization experiments. In addition, many such lines actively shed EB virus under certain culture conditions. It is thus highly suggestive that the cells of these lines are viral transformants. The association of EB virus with the Burkitt lymphoma, nasopharyngeal carcinoma, and infectious mononucleosis is now well established, though it has not been proven that EB virus is etiologic in these diseases. Furthermore, there is increasing experimental evidence for the oncogenicity of established lymphocyte lines, and it is for this reason in particular that we regard the inclusion of a chapter on the cytogenetics of lines that contain EB virus (i.e., are EB virus-positive) to be relevant to a book on chromosomes and cancer. It may well be that the EB virus and its infection and transformation of human lymphoid cells will provide a model for the study of the effects of other oncogenic DNA viruses on mammalian cells.

When fibroblasts are transformed by oncogenic viruses, there is a highly predictable loss of contact inhibition, change in cellular morphology, and a marked increase in aneuploidy. The metabolism of the transformed cells differs from that of the nontransformed fibroblasts, and there is a general movement away from the phenotype and genotype of the donor. The lymphocyte in culture is different. Though the lymphocytes of established lines may be thought of as viral transformants, the cells generally retain the chromosome complement of the donor; and, metabolically, the cells reflect the genotype of the donor. Furthermore, despite a morphologic dedifferentiation into blast-like cells (Fig. 1), the lymphocytes in long-term cultures are functionally highly differentiated, retaining their surface receptors (HL-A, complement, immunoglobulin) and producing factors required in both cell-mediated and humoral immunity. It should be stressed that these immunologic and metabolic functions, which are the result of the expression of the donor genome, are maintained despite a seemingly infinite capacity for cellular proliferation in vitro.

In numerous ways, then, the lymphocytes in established lines represent a cross between homoploidy and heteroploidy, a cross between non-neoplastic and neoplastic behavior. We shall here review the evidence for the role of EB virus in the establishment of these lines; the evidence for the oncogenicity of the cells in experimental animals; and, especially, the cytogenetic findings in lymphocyte lines established in a variety of ways from normal and abnormal donors. (Most of these cytogenetic studies were done prior to the arrival of the banding techniques, and, therefore, we are here to consider primarily gross changes in chromosome number or structure.)

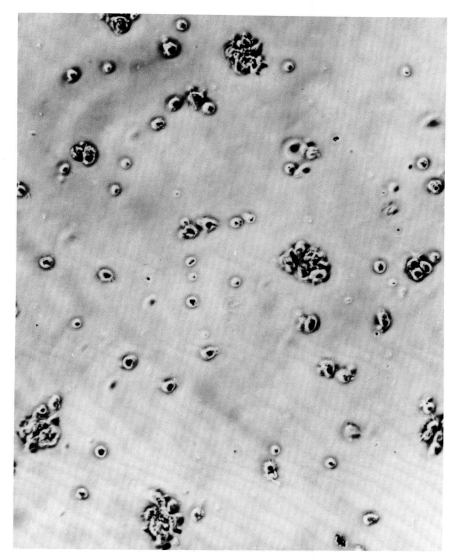

Figure 1. Lymphocytes growing in continuous suspension culture.

ESTABLISHMENT OF LYMPHOCYTE CELL LINES AND THE PRESENCE OF THE EB VIRUS

Most lymphocyte cell lines (LCLs) are initiated from leukocytes of the peripheral blood, though spleen, lymph node, and bone marrow may also be used. Belpomme et al. [5] state that, with few exceptions, no unique differences exist between the morphology of LCLs established from normal donors and those established from patients with hematopoietic disease, regardless of the tissue of origin or the benign or malignant nature of the tissue. Fahey et al. [22] report that in LCLs established by them 85% of the cells resemble lymphoblasts, 10% resemble small and medium-sized lymphocytes, and 5% are multinucleated cells. Grouchy et al. [36] have demonstrated the persistence of macrophages in LCLs as long as two years post-establishment. Rosenfeld et al. [91, 92] presented data on kinetics and mathematical analysis of the establishment of LCLs by the method of Moore et al. [70]. This method consists of the incubation of lymphocytes in mass culture without the addition of phytohemagglutinin (PHA) or EB virus. Establishment through in vitro mutation was considered improbable, since higher mutation rates than observed in vivo would be necessary, and the growth curves during establishment did not fit a mutation hypothesis. The data also did not fit a theory of clonal origin. The data of Moore et al. [73] and our own experience on establishment of mosaic cell lines do not rule out the clonal origin of the lines, but suggest rather that the lines are multiclonal in origin.

Early attempts to establish lymphocytes in long-term culture stressed the importance of initial higher leukocyte densities and an optimum feeding schedule. Osgood and Brooke [85] expressed the view that optimum culturing conditions had to be determined for each donor by means of a "gradient factor" that was the product of cellular density, expressed in thousands of cells per cubic millimeter and the depth of the cells, in centimeters, below the medium surface. By their methods they initiated cultures at 1 to 4×10^6 cells/ml and were able to maintain three cultures obtained from peripheral blood leukocytes of leukemic patients for 6 to 12 months. Iwakata and Grace [48] seeded 65×10^6 leukocytes from a patient with acute myelogenous leukemia onto an X-irradiated feeder layer of human embryonic lung cells (WI-38) to obtain a LCL at 46 days which no longer required the feeder fibroblasts for growth. Foley et al. [23] and Armstrong [2], using high initial cell densities, 2 to 15×10^6 cells/ml, obtained LCLs from buffy-coat cells of acute leukemics. Clarkson [12] established seven LCLs from leukemic blood by incubating leukocytes at high cell densities with and without X-irradiated WI-38 feeder fibroblasts.

In their early work on establishment of LCLs from patients with leukemias and lymphomas, Moore and co-workers [69, 70] optimized the culture conditions for maintenance of such lines. They modified McCoy's 5A medium, for example, by increasing the arginine concentration and developed medium 1640 of the Roswell Park Memorial Institute (RPMI 1640), which has proven to be a consistently effective growth medium for mammalian lymphocytes. Moore's group further found that fetal calf serum (FCS) at 20 to 30% was superior to horse and human serum for maintenance of the lines. From leukemic bloods, they initiated cultures at 4 to 5 × 10^6 leukocytes/ml, with adjustment of the pH of the medium to 7.1 to 7.3. The time of feeding of cells was determined by changes in pH, cell density, and even by time between changes to resupply labile nutrients in slowly metabolizing cultures. Viable cell density, as determined by trypan blue exclusion, was maintained at a minimum of 5 × 10^5 cells/ml by combining cultures or reducing the volume of medium. Using these techniques Moore et al. [71] established two LCLs from normal donors and, subsequently, LCLs from patients with solid malignant tumors [30], from patients with sex chromosome anomalies [74], from patients with nonmalignant diseases [77], and from a normal donor with a mosaic (normal/normal plus minute marker) karyotype [73].

Other workers used basically Moore's principles of high cell density at initiation, maintenance of a minimum viable cell density, timely nutrient replacement, and pH maintenance, and established a number of LCLs from normal donors [25], from patients with infectious hepatitis [100], and from leukocytes stored for two months at −70° in dimethyl sulfoxide and derived from a patient with acute leukemia [4].

Some workers began to use smaller numbers of cells to initiate LCLs from patients with other diseases. Pope [87] and Glade et al. [32] established lines from infectious mononucleosis patients, again by simple incubation of leukocytes into an appropriate medium. Other investigators, using similar techniques, established lymphocyte lines from patients with the Chediak-Higashi syndrome [6], myelofibrosis [98], and viral hepatitis [31]. Lines were also established in this way from normal donors [9].

While numerous LCLs were established, then, without addition of external agents, it became clear that the efficiency of establishment of lines *from certain kinds of donors* was not very high. The ease of establishment of leukemic and infectious mononucleosis lines by incubation alone did not hold, for example, when the donor was a clinically normal person, had an inborn metabolic error, or had a primary immunologic deficiency disorder. One attempt to improve the efficiency of establishing lines from such patients was made by Broder et al. [7], who used a highly purified form of PHA. This PHA (prepared by Burroughs Wellcome Co.) was added at the initiation of culture, and, in some cases, also at weekly intervals until the time of establishment. While this approach improved the efficiency of establishment somewhat, it still did not result in a highly consistent establishment of LCLs.

Before the early successes cited above in establishing LCLs, Pulvertaft [90] and Epstein and Barr [17] obtained long-term cultures of lymphoblasts from tissue of the Burkitt lymphoma. These cultures were subsequently found to harbor a herpeslike virus [18] which was to become known as the Epstein-Barr virus. At present, there are few if any LCLs which do not harbor the EB viral genome [55]. It is now believed, therefore, that the "spontaneous" establishment of LCLs is the result of transformation by EB virus. With 70 to 85% of the normal adult population having antibody to EB virus [24, 61], it is likely that stimulation with PHA and use of fibroblast feeder layers has little if any effect on increasing the efficiency of establishment of LCLs. Cultures from EB virus-positive individuals probably have a significantly increased likelihood of establishment when compared with cultures from EB virus-negative persons.

Numerous experiments on this issue tend to support these views. Henle et al. [40] co-cultivated leukocytes from the peripheral blood of female infants with X-irradiated cells from two male Burkitt lymphoma lines: one line, Jijoye, was known to be EB virus-positive, the other, Raji, was EB virus-negative by electron microscopy. Donor and X-irradiated cells were cultured alone and together on WI-38 fibroblasts. Only those cultures of donor lymphocytes plus X-irradiated Jijoye cells established LCLs.

Pope et al. [88, 89] successfully established LCLs from fetal leukocytes, obtained from thymus, bone marrow, and spleen, by co-cultivation with filtrates from an EB virus-positive LCL established from a leukemic donor. Filtrates of the Raji Burkitt lymphoma cell line (EB virus-negative) were not effective.

Gerber et al. [28] obtained leukocytes from a healthy adult who showed no antibodies to EB virus by complement-fixation testing. Three attempts to establish a LCL were unsuccessful. Leukocytes were then cultivated in five systems:

(1) Donor cells + culture medium.
(2) Donor cells + Raji culture medium (EB virus-negative).
(3) Donor cells + EB virus neutralized with EB virus-positive serum.
(4) Donor cells + EB virus treated with EB virus-negative serum.
(5) Donor cells + 400 times concentrated EB virus.

Only the culture from system 4 survived to establish at 24 days. Culture

5 degenerated completely at 18 days, and the remainder of the cultures were lost at 40 to 45 days. A repeat experiment using the same donor cells with 100-times-concentrated EB virus yielded a second LCL. In a later experiment, Gerber and Hoyer [29] were to show that EB-virus infection of human leukocytes results in cellular DNA synthesis demonstrable by incorporation of ^{3}H-thymidine.

In our laboratory at the University of Michigan, we have used a lysate method for establishing LCLs since 1969 [10, 11]. In essence, lysates of LCLs from normal donors are prepared by freezing and thawing. From 10 ml of blood with a normal white cell count, we separate leukocytes for inoculation of four flasks, with a lysate of cells of the opposite sex being added in 10 ml of RPMI 1640, 20% fetal calf serum, and antibiotics. Establishment of LCLs occurs at 4 to 6 weeks, in 75 to 80% of cases. We have successfully established cultures from a wide variety of donors of all ages, from normal donors to patients with inborn metabolic errors, immunodeficiency diseases, and neoplastic disease of the lymphoid tissue. All of our lines that have been tested are EB virus-positive, as demonstrated by immunofluorescence, and we, therefore, assume that the EB virus in these lysates is the agent that induces transformation and ultimate establishment of the lines.

Since most LCLs are immunoglobulin-producing, it is reasonable to believe that bursa-dependent lymphocytes (BL) comprise a major population of lymphocytes in the lines. Whether thymus-dependent lymphocytes (TL) are also present is not clear, but the LCLs do produce many of the factors (MIF, macrophage migration-inhibition factor; lymphotoxin; BF, blastogenic factor) involved in cell-mediated immunity. The relative nonspecificity of the assays for these factors causes us to hesitate to conclude that the LCLs are a mixture of T and B lymphocytes. Absence of a specific antigen for human T cells, like the theta antigen of mouse T cells, precludes a definitive statement on this point at the present time.

That EB virus is the transforming agent in the establishment of most lymphoid lines is further verified by the technique used by Baumal et al. [3]. Leukocytes were obtained from peripheral blood of healthy donors demonstrating a delayed skin reaction to tuberculin (PPD) or to streptokinase-streptodornase (SKSD). Cells were incubated for 72 hours with or without PPD or SKSD. The cells were collected by centrifugation and EB virus was added for 1 hour at 37°C, after which time the mixture was diluted with culture medium. After 6 weeks, LCLs were obtained only from cells stimulated with appropriate antigen *and* EB virus. Use of specific antigen alone has thus far not resulted in establishment of LCLs that produce specific antibody to the antigen. Furthermore, specific antibody production has not been demonstrated by any LCL, regardless of the method of establishment, though immunoglobulin production is a more or less consistent finding.

In summary, at this time the EB virus appears to play a primary role in the establishment of LCLs, regardless of the status of the donor. The efficiency with which added EB virus results in establishment of these lines and its demonstration in virtually all LCLs point strongly to this agent as causative in the induction of the transformation from small lymphocyte to lymphoblast.

LYMPHOMATOUS DISEASE AND THE EB VIRUS

Since the discovery of the Epstein-Barr virus in cultures of Burkitt lymphoma cells [18], efforts have been directed toward defining the role of this agent in the pathogenesis of the disease. More recently, and as described above, the presence of EB virus has been demonstrated in cultured lymphocytes from patients with a variety of neoplastic diseases (such as leukemia, nasopharyngeal carcinoma, and Hodgkin's lymphoma), in infectious mononucleosis, and even in cell lines established from normal individuals. As has been mentioned, Henle and Henle [40] were the first to propose that EB virus may be the agent which stimulates continuous proliferation of human leukocytes in vitro.

Much of the early interest in the EB virus centered around the seroepidemiological evidence supporting a viral etiology of the Burkitt lymphoma. The restricted geographical distribution of this disease indicated an apparent dependence on climatic factors. The tumor occurs primarily, though not exclusively, in Africa, in areas of low altitude and high humidity [8]. The time and space clustering of the tumor showing epidemic "drift" is characteristic of infectious conditions [87]. Restriction of the disease to the young, except among migrants from low-risk areas to high-risk areas, suggested that these child and adult migrants might be lacking an immunity that is present in adults who reside in the high risk areas.

Attempts to demonstrate directly the presence of a virus in biopsies of Burkitt lymphomas by the use of electron microscopy have failed [16]. However, in 1964, Epstein and Barr [17] reported the in vitro cultivation of lymphocytes from patients with the Burkitt lymphoma. Upon examination of the cells by electron microscopy, a small number (1 to 2%) of the cells were found to harbor herpes-like viruses [18]. Henle and Henle [42] suggested that the reason EB virus has been observed by electron microscopy in cell cultures of Burkitt tumors, but as yet not in tumor biopsies, might be the suppression of the infection in vivo by EB viral antibodies.

Upon cultivation of the cells, however, the virus might no longer be restrained.

Immunofluorescence studies indicate that the EB viruses from different Burkitt lymphoma cell lines are identical and that the virus is distinct from other known human herpesviruses [20]. Immature virus particles can be seen in both the nucleus and the cytoplasm. They have a diameter of approximately 75 to 80 mμ. The particles mature by budding at cellular membranes and thus acquire an additional membrane. These mature virus particles are polygonal, mostly hexagonal, and their diameter is about 115 to 120 mμ [35] (Fig. 2). Similar virus particles have since been observed by electron microscopy in lymphocyte lines established from persons with various types of leukemia [48, 110], from patients with various types of cancer [50], from patients with infectious mononucleosis [77], and from normal individuals [25, 71]. Since the discovery of the virus in cultured lymphocytes of diverse origins, seroepidemiological studies have been carried out to assess the relevance of EB virus in the etiology of both benign and malignant leukoproliferative diseases of man.

Klein et al. [51] reported the presence of an antigen common to Burkitt cells which was detected by the membrane-immunofluorescence test on living, target cells. They found that the presence of the membrane antigen appeared to depend upon the presence of EB virus [52]. Cultured blast-like cells derived from patients with infectious mononucleosis possessed membrane antigens similar, possibly identical, to those found on the Burkitt lymphoma cells obtained from biopsies or from continuous cultures [53]. Henle and Henle [39] used the indirect immunofluorescence test on acetone-fixed Burkitt cells to detect the presence of EB viral antigen. Klein et al. [54] suggested that the antigens involved in the two types of immunofluorescence tests are distinct, since the number of cells showing membrane immunofluorescence generally exceeded the number containing EB viral antigen by a factor of 10 or more. Further, sera were found that were highly reactive in the immunofluorescence test but had low titers against EB virus; and the reverse was also true.

Henle and Henle [39] employed the indirect immunofluorescence technique on acetone-fixed Burkitt cells to detect humoral antibodies against EB viral antigens in the sera of Burkitt lymphoma patients. They also tested sera from American controls, including healthy donors and patients suffering from various diseases. All of the sera from patients with the Burkitt lymphoma gave positive reactions. The Henles observed that immunofluorescence was also produced with many sera from the Americans, the incidence increasing from about 30% in children to 90% in adults. Similar results were obtained by Levy and Henle [61] and Moore [69]. Levy and Henle reported that 100% of the sera from patients with the Burkitt lym-

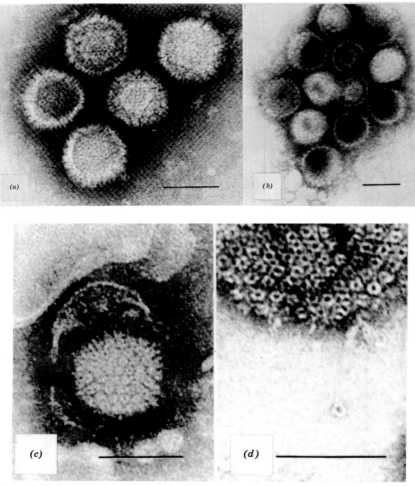

Figure 2. Electron micrographs of negatively stained virus particles in the P3HR-1 Burkitt lymphoma cell line. The bar in each figure represents 1000 Å. (From [109]; reprinted by permission of the American Society for Microbiology.)

2a,b: Two groups of unenveloped virus particles of different appearance.

2c: Full capsid enclosed in an envelope; hexagonal packing of capsomeres is seen on the surface of capsid.

2d,e: Two disintegrated capsids; polygonal cross section of capsomeres and hexagonal packing of them are clearly visible. A free-lying capsomere released by disintegration of capsid is seen in the lower part of d.

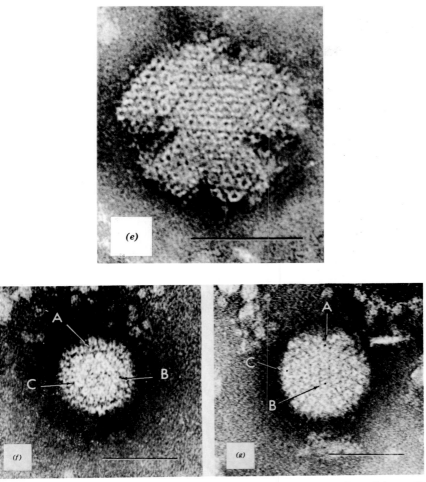

2f,g: Two capsids, each showing an equilateral triangular facet (indicated by A, B, and C) with five capsomeres on its base (A-B). Four capsomeres (A and B in each figure) are seen surrounded by five neighbors and situated on an axis of fivefold symmetry, respectively.

phoma gave positive results in the immunofluorescence test, whereas only 54% of the sera from African control children were positive. Also, they noted a difference in the intensity of the reactions between the sera from Burkitt patients and controls, the control group giving much less intense reactions, presumably indicative of a lower antibody titer. The percentage of fluorescent cells in the continuous cultures correlated well with the number of cells containing viral particles as determined by electron microscopy [39]. The percentage of cells which fluoresced or were observed to harbor EB virus usually ranged from 0.5 to 10%, depending upon the line examined. When the fluorescing cells were isolated and examined under the electron microscope, they were found to harbor EB virus, whereas the nonfluorescing cells did not [112].

Gerber [26] studied complement-fixing antibodies to EB virus and showed that nearly all of the sera from newborns contained such antibodies (presumably of maternal origin); about 30% of the children ages 1 to 2 years old, and about 90% of the adult population also had antibodies to EB virus.

Another method employed to test for the presence of EB virus in cells is the DNA-DNA hybridization technique. Zur Hausen and Schulte-Holthausen [114] studied the homology between the DNA from purified EB virus and the nucleic acid from a nonvirus-producing Burkitt line (Raji). When the Raji line was examined by the immunofluorescent technique and by electron microscopy, it was reported to be EB virus-negative [19]. However, using DNA-DNA hybridization, zur Hausen calculated that Raji cells contain an average of six EB virus genome equivalents per cell. Furthermore, hybridization of viral DNA with Raji cellular RNA indicated that the viral genome was being transcribed.

Nonoyama and Pagano [81] employed the DNA–RNA hybridization technique (cellular DNA: viral complementary RNA) and estimated that there were approximately 65 EB viral genome equivalents per Raji cell. The difference in these results may be due to the low specific radioactivity of ^{3}H–DNA as compared to ^{3}H–cRNA. Nonoyama and Pagano performed DNA–RNA hybridization on three established human leukocyte lines not originating from the Burkitt lymphoma. All were negative for EB virus by immunofluorescence, but by the hybridization technique the lines were found to contain EB viral DNA, ranging from 45 to 100 genome equivalents per cell. In fact, no human lymphoblastoid lines have been found yet which do not contain nucleic acid sequences that are homologous with EB virus [55]. One must keep in mind, however, that the low amount of hybridization in some lymphocyte lines may merely reflect the existence of sequences of cellular DNA similar, and not identical, to EB viral DNA.

It has been suggested that lymphoblastoid cells cannot grow in continuous culture unless they carry the EB viral genome. It may be that the viral genome is present in all cells of a line, though only a small proportion of the cells are detected to have EB viral antigens. This hypothesis is supported by the work of Mauer et al. [66], who studied the incidence of EB virus-containing cells in clones of several Burkitt lymphoma lines. Lines with a low incidence of EB virus-containing cells (less than one fluorescent cell per thousand) could produce virus-containing clones. Recloning of virus-negative clones gave rise to secondary clones that were virus-producing. It was concluded that the "virus-negative" cell lines actually do contain EB virus in some sort of carrier state (an episome?), which is maintained by vertical transmission of the virus-producing material. This is in accord with the results of DNA–DNA hybridization which detected viral genome even in "virus-negative" lines.

Using DNA–DNA hybridization, EB viral DNA has now been detected in biopsy material obtained from patients with the Burkitt lymphoma and those with nasopharyngeal carcinoma (NPC) [115]. The DNA derived from biopsies of these tumors annealed with EB viral DNA to a greater extent than the DNA from malignant tumors of other types. Previously, Old et al. [83] had tested sera for precipitating antibody to EB virus by the double-diffusion method and had found that the incidence of positive reactions was high in two groups of patients—those with the Burkitt lymphoma and those with NPC. Klein and co-workers found that many African and Chinese patients with NPC had high titers of antibodies ($\geqslant$1:160) to the membrane antigen of EB virus [55, 96], and Henle et al. found high titers of anti-EB virus in NPC patients [43]. de-Thé et al. [15] detected the presence of herpes-type viral particles in cultured lymphoblastoid cells derived from a patient with NPC. The above findings suggest a possible serological association between EB virus and NPC.

Epidemiological and immunological evidence has also implicated EB virus as a causal agent in heterophile-positive infectious mononucleosis (IM). A single attack of IM usually confers lasting immunity. Henle et al. [41] first reported a relationship between EB virus and IM when they noticed that an immune response to EB virus develops during the course of the illness. They were able to grow the leukocytes from a patient with IM on a monolayer of human diploid cells; attempts to do this before the onset of the disease had been unsuccessful. Also, before the development of IM, the patient's serum had been negative for antibodies to EB virus; however, shortly after the patient contracted the disease, the plasma was shown to have a high titer of anti-EB virus. Niederman et al. [80] studied 29 patients with IM in each of whom the pre-illness serum was negative for anti-EB virus. In all the patients, antibodies to EB virus appeared early in the course of the disease. Gerber et al. [27] showed that complement-fixing antibodies to herpes-like virus were absent in pre-infectious IM sera but that the antibodies develop during the course of the disease. Moses et al. [77] detected the presence of herpes-like viruses in cell lines established from individuals with IM.

If EB virus is etiologic in IM, the disease apparently results only when non-immune or possibly partially immune older children or young adults are infected. Infection in young children is inapparent, though antibodies do appear. Epidemiological analysis by Evans et al. [21] showed that susceptibility to IM was correlated with the absence of antibodies to EB virus.

While EB virus may be the causative agent in IM, it is also possible that the virus is simply a passenger virus that is activated nonspecifically during the course of the disease, thus resulting in the appearance of antibodies to EB virus. However, the probable etiological relationship of EB virus to IM was solidified when Grace et al. [33] demonstrated that human volunteers who were injected with herpes-like virus isolated from Burkitt cells developed IM.

EB virus has also been suggested as the causal agent in Hodgkin's disease. Epidemiological data link some cases of Hodgkin's disease with a virus. It has been observed that there is some familial aggregation of this disease [65]. It has also been observed that there is a seasonal variation in the onset of the disease [14]. Stewart et al. found herpes-type virus in a lymphocyte line from a patient with Hodgkin's disease [102], and high antibody titers to EB virus have been observed in patients with certain forms of the disease [60].

The role of EB virus in the etiology of any human malignancy or disease is far from established. With a suspected human tumor virus, it is difficult to devise experiments which show *conclusively* that the virus is, in fact, etiologically related to a particular malignant disease.

It is possible that EB virus is the etiological agent of infectious mononucleosis and perhaps a co-carcinogen with some other factor in the Burkitt lymphoma. Alternatively, EB virus may simply be a passenger virus with an attraction for poorly differentiated lymphoblastoid cells. Henle et al. [40] suggested that EB virus may be responsible for the long-term in vitro proliferative potential of the circulating cells. It is also possible that the agent is present in the host cells and that the proliferation in vitro is caused by another stimulus that results in the concomitant replication of the EB virus. But the presence of EB virus in continuously cultivated lymphocyte lines clearly is not in doubt, whatever the role of the virus in human disease.

ONCOGENICITY OF LYMPHOCYTE CELL LINES

The experiments of Adams and colleagues [1] on the heterotransplantation of LCLs has involved pathologic, karyotypic, and immunologic characterization of tumors produced in immunologically suppressed, neonatal hamsters. Experiments of these workers focused on transplantation of lines established both from patients with malignant, lymphomatous diseases (such as lymphosarcoma, lymphoma, and lymphatic leukemia) and from normal individuals. The LCLs used in these experiments were established from lymph node, spleen, peripheral blood, or bone marrow. Cells were given intraperitoneally or subcutaneously to neonatal golden Syrian hamsters suppressed with antilymphocyte serum (ALS). The lines were serially passed, and evidence of both primary site and metastatic tumors was found, regardless of the normality or abnormality of the donor. Thus lymphoid cells established from normal persons were equally capable of forming tumors under these experimental conditions, as were lines established from patients with Hodgkin's lymphoma, Letterer-Siwe's disease, or infectious mononucleosis. Adams described the tumors as "progressively growing, fatal, serially transplantable, locally invasive, and metastatic." The proportion of injected hamsters that developed tumors ranged from 37 to over 80%.

Furthermore, many of the transplanted human lines were immunoglobulin-producing in vitro, and human immunoglobulins were detectable by immunodiffusion in the serum of the injected hamsters. Such tumors have, therefore, been classified by Adams et al. [1] as "malignant immunoblastomata." Experiments in our own laboratory have been less conclusive. We have used a different strain of golden Syrian hamster, and have injected over 200 of them with IgG- and IgM-producing normal lymphocyte lines derived from normal donors. Less than 5% of these hamsters had detectable levels of human immunoglobulin up to 41 days post-injection. We did not attempt, however, to correlate macro- or microscopically visible tumor formation with these transplants, so that we have as yet no evidence for the persistence of these transplanted cells in the recipients.

Schneider [95] has reviewed the potential biohazards of established lymphocyte lines. He points out quite properly that the large quantities of cells grown by some investigators might result in release of considerable numbers of viral particles. Use of mutagenic agents, such as fluorodeoxyuridine (FUDR), might well increase viral production and release. Lastly, cross-contamination of cultures suggests that the cells may be passed by aerosol spread.

From the evidence of the oncogenicity of the lines in the heterotransplantation experiments, and from the theoretical reasons to think that virus release may be enhanced under certain in vitro conditions, it is clear that reasonable precautions must be taken in the handling of lymphocyte lines. In our view, these cells should be treated as active virus. They should be passaged only in *vertical* laminar flow hoods, for example, with no mouth pipetting of media or cells, and with caution being used in spills and medium disposal.

CYTOGENETICS OF LYMPHOCYTE CELL LINES FROM NORMAL DONORS

The general experience with LCLs established from normal donors is that a clear majority of cells of most lines accurately reflect the karyotype of the donor. This is true regardless of the method used for establishment of the lines.

As seen in Table 1, our data strongly support this concept. In terms

Table 1. Cytogenetic Analysis of Normal Lymphocytic Cell Lines[*]

Lymphocyte Cell Line	Date Established	Percentage Cells with Normal Karyotype	Percentage Cells with Aneuploidy[†]	Percentage Cells with Polyploidy[‡]
UM-3	2-17-70	73–93	0–27	0–10
UM-4	2-17-70	86	10	2
UM-5	3-12-70	86	2	12
UM-6	3-15-70	82	12	6
UM-8	3-24-70	77–97	13	0–10
UM-9	4-20-70	96	2	2
UM-12	5-16-70	87–97	3	5–10
UM-13	5-23-70	90	3	7
UM-14	5-29-70	80–97	3	7–17
UM-15	5-30-70	80–97	2–3	3–20
UM-16	6-1-70	90–97	3	3–7
UM-17	6-7-70	93–97	3	3–7
UM-19	6-23-70	93–97	3–7	1
UM-26	11-9-70	83–87	3–10	0–1
UM-27	11-18-70	90	3	7
UM-39	4-5-71	97	3	0
UM-54	3-23-72	97	0	3
UM-56	3-15-72	100	0	0
UM-61	9-18-72	100	0	0

[*] Lines screened serially once every 3 to 6 months.
[†] The majority were hypodiploid.
[‡] Virtually all were tetraploid.

of chromosome number, among the 19 normal lines followed serially by us for up to three years, the percentage of cells with the diploid karyotype ranged from 73 to 100. As in short-term (48- to 72-hour) cultures of PHA-stimulated cells, the most common aneuploid cells were hypodiploid. In long-term cultures, polyploidy was seen in up to 20% of cells, with an average of 2 to 8%.

Occasionally, as illustrated by UM-1 (Fig. 3), pseudodiploidy was seen. UM-1 was established in 1969 by use of a lysate of an established infectious mononucleosis line. A small inversion of a member of group C resulting in a chromosome resembling a No. 3 first appeared after six months of continuous culture in what had been a cytogenetically normal line at the time of establishment. The inversion-containing cells increased to 100% by one year of culture, and this karyotype has persisted.

Ten clonal sublines were established from a LCL derived from a patient with hypoimmunoglobulin A. This patient had a 46,XX,rE karyotype (i.e., a ring chromosome in group E), both in short-term cultured cells and in all 30 cells examined at the time of establishment of the cell line. Seven of the clones maintained the same karyotype, but one became polyploid in 60% of its cells, with duplication of the ring, while two clones had 16 to 20% of cells that lost the ring E. Thus, the instability of the ring chromosome in some of the cells of these clones became apparent on continuous cultivation.

Except in the case of mosaicism in the initial cell inoculum, when selection appears to occur in the cultured cells (to be discussed later), the karyotype of a human lymphocyte line will accurately reflect the karyotype of the donor. The instances of an altered karyotype in the line, as our UM-1, are uncommon, at least in studies using conventional cytogenetic techniques (i.e., without banding).

While work is now well under way in a number of laboratories on G-, Q-, and C-banding of LCL chromosomes, we have elected here not to review these findings, since this approach is as yet inconclusive on the frequency with which variants or true polymorphisms are seen. However, such data will undoubtedly reveal more variation than is detectable with the more standard methods. For example, O'Neill and Miles [84] stained cells of five LCLs for constitutive heterochromatin. Twenty percent of cells of line LK60 had an enlarged secondary constriction on one of the No. 1 homologues. Lines NC37 and LK60 were polymorphic for the size of the C-bands on chromosome No. 1. RPMI 6410 cells showed two chromosomes with double C-bands, perhaps suggestive of dicentric formation. One approach that must be followed in such banding studies of the cells of lymphocyte lines is banding of the short-term cultured lymphocytes of the donor. Only in this way may correlations be made

Figure 3. Pseudodiploid karyotype of the UM-1 lymphocyte cell line from a normal male donor. The marker chromosome (arrow) originated in culture and is now present in 100% of the cells.

between variations arising de novo in the established lines and variants that were present in the cells before culture.

It had previously been demonstrated that certain viruses, both oncogenic and non-oncogenic, are capable of inducing chromosomal lesions in apparently normal cells of varoius types. Moorhead and Saksela [76] observed nonrandom chromosomal aberrations in SV40-transformed human cells. Also, the damage caused to chromosomes in vitro by herpes simplex virus has been thought to affect certain chromosomes and chromosomal regions more often than others [37]. Zur Hausen [113] observed nonrandom chromosomal breaks in chromosomes No. 1 and No. 17 in human embryonic kidney cells infected with adenovirus type 12. Other viruses which cause chromosomal damage have been discussed in detail in two recent review articles [38, 79].

A brief statement regarding cells with isolated breaks and rearrangements is in order. While all of our lymphoid lines may be EB virus-producing, the proportion of cells with chromatid- or chromosome-type breaks rarely exceeds the percentages found in short-term cultured cells. We find no increases in these types of aberrations, which is perhaps surprising in the light of the increase in polyploidy, especially tetraploidy, when comparisons are made with short-term cultures. Again, the banding techniques may reveal more subtle types of rearrangements.

CYTOGENETICS OF CELL LINES FROM ABNORMAL DONORS

Most of the interest in the cytogenetics of lymphocyte lines has revolved about the findings in lines established from patients with hematologic disorders and neoplastic disease. As indicated above, most human LCLs, unlike cell lines of epithelial or fibroblastic origni, maintain a diploid or pseudodiploid character even after a prolonged period of time in vitro. In this section we will describe some of the karyotypic changes observed in lymphocyte cultures established from abnormal donors and discuss the theoretical implications of the chromosomal changes.

Abnormal Chromosomal Constitution

LCLs from individuals with abnormal chromosome constitutions, but without malignant disease, have also been established and karyotyped after different times in culture. Moore et al. [73] established an LCL from an individual with 46 normal chromosomes plus one minute marker chromosome. The minute marker persisted in the cultured cells that were examined serially for two years after establishment. Moore et al. [74] established LCLs from patients with 47,XYY and 47,XXY chromosome constitutions. The extra chromosome was still detected in almost all of the cells after 10 to 11 months in culture. In our own laboratory, as mentioned earlier, we established a LCL from a patient with hypogammaglobulinemia A whose two-day culture had shown 46,XX,rE in 100% of the cells. This ring chromosome has been retained in the long-term cultures for 21 months now in the parental line, though some of the clones have partially lost it. It might be anticipated that cultures beginning with cells with unbalanced chromosome constitutions might be more apt to undergo significant change in vitro. However, in the above four cases, the lines have maintained the donors' chromosome constitution for long periods of time in culture.

Our experience with mosaic cell lines is revealing with respect to selection. As seen in Table 2, we have serially harvested cells of lines from mosaic patients, two with Down's syndrome and two with Turner's syndrome. Both Down's syndrome patients had a trisomic line and a normal line. In one case, on two-day culture, 36% of cells had a 47,XX,+G karyotype, and 64% had 46,XX. The first cytogenetic analysis, shortly after establishment of the line, showed only 12% of cells to be trisomic, and three months later, no trisomic cells were seen. The other mosaic had, on two-day culture, 89% of cells with a 47,XY,+G karyotype and 6% 46,XY. On establishment, 93% of cells were trisomic; and on subsequent analyses, all cells have been trisomic, with an additional F trisomy (i.e., double trisomy) appearing in 98% of the analyzed cells.

The first Turner's syndrome patient, age 19, had 50% of cells 45,X and 50% 46,XY in short-term culture. On establishment, and during the first 4 months thereafter, 80% of cells were 45,X, and within 18 months, no XY cells were seen. The second patient, age 16, had a 45,X/46,XrX karyotype, with 67% of cells 45,X, and 28% 46,XrX in short-term culture. On establishment, 95% of cells were 45,X; no XrX cells (i.e., with the ring chromosome) were present.

These findings argue strongly for a high degree of selection in these LCLs, as in other homoploid and heteroploid lines of diverse origins. It is clear that there is no ready way to predict in what direction the selection will take place after establishment, and it is perhaps still more important that a high degree of selection may take place even before the line is established. The argument that each lymphocyte line may be composed of a small number of clones is supported by these cytogenetic findings.

In the largest series of lines established from aneuploid donors, Woods

Table 2. Cytogenetic Analysis of Mosaic Donors

Diagnosis of Donor	2-Day Culture Results	Time after Establishment of Line	Chromosome Findings in Cell Lines	
Down's syndrome (6-year-old girl)	36% 47,XX,+G 64% 46,XX	1 week 5 weeks 3 months	12% 47,XX,+G 4% 47,XX,+G 0% 47,XX,+G	86% 46,XX 86% 46,XX 93% 46,XX
Down's syndrome (4-month-old boy)	89% 47,XY,+G 6% 46,XY	5 weeks 12 months	93% 47,XY,+G 98% 48,XY,+G,+F 2% 47,XY,+G	7% 46,XY
Turner's syndrome (19-year-old girl)	50% 45,X 50% 46,XY	2 weeks 4 weeks 11 weeks 4 months 5 months 8 months 18 months	78% 45,X 78% 45,X 73% 45,X 83% 45,X 90% 45,X 90% 45,X 97% 45,X	20% 46,XY 22% 46,XY 3% 46,XY 7% 46,XY 7% 46,XY 10% 46,XY 0% 46,XY
Turner's syndrome (19-year-old girl)	67% 45,X 28% 46,XrX	3 weeks	95% 45,X 0% 46,XrX	

et al. [108] have reported on the cytogenetic findings in lymphocyte lines established from 18 patients with trisomy 21. Fifteen of these lines remained consistently trisomic. Two became mosaic (46/47; 47/48), and one became 100% polyploid. Two lines we established from patients with nonmosaic sex chromosomal aneuploidy revealed no significant deviation from the donor karyotype: from 86 to 100% of cells in both of these lines retained the original chromosome constitution of the donor.

Infectious Mononucleosis

Continuous LCLs have also been established from persons with IM [86]. Tomkins [103] karyotyped three IM lines (QIMR–STE, QIMR–GOE and QIMR–DAR) at 118, 131, and 152 days, respectively, after establishment and observed that all of the lines consisted of a majority of cells with 46 chromosomes. Kohn et al. [57] obtained similar results on three IM lines they examined. Huang et al. [46] karyotyped 12 IM lines that had been in culture for 6 to 12 months and found that they were all normal diploid lines with the exception of one line in which 14% of the cells contained 47 chromosomes, the extra chromosome usually being one of group C.

Other investigators have observed karyotypic evolution in various IM lines. Steel et al. [99] examined eight lines from patients with IM and noted that one line (GOL_1), which was originally diploid, had become pseudotetraploid after 10 months in culture. A subline of GOL, which initially had a diploid karyotype, showed an increasing number of cells that were 46XX,-C,+abn, Dg+ with increasing time in culture (0% at 7 months; 78% at 11½ months; 96% at 12½ months; and 100% at 13½ months). Macek et al. [63, 64] observed a similar karyotypic evolution in some of the IM lines they examined. One line, IM39 subline A, underwent a shift in the modal chromosome number from diploidy to hyperdiploidy with 47 chromosomes, as it was examined between passages 12 and 249. IM39 subline B, however, was examined at passage 224, and 81% of the cells were hypotetraploid. Complete heteroploidy also developed in four other IM lines Macek examined. In two of these, IM529 (at the 571st passage) and IM566 (at the 688th passage), 73 and 95% of the cells, respectively, were hypotetraploid. Other IM lines were predominantly diploid.

The presence of marker chromosomes was frequently seen in these lines and sometimes, though not always, the percentage of cells with the markers was seen to increase with time in culture. Some of the lines showed an increase in the proportion of cells with an extra group C chromosome with an increased time in culture. As in the Burkitt lines, the group C marker has been observed in some of these lines [46, 57, 63]. However, Steel et al. [99] did not observe it in eight IM lines initiated in his laboratory, and Tomkins [103] did not detect it in the three lines he examined. Here again there seems to be no apparent association between the group C marker and the presence of EB virus.

An extra group C chromosome has been reported in various other LCLs by other investigators. Tomkins [103] observed an extra group C chromosome in some cells of two Burkitt lines (GOR and AMB/1), an IM line (DAR), and a leukemic line (WIL). Toshima et al. [104] reported that the Jijoye Burkitt-lymphoma line had a mode of 47 chromosomes with an extra group C chromosome. The P3HR-1 subline of Jijoye was found to be trisomic for a group C chromosome, probably the No. 10 chromosome [59]. Kurita also noted an extra No. 10 chromosome in two lines derived from nine human-embryo cells exposed to human leukemic cells in vitro. An extra group C chromosome has also been seen in the EB_1, EB_2, and SL_1 Burkitt lines [56] and in lines from normal individuals [28, 44].

These findings suggest a distinct, but not invariable, tendency toward an excess of group C chromosomes in LCLs, particularly in those of Burkitt origin. The evolution of an aneuploid line from the diploid state may be due to a selective advantage of those cells with the abnormal karyotype. The mechanism by which such chromosome aberrations confer an advantage upon these cells is unknown at the present time.

Burkitt Lymphoma

The karyotype of cultured Burkitt lymphoma cells has been described in several laboratories. Most investigators have found the lines to be predominantly diploid or pseudodiploid, though deviations from normalcy have occasionally been observed [78]. Kohn et al. [56] examined five Burkitt lines that were EB virus-positive (EB_1, EB_2, EB_3, Jijoye, and SL_1) and found what was considered to be a specific marker chromosome in all lines except SL_1. This marker was a prominent subterminal secondary constriction on the long arm of the chromosome No. 10. Other investigators have also noted the group C marker chromosome in Burkitt LCLs [45, 59, 67, 68, 103] (Fig. 4). Henle et al. [40] reported that when leukocytes from the peripheral blood of normal donors were established by co-cultivation with X-irradiated cultured Burkitt cells (Jijoye), between 5 and 16% of the established donor cells contained the group C marker chromosome after three to eight months of continuous growth. It was thought that the appearance of the marker in the donor cells was related to their exposure to the X-irradiated Jijoye cells and that perhaps the

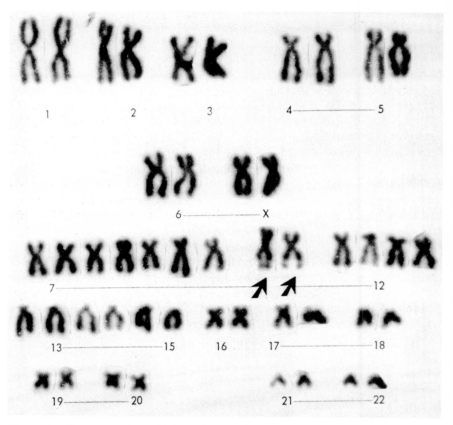

Figure 4. Karyotype of a cell from the EB$_2$ line containing 47 chromosomes, with one extra group C chromosome and two marker chromosomes in group C (from [56]).

marker in the Burkitt lines was due to the presence of EB virus. Kohn [56] postulated that the EB virus is able to attack a specific locus on the chromosome No. 10 and in some way alter it.

Though there have been reports from several different laboratories of the presence of the C marker in different Burkitt LCLs, the overall data are conflicting. Huang et al. [46] karyotyped four Burkitt-lymphoma LCLs (MOB 12, 14, 16, and 17), which were found by immunofluorescence to be EB virus-positive, and yet they found no evidence of the group C marker in any of them. Tomkins [103] observed the group C marker at a low frequency in one Burkitt LCL (QIMR-AMB/2) but not in two others (QIMR-GOR and QIMR-AMB/1). Ikeuchi et al. [47] studied a

Burkitt LCL (B46M) and 10 cloned sublines and found the group C marker to be present in only one of the sublines. Huang et al. [46] suggest that there is no apparent association between the presence of EB virus and the group C marker chromosome. Tough et al. [105] discussed these conflicting reports and proposed that the frequency of chromosomal changes observed in the lines may be a reflection of the length of time the cells have been in culture and may also depend on the varying culture conditions in different laboratories.

The time in culture and the culture conditions may also affect the degree of ploidy seen in Burkitt LCLs. For example, Stewart et al. [101] examined the EB_2 line after it had been in culture for several months and found the modal number of chromosomes to be 46 (98% of the cells), though there was a high percentage of cells with breaks (23%). Kohn [56] found the line to be diploid but observed an increase in the proportion of cells with 47 chromosomes with increasing time in culture. However, Cooper et al. [13] observed that the EB_2 line in their laboratory underwent marked alteration to heteroploidy with most of the cells having 86 to 88 chromosomes. Tough [105] has constructed a chart of the karyotypes of different Burkitt LCLs indicating the laboratory in which the lines were examined and the duration of time in culture when karyotype analysis was performed. The degree of polyploidy, the percentage of cells with the group C marker, and the percentage of cells with other markers varied considerably from laboratory to laboratory, even when the analyses were done after the cells had been in culture for approximately the same time.

An important question with respect to these extensively studied Burkitt LCLs is whether the chromosome aberrations observed reflect the true in vivo chromosome constitution of the tumor cells or whether they were generated in vitro. Jacobs et al. [49] examined short-term cultures of peripheral blood and direct biopsies of tumor tissues from 19 patients with the Burkitt lymphoma. The finding of chromosomal abnormalities in the tumor biopsies of six untreated cases indicated that some of these tumors do, indeed, have cytological changes before cultivation—not a surprising finding. However, the absence of chromosomal abnormalities in four untreated cases indicates that these tumors are not always associated with chromosomal aberrations. None of the blood or bone-marrow preparations showed similar chromosome abnormalities, or the karyotypic changes were restricted to the tumor cells in these cases. On the other hand, the SL_1 Burkitt line was observed to retain two marker chromosomes that were originally present in the direct tumor preparations [101]. Also, Gripenberg et al. [34] examined a case of the Burkitt lymphoma with bilateral tumors and found identical marker chromosomes in direct preparations of tumor-biopsy material and in the established cell lines that had been in culture for six months.

It has, therefore, not been determined whether many of the chromosomal changes that are observed in the different Burkitt LCLs were already represented in low frequency by some of the cells of the initial population or whether they are induced in vitro by culture conditions. It is likely that at least some of the chromosomal abnormalities in cultured Burkitt cells are the result of continuous propagation in culture, while a selective advantage may also exist for some cells with an initially abnormal karyotype.

Malignant Lymphoma other than Burkitt

Cell lines from malignant lymphomas other than the Burkitt have also been established. Saksela and Pontén [93] followed the karotypic evolution of two lymphoid lines (61M and 129M) derived from patients with Hodgkin's disease. They found that with an increased time in culture, an increasing, though small, fraction of the cells had abnormal chromosomes. The changes were clonal in character, and the pattern of changes suggested the presence of competing clonal cell populations. Spiers and Baikie [97] examined the karyotypes of short-term cultures of lymph-node material from several patients with Hodgkin's disease and found numerous chromosomal anomalies. It is not known whether the cell lines examined by Saksela and Pontén represented clones that were selectively grown from cells present in the lymph nodes in vivo or whether the chromosomal changes evolved in vitro—again, the Burkitt lymphoma dilemma.

Leukemia

Lymphoblastoid cell lines have also been established from patients with different types of leukemia. These lines frequently maintain a diploid karyotype after continuous growth in culture, though some lines do undergo karyotypic evolution. Zur Hausen [111] studied seven established cell lines from patients with leukemia (one with monocytic leukemia, three with myeloblastic leukemia, and three with lymphocytic leukemia). All lines had predominantly diploid chromosome numbers, though one of the myeloblastic lines contained some marker chromosomes. In all cell lines, prominent secondary constrictions were found in the centromeric regions of chromosomes Nos. 1 and 16. Krishnan and Raychaudhuri [58] observed that the CCRF-SB line established from the peripheral blood of a patient with acute lymphoblastic leukemia maintained a diploid karyotype for over three years in culture. Two chronic lymphocytic leukemia lines established in our laboratory (UM-57 and UM-58) have normal karyotypes.

Clarkson et al. [12] obtained similar results on seven acute leukemia lines that had been in culture for 6 to 18 months. However, one line (SK-L4) showed slight karyotypic variation from the beginning, and another line (SK-L3) developed into a mosaic line after 14 months in culture, with 50% of the cells having 47 chromosomes including an extra member of group C. Macek et al. [63] examined three acute leukemia lines: he found one (L77) to have a predominantly normal-diploid karyotype after 50 passages in culture; one (L48) had two cell populations (one normal-diploid and one pseudodiploid) after 127 passages; and one (L56) demonstrated a shift to hyperdiploidy, with one to two extra group C chromosomes, after 23 passages. Steel et al. [99] describe the karyotypic evolution they observed in four established leukemia lines but found that there is no evidence for the recurrence of any specific chromosomal aberration in the different cell lines. Tomkins [103] karyotyped a lymphoblastic leukemia line (QIMR-FIN) that had been in culture for 31 weeks and a myeloblastic leukemia line (QIMR-WIL) in culture for 104 weeks. The cells of QIMR-FIN were predominantly diploid. In QIMR-WIL, 64% of the cells had 46 chromosomes, and 31% had 47 chromosomes. In our laboratory we established a line (UM-51) from a patient with acute undifferentiated leukemia. Of 48 metaphases examined in a two-day, PHA-stimulated culture, seven were 46,XY and 41 were 46,XY,-D,+Mar karyotype. The marker was lost on establishment.

How do these in vitro karyotypes compare with the karyotypes from short-term cultures from leukemic patients? Cytogenetic abnormalities frequently occur in the neoplastic cells of leukemic patients, but their significance is difficult to assess. Sandberg et al. [94] compared the chromosome constitution of marrow cells in 113 cases of *acute myeloblastic* leukemia and 106 cases of *acute lymphoblastic* leukemia. They found aneuploidy in approximately 50% of all cases and diploidy in the remaining 50%. There was variability from case to case, and the only consistent chromosomal change was group C trisomy, which occurred in a small number of cases. Because of the great variability observed, Sandberg suggested that the chromosomal changes may be secondary to the leukemic state, rather than causative.

Whang-Peng et al. [107] examined the chromosomes of the bone marrow and the peripheral blood of 45 patients with *acute lymphocytic* leukemia, and they also commented on the great variability of the chromosome constitutions, concluding that there were no unique abnormalities associated with acute lymphocytic leukemia. Nowell and Hungerford [82] reported that they found no consistent abnormality in chromosome number or morphology in three patients with *chronic lymphocytic* leukemia. Trujillo et al. [106] compared the karyotype of a lymph node biopsy from a patient with chronic lymphocytic leukemia with the karyotype of short-term cultures of peripheral white blood cells and of a continuous lymphocyte culture established from the same patient. The lymph-node biopsy and the short-term culture revealed a normal chromosome constitution. However, the continuous culture showed a shift to increased aneuploidy at about six months, with hyperdiploid and pseudotetraploid cells increasing. Later, the pseudotetraploid cells became predominant.

Moore [70] examined four lines established from a chronic myelocytic leukemia (CML) patient in which about half of the metaphases from bone-marrow biopsies contained the Ph^1 chromosome. None of the cells from the lymphocyte cultures were seen to have the Ph^1 chromosome. Similarly, Miles [68] did not observe the Ph^1 chromosome in two lines he examined which were established from patients with CML. In our laboratory, we established a line from the bone marrow of a patient with CML (UM-22). The karyotype of the uncultured marrow cells was 46,XY, and six out of seven metaphases contained the Ph^1 chromosome. However, the Ph^1 chromosome was not seen in any cells of the established cultures, raising the question of whether we had established the leukemic myeloblasts or bone-marrow lymphocytes in the line.

Lucas et al. [62] reported the establishment of a LCL (M-2) from a patient with CML. The karyotypes from the patient's bone marrow, short-term blood cultures, and the M-2 line were compared. One hundred percent of the bone-marrow cells and the cells from the one-day culture contained the Ph^1 chromosome. Seventy percent of the cells from a three-day-old culture and 59% of the cells from a 113-day-old culture contained the Ph^1 chromosome. However, when Moore et al. [72] studied this line several months later, they were unable to detect the presence of the Ph^1.

SUMMARY

In general, when a lymphocyte cell line (LCL) is established from a nonmosaic donor whose karyotype is diploid or aneuploid, the karyotype of the donor is retained. When the donor has a hematologic disorder, like infectious mononucleosis, or a frankly neoplastic disease, like the Burkitt lymphoma or one of the leukemias, the chromosomal variation is considerably increased. Since most LCLs, regardless of origin, contain EB virus genome, these cytogenetic findings may represent a differential sensitivity to the effects of the EB virus.

We have discussed the role that EB virus might play in the transformation of lymphocytes in vitro. The heterotransplantation experiments suggest the oncogenic potential these cells may possess. It would be of interest to determine whether EB virus causes any consistent subtle chromosomal aberration (since no consistent major chromosomal aberration has been observed). It is now thought that EB virus confers on human lymphocytes the property of sustained proliferation in vitro.

Acknowledgements.—This work was supported by a Program Project Grant from the National Institute of General Medical Sciences (NIH-1-PO1-GM-15419-06).

LITERATURE CITED

1. ADAMS RA, FOLEY GE, FARBER S, HELLERSTEIN EE, POTHIER L: Oncogenicity of cell cultures isolated from normal subjects and from patients with infectious mononucleosis, Letterer-Siwe's disease or Hodgkin's disease. In Long-Term Lymphocyte Cultures in Human Genetics (Smith GF, Bloom AD, eds), New York, The National Foundation–March of Dimes, 1973.
2. ARMSTRONG D: Serial cultivation of human leukemic cells. *Proc Soc Exp Biol Med* 122:475–481, 1966.
3. BAUMAL R, BLOOM B, SCHARFF MD: Induction of long-term lymphocyte lines from delayed hypersensitive human donors using specific antigen plus Epstein-Barr virus. *Nature New Biol* 230:20–21, 1971.
4. BELPOMME D, SEMAN G, DORE JF, VENUAT AM, BERUMEN L, LE BORGNE C, DE KAOVEL, MATHE G: Established cell line obtained from frozen human lymphoblastic leukemia cells: morphological and immunological comparison with the patients' fresh cells. *Eur J Cancer* 5:55–59, 1969.
5. BELPOMME D, MINOWADA J, MOORE GE: Are some human lymphoblastoid cell lines established from leukemic tissues actually derived from normal leukocytes? *Cancer* 30:282–287, 1972.
6. BLUME RS, GLADE PR, GRALNICK HR, CHESSIN N, HAASE AT, WOLF SM: The Chediak-Higashi syndrome: continuous suspension cultures derived from peripheral blood. *Blood* 33:821–832, 1969.
7. BRODER SW, GLADE PR, HIRSCHHORN K: Establishment of long-term lines from small aliquots of normal lymphocytes. *Blood* 35:539–542, 1970.
8. BURKITT D: Determining the climatic limitations of a children's cancer common in Africa. *Brit Med J* 2:1019–1023, 1962.
9. CHANG RS: The loss of growth vitality of human lymphoid cell lines derived from healthy adults. *Proc Soc Exp Biol Med* 135:212–215, 1970.
10. CHOI KW, BLOOM AD: Cloning human lymphocytes in vitro. *Nature* 227:171–172, 1970.
11. CHOI KW, BLOOM AD: Biochemically marked lymphocytoid lines: establishment of Lesch-Nyhan cells. *Science* 2:89–90, 1970.
12. CLARKSON BA, STRIFE A, HARVEN E DE: Continuous culture of seven new cell lines (SK-L1 to 7) from patients with acute leukemia. *Cancer* 20:926–947, 1967.

13. Cooper TH, Hughes DT, Topping NE: Kinetics and chromosome analyses of tissue culture lines derived from Burkitt lymphomata. Brit J Cancer 20:102–113, 1966.
14. Cridland MD: Seasonal incidence of clinical onset of Hodgkins disease. Brit Med J II:621–623, 1961.
15. de The G, Ambrosioni JC, Ho HC, Kwan HC: Lymphoblastoid transformation and presence of herpes-type viral particles in a Chinese nasopharyngeal tumor cultured in vitro. Nature 221:770–771, 1969.
16. Epstein MA, Herdson PB: Cellular degeneration associated with characteristic nuclear fine structural changes in the cells from two cases of Burkitt's malignant lymphoma syndrome. Brit J Cancer 17:56–58, 1963.
17. Epstein MA, Barr YM: Cultivation in vitro of human lymphoblasts from Burkitt's malignant lymphoma. Lancet 1:252–253, 1964.
18. Epstein MA, Achong BG, Barr YM: Virus particles in cultured lymphoblasts from Burkitt's lymphoma. Lancet 1:702–703, 1964.
19. Epstein MA, Achong BG, Barr YM, Zajac B, Henle G, Henle W: Morphological and virological investigations on cultured Burkitt tumor lymphoblasts (strain Raji). J Nat Cancer Inst 37:547–559, 1967.
20. Epstein MA, Achong BG: Observations on the nature of the herpes-type virus in cultured Burkitt lymphoblasts using a specific immunofluorescence test. J Nat Cancer Inst 40:609–621, 1968.
21. Evans AS, Niederman JC, McCollum RW: Seroepidemiologic studies of infectious mononucleosis with EB virus. New Engl J Med 279:1121–1127, 1968.
22. Fahey JL, Buell DN, Sox HC: Proliferation and differentiation of lymphoid cells: studies with human lymphoid cell lines and immunoglobulin synthesis. Ann NY Acad Sci 190:221–234, 1971.
23. Foley GE, Lazarus H, Farber S, Uzman BG, Boone BA, McCarthy RE: Continuous culture of human lymphoblasts from peripheral blood of a child with acute leukemia. Cancer 18:522–529, 1965.
24. Gerber P, Birch SM: Complement-fixing antibodies in sera of human and non-human primates to viral antigens derived from Burkitt's lymphoma cells. Proc Nat Acad Sci (US) 58:478–484, 1967.
25. Gerber P, Monroe JH: Studies on leukocytes growing in continuous culture derived from normal human donors. J Nat Cancer Inst 40:855–866, 1968.
26. Gerber P, Rosenblum EN: The incidence of complement-fixing antibodies to herpes simplex and herpes-like viruses in man and Rhesus monkeys. Proc Soc Exp Biol Med 128:541–546, 1968.
27. Gerber P, Hamre D, Moy RA, Rosenblum EN: Infectious mononucleosis: complement-fixing antibodies to herpes-like virus associated with Burkitt lymphoma. Science 161:173–175, 1968.
28. Gerber P, Whang-Peng J, Monroe JH: Transformation and chromosome change induced by Epstein-Barr virus in normal human leukocyte cultures. Proc Nat Acad Sci (US) 63:740–747, 1969.
29. Gerber P, Hoyer BH: Induction of cellular DNA synthesis in human leukocytes by Epstein-Barr virus. Nature 231:46–47, 1971.
30. Gerner RE, Moore GE: Establishment of leukocyte cultures from patients with solid malignant tumors. Surg Forum 76–77, 1967.

31. GLADE PR, HIRSHAUT V, DOUGLAS SD, HIRSCHHORN K: Lymphoid suspension cultures from patients with viral hepatitis. *Lancet* 2:1273–1275, 1968.
32. GLADE PR, KASEL JA, MOSES HL, WHANG-PENG J, HOFFMAN PF, KAMMERMEYER JK, CHESSIN LN: Infectious mononucleosis: continuous suspension culture of peripheral blood leukocytes. *Nature* 217:564–565, 1968.
33. GRACE JT, BLAKESLEE J, JONES R: Induction of infectious mononucleosis in man by the herpes-type virus (HTV) in Burkitt lymphoma cells in tissue culture. *Proc Am Assoc Cancer Res* 10:31, 1969.
34. GRIPENBERG U, LEVAN A, CLIFFORD P: Chromosomes in Burkitt lymphomas. I. Serial studies in a case with bilateral tumors showing different chromosomal stemlines. *Int J Cancer* 4:334–349, 1969.
35. GROSS L: Oncogenic Viruses, 2nd ed, Headington Hill Hall, Oxford, Pergamon Press, p 830, 1970.
36. GROUCHY J DE: From a film presentation at the Fourth International Congress of Human Genetics, Paris, September 6–11, 1971.
37. HAMPAR B, ELLISON SA: Cellular alterations in the MCH line of Chinese hamster cells following infection with herpes simplex virus. *Proc Nat Acad Sci (US)* 49:474–480, 1963.
38. HARNDEN DG: Viruses, chromosomes and cancer: the interaction between viruses and chromosomes. This volume, 151–190, 1974.
39. HENLE G, HENLE W: Immunofluorescence in cells derived from Burkitt's lymphoma. *J Bacteriol* 91:1248–1256, 1966.
40. HENLE W, DIEHL V, KOHN G, ZUR HAUSEN H, HENLE G: Herpes-type virus and chromosome marker in normal leukocytes after growth with irradiated Burkitt cells. *Science* 157:1064–1065, 1967.
41. HENLE G, HENLE W, DIEHL V: Relation of Burkitt's tumor-associated herpes-type virus to infectious mononucleosis. *Proc Nat Acad Sci (US)* 59:94–101, 1968.
42. HENLE G, HENLE W: Studies on cell lines derived from Burkitt's lymphoma. *Trans NY Acad Sci* 29:71–79, 1969.
43. HENLE W, HENLE G, HO H, BURTIN P, CACHIN Y, CLIFFORD P, SCHRYVER A DE, DE THE G, DIEHL V, KLEIN G: Antibodies to Epstein-Barr virus in nasopharyngeal carcinoma, other head and neck neoplasms, and control groups. *J Nat Cancer Inst* 44:225–231, 1970.
44. HUANG CC, MOORE GE: Chromosomes of 14 hematopoietic cell lines derived from peripheral blood of persons with and without chromosome anomalies. *J Nat Cancer Inst* 45:1119–1128, 1969.
45. HUANG CC, IMAMURA T, MOORE GE: Chromosomes and cloning efficiencies of hematopoietic cell lines derived from patients with leukemia, melanoma, myeloma and Burkitt lymphoma. *J Nat Cancer Inst* 43:1129–1146, 1969.
46. HUANG CC, MINOWADA J, SMITH RT, OSUNKOYA BO: Re-evaluation of relationship between C chromosome marker and Epstein-Barr virus: chromosome and immunofluorescence analyses of 16 human hematopoietic cell lines. *J Nat Cancer Inst* 45:815–829, 1970.
47. IKEUCHI T, MINOWADA J, SANDBERG AA: Chromosomal variability in ten cloned sublines of a newly established Burkitt's lymphoma cell line. *Cancer* 28:499–512, 1971.

48. IWAKATA S, GRACE JT JR: Cultivation *in vitro* of myeloblasts from human leukemia. *NY State J Med* 64:2279–2282, 1964.
49. JACOBS PA, TOUGH IM, WRIGHT DH: Cytogenetic studies in Burkitt's lymphoma. *Lancet* 2:1144–1146, 1963.
50. JENSEN EM, KORAL W, DITTMAN SL, MEDREK TJ: Virus containing lymphocyte cultures from cancer patients. *J Nat Cancer Inst* 39:745–754, 1967.
51. KLEIN G, CLIFFORD P, KLEIN E, STJERNSWARD J: Search for tumor-specific immune reactions in Burkitt lymphoma patients by the membrane immunofluorescence reaction. *Proc Nat Acad Sci (US)* 55:1628–1635, 1966.
52. KLEIN G, PEARSON G, NADKARNI JS, NADKARNI JJ, KLEIN E, HENLE G, HENLE W, CLIFFORD P: Relation between Epstein-Barr viral and cell membrane immunofluorescence of Burkitt tumor cells. *J Exp Med* 128:1011–1020, 1968.
53. KLEIN G, PEARSON G, HENLE G, HENLE W, DIEHL V, NIEDERMAN JC: Relation between Epstein-Barr viral and cell membrane immunofluorescence in Burkitt tumor cells. *J Exp Med* 128:1021–1030, 1968.
54. KLEIN G, PEARSON G, HENLE G, HENLE W, GOLDSTEIN G, CLIFFORD P: Relation between Epstein-Barr viral and cell membrane immunofluorescence in Burkitt tumor cells. *J Exp Med* 129:697–705, 1969.
55. KLEIN G: Herpes virus and oncogenesis. *Proc Nat Acad Sci (US)* 69:1056–1064, 1972.
56. KOHN G, MELLMAN WJ, MOORHEAD PS, LOFTUS J, HENLE G: Involvement of C-group chromosomes in five Burkitt lymphoma cell lines. *J Nat Cancer Inst* 38:209–222, 1967.
57. KOHN G, DIEHL V, MELLMAN WJ, HENLE W, HENLE G: C-group chromosome marker in long-term leukocyte cultures. *J Nat Cancer Inst* 41:795–804, 1968.
58. KRISHAN A, RAYCHAUDHURI R: Chromosome studies of cell lines and tumors derived from a single specimen of human leukemic blood by cell culture and heterotransplantation. *Cancer Res* 30:2012–2016, 1970.
59. KURITA Y, OSATO T, ITO Y: Studies on chromosomes of three human cell lines harboring the EB virus particles. *J Nat Cancer Inst* 41:1355–1366, 1968.
60. LEVINE PH, ABLASHI DV, BERARD CW, CARBONG PP, WAGGONER DE, MALAN L: Elevated antibody titers to Epstein-Barr virus in Hodgkin's disease. *Cancer* 27:416–421, 1971.
61. LEVY JA, HENLE G: Indirect immunofluorescence tests with sera from African children and cultured Burkitt lymphoma cells. *J Bacteriol* 92:275–276, 1966.
62. LUCAS LS, WHANG JJK, TIJOR JH, MANAKER RA, ZEVE YH: Continuous cell culture from a patient with chronic myelogenous leukemia. I. Propagation and presence of Philadelphia chromosome. *J Nat Cancer Inst* 37:753–756, 1966.
63. MACEK M, SEIDEL EH, LEWIS RT, BRUNSCHWIG JP, WIMBERLY I, BENYESH-MELNICK M: Cytogenetic studies of EB virus-positive and EB virus-negative lymphoblastoid cell lines. *Cancer Res* 31:308–321, 1971.
64. MACEK M, BENYESH-MELNICK M: Chromosomal analysis of lymphoblastoid cell lines from patients with leukemia, infectious mononucleosis, or Burkitt lymphoma. *Neoplasma* 19:51–56, 1972.
65. MACMAHON B: Epidemiology of Hodgkin's disease. *Cancer Res* 26:1189–1200, 1966.
66. MAUER BA, IMAMURA T, WILBERT SM: Incidence of EB virus-containing cells

in primary and secondary clones of several Burkitt lymphoma cell lines. *Cancer Res* 30:2870–2875, 1970.
67. MILES CP, O'NEILL F: Chromosome studies of 8 *in vitro* lines of Burkitt's lymphoma. *Cancer Res* 27:392–402, 1967.
68. MILES CP, O'NEILL F, ARMSTRONG D, CLARKSON B, KEANE J: Chromosome patterns of human leukocyte established cell lines. *Cancer Res* 28:481–490, 1968.
69. MOORE GE, GRACE JT, CITRON P, GERNER R, BURNS A: Leukocyte cultures of patients with leukemia and lymphomas. *NY State J Med* 66:2757–2764, 1966.
70. MOORE GE, ITO E, ULRICH K, SANDBERG AA: Culture of human leukemia cells. *Cancer* 19:713–723, 1966.
71. MOORE GE, GERNER RE, FRANKLIN HA: Culture of normal human leukocytes. *JAMA* 199:519–524, 1967.
72. MOORE GE, KITAMURA H, TOSHIMA S: Morphology of cultured hematopoietic cells. *Cancer* 22:245–267, 1968.
73. MOORE GE, FJELDE A, HUANG CC: Established hyperdiploid hematopoietic cell line with a minute marker chromosome persisting both in culture and in the "normal" donor. *Cytogenetics* 8:332–336, 1969.
74. MOORE GE, PORTER IH, HUANG CC: Lymphocytoid lines from persons with sex chromosome anomalies. *Science* 163:1453–1454, 1969.
75. MOORE GE: Lymphoblastoid cell lines from normal persons and those with non-malignant diseases. *J Surg Res* 9:139–141, 1969.
76. MOORHEAD PS, SAKSELA E: Non-random chromosomal aberrations in SV_{40}-transformed human cells. *J Cell Comp Physiol* 62:57–84, 1963.
77. MOSES HL, GLADE PR, KASEL JA, ROSENTHAL AS, HIRSHAUT Y, CHESSIN LN: Infectious mononucleosis: detection of herpes-like virus and reticular aggregates of small cytoplasmic particles in continuous lymphoid cell lines derived from peripheral blood. *Proc Nat Acad Sci (US)* 60:489–496, 1968.
78. NADKARNI JS, NADKARNI JJ, CLIFFORD P, MANOLOV G, FENYO EM, KLEIN E: Characteristics of new cell lines derived from Burkitt lymphomas. *Cancer* 23:64–79, 1969.
79. NICHOLS WW: Virus-induced chromosome abnormalities. *Ann Rev Microbiol* 24:479–500, 1970.
80. NIEDERMAN JC, MCCOLLUM RW, HENLE G, HENLE W: Infectious mononucleosis: clinical manifestations in relation to EB virus antibodies. *JAMA* 203:139–143, 1968.
81. NONOYAMA M, PAGANO JS: Detection of Epstein-Barr viral genome in non-productive cells. *Nature New Biol* 233:103–106, 1971.
82. NOWELL PC, HUNGERFORD DA: Chromosome changes in human leukemia and a tentative assessment of their significance. *Ann NY Acad Sci* 113:654–662, 1964.
83. OLD LJ, BOYSE EA, OETTGEN HF, HARVEN E DE, GEERING G, WILLIAMSON B, CLIFFORD P: Precipitating antibody in human serum to an antigen present in cultured Burkitt's lymphoma cells. *Proc Nat Acad Sci (US)* 56:1699–1704, 1966.
84. O'NEILL FJ, MILES CP: C-Band patterns of human lymphoblastoid cell lines. *In* Eleventh Annual Mammalian Cell Genetics Conference, Sarasota, Fla, January 8–10, 1973.

85. Osgood EE, Brooke JH: Continuous tissue culture of leukocytes from human leukemic bloods by application of "gradient" principles. *Blood* 10:1010–1022, 1955.
86. Pike MC, Williams EH, Wright B: Burkitt's tumour in the West Nile district of Uganda 1961–1965. *Brit Med J* 2:395–399, 1967.
87. Pope JH: Establishment of cell lines from peripheral leukocytes in infectious mononucleosis. *Nature* 216:810–811, 1967.
88. Pope JH, Horne MK, Scott W: Transformation of foetal human leukocytes *in vitro* by filtrates of a human leukemic cell line containing herpes-like virus. *Int J Cancer* 3:857–866, 1968.
89. Pope JL, Horne MK, Scott W: Identification of the filtrable leukocyte-transforming factor of QIMR-WIL cells as herpes-like virus. *Int J Cancer* 4:255–260, 1969.
90. Pulvertaft RJV: Cytology of Burkitt's tumor (African lymphoma). *Lancet* 1: 238–240, 1964.
91. Rosenfeld C, Maciera-Coelho A, Venuat AM, Jasmin C, Tuan TQ: Kinetics of the establishment of human peripheral blood cultures. *J Nat Cancer Inst* 43:581–595, 1969.
92. Rosenfeld C, Macieira-Coelho A: Mathematical analysis of the establishment of human peripheral blood cell lines. *J Nat Cancer Inst* 43:597–602, 1969.
93. Saksela E, Ponten J: Chromosomal changes of immunoglobulin-producing cell lines from human lymph nodes with and without lymphoma. *J Nat Cancer Inst* 41:359–372, 1968.
94. Sandberg A, Takagi N, Sofuni T, Crosswhite LH: Chromosomes and causation of human cancer and leukemia. V. Karyotypic aspects of acute leukemia. *Cancer* 22:1268–1282, 1968.
95. Schneider JA: Biohazards of long-term lymphocyte cultures. In Long-Term Lymphocyte Cultures in Human Genetics (Smith GF, Bloom AD, eds), New York, The National Foundation–March of Dimes, 1973.
96. Schryver A de, Friberg S, Klein G, Henle W, Henle G, de The G, Clifford P, Ho HC: Epstein-Barr virus associated antibody patterns in carcinoma of the post-nasal space. *Clin Exp Immunol* 5:443–459, 1969.
97. Spiers ASD, Baikie AG: Cytogenetic studies in the malignant lymphomas and related neoplasms. *Cancer* 22:193–217, 1968.
98. Steel CM, Edmond E: Human lymphoblastoid cell lines. I. Culture methods and examination for Epstein-Barr virus. *J Nat Cancer Inst* 47:1193–1201, 1971.
99. Steel CM, McBeath S, O'Riordan ML: Human lymphoblastoid cell lines. II. Cytogenetic studies. *J Nat Cancer Inst* 47:1203–1214, 1971.
100. Stevens DP, Barker LF, Fike R, Hopps HE, Meyer HM Jr: Lymphoblastoid cell cultures from patients with infectious hepatitis. *Proc Soc Exp Biol Med* 132:1042–1046, 1969.
101. Stewart SE, Lovelace E, Whang JJ: Burkitt tumor: tissue culture, cytogenetic and virus studies. *J Nat Cancer Inst* 34:319–322, 1965.
102. Stewart SE, Mitchell EZ, Whang JJ, Dunlop WR, Ben T, Nomura S: Viruses in human tumors. I. Hodgkin's disease. *J Nat Cancer Inst* 43:1–14, 1969.

103. TOMKINS GA: Chromosome studies on cultured lymphoblast cell lines from cases of New Guinea Burkitt lymphoma, myeloblastic, and lymphoblastic leukaemia and infectious mononucleosis. *Int J Cancer* 3:644–653, 1968.
104. TOSHIMA S, TAKAGI N, MINOWADA J, MOORE GE, SANDBERG AA: Electron microscopic and cytogenetic studies of cells derived from Burkitt's lymphoma. *Cancer Res* 27:753–771, 1967.
105. TOUGH IM, HARNDEN DG: Chromosome markers in cultured cells from Burkitt's lymphoma. *Eur J Cancer* 4:637–646, 1968.
106. TRUJILLO JM, BUTLER JJ, AHEARN MJ, SHULLENBERGER CC, LIST-YOUNG B, GOTT C, ANSTALL HB, SHIVELY JA: Long-term culture of lymphnode tissue from a patient with lymphocytic lymphoma. II. Preliminary ultrastructural, immunofluorescence and cytogenetic studies. *Cancer* 20:215–224, 1967.
107. WHANG-PENG J, FREIREICH EJ, OPPENHEIM JJ, FREI E III, TJIO JH: Cytogenetic studies in 45 patients with acute lymphocytic leukemia. *J Nat Cancer Inst* 42:881–897, 1969.
108. WOODS LK, MOORE GE, BAINBRIDGE CJ, HUANG CC, HUNZELLA C, QUINN LA: Lymphoid cell lines established from peripheral blood of persons with Down's syndrome. *NY State J Med*, in press, 1974.
109. YAMAGUCHI J, HINUMA Y, GRACE JT: Structure of virus particles extracted from a Burkitt lymphoma cell line. *J Virol* 1:640–642, 1967.
110. ZEVE VH, LUCAS LS, MANAKER RA: Continuous cell culture from a patient with chronic myelogenous leukemia. II. Detection of a herpes-like virus by electron microscopy. *J Nat Cancer Inst* 37:761–774, 1966.
111. ZUR HAUSEN H: Chromosomal changes of similar nature in seven established cell lines derived from the peripheral blood of patients with leukemia. *J Nat Cancer Inst* 38:683–696, 1967.
112. ZUR HAUSEN H, HENLE W, HUMMELER K, DIEHL V, HENLE G: Comparative study of cultured Burkitt tumor cells by immunofluorescence, autoradiography, and electron microscopy. *J Virol* 1:830–837, 1967.
113. ZUR HAUSEN H: Introduction of specific chromosomal aberrations by adenovirus type 12 in human embryonic kidney cells. *J Virol* 1:1174–1185, 1967.
114. ZUR HAUSEN H, SCHULTE-HOLTHAUSEN H: Presence of EB virus nucleic acid homology in a "virus-free" line of Burkitt tumor cells. *Nature* 227:245–248, 1970.
115. ZUR HAUSEN H, SCHULTE-HOLTHAUSEN H, KLEIN G, HENLE W, HENLE G, CLIFFORD P, SANTESSON L: EBV DNA in biopsies of Burkitt tumors and anaplastic carcinomas of the nasopharynx. *Nature* 228:1056–1058, 1970.

BLOOM'S SYNDROME. II. THE PROTOTYPE OF HUMAN GENETIC DISORDERS PREDISPOSING TO CHROMOSOME INSTABILITY AND CANCER

JAMES GERMAN
The New York Blood Center, New York, New York

Introduction 601
Definition of Bloom's Syndrome 602
    The Clinical Entity 603
    The Genetics 605
    The Cytogenetics 607
    Incidence of Cancer 612
Discussion 613
Conclusion 615
Acknowledgements 616
Literature Cited 616

INTRODUCTION

Homozygosity for the gene bl results in Bloom's syndrome, a rare genetic disorder the cardinal clinical features of which are stunted growth and a sun-sensitive, telangiectatic erythema of the face [3, 4, 9]. This human gene can be grouped with certain rare genes known in nonhuman species that fall into two categories: those that have been demonstrated to exert an effect, directly or indirectly, on the genetic material itself (i.e., on the behavior or function of the chromosomes at some stage in the cell cycle); and those that predispose the affected animal to cancer.

In the first category are many examples among both plants and animals [6, 22]. A familiar one is *sticky chromosomes, st,* in maize [2], an appropriately named gene resulting in chromosome disruption and aberrant segregation of chromosomes at cell division, especially at meiosis. In *Drosophila,* a number of genes have been shown to affect certain steps in the meiotic process such as pairing, recombination, or disjunction of the chromosomes [5, 15, 23]. An example of a gene in the second category is *viable yellow,* A^{vy}, in the mouse; when it is present, cancer usually develops, regardless of whether the host is a member of a low- or a high-cancer strain [17, 27].

The activity of such genes is of obvious interest not only in the study of chromosomal and cellular disorders in general but also in the study of factors responsible for induction of cancer. Only one of several interesting human genes now recognized [10, 11], *bl* may belong to either, and probably to both, of the categories just mentioned. In the homozygous state, *bl* results in an increased disruption and rearrangement of chromosomes (referred to here as chromosome instability), as well as an increased risk of developing cancer [9, 12, 14].

In man, Bloom's syndrome can be considered a prototype of such genetic disorders, which is why I have included a concise summary of certain of its features in this volume devoted to a review of the various cytogenetic aspects of cancer. A long-term program of surveillance for cancer among a cohort of 50 affected persons is presently in progress and will also be described.

DEFINITION OF BLOOM'S SYNDROME

Many genetic disorders of man are now known, several thousand of them examples of simple inheritance—recessive, dominant, or X-linked—as listed in McKusick's most recent catalogue, "Mendelian Inheritance in Man" [19]. Although it is not difficult to find something of interest for study in most of the thousands of inherited conditions known, certain of them take on particular interest at certain times, depending, for instance, on whether they constitute major health problems or whether they promise to provide new understanding of important *normal* life processes, especially those important to health. Bloom's syndrome does not qualify as a major health problem: it would hardly be worth a large expenditure of time and money when only two to three affected persons are recognized each year. But study of this syndrome, at least by one or a few laboratories, is certainly justified because it may provide insight into life processes in general—for several reasons.

First, Bloom's syndrome exhibits a simple recessive transmission, which indicates that a single gene, and a single enzyme, is affected. The major manifestation of the gene, when homozygous, is growth retardation, so that some interesting aspects of normal growth regulation will undoubtedly be clarified as the syndrome is more clearly defined. The question seems simple: at the end of a certain growth period, why do the bodies of normal individuals contain n cells, whereas those of individuals homozygous for the gene bl contain only $3/4n$, $2/3n$, or even fewer.

A second feature of interest is excessive chromosome instability ("breakage"). This suggests that the affected enzyme exerts its effect, directly or indirectly, on the genetic material itself—in this case, the chromosomes. Definition of the gene's effect can scarcely fail to be of interest, especially if the affected gene proves to be a mutant allele at a locus coding for some protein of major importance in the nucleus.

The feature of greatest interest to us now, however, is that persons with Bloom's syndrome are at an increased—a greatly increased—risk of cancer. It therefore seems reasonable to assume that identification of the enzyme affected, and accumulation of information about its defective function, may contribute to an understanding of conversion of a normal to a malignant cell.

The Clinical Entity

Born at term, the affected homozygote is well-developed and normally mature, although he is extraordinarily small: boys' birth weights average 2,094 g and girls', 1,841 g [9]. Health is good, although in most cases there is a predisposition to life-threatening infections of the respiratory or gastrointestinal tracts, ameliorating with increasing age. The advent and widespread use of antibiotic therapy after World War II greatly increased the chance of survival of Bloom's syndrome patients, and only then was the syndrome discovered and first reported [3]. The number of recognized cases has risen noticeably since then (Fig. 1). An immunopathy is present but as yet has been poorly investigated; it is characterized by a weak, or absent, delayed-hypersensitivity response and by subnormal concentrations of one or several of the circulating immunoglobulins.

In general, postnatal physical and mental development are normal, except that the body length is far below the normal range (Fig. 2). The growth curve indicates that the rate of growth is approximately normal, however; the body retains its normal proportions, except that the head is narrow and long, and sometimes disproportionately small. Adult height is (with one exception) less than 5 feet. In infancy and childhood, little

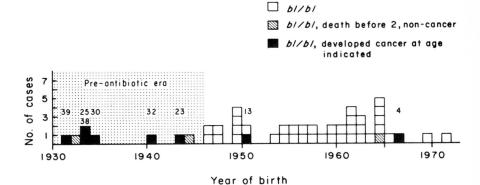

Figure 1. Recognized cases of Bloom's syndrome by years of birth. Those who died before age 2 (hatched areas) are not members of the cohort of 50 which is under surveillance for cancer (see text). For each of those in whom cancer has been detected (solid areas), the age of onset of symptomatic cancer is indicated directly on the figure.

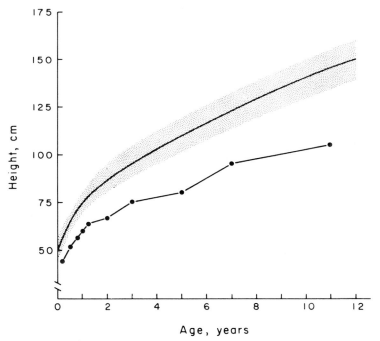

Figure 2. Typical growth in Bloom's syndrome. Patient 32(Mi.Ko.) (dots), showing that both intrauterine and postnatal retardation occurred. This boy weighed 1302 g when born at term. At present age of 10, patient's height is 106.4 cm, weight 15.5 kg. Stippled area indicates the range for normal children (mean ±2 S.D.).

subcutaneous fat accumulates which, together with the short stature, gives these individuals a delicate appearance. The facies is characteristic also (Fig. 3): narrow, with a prominent nose, relatively hypoplastic malar areas, and a retruded mandible.

The skin seems normal at birth, but during infancy or early childhood a telangiectatic lesion appears in a butterfly distribution on the face; occasionally the ears or the dorsa of the hands are also affected. The lips and lower eyelids may show crusting, bullae, and scarring. The lesion is accentuated by exposure to sunlight and so usually worsens in summertime. Severity varies from a minimal and almost asymptomatic telangiectasia to an ugly, brilliant redness which disfigures the face; males as a group are more severely affected than females. It is probable, in fact, that many individuals with a minimal lesion have been misdiagnosed as "primordial dwarfism" rather than as Bloom's syndrome, because dermatologists, until recently those most aware of this rare condition, would not have been consulted.

Sexual development appears generally to be normal; at puberty, girls begin menstruating regularly, and virilization in males is normal. However, of the three with Bloom's syndrome who have married (all men), none has had children; one of these was evaluated for fertility and reportedly produced "no active sperm." The testes of all the males I have examined have been disproportionately small, which suggests that meiosis is disturbed.

The Genetics

Formal genetic analysis of Bloom's syndrome indicates autosomal recessive inheritance [9]. The affected homozygote may be described by bl/bl, the heterozygote (normally developed) by $bl/+$. The sex ratio is at present 1.61 (M/F); this might be explained by the fact that the skin lesion is much less prominent among females, which may result in misdiagnoses, as mentioned above. Carriers of the gene are exceedingly rare, but among Ashkenazik Jews they are much less rare [9]. The gene present in persons living today, at least in the Ashkenazim, is derived from an area of Eastern Europe lying between Warsaw and Krakow on the east and Kiev and Chernovtsy on the west, i.e., the area of southeastern Poland and southwestern Ukraine. The 49 homozygous affected persons recognized so far are indicated in Table 1. Twenty-seven of them were Jewish, all of whom were Ashkenazik. Thirty-one of the 49 were living in North or Meso-America when recognized; 10 were in continental Europe (only 1 was Jewish); and 7 were in Asia Minor (all in Israel).

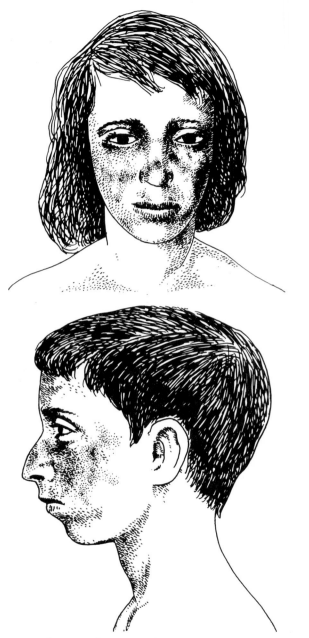

Figure 3. Composite drawings from several photographs of children with Bloom's syndrome, to demonstrate the characteristic head configuration and facies (drawings by Bunji Tagawa; reproduced by permission of *Hospital Practice*; see ref. [13]). See ref. [10] for photographs of affected persons.

(One was born in Malaysia while his European parents were there in foreign service; his affected sib was born subsequently in Germany.)

The Cytogenetics

Short-term cultures of lymphocytes from the blood of Bloom's syndrome patients display an increased incidence of metaphases with various types of chromosome rearrangements [14]; these are rarer in long-term cultures of dividing dermal fibroblasts from these patients but clearly are more frequent than in control cultures (i.e., cultures of cells neither hetero- nor homozygous for bl [13]).

The most characteristic of these aberrations is present in 0.5 to 14.0% of all dividing PHA-stimulated lymphocytes. It is a quadriradial configuration (Qr), more specifically described as a symmetrical Qr with centromeres in opposite arms of the figure (Fig. 4; ref. [8]). This rearrangement occurs before onset of mitosis and results from exchange of chromatid segments of two homologous chromosomes (Table 2), apparently at the same point in each of the affected chromatids; the preferred point of exchange appears to be in the centromeric region. Certain chromosomes are much more commonly involved in the formation of the Qr than are others: for example, the Nos. 1 as opposed to the Nos. 2 (which are chromosomes of similar length) (Table 2).

This Qr is so characteristic of Bloom's syndrome and is so rarely encountered in cells from normal persons that we do not make the diagnosis in its absence. My analysis in 1969 [9] of the first 27 recognized cases of Bloom's syndrome revealed that each person who had all of the clinical criteria of the syndrome, and whose cells had been subjected to cytogenetic study, displayed the chromosome instability, and that all but one of these patients had at least one Qr among a reasonable number (100 to 200) of cells. It was at that point in the survey that Dr. Bloom and I added the Qr to sun-sensitive telangiectatic erythema and stunted growth as a cardinal feature of the syndrome.

Other aberrations are also increased in frequency: Qr's resulting from exchanges at nonhomologous sites; separation ("breakage") near the centromere or elsewhere, with or without sister-chromatid reunion; acentric segments of paired sister chromatids; dicentric chromosomes; abnormal (rearranged) monocentric chromosomes. Chromatid and isochromatid gaps and breaks also occur, although they are not consistently increased in number. At anaphase and telophase, the increased chromosome disruption and rearrangement is represented by increased numbers of chromatin bridges or by lagging or displaced chromatin fragments. At telophase and interphase, the consequence of the chromosome instability

Table 1. Present Status, Surveillance-for-Cancer Program in Families Transmitting the Gene bl*

Case Identification†	Year of Birth/Death	Cancer in Family, by Genotype		
		bl/bl (patient)	$bl/+$ (parents)	Possible $bl/+$ (other relatives)
1 (Ge.So.)	1932/1932	Acute leukemia (25)	—	—
24 (Ro.So.)	1933/1958	—	—	PFS, abdominal cancer (53)
2 (Su.Bu.)	1950/1963	Acute leukemia (13)	—	PFF, lung cancer (78)
				PMF, prostate cancer (75)
3 (Ho.Co.)	1949	—	—	—
4 (Ge.Ho.)	1933	Adenocarcinoma, sigmoid (37)	M, colon cancer (50); stomach cancer (56)	—
5 (Ja.Oa.)	1934/1966	Squamous cell carcinoma, base of tongue (30)	—	PB, benign kidney tumor
				PFS, breast cancer
				PMB, leukemia
6 (De.Sou.)	1947	—	—	PFF, prostate cancer (82)
				PFB, metastatic cancer (22)
7 (Ro.Ta.)	1946	—	—	—
8 (Ka.¹)	1957	—	—	—
9 (Ka.²)	1949	—	—	—
10 (Gr.St.)	1950	—	—	—
11 (Ia.Th.)	1949	—	—	PMF, intestinal cancer
12 (De.Th.)	1953	—	—	
13 (De.Si.)	1954	—	—	
14 (Le.Si.)	1955	—	—	
15 (Ma.Ro.)	1961	—	—	—
16 (Et.Fi.)	1961	—	—	—
17 (Ch.Sm.)	1957	—	—	—
18 (La.Sm.)	1960	—	—	PFM, stomach cancer
19 (Sy.Tik.)	1943/1966	Acute leukemia (23)	—	PFS, stomach cancer
21 (La.¹)	1960	—	—	—
22 (El.Ha.)	1956	—	—	—
23 (Na.Ha.)	1959	—	—	—

⌈25	(St.Ti.)	1963	—	PFFF, prostate cancer (70)
⌊26	(Sa.Ti.)	1964	—	PFFM, breast cancer
27	(Ly.Se.)	1964	—	—
⌈28	(Cas.[1])	1961‡	—	—
⌊29	(Cas.[2])	1962‡	—	—
30	(Ke.[1])	1964	—	—
31	(Ca.D.)	1955	—	PMF, stomach cancer
32	(Mi.Ko.)	1962	—	—
33	(Be.Be.)	1954	—	PFM, leukemia
34	(Al.St.)	1946	—	—
35	(Ev.Ne.)	1965	—	
36	(St.Ba.)	1964	—	—
⌈37	(And.Ca.)	1966/1971	Acute leukemia (4)	—
⌊38	(Ant.Ca.)	1964/1964	—	—
39	(St.Pe.)	1931/1971	Squamous cell carcinoma, esophagus (39); adenocarcinoma, sigmoid (39)	F, myeloid metaplasia; myelofibrosis (74)
40	(Do.Roe.)	1962	—	PMM, lung cancer
41	(Zv.Ku.)	1947	—	—
⌈42	(Ra.Fr.)	1949	—	M, breast cancer
⌊43	(Ka.Fr.)	1944/1945	—	M, breast cancer
44	(Ab.Ru.)	1959	—	—
45	(Zv.Sh.)	1969	—	—
46	(Ma.Pr.)	1956	—	—
47	(Ar.Smi.)	1958	—	—
48	(Ma.Sa.)	1961	—	—
49	(Da.Ben.)	1940/1973	Pleomorphic reticulum cell sarcoma (31)	M, melanoma (55)	PFM, liver cancer (50)
50	(Je.Bl.)	1971	—	PFMF, "cancer" PFFM, stomach cancer PFMM, breast cancer PMMF, lung cancer (80)

* Blanks = no information; — = none known; P = patient('s), F = father('s), M = mother('s), S = sister('s), B = brother('s). Numbers in parentheses indicate ages at symptomatic onset of cancer.
† Affected sibs are bracketed; see footnote, page 612 for explanation of numbers and initials.
‡ Approximate birth dates; no information after ages 4 and 2.

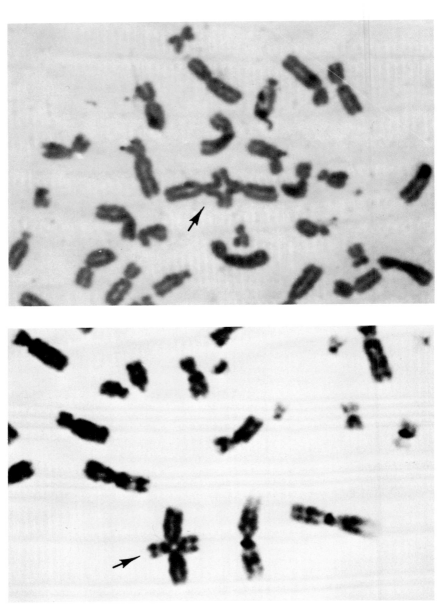

Figure 4. Chromosomes from two dividing lymphocytes, showing quadriradial configurations (Qr's) of the type characteristic of Bloom's syndrome. (a) The homologues of pair No. 1 have undergone chromatid exchange at the same, or approximately the same, points near the middle of the long arm. Orcein stain. (b) The Nos. 6 have undergone exchange near or at the centromeric regions. Giemsa stain for G-bands.

Table 2. Analysis of 99 Cells, Indicating the Two Chromosomes Forming the Qr*

One Chromosome	The Other Chromosome									
	1	2	3	B	C	D	16	17 or 18	F	G
1	17	0	0	0	0	0	0	1	0	0
2		0	0	0	0	0	0	0	0	0
3			4	0	1	0	0	0	0	0
B				5	2	0	0	0	0	0
C					25	0	0	1	1	0
D						9	1	1	0	1
16							5	0	0	0
17 or 18								9	0	0
F									12	0
G†										4

* Qr = quadriradial configuration. Chromosomes identified by number (1-3, 16-18) or group (B-D, F-G). These Qr's were discovered in the course of an examination of 4000 PHA-stimulated lymphocytes from 18 different persons (unpublished data, with L. P. Crippa). The cells were not selected, and the analysis comprises the first 101 cells discovered in this laboratory to contain a Qr. Two cells could not be analyzed because of poor morphology.
† The Y was not seen in Qr formation.

is an increased number of cells with distorted nuclei and micronuclei [13].

Thus, among the cells in cultures, increased numbers, in comparison to normals, of cells with chromosome mutations of various types, producing deletion, duplication, and repositioning of segments. The question inevitably arises whether such changes could somehow be in vitro-induced, not reflecting the true in vivo picture. It is probable that chromosome instability is also present in vivo, because I have found exchange figures and dicentric chromosomes (but, as yet, none of the characteristic, symmetrical Qr's) in dividing bone-marrow cells aspirated directly from one of the two patients whose marrow I have been able to examine.

Incidence of Cancer

In his original paper in 1954, Bloom described three individuals with this syndrome, a young man, a boy, and a young girl [3]. In 1960 Dr. Bloom referred the girl to me for karyotype study. The chromosome instability was discovered and, because of it, I followed her closely, confirming the finding repeatedly. She remained healthy until 1963, when she developed weakness and pallor and died six weeks later from acute leukemia [24]. In discussion with Dr. Bloom, I learned that the young man in his report had died of acute leukemia also, in 1958. In 1963 only 14 persons with Bloom's syndrome had been recognized. Two instances of acute leukemia, itself a rare disease, seemed therefore of real significance, and we began to collect clinical data and to make cytogenetic studies of each possibly affected person about whom we could learn. By assiduous and conservative application of Dr. Bloom's original clinical criteria for diagnosis [4], we have been able to collect a "pure sample" of almost 50 homozygous-affected individuals (Table 1)*, and we propose to follow them all during coming years.

What may now be said about cancer in Bloom's syndrome, about its incidence and site 20 years after the syndrome was first described and 10 years after serious efforts were begun to detect cases throughout the world and to make surveillance for cancer? The 49 recognized cases are listed in Table 1 and depicted in Figure 1. Of these, 46 survived till the age of 2 years. We are in contact with the patients, their families, or their physicians in all but two of the recognized cases. These 44 will be members of our "cohort" of affected persons to be followed during the coming years, the total number in the cohort to be brought to 50.

To date, eight of the homozygotes in our cohort have developed cancer. Four of these cancers were acute leukemia [21, 24]; classification of acute leukemia is difficult, but none of these four was lymphocytic in type. [These persons are identified in Table 1 as 1(Ge.So.), 2(Su.Bu), 19(Sy.Tik.), and 37(And.Ca.).] The deaths occurred at ages 25, 13, 23, and 55. Four other cancers have originated in the alimentary canal, each in a person past 30: a squamous cell carcinoma developed at the base of the tongue in a 30-year-old man, 5(Ja.Oa.); local extension of the tumor led to his death at 32. An adenocarcinoma of the sigmoid colon developed in a 37-year-old man, 4(Ge.Ho.), but early diagnosis (as direct result of the surveillance program) permitted prompt and apparently successful surgical excision. A 39-year-old man, 39(St.Pe), developed a squamous cell carcinoma of the upper esophagus and died from it; shortly after his admission to our hospital for unexplained dysphagia, he was found to have an additional but asymptomatic cancer, an adenocarcinoma of the sigmoid colon, which was apparently successfully excised. A 31-year-old man, 49(Da.Ben.), developed lymphadenopathy and ascites and died at 32 of reticulum cell sarcoma.

---

* Each person with Bloom's syndrome is identified in publications from my laboratory by official number and initials as used in Table 1 (and first used in reference [9]). The number indicates the approximate order of recognition; the initials are derived from the person's first and family names (or sometimes from the name of the author first reporting the case).

Thus 8 of the 44 persons in our cohort, approximately 1 in 6, developed primary cancer, and one of the 8 developed primary carcinomas in two different sites. The average age of all persons alive in 1973 with Bloom's syndrome is only 16.6 years. Of the 16 who have reached age 20, six have developed cancer; all four of those who have reached 30 developed cancer (Fig. 1). Of the 8 with cancer, 6 were Jewish; in the entire cohort of 44, 26 are Jewish. Six of the 8 with cancer were male; in the cohort of 44, 25 are male.

Are heterozygotes ($bl/+$) at increased risk of developing cancer? The answer to this important question is not yet available. I prepared Table 1 from the information in my files collected from the families of bl/bl individuals during ordinary pedigree-taking. It is, therefore, a minimum estimate of the incidence of cancer in heterozygotes, because my exhaustive search for cancer among the relatives of persons with Bloom's syndrome, perhaps along the lines used by Swift in Fanconi's anemia [26], remains undone. Three, or four, parents (who necessarily are $bl/+$) are known to have had cancer; of these, one of them has had multiple primary tumors and one has the "premalignant" myelofibrosis. Nine grandparents are reported to have had cancer.

DISCUSSION

Because many other authors in this volume have written on the significance of chromosome change in the etiology of cancer, I shall do no more here than summarize what I have suggested elsewhere [11] to be the significance of the association between the chromosome instability and the high incidence of cancer in Bloom's syndrome.

Groups of persons predisposed to cancer as result of excessive exposure to ionizing radiation can be shown to have chromosome damage and increased numbers of chromosome rearrangements in their somatic cells [1, 7]. (Some of these changes then appear in the chromosome complement of in vivo clones of cells which, however, express none of the features of malignancy.) The sequence of events is (a) irradiation over a brief period of time, with mutation of the genetic material of many somatic cells; (b) a lag period of months to years; and then, (c) clinical cancer. In persons with the genetic constitution bl/bl, the sequence of events appears to be similar (if the assumption is made that the chromosome instability occurs in vivo as well as in vitro): (a) mutation of the genetic material in all proliferating tissues, probably repeatedly from conception, through embryonic life, and in the adult predominantly in the tissues with highest mitotic indices (e.g., the lymphoid tissues, bone marrow, the mucosa of the alimentary tract, and the skin); (b) a lag period of years; and then, (c) clinical cancer. Of course, in neither group can it be stated whether an interplay of mutated cells with some ubiquitous endogenous or exogenous virus is of importance, although the possibility exists.

Thus the gene *bl* appears to accomplish what environmental mutagens accomplish. Somatic tissues accumulate increased numbers of cells with aberrant chromosome complements, cells in which rearrangements of various types lead to segmental deficiencies, duplications, new positioning of genes through chromosome translocations, or homozygosity for genes on chromosome arms distal to the points of exchanges of the type signified by the symmetrical Qr. In an as yet undefined way, a background of chromosome instability—whether environmentally or genetically produced—appears to predispose to the emergence of a cancer which usually, at least in the human, as I have shown elsewhere [11], begins as a clone of cells derived from one having a chromosome mutation. Such observations point to chromosome mutation, whatever its cause, as an event of importance in the majority of human cancers.

I have advanced the argument elsewhere [11], and earlier in this paper, that the gene *bl* may be grouped with other rare human genes each of which, when homozygous, produces a different and distinctive phenotypic effect, causes chromosome instability, and predisposes to cancer. The clinical conditions produced by three other such genes are Fanconi's anemia, the Louis-Bar or ataxia-telangiectasia syndrome, and xeroderma pigmentosum. Still other genes with these qualities will certainly be recognized in the future. Some of them may not affect an entire spectrum of dividing cell types as *bl* seems to do but may affect only certain types. Others may lead to chromosome instability only under certain conditions; for example, cells homozygous for the gene for xeroderma pigmentosum, which have therefore a severely limited ability to excise UV-induced dimers [25], do not have increased numbers of breaks and rearrangements in the chromosomes under ordinary conditions of tissue culture, but are reported to have significantly more of such changes than other cells after UV irradiation [20]. Among individuals with xeroderma pigmentosum, there is a greatly increased incidence of cancer, mostly in those areas of the skin regularly exposed to direct sunlight (and therefore to UV radiation).

Some of the genes that share this ability to increase the mutation rate in somatic cells (e.g., as evidenced by visible chromosome rearrangements) may be expected to code for proteins that are of direct importance in chromosome activity, a DNA polymerase, for example. Others may code for proteins that act at a distance from the chromosome and only indirectly lead to a metabolic imbalance in cells of a given type and predispose to chromosome instability therein. (We have studied a girl homozygous for one such gene [18]; she had a severe megaloblastic anemia because of an inability to transport folic acid across the intestinal mucosa, so that maturation of her bone-marrow cells was arrested. The chromosomes in her dividing marrow cells appeared severely damaged; however, parenteral administration of folic acid promptly reversed the anemia, and the chromosome abnormalities disappeared.) It may be anticipated that elucidation of the biochemical disturbances associated with this class of genes will be useful and interesting, whether the genetic material is affected directly or indirectly. Those genes whose action directly concerns the genetic material itself (of which *bl* may be an example) may prove to be of particular value as experimental tools in cell biology.

It is important to determine which of the genes in this class predispose to cancer when they are in the heterozygous state (but produce no recognizable phenotype). The gene for Fanconi's anemia appears to do so [26]; *bl* remains to be studied. There may be others which do so when heterozygous but which will be difficult to recognize because of homozygous lethality. The generality of cancer appears to be polygenically determined [16, 17]; genes such as *bl* could be members of the pool of genes responsible for this inheritance. On the other hand, they could be more analogous to the murine gene A^{vy} which, when present in a genome, exerts a strongly pro-cancer effect. Although genes such as *bl* and those for Fanconi's anemia and xeroderma pigmentosum are rare individually, they are not rare as a pro-cancer *class* of genes. I have data (unpublished) showing that more than 1 Ashkenazik Jew in 200 carries *bl*. The frequency of persons in the general population who are carriers of some one gene of this class actually may prove to be not much different from the frequency of cancer itself, which again would point to a role for chromosome instability as a factor in the etiology of cancer.

CONCLUSION

Bloom's syndrome can be viewed as the prototype of a class of infrequently occurring genetic disorders which share two features: chromosome instability and a predisposition to cancer. Several disorders in this class, each transmitted as an autosomal recessive trait, are already recognized, and others probably will be recognized in the future.

It may be anticipated that at least some of the genes associated with this class of disorders will exert an effect also in the heterozygote, including predisposing to cancer. Some such genes may exert a weakly pro-cancer effect and may be of importance in the polygenic inheritance of human cancer, whereas others may be strongly pro-cancer, in which case a pattern of dominantly inherited cancer might emerge.

The rarity of these genes individually may at first seem to detract from their importance. However, as a *class* of pro-cancer genes, they, and how they act, should be viewed, along with the accepted environmental oncogenic factors, as of potential importance in the etiology of human cancer. Finding out how such genes affect the genetic material, in diverse ways, either directly or indirectly, doubtless will be useful and interesting in cell biology and, therefore, in the study of cancer.

Acknowledgements.–Since 1960, Dr. David Bloom has provided stimulation, encouragement, and sound advice in my study of the syndrome that appropriately bears his name. I am pleased to acknowledge his close collaboration and friendly help. I also acknowledge, with gratitude, research funds from the American Cancer Society since 1968, and grants HD 04134 and HL 09011 from the National Institutes of Health.

LITERATURE CITED

1. Awa AA: Cytogenetic and oncogenic effects of the ionizing radiations of the atomic bombs. This volume, pp. 637–674, 1974.
2. Beadle GW: A gene for sticky chromosomes in *Zea mays*. *Z Indukt Abstammungs- Verebungsl* 63:195–217, 1933.
3. Bloom D: Congenital telangiectatic erythema resembling lupus erythematosus in dwarfs. *Am J Dis Child* 88:754–758, 1954.
4. Bloom D: The syndrome of congenital telangiectatic erythema and stunted growth: observations and studies. *J Pediatr* 68:103–113, 1966.
5. Bridges CB: Elimination of chromosomes due to a mutant (Minute-m) in *Drosophila melanogaster*. *Proc Nat Acad Sci (US)* 11:701–706, 1925.
6. Chaganti RSK: Cytogenetic Studies of Maize–Trypsicum Hybrids and Their Derivatives. Boston, Bussey Institution of Harvard University, 74–83, 1965.
7. Evans HJ, Court Brown WM, McLean AS (eds): Human Radiation Cytogenetics, Amsterdam, North Holland Publishing Company, 1967.
8. German J: Cytological evidence for crossing-over in vitro in human lymphoid cells. *Science* 144:298–301, 1964.
9. German J: Bloom's syndrome. I. Genetical and clinical observations in the first twenty-seven patients. *Am J Hum Genet* 21:196–227, 1969.
10. German J: Chromosomal breakage syndromes. *Birth Defects: Original Article Ser* 5(5):117–131, 1969.
11. German J: Genes which increase chromosomal instability in somatic cells and predispose to cancer. *In* Progress in Medical Genetics (Steinberg AG, Bearn AG, eds), vol VIII, New York, Grune & Stratton, 61–101, 1972.
12. German J: Oncogenic implications of chromosomal instability. *Hosp Pract* 8:93–104, 1973.
13. German J, Crippa LP: Chromosomal breakage in diploid cell lines from Bloom's syndrome and Fanconi's anemia. *Ann Génét* 9:143–154, 1966.

14. GERMAN J, ARCHIBALD R, BLOOM D: Chromosomal breakage in a rare and probably genetically determined syndrome of man. Science 148:506–507, 1965.
15. GOWEN MS, GOWEN GW: Complete linkage in Drosophila melanogaster. Am Naturalist 56:286–288, 1922.
16. GRÜNEBERG H: The genetics of cancer. In The Genetics of the Mouse, 2nd ed, The Hague, Martinus Nijhoff, 435–471, 1952.
17. HESTON WE: Genetics of neoplasia. In Methodology in Mammalian Genetics (Burdette WJ, ed), San Francisco, Holden-Day, 247–268, 1963.
18. LANZKOWSKY P, ERLANDSON MF, BEZAN AI: Isolated defect of folic acid absorption associated with mental retardation and cerebral calcification. Blood 34: 452–465, 1969.
19. MCKUSICK VA: Mendelian Inheritance in Man, 3rd ed, Baltimore, Johns Hopkins Press, 1971.
20. PARRINGTON JM, DELHANTY JDA, BADEN HP: Unscheduled DNA synthesis, u.v.-induced chromosome aberrations and SV_{40} transformation in cultured cells from xeroderma pigmentosum. Ann Hum Genet 35:149–160, 1971.
21. PFEIFFER RA, KIM A: Leukemia in patients with a "chromosomal breakage syndrome." Excerpta Med Int Congr Ser 233:142, 1971.
22. RIIS H: Genotypic control of chromosome form and behavior. Botan Rev 27:288–318, 1961.
23. SANDLER L, LINDSLEY DL, NICOLETTI B, TRIPPA G: Mutants affecting meiosis in natural populations of Drosophila melanogaster. Genetics 60:525–558, 1968.
24. SAWITSKY A, BLOOM D, GERMAN J: Chromosomal breakage and acute leukemia in congenital telangiectatic erythema and stunted growth. Ann Int Med 65:487–495, 1966.
25. SETLOW RB, REGAN JD, GERMAN J, CARRIER WL: Evidence that xeroderma pigmentosum cells do not perform the first step in the repair of ultraviolet damage to their DNA. Proc Nat Acad Sci (US) 64:1035–1041, 1969.
26. SWIFT M: Fanconi's anemia in the genetics of neoplasia. Nature 230:370–373, 1971.
27. VLAHAKIS G, HESTON WE, SMITH GH: Strain C3H-A^{vy} fB mice: ninety percent incidence of mammary tumors transmitted by either parent. Science 170:185–187, 1970.

ATAXIA TELANGIECTASIA SYNDROME: CYTOGENETIC AND CANCER ASPECTS

D. G. HARNDEN

Department of Cancer Studies, University of Birmingham, Birmingham, England

Clinical Features	620
Occurrence of Cancer	621
Chromosome Abnormalities	623
Abnormalities in Lymphocytes	623
Abnormalities in Fibroblasts	631
Comments on the Cytogenetics	632
Acknowledgements	633
Literature Cited	633

The ataxia telangiectasia syndrome (AT) was first described clearly in 1941 by Louis-Bar [31] and is alternatively known as the Louis-Bar syndrome. A previous report in 1926 [52] had described a familial syndrome with many of the features of AT, but it was not clear that the neurological disorder in the cases described was cerebellar in origin. Further cases were not reported until 1957, when several independent reports appeared almost simultaneously [3, 10, 52, 58]. Many other reports have now appeared, but the syndrome is relatively rare. In 1963 Boder and Sedgwick [7] reviewed 101 cases, and the number of new cases reported in the literature since then brings the total reported to approximately 250. The syndrome is inherited in an autosomal recessive manner [33, 34, 53] and the basic karyotype is normal [60]. Much interest has centered on the immunological deficiency found in this syndrome and its possible relationship to an apparently increased incidence of malignant disease [41]. Little work has been done on the cytogenetics, but the present short review of the literature and presentation of data from four new cases strongly suggests that further study will yield much of interest.

CLINICAL FEATURES

The clinical features have been described in detail elsewhere [6, 7, 25]. The two constant features are progressive cerebellar ataxia and oculocutaneous telangiectasia. The ataxia usually appears first and is typically recognized about age 12 to 14 months when the child begins to walk. The ataxia is severe and progressive, and the patient is usually confined to a wheelchair before adolescence. In cases that have been autopsied there has been a uniform cerebellar cortical atrophy involving both the Purkinje cell layer and the internal granular layer, with demyelination of the posterior columns and dorsal spinocerebellar tracts of the spinal cord [51]. Abnormalities of the cerebrum or brain stem are not usually reported.

The telangiectases become evident usually between the third and the sixth year, but there is considerable variation in both age of onset and the extent to which the telangiectasia progresses. Characteristically, it is first recognized on the exposed bulbar conjunctivae but later spreads in a symmetrical fashion to the eyelids, the external ear, the face, the neck, the antecubital and popliteal spaces, and less often to the palate and to the backs of the hands and feet [10, 16]. Other neurological features that may appear in a relatively high proportion of cases are: diminished or absent deep reflexes, flexor or equivocal plantar responses, choreoathetosis, peculiar eye movements, dysarthric speech, drooling, characteristic posturing, and a sad, masklike face. The incidence of sinopulmonary infections is high, and in many cases an early death is the result of such infection, although McKusick [33] has recorded one patient aged 37 and another aged 41. Retardation of growth has been reported and also mental deficiency, though some authors believe that poor performance in formal intelligence testing is due to motor impairment rather than to mental defect [27]. Changes in hair and skin, particularly the appearances of cafe-au-lait spots, have been described by several authors, and it has been suggested that premature aging is one of the characteristic features [44].

Various associated immunological deficiencies have been reported in detail by many authors and are reviewed by Leveque et al. [29] Naspitz et al. [39], and Peterson and Good [41]. An unusual finding has been the lack of uniformity from case to case. The commonest defect is a low level or complete absence of IgA [13], but this is not found in all cases. Deficiency of IgE has also been reported [1, 4], but in fewer cases. Levels of IgG and IgM are most often normal, but there have been reports of abnormalities. The response of peripheral blood lymphocytes to stimulation with phytohemagglutinin (PHA) and other mitogens is often poor [32], but again this is not a constant feature [47]; in the cases we have studied, the PHA response, though slightly depressed, was quite adequate for making chromosome preparations of good quality. Taken together with the frequent reports of poor or absent delayed-hypersensitivity responses and poor homograft rejection [39], this suggested a basic deficiency of thymus-derived lymphocytes. Complete or partial absence of thymus tissue [6, 42] and also abnormalities of other lymphoid tissues, such as tonsils and lymph nodes [40], have been found at autopsy. Abnormalities of the ovaries have been described in some cases [11, 12, 37], particularly ovarian follicular agenesis, which has sometimes been associated with poor development of secondary sexual characteristics.

Thus the mode of AT inheritance seems to suggest that a basic lesion underlies the range of abnormalities described, but so far there is no indication of the nature of this defect. As Tadjoedin and Fraser [53] point out, the inconsistency of the IgA findings seems to rule this out as the primary defect. However, it is likely that there is a strong association between the immunological deficiencies and the frequent occurrence of sinopulmonary infections, and it has been suggested that the immunological abnormalities and the increased incidence of malignant disease are also closely connected [42].

OCCURRENCE OF CANCER

It has been recognized for a decade that there is a close association between AT and malignant diseases of the reticuloendothelial system [7, 42]. More recently, it has become apparent that other malignancies may also occur with an increased frequency in these patients (Table 1). Data on some of the patients is lacking, since they have been reported at second hand by other authors, but only cases confidently said to be AT are included; for example, two cases of acute lymphocytic leukemia in children with ataxia mentioned by Miller [35] have been omitted because there was no indication that the children had signs of telangiectasia. It is not meaningful to give incidence figures, since cases with malignancies are now more likely to be reported; but there can be little doubt that there is a very large increase in the incidence of lymphoid neoplasms. It is also probable, since many of these cases die at an early age from respiratory-tract infections, that the incidence among those at risk is very high indeed if they live long enough. The occurrence of other neoplasms, possibly at an older age, also seems higher than would be expected by chance in such a small group of individuals in this age group.

It could be argued that the occurrence of the lymphoid neoplasms is

Table 1. Malignancies in Patients with Ataxia Telangiectasia

Tumor Type	Age	Sex	Comment	Reference
Reticulum cell sarcoma	—	—		7
	5½	M	Described as generalized reticulosis but in a later paper classified RCS (Peterson et al. [40])	42
	—	—	No details. Szanto*	
	—	—	Reported at conference; no details available. Murphy and O'Neal (1965)†	
	2½	F		12
Hodgkin's disease	—	—		7
	—	—	No details. Szanto*	
Lymphoma	9	F		45
	—	—	No details. Landing†	
Lymphosarcoma	8	F	One of cases reported in detail by Boder and Sedgwick [6]	7
	9	M	First suggestion of increased incidence of neoplasia	57
	9½	F		40
	8	M		50
	14	M	} Brothers	8
	—	M		
	11	M	Death certificate data but said to be previously unreported	36
Lymphoid tumor, type unspecified	—	—	No details. Szanto‡	
γM monoclonal gammopathy	9	F	Considered malignant by authors	9
Acute lymphatic leukemia	9	M		54
	5	F		24
	6	M		24
	3	M		28
	15	M		28
Dysgerminoma, bilateral	17	F	Considered malignant by authors	11
Colloid carcinoma of pylorus	16	—		48
Adenocarcinoma of stomach	21	F	} Sisters	22
Adenocarcinoma of stomach	19	F		
Cerebellar medulloblastoma	13	F		49
Frontal cystic mixed glioma	19	M		59

* One of three sibs quoted by Peterson et al. [40].
† Quoted by Miller [35].
‡ One of three sibs quoted by Peterson et al. [42].

one feature of a more generalized disturbance of the lymphoid system. The occurrence of the other neoplasms suggests either a failure in immunological surveillance or the existence of an unknown factor making more probable the initial event that changes a normal cell into a potentially malignant cell. The high incidence of neoplasms in immunologically deficient individuals is now well documented [15, 56]. However, the additional possibility that in some individuals malignant cells may arise more often than in others has only recently received much attention. In most instances, we do not yet know the nature of this event, but it seems inescapable that an alteration in either the structure or the regulation of the cell genome is an integral part of the process. The occurrence of chromosomal abnormalities in genetically determined syndromes in association with an increased incidence of malignant disease is therefore of considerable interest [18].

CHROMOSOME ABNORMALITIES

Surprisingly few cytogenetic studies have been carried out on patients with AT, and only one or two have been reported in sufficient detail to determine whether or not spontaneous chromosome damage is a feature of the syndrome. All but one of the cytogenetic studies so far reported deal with phytohemagglutinin-stimulated peripheral blood lymphocytes.

Abnormalities in Lymphocytes

Early studies of single cases by Zellweger and Khalifeh [60], Young et al. [59], Utian and Plit [55], and Rosenthal et al. [45], and a report on two cases by Schuster et al. [49] indicated that the karyotype was normal but gave no details; it is possible that only a small number of cells were examined. These reports of normal karyotypes in AT may have deterred other investigators. Similarly, reports [32] of impaired lymphocyte transformation in AT have discouraged investigation, and there seems little doubt that such impairment has made cytogenetic investigations much more difficult. I have been able to find in the literature only 10 reports, dealing with 18 cases, which give any details of counts or analyses [17, 21–24, 30, 43, 46, 48, and Hecht et al., 1972, personal communication]. The results on the lymphocytes of these patients are summarized in Table 2.

Schmid and Jerusalem [46] found no evidence of abnormality in the lymphocytes of relatives. They found 0/50, 2/58, and 1/56 cells with breaks in the father, mother, and sister, respectively, of their two cases and considered them not to differ from the controls.

Two further families have now been studied in detail and will be reported fully elsewhere, but a summary of the information available on the cytogenetics of these cases is presented here (Table 3). One family, a local one, was studied in our laboratory in Birmingham, while the second, a Maltese family, was studied at the Galton Laboratory, University Col-

Table 2. Cytogenetic Findings on Lymphocytes from Patients with Ataxia Telangiectasia[*]

Case Identification and Reference	Age	Sex	Abnormal Cytogenetic Findings	Abnormal Cells No./Total Examined	Percentage	Clone Present (+)
Hecht et al. [23, 24, and unpublished data]	18–23	M	Breaks (15) at first observation	15/60	25	—
			Decreasing breaks with time, increasing clone with time	1570/2676	59	+
Gropp and Flatz [21]						
U.K.-B.	7	F	Gaps or fragments (3), fragments and exchange figure (1)	4/12	33	—
C.K.-B.	3	F	Breaks and gaps (2), exchange (1)	3/12	25	—
Haerer et al. [22]	19	F	47 chromosomes (2), fragments (1)	3/11	27	—
German [17]	6	F	Fragment (1), exchange (1)	2/30	7	—
Pfeiffer [43]						
A.Wi.	5	M	Gaps or breaks (7), fragments (2), exchange (1)	10/80	12.5	—
M.Wi.	7	M	Gaps or breaks (4), fragments (3), Dq+ (1), exchange (1)	9/60	15	—
H.Ma.	3	F	Filamental bridges between Dq	1/40	2.5	—
U.Ma.	6	F	Nil	0/40	0	—
P.Re.	16	M	Fragments (2), dicentrics (3), Dq+ (5)	10/107	11	+
W.Mu.	15	M	Marker Dq+ (36), dicentrics (>10), polyploidy	46/55	84	+
Lisker and Cobo [30]	8	F	Gaps or breaks (22) (1 or 2 per cell, mostly chromatid breaks)	22/60	37	—
Schuler et al. [48]						
Zs.Sz.	14	—	Gaps or breaks (4)	4/38	10	—
D.L.	10	—	Gap (1)	1/30	3	—
P.J.	10	—	Nil	0/27	0	—
H.L.	11	—	Gap (1)	1/15	6.6	—
Schmid and Jerusalem [46]						
B.I.	17	M	Break (1), dicentrics (6), Dq+ (89)	96/107	>83	+
	12	M	Break (1)	1/15	6.6	—

[*] Data were obtained by mitogen-stimulation of peripheral lymphocytes.

Table 3. Cytogenetic Findings on Lymphocytes from Two Families with Ataxia Telangiectasia

Case Identification		Age	Sex	Chromatid Aberrations only		Chromosome Aberrations[*]		Percentage of Abnormal Cells[†]	Clone Present (+)
				Gaps and Breaks	Exchanges	Unstable	Stable		
Birmingham[‡]									
J.G.	AT	25	F	1	—	18	84	85	+
P.G.	AT	16	F	8	—	5	7	19	—
S.G.	brother	17	M	7	—	—	1	8	—
M.G.	brother	23	M	8	—	—	—	8	—
E.G.	mother	—	F	3	—	1	—	4	—
J.G.	father	—	M	1	—	3	—	4	—
Galton Laboratory[§]									
M.S.	AT	12	F	3	—	7	10	20	—
J.S.	AT	9	F	7	—	9	10	26	—
C.S.	mother	—	F	1	—	3	1	5	—
J.S.	father	—	M	—	—	1	1	2	—

[*] Unstable aberrations are those that will tend to undergo further change at the next mitosis (e.g., dicentrics, fragments and rings). Stable aberrations are abnormal monocentric chromosomes.
[†] 100 cells were analyzed per patient.
[‡] Data from D. G. Harnden, Department of Cancer Studies, University of Birmingham.
[§] Previously unpublished data by courtesy of Dr. J. Parrington and Dr. J. Delhanty of the Galton Laboratory, University College, London.

lege, London; I am grateful to Dr. Jennifer Parrington and Dr. Joy Delhanty for allowing me to quote their unpublished results.

In the Birmingham family, the two patients with AT showed chromosome abnormalities in excess of normal levels. In the younger sister, the number of cells with gaps and breaks was only moderately elevated and, indeed, no higher than the number found in her normal brothers. The other abnormal findings were 2 cells with dicentric chromosomes, 3 with fragments, and 7 with marker chromosomes or an abnormal analysis (Fig. 1), which far exceeded our normal level. The older sister had only

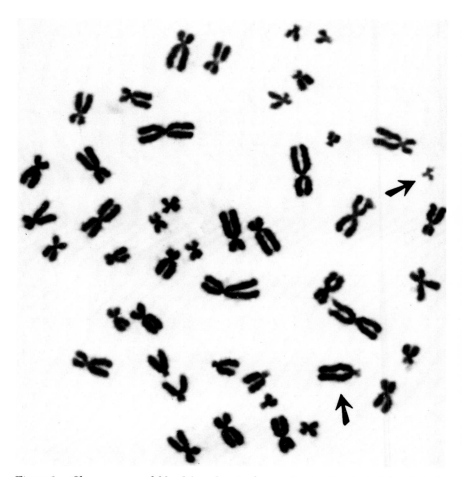

Figure 1. Chromosomes of blood lymphocyte from patient with ataxia telangiectasia (P.G.) showing an apparent balanced translocation between two group D chromosomes (arrows). (Also, see Table 3).

a single cell with a chromatid gap. The striking finding was a clone of cells whose basic karyotype was 46,XX,t(X;14)(q27;q12) and which comprised 73% of the total lymphocytes examined (Fig. 2). There were in addition 13 cells with dicentrics, 5 with fragments, and 20 cells that had stable rearrangements of the chromosomes, either alone or with fragments, dicentrics, or the clonal rearrangement (Fig. 3). A preliminary analysis of the breakage points suggested that the exchange points were not random but that the chromosomes of groups B, D, and G were involved more often than would be expected on the basis of a random distribution. The finding of a single dicentric in the father was slightly

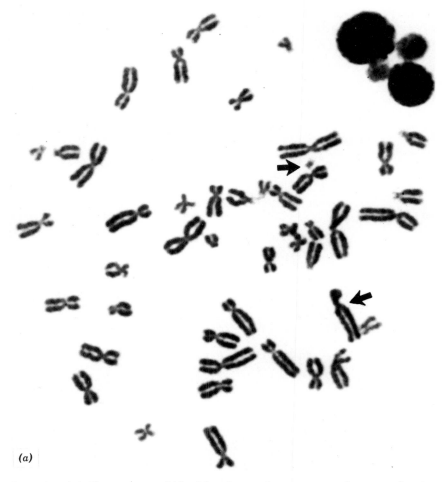

(a)

Figure 2. (a) Chromosomes of blood lymphocyte from patient with ataxia telangiectasia (J.G.) showing clonal translocation 46,XX,t(X;14)(q27;q12). (Also see Tables 3 and 4.)

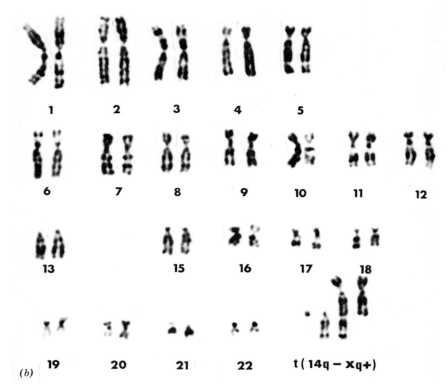

Figure 2. (b) Karyotype of lymphocytic clone cell (from J.G.) stained with modified Giemsa to show banding patterns. (Also, see Tables 3 and 4.)

unusual, but the other abnormalities in the parents were within control levels.

In the family studied by Parrington and Delhanty, the number of cells with abnormalities in the parents was given within control levels, and even in the patients themselves the number of cells with gaps and breaks was not elevated. As in the Birmingham family, however, there was an excess of cells with fragments, dicentrics, and chromosome exchanges. If the clone cells in J.G. were omitted the percentages of abnormal cells in the four cases would be: 29 (J.G.), 19 (P.G.), 20 (M.S.), and 26 (J.S.).

When one considers these data together with the previously published data, several interesting features emerge. As already indicated by Schmid and Jerusalem [46], the occurrence of chromatid gaps and breaks does

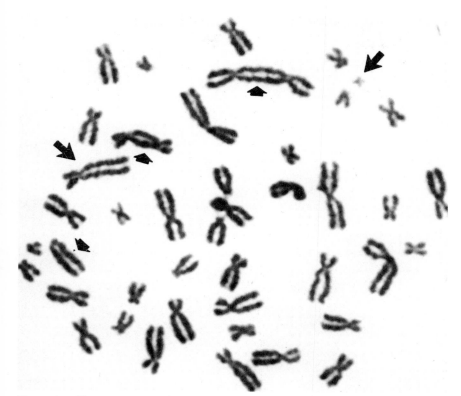

Figure 3. Chromosomes of blood lymphocyte from patient with ataxia telangiectasia (J.G.) showing clonal abnormality (long arrows) together with two dicentric and an abnormal acrocentric chromosome (short arrows). (Also, see Tables 3 and 4.)

not appear to be a regular feature in this syndrome. Of those studied on a large number of cells, only two cases [23, 24, 30] showed a significant increase in gaps and breaks The first two cases of Pfeiffer [43] did seem to show an increase, but the author considered that the number was not excessive when compared with his own controls, who had received comparable amounts of diagnostic irradiation. Hecht et al. (1972 personal communication), however, reported the finding of increased chromosome breakage in six cases of AT in addition to the subject originally reported by them.

Much more significant is the increase in the incidence of chromosome rearrangements leading to the formation of dicentric or abnormal monocentric chromosomes. It is not always clear from the way in which data are presented whether or not cells have been fully analyzed, so that dogmatic statements cannot be made. However, it seems likely that when adequate numbers of cells are fully analyzed, rearrangements will be found in 10 to 30% of the cells. It is possible that gaps and breaks are commoner in younger patients, and rearrangements in older ones. The most interesting feature, however, is the emergence of *clones* of cytogenetically abnormal lymphocytes (Table 4), and it is again worth noting that these have been found in the older patients. The suggestion above that the exchange points are not randomly distributed is confirmed by the fact that all six clones (Table 4) involved a group D chromosome, and in two cases, at any rate, the chromosome in question was a No. 14. In the Hecht case the exchange points were in the proximal third of one chromosome No. 14 and in the telomeric region of the other. The location of the exchange points could not be ascertained in the two Pfeiffer cases, while in the Schmid and Jerusalem case there was only one obviously abnormal chromosome, which makes it impossible to identify the exchange point until the banding of the chromosomes is studied. In our case, J.G., the exchange point was in the long arm near the centromere of the chromosome No. 14 and near terminal in the long arms of the X.

It is clear that in all six instances the clone was very large and constituted the majority of the lymphocyte population. Our case has been found to be hematologically normal on two occasions after the recognition of

Table 4. CLONES DETECTED IN LYMPHOCYTE CULTURES FROM PATIENTS WITH ATAXIA TELANGIECTASIA

Case Identification (Reference)	Age	Sex	Karyotype of Clone	Proportion of Cells with the Clonal Karyotype
Hecht and Case [23] and Hecht et al. (unpublished data, 1972)	18–23	M	46,XY,t(14q+;14q−)	2–3%; 75%*
Goodman et al. [20]	—	—	Group D translocation	All cells examined
Pfeiffer [43]				
P.Re.	15	M	46,XY,D−,Dq+	11 of 20†
W.Mu.	14	M	46,XY,Dq+	36 of 55
Schmid and Jerusalem [46]				
B.I.	16	M	46,XY,Dq+	89 of 107
Harnden (1974)‡				
J.G.	25	F	46,XX,t(X;14)(q27;q12)	73 of 100

* Second observation made 4½ years later.
† Cells fully analyzed had related, but varied, karyotypes.
‡ Present paper (also, see Figs. 2a, 2b, and 3).

the clone, and there has been no report of hematological abnormality in the other cases. Schmid and Jerusalem [46] speculated that the abnormal clone cells were persisting because of a defect of the immune rejection mechanism. This does not seem to explain their apparent advantage, and one must agree with Hecht and Case who talk of a nonmalignant clone of lymphocytes with a proliferative advantage. This concept is, however, difficult to accept unless it is a step toward malignancy. One other possible way of looking at it, since one of the features of the syndrome is that the lymphocytes are basically defective, might be that the clone cells, though not malignant, may have an advantage over the other, defective, lymphocytes. Such a situation could be directly comparable to that described by Ford et al. [14] in chimeric mice in which, following the rejection of a marrow homograft by mice recovering from radiation-induced immunological incompetence, the entire reticuloendothelial system is repopulated by the progeny of a very small number of cells cytogenetically marked by radiation. The finding of Hecht et al. (F. Hecht, B. K. McCaw, and R. D. Koler, unpublished data, 1972) of a lower rate of chromosome breakage in the clone cells as compared with the nonclone cells of the same patient would be consistent with such an idea.

Abnormalities in Fibroblasts

The comment by Hecht et al. that the chromosomes of fibroblasts cultured from an AT patient were normal, was the only mention of such a study, perhaps because cells from these patients are not easy to grow (see Bloom's syndrome, ref. [18]). We have found that the initial growth from the explant is very slow and, even when established, AT cells are more difficult to maintain than normal cells. However, Parrington and Delhanty (personal communication, 1972) have examined fibroblasts from one of their patients and her parents, while we have studied two patients and their parents. Fibroblasts from Parrington and Delhanty's case showed an increased incidence of fragments and dicentrics. Fibroblasts of one of the Birmingham cases (J.G.) were found to have a translocation affecting a group D and a group F chromosome in all the cells examined; numerous other abnormalities were also found, most of them chromosome exchanges, although some chromatid gaps, breaks, and one chromatid exchange were also detected. Only a few cells from the other case have been examined so far, but it is clear that chromosome and chromatid gaps and breaks are common, and that some chromosome exchanges also occur. It is possibly even more interesting that Parrington and Delhanty found one out of 50 cells cultured from the mother to have a ring chromosome, and 13 out of 50 to have an extra group F chromosome. The father was normal. Preliminary observations showed fibroblasts from the father of the Birmingham cases to be normal, but the mother's fibroblasts to have a high incidence of abnormalities, most of which were chromosome exchanges. These studies have not yet been completed but do suggest that, as in Bloom's syndrome [18] and Fanconi's anemia [19], the chromosomal instability is not confined to the lymphoid cells.

COMMENTS ON THE CYTOGENETICS

It is of interest to know whether or not these abnormalities occur in vivo or are a chromosomal instability only manifested in vitro. We have no evidence yet that the clone of fibroblasts derived from the AT patient was present in vivo, since we only had observations on a single culture. The lymphoid clone in J.G. was seen in two independent samples, and the multiple observations of Hecht et al. (unpublished data, 1972) left no doubt that lymphoid cell clones did occur in vivo. Direct observations are only possible on marrow cells, but Hecht et al. [24, and unpublished data, 1972] reported the marrow of their case to be normal, while Lisker and Cobo [30] found 4 out of 12 cells examined to have chromatid breaks. Clearly more observations are required.

In view of the small number of cases fully investigated, it is really too early to attempt to draw firm conclusions about the significance of chromosome abnormalities seen in cases of AT, but a few comments can be made. There is no doubt that chromosome abnormalities are a feature of both lymphoid cells and fibroblasts from patients with AT. Some cases, but not all, have a high incidence of chromosome breakage. Chromosome rearrangements occur in most of the well-studied cases, and marker chromosomes derived through rearrangement may be present in clones of cells. There is a striking absence of the quadriradial configurations which are such a common occurrence in Bloom's syndrome [18]. The occurrence of clones of lymphocytes has not been reported in either Bloom's syndrome or in Fanconi's anemia, and the relationship between the clones in AT and the lymphoid neoplasms is obscure. The only report of cytogenetic studies on malignant cells in AT comes from Lampert [28], who carried out a direct examination of the leukemic cells on one of his patients. In four out of nine cells examined there was one deleted group G chromosome that was said to resemble an acentric fragment.

As in Bloom's syndrome, the distribution of breakage and reunion points in AT is not random. The high frequency of chromatid exchanges between homologues, characteristic of Bloom's syndrome, is not found, and more work must be done before the pattern of specificity in AT can be defined precisely. It is of interest, however, that the group B, D, and G chromosomes appear to be involved more frequently in rearrangements. Atkin and Baker [2] have found that the chromosomes of these groups are often underrepresented in malignant cells, and if it can be shown using banding techniques that this is a consequence of their more frequent involvement in rearrangements, this will make the association between chromosomal rearrangements and malignant disease in AT all the more interesting.

Another possible approach to a study of the relationship of the chromosome abnormalities to the malignancies in these cases is to examine the in vitro transformation of the cells by oncogenic agents. There are now several reports of an increased transformation rate by SV40 virus of fibroblasts from genetically abnormal individuals (e.g., ref. [38]). The only report published so far, however, on fibroblasts from patients with AT suggests they are not unusually susceptible to transformation by SV40 [26]. This report, however, deals with only a small number of observations on two cases, and observations on other patients, especially those with cytogenetic abnormalities in the fibroblasts, would be of interest.

Acknowledgements.–I am grateful to Miss Val Tryner and Miss Jennifer Oxford for their cytogenetic analyses of the Birmingham family; to Dr. A. C. Kendall for permission to quote details of his cases; to Dr. F. Hecht, Mrs. B. McCaw, Dr. Jennifer Parrington, and Dr. Joy Delhanty for permission to quote their unpublished results; to Mrs. Carol Melville and Mr. Christopher Richardson for their technical assistance. This work was carried out with grants from the Cancer Research Campaign and the Medical Research Council.

Addendum.–Since writing this paper several important advances have occurred in this field. The occurrence of a high frequency of neoplasms in AT patients has been confirmed by extensive data from the Immunodeficiency-Cancer Registry.[a] Hecht et al.[b] have published details of the case quoted above and report two further cases with lymphocyte clones involving chromosome 14. Our patient P.G. has now developed a lymphocyte clone with a 46,XX, t(14;14)(q12;q31) karyotype, apparently identical to the clone reported by Hecht et al. (See Figure 1 which, by chance, shows the one cell in 100 with this karyotype seen in our first observation on P.G.) The specific involvement of chromosome 14, especially at band 14q12, now seems quite clear.

[a] KERSEY JH, SPECTOR BD, GOOD RA: Primary immunodeficiency diseases and cancer: The immunodeficiency-cancer registry. *Int J Cancer* 12:333–347, 1973.

[b] HECHT F, McCAW BK, KOLER RD: Ataxia telangiectasia–clonal growth of translocation lymphocytes. *New Eng J Med* 289:286–291, 1973.

LITERATURE CITED

1. AMMANN AJ, CAIN WA, ISHIZAKA K, HONG R, GOOD RA: Immunoglobulin E deficiency in ataxia-telangiectasia. *New Engl J Med* 281:469–472, 1969.
2. ATKIN NB, BAKER MC: Possible differences between the karyotype of preinvasive and malignant tumours. *Brit J Cancer* 23:329–336, 1969.

3. BIEMOND A: Paleocerebellar atrophy with extrapyramidal manifestations in association with bronchiectases and telangiectases of the conjunctive bulbi as a familial syndrome. In Proceedings of the First International Congress on the Neurological Sciences (van Bogaert L, Radermacker J, eds), vol. 4, New York, Pergamon Press, 202–208, 1957.
4. BIGGAR D, LAPOINTE N, ISHIZAKA K, MEUWISSEN H, GOOD RA, FROMMEL D: IgE in ataxia-telangiectasia and family members. *Lancet* 2:1089, 1970.
5. BODER E, SEDGWICK RP: Ataxia telangiectasia. A familial syndrome of progressive cerebellar ataxia oculocutaneous telangiectasia and frequent pulmonary infection. A preliminary report on seven children, an autopsy and a case history. *Med Bull Univ South Calif* 9:15–28, 1957.
6. BODER E, SEDGWICK RP: Ataxia telangiectasia. A familial syndrome of progressive cerebellar ataxia, oculocutaneous telangiectasia and frequent pulmonary infection. *Pediatrics* 21:526–554, 1958.
7. BODER E, SEDGWICK RP: Ataxia telangiectasia. A review of 101 cases. *Little Club Clin Develop Med* 8:110–118, 1963.
8. CASTAIGNE P, CAMBIER J, BRUNET P: Ataxie télangiectasies, désordres immunitaires, lymphosarcomatose terminale chez deux frères. *Presse Med* 77:347–348, 1969.
9. CAWLEY LP, SCHENKEN JR: Monoclonal hypergammaglobulinaemia of the γM type in a nine year old girl with ataxia telangiectasia. *Am J Clin Pathol* 54: 790–801, 1970.
10. CENTERWALL WR, MILLER MM: Ataxia, telangiectasia, and sinopulmonary infections. A syndrome of slowly progressive deterioration in childhood. *Am J Dis Child* 95:385–396, 1958.
11. DUNN HG, MEUWISSEN H, LIVINGSTONE CS, PUMP KK: Ataxia telangiectasia. *Can Med Assoc J* 91:1106–1118, 1964.
12. FEIGIN RD, VIETTI TJ, WYATT RG, KAUFMAN DG, SMITH CH: Ataxia telangiectasia with granulocytopenia. *J Pediatr* 77:431–438, 1970.
13. FIREMAN P, BOESMAN M, GITLIN D: Ataxia telangiectasia. A dysgammaglobulinaemia with deficient γ,A (β_2A)—globulin. *Lancet* 1:1193–1195, 1964.
14. FORD CE, MICKLEM HS, GRAY SM: Evidence for selective proliferation of reticular cell clones in heavily irradiated mice. *Brit J Radiol* 32:280, 1959.
15. FUDENBERG HH: Immunologic deficiency, autoimmune disease and lymphoma, observations, implications and speculations. *Arthritis Rheum* 9:464–472, 1966.
16. GELLIS SS, FEINGOLD M: Ataxia telangiectasia. *Am J Dis Child* 117:317–318, 1969.
17. GERMAN J: Chromosomal breakage syndromes. *Birth Defects: Original Article Ser* 5(5):117–131, 1969.
18. GERMAN J: Genes which increase chromosomal instability in somatic cells and predispose to cancer. In Progress in Medical Genetics (Steinberg AG, Bearn AG, eds), vol 8, New York, Grune & Stratton, 61–101, 1972.
19. GERMAN J, CRIPPA LP: Chromosomal breakage in diploid cell lines from Bloom's syndrome and Fanconi's anaemia. *Ann Genet* 9:143–154, 1966.
20. GOODMAN WN, COOPER WC, KESSLER GB, FISHER MS, GARDNER MB: Ataxia telangiectasia. Report of 2 cases in siblings presenting picture of progressive spinal muscular atrophy. *Bull Los Ang Neurol Soc* 34:1–22, 1969.

21. GROPP A, FLATZ G: Chromosome breakage and blastic transformation of lymphocytes in ataxia telangiectasia. *Humangenetik* 5:77–79, 1967.
22. HAERER AF, JACKSON JF, EVERS CG: Ataxia telangiectasia with gastric adenocarcinoma. *J Am Med Assoc* 210:1884–1887, 1969.
23. HECHT F, CASE M: Emergence of a clone of lymphocytes in ataxia telangiectasia. *In* Program of Annual Meeting of the American Society on Human Genetics, San Francisco (abstract), October 1–4, 1969.
24. HECHT F, KOLER RD, RIGAS DA, DAHNKE GS, CASE MP, TISDALE V, MILLER RW: Leukaemia and lymphocytes in ataxia telangiectasia. *Lancet* 2:1193, 1966.
25. KARPATI G, EISEN AH, ANDERMANN F, BACAL HL, ROBB P: Ataxia telangiectasia. *Am J Dis Child* 110:51–63, 1965.
26. KERSEY JH, GATTI RA, GOOD RA, AARONSON SA, TODARO GJ: Susceptibility of cells from patients with primary immunodeficiency diseases to transformation by simian virus 40. *Proc Nat Acad Sci (US)* 69:980–982, 1972.
27. KOREIN J, STEINMAN PA, SENZ EH: Ataxia telangiectasia—report of a case and review of the literature. *Arch Neurol* 4:272–280, 1961.
28. LAMPERT F: Akute lymphoblastische Leukamie bei Geschwistern mit progressive Kleinhirn-ataxie (Louis-Bar syndrome). *Deut Med Wochenschr* 94:217–220, 1969.
29. LEVEQUE B, DEBAUCHEZ CL, DESBOIS J-C, FEINGOLD J, BARBE J, MARIE J: Les anomalies immunologiques et lymphocytaires dans le syndrome d'ataxie télangiectasie. *Sem Hop, Paris* 42:2709–2717, 1966.
30. LISKER R, COBO A: Chromosome breakage in ataxia telangiectasia. *Lancet* 1:618, 1970.
31. LOUIS-BAR D: Sur un syndrome progressif comprenant des télangiectasies capillaire cutanées et conjonctivales symmetriques, à diposition naevoide et des troubles cérébelleux. *Confin Neurol (Basel)* 4:32–42, 1941.
32. MACFARLIN DE, OPPENHEIM JJ: Impaired lymphocyte transformation in ataxia telangiectasia in part due to a plasma inhibitory factor. *J Immunol* 103:1212–1222, 1969.
33. MCKUSICK VA: Mendelian Inheritance in Man, 3rd ed, Baltimore and London, Johns Hopkins Press, 1971.
34. MCKUSICK VA, GROSS HE: Ataxia telangiectasia and Swiss type agammaglobulinaemia. Two genetic disorders of the immune mechanism in related Amish sibships. *J Am Med Assoc* 195:739–745, 1966.
35. MILLER RW: Relation between cancer and congenital defects in man. *New Engl J Med* 275:87–93, 1966.
36. MILLER RW: Childhood cancer and congenital defects. A study of U.S. death certificates during the period 1960–1966. *Pediatr Res* 3:389–397, 1969.
37. MILLER ME, CHATTEN J: Ovarian changes in ataxia telangiectasia. *Acta Paediatr Scand* 56:599–561, 1967.
38. MUKERJEE D, BOWEN JM, TRUJILLO JM, CORK A: Increased susceptibility of cells from cancer patients with XY gonadal dysgenesis to simian papovavirus 40 transformation. *Cancer Res* 32:1518–1520, 1972.
39. NASPITZ CK, EISEN AH, RICHTER M: DNA synthesis in vitro in leukocytes from patients with ataxia telangiectasia. *Int Arch Allergy* 33:217–226, 1968.

40. PETERSON RDA, COOPER MD, GOOD RA: Lymphoid tissue abnormalities associated with ataxia telangiectasia. Am J Med 41:342–359, 1966.
41. PETERSON RDA, GOOD RA: Ataxia telangiectasia. In Immunological Deficiency Diseases in Man. Birth Defects: Original Article Ser 4(1):370–377, 1969.
42. PETERSON RDA, KELLY WD, GOOD RA: Ataxia telangiectasia—its association with a defective thymus, immunological deficiency disease, and malignancy. Lancet 1:1189–1193, 1964.
43. PFEIFFER RA: Chromosomal abnormalities in ataxia telangiectasia (Louis-Bar syndrome). Humangenetik 8(4):302–306, 1970.
44. REYE C: Ataxia telangiectasia. Am J Dis Child 99:238–247, 1960.
45. ROSENTHAL IR, MAKOWITZ AD, MEDENIS R: Immunologic incompetence in ataxia telangiectasia. Am J Dis Child 110:69–75, 1965.
46. SCHMID W, JERUSALEM F: Cytogenetic findings in two brothers with ataxia telangiectasia (Louis-Bar syndrome). Arch Genet 45:49–52, 1972.
47. SCHULER D, GACS G, SCHONGUT L, CSERHATI E: Lymphoblastic transformation in ataxia telangiectasia. Lancet 2:753–754, 1966.
48. SCHULER D, SCHONGUT L, CSERHATI E, SIEGLER J, GACS G: Lymphoblastic transformation, chromosome pattern and delayed type skin reaction in ataxia telangiectasia. Acta Paediatr Scand 60:66–72, 1972.
49. SCHUSTER J, HART Z, STIMSON CW, BROUGHT AJ, POULIK MD: Ataxia telangiectasia with cerebellar tumour. Paediatrics 37:776–786, 1966.
50. SMEBY B: Ataxia telangiectasia. Acta Paediatr Scand 55:239–243, 1966.
51. SOLITARE GB, LOPEZ VF: Louis-Bar's syndrome (ataxia telangiectasia). Neuropathologic observations. Neurology 17:23–31, 1967.
52. SYLLABA L, HENNER K: Contribution à l'indépendence de l'athétose double idiopathique et congénitale. Rev Neurol 33(1):541–562, 1926.
53. TADJOEDIN HK, FRASER FC: Heredity of ataxia telangiectasia (Louis-Bar syndrome). Am J Dis Child 110:64–68, 1965.
54. TALEB N, TOHME S, GHOSTINE S, BARMADA B, NAHAS S: Association d'une ataxie-télangiectasie avec une leucémie aiguë lymphoblastique. Presse Med 77:345–347, 1969.
55. UTIAN HL, PLIT M: Ataxia telangiectasia. J Neurol Neurosurg Psychiatr 27:38–40, 1964.
56. WALDER BK, ROBERTSON MR, JEREMY D: Skin cancer and immunosuppression. Lancet 2:1282–1283, 1971.
57. WAROT P, DUPUIS C, WALBAUM R, BONIFACE L: Le syndrome d'ataxia télangiectasies de Louis-Bar. A propos de deux observations dans une fratrie. Pédiatrie, 18:340–345, 1963.
58. WELLS CE, SHY GM: Progressive familial choreoathetosis with cutaneous telangiectasia. J Neurol Neurosurg Psychiatr 20:98–104, 1957.
59. YOUNG RR, AUSTEN KF, MOSER HW: Abnormalities of serum γ-IA globulin and ataxia telangiectasia. Medicine 43:423–433, 1964.
60. ZELLWEGER H, KHALIFEH RR: Ataxia telangiectasia. Report of two cases. Helv Pediatr Acta 18:267–279, 1963.

CYTOGENETIC AND ONCOGENIC EFFECTS OF THE IONIZING RADIATIONS OF THE ATOMIC BOMBS

AKIO A. AWA

Cytogenetics Section, Department of Clinical Laboratories, Atomic Bomb Casualty Commission, and Hiroshima Branch Laboratory, Japanese National Institute of Health, Ministry of Health and Welfare, Hiroshima, Japan

Introduction 637
Mortality Due to Leukemia and Cancer Among Atom-bomb Survivors 639
   Leukemia 640
   Cancers Other than Leukemia 642
   Benign or Unspecified Neoplasms 644
   Summary 644
Radiation-induced Chromosome Aberrations in Human Somatic Cells 645
   Methods of Study 645
   Relationship between Absorbed Dose of Radiation and Yield of
      Chromosome Aberrations 647
   Sources of Whole-body Irradiation Data 647
   Detailed Cytogenetic Studies of Atom-bomb Survivors 648
      Earlier Studies Using Peripheral Lymphocytes 648
      Current Studies Using Peripheral Lymphocytes 649
      Use of Bone Marrow Samples 660
      Leukemia and Related Disorders 661
   Evidence for In Vivo Clones 662
      Lymphocyte Clones 662
      Bone Marrow Clones 669
Concluding Remarks 669
Literature Cited 672

INTRODUCTION

Several observations concerning the effects of irradiation have now been firmly established in a number of independent investigations: (1)

Ionizing radiations induce a high incidence of malignant neoplasms. (2) Ionizing radiations induce damage in somatic cell chromosomes; breakage and re-union of chromosomes or DNA strands in somatic and germ cells result in abnormal morphology of the chromosomes, and possibly in cell death. (3) Cancer cells often show abnormalities in both chromosome structure and number. (4) Both the incidence of cancer and the yield of aberrant chromosomes produced by radiation exposure are related to the exposure dose. The interrelationships of these observations will serve as the basis of this paper.

There is now an extensive literature describing the somatic and genetic effects of ionizing radiation on experimental animals, and the data seem to be pertinent to man when he is exposed to high doses of radiation. Among the somatic effects thus far demonstrated in man, the carcinogenic effects perhaps are of most significance clinically.

Chromosome studies on experimental animal tumors were initiated in the early 1950s. Makino and his associates (see review by Makino, 1957 [37]), and Hauschka and Levan [19–21] studying transplantable tumors in rats and mice, established almost simultaneously the "stemline-cell" concept of cancer. This concept was based on the observations that (1) the chromosome constitution of tumor cells often differs from that of normal cells of the species; (2) each tumor has a characteristic chromosome pattern in both morphology and number, and this is true even for different tumors developing in the same tissue or organ; and (3) such a specific karyotypic alteration can be transmitted, without any notable change, through serial transplantations of the tumor.

Further supporting evidence for this stemline concept comes from observations that the same abnormal chromosome constitution of the primary tumor can be demonstrated in metastatic tumors; also, after suppression of a tumor by chemotherapy, the tumor cells present in relapse have the same karyotype as the original neoplasm. This karyotypic uniformity in the cells of a given tumor usually appears to evolve, presumably the result of mutation(s) and selection under the pressure of different host factors. The consistent occurrence of a specific anomaly, such as the Ph^1 chromosome in chronic granulocytic leukemia in man, represents an exception in which we have the presence of the same chromosomal abnormality in a specific malignancy rather than extensive variation from case to case (see review by Sandberg and Hossfeld, 1970 [44]).

A variety of chromosome abnormalities have been reported to occur in tumor cells of both experimental animals and man, but it still remains unclear whether these chromosome irregularities are the *cause* or the *consequence* of cancer development. The findings cited above lead us to believe that radiation-induced aberrations in the chromosomes of somatic cells may be related to, or the cause of, chromosome abnormalities in cancer cells. Unfortunately, there is neither direct nor indirect evidence for this assumption, and cytogenetic evidence relating cancer development to radiation effects in man is not extensive.

With the above considerations in mind, I shall describe here studies made by the Atomic Bomb Casualty Commission (ABCC), in collaboration with the Japanese National Institute of Health (JNIH), in two major areas of investigation: (1) mortality due to cancer and leukemia among atom-bomb survivors of Hiroshima and Nagasaki [29] and (2) the late effects of atom-bomb radiation on somatic chromosomes of the survivors, with special reference to the in vivo formation of clones of cells with radiation-induced chromosome aberrations. The possible implications of induced somatic-cell chromosome aberrations in irradiated persons will be discussed in relation to the etiology of malignant diseases. Relevant findings of other irradiated populations will also be considered.

MORTALITY DUE TO LEUKEMIA AND CANCER AMONG ATOM–BOMB SURVIVORS

The population studied consisted of about 109,000 persons in Hiroshima and Nagasaki, referred to as the JNIH–ABCC Life Span Study Sample (Extended). The cohort included both survivors of the bombings and, as one comparison group, persons who were not in the cities. All survivors, and the majority of the not-in-city groups, were chosen from listings dated October 1, 1950 (Table 1). Special arrangements were made by the Ministry of Health and Welfare whereby, for the purposes of this collaborative study, the Hiroshima and Nagasaki Branch Laboratories of JNIH obtained transcripts of death certificates for deaths that occurred in members of the study cohort, anywhere in Japan. Virtually complete records of mortality rates were obtained by means of the Japanese family registration system. The causes of death were obtained from the death certificate transcripts and were coded according to the International Classification of Diseases (ICD). It has been shown that in these two cities death certification of cancer as the underlying cause of death is about 90% accurate.

The dosimetry system employed at ABCC was devised by collaborators at Oak Ridge National Laboratory [1]. Hashizume and his colleagues [18], of the National Institute of Radiological Sciences of Japan, have obtained experimental estimates of the γ-radiation and neutron air-dose curves that agree well with the theoretically derived air-dose curves employed

Table 1. HIROSHIMA AND NAGASAKI POPULATIONS UNDER STUDY*

Parameter of Study	Numbers of Subjects							
	Calculated Radiation Dose (rad)†							
	None‡	0–9	10–49	50–99	100–199	200+	Un-known	Totals
By city								
Hiroshima	20,176	43,730	10,707	2,665	1,677	1,460	1,670	82,085
Nagasaki	6,347	11,404	3,700	1,231	1,229	1,310	1,461	26,682
By sex								
Male	11,143	23,176	5,664	1,533	1,282	1,267	1,427	45,492
Female	15,380	31,958	8,743	2,363	1,624	1,503	1,704	63,275
By age§								
0–9	5,015	10,759	2,992	695	422	398	302	20,583
10–19	5,978	11,815	2,626	772	795	826	1,117	23,929
20–34	5,671	10,836	2,801	798	599	633	775	22,113
35–49	6,161	12,664	3,570	974	695	618	583	25,265
50+	3,698	9,060	2,418	657	395	295	354	16,877
Totals	26,523	55,134	14,407	3,896	2,906	2,770	3,131	108,767

* Referred to as the JNIH–ABCC Life Span Study Sample (Extended).
† Estimates made using the T-65 dosimetry system [38].
‡ Residents not in the city at the time of bombing.
§ Age at the time of bombing.

in the dosimetry system. The dose estimates are designated T-65 to distinguish them from earlier estimates resulting from the cruder T-57 dosimetry system. The T-65 dosimetry provides independent estimates in rads of the γ- and fast-collision neutron doses, for individual survivors; these estimates are based on shielding information obtained from survivors by field interview [38]. For 3% of the survivors, dose estimates are not available, either because the survivor died or moved away from the cities before he could be interviewed or because he was located in a complex shielding situation for which no satisfactory method of estimating the attenuation due to shielding has yet been devised.

Mortality Due to Leukemia

It has long been known that mortality rates due to leukemia were sharply elevated among survivors of the atomic bombs, and especially among those who received high radiation doses.

As shown in Table 2, 147 deaths attributed to leukemia occurred in the cohort during 1950–1970, in contrast to the 55.1 that might have been expected at Japanese national leukemia rates. For the not-in-city group,

Table 2. Mortality from Leukemia (1950–1970) Correlated with Radiation*

Parameter of Study	Total Number of Deaths (mortality ratio)† Estimated Radiation Dose (rad)‡							
	None§	0–9	10–49	50–99	100–199	200+	Unknown	Totals
By city								
Hiroshima	10 (1.0)	34 (1.5)	17 (3.1)ᵃ	7 (5.1)ᵇ	10 (11.3)ᶜ	27 (35.5)ᶜ	5 (5.9)	110 (2.6)
Nagasaki	3 (0.9)	11 (2.0)	2 (1.1)	0 (0.0)	3 (4.8)	15 (22.5)ᶜ	3 (4.1)	37 (2.8)
By age‖								
0–9	0 (0.0)	7 (1.5)	3 (2.4)	1 (3.4)	4 (22.7)ᶜ	11 (66.5)ᶜ	0 (0.0)	26 (3.0)
10–19	5 (1.9)	4 (0.7)	6 (5.1)	1 (2.9)	3	6 (16.1)ᶜ	3 (6.1)	28 (2.6)
20–34	2 (0.6)	8 (1.3)	3 (2.0)	1 (2.3)	3 (8.4)ᵇ	9	3 (7.3)	29 (2.4)
35–49	3	19 (2.2)	4 (1.6)	2 (3.0)	1 (9.2)ᵃ	10 (26.2)ᶜ	1 (2.4)	40 (2.3)
50+	3 (2.3)	7 (2.1)	3 (3.4)	2 (7.9)	2 (13.5)	6 (23.6)ᶜ	1 (7.5)	24 (3.9)
By time period								
1950–1954	1 (0.8)	10 (2.7)	1 (1.0)	3 (11.1)	5 (25.0)ᶜ	15 (48.9)ᶜ	6 (29.6)	41 (5.9)
1955–1959	1 (0.3)ᵃ	16 (2.5)	7 (4.2)	2 (4.4)	5 (14.7)ᵇ	13 (79.1)ᶜ	0 (0.0)	44 (3.5)
1960–1964	3 (0.8)	5 (0.7)	7 (3.5)	0 (0.0)	3 (7.4)ᵇ	6 (40.2)ᶜ	1 (2.3)	25 (1.7)
1965–1970	8 (1.6)	14 (1.4)	4 (1.5)	2 (2.8)	0 (0.0)	8 (15.5)ᶜ	1 (1.7)	37 (1.8)
Totals	13 (1.0)	45 (1.6)	19 (2.6)	7 (3.5)	13 (8.6)ᶜ	42 (15.1)ᶜ	8 (5.1)	147 (2.7)

* Modified from Jablon and Kato [29].
† Mortality ratio = observed mortality/expected mortality.
Significance levels: ᵃ = $.01 < P \leq .05$; ᵇ = $.001 < P \leq .01$; ᶜ = $P \leq .001$.
‡ T-65 dosimetry system [38].
§ Not in city at time of bombing.
‖ Age at time of bombing.

observation and expectation for that time period agreed almost perfectly: 13 deaths vs. 13 expected—a mortality ratio of 1.0, where the ratio is defined as the mortality observed/mortality expected. However, in every group of survivors who had received the same radiation dose (including even those whose doses were in the range of 0–9 rad), the mortality ratio exceeded unity, and, among those with doses of 200+ rad, the number of deaths was nearly 30 times expectation.

The mortality ratios have, in general, declined over the years. However, in the 200+ rad group, the ratio still remained high, being 15 times expectation from 1960 to 1970. Those who were 0 to 9 years old at the time of bombing had far greater mortality ratios (about 66 times expectation) than those who were older (16 to 49 times expectation). After declining from ages 0–9 to 10–19, the ratios again rose from ages 35–50 and over. Although the number of cases is not large, it appears that those under 10 years of age or over 50 at exposure were most sensitive to the leukemogenic effects of radiation.

At all dose levels above 10 rad, the mortality ratios were much higher in Hiroshima than in Nagasaki. Ishimaru and colleagues [28] have tentatively implicated the larger neutron component of total dose in Hiroshima as being a possible reason for this discrepancy.

Mortality Due to Cancers other than Leukemia

It is evident from Table 3 that the data for survivors receiving radiation doses over 200 rad revealed a significantly elevated mortality ratio due to malignancies other than leukemia. It is possibly of interest that among those less than 10 years old at the time of the bomb the mortality ratio was very high for the high-dose groups and that most of the cases occurred after 1960. In this age group, the mortality ratio for those not in the city (controls) was also high, but all other exposure groups below 100 rad had a lower mortality than would be expected at national rates. The data for males and for females are similar. However, among those receiving an estimated dose of 200+ rad, the mortality ratio was much higher for the population in Hiroshima than for that in Nagasaki, although in both cities the ratio was significantly elevated.

As mentioned previously, death certification of cancer as the cause of death is about 90% accurate, although the accuracy of designation of tumors in particular sites is much less reliable. Relative risks for the 200+ rad groups were compared with the 0–9 rad group, for various tumors, and there is considerable variation in the relative risks for different tumors. For a number of them the probabilities are quite low, includ-

Table 3. Mortality from Malignant Neoplasms other than Leukemia (1950–1970) Correlated with Radiation Dosage[*]

Parameter of Study	Total Number of Deaths (mortality ratio)[†]							
	Estimated Radiation Dose (rad)[‡]							
	None	0–9	10–49	50–99	100–199	200+	Unknown	Totals
By city								
Hiroshima	650	1672	443	126	81	84	73	3129
	(1.0)[a]	(1.1)	(1.1)	(1.2)	(1.3)	(1.7)[c]	(1.5)	(1.1)
Nagasaki	189	318	118	40	25	43	38	771
	(1.1)	(1.0)	(1.1)	(1.1)	(0.8)	(1.4)[a]	(1.1)	(1.1)
By sex								
Male	437	1053	254	77	60	68	66	2015
	(1.0)[a]	(1.2)	(1.1)	(1.1)	(1.1)	(1.5)[a]	(1.2)	(1.1)
Female	402	937	307	89	46	59	45	1885
	(1.0)	(1.0)	(1.2)[a]	(1.3)[a]	(1.1)	(1.7)[c]	(1.7)	(1.1)
By age§								
0–9	10	6	0	0	1	6	2	25
	(1.8)[a]	(0.5)	(0.0)	(0.0)	(2.0)	(13.5)[c]	(5.5)	(1.1)
10–19	32	49	11	4	2	9	3	110
	(1.3)	(1.0)	(1.0)	(1.2)	(0.6)	(2.5)[a]	(0.6)	(1.1)
20–34	96	197	55	16	10	21	13	408
	(0.9)	(1.0)	(1.0)	(1.1)	(1.0)	(1.9)[b]	(1.1)	(1.0)
35–49	383	806	253	75	43	55	44	1659
	(1.0)	(1.0)	(1.2)[a]	(1.2)	(1.0)	(1.5)[b]	(1.2)	(1.1)
50+	318	932	242	71	50	36	49	1698
	(1.0)[b]	(1.2)	(1.1)	(1.2)	(1.5)	(1.3)	(1.7)	(1.2)
By time period								
1950–1954	65	291	85	95	18	18	37	539
	(0.9)	(1.0)	(1.1)	(1.1)	(1.3)	(1.6)[a]	(3.1)	(1.1)
1955–1959	194	474	114	37	20	22	16	877
	(1.0)	(1.1)	(1.0)	(1.2)	(1.0)	(1.3)	(0.9)	(1.1)
1960–1964	237	499	159	36	37	26	22	1016
	(1.0)	(1.0)	(1.2)	(1.0)	(1.5)[a]	(1.2)	(1.0)	(1.1)
1965–1970	343	726	203	68	31	61	36	1468
	(1.1)	(1.1)	(1.1)	(1.4)[a]	(0.9)	(2.0)[c]	(1.1)	(1.1)
Totals	839	1990	561	166	106	127	111	3900
	(1.0)	(1.1)	(1.1)	(1.2)	(1.1)	(1.6)[c]	(1.3)	(1.1)

[*] Modified from Jablon and Kato [29].
[†] Mortality ratio = observed mortality/expected mortality. Significance levels: [a] = $.01 < P \leq .05$; [b] = $.001 < P \leq .01$; [c] = $P \leq .001$.
[‡] T-65 dosimetry system [38].
[§] Age at time of bombing.

ing gastrointestinal cancer (other than stomach cancer), lung cancer, and others.

Mortality Due to Benign or Unspecified Neoplasms

Mortality attributed to benign neoplasms (or those unspecified on the certificates as to malignancy) in the 200+ rad group was elevated significantly for the whole period, specifically in 1965–1970 when the mortality ratio reached 4.7. There was no significant evidence of increased risk for survivors whose doses were less than 200 rad. There was also no evidence that those who were children when heavily irradiated suffered especially from benign neoplasms. The radiation effect was more marked in females (mortality ratio 3.8) than in males (mortality ratio 2.0).

Summary of Mortality

In summary, the causes of death strongly related to high radiation dose are leukemia and other cancers. Both leukemia and other cancers show one common feature: those who were under 10 years of age at the time of bombing were at the highest risk. However, this was not so for benign neoplasms.

It is apparent that leukemia rates among the survivors have declined over the 20-year period of this study. On the other hand, among survivors whose doses exceeded 200 rad, excessive numbers of deaths were still occurring during the interval 1965–1970, or 20 to 25 years after exposure. The latent period for radiation leukemogenesis varies widely, from less than 5 years to more than 20 years.

In Hiroshima the mortality ratios for cancer other than leukemia rose constantly during the 20-year period, and the rate of increase was accelerated during the last 6 years. In Nagasaki, on the other hand, the ratios declined from a moderately high level in the early period, reaching a minimum in 1960–1964, then rose sharply in 1965–1970. There is no reasonable explanation for the unusually high ratio in Nagasaki in 1950–1954, but the data from the two cities are in agreement in that during the most recent period, 1965–1970, cancer rates were increasing steeply in survivors who had high radiation doses. If the early elevation in Nagasaki is dismissed, it can be concluded that the latent period for death from radiation-induced cancers other than leukemia, at the dose levels experienced by the heavily irradiated survivors, is generally of the order of 20 years or more.

RADIATION-INDUCED CHROMOSOME ABERRATIONS IN HUMAN SOMATIC CELLS

It has long been known that ionizing radiations induce breakage and rearrangement of the chromosomes in somatic as well as germ cells of plants and animals. The incidence of such abnormalities has been proved to be proportional to the dose administered and to the dose rate.

Methods of Study

In the late 1950s, routine cytogenetic studies of mammalian chromosomes became feasible, along with the development of easier tissue culture methods for mammalian cells in vitro. Moreover, in 1960 a simple and reliable technique was developed for obtaining preparations of mitotic cells from cultured peripheral blood leukocytes by use of mitogenic agents [39]. It then became possible to examine the response of human chromosomes in somatic cells exposed to ionizing radiations in vivo and in vitro.

Lymphocytes from peripheral blood can enter into mitosis as early as 36 hours after culture initiation in the presence of phytohemagglutinin (PHA), a mitogen extracted from red kidney bean, and many mitotic cells can be seen at about 48 hours. A second peak of cells in mitosis is observed at 72 hours. Many laboratories in which human cytogenetic studies are carried out have cultured peripheral blood leukocytes for 72 hours to obtain the maximum number of cells in division. There is, however, good reason to believe that a majority of the mitotic cells cultured for 72 hours may be in their second or third in vitro cell division.

It has been suggested that the yield of some types of chromosome aberrations induced by radiation is influenced by increasing culture time. For instance, cells observed after 72 hours in culture were found to contain fewer aberrations, such as dicentrics, rings, or acentric fragments, than cells from the same blood sample cultured for 48 hours. Furthermore, there was a high frequency of tetraploid cells with duplicated aberrations. Such tetraploid cells with duplicated aberrations were rare or absent in 48-hour cultures [12, 25]. It has, therefore, been recommended that consideration be given to the influence of culture-sampling time on the yield of radiation-induced chromosome aberrations, if the frequency of aberrations obtained from blood cultures is to be used for biological dosimetry.

The lymphocytes in the circulating blood are now believed to be quite homogeneous in the sense that they are uniformly in the G_1 stage, the pre-DNA-synthesis period. In cultures containing PHA and incubated at

37°C, the small lymphocytes are stimulated to undergo DNA synthesis almost synchronously to reach their first in vitro mitosis. Consequently, when the lymphocytes are exposed to ionizing radiations either in vivo or in vitro before initiation of cultures, the cells are uniformly in their G_1 phase, when the chromosomes are single stranded. When these cells are observed at the next metaphase, the chromatids of a chromosome are broken or exchanged at the same locus. This type of abnormality is called "chromosome-type aberration," in contrast to "chromatid-type aberration," which can be induced when the cells are exposed to irradiation at interphase during or after DNA synthesis. Therefore, chromosome-type aberrations seen in cultured human lymphocytes are more generally considered to be significant in evaluating the relationship between yield of aberrations and radiation dose than chromatid-type aberrations, which are thought to be of little importance in this respect.

The chromosome-type aberrations are usually divided into two major types: the *simple deletion*, which is the result of a single break in the chromosome, and the *exchange*, which involves at least two breaks and an exchange of broken parts either between different chromosomes (i.e., interchanges, such as dicentrics and reciprocal translocations) or between different parts of the same chromosome (i.e., intrachanges, such as rings and paracentric or pericentric inversions).

Among chromosome-type aberrations, asymmetric (often called unstable-type) aberrations such as dicentrics, rings, and terminal deletions are easily identifiable on the basis of their distinctive morphology. It is more difficult to identify reciprocal translocations and inversions, referred to as symmetric exchange (stable-type) aberrations, and their detection varies significantly among laboratories. Furthermore, paracentric inversions (intra-arm exchanges of chromosome parts) do not result in an altered morphology of the affected chromosome. The same is true for reciprocal translocations, as well as for pericentric inversions, if the exchange occurs between points equidistant from the centromeres of two chromosomes of similar shape and size. These aberrations are, therefore, practically impossible to detect with classical methods, even by detailed karyotype analysis. It is considered to be highly reliable, then, to use only asymmetric aberrations such as dicentrics, rings, and acentric fragments as indicators for biological dosimetry.

As described previously, some of the cells with such asymmetric aberrations as dicentrics and rings cannot proceed through a normal mitosis. For example, bridge formation and lagging of chromatids can be seen at anaphase of cells with those aberrations, and such mechanical disturbances of mitosis may eventually result in cell death. It is possible that if the chromosome aberrations of the asymmetric exchange type are examined in irradiated persons at long intervals after exposure, the observed frequency will be lower than immediately after exposure to irradiation. As long as the cells are examined immediately after in vivo or in vitro radiation exposure, asymmetric aberrations can be readily identified, with few discrepancies in interpretation between observers, and thus are quite useful as indicators of the absorbed dose in man.

Relationship between Absorbed Dose of Radiation and Yield of Chromosome Aberrations

Extensive data from in vitro irradiation studies on cultured lymphocytes have been accumulated, and it is now widely accepted that the yield of chromosome aberrations, expressed as "number of dicentrics and rings per cell," is closely related to the absorbed dose (for details, see United Nations Report, 1969 [43]).

In contrast to the results of in vitro radiation experiments, very limited information on the relationship of in vivo dose to chromosome aberration has been obtained, because there are a variety of factors that make the estimation of biological doses more complicated. One of the reasons for this is that with the passage of time after irradiation, it becomes more difficult to obtain the true frequency of induced, but unstable, types of aberrations.

Despite this disadvantage in deriving dose estimates by scoring the number of chromosome aberrations in these irradiated individuals, attempts have been made by some investigators to correlate the yield of chromosome aberrations with the absorbed dose. Since the original work of Tough et al. [46], who observed gross chromosome damage in cells from blood cultures of two patients after X-ray therapy for ankylosing spondylitis, it has become increasingly evident that there is a general correlation between an increase in frequency of cells with induced chromosome aberrations and an increase in radiation dose, even many years after exposure (for details, see United Nations Report, 1969 [43]).

Sources of Whole-body Irradiation Data

Numerous studies of chromosome aberrations in cultured lymphocytes have now been made from persons exposed to (1) clinical, (2) occupational, (3) accidental, and (4) nuclear-explosion irradiations. The persistence of lymphocytes with radiation-induced chromosome aberrations has been well documented by repeated examinations of those persons for many years after radiation exposure. However, there are a limited number of papers reporting cytogenetic studies of persons exposed to high doses of total-body irradiation. Studies in this category include those done on: (1) humans who have received accidental exposure to irradiation ([4–6, 16, 17, 34, 35]; and others), (2) those who received internal irradiation by Thorotrast injections [24, 11], (3) Japanese fishermen exposed to fallout radiation by the 1954 nuclear test [24–27], (4) Marshallese exposed to the same source of irradiation as the Japanese fishermen [36], and (5) atom-bomb survivors of Hiroshima and Nagasaki in 1945 ([2, 7–9, 23–25, 45] and others).

Among these irradiated subjects, the atom-bomb survivors of Hiroshima and Nagasaki have been extensively studied from many points of view, and it is worth describing here in some detail the cytogenetic findings in these people.

Detailed Cytogenetic Studies of Atom-bomb Survivors

Earlier Studies Using Peripheral Lymphocytes.—Ishihara and Kumatori [24] reported the results of cytogenetic examinations of six atom-bomb survivors (two in Hiroshima and four in Nagasaki), in whom an increased frequency of induced chromosome aberrations of both stable and unstable types was demonstrated. Similar results were also obtained by Iseki [23], who studied 25 exposed persons in Nagasaki.

Twenty-two years after atom-bomb exposure, 51 Hiroshima survivors and 11 controls were studied by Sasaki and Miyata [45]. They attempted to correlate the yield of residual chromosome aberrations (as an indicator of biological dose) with the physical dose estimate. Based on the assumption that the proportion of peripheral blood lymphocytes with stable chromosome aberrations observed many years after an irradiation exposure remains unchanged from the proportion observed shortly after exposure, they were able to obtain fairly reasonable individual dose estimates by relating the proportion of cells with stable aberrations to distance from the hypocenter. On the other hand, they also attempted to estimate individual doses of exposed survivors according to the frequency of asymmetric (or unstable) chromosome aberrations, in terms of the number of dicentrics and rings per cell. These aberrations, which have been widely used in short-term in vitro radiation experiments, would have been largely eliminated in vivo over the years after exposure, so that the total frequencies of asymmetric aberrations in persons at different distances from the hypocenter were not considered by Sasaki and Miyata to offer a reliable basis for estimating absorbed dose.

Bloom et al. [7–9], using 3-day cultures of blood samples, examined younger (less than 30 years of age at the time of bombing) and older (over 30 years of age at that time) survivors of Hiroshima and Nagasaki; these were Adult Health Study participants at ABCC. The investigators found that cells with chromosome aberrations induced by irradiation were significantly increased in frequency in the exposed subjects and that the frequency of persons carrying chromosome aberrations was definitely higher among the exposed than among the control subjects in both cities. Predominance of cells with translocations was noted among older survivors. Honda et al. [22] demonstrated similar cytogenetic evidence in cultured lymphocytes obtained from atom-bomb survivors of Hiroshima.

Bloom et al. [10] further studied subjects who were in utero and whose mothers were exposed to more than 100 rad at the time of the bombings at Hiroshima and Nagasaki. Although it was difficult to estimate the doses absorbed by the subjects, since there was attenuation by maternal tissues, a significant difference was demonstrated between the 38 exposure subjects and the 48 controls in the frequency of cells with complex chromosome rearrangements. A suggestion of a positive dose-response relationship was present, though the increase of aberrations with increased doses was not statistically significant.

Current Studies Using Peripheral Lymphocytes. Since the data reported by Bloom et al. [7–9] were derived from 3-day cultures of blood samples, we have, for reasons given above, reexamined the adult atom-bomb survivors of both cities, using a 2-day-culture method. The results of the study appear in the following sections.

The individuals selected for cytogenetic studies were chosen on the basis of tentative dose estimates [1, 38] assigned to participants in the ABCC–JNIH Adult Health Study. This population sample comprised atom-bomb survivors as well as nonexposed control subjects living in Hiroshima and Nagasaki [3]. To date, the number of persons examined cytogenetically totals 130 in Hiroshima and 113 in Nagasaki, both males and females, whose estimated doses were greater than 100 rads (referred to here as the exposed group). For comparison, 134 individuals were selected in Hiroshima, and 79 in Nagaskai (herein referred to as the control group), whose estimated doses were less than 1 rad (Table 4).

Peripheral blood obtained from each subject was cultured for 54 hours; therefore, most of the observed metaphases were probably in their first in vitro cell division. Microscopic slides were prepared routinely by an air-drying method. An attempt was made to examine 100 metaphases in each case, but there were a few cases with poor proliferation in which 30 metaphases was the minimum number accepted for the present

Table 4. Cytogenetic Aberrations Correlated with Radiation Dosage

Radiation Dose (rad)		Number of Subjects	Number of Cells (and %) with Aberrations*				Cells Examined	
Range	Mean		dic+r	ace	t+inv	del	Total	With Aberrations (%)
Hiroshima								
Controls†	—	134	24 (0.2)	28 (0.2)	80 (0.6)	12 (0.1)	13,289	144 (1.1)
100–199	145	59	24 (0.4)	15 (0.3)	199 (3.4)	20 (0.3)	5,859	258 (4.4)
200–299	243	36	20 (0.6)	7 (0.2)	226 (6.3)	23 (0.6)	3,574	276 (7.7)
300–399	353	18	18 (1.0)	4 (0.2)	157 (7.0)	12 (0.7)	1,750	191 (10.9)
400–499	442	11	8 (0.8)	4 (0.4)	117 (10.9)	10 (0.9)	1,069	139 (13.0)
500+	717	6	4 (0.7)	4 (0.7)	110 (18.3)	6 (1.0)	600	124 (20.7)
Totals	252	130	74 (0.5)	34 (0.3)	809 (6.3)	71 (0.6)	12,852	988 (7.7)
Nagasaki								
Controls†	—	79	17 (0.2)	17 (0.2)	25 (0.3)	10 (0.1)	7,418	69 (0.9)
100–199	142	47	11 (0.3)	11 (0.3)	33 (0.8)	6 (0.1)	4,423	61 (1.4)
200–299	247	40	9 (0.2)	14 (0.4)	66 (1.7)	14 (0.4)	3,791	103 (2.7)
300–399	350	12	3 (0.3)	3 (0.3)	12 (1.1)	1 (0.1)	1,051	19 (1.8)
400–499	436	10	4 (0.5)	3 (0.4)	46 (5.4)	8 (0.9)	847	61 (7.2)
500+	636	4	1 (0.3)	1 (0.3)	46 (12.6)	0	366	48 (13.1)
Totals	245	113	28 (0.3)	32 (0.3)	203 (1.9)	29 (0.3)	10,478	292 (2.8)

* *Abbreviations:* dic=dicentric (and polycentric); r=ring; ace=acentric fragment; t=reciprocal translocation; inv=pericentric inversion; del=deletion.
† Control refers to those who received less than 1 rad.

analysis. The chromosomes were analyzed directly under the microscope. All of the cells in which definite or suspected structural aberrations were detected by direct microscopy were then photographed for karyotype analysis. All of these aberrant cells were karyotyped to identify which chromosomes were involved in the formation of exchange aberrations. Representative chromosome aberrations are shown in Figs. 1 to 8.

The cytogenetic findings are given in Table 4. In the data from both cities, there are striking differences between the control and exposed groups in the frequencies of cells with aberrations: while 1% of cells from controls had one or more chromosome aberrations, 7.7% of cells from heavily exposed Hiroshima survivors and 2.8% of cells from Nagasaki survivors were found to be abnormal. Although the distributions of percentage aberrations for individuals within different exposure groups overlapped considerably, the average percentage of cells with aberrations increased as the dose increased.

The frequencies of cells with unstable acentric fragments did not always increase with increasing dose except among the higher dose groups in Hiroshima. This suggests that an unknown proportion of aberrations classified as acentric fragments might arise from spontaneous terminal deletion and may not be related to radiation exposure.

Of the aberrant cells scored, the preponderance of cells with reciprocal translocations and pericentric inversions, or stable-type aberrations, was striking in all dose groups, and these aberrant cells constituted the major evidence of a dose–aberration relationship. In contrast to cells with stable-type aberrations, the cells with dicentrics and rings, or unstable-type aberrations, were less frequent, by an order of magnitude. There was, however, even in this class of aberrations, a suggestive increase with increasing dose. There was no evidence suggesting any difference in the aberrant cell frequencies by age.

The present findings demonstrate several important features. The predominance of stable-type aberrations over the unstable ones suggests that cells with unstable aberrations probably would have been eliminated from the in vivo lymphocyte population by mitotic events after atom-bomb irradiation, if we assume that the proportion of peripheral blood lymphocytes with stable chromosome rearrangements remains almost unchanged and that both stable and unstable aberrations had an equal probability of being induced by irradiation. Our data, together with those in other reports [40], indicate that the elimination of cells with unstable changes appears to be almost proportional to the increasing dose, but that nonetheless a dose–aberration relationship still holds for those remaining cells with dicentrics and rings.

As described in the preceding section, the use of cells with stable-

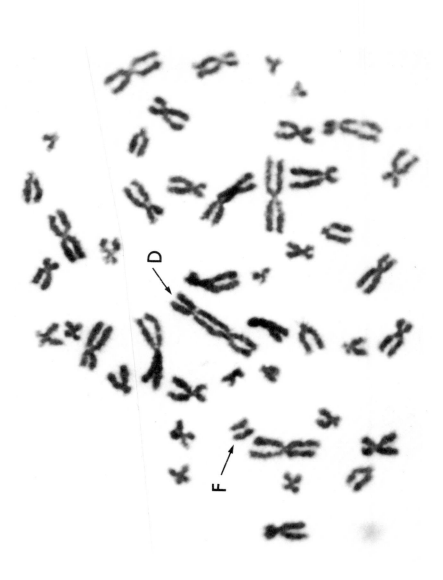

Figure 1. Metaphase from 2-day lymphocyte culture. Dicentric chromosome (D) with accompanying acentric fragment (F).

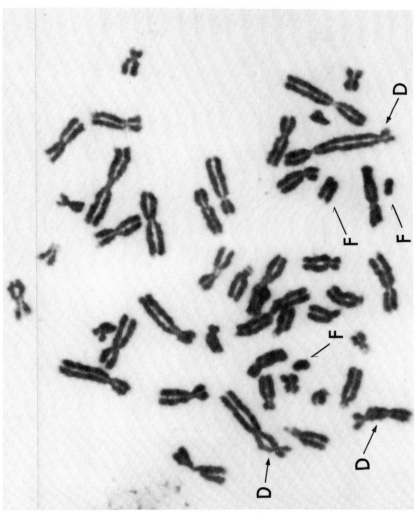

Figure 2. Metaphase showing three dicentrics (D) with three accompanying fragments (F).

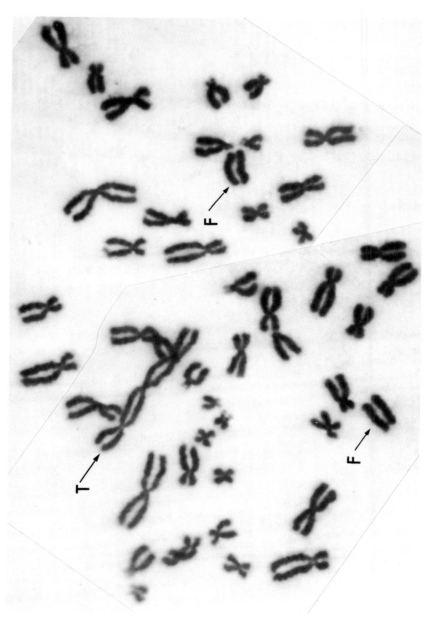

Figure 3. Metaphase with tricentric chromosome (T) and two accompanying fragments (F).

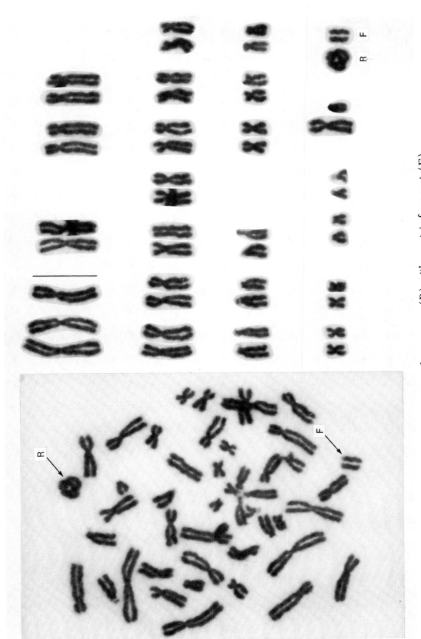

Figure 4. Metaphase and its karyotype showing ring chromosome (R) with acentric fragment (F).

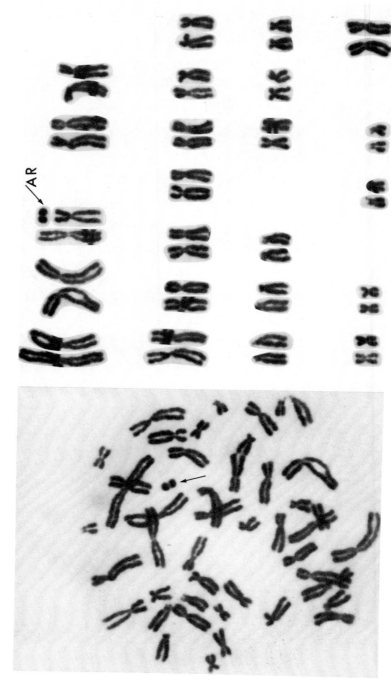

Figure 5. Metaphase and its karyotype showing acentric ring (AR), presumably the result of interstitial deletion from chromosome No. 3.

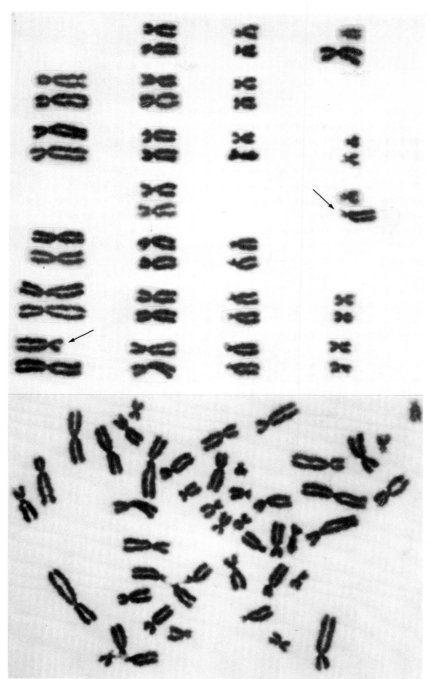

Figure 6. Metaphase and its karyotype showing a reciprocal translocation, t(1q−;Gq+).

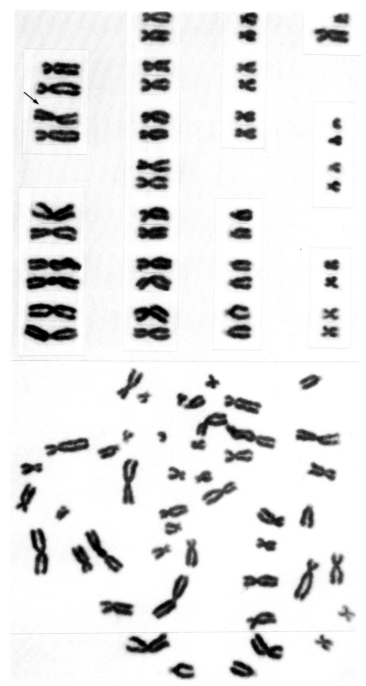

Figure 7. Metaphase and its karyotype showing pericentric inversion, inv(Bp+q−).

658

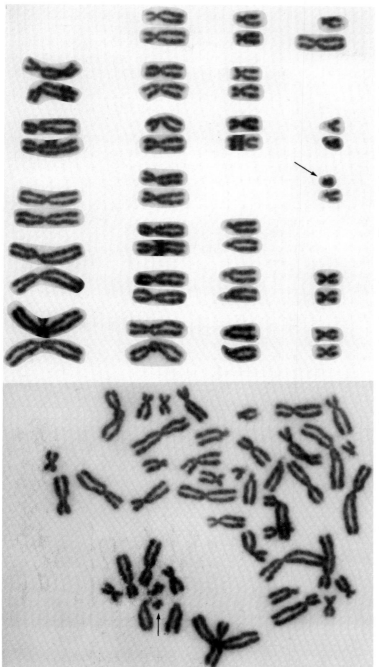

Figure 8. Metaphase and its karyotype showing a deletion, Gq−. This abnormality resembles morphologically the Ph¹ chromosome.

type aberrations formerly were considered to be rather unreliable for the evaluation of the dose–aberration relation. However, the present results seem to support the evidence presented by Sasaki and Miyata [45] that stable-type aberrations are indeed useful for dosimetry work. In fact, the frequency of cells with stable-type aberrations may be a more sensitive indicator than the frequency of unstable aberrations, particularly in those who received ionizing radiation *many years* before cytogenetic examination, and especially if a system with standardized criteria for determination of chromosome aberrations were to be established. However, although a dose–aberration response relationship is now demonstrable among the individuals exposed to atom-bomb radiation, it is still extremely difficult to use the frequency of chromosome aberrations found in any *one* irradiated individual as a biological dosimeter to estimate the amount of irradiation received.

Recent preliminary analysis in this laboratory on atom-bomb survivors who were in the low-dose group (1 to 99 rad) has confirmed the evidence for the dose–aberration response in both cities (unpublished data from the cytogenetics laboratory, ABCC). A comparison of the data from Hiroshima and Nagasaki shows a difference between the two cities in the frequencies of aberrant cells in each dose group. Furthermore, the increased frequency in Hiroshima appears to be almost linear with dose, while it seemed to be somewhat exponential in Nagasaki. It is not clear at this moment whether the difference between Hiroshima and Nagasaki is real or not. In order to confirm more objectively this observation, the frequencies of dicentrics and rings are compared for the two cities. It is apparent from the results shown in Table 4 that the frequency of these aberrations is always higher in Hiroshima than in Nagasaki, while there is no difference in the control values in both of the two cities. This suggests that the differences between the two cities may be ascribable in part to the differences in the radiation spectrum: the ratio of neutrons to γ-rays was comparatively high in Hiroshima, while γ-rays contributed almost the total dose in Nagasaki [1].

Use of Bone Marrow Samples. There have been a few reports on studies of bone marrow chromosomes obtained at biopsy from persons exposed to whole-body irradiation. After the work of Bender and Gooch [4, 5], Goh [16, 17] undertook a follow-up study of chromosomes in direct bone marrow preparations and cultured blood leukocytes from six patients, seven years after accidental total-body exposure to mixed fast neutrons and γ-rays. He reported the presence of residual chromosome aberrations, presumably in the myeloid cells, of the bone marrow specimens.

Ishihara and Kumatori [27] have continued cytogenetic examinations of direct preparations from bone marrow cells of the aforementioned Japanese fishermen. Of the 11 individuals studied, 5 have been found to have chromosome aberrations of the stable type. Despite the finding of cultured lymphocytes with unstable-type aberrations in all of these fishermen, no such cells were found in their bone marrow specimens.

In his initial studies of bone marrow cells from 47 atom-bomb survivors of Hiroshima, Kamada [30, 31] failed to demonstrate the presence of cells with radiation-induced aberrations. These persons were exposed at various distances from the hypocenter. Subsequently, however, Kamada et al. [32, 33] reported structurally aberrant chromosomes in bone marrow cells from proximally exposed atom-bomb survivors of Hiroshima.

It is interesting to note here that each of the above three research groups described the occurrence of cells with stable-type aberrations only. This could be explained by the more rapid turnover of actively proliferating bone marrow cells than of circulating lymphocytes; most lymphocytes of the peripheral blood are nondividing at any given time. Consequently, only cells with stable-type aberrations (i.e., those causing no mechanical disturbance at mitosis), could have survived in the marrow; those carrying unstable-type aberrations should have been eliminated by successive cell divisions in the actively proliferating bone marrow.

Of further interest is the occasional finding among the aberrations detected in bone marrow cells of an abnormally small group G chromosome, morphologically similar to the Ph^1 chromosome which is characteristic of chronic granulocytic leukemia in man.

Leukemia and Related Disorders. Kamada and his associates [30–33], using bone marrow specimens, have extensively studied chromosome constitutions in cases with leukemia and in preleukemic hematological disorders found among atom-bomb survivors of Hiroshima. In some of these leukemic cases, retrospective chromosome analysis was possible. The results of their findings are summarized below.

Of the 18 cases with *chronic granulocytic leukemia (CGL)*, seven were exposed to the atomic bomb within 2 km of the hypocenter. All of them had the Ph^1 chromosome characteristic of this leukemia. Other than the presence of the Ph^1, there was no specific chromosome aberration that might be related to atom-bomb radiation exposure. Chromosome analysis from peripheral blood cultures with PHA was also carried out in five of the seven exposed CGL cases: two of the five showed structural chromosome rearrangements such as dicentrics or translocation, or both, in the blood, and a third showed terminal deletions (acentric fragments) in bone marrow cells. These abnormalities are presumed to be related not to the leukemia but to the radiation exposure.

Of 36 individuals in Hiroshima having acute leukemia, 25 were diagnosed as the *acute granulocytic type*. Ten of the 36 were found to be exposed atom-bomb survivors; in 7 of these, abnormal karyotypes were noted with modal chromosome numbers ranging from 45 to 48. Kamada concluded that although abnormal karyotypes specific to each leukemia case were demonstrated, no consistent abnormal patterns related either directly or indirectly to atom-bomb irradiation were noted. Chromosome studies of these subjects done before the onset of leukemia failed to demonstrate any specific pattern of chromosome abnormality associated with radiation exposure; this had also been found true in examination of individuals with anemia, monocytosis, and prolonged leukopenia and possible preleukemic conditions.

Recently, Shimba and Thompson (H. Shimba and L. R. Thompson, Hiroshima ABCC, personal communication) examined an acute leukemia case by using cultured bone marrow samples obtained at autopsy. The modal chromosome number was found to be 74 to 75 with several marker chromosomes. Peripheral blood taken at the same time was also successfully cultured in the presence of PHA. Some cultured blood cells showed karyotypic changes similar to those observed in the leukemic bone marrow cells. The subject had been exposed heavily to atom-bomb irradiation in Hiroshima. However, several years before the onset of leukemia, she had had surgery because of breast cancer, following which she had been treated with more than 13,000 rad of local ^{60}Co radiation. Lymphocyte chromosomes had been examined twice, at 6 and at 36 months after radiotherapy, and on each occasion a majority of the cells showed both stable- and unstable-type aberrations. There was no evidence on those two occasions of the leukemic cell line with its modal chromosome number in the hypertriploid range, nor was there any evidence in cultured lymphocytes of stable-type aberrations that might have been related to the marker chromosomes of this leukemic cell line. Therefore, it seems likely, as suggested by Kamada et al. [31–33], that the chromosome abnormalities observed in this case of leukemia have no relationship to the aberrations in this same case, which were induced earlier by exposure to atom-bomb or ^{60}Co-irradiation.

Evidence for In Vivo Clones

Lymphocyte Clones. Among the cells having stable-type rearrangements of the chromosomes induced by irradiation, two or more cells with cytogenetically identical aberrations may occasionally be found in a blood culture. If the metaphases of such cells are examined at their first cell division in vitro, then identical cells may be considered to have been members of an in vivo clone. This means that in vivo propagation of cells from a progenitor cell with stable, radiation-induced rearrangements may form a clone, each cell of which has an identical abnormal karyotype. Since the detection of cells with stable rearrangements is tedious technically, it is not simple to identify a clone and only detailed karyotype analysis makes it possible.

Buckton et al. [11] considered three major factors in evaluating the cytogenetic evidence for the existence of a lymphocyte clone. We will cite the essence of their discussion directly:

> The first was the number of like cells observed in a blood culture, harvested when most cells are known to be in their first division in culture. Two like cells were considered to be insufficient evidence for the existence of a clone since they could conceivably have arisen as the result of a division in culture or as the result of separate events in vivo.
>
> The second consideration was the nature of rearrangements by which the cells were identified. For example, aberrations which consist of a deletion of a medium chromosome are most likely to have arisen as separate events since there are 15 or 16 morphologically similar chromosomes in this group.
>
> Lastly, consideration was given to the complexity of the rearrangements. Only those rearrangements which are the result of at least two breaks are acceptable as radiation-induced. The more complex the rearrangements the less likely it is that the cells arose as separate events in vivo and the more likely it is that they are derived from a common radiation-changed progenitor cell.

Ishihara and Kumatori [25] demonstrated several different lymphocyte clones collected from four different cultures in a patient who received Thorotrast years before. Buckton et al. [11] examined 36 patients who had received Thorotrast by intra-arterial injection from 11 to 31 years previously, and they found five individuals in whom two or more cells carried apparently identical structural rearrangements in blood cultures. Because of the above limitation in defining the clone, they conservatively concluded that in only two patients were clonal cells found on separate occasions.

In the course of studies of the atom-bomb survivors of Hiroshima and Nagasaki, Bloom et al. [8, 9] and Honda et al. [22] reported possible in vivo lymphocyte clones with identical karyotype abnormality in the heavily exposed survivors, although the earlier study was based on 3-day-culture blood samples. Of the 51 atom-bomb survivors of Hiroshima studied by Sasaki and Miyata [45], seven heavily exposed cases showed two or more cells with an identical karyotype, providing virtually unequivocal evidence of clonal proliferation of lymphocytes in vivo. With some exceptions, chromosome aberrations detected in the reported clones of lymphocytes were uniformly reciprocal translocations and pericentric inversions of the balanced type, with deletions occasionally also noted.

In our previous report [2] we described 14 cases of heavily exposed atom-bomb survivors of Hiroshima and Nagasaki carrying in vivo lymphocyte clones, whereas no clone was found in the 213 control persons in both cities. Based on newly selected, atom-bomb survivors of Hiroshima, most of whom were already reported in our previous study [2], we have examined 388 exposed survivors as well as 268 controls who were either not in the city at the time of bombing or whose doses were estimated to be less than 1 rad. As shown in Table 5, there was no clone-former in the control group, while 36 cases were found to have clones in those exposed, using, in this instance, the definition that two or more cells with an identical karyotypic abnormality in 2-day-culture preparations signify the presence of an in vivo clone. The frequency of clone-formers increased with increasing exposure doses.

An apparently identical chromosome aberration detectable in two cells in the same specimen could be fortuitous, or it could be that the resemblance was merely superficial. Therefore, *three or more* cells with an identical abnormality is a more conservative definition of a clone. Looked at in this way, and as given in the right-hand column of Table 5, the results showed again that the frequency of clone-formers in-

Table 5. Clone Formation Correlated with Radiation Dosage*

Radiation Dose (range in rad)	Number of Subjects Examined			
	Total	Clone-Formers	Percentage	With 3 or more Cells per Clone
Controls†	268	0	(0)	0
1–99	69	1	(1.4)	0
100–199	128	6	(4.7)	0
200–299	77	5	(6.5)	1
300–399	43	6	(14.0)	2
400+	71	18	(25.4)	7
Total exposed	388	36	(9.3)	10

* All subjects were from Hiroshima.
† Not in city at time of bomb or received less than 1 rad.

creased almost exponentially with increasing dose. In general, the frequency of aberrant cells with structural rearrangements was remarkably higher in the clone-formers than in the other exposed persons within each dose group [2]. Structural aberrations in the complements of each clone were of the stable type, and none of the unstable type aberrations was observed. Six of the 36 clones showed a terminally deleted chromosome, characterized by the loss of chromosome material. There were another three cases in whom more than 10% of the observed cells formed a clone (Table 6, Figs. 9 to 11). Although all of these cases were exposed heavily to atom-bomb radiation, they appear to be healthy at present.

Thus the existence of in vivo lymphocyte clones in irradiated persons is clearly demonstrated. Since heavily exposed atom-bomb survivors had received acute whole-body irradiation at the time of the bombing, repopulation of blood cells should have occurred in the recovery phase after the depression in the number of circulating blood cells as an acute radiation symptom. The cells contributing principally to regrowth of blood cells would be the precursor cells with either normal or abnormal, but stable, chromosome constitutions. These latter aberrations are assumed to cause no significant mechanical disturbance in mitotic events. If the lymphoid tissues were severely damaged by irradiation, a subsequent and continuing production of clonal cells carrying a new and consistent chromosome abnormality would be expected in the circulating blood cells. The three cases just described who have an unusually high proportion of circulating clonal cells favor this view (Table 6).

Six of the clone-formers found in our survey may be of particular interest because their clones were characterized by a deficiency of chromosome material. While this type of abnormality very well might have

Table 6. Cytogenetic Data from Three Heavily Irradiated Subjects Who Formed Multiple Clonal Cells[*]

Subjects		Radiation Dose (rad)	Cells Examined			Karyotype of Clone
Sex	Age[†]		Total	Clonal Cells	Other Aberrations[‡]	
Male	40	284	70	8	10	$t(2q-; Cq+)$
Female	51	503	100	10	30	$t(Cp+; Cp-)$ $t(Bq-; Cq+)$ $Gq-$
Female	66	892	100	16	23	$t(Cp+; Dq-)$

[*] Detailed cytogenetic studies of selected subjects from those described in Table 5.
[†] Age at examination.
[‡] Dicentrics, rings, acentric fragments, translocations, inversions.

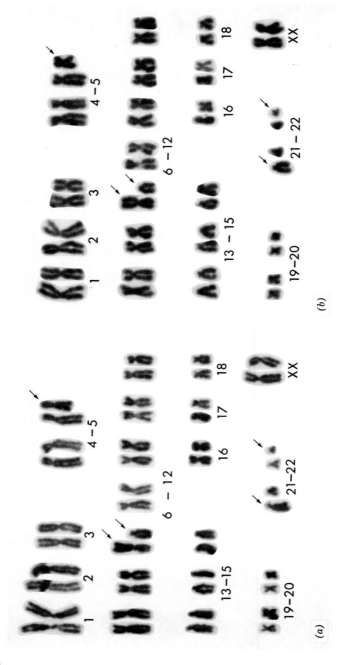

Figure 9. Karyotypes of clonal cells from case 1. Aberrations are identified as t(Bq−;Gq+), t(Cp+;Cp−), and Gq−.

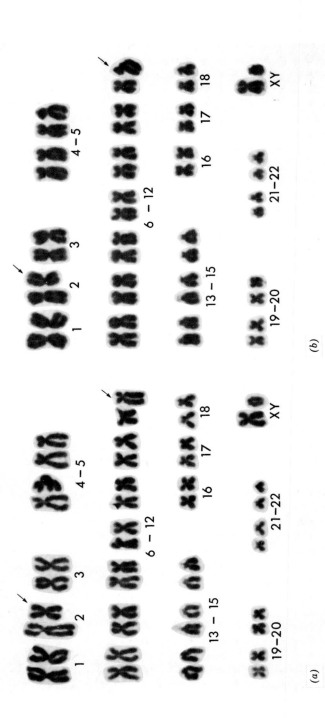

Figure 10. Karyotypes of clonal cells from case 2, showing t(2q−;Cq+). The presence of this clone has recently been confirmed using the Giemsa G-band method.

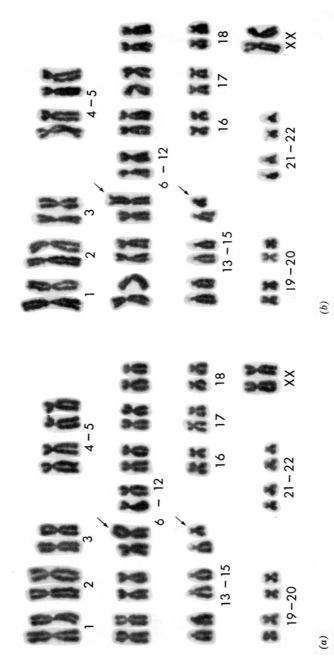

Figure 11. Karyotypes of clonal cells from case 3, showing t(Cp+;Dq−).

resulted in cell death due to the genetic deficiency, it seemed in these instances to have had some kind of selective advantage in vivo, which resulted in clone formation.

Bone Marrow Clones. Clones of bone marrow cells with rearranged chromosomes were also identified by Ishihara and Kumatori [27] in the irradiated Japanese fishermen, and by Kamada and his colleagues [32, 33] in atom-bomb survivors of Hiroshima. Similar to aberrations in clones of lymphocytes, those in bone-marrow clones were consistently of the stable type: reciprocal translocations, inversions, and deletions.

The frequency of clones or clone-bearers appears to be higher in the bone marrow samples than in the cultures of lymphocytes. This may be due partly to the faster turnover of cells in the bone marrow than in the circulating lymphocyte pool. Thus the chance of forming aberrant clones would be higher in bone marrow than in the peripheral blood, if the affected cells are viable and capable of surviving mitosis.

Another interesting finding in the bone marrow clones is that there were at least three cases among Hiroshima atom-bomb survivors [33] and one in irradiated fishermen [27] in whom an abnormally small group G chromosome was the principal component of the clone. In still another Hiroshima case, a second abnormality was also detected in addition to the presence of the small group G chromosome.

CONCLUDING REMARKS

Results of population surveillance undertaken at ABCC have been described. Mortality due to cancer and leukemia is now known to be higher among the heavily exposed atom-bomb survivors of Hiroshima and Nagasaki, and a higher risk of malignant neoplasms is expected among the younger persons who were exposed at 0 to 9 years of age. From the cytogenetic findings, we concluded not only that induced chromosome aberrations have persisted in the peripheral lymphocytes of those exposed more than 20 years after irradiation but also that the frequency of such aberrant cells is correlated with the radiation dose absorbed (i.e., increases with an increase in the estimated dose).

Nevertheless, it is not known whether these two effects of irradiation (cancer induction and chromosome aberration) are in fact related or not, since some discrepancies exist. For example, age appears to be an important factor in mortality due to cancer, whereas there is no evidence of an age-related difference in the frequency of aberrant cells among cultured lymphocytes. This would seem to imply that when younger people are exposed to a large dose of irradiation they have a higher susceptibility to radiation-induced tumor induction (i.e., at the systemic tissue or organ levels), but not to chromosome changes in the somatic cell. However, there may be completely different mechanisms that determine the fate of tumor cells other than those associated with radiation-induced chromosome abnormalities in lymphoid cells. Although there is a sizable literature concerning somatic mutations, including chromosome rearrangements, as a possible cause of cancer (see review by German, 1972 [15]), we are not yet in a position to conclude that there is a predisposition to malignant transformation in cells with chromosomal abnormality. There are still a number of issues that require clarification.

In order to correlate these two radiation effects, we must look for several missing links. First, we should determine whether the chromosome aberrations induced by irradiation in lymphocytes, for example, can be related to those commonly found in cancer cells. We must learn the biological and clinical significance of persistent chromosomal aberrations in circulating lymphocytes. Since it is established that there is no specific, radiation-induced aberration involving particular chromosomes, the probability of finding a given chromosome in an exchange formation, for instance, is in general proportional to its length. This is especially true for dicentrics [41; Awa et al., unpublished]. Thus it is impossible to predict which type of chromosome aberration will form and whether it is likely to cause the cell to transform.

Second, radiation-induced chromosome aberrations of the stable type seen in lymphocytes seem to be basically different from those seen in malignant cells. Most radiation-exposed lymphocytes remain nearly diploid, and, in general, there is neither significant loss nor gain of chromosome material in a given aberrant cell; that is, many such cells are quasidiploid. On the other hand, most abnormalities in cancer cells are of the aneuploid or even heteroploid types, and abnormal marker chromosomes appear to be derived through complex rearrangements. It is thus extremely difficult to determine just which chromosomes are involved in the formation of a rearranged marker in cancer cells.

Clones of lymphocytes with radiation-induced chromosome aberrations may be one of the important missing links to correlate radiation exposure with the development of neoplasms. The clonal proliferation appears to be similar in many respects to that of malignant cells. Perhaps what most distinguishes cancer cells from clonal lymphocytes is that the lymphoid clones do not seem to produce any deleterious effect upon the clone-bearer. This may be because the capacity for proliferation of lymphoid clones, unlike that of tumor cells, is somewhat limited and under the normal regulatory control.

Some of the clones carrying chromosome, as well as gene, mutations may undergo transformation toward malignancy if other oncogenic stimulants are superimposed on them. The stimulants here considered may be oncogenic viruses, physicochemical mutagens, or physiological imbalance in the host environment. Thus it seems particularly important to discover whether there is any correlation between the lymphocyte clones and subsequent development of malignant lymphoid tissue. No positive evidence for this has been reported to date and, because of the small numbers of clone-formers available, it seems unlikely that we will be able to confirm (or refute) this possibility in the near future.

There is evidence also of clonal proliferation of radiation-induced aberrant cells in the bone marrow of irradiated persons, and it undoubtedly occurs in other tissues as well [13, 64]. However, it might be expected that clones in the bone marrow and in other tissues would be different from those found in a well-differentiated, mature lymphocyte population because the turnover rate of cells would be so different. The risk of malignant transformation among clonal cells in irradiated tissues other than lymphocytes may be high—but, again—this will be difficult to prove.

Of special interest among persons receiving total-body exposure is the existence of bone marrow clones with a small group G chromosome resembling the Ph^1 chromosome of chronic granulocytic leukemia. However, it is premature to consider this particular abnormality as a possible cause of chronic myelogenous leukemia.

Nowell et al. [42], who were the first to detect bone marrow clones in therapeutically irradiated persons with myeloproliferative disorders, drew the following conclusions at that time. (1) Clones of cells with the same chromosome aberration might not be the cause of the neoplasms, since such aberrant clones have been found to persist in hemopoietic tissues free of leukemia long after irradiation. (2) The chromosome changes in radiation-induced neoplasms are not uniform from case to case. Kamada et al. [33] also found no common abnormal features of chromosomes among leukemic, preleukemic, and healthy bone-marrow cells obtained from atom-bomb survivors of Hiroshima.

None of these considerations provides more than circumstantial evidence for radiation-induced chromosome changes as the mechanism responsible for radiation-induced malignancy, although both events occur, and both can be correlated with the radiation dose absorbed. Further cytogenetic, immunological, and biochemical studies are necessary to determine the significance of cells with radiation-induced chromosome abnormality, and, in particular, the significance of persistent clonal cells, as they may relate to the etiology of malignant neoplasms (and other diseases) developing in man after radiation exposure.

LITERATURE CITED

1. Auxier JA, Cheka JS, Haywood FF, Jones TD, Thorngate JH: Free-field radiation-dose distributions from the Hiroshima and Nagasaki bombings. *Health Phys* 12:425–429, 1966.
2. Awa AA, Neriishi S, Honda T, Yoshida MC, Sofuni T, Matsui T: Chromosome-aberration frequency in cultured blood-cells in relation to radiation dose of A-bomb survivors. *Lancet* 2:903–905, 1971.
3. Belsky JL, Tachikawa K, Jablon S: ABCC-JNIH Adult Health Study, Report 5. Results of the first five examination cycles, 1958–68. Hiroshima and Nagasaki. ABCC Tech Rep 9–71, 1971.
4. Bender MA, Gooch PC: Persistent chromosome aberrations in irradiated human subjects. *Radiat Res* 16:44–53, 1962.
5. Bender MA, Gooch PC: Persistent chromosome aberrations in irradiated human subjects. II. Three and one-half year investigation. *Radiat Res* 18:389–396, 1963.
6. Bender MA, Gooch PC: Somatic chromosome aberrations induced by human whole-body irradiation: the "Recuplex" criticality accident. *Radiat Res* 29: 568–582, 1966.
7. Bloom AD, Neriishi S, Kamada N, Iseki T, Keehn R: Cytogenetic investigation of survivors of the atomic bombings of Hiroshima and Nagasaki. *Lancet* 2:672–674, 1966.
8. Bloom AD, Neriishi S, Kamada N, Iseki T: Leukocyte chromosome studies of adult and in-utero exposed survivors of Hiroshima and Nagasaki. In Human Radiation Cytogenetics (Evans HJ, Court Brown WM, McLean AS, eds), Amsterdam, North Holland Publishing Company, 136–143, 1967.
9. Bloom AD, Neriishi S, Awa AA, Honda T, Archer PG: Chromosome aberrations in leukocytes of older survivors of the atomic bombings of Hiroshima and Nagasaki. *Lancet* 2:802–805, 1967.
10. Bloom AD, Neriishi S, Archer PG: Cytogenetics of the in-utero exposed of Hiroshima and Nagasaki. *Lancet* 2:10-12, 1968.
11. Buckton KE, Langlands AO, Woodcock GE: Cytogenetic changes following Thorotrast administration. *Int J Rad Biol* 12:565–577, 1967.
12. Buckton KE, Pike MC: Time in culture—an important variable in studying in vivo radiation-induced chromosome damage in man. *Int J Rad Biol* 8:439–452, 1964.
13. Dekaban A: Persisting clone of cells with an abnormal chromosome in a woman previously irradiated. *J Nucl Med* 6:740–746, 1965.
14. Engel E, Flexner JM, Engel de Montmollin ML, Frank HE: Blood and skin chromosome alterations of a clonal type in a leukemic man previously irradiated for a lung carcinoma. *Cytogenetics* 3:228–251, 1964.
15. German J: Genes which increase chromosomal instability in somatic cells and predispose to cancer. In Progress in Medical Genetics (Steinberg AG, Bearn AG, eds) vol 8, New York, Grune & Stratton, 61 pp, 1972.
16. Goh K: Smaller G chromosome in irradiated man. *Lancet* 1:659–660, 1966.
17. Goh K: Total body irradiation and human chromosomes: cytogenetic studies of the peripheral blood and bone marrow leukocytes seven years after total body irradiation. *Radiat Res* 35:155–170, 1968.

18. HASHIZUME T, MARUYAMA T, SHIRAGAI A, TANAKA E, IZAWA M, KAWAMURA S, NAGAOKA S: Estimation of the air dose from the atomic bombs in Hiroshima and Nagasaki. *Health Phys* 13:149–161, 1967.
19. HAUSCHKA TS: The chromosomes in ontogeny and oncogeny. *Cancer Res* 21: 957–974, 1961.
20. HAUSCHKA TS, LEVAN A: Characterization of five ascites tumors with respect to chromosome ploidy. *Anat Rec* 111:4671, 1951.
21. HAUSCHKA TS, LEVAN A: Inverse relationship between chromosome ploidy and host-specificity of sixteen transplantable tumors. *Exp Cell Res* 4:457–467, 1953.
22. HONDA T, KAMADA N, BLOOM AD: Chromosome aberrations and culture time. *Cytogenetics* 8:117–124, 1969.
23. ISEKI T: Cytogenetic studies in atomic bomb survivors. *J Jap Soc Int Med* 55: 76–83, 1966.
24. ISHIHARA T, KUMATORI T: Chromosome aberrations in human leukocytes irradiated *in vivo* and *in vitro*. *Acta Haematol Jap* 28:291–307, 1965.
25. ISHIHARA T, KUMATORI T: Polyploid cells in human leukocytes following *in vivo* and *in vitro* irradiation. *Cytologia* 31:59–68, 1966.
26. ISHIHARA T, KUMATORI T: Chromosome studies on Japanese exposed to radiation resulting from nuclear bomb explosions. In Human Radiation Cytogenetics (Evans HJ, Court Brown WM, McLean AS, eds), Amsterdam, North Holland Publishing Company, 144–166, 1967.
27. ISHIHARA T, KUMATORI T: Cytogenetic studies on fishermen exposed to fallout radiation in 1954. *Jap J Genet* 44 (Suppl 1):242–251, 1969.
28. ISHIMARU T, HOSHINO T, ICHIMARU M, OKADA H, TOMIYASU T: Leukemia in atomic bomb survivors, Hiroshima and Nagasaki, 1 October 1950–30 September 1966. *Radiat Res* 45:216–233, 1971.
29. JABLON S, KATO H: Studies of the mortality of A-bomb survivors. 5. Radiation dose and mortality, 1950–1970. *Radiat Res* 50:649–698, 1972.
30. KAMADA N: The effects of radiation on chromosomes of bone marrow cells. II. Studies on bone marrow chromosomes of atomic bomb survivors in Hiroshima. *Acta Haematol Jap* 32:236–248, 1969.
31. KAMADA N: The effects of radiation on chromosomes of bone marrow cells. III. Cytogenetic studies on leukemia in atomic bomb survivors. *Acta Haematol Jap* 32:249–274, 1969.
32. KAMADA N, TSUCHIMOTO T, UCHINO H: Smaller G chromosomes in the bone-marrow cells of heavily irradiated atomic bomb survivors. *Lancet* 2:880–881, 1970.
33. KAMADA N, UCHINO H: Preleukemic state in atomic bomb survivors, with special reference to retro- and prospective analysis. *Jap J Clin Hematol* 13:313–332, 1972.
34. KELLY S, BROWN CD: Chromosome aberrations as a biological dosimeter. *Am J Publ Health* 55:1419–1429, 1965.
35. LEJEUNE J, BERGER R, LEVY CI: Analyse caryotypique de quatre cas d'irradiation accidentelle. *Ann Genet* 10:118–123, 1967.
36. LISCO H, CONARD A: Chromosome studies on Marshall Islanders exposed to fallout radiation. *Science* 157:445–447, 1967.

37. MAKINO S: The chromosome cytology of the ascites tumors of rats, with special reference to the concept of the stemline cell. *Int Rev Cytol* 6:25–84, 1957.
38. MILTON RC, SHOHOJI T: Tentative 1965 radiation dose estimation for atomic bomb survivors, Hiroshima and Nagasaki. ABCC Tech Rep 1–68, 1968.
39. MOORHEAD PS, NOWELL PC, MELLMAN WJ, BATTIPS DM, HUNGERFORD DA: Chromosome preparations of leukocytes cultured from human peripheral blood. *Exp Cell Res* 20:613–616, 1960.
40. NORMAN A, SASAKI MS: Chromosome-exchange aberrations in human lymphocytes, *Int J Rad Biol* 11:321–328, 1966.
41. NORMAN A, SASAKI MS, OTTOMAN RE, FINGERHUT AG: Elimination of chromosome aberrations from human lymphocytes. *Blood* 27:706–714, 1966.
42. NOWELL PC, HUNGERFORD DA, COLE LJ: Chromosome changes following irradiation in mammals. *Ann NY Acad Sci* 114:252–258, 1964.
43. Report of the United Nations Scientific Committee on the Effects of Atomic Radiation. Annex C. Radiation-induced chromosome aberrations in human cells. General Assembly Official Records: Twenty-fourth Session, Suppl 13 (A/7613), United Nations, New York, 1969.
44. SANDBERG AA, HOSSFELD DK: Chromosomal abnormalities in human neoplasia. *Ann Rev Med* 21:379–408, 1970.
45. SASAKI MS, MIYATA H: Biological dosimetry in atomic bomb survivors. *Nature* 220:1189–1193, 1968.
46. TOUGH IM, BUCKTON KE, BAIKIE AG, COURT BROWN WM: X-ray-induced chromosome damage in man. *Lancet* 2:849–851, 1960.

THE ROUS SARCOMA VIRUS STORY: CYTOGENETICS OF TUMORS INDUCED BY RSV

FELIX MITELMAN

Department of Internal Medicine A, University Hospital of Lund, Lund, Sweden

---

Introduction 675
Primary Sarcomas 676
    Karyotypic Evolution 676
    Different Tumor Regions 684
    Latent Period and Age of Tumor 684
Metastatic Sarcomas 685
Chromosomes and Histopathology 687
Conclusion 688
Acknowledgements 690
Literature Cited 691

INTRODUCTION

The Rous sarcoma virus (RSV), originally recovered from a chicken sarcoma, induces tumors only in fowl. However, other strains of this virus, including the Schmidt-Ruppin strain (RSV-SR), have been developed which will produce tumors in fowl as well as in a wide variety of mammals. The pathology and epidemiology of the RSV-SR has been studied extensively by Prof. C. G. Ahlström and his co-workers at the Institute of Pathology, Lund [1–8].

In the beginning of the 1960s, a broad collaborative investigation concerning the chromosomes of RSV-SR tumors in various rodent species was initiated by Prof. C. G. Ahlström and Prof. A. Levan at the Institutes of Pathology and Genetics. Because of the wide variety of mammals in which the virus produces tumors, it was regarded as an excellent model system for the study of the relationship of viruses, chromosomes, and carcinogenesis. An experimental system could thus be utilized in which as many factors as possible were controllable, increasing the chances to detect possible chromosome patterns.

The present communication is a report on some of the relevant results on the effects of the vrius on chromosomes in vivo, as studied in cells from primary, metastatic, and transplanted tumors induced in the mouse, the Chinese hamster, and the rat. Another part of this collaborative study, the action of RSV in vitro on chromosomes of normal cells, has been done by Dr. Warren W. Nichols at the South Jersey Medical Research Foundation, Camden, and was reported by Nichols [36, 37] and by Nichols et al. [38].

PRIMARY SARCOMAS

The SR strain of Rous sarcoma virus used in all experiments below was originally obtained in 1960 from Dr. Schmidt-Ruppin, Frankfurt am Main Hoechst. Cell-free virus suspensions were prepared from homogenized chicken sarcoma by one standard centrifugation followed by repeated ultracentrifugations. The virus suspensions were pooled and kept in saline at −70°C until used. In part of the experiments a crude cell suspension of chicken sarcomas induced by the virus was used as a virus source. It was prepared by filtering a 1:5 suspension of finely minced chicken sarcoma tissue through gauze. Sarcomas were produced by a single injection subcutaneously of the RSV-SR into the thigh of newborn or young animals. For details concerning virus preparation and tumor induction, see the papers by Ahlström and co-workers cited above.

Karyotypic Evolution

The most characteristic cytogenetic feature of tumors is chromosomal variability. Usually a variety of karyotypes are present at every moment. This variation is never haphazard, however, but gathers around stemlines and sidelines. As used by our group [12, 24, 29] a *stemline* is defined as the most frequent karyotype of a tumor cell population, a *sideline* is defined as other karyotypes of the population present in a frequency of 10% or more, and *variant cells* are defined as all other karyotypes present in a frequency of less than 10%. By selective pressures, a dynamic equilibrium is maintained between these cell populations. Any change in environment may, however, upset the equilibrium, causing shifts of the stemline karyotype. It has been pointed out by Levan [20] that the deviations from the original karyotype of the species found in all advanced tumors may be regarded as the result of a genetic evolution, each tumor stemline representing a "survival of the fittest." The most viable karyotype prevails at all times. To quote Levan: "While the natural evolution of species proceeds at the level of organisms, the cancer evolution proceeds at the level of cells."

This interpretation of the cytogenetic events in neoplasms was mainly derived from studies of transplantable ascitic tumors and to some extent from studies on human leukemias; for review see Levan, 1959 [17] and de Grouchy et al., 1966 [11], respectively. Within the Rous system it has been ascertained that this concept applies to solid tumors as well. Systematic investigations of early in vivo passages of primary RSV-induced sarcomas have further permitted a detailed characterization of the early evolution in tumors that still have a completely normal diploid stemline. Such studies were done in the rat [13, 18, 28, 31, 36] and the mouse [26, 27]; in both materials the following steps of karyotypic progression could be visualized: (1) the appearance of single heteroploid variant cells;* (2) the development of a sideline from one of these cells; (3) replacement of the normal diploid stemline by the sideline, either directly or after modifications of the sideline; (4) modifications of the heteroploid stemline, including shifts of stemline category (e.g., changes from a diploid to a tetraploid level).

These results have thus clearly demonstrated that chromosome changes, numerical as well as structural, are an integral part of the evolution of tumors. However, few if any conclusions can be drawn as to the role of chromosome changes in the etiology of neoplasia. This can only be done when a large number of early stages of primary tumors of clear-cut etiology have been investigated. The Rous tumors of the different rodent species have made such a study possible, and the results described below are the first contribution to this field.

Primary, RSV-induced sarcomas have been studied in detail in the mouse [24], the Chinese hamster [12], and the rat [18, 29, 33, 36, and Ahlström and Levan, unpublished]. Figure 1 summarizes diagrammat-

---

* In the present paper, as in other papers by our group, heteroploidy is defined as all karyotypes except the normal one of the species. In this extended definition, "pseudodiploidy"—including both diploidy with abnormal representation of morphologically normal chromosomes ("quasidiploidy" according to Yerganian et al. [45]) and diploidy with structurally changed chromosomes—is also included in the concept of heteroploidy.

ically the chromosome numbers of stemlines and sidelines in 91 tumors of the mouse, 42 tumors of the Chinese hamster, and 80 tumors of the rat. As can be seen, the general numerical patterns show striking similarities among the three species. Thus in materials from all three sources the normal diploid chromosome number dominates, and the most important chromosomal deviation is hyperdiploidy (gain of one or several chromosomes in the diploid region); the trisomic number (the normal diploid number plus one chromosome) is second in frequency to the diploid number, and hypodiploidy (loss of one or several chromosomes in the diploid region) and polyploidy are of minor importance.

In the Rous tumors of the mouse it was impossible, because of the

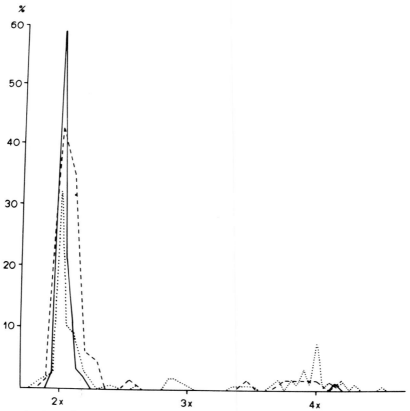

Figure 1. Distributions of chromosome numbers of stemlines and sidelines in primary RSV-SR-induced sarcomas; comparison between the rat (solid curve), the mouse (dotted curve), and the Chinese hamster (dashed curve) (from [29]).

uniform chromosome morphology, to determine whether any specific chromosome or chromosomes were preferentially involved in the heteroploid evolution. In the Chinese hamster, however, where the diploid chromosome number is 22 and considerable morphologic chromosome variation is present, each stemline and each sideline karyotype could be characterized in relation to the normal diploid or tetraploid karyotype by gains or losses of individual chromosomes or by chromosomal rearrangements. In the entire material of the 42 Chinese hamster tumors, 42 stemlines and 26 sidelines occurred. Table 1 shows the distribution in these 68 stemlines and sidelines of the 25 different karyotypes found. Out of 32 hyperdiploid stemlines and sidelines (cells with 23 to 28 chromosomes) 28 had one or more additional chromosomes of pairs Nos. 5, 6, or 10. In 20 of the 25 stemlines and sidelines with a chromosome number of 23, the addition of a chromosome of one of these pairs was the only abnormality. It is noteworthy that this was the first obvious nonrandom karyotypic pattern ever to be observed in experimental neoplasms.

In the rat, the RSV-induced sarcomas manifested one of the most striking nonrandom patterns hitherto observed. Table 2 summarizes the karyotypes of stemlines and sidelines in 80 prmiary sarcomas studied in our laboratory during the last 10 years. The diploid chromosome number of the rat is 42, and the system of classification of the rat chromosomes follows Tjio and Levan [44]; the position of the centromere is given as terminal (t), subterminal (st), or median (m), and the subscript gives the size classification, with 1 the largest ([21]; Fig. 2). As can be seen from Table 2, 43 out of 58 deviating stemlines and sidelines displayed the same apparent karyotypic feature (i.e., one or more extra t chromosomes). As in the Chinese hamster, this was most clearly seen among the trisomic stemlines and sidelines (i.e., those with 43 chromosomes), in which one extra t chromosome was the only change from normality in 24 out of 28 karyotypes. In the remaining 19 stemlines and sidelines with extra t chromosomes, additional changes had occurred. The most frequent one, a gain of one st_3 chromosome, was observed in 14 of these karyotypes (Fig. 2, upper row). Altogether, one extra t or st_3 chromosome, or both, were found in almost 80% of all heteroploid stemlines and sidelines, and also in about 70% of all variant cells. These chromosomal deviations are highly significant when tested statistically according to the method described by Levan [19].

There were many indications that the karyotypic changes were stepwise and that the gain of the t chromosome preceded that of the st_3 chromosome. Thus the dominating variant cell in the sarcomas with diploid stemlines without sidelines had 43 chromosomes, and three-

Table 1. Gains and Losses of Individual Chromosomes in Stemlines and Sidelines of 42 Primary RSV-induced Sarcomas of the Chinese Hamster[*]

Karyotype No.	Number of Chromosomes per Cell	Big m 1	Big m 2	Medium m-sm X	Medium m-sm Y	Medium m-sm 4	Medium m-sm 5	Medium st 6	Medium st 7	Medium st 8	Small m 9	Small m 10	Small m 11	Markers	S	s	S+s
1	21						−1									1	1
2	22														22	5	27
3			−1													1	1
4						+1				−1				+1		1	1
5	23			+1											1		1
6					+1											1	1
7						−1									1		1
8							+1					+1		+1		1	1
9								+1							2	4	6
10												+1			6	3	9
11														+1	1	4	5
12	24					+1	+1								1		1
13							+1			+1					1		1
14												+1	+1		1		1
15														+2		1	1
16	25					+1	+1	+1									
17							+1			+1	+1		+1			1	1
18							+1				+1	+1				1	1
19	28						+1		+1	+1	+1	+1	+1		1		1
20	37†	−2	−1	−1	−1	−1		−2	−2	−2		+1		+4			
21	41	−1	−1				−1								1		1
22	43	−1												+1	1		1
23		−2												+6	1		1
24		−1	−1					−1	−2	−2					1		1
25	46					+2	−1	−1				+1	+1		1		1
Total															42	26	68

[*] Modified from Kato [12]. The position of the centromere is indicated by (m) median, (sm) submedian, or (st) subterminal. See Levan et al. [21]. S=stemline, s=sideline.
† Cells containing 37 to 46 chromosomes are compared with the normal tetraploid karyotype.

TABLE — KARYOTYPES AND SIDELINES OF 80 PRIMARY RSV-INDUCED SARCOMAS OF THE RAT*

Karyotype No.	Number of Chromosomes per Cell	m 1-7	st 1	st 3	st 5	t 2-9	Markers	S	s	S+s
1	38	-2	-1			-1		1		1
2	41	-1						1	1	2
3	41								1	1
4	42	-1		+1		-2	+1	1		1
5	42	-2				+1			1	1
6	42	-2					+2		1	1
7	42		-1				+1			1
8	42						+1	1	2	3
9	42					-1	+1	66	6	72
10	43					+1		4	20	24
11	43						+1	1	3	4
12	44	-1				+1	+2		1	1
13	44			+1		+1		1	6	7
14	44			+1			+1		1	1
15	44					+2			3	3
16	45	+1		+1		+1			1	1
17	45			+1		+2		1	1	2
18	45				+1	+2			1	1
19	46			+1	+1	+1	+1	1		1
20	46			+1		+3			1	1
21	87†					+3		1		1
Total								80	50	130

* Modified from Mitelman [29]. The position of the centromere is indicated by (t) terminal, (st) subterminal, or (m) median, and the subscript gives the size classification, with 1 the largest. See Levan et al. [21]. S=stemline, s=sideline.
† Compared with the normal tetraploid karyotype.

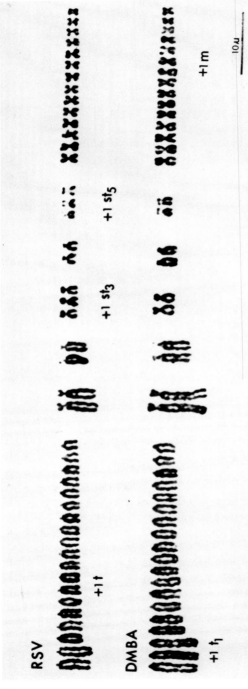

Figure 2. Deviations from the normal (female) rat karyotype characteristic of sarcomas induced by RSV (upper row) and DMBA (lower row). (See footnote *, Table 2. From [35].)

fourths of these cells differed from the normal by having only an extra t chromosome. This particular karyotype—one extra t plus 42 apparently normal rat chromosomes—dominated all sidelines of the diploid sarcomas, and it was very striking that among the 10 sidelines that were earliest to appear, 8 had this karyotype (see "Metastatic Sarcomas"). With increasing age of the tumors, karyotypes with one extra st_3 chromosome increased in prominence and gradually outgrew the trisomics.

In order to test the hypothesis that sequential karyotypic changes occur and to determine whether additional steps could be discovered, we used the following experimental system [31]. Two primary rat sarcomas with a normal diploid stemline and lacking variant cells were selected. From each tumor, five and six different series of transplant passages were established, respectively; the chromosomes were studied intermittently in each series during approximately 20 successive transplantations. The results clearly confirmed the sequential karyotypic evolution, with additions in turn of one t and then one st_3 chromosome. In addition, a third step—a gain of one st_5 chromosome—could now be demonstrated as a regular part of the chromosome variation (Fig. 2, upper row). From this point on, the karyotypic changes observed manifested no obvious pattern. Analyses using fluorescence and Giemsa banding, not yet used in this experimental system, are required to determine whether later stages of tumor development also are characterized by nonrandom karyotypic patterns which, because of superimposed secondary changes, cannot be studied by means of conventional staining methods.

Another interesting question related to the present discussion is to what extent the initial karyotypic changes in each tumor are predetermined. In the experiment cited above it was found that in all the different series of passage established from the same primary sarcoma, the earliest deviating cells had exactly the same karyotype—an unexpected and rather remarkable observation. Two possible explanations were offered [31]: (1) The deviating cells originated de novo in the different series. The initial step in the heteroploid evolution of each tumor would then be predetermined with a specificity comprising not only the numerical direction but also the first karyotypic change. (2) The deviating cells were present in the primary sarcoma but not detected due to the restricted number of analyzed cells. By a selective advantage of the particular karyotype, these cells gradually increased in number and were observed as single deviating cells in later passages of each different series.

Neither of these two hypotheses has been proven so far. The likelihood, however, that the sequence of events producing a conspicuous marker in several series of a sarcoma would occur repeatedly and independently in different cells must be considered as very small. Accordingly, an explanation in which such a mechanism is a prerequisite was regarded as less probable. The question is of considerable theoretical importance and is studied further in our laboratory.

Different Tumor Regions

Different tumor regions of the same primary sarcoma have been studied in the mouse [24] and the rat [18, 29]. In the mouse, five sarcomas were analyzed in two or three different regions: four tumors manifested the same chromosomal picture in the different parts, whereas one sarcoma had two different tetraploid stemlines in the two parts studied. In the rat, 12 primary sarcomas were successfully analyzed in samples from two or three different regions: 9 sarcomas had identical stemlines in the different regions, in one case including a similar marker. Three sarcomas had different stemlines in different regions; in two of them, the stemlines had closely related karyotypes. With the dynamic equilibria among different elements of tumor cell populations, such differences as were occasionally observed are only to be expected during the sequential chromosomal changes, and none of the differences among samples from the same tumor indicated necessarily a polyclonal origin. On the contrary, the presence of a characteristic marker chromosome in two samples from different regions of the same rat tumor (mentioned above) is a good indication of a common clonal origin of both parts of the tumor, supporting the concept that progression is clonal (i.e., originating from a single mutant cell).

The study of different tumor regions led to another interesting observation. The necrotic and hemorrhagic centers of three rat sarcomas were successfully analyzed. In all three, related hypodiploid karotypes were present; these differed from the normal rat karyotype by loss of one or two m chromosomes and were found to constitute a stemline in two of the tumors and a sideline in the third. It might be that the particular internal environment of tumors favors the selection of a specific mutant karyotype. The findings may be incidental, but the loss of this particular chromosome, in a particular milieu in the interior of a tumor, may be comparable to the preferential loss of one specific chromosome previously observed in a transplanted rat Rous sarcoma after environmental changes [28].

Latent Period and Age of Tumor

The relationship between the latent period of the primary tumor (i.e., the period of time elapsing between injection of the virus and appearance of tumor) and the chromosomal constitution, has long been a matter for debate. The relationships, short latent period and diploidy and long latent period and polyploidy, have been proposed by some authors, but others have failed to find any such relationship (for reference, see [29]). Among the RSV-induced primary sarcomas of the mouse, the Chinese hamster, and the rat, no correlation between the duration of the latent period and the chromosomal constitution of the tumors could be discerned. In the mouse [24], the shortest latent period was found in tumors with pseudodiploid stemlines and the longest, in hypodiploid tumors. Among tumors with a normal diploid stemline, those without sidelines appeared earlier than those with sidelines, the former having about the same latent period as the tetraploid tumors. In the Chinese hamster [12], tumors with normal diploid stemlines with and without sidelines had a slightly shorter latent period than tetraploid tumors. In the rat [29], the mean latent period was roughly the same in diploid tumors with and without sidelines and in heteroploid tumors.

By relating the age of the tumor to the chromosomal picture, however, some interesting observations were made. In all three species, it could be demonstrated that the proportion of normal diploid cells decreased with increasing age of the tumor. Furthermore, during the earliest stages the tumors not only had normal diploid stemlines but they consisted exclusively of normal diploid cells. As cells deviating in karyotype from normal began to appear and gradually to form sidelines, it followed that the proportion of normal cells decreased. In primary tumors from all three species, this correlation was found to be nonlinear with a period of unusually rapid loss of the normal diploid cells. This nonlinear change was confirmed by transplantation studies [31] in which each individual tumor could be followed through different periods with samples taken at frequent intervals (Fig. 3). Evidently, once the evolution has reached a critical point, the diploid cells are lost at an accelerated rate.

METASTATIC SARCOMAS

It has been possible to analyze the chromosomes of 20 metastatic RSV-induced rat tumors together with their corresponding primary sarcomas [30, 33]. All the metastases were to either regional or para-aortic lymph nodes. In 14 animals, one metastasis was studied, and in three animals two different metastases of the same primary tumor could be analyzed. The results demonstrated a close karyotypic relationship between the primary tumors and their metastases, on the one hand, and among the different metastases of the same primary tumor, on the other. In three cases, similar marker chromosomes were present in both sites. Hetero-

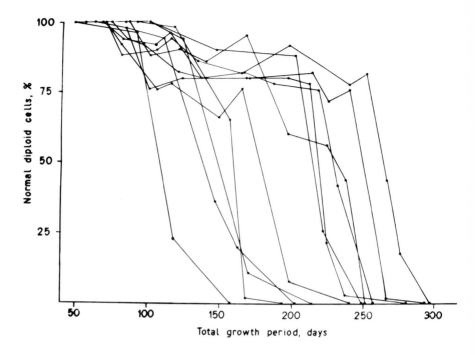

Figure 3. Relation between the frequency of normal diploid cells and age of tumor in 11 series of transplant passages of RSV-induced rat sarcomas initially comprised of normal diploid cells (from [31]).

ploidy of the primary tumor was not a prerequisite for the spreading process, since apparently normal diploid cells were capable of metastasizing. Not only stemline or sideline cells but also variant cells were potential stem cells for metastatic tumors. There was no selection advantage for polyploid cells as has been proposed by some workers (for review, see [30]). Whether or not diploid–near-diploid cells with any particular chromosome constitution were favored could not be determined. However, the general pattern of the karyotypic evolution in the metastases was the same as in primary Rous sarcomas of the rat. Thus hyperdiploidy was the most important chromosomal pathway, the second most important being pseudodiploidy; hypodiploidy and polyploidy were of little or no importance.

Each primary tumor and its metastases were processed on the same occasion. Yet 60% of the metastatic tumors in the lymph nodes had heteroploid stemlines, as compared with approximately 20% in the primary site. Approximately 40% of the primary tumors with a normal diploid stemline had sidelines, whereas 75% of the metastases had sidelines. It thus appears that the sequential chromosomal change is accelerated in the lymph node metastases. The specific chromosome losses, gains and structural alterations were also compatible with an acceleration of the nonrandom, stepwise changes observed in the primary tumors. In both the primaries and metastases one extra t, st_3, or both, were observed in approximately 75% of the deviating karyotypes, but their distribution was different. The trisomic karyotypes with one extra t chromosome, which dominated the deviating stemlines and sidelines and variant cells of the primaries, were only seen in about 10% of the metastatic heteroploid stemlines and sidelines, whereas the frequency of stemlines and sidelines with one extra st_3 had increased from 24% to 68%. In addition to the extra t and st_3, a threefold increase of stemlines and sidelines possessing one extra st_5 chromosome was found in the metastatic tumors (see p. 683). The accelerated progression in the metastases was further emphasized by the considerable increase in numbers of structural rearrangements: thus the frequency of marker chromosomes among the deviating stemlines and sidelines was almost twice that of the primary tumors (45 and 24%, respectively). As for all virus-induced tumors, it might be argued that the metastases represent inductions of new tumors. Though this possibility cannot be excluded wholly, Ahlström and Jonsson [3] found it unlikely for several reasons. Certainly, the generally close karyotypic relationships, and particularly the finding of the three cases with similar markers in both the primary and the secondary lesions, just discussed, seem to support the concept of metastasization rather than repeated primary inductions by the virus. Unfortunately, no blood-borne metastases could be studied because these were usually of small size and yielded poor preparations. Hence it cannot be stated whether an acceleration of the chromosomal changes typical for primary tumors applies to all types of metastatic growth.

CHROMOSOMES AND HISTOPATHOLOGY

Since chromosome variation between cells means differences in genetic constitution, it was reasonable a priori to assume that the chromosome variation, which is so characteristic of cancer development and progression, was reflected in phenotypic differences (i.e., histologic parameters). Any correlation between chromosomal constitutions and histologic characteristics would be of theoretical significance and, if applied to human tumors, might become of clinical interest. Thus it is known that chronic myelogenous leukemia, lacking the Philadelphia chromosome (Ph^1), is a clinically recognizable entity with a prognosis different from the cases with the Ph¹ chromosome [10, 43]. Systematic investigations of the chromosomes and the histology of a large number of primary tumors with uniform etiology ought to be rewarding, and we made this a part of the planned collaborative study of our group. All together, 453 Rous sarcomas that were subjected to chromosome analysis have also been examined histologically. This included: 92 primary and 115 transplanted mouse tumors [25, 27], 42 primary Chinese hamster tumors [12], and 63 primary, 121 transplanted, and 20 metastatic rat tumors [29–31].

All these tumors were identified as fibrosarcomas of varying degrees of maturity. In the three species, roughly the same histological classes could be distinguished: highly differentiated fibrosarcomas, spindle-cell sarcomas, and anaplastic sarcomas. This classification was based on the degree of cellular differentiation and the amount of collagen produced. Other histologic features proved less reliable, because they lacked sufficient specificity; small necrotic and hemorrhagic areas were found in all types, and no sure difference was observed as regards the mitotic rate. When uncertainty of classification arose, it always concerned neighboring classes; in all three species such tumors were assigned to the less-differentiated class. Histologic classifications were made without knowledge of the chromosome data.

There were no specific chromosomal changes related to histopathologic tumor types except for a tendency for the highly differentiated fibrosarcomas of the Chinese hamster to be associated with trisomy of chromosome No. 6. The same trends, however, were found in primary, metastatic, and transplanted sarcomas: tumors with a normal diploid stemline, lacking sidelines, were highly differentiated fibrosarcomas; with the appearance of deviating sidelines, the histologic picture changed to that of a spindle-cell sarcoma, and heteroploid stemlines—irrespective of specific chromosomal alterations—were associated with a change to an anaplastic sarcoma. It should be noted that the correlation was by no means an absolute one, and in many instances during transplantations in vivo the change in histologic maturity preceded or lagged behind the chromosomal changes by one or several passages. The trend, however, was obvious: the sequential chromosomal change from normal diploidy to heteroploidy was associated with the apparent state of differentiation as determined histologically.

CONCLUSIONS

The results presented above, based on the most comprehensive chromosome analyses existing of primary tumors having the same etiology, have demonstrated striking similarities in the general numerical pattern of the chromosomes in RSV-SR-induced sarcomas from different species.

The observations strongly suggest that all of these neoplasms are initially comprised of normal diploid cells and that, with increasing age of the tumor, these cells decrease in number at an accelerated rate. Evidently, after malignant conversion of a normal cell, the normal diploid karyotype is no longer the ideal vehicle for the genetic material, or perhaps the malignant cell can tolerate chromosomal change more readily than the normal cell. In either case a main difference between normal cells and tumor cells is this chromosomal variability. This variability gradually leads to the appearance of karyotypes deviating from normal in sidelines that then outgrow the normal stemline, either directly or after modification. The initial deviation from normality seemingly was the formation of a trisomic karyotype, judging from the observation that trisomy was the dominating abnormality in all three species. In the Chinese hamster and the rat—the two species with chromosome complements which permit detailed karyotype analyses—this chromosome variation was nonrandom. In the Chinese hamster, the variation preferentially affected chromosome pairs Nos. 5, 6, and 10; in the rat, the characteristic change was a three-step evolution, with additions in turn of one medium-sized t, one st_3, and one st_5 chromosome (Fig. 2, upper row).

Some intriguing conclusions can be drawn when these results are considered together with those obtained from recent studies of neoplasms induced in the Chinese hamster and the rat by 7,12-dimethylbenz(α)-anthracene (DMBA) [35]. In the Chinese hamster, five out of six primary sarcomas induced by DMBA showed a preferential involvement of the smallest chromosome of the karyotype (i.e., a No. 11 [34])—a pattern quite different from that seen in the RSV-induced tumors. This was found in four near-diploid tumors and one near-triploid tumor. (One diploid tumor showed no deviation from the normal karyotype.) Chromosome analyses of 12 primary DMBA-induced sarcomas in the rat have also revealed a karyotypic pattern completely different from that of RSV-induced sarcomas of the same inbred strain [32]: trisomy of the longest terminal chromosome, t_1, was seen in 10 out of 12 DMBA-induced sarcomas, and trisomy of one m chromosome was present in seven of these (Fig. 2, lower row). The chromosomes most often involved in the evolution of the RSV tumors were participating only exceptionally in the variation of the DMBA tumors. It is of great interest that the particular pattern found in these solid DMBA-induced tumors was also a characteristic feature of primary rat leukemias induced by pulse doses of DMBA [14, 15, 39, 42]: about 30% of the leukemic rats had stemlines with trisomy t_1, and the extra t_1 was also present in about 20% of the sidelines and variant cells. The same abnormality has also been reported in rat leukemias induced by the two closely related hydrocarbon carcinogens 6,8,12- and 7,8,12-trimethylbenz(α)anthracene [41]. Furthermore, it has recently been demonstrated that, after intravenous administration of DMBA, the breaks produced in chromosomes of rat bone marrow cells are predominantly localized to the t_1 chromosomes [16, 39], with a vulnerability specifically of two regions of this chromosome [40].

Another indication of possible specificity of altered genetic expression in tumor cells is seen in recent isozyme studies. Levan et al. [22, 23] studied the isozyme patterns of nine enzyme systems in 50 RSV-SR-induced and 25 DMBA-induced sarcomas in inbred rats and compared the patterns obtained to those found in the normal tissue of origin (i.e., subcutaneous connective tissue). No consistent differences were found among or between the different RSV and DMBA tumors, but similar qualitative differences were apparent in the esterase patterns between the tumors and their tissue of origin. Thus the shifts in enzyme production due to developing malignancy must take similar ways in the two different tumors. It is noteworthy that all tumors differed from adult connective tissue in two or three C-zone esterase bands but agreed with placenta and umbilical-cord tissues and in part with embryonic skin tissue [22]. This suggests the derepression of fetal genes along with the "dedifferentiation" in tumors [9].

The results seem to indicate a specificity in the participation of the genetic apparatus of the host cell in the early stages of oncogenesis. Nichols [36, 37] proposed that the visible chromosomal mutations found within a tumor may be indicators of subvisible gene mutations and that these gene mutations may frequently be the most important initiator of change of a normal cell to the malignant state. This was based primarily on two points; the first is the high correlation between the ability of an agent to produce chromosomal mutations and its ability to produce gene mutations. The second is the finding of early tumors with apparently normal chromosomes. If this is true, it appears that, with different oncogenic agents, essentially different molecular pathways must be involved. The details of the karyotypic evolution seem to be determined by the inducing agent in a similar way in different tissues. This suggests, furthermore, that tumors within a well-defined histopathologic category could actually consist of cytogenetically different subgroups, each determined by its *mode of induction*. The apparent lack of consistent karyotypic abnormalities in most malignancies might thus be at least partly explained by tumors in the same histopathological category being produced by different etiological agents.

Acknowledgements.–I wish to thank Drs. C. G. Ahlström, A. Levan, and Warren W. Nichols for constructive criticism. This work was supported by grant No. 622-B72-01X from the Swedish Cancer Society.

LITERATURE CITED

1. AHLSTRÖM CG: Neoplasms in mammals induced by Rous chicken sarcoma material. Nat Cancer Inst Monogr 17:299–319, 1964.
2. AHLSTRÖM CG, FORSBY N: Sarcomas in hamsters after injection with Rous chicken tumor material. J Exp Med 115:839–852, 1962.
3. AHLSTRÖM CG, JONSSON N: Induction of sarcoma in rats by a variant of Rous virus. Acta Pathol Microbiol Scand 54:145–172, 1962.
4. AHLSTRÖM CG, MARK J: The reaction of rabbits to Rous sarcoma virus. Int J Cancer 1:51–60, 1966.
5. AHLSTRÖM CG, BERGMAN S, EHRENBERG B: Neoplasms in guinea pigs induced by an agent in Rous chicken sarcoma. Acta Pathol Microbiol Scand 58:177–190, 1963.
6. AHLSTRÖM CG, BERGMAN S, FORSBY N, JONSSON N: Rous sarcoma in mammals. Acta Union Int contre Cancer 19:294–298, 1963.
7. AHLSTRÖM CG, JONSSON N, FORSBY N: Rous sarcoma in mammals. Acta Pathol Microbiol Scand Suppl 154:127–129, 1962.
8. AHLSTRÖM CG, KATO R, LEVAN A: Rous sarcoma in Chinese hamsters. Science 144:1232–1233, 1964.
9. ALEXANDER P: Foetal "antigens" in cancer. Nature 235:137–140, 1972.
10. EZDINLI EZ, SOKAL JE, CROSSWHITE L, SANDBERG AA: Philadelphia-chromosome-positive and -negative chronic myelocytic leukemia. Ann Int Med 72:175–182, 1970.
11. GROUCHY J DE, NAVA C DE, CANTU J-M, BILSKI-PASQUIER G, BOUSSER J: Models for clonal evolutions: a study of chronic myelogenous leukemia. Am J Hum Genet 18:485–503, 1966.
12. KATO R: The chromosomes of forty-two primary Rous sarcomas of the Chinese hamster. Hereditas 59:63–119, 1968.
13. KATO R, LEVAN A, NICHOLS WW: Chromosomal changes in Rous sarcomas in rodents. In Unio Nord Contra Cancrum: Symposium on Virus and Cancer, Sweden, 75–94, 1964.
14. KURITA Y, SUGIYAMA T, NISHIZUKA Y: Cytogenetic studies on rat leukemia induced by pulse doses of 7,12-dimethylbenz(α)anthracene. Cancer Res 28:1738–1752, 1968.
15. KURITA Y, SUGIYAMA T, NISHIZUKA Y: Cytogenetic analysis of cell populations in rat leukemia induced by pulse doses of 7,12-dimethylbenz(α)anthracene, Gann 60:529–535, 1969.
16. KURITA Y, SUGIYAMA T, NISHIZUKA Y: Chromosome aberrations in rat bone marrow cells by 7,12-dimethylbenz(α)anthracene. J Nat Cancer Inst 43:635–641, 1969.
17. LEVAN A: Relation of chromosome status to the origin and progression of tumors: the evidence of chromosome numbers. Ann Symp Fundam Cancer Res 13:151–182, 1959.
18. LEVAN A: Preliminary chromosome data on Rous sarcoma in rats. In First Scandinavian Symposium on Carcinogenesis, Oslo, 1961.
19. LEVAN A: Non-random representation of chromosome types in human tumor stemlines. Hereditas 55:28–38, 1966.
20. LEVAN A: Some current problems of cancer cytogenetics. Hereditas 57:343–355, 1967.

21. LEVAN A, FREDGA K, SANDBERG AA: Nomenclature for centromeric position on chromosomes. *Hereditas* 52:201–220, 1964.
22. LEVAN G, MITELMAN F, MARK J: Isozymes of experimentally induced primary sarcomas in the rat. *Hereditas* 73:318–321, 1973.
23. LEVAN G, MITELMAN F, NICHOLS WW, BECKMAN G, BECKMAN L: Isozyme patterns in Rous sarcoma virus-induced tumors in the rat. *Hereditas* 68:143–150, 1971.
24. MARK J: Chromosomal analysis of ninety-one primary Rous sarcomas in the mouse. *Hereditas* 57:23–82, 1967.
25. MARK J: Relationships of chromosomal and pathological findings in Rous sarcoma virus-induced tumors in the mouse. *Int J Cancer* 3:663–676, 1968.
26. MARK J: Rous sarcomas in mice: the chromosomal progression in primary tumours. *Eur J Cancer* 5:307–315, 1969.
27. MARK J: Rous sarcomas in mice: the chromosomal progression during early in vivo transplantation. *Hereditas* 65:59–82, 1970.
28. MITELMAN F: Preferential chromosome loss in a Rous rat sarcoma in response to environmental changes. *Hereditas* 54:202–212, 1965.
29. MITELMAN F: The chromosomes of fifty primary Rous rat sarcomas. *Hereditas* 69:155–186, 1971.
30. MITELMAN F: Comparative chromosome analysis of primary and metastatic Rous sarcomas in rats. *Hereditas* 70:1–14, 1972.
31. MITELMAN F: Predetermined sequential chromosome changes in serial transplantation of Rous rat sarcomas. *Acta Pathol Microbiol Scand* Sect A 80:313–328, 1972.
32. MITELMAN F, LEVAN G: The chromosomes of primary 7,12-dimethylbenz(α)-anthracene-induced rat sarcomas. *Hereditas* 71:325–334, 1972.
33. MITELMAN F, MARK J: Chromosomal analysis of primary and metastatic Rous sarcomas in the rat. *Hereditas* 65:227–235, 1970.
34. MITELMAN F, MARK J, LEVAN G: The chromosomes of six primary sarcomas induced in the Chinese hamster by 7,12-dimethylbenz(α)anthracene. *Hereditas* 72:311–318, 1972.
35. MITELMAN F, MARK J, LEVAN G, LEVAN A: Tumor etiology and chromosome pattern. *Science* 176:1340–1341, 1972.
36. NICHOLS WW: Relationships of viruses, chromosomes and carcinogenesis. *Hereditas* 50:53–80, 1963.
37. NICHOLS WW: Studies on the role of viruses in somatic mutation. *Hereditas* 55:1–27, 1966.
38. NICHOLS WW, LEVAN A, CORIELL LL, GOLDNER H, AHLSTRÖM CG: In vitro chromosome abnormalities in human leukocytes associated with Schmidt-Ruppin Rous sarcoma virus. *Science* 146:248–250, 1964.
39. REES ED, MAJUMDAR SK, SHUCK A: Changes in chromosomes of bone marrow after intravenous injections of 7,12-dimethylbenz(α)anthracene and related compounds. *Proc Nat Acad Sci (US)* 66:1228–1235, 1970.
40. SUGIYAMA T: Specific vulnerability of the largest telocentric chromosome of rat bone marrow cells to 7,12-dimethylbenz(α)anthracene. *J Nat Cancer Inst* 47:1267–1275, 1971.

41. Sugiyama T, Brillantes FP: Cytogenetic studies of leukemia induced by 6,8,12- and 7,8,12-trimethylbenz(α)anthracene. *J Exp Med* 131:331–341, 1970.
42. Sugiyama T, Kurita Y, Nishizuka Y: Chromosome abnormality in rat leukemia induced by 7,12-dimethylbenz(α)anthracene. *Science* 158:1058–1059, 1967.
43. Tjio JH, Carbone PP, Whang J, Frei E III: The Philadelphia chromosome and chronic myelogenous leukemia. *J Nat Cancer Inst* 36:567–584, 1966.
44. Tjio JH, Levan A: Comparative idiogram analysis of the rat and the Yoshida rat sarcoma. *Hereditas* 42:218–234, 1956.
45. Yerganian G, Kato R, Leonard MJ, Gagnon HJ, Grodzins LA: Sex chromosomes in malignancy, transplantability of growths, and aberrant sex determination. *In* Cell Physiology of Neoplasia. Austin, University of Texas Press, **49–96**, 1960.

AUTHOR INDEX

Numbers in parentheses are reference numbers and show that an author's work is referred to although his name is not mentioned in the text.
Numbers in *italics* indicate the pages on which the full references appear.

Aaronson, S. A., 65(69), 66(1), 68(1), 72(69), 157(159), 165(1), 167(1), 180(1), 189(159), 633(26), 635(26)
Abbatt, J. D., 78(7), 92(7)
Abbot, J. A., 464(132), 477(132)
Abbott, C. R., 470(1)
Ablashi, D. V., 579(60), 596(60)
Achong, B. G., 159(107), 186(107), 570(18), 572(18), 573(20), 576(19), 594(18,19,20)
Ada, G. L., 28(38), 37(38)
Adam, A., 320(17), 332(17)
Adam, M., 399(1), 411(1)
Adams, A., 203(1), 211(59), 212(59), 222(59), 228(1), 231(59), 460(2), 464(60), 465(61,84), 470(2), 473(60,61), 474(84)
Adams, R. A., 579(1), 593(1)
Aderca, I., 167(109), 186(109)
Agnish, N. D., 550(1), 555(1)
Ahearn, M. J., 592(106), 599(106)
Ahlstrom, C. G., 171(2), 172(117), 180(2), 186(117), 675(1,2,3,4,5,6,7,8), 676(38), 687(3), 691(1,2,3,4,5,6,7,8), 692(38)
Aikat, B. K., 399(57), 414(57), 466(38), 472(38)
Ajello, P., 366(1), 367(1)
Alcini, E., 384(138), 385(138), 387(138), 418(138)
Alexander, D. S., 161(27), 168(27), 169(27), 181(27)

Alexander, P., 27(1), 35(1), 100(70), 130(70), 208(2,96,97), 228(2), 233(96,97), 690(9), 691(9)
Alfaro-Kofman, S. A., 169(92), 185(92), 394(166), 419(166)
Alhrecht, A. M., 554(125), 561(125)
Allderdice, P. W., 138(24), 139(24), 150(24), 525(96), 539(2), 542(96), 555(2), 560(96)
Allendorf, F. W., 89(20), 93(20)
Allison, A. C., 174(3), 180(3), 532(3,4), 555(3,4)
Alter, A., 461(3), 470(3)
Aly, F. W., 399(183), 420(183)
Ambrose, K. R., 52(2), 68(2)
Ambrosioni, J. C., 577(15), 594(15)
Amman, A. J., 465(141), 477(141), 620(1), 633(1)
Amos, D. B., 523(126), 561(126)
Anagnostakis, D. E., 168(94), 170(93), 185(93,94)
Andermann, F., 620(25), 635(25)
Anderson, D. E., 138(21), 147(21), 150(21)
Anderson, N. G., 26(2), 29(24), 35(2), 36(24), 52(2), 68(2)
Angel, T., 60(3), 68(3)
Angioni, G., 315(15), 316(15,18), 332(15,18)
Anstall, H. B., 592(106), 599(106)
Archer, P. G., 648(9), 649(9), 663(9), 672(9,10)

Archibald, R., 304(26), *306*(26), 602 (14), 607(14), *617*(14)
Argiolas, N., 315(15), 316(15,18), *332* (15,18)
Arias-Bernal, L., 391(2), *411*(2)
Armitage, P., 32(3), *35*(3)
Armstrong, D., 162(100), *185*(100), 568(2), *593*(2)
Armstrong, G., 170(167), *189*(167)
Arvan, D. A., 385(61), 387(61), *414* (61)
Ashley, D. J. B., 319(1), *331*(1)
Aspiras, L., 65(61), *71*(61)
Asquith, P., 458(147), *477*(147)
Atkin, N. B., 78(1), 87(1), *92*(1), 293 (1), *305*(1), 377(6), 378(20), 379(4, 12,13,18,22,43), 382(7,8), 384(7), 385(7), 387(7,28), 390(9,13,15,43), 391(17), 394(12,15,76), 395(3,8), 398(10), 399(15,19,27), 402(5,7,10, 21,28), 403(15), 405(16,22), 406(7, 14), 407(4,8,9,10,11,18,43), *411*(3, 4,5,6,7,8,9,10,11,12), *412*(13,14,15, 16,17,18,19,20,21,22,27,28), *413* (43), *415*(76), 430(3), 431(10), 433 (6,8,9,11,13), 434(1,4,5,8,10), 437 (7), 438(3), 439(9), 442(2,3,13,14), *446*(1,2,3,4,5,6,7,8,9,10), *447*(11, 12,13,14), 459(4), 465(13), 466 (13), *470*(4), *471*(13), 633(2), *633* (2)
Atkins, L., 295(2), *305*(2), 462(5), 463(5), 470(5)
Atwood, K. C., 123(1), 124(1,127), *126*(1), *133*(127), 205(177), *237* (177)
Auersperg, N., 269(2), *284*(2), 430 (18), 433(17), 434(17,86), 439(16), 440(16), *447*(15,16,17,18), *450*(86)
Aula, P., 161(4,5), 168(5,25,116,118), 169(5,7,25,136), 174(6), 178(118), *180*(4,5,6,7), *181*(25), *186*(116, 118), *187*(136)
Aursperg, N., 164(44), *182*(44)
Austen, K. F., 623(59), *636*(59)
Austin, B., 60(3,21), *68*(3), *69*(21)
Austin, B. M., 28(38), *37*(38)
Austin, G., *447*(15)
Austin, J. B., 158(45), *182*(45)
Auxier, J. A., 639(1), 649(1), 660(1), *672*(1)
Avelino, E., 136(8), *149*(8)
Avila, L., 158(149), *188*(149)
Awa, AA., 224(3), *228*(3), 344(2), 368(2), 613(1), *616*(1), 648(2,9), 649(9), 663(9), 664(2), 665(2), *672* (2,9)
Awano, I., 384(23), *412*(23)
Aya, T., 160(10,87), 169(138), 170(8, 9,75,87), 174(138), *180*(8,9,10), *184*(75,87), *187*(138)
Ayraud, N., 401(24,153), *412*(24), *419*(153)
Azarnia, R., 546(6), 549(5,6), *555*(5, 6)
Azyma, M., 337(93,94), 338(94), 340 (93,94), 341(94), 343(94), 345(94), 363(94), *372*(93,94)

Bacal, H. L., 620(25), *635*(25)
Back, F., 290(96), 295(96), *310*(96), 460(187), *479*(187)
Baden, H. P., 210(121), *234*(121), 614 (20), *617*(20)
Bader, S., 391(25), *412*(25)
Baguena-Candela, R., 295(23), *306* (23), 390(71), *414*(71), 463(6), *470* (6)
Baikie, A. G., 83(2), *92*(2), 289(99), 303(3), 304(3), *305*(3), *310*(99), 362(3), *368*(3), 399(190,191,192), *420*(190), *421*(191,192), 456(9), 457(7,8,9,10,11,190,192), 458(190, 192), 461(7), 461(191), 462(190), 463(11,12), 466(184,185), 468 (146), *470*(7), *471*(8,9,10,11,12), *477*(146), *479*(184,185), *480*(190, 191,192), 590(97), *598*(97), 647 (46), *674*(46)
Bain, A. D., 399(160), *419*(160)
Bainbridge, C. J., 586(108), *599*(108)
Bajer, A., 122(2), *126*(2)
Bajerska, A., 211(4), 217(4), *228*(4)
Bakay, B., 316(41), *334*(41)
Baker, D. G., 100(3), *126*(3)
Baker, M. C., 379(13,22), 387(28), 390(13,15), 391(17), 394(15,76), 395(26), 399(15,27), 402(21,28), 403(15), 405(16,22), 406(14,26), *412*(13,14,15,16,17,21,22,26,27,28),

AUTHOR INDEX

415(76), 428(12), 431(10), 433(8,9, 11,13), 434(8,10), 437(7), 439(9), 442(13,14), *446*(7,8,9,10), *447*(11, 12,13,14), 465(13), 466(13), *471* (13), 633(2), *633*(2)
Baldwin, R. W., 26(4,5), *35*(4,5)
Ballard, P. L., 549(7), *555*(7)
Ballieux, R. E., 23(26), *36*(26)
Baltimore, D., 28(6), *35*(6), 49(5), *68* (5), 80(3), *92*(3), 526(8), 530(8), *555*(8)
Baltzer, Fritz., 4(1), *19*
Baluda, M. A., 64(6), *68*(6)
Baluda, M. C., 317(50), *334*(50)
Banerjee, A. R., 465(14), *471*(14)
Banerjee, K. B., 461(15), *471*(15)
Banfield, W. G., 363(8,9), *368*(8,9)
Barbe, J., 620(29), *635*(29)
Barcinski, M. A., 100(4), *126*(4)
Barker, C. R., 26(4,5), *35*(4,5)
Barker, L. F., 569(100), *598*(100)
Barmada, B., *636*(54)
Barr, Y. M., 570(17,18), 572(17,18), 576(19), *594*(17,18,19)
Barren, B. A., 240(44), *262*(44)
Barski, G., 157(11), *180*(11), 343(4), 356(4), 363(4), *368*(4), 522(9,12, 13), 523(10,11,13), 525(9), 527 (11), 533(10,13), 535(10), 538(15), 540(10), 541(16), *555*(9,10,11,12, 13), 556(15,16)
Bartsch, H. D., 158(12), 170(12), *180* (12)
Bashore, R. A., 304(24), *306*(24)
Basilico, C., 48(7), 56(124), *69*(7), *75* (124), 530(14), *556*(14)
Batchelor, A. L., 213(149,150), *236* (149,150)
Bateman, A. J., 227(5), *228*(5)
Battips, D. M., 192(103), 211(103), *233*(103), 645(39), *674*(39)
Baughman, F., 170(13), *180*(13)
Bauke, J., 290(4), *305*(4), 399(29), *412*(29), 457(18), 460(16), 466(17), *471*(16,17,18)
Baumal, R., 571(3), *593*(3)
Banmer, A., 464(45), *472*(45)
Baumiller, R. C., 153(14), *180*(14)
Bayreuther, K., 163(15), 167(15), *180* (15)

Beadle, G. W., 602(2), *616*(2)
Becak, M. L., 464(19), *471*(19)
Becak, W., 464(19), *471*(19)
Becker, S., *479*(188)
Beckman, G., 690(23), *692*(23)
Beckman, L., 690(23), *692*(23)
Beckwith, J. B., 30(7), *35*(7)
Beebe, S. P., 337(5), *368*(5)
Beechey, C. V., 226(152), *236*(152)
Beermann, W., 59(8), *69*(8)
Begin, P., 304(43), *307*(43)
Belehradek, J., 538(15), *556*(15)
Bell, A. G., 100(3), *126*(3)
Bell, W. R., 292(14), *306*(14), 468 (49), *473*(49)
Belpomme, D., 568(5), 569(4), *593*(4, 5)
Belsky, J. L., 649(3), *672*(3)
Ben, T., 578(102), *598*(102)
Ben-Bassat, H., 548(71), *558*(71)
Benda, C. E., 170(13), *180*(13)
Bender, M. A., 100(4,5,48), *126*(4), *127*(5), *129*(48), 203(6,7,9), 211(6), 212(6), 224(10), *228*(6,7,8,9,10), 384(30), *412*(30), 648(4,5,6), 660 (4,5), *672*(4,5,6)
Benditt, E. P., 330(2), *332*(2)
Benditt, J. M., 330(2), 332(2)
Benedict, W. F., 387(31), 390(31,165), 391(32), 398(31), 409(165), *412* (31,32), *419*(165), 507(14), 508(14), *516*(1), *517*(14)
Ben Haim, A., 219(66), *231*(66)
Benirschke, K., 372(98), 394(76), 415 (76), 465(20), 466(20), *471*(20)
Bennett, B., 27(8), *35*(8)
Bennett, J. M., 459(91), 463(91), *475* (91)
Benyesh-Melnick, M., 161(16), *181* (16), 586(63,64), 587(63), 591(63), *596*(63,64)
Berard, C. W., 578(60), *596*(60)
Berebbi, M., 541(16), 548(37), *556* (16,17), *557*(37)
Berger, C. A., 244(1), *260*(1)
Berger, R., 256(16), *260*(16), 288(62), 293(62,64), 302(66), *308*(62,64,66), 377(35,36), 382(34), 385(34), 390 (34,114), 394(38), 398(34,39), 400 (34), 401(34), 402(115), 409(33,37),

413(33,34,35,36,37,38,39), 417 (114,115), 458(21), *471*(21), 648 (35), *673*(35)
Bergin, J. W., 292(58), *308*(58)
Bergman, S., 675(516), *691*(5,6)
Bernadou, A., 302(38), *307*(38)
Bernard, J., 159(156), 160(156), *188* (156)
Bernhard, W., 530(18), *556*(18)
Berry, E. W., 272(17), *285*(17), 459 (91), 463(91), *475*(91)
Berumen, L., 569(4), *593*(4)
Berwald, Y., 65(10), *69*(10)
Bettinger, H. F., 424(19), *447*(19)
Betty, A. V., 100(93), *131*(93)
Beutler, E., 326(3), *332*(3)
Bezan, A. I., 615(18), *617*(18)
Bick, Y. A. E., 203(11), *228*(11)
Biedler, J. L., 554(125), *561*(125)
Biemond, A., 619(3), *634*(3)
Bierme, R., 290(89), *310*(89)
Biesele, J. J., 170(41), *182*(41)
Biggar, D., 620(4), *634*(4)
Biggs, P. M., 159(17), *181*(17)
Billardan, C., 523(11), 527(11), *555* (11)
Billen, D., 363(21), *368*(21)
Bilski-Pasquier, G., 289(35), 290(32, 73), 292(30,73), 293(31,33), 294 (31,33), 295(41), 298(41), 302(38), *307*(30,31,32,33,35,38,41), *309* (73), 402(88), *415*(88), 458(39,41, 42), 459(40), *472*(39,40,41,42), 677 (11), *691*(11)
Birbeck, M., 293(61), *308*(61)
Birbeck, M. S. C., 462(125), *476*(125)
Birch, S. M., 570(24), *594*(24)
Biryulina, T. I., 530(121), *561*(121)
Bischoff, R., 62(11,12), *69*(11,12)
Bishop, C. J., 119(6), *127*(6)
Bjerknes, R., 22(51), *38*(51)
Blakeslee, J., 578(33), *595*(33)
Black, P. H., 158(45), *182*(45), 167 (29,30), 178(30), *181*(29,30), 529 (25), 542(111), 546(95), 551(111), *556*(25), *560*(95), *561*(111)
Blaisdell, R. K., 293(87), *310*(87), 462 (168), 463(168), *479*(168)
Blank, M. H., 441(80), 450(80)
Blatnik, D., 168(130), *187*(130)

Blattner, R. J., 462(22), 463(23), *471* (22,23)
Block, J. B., 457(31), *472*(31)
Bloom, A., 399(40), *413*(40)
Bloom, A. D., 571(10,11), *593*(10,11), 648(7,8,9), 649(7,8,9,22), 663(8,9, 22), *672*(7,8,9,10), *673*(22)
Bloom, B., 571(3), *593*(3)
Bloom, B. R., 27(8), *35*(8)
Bloom, D., 78(48), *94*(48), 304(26), *306*(26), 601(3,4), 602(14), 603(3), 607(14), 612(3,4,24), *616*(3,4), *617* (14,24)
Bloom, F., 337(6), *368*(6)
Bloom, G. E., 457(24), *471*(24)
Bloom, S. E., 158(18), 161(18), *181* (18)
Blume, R. S., 569(6), *593*(6)
Bocharov, E. F., 159(134), 160(134), *187*(134)
Bochkov, N. P., 212(157), 213(157), 214(157), *236*(157)
Bockelmann, W., 456(199), *480*(199)
Boddington, M. M., 387(198), 390 (197,198), 398(198), 399(196), 401 (197), *421*(196,197,198), 429(70), 435(70), 436(20), 439(20), *447*(20), *450*(70), 466(186), *479*(186)
Boder, E., 619(7), 620(6,7), 621(6,7), 622(6), *634*(5,6,7)
Bodycote, D. J., 207(75,77), *232*(75, 77)
Bodycote, J., 105(52), 107(52), 109 (52), *129*(52)
Boesman, M., 620(13), *634*(13)
Boiron, M., 159(19,156), 160(156), 168(157), *181*(19), *188*(156,157)
Bombik, B. M., 549(22), *556*(22)
Boniface, L., *636*(57)
Bonnette, J., 304(43), *307*(43)
Boon, T., 57(13,14), *69*(13,14)
Boone, B. A., 568(23), *594*(23)
Bootsma, D., 210(12), *228*(12), 543 (38), *557*(38)
Borch, S., von der., 390(41), *413*(41)
Borek, C., 65(15), *69*(15), 549(19), *556*(19)
Borgers, W. H., 462(25), *471* (25)
Borgne, Le C., 568(4), *593*(4)

AUTHOR INDEX 699

Borsos, T., 26(52,53), 27(14,29,30), 36(14,29,30), *38*(52,53)
Boskovic, D., 458(164), 468(164), *478* (164)
Bottura, C., 295(5), 399(42,48), *305* (5), *413*(42,48), 465(26), *471*(26)
Boué, A., 168(20), 170(21,22), *181* (20,21,22)
Boué, J. G., 168(20), 170(21,22,26), *181*(20,21,22,26)
Bouroncle, B. S., 460(182), 464(183), *479*(183)
Bousser, J., 289(35), 290(32,73), 292 (73), 295(37,41), 298(41), 402(88), *307*(32,35,37,41,43), *309*(73), *415* (88), 458(41,42), 459(40), 463(43), 464(27), *471*(27), *472*(40,41,42,43), 677(11), *691*(11)
Boveri, Marcella, O'Grady, *see* O'Grady, Boveri, Marcella
Boveri, Margret., 5
Boveri, T., 8(6), 9(2,6,8), 10(2,3,4,5, 6), 13(7), *19*, 78(4), *92*(4), 361(7), *368*(7)
Bowen, J. M., 633(38), *635*(38)
Bowen, P., 463(28), *472*(28)
Bowery, C. E., 425(71), 429(21), 435 (71), 438(71), 439(71), 440(71), 441(71), 442(71), 443(71), 444(21), *447*(21), *450*(71)
Boyce, R. P., 209(13), *228*(13)
Boyd, E., 459(29), 463(29), *472*(29)
Boyd, J. T., 219(14), *228*(14)
Boyse, E. A., 80(52), *94*(52), 303(78), *309*(78), 577(83), *597*(83)
Bradt, C., 155(120), *186*(120)
Brady, R. O., 546(31), *556*(31)
Brailovsky, C., 157(23), *181*(23)
Brandao, H. J. S., 379(18,43), 390 (43), 407(18,43), *412*(18), *413*(43), 430(22), 440(22), 441(22), *447*(22)
Braun, A. C., 280(28), 282(28), 283 (28), *285*(28)
Braun, H. J., 399(183), *420*(183)
Brecher, G., 289(105), *310*(105), 458 (196), 459(196), *480*(196)
Breckenridge, B. Mcl., 549(22), *556* (22)
Breckton, G., 226(49), *230*(49)
Breg, W. R., 525(96), 542(96), *560*(96)

Bregula, U., 532(82), 536(82), 537(20, 82), 538(82), 539(82), 543(57,82), 545(57,82), *556*(20), *558*(57), *559* (82)
Breinl, H., 439(24), *447*(24)
Brewen, J. G., 100(9), 105(9), 106(8, 9), 107(9), 122(114), 125(7,84), *127*(7,8,9), *131*(84), *132*(114), 203 (6,16), 207(15,17), 211(6), 212(6), 213(19), 218(20), 220(18,124), 221 (16,21), 224(22), *228*(6,15), *229*(16, 17,18,19,20,21,22), 234(124)
Bridges, B. A., 66(16), *69*(16)
Bridges, C. B., 602(5), *616*(5)
Brigato, G., 390(44), *41*(44)
Brill, A. M., 461(95), *475*(95)
Brillantes, F. P., 690(41), *693*(41)
Brindley, D. C., 363(8), *368*(8)
Brinkley, B. R., 109(10), *127*(10), 461 (142), *477*(142)
Brittin, G. M., 463(70), *474*(70)
Brock, R. D., 100(9), 105(9), 106(9), 107(9), *127*(9), 207(17), *229*(17)
Broder, S. W., 569(7), *593*(7)
Brodey, R. S., 303(78), *309*(78)
Brøgger, A., 109(11), *127*(11)
Brooke, J. H., 568(85), *597*(85)
Brooks, A. L., 119(12), *127*(12)
Brooks, S. E. H., 399(1), *411*(1)
Brought, A. J., 623(49), *636*(49)
Broustet, A., 295(6), *305*(6)
Brown, C. D., 387(31), 390(31,165), 398(31), 409(165), *412*(31), *419* (165), 507(14), 508(14), 517(14), 648(34), *673*(34)
Brown, D. D., *516*(1)
Brown, J. K., 203(11), *228*(11)
Brown, M., 529(77), *559*(77)
Brownhill, L., 465(20), 466(20), *471* (20)
Bru., 337(54), *370*(54)
Brumfield, R. T., 118(13), *127*(13)
Brunet, P., *634*(8)
Brunschwig, J. P., 586(63), 587(63), 591(63), *596*(63)
Brussieu, J., 304(43), *307*(43)
Bryan, E., 158(43), *182*(43)
Bryans, A. M., 161(27), 168(27), 169 (27), *181*(27)
Bryant, J. I., 314(12), 327(12), *332*(12)

Buck, C. A., 546(139), *562*(139)
Buckton, D. E., 100(14), *127*(14)
Buckton, K. E., 83(2), *92*(2), 100(67), 123(67), *130*(67), 136(11,26), 137(3), *149*(3,11), *150*(26), 212(24), 213(24), 219(23), 220(24), 224(25), *229*(23,24,25), 456(9), 457(9,190), 461(191), 462(190), *471*(9), *480*(190,191), 645(12), 647(46), 648(11), 663(11), *672*(11,12), *674*(46)
Buell, D. N., 568(22), *594*(22)
Bugszewski, C., 164(44), *182*(44)
Buhler, W. L., 318(4), *332*(4)
Buist, D. G. P., 459(102), *475*(102)
Bulkin, W., 387(58), *414*(58)
Bulme, B., 163(47), *182*(47)
Bunker, M. C., 225(73), *232*(73)
Burch, P. R. J., 31(9), *35*(9), 469(30), *472*(30)
Burdette, W. J., 153(24), *181*(24)
Burger, M. M., 56(36), *70*(36), 546(21), 548(23,24,41), 549(22), *556*(21,22,23,24), *557*(41)
Burk, R. R., 54(113), *74*(113)
Burkitt, D., 572(8), *594*(8)
Burmeister, P., 460(143), *477*(143)
Burnet, F. M., 21(10,11), 25(13), 30(12), *36*(10,11,12,13)
Burns, A., 569(69), 573(69), *597*(69)
Burns, W. H., 529(25), *556*(25)
Burny, A., 49(104,105), *74*(104,105)
Burtin, P., 577(43), *595*(43)
Bushong, S. C., 215(26), *229*(26)
Butcher, R. W., 82(46), *94*(46)
Butel, J. S., 47(17), 48(17), *69*(17), 167(110), *186*(110)
Butler, J. J., 592(106), *599*(106)
Buttner, H. H., 400(213), *422*(213)

Cachin, Y., 578(43), *595*(43)
Cailie, B., 293(64), *308*(64)
Caille, B., 458(127), *476*(127)
Cailleau, R., 289(51), *308*(51)
Cain, W. A., 620(1), *633*(1)
Caldwell, I., 100(70), *130*(70), 208(97), *233*(97)
Callan, H. G., 59(18), *69*(18)
Camaeho, de Osorio, O., 395(161), 400(162), *419*(161,162)
Cambier, J., *634*(8)

Cameron, C., 273(20), *285*(20)
Cameron, C. B., 463(145), *477*(145)
Campbell, A. M., 42(19), *69*(19)
Campbell, B., 316(19), *332*(19)
Campbell, B. K., 329(21), *333*(21)
Campbell, H. A., 303(78), *309*(78)
Camus, J. P., 304(16), *306*(16)
Canellos, G. P., 324(5), *332*(5), 457(197), 460(197), *480*(197)
Cangella, K., 169(136), *187*(136)
Cantell, K., 168(25), 169(25), *181*(25)
Cantu, J. M., 289(35), *307*(35), 458(41), *472*(41), 677(11), *691*(11)
Carbone, P. P., 289(25,98,105), 292(14), *306*(14,25), *310*(98,105), 457(31,189,197), 458(196), 459(196), 460(189,197), 468(49), *472*(31), *473*(49), *480*(189,196,197), 688(43), *693*(43)
Carbong, P. P., 578(60), *596*(60)
Carlson, P. S., 60(20), *69*(20)
Carrier, W. L., 78(50), *94*(50), 209(155), 210(156,164), *236*(155,156), *237*(164), 537(117), 542(117), *561*(117), 614(25), *617*(25)
Carswell, E., 523(11), 527(11), *555*(11)
Carswell, E. A., 26(39), *37*(39)
Carter, C. O., 462(97), *475*(97)
Carter, R. L., 31(21), *36*(21)
Case, M., 623(23), 624(23), 629(23), 630(23), *635*(23)
Case, M. P., 78(19), *93*(19), 461(93), *475*(93), 623(24), 624(24), 629(24), 632(24), *635*(24)
Caspersson, T., 136(1,2), 137(3), *149*(1,2,3), 289(7), *305*(7), 323(6), *332*(6)
Cassingena, R., 157(23), *181*(23)
Castaigne, P., *634*(8)
Castoldi, G., 462(32), *472*(32)
Castoldi, G. L., 390(46), 399(45), *413*(45,46), 466(202), *480*(202)
Castro-Sierra, E., 292(8), *305*(8), 458(33), *472*(33)
Catcheside, D. G., 60(3,21), 68(3), *69*(21), 100(16), 106(16), 119(15,16), *127*(15,16), 205(93), *233*(93)
Cawein, M., 289(9), 304(9), *305*(9)
Cawley, L. P., *634*(9)

Cecco, L., de., 390(47,172), 395(47), 413(47), 420(172)
Cellier, K. M., 433(41), 439(41), 444 (41), 448(41)
Cenani, A., 404(163), 419(163)
Cerna, H., 56(65,66), 72(65,66)
Ceuterwall, W. R., 619(10), 620(10), 634(10)
Chaganti, R. S. K., 602(6), 616(6)
Chandley, A. C., 225(90), 233(90)
Chang, R. S., 569(9), 593(9)
Chang, T. H., 170(26), 181(26)
Chapelle, A. De La., 137(10), 149(10), 293(10), 305(10), 467(178), 479(178)
Chaplain, M., 552(146), 563(146)
Chardonnet, Y., 163(129), 187(129)
Charles, D. R., 100(72), 130(72)
Charney, E., 462(98), 475(98)
Chasin, L., 524(75), 559(75)
Chatten, J., 621(37), 635(37)
Cheka, J. S., 639(1), 649(1), 660(1), 672(1)
Chen, S., 315(22), 333(22)
Chen, T. R., 140(4), 144(5), 149(4,5), 380(181), 420(181), 494(14), 495(14)
Chessin, L. N., 569(32,77), 573(77), 578(77), 595(32), 597(77)
Chessin, N., 569(6), 593(6)
Chicago, Conference, 499(2), 516(2)
Childs, B., 315(9), 332(9)
Chio, K. W., 571(10,11), 593(10,11)
Cho, S. S., 164(174), 189(174)
Chretien, P., 398(214), 422(214)
Chu, E. H. Y., 106(17), 127(17), 203(28), 210(27,164), 229(27,28), 237(164)
Chu, E. W., 324(7), 332(7)
Chubb, R., 80(42), 94(42)
Chudina, A. P., 222(29), 229(29)
Chun, T., 161(27), 168(27), 169(27), 181(27)
Churchill, W. H., Jr., 26(52), 27(14, 30), 36(14,30), 38(52)
Cipani, F., 390(44), 413(44)
Citron, P., 569(69), 573(69), 597(69)
Clark, H. F., 157(28), 181(28)
Clark, R., 385(123), 387(123), 400(123), 417(123)

Clarke, C. H., 29(19), 36(19)
Clarke, C. M., 387(198), 390(198), 398(198), 421(198), 429(70), 435(70), 450(70), 463(68), 474(68)
Clarke, M., 289(22), 306(22)
Clarkson, B., 162(100), 185(100)
Clarkson, B. A., 568(12), 591(12), 593(12)
Clarkson, J. M., 209(30), 229(30)
Cleaver, J. E., 78(5), 92(5), 209(117, 118), 210(31,32,33,34), 229(31,32, 33,34), 234(117,118)
Clein, G. P., 399(184), 420(184), 458(34), 466(181), 472(34), 479(181)
Clemenger, J. F., 219(35), 230(35)
Clifford, P., 24(17,35), 36(17), 37(35), 324(11), 332(11), 407(134), 418(134), 454(59), 473(59), 573(51,52, 54), 577(43,83,96), 587(78), 589(34), 595(34,43), 596(51,52,54), 597(78,83), 598(96)
Cobo, A., 623(30), 624(30), 629(30), 635(30)
Cochran, A., 551(153), 552(153), 553(153), 563(153)
Coffin, J. M., 530(27), 556(27)
Coggin, J. H., 26(2), 35(2), 52(2), 68(2)
Cohen, J. A., 210(12), 228(12)
Cohen, M. M., 108(18), 127(18), 157(28), 181(28)
Cole, A., 106(19), 127(19), 208(38), 230(38)
Cole, L., 271(15), 285(15)
Cole, L. J., 671(42), 674(42)
Coleman, D. V., 163(47), 182(47)
Collins, Z., 326(3), 332(3)
Colombies, P., 290(89), 310(89)
Comings, D. E., 105(23,24,25), 113(20,22,23,24,25,26,27), 114(20,23, 24,25), 117(21), 121(22), 122(21), 127(20,21), 128(22,23,24,25,26,27), 136(7,8), 148(6), 149(6,7,8), 208(36), 230(36)
Conard, A., 648(36), 673(36)
Condamine, H., 328(8), 332(8)
Conen, P. E., 290(20), 306(20), 458(54), 463(55), 467(35), 472(35), 473(54,55)
Conger, A. D., 100(28), 106(28,30),

124(29), *128*(28,29,30), 214(37), *230*(37)
Connor, J. D., 316(41), *334*(41)
Conrad, M., 292(58), *308*(58)
Constantinides, S. M., 549(109), *561* (109)
Cook, P. R., *560*(97)
Cookson, M. J., 65(22), *69*(22)
Coon, H. C., 525(28), *556*(28)
Cooper, E. H., 378(193), 395(193), 398(193), *421*(193)
Cooper, H. L., 167(29,30), 178(30), *181*(29,30), 363(9), *368*(9)
Cooper, J. E. K., 67(23), *69*(23), 155 (32), 158(31), *181*(31,32)
Cooper, L. Z., 170(121), *186*(121)
Cooper, M. D., 621(40), 622(40), *636* (40)
Cooper, T. H., 589(13), *594*(13)
Cooper, W. C., 630(20), *634*(20)
Corey, M. J., 269(2), *284*(2), 439(16), 440(16), *447*(15,16)
Coriell, L. L., 172(117), *186*(117), 676(38), *692*(38)
Cork, A., 210(81), *232*(81), 300(101), *310*(101), 633(38), *635*(38)
Cornefert, F. R., 157(11), *180*(11), 522(12,13), 523(10,13), 533(10, 13), 535(10), 540(10), *555*(10,12, 13)
Cornefert-Jensen, F. R., 343(4), 356 (4), 363(4), *368*(4), 532(29), *556* (29)
Corneo, G., 176(65), *183*(65)
Correns, C., 7
Corry, P. M., 106(19), *127*(19), 208 (38), *230*(38)
Court Brown, M. W., 457(11), 463(11), *471*(11)
Court Brown, W. M., 78(7,8), 83(2), *92*(2,7,8), 219(14), 224(25), *228* (14), *229*(25), 289(99), *310*(99), 362(3), *368*(3), 456(9), 457(9,11, 36,190,192), 458(190,192), 461 (191), 462(190), 463(11), *471*(191), *472*(36), *480*(190,191,192), 613(7), *616*(7), 647(46), *674*(46)
Coutinho, V., 399(48), *413*(48)
Cowdell, R. H., 425(71), 435(71,72), 438(71), 439(71), 440(71), 441(71), 442(71), 443(71), *450*(71,72)
Cox, D., 400(49), 401(50,215), 402 (51), *413*(49,50,51), *422*(215), 487 (1), *494*(1)
Craddock, C. G., 220(170), 237(170), 293(107), *311*(107),
Craig, D., 271(15), *285*(15)
Craig-Holmes, A. P., 135(9), *149*(9)
Crawford, L. V., 48(24), 52(24), 54 (24,113), *69*(24)
Crawfurd, M., 457(37), *472*(37)
Creasy, B., 64(88), *73*(88)
Crick, F. H., 153(33), *182*(33)
Cridland, M. D., 578(14), *594*(14)
Crippa, L. P., 304(27), *306*(27), 606 (13), 607(13), 611(13), *617*(13), 632(19), *634*(19)
Crippa, M., 119(74), *130*(74)
Croce, C. M., 525(30), *556*(30)
Croll, P. M., 542(54), *558*(54)
Crookston, J. H., 290(20), *306*(20), 458(54), 463(55), *473*(54,55)
Crosby, W. H., 462(32), *472*(32)
Crosswhite, L., 688(10), *691*(10)
Crosswhite, L. H., 78(47), *94*(47), 289 (90), 290(90,92), 293(91), 295(90), *310*(90,91,92), 399(175), *420*(175), 457(56), 459(171), 461(56), 463 (171,172,175), 466(173), 467(172), *473*(56), *479*(171,172,173,175), 591(94), 598(94)
Crouse, H. V., 122(31,32), *128*(31,32)
Crow, R. S., 399(121), *417*(121)
Cserhati, E., 621(47), 623(48), 624 (48), *636*(47,48)
Culp, L. A., 546(95), *560*(95)
Cumar, F. A., 546(31), *556*(31)
Curcio, S., 390(53,55), 391(54), 400 (52), *413*(52,53,54), *414*(55)
Curtis, A. C., 31(15), *36*(15)
Curtis, H. J., 124(29), *128*(29)
Custer, R. P., 328(8,36), 329(36), *332* (8), *333*(36)
Czuppon, A. B., 85(44), *94*(44)

Dahnke, G. S., 78(19), *93*(19), 461 (93), *475*(93), 623(24), 624(24), 629(24), 632(24), *635*(24)
Dalla Piccola, B., 384(56), *414* (56)

AUTHOR INDEX 707

Galton, D. A. G., 293(60), *308*(60), 459(123), 468(146), 469(123), *476* (123), *477*(146)
Galton, M., 394(76), *415*(76)
Gandini, E., 315(15), 316(15,18,19), *332*(15,18,19)
Ganner, E., 399(77), *415*(77)
Ganshow, R., 523(51), *558*(51)
Gant, N., 329(21), *333*(21)
Garcia, I. M., 64(96), *73*(96)
Garcia-Benitez, C., 125(45), *129*(45)
Gardner, L. I., 399(100), *416*(100)
Gardner, M. B., 630(20), *634*(20)
Gardner, S. D., 163(47), *182*(47)
Gars, U., 543(81), *559*(81)
Garson, O. M., 457(10), 459(102), *471* (10), *475*(102)
Gart, J. J., 289(106), *311*(106), 457 (198), 463(198), 466(198), *480* (198)
Gartler, S. M., 24(16,17), 24(31), *36* (16,17), *37*(31), 139(15), *149*(15), 281(29), *285*(29), 314(20,39), 315 (15,16,22,26), 316(15,18,19,28), 317(26,27), 318(26), 320(13,17), 323(10), 324(11), 329(21), *332*(10, 11,13,15,16,17,18,19), *333*(20,21, 22,26,27,28,39), 430(28), *447*(28), 454(59), 458(58), *473*(58,59)
Garvey, M. G., 466(202), *480*(202)
Gatti, R. A., 65(69), *72*(69), 633(26), *635*(26)
Gavosto, F., 290(82), *309*(82), 458 (156), 460(156), 461(73), *474*(73), *478*(156)
Gavrilov, V. I., 161(37), *182*(37)
Gay, H., 257(48), *262*(48)
Gazzolo, L., 163(129), *187*(129)
Geering, G., 577(83), *597*(83)
Geitler, L., 241(8), 244(7,8), *260*(7,8)
Gelb, L. D., 47(42), *70*(42)
Geller, W., 399(145), *418*(145)
Gellis, S. S., 620(16), *634*(16)
Genes, I. S., 387(78), 390(78), 398 (78), *415*(78)
Genozian, N., 220(18), *229*(18)
George, K. P., 136(16), *149*(16)
George, S. L., 467(89), *475*(89)
Gerald, P. S., 457(24), *471* (24)

Gerber, P., 162(171), *189*(171), 528 (52), *558*(52), 569(25), 570(24,28), 571(29), 573(25), 576(26), 578(27), 587(28), *594*(24,25,26,27,28,29)
Gerli, M., 390(73,130), 391(74), *415* (73,74), *417*(130)
German, J., 18(9), *19*, 59(43), 66(44), 67(43,44,45), 70(43), 71(44,45), 78 (12,48,50), 88(11), *92*(11,12), *94* (48,50), 147(17), *149*(17), 210 (156), *236*(156), 249(9), *260*(9), 269(1), 271(1), 277(1), 279(1), *284* (1), 289(11), 304(11,26,27,28), *305* (11), *306*(26,27,28), 409(79), *415* (79), 464(44,74), *472*(44), 474(74), 537(117), 542(117), *561*(117), 601 (9), 602(9,10,11,12,14), 603(9), 606 (9,10,13), 607(8,9,13,14), 611(13), 612(24), 613(11), 614(11,25), *616* (8,9,10,11), *617*(12,13,14,24,25), 623(17,18), 624(17), 631(18), 632 (18,19), *634*(17,18,19), 670(15), *672*(15)
Gerner, R., 569(69), 573(69), *597*(69)
Gerner, R. E., 569(30,71), 573(71), *594*(30), *597*(71)
Gershon, D., 56(46), *71*(46), 523(53), 543(53), *558*(53)
Gershon, R. H., 31(21), *36*(21)
Gey, G. O., 79(13), *92*(13)
Ghose, J., 461(95), *475*(95)
Ghostine, S., *636*(54)
Giannelli, F., 542(54), *558*(54)
Gibbs, W. H., 391(1), *411*(1)
Giblett, E. R., 315(22), *333*(22)
Gibson, D. A., 218(70), *232*(70)
Gilden, R. V., 154(78,80), *184*(78,80)
Giles, N. H., 99(46), 100(72,93), 106 (17), *127*(17), *129*(46), *130*(72), *131*(93), 205(71), *232*(71), 203(28), *229*(28)
Gilgenkrantz, S., 290(97), *310*(97)
Gillies, N. E., 203(168), 212(168), 216 (168), *237*(168)
Gilman, J. G., 316(51), *334*(51)
Gilmore, C. E., 159(63), *183*(63)
Gilmore, V. H., 344(63), *371* (63)
Gimemez-Martin, G., 122(47), *129* (47)

Girardi, A. J., 165(48), 166(48), 178 (48), *182*(48)
Gitlin, D., 620(13), *634*(13)
Glade, P. R., 569(6,7,31,32,77), 573 (77), 578(77), *593*(6,7), *595*(31,32), *597*(77)
Glaser, R., 528(55), 530(55), *558*(55)
Glaves, D., 26(5), *35*(5)
Glazer, D., 552(98), *560*(98)
Glomset, J. A., 330(44), 331(44), *334* (44)
Gmyrek, D., 464(75), *474*(75)
Gofman, J. W., 384(147), *418*(147)
Goh, K., 648(16,17), 660(16,17), *672* (16,17)
Goh, K. O., 289(29), *306*(29), 398 (80), 399(80), *415*(80), 456(79,80), 457(76,78), 460(77), 461(80), 465 (80), *474*(76,77,78,79,80)
Goldberg, A. R., 548(23), *556*(23)
Goldner, H., 172(117), *186*(117), 676 (38), *692*(38)
Goldschmidt, Richard, 4(10), 5(10), 7,9(10), *19*
Goldstein, G., 573(54), *596*(54)
Goldstein, S., 543(56), *558*(56)
Golub, E., 219(66), *231*(66), 390(66), 394(65), 395(66), *414*(65,66), 447 (25,26)
Gonczol, E., 160(39), *182*(39)
Gonzalez-Fernandez, A., 122(47), *129* (47), 257(10), *260*(10)
Gooch, P. C., 100(5,48), *127*(5), *129* (48), 203(7,9), 224(10), 228(7,8,9, 10), 648(4,5,6), 660(4,5), *672*(4,5, 6)
Good, R. A., 52(47), 65(69), *71*(47), *72*(69), 619(41), 620(1,4,41), 621 (40,42), 622(40,42), 633(26), *633* (1), *634*(4), *635*(26), *636*(40,41,42)
Goodheart, C., 155(120), *186*(120)
Goodlin, R. C., 390(81), *415*(81), *447* (29)
Goodman, W. N., 630(20), *634*(20)
Goodpasture, E. W., 340(10), 366(10), *368*(10)
Gordon, M., 387(64), *414*(64)
Gore, H., 314(23), 322(23), *333*(23)
Gorman, L. Z., 292(8), *305*(8), 458 (33), *472*(33)

Gott, C., 593(106), *600*(106)
Goulian, M., 295(2), *305*(2), 462(5), 463(5), *470*(5)
Govan, A. D. T., 424(30), *447*(30)
Gowen, G. W., 602(15), *617*(15)
Grace, J. T., 569(69), 573(69), 575 (109), 578(33), *595*(33), *597*(69), *599*(109)
Grace, J. T., Jr., 568(48), 573(48), 586 (48), *595*(48)
Grado, F. D., 464(46), *473*(46)
Gralnick, H. R., 569(6), *593*(6)
Granberg, I., 401(82), *415*(82), 433 (31), 434(31), 436(31), 438(31), 439(31,32), 444(31), 445(31), *448* (31,32), 455(81), 470(81), *474*(81)
Grant, C. J., 119(49), *129*(49)
Gray, H., 214(41), 217(41), *230*(41)
Gray, S. M., 631(14), *634*(14)
Green, H., 53(85), 57(120), 66(121), *73*(85), *75*(120,121), 524(87), 525 (149), 529(150), 530(14), 547(107), *556*(14), *559*(87), *560*(107), *562* (133), *563*(149,150)
Green, M., 49(49), 53(48), *71*(48,49), 154(78), 184(78)
Greisen, O., 387(83), *415*(93)
Grewal, M. S., 138(24), 139(24), *150* (24)
Griffen, A. B., 225(72,73), *232*(72,73)
Gripenberg, U., 161(49), 168(49), 169 (49), *182*(49), 589(34), *595*(34)
Griscelli, C., 289(85), *309*(85)
Grodzins, L. A., 677(45), *693*(45)
Grofova, M., 172(144), *188*(144)
Gropp, A., 78(14), *92*(14), 461(82,83), 466(107), *474*(82,83), *476*(107), 623(21), 624(21), *635*(21)
Gropp, H., 390(84), *415*(84)
Gross, H. E., 619(34), *635*(34)
Gross, L., 573(35), *595*(35)
Groth, C. G., 30(46), *37*(46)
Grouchy, J. de., 137(18), *149*(18), 161 (36), 168(36), 170(36), *182*(36), 289(35,40), 290(32,73), 292(30,73), 293(31,33), 294(31,33), 295(37,41), 298(41), 300(74,102), 302(38), 304 (16,36,39,40,42,43), *306*(16), *307* (30,31,32,33,34,35,36,37,38,39,40, 41,42,43), *309*(73,74), *310*(102),

390(86,87), 402(88), 404(152), 409 (85), *415*(85,86,87,88), *419*(152), 458(39,41,42), 459(40), 463(43), *472*(39,40,41,42,43), 568(36), *595* (36), 677(11), *691*(11)
Grover, P. L., 65(22,61), *69*(22), *71* (61)
Gruenwald, H., 290(44), *307*(44)
Grundner, G., 531(47), 543(57), 544 (47), 545(57), *557*(47), *558*(57)
Gruneberg, H., 615(16), *617*(16)
Guillan, B. A., 399(89), *415*(89)
Gunz, F. W., 458(159), 462(159), 464 (60), 465(84), *473*(60), *474*(84), *478*(159)
Gustafsson, T., 399(68), 410(68), *414* (68)
Gwynn, R. J. R., 343(85), 359(85), *372*(85)

Haase, A. T., 569(6), *593*(6)
Habel, K., 28(22), *36*(22)
Habermehl, K. O., 158(12), 170(12), *180*(12)
Haemmerli, G., 248(45), *262*(45)
Haerer, A. F., 623(22), 624(22), *635* (22)
Hahn, E. C., 48(50), *71*(50)
Haines, M., 288(62), 293(62), *308*(62)
Haines, R. M., 424(30), *447*(30)
Hakansson, C. H., 137(12,13), 399(67, 68,69), 402(67,69), 410(67,68,69), *149*(12,13), *414*(67,68,69), 455(65), 466(64,65), 469(65), *473*(64,65), 507(3), *516*(3)
Halgrimson, C. G., 30(41,46), *37*(41, 46)
Hall, B., 168(115), *186*(115), 463 (129), *477*(129)
Halliday, W. J., 27(23), *36*(23)
Hamerton, J. L., 361(13), *368*(13)
Hammack, W. J., 467(133), *477*(133)
Hammouda, F., 290(45), *307*(45), 457 (90), 458(85), 460(85), 465(90), *474*(85), *475*(90)
Hampar, B., 159(50,51), 176(51), *182* (50,51), 583(37), *595*(37)
Hampe, A., 159(19), *181*(19)
Hampel, K. E., 290(46), *307*(46), 457 (87), 458(86), *475*(86,87)

Hamre, D., 578(27), *594*(27)
Han, T., 290(48), *308*(48), 457(99), 462(99), *475*(99)
Hanafusa, H., 53(68), 64(90), *72*(68), *73*(90)
Hanafusa, T., 64(90), *73*(90)
Hanawalt, P. C., 209(122), *234*(122)
Hanchard, B., 399(1), *411*(1)
Handler, A. H., 550(42), *557*(42)
Hansen-Melander, E., 362(14), *368* (14), 390(90), *415*(90), 523(34), *557*(34)
Hansjoerg, S., 461(15), *471*(15)
Haque, 105(50), *129*(50)
Hardisty, R. M., 457(88), 467(136), *475*(88), *477*(136)
Hare, C. D., 343(96), 352(96), 356 (92), 359(96), 363(96), *372*(96)
Hare, D., 279(27), *285*(27)
Hare, J. D., 56(89), *73*(89)
Hare, W. C. D., 532(29), *556*(29)
Harel, P., 168(157), *188*(157)
Harnden, D. G., 83(2), *92*(2), 161(52), 162(160), 168(52), 170(52), *183* (52,53), *189*(160), 362(3), *368*(3), 409(91), *415*(91), 456(9), 457(9, 190), 461(191), 462(190), *471*(9), *480*(190,191), 526(58), *558*(58), 583(38), 589(105), *595*(38), *599* (105)
Harrell, B. W., 29(24), *36*(24)
Harris, H., 47(51), 53(51,52,135), 66 (51), *71*(51,52), 76(135), 83(15), 88(15), *92*(15), 248(62), *263*(62), 524(62,63), 525(43,59,60), 531(47), 532(61,82,156), 534(59,60,61), 535 (59,60,61), 536(61,82), 537(20,82, 155), 538(61,82,155), 539(2,61,82, 156), 540(60,156), 542(154), 543 (57,61,81,82,83), 544(47,61,81,83), 545(57,61,81,82,84,155), 546(60, 81), 551(153), 552(153,154), 553 (153), *555*(2), *556*(20), *557*(43,47), *558*(57,59,60,61,62,63), *559*(81,82, 83,84), *563*(153,154,155,156)
Harris, M., 85(16), *92*(16), 554(64), *558*(64)
Harris, R. J. C., 172(139), *187* (139)
Harris, W.

Hart, J. S., 300(101), *310*(101), 467 (89), *475*(89)
Hart, Z., 623(49), *636*(49)
Hartley, B., 122(60), *130*(60), 194 (88), *233*(88)
Hartley, J. W., 64(92), 73(92), 527 (112), *561*(112)
Hartmann, L., 295(6), *305*(6)
Hartmann, Max., 7
Hartmann, R. C., 295(18), 304(18), *306*(18), 458(52), *473*(52)
Hartnett, E. M., 170(98), *185*(98)
Harven, E. de., 568(12), 577(83), 591 (12), *593*(12), *597*(83)
Harvey, P. W., 401(151), *418*(151), 433(51), *449*(51)
Hashizume, T., 639(18), *673*(18)
Hastings, P. J., 60(53), *71*(53)
Hataya, M., 337(15,88), *368*(15), *372* (88)
Hauschka, T. S., 78(26,47), 79(17), *92* (17), *93*(26), *94*(47), 241(19), 244 (19), 254(19), 255(20), 256(19), *260*(19,20), 274(25), 275(25), *285* (25), 289(90), 290(90), *310*(90), 362(16), 366(16,17), *368*(16,17), 463(174), *479*(174), 523(126), *561* (126), 638(19,20,21), *673*(19,20,21)
Hausen, P., 56(46), *71*(46)
Haust, M. D., 161(27), 168(27), 169 (27), *181*(27)
Hay, J., 54(113), *74*(113)
Hayes, W., 46(54), *71*(54)
Hayflick, L., 78(18), *92*(18), 139(19), *149*(19)
Hayhoe, F. G. J., 290(45), *307*(45), 457(90), 458(85), 460(85), 465(90), *474*(85), *475*(90)
Haywood, F. F., 639(1), 649(1), 660 (1), *672*(1)
Hazlett, B., 463(55), *473*(55)
Heath, C. W., 290(47), *308*(47), 459 (91), 462(92), 463(91), *475*(91,92)
Hecht, F., 78(19), *93*(19), 461(93), *475*(93), 623(23,24), 624(23,24), 629(23,24), 630(23), 632(24), *635* (23,24)
Heckaman, J. H., 31(15), *36*(15)
Heckman, J. R., 89(20), *93*(20)
Heddle, J. A., 105(52), 107(52,53), 109(52), 119(54), 122(51), *129*(51, 52,53,54), 196(74), 200(75), 207 (75,77), 212(76), *232*(74,75,76,77)
Heidelberger, C., 65(61), *71*(61)
Heidrick, M. L., 549(65), *558*(65)
Hellerstein, E. E., 579(1), *593*(1)
Hellstrom, I., 164(54,55), *183*(54,55)
Hellstrom, K. E., 30(25), *36*(25), 164 (54,55), *183*(54,55)
Hemel, J., 458(94), *475*(94)
Henderson, E. S., 289(106), *311*(106), 457(198), 463(198), 466(198), *480* (198)
Heneen, W. K., 172(119), *186*(119), 248(11), *260*(11), 174(56), *183*(56), 444(33), *448*(33)
Henle, G., 162(72), 163(57), *183*(57), *184*(72), 466(117), *476*(117), 570 (40,61), 572(40,42), 573(39,52,53, 54,61), 576(19,39,112), 577(41,43, 80,96,115), 578(40), 586(57), 587 (40,56,57), 588(56), 589(56), *594* (19), *595*(39,40,41,42,43), *596*(52, 53,54,56,57,61), *597*(80), *598*(96), 599(112,115)
Henle, W., 159(168), 160(168), 163 (57), *183*(57), *189*(168), 570(40), 572(40,42), 573(39,52,53,54), 576 (19,39,112), 577(41,43,80,96,115), 578(40), 586(57), 587(40,57), *594* (19), *595*(39,40,41,42,43), *596*(52, 53,54,57), *597*(80), *598*(96), 599 (112,115)
Henner, K., 619(52), *636*(52)
Hennessen, W., 169(95), *185*(95)
Henry, J., 462(119), *476*(119)
Henry, P. H., 158(45), *182*(45)
Hensen, S., 461(83), *474*(83)
Herdson, P. B., 572(16), *594*(16)
Hertig, A. T., 314(23), 322(23), *333* (23)
Hertwig, O., 15(12), *20*
Hertwig, Richard., 4, 10(12), 12, 15 (12), *20*
Hervei, S., 168(140), *187*(140)
Hess, O., 59(55), *71*(55)
Heston, W. E., 602(17,27), 615(17), *617*(17,27)
Heysell, R., 461(95), *475*(95)
Higashino, S., 549(19), *556*(19)

AUTHOR INDEX 711

Higgins, L. C., Jr., 462(104), 475(104)
Higuera-Ballesteros, F. J., 169(92), 185(92)
Hill, M., 530(66), 558(66)
Hillova, J., 530(66), 558(66)
Hilton, H. B., 467(96), 475(96)
Hinuma, Y., 575(109), 599(109)
Hirai, K., 63(56,57), 71(56,57)
Hirayama, A., 170(9), 180(9)
Hirono, I., 248(12), 260(12), 345(18), 368(18)
Hirsch, B., 170(13), 180(13), 271(14), 285(14)
Hirschhorn, K., 164(173), 170(121), 186(121), 189(173), 458(94), 475(94), 569(7,31), 593(7), 595(31)
Hirshaut, Y., 569(31,77), 573(77), 578(77), 595(31), 597(77)
Hitotsumachi, S., 54(58), 65(58), 71(58), 410(92), 415(92), 547(67,68), 558(67,68)
Ho, H., 577(43), 595(43)
Ho, H. C., 577(15,96), 594(15), 598(96)
Ho, T., 164(174), 189(174)
Hobbs, J., 24(37), 37(37), 324(37), 333(37)
Hobson, B. M., 399(160), 419(160)
Hocker, E. V., 399(89), 415(89)
Hoffman, P. F., 569(32), 595(32)
Holdsworth, R. N., 290(48), 308(48), 457(99), 462(99), 475(99), 490(13), 495(13)
Holland, J. F., 461(15), 471(15)
Holland, S. N., 462(97), 475(97)
Holliday, R., 58(59), 71(59)
Holtzer, H., 62(11,12), 69(11,12)
Holzner, J. H., 390(66), 395(66), 414(66), 447(25,26)
Homer, J. W., 465(63), 473(63)
Homma, M., 169(58), 183(58)
Honda, F., 462(98), 475(98)
Honda, T., 648(2,9), 649(9,22), 663(9,22), 664(2), 665(2), 672(2,9), 673(22)
Hong, R., 465(141), 477(141), 620(1), 633(1)
Honjo, H., 337(29), 369(29)
Hopps, H. E., 569(100), 598(100)
Hori, S. H., 274(21), 285(21)

Horikawa, M., 100(55,56), 106(56), 129(55,56)
Horne, M. K., 570(88,89), 598(88,89)
Hosaka, T., 337(93), 340(93), 363(93), 372(93)
Hoshino, T., 461(95), 475(95), 642(28), 673(28)
Hoskins, J. M., 170(26), 181(26)
Hossfeld, D. K., 269(3), 271(3), 272(3), 273(3), 274(3), 279(3), 284(3), 290(48,92), 308(48), 310(92), 409(173), 420(173), 457(99), 459(171), 460(170), 462(99,170), 463(171), 464(170), 475(99), 479(170,171), 507(16), 517(16), 638(44), 674(44)
Hotta, Y., 61(60,63,108), 71(60,63), 74(108), 225(78), 232(78)
Housset, E., 304(16), 306(16)
Houston, E. W., 274(22), 285(22), 466(100), 475(100)
Howard, A., 194(79), 232(79)
Howard-Flanders, P., 66(93,94), 73(93,94), 209(13), 210(131,132), 228(13), 235(131,132)
Hoyer, B. H., 571(29), 594(29)
Hsu, T. C., 100(57), 119(57), 121(83), 129(57), 131(83), 135(20), 136(33), 150(20,33), 159(131), 160(59,150), 161(16), 181(16), 183(59), 187(131), 188(150), 203(80), 232(80), 240(32), 244(22), 248(32), 250(13), 253(13), 256(21), 260(13,21), 261(22,32), 344(19), 363(20,21), 368(19,20,21), 523(69), 558(69)
Huang, C. C., 155(60), 162(61), 183(60,61), 568(73), 569(73,74), 583(73), 584(74), 586(46,108), 587(44,45,46), 588(46), 590(46), 595(44,45,46), 597(73,74), 599(108)
Huberman, E., 65(61), 71(61)
Huckle, J., 66(16), 69(16)
Huebner, R. J., 64(62,88,96,122), 71(62), 73(88,96), 75(122), 80(21), 93(21), 154(62,79,80), 158(45,62), 182(45), 183(62), 184(79,80), 526(70,134), 531(114), 558(70), 561(114), 562(134)
Hughes, D. T., 395(212), 422(212), 433(82), 438(82), 450(82,83), 589(13), 594(13)

Hughes, W. L., 218(162), *237*(162)
Hugosson, R., 487(15), *495*(15)
Hummeler, K., 576(112), *599*(112)
Humphrey, R. M., 100(57), 119(33, 57), *128*(33), *129*(57), 203(42,80), 208(82), 210(81), 218(43,45), *230* (42,43,45), *232*(80,81,82)
Humphrey, R. R., 247(5), *260*(5)
Hungerford, D. A., 83(37), *93*(37), 163(85), 178(85), *184*(85), 192 (103), 211(103), *233*(103), 269(5), *284*(5), 289(76), 293(49,50), 295 (76), *308*(49,50), *309*(76), 323(40), *334*(40), 362(58), *370*(57,58), 378 (155), *419*(155), 456(148,149,150), 460(152), 462(151), 463(153), 464 (101), 465(152), 475(101), *477* (148), *478*(149,150,151,152), 591 (82), *597*(82), 645(39), 671(42), *674*(39,42)
Hunt, R. D., 159(63), *183*(63)
Hunzella, C., 587(108), *600*(108)
Hurdle, A. D., 459(102), *475*(102)
Hutchison, H. E., 459(29), 463(29), *472*(29)
Hutchison, H. T., 316(19), *332*(19)

Ichigi, H., 337(15), *368*(15)
Ichikawa, K., 337(22), 338(22), *368* (22)
Ichimaru, M., 642(28), *673*(28)
Ida, N., 172(64), *183*(64)
Ikeuch, T., 169(138), 174(138), *187* (138), 586(47), 588(47), 595(47)
Imamaki, K., 337(23), 338(23), *369* (23)
Imamura, T., 577(66), 587(45), *595* (45), *596*(66)
Imhof, J. W., 23(26), *36*(26)
Inbar, M., 548(71,72), *558*(71), *559* (72)
Inui, N., 360(60), *370*(60)
Irwin, L. E., 326(3), *332*(3)
Isaacs, J. J., 159(101), *185*(101)
Iseki, T., 648(7,8,23), 649(7,8), 663 (8), *672*(7,8) *673*(23)
Ishida, N., 169(58), *183*(58)
Ishihara, H., 336(44), 344(44), 362 (44), *370*(44)
Ishihara, T., 78(47), *94*(47), 219(83), 232(83), 362(24), *369*(24), 382(95), 385(93,94,95,96,176), 387(95), 390 (94,95,96,126,209), 398(96,126), 399(126,175), 400(126), *415*(93), *416*(94,95,96), *417*(126), *420*(175, 176), *421*(209), 429(46), *448*(46), 463(172,174), 466(173), 467(172), *479*(172,173,174), 645(25), 648(24, 25,26,27), 661(27), 663(25), 669 (27), *673*(24,25,26,27)
Ishihara, T. C., 289(90), 290(90), 293 (91), 295(90), *310*(90,91)
Ishii, S., 401(97), *416*(97)
Ishimaru, T., 642(28), *673*(28)
Ishizaka, K., 620(1,4), *633*(1), *634*(4)
Ising, U., 336(25), *369*(25), 385(98), 387(98), *416*(98)
Islas, G. M., 398(167), *419*(167)
Israels, M. C. G., 293(15), *306*(15), 462(51), 466(50), *473*(50,51)
Isselbacher, K. J., 546(73), *559*(73)
Ito, E., 568(70), 569(70), 592(70), *597*(70)
Ito, H., 226(106), *234*(106)
Ito, M., 61(63), 71(63), 225(78), *232* (78)
Ito, Y., 587(59), *596*(59)
Itzhaki, R., 208(2), *228*(2)
Iwakata, S., 568(48), 573(48), 586 (48), *595*(48)
Iwanaga, K., 251(14), 252(14), *260* (14)
Izawa, M., 639(18), *673*(18)

Jablon, S., 639(29), 641(29), 643(29), 649(3), *672*(3), *673*(29)
Jackson, J. F., 390(99), *416*(99), 462 (104), *475*(104), 623(22), 624(22), *635*(22)
Jacob, F., 81(22), *93*(22)
Jacobasch, G., 464(75), *474*(75)
Jacobs, E. M., 289(51), *308*(51)
Jacobs, P. A., 83(2), *92*(2), 289(99), *310*(99), 362(3), *368*(3), 456(9), 457(9,11,192), 458(192), 463(11, 12), 466(103), *471*(9,11,12), *475* (103), *480*(192), 589(49), *596*(49)
Jacobson, C. B., 292(58), *308*(58)
Jacobson, L. O., 293(87), *310*(87), 462(168), 463(168), *479*(168)

AUTHOR INDEX 713

Jacquemont, M., 163(129), *187*(129)
Jagiello, G. M., 442(37), *448*(37)
Jankic, M., 458(164), 468(164), *478* (164)
Janzen, H. W., 551(74), *559*(74)
Jasmin, C., 568(91), *598*(91)
Jemilev, Z. A., 203(84,85), *232*(84, 85)
Jenkins, D. E., 293(17), *306*(17)
Jensen, E. M., 573(50), *596*(50)
Jensen, F., 157(28), 163(73), 165(73), 166(73), *181*(28), *184*(73)
Jensen, F. C., 528(86), 529(85), *559* (85,86)
Jensen, M. K., 462(106), *476*(106)
Jeremy, D., 30(49), 37(49), 622(56), *636*(56)
Jerusalem, F., 623(46), 624(46), 628 (46), 630(46), 631(46), *636*(46)
Johnson, L. D., 314(23), 322(23), *333* (23)
Johnson, T., 55(33), *70*(33)
Johnston, A. W., 462(105), *475*(105)
Jones, B. A., 218(43), *230*(43)
Jones, C., 336(74), 337(74), 338(74), 363(74), *371*(74)
Jones, H. J., 439(88), *450*(88)
Jones, H. W., 322(43), *334*(43), 430 (53), 433(36), 434(53,65), 436(36), 438(36), 439(34,35), 440(36), 442 (34), *448*(34,35,36), *449*(53,65)
Jones, H. W., Jr., 390(204), 391(2), 394(204), 395(102,208), 398(105), 400(105,221), *411*(2), *416*(102,105), *421*(204,208), *422*(221)
Jones, K. P., 220(124), *234*(124)
Jones, K. W., 51(79), 67(79), 72(79), 176(65), *183*(65)
Jones, R., 578(33), *595*(33)
Jones, T. C., 159(63), *183*(63), 344 (63), *371*(63)
Jones, T. D., 639(1), 649(1), 660(1), *672*(1)
Jonsson, N., 675(3,6,7), 687(3), *691* (3,6,7)
Jordan, J. A., 158(145), *188*(145)
Judd, B. H., 59(64), *72*(64)
Jullien, P. M., 523(11), 527(11), *555* (11)

Kaalund-Jorgensen, J., 336(26), *369* (26)
Kaczala, O. A., 387(64), *414*(64)
Kafer, E., 59(86), *73*(86)
Kajii, T., 399(100), *416*(100)
Kalicanin, P., 458(164), 468(164), *478* (164)
Kalnins, V. I., 155(84), *184*(84)
Kamada, N., 290(52), *308*(52), 648 (7,8), 649(7,8,22), 661(30,31,32, 33), 662(31,32,33), 663(8,22), 669 (33), 671(33), *672*(7,8), *673*(22,30, 31,32,33)
Kaminetzky, H. A., 442(37), *448*(37)
Kammer-Meyer, J. K., 569(32), 595 (32)
Kanazir, D. T., 208(86), *232*(86)
Kang, Y. S., 203(87), 221(87), *233* (87), 395(106), *416*(106)
Kano, K., 250(25), *261*(25)
Kansow, U., 466(107), *476*(107)
Kao, F. T., 524(75), *559*(75)
Kaplan, H. S., 64(76), 72(76), 106 (58), *129*(58)
Karasawa, T., 343(30), 344(30), 352 (30), 356(30), 363(30), *369*(30)
Karazas, N. V., 163(143), *188*(143), 530(121), *561*(121)
Kara, J., 56(65,66), 72(65,66)
Karlson, A. G., 337(27), 341(27), *369* (27)
Karpati, G., 620(25), *635*(25)
Kasel, J. A., 569(32,77), 573(77), 578 (77), *595*(32), *597*(77)
Kastenbaum, M. A., 384(30), *412*(30)
Katayama, K. P., 390(103,104,204, 205), 394(204), 395(102), 398 (105), 400(101,105), *416*(101,102, 103,104,105), *421*(204,205), 433 (36), 434(38), 436(36), 438(36), 439(35,38), 440(36), 442(38), *448* (35,36,38)
Kato, H., 174(68), *183*(68), 639(29), 641(29), 643(29), *673*(29)
Kato, R., 67(67), 72(67), 171(2,66, 67), 179(67), *180*(2), 183(66,67), 675(8), 676(12), 677(12,13,45), 680 (12), 685(12), *691*(8,12,13), *693* (45)
Katseoman, A. E., 461(169), *479*(169)

Katz, E., 81(55), *94*(55)
Kauer, G. L., Jr., 462(108), *476*(108)
Kaufman, D. G., 621(12), *634*(12)
Kaufman, T. C., 59(64), 72(64)
Kaufmann, B. P., 257(48), *262*(48)
Kawai, S., 53(68), *72*(68)
Kawamura, S., 639(18), *673*(18)
Kay, H., 273(20), *285*(20), 293(61), *308*(61)
Kay, H. E. M., 293(53), 295(53), *308* (53), 459(109), 462(125), 463(109, 126,145), *476*(109,125,126), *477* (145)
Keane, J., 162(100), *185*(100)
Keane, J. F., 428(49), *449*(49)
Keehn, R., 648(7), 649(7), *672*(7)
Keele, D. K., 316(41), *334*(41)
Keijzer, W., 543(38), *557*(38)
Keir, H. M., 54(113), 74(113)
Kellerer, A. M., 325(45), *334*(45)
Kelley, W. N., 85(49), *94*(49)
Kelly, S., 648(34), *673*(34)
Kelly, W. D., 621(42), 622(42), *636* (42)
Kemp, N. H., 289(54), *308*(54), 457 (110), 458(110), 462(111), *476* (110,111)
Kenis, Y., 290(55), *308*(55), 458(112), 462(112), *476*(112)
Kermarec, J., 401(24), *412*(24)
Kern, J., 154(80), *184*(80)
Kerr, M. G., 391(168), *419*(168)
Kersey, J. H., 65(69), 72(69), 633(26), *635*(26)
Kessler, G. B., 630(20), *634*(20)
Keydar, J., 49(104,105), 74(104,105)
Khalifeh, R. R., 619(60), 623(60), *636* (60)
Khan, M. H., 463(113), *476*(113)
Kibrick, S., 550(42), *557*(42)
Kihlman, B. A., 108(60), *130*(60), 194 (88), *233*(88)
Kikuchi, Y., 382(95), 385(95), 387 (95), 390(95), 399(175), *416*(95), *420*(175), 463(172), 466(173), *479* (172,173)
Kim, A., 612(21), *617*(21)
Kim, J., 462(25), *471*(25)
Kim, S. R., 395(106), *416*(106)
Kim, S. W., 395(106), *416*(106)
Kim, U., 31(27), *36*(27)
Kimishige, I., 465(141), *477*(141)
King, E. D., 122(103), *132*(103)
King, N. W., 159(63), *183*(63)
King, T., 283(31), *285*(31)
Kiossoglou, K. A., 170(93), 168(94), *185*(93,94), 289(11), 290(44,56), 290(56), 293(56,57), 295(56), 304 (11), *305*(11), *307*(44), *308*(56,57), 457(116), 462(114,115,116), 463 (114,115), 464(44), *472*(44), *476* (114,115,116)
Kirby-smith, J. S., 106(62), *130*(62)
Kirchner, M., 168(140), *187*(140)
Kirkland, J. A., 390(41), 395(199), *413*(41), *421*(199), 433(40,41), 436 (73), 437(40), 438(40), 439(41,84), 440(39,84), 441(40), 444(41), *448* (39,40,41), *450*(73,84)
Kit, S., 56(70), 57(70), 58(70), 72 (70), 157(69), *184*(69), 527(78), 528(79,151), 529(77), 537(80), *559* (77,78,79,80), *563*(151)
Kitamura, H., 592(72), *597*(72)
Klatt, O., 363(20), *368*(20)
Klatt, R. W., 292(58), *308*(58)
Klease, M. P., 461(83), *474*(83)
Klein, E., 88(23), *93*(23,24), 531(47), 543(57), 544(47), 545(57), *557*(47), *558*(57), 573(51,52), 587(78), *596* (51,52), *597*(78)
Klein, G., 24(17), *36*(17), 48(72), 49 (72), 52(71), 53(52,135), 62(71), *71* (52), *72*(71,72), *76*(135), 83(15), 88 (15,24), *92*(15), *93*(24), 158(71), 159(71), 159(70), *184*(70,71), 248 (62), *263*(62), 324(11), *332*(11), 361(28), *369*(28), 404(217), 410 (132,133), *418*(132,133), *422*(217), 454(59), 466(134), *473*(59), *477* (134), 531(47), 532(61,82,156), 534 (61), 535(61), 536(61,82), 537(20, 82,155), 538(61,82,155), 539(2,61, 82,156), 540(156), 542(154), 543 (57,61,81,82,83), 544(47,61,81,83), 545(57,61,81,82,84,155), 546(81), 551(153), 552(153,154), 553(153), *555*(2,57), *556*(20), *557*(47), *558* (61), *559*(81,82,83,84), *563*(153, 154,155,156), 570(55), 573(51,52,

53,54), 576(55), 577(43,55,96,115), 595(43), 596(51,52,53,54,55), 598 (96), 599(115)
Kleinman, L., 550(42), 557(42)
Klemment, V., 530(128), 562(128)
Klinger, H. P., 399(19), 412(19)
Knorr-Gartner, H., 439(24), 447(24)
Knospe, W. H., 292(58), 308(58)
Knowles, B. B., 529(85), 559(85)
Knudson, A. G., 33(28), 36(28), 138 (21), 147(21), 150(21), 320(24,25), 321(24), 333(24,25)
Knutsen, T., 162(171), 189(171), 289 (106), 311(106), 398(214), 422 (214), 457(198), 463(198), 466 (198), 480(198)
Koernicke, M., 192(89), 233(89)
Kofman-Alfaro, S., 225(90), 233(90)
Kohn, G., 162(72), 184(72), 163(57), 183(57), 466(117), 476(117), 570 (40), 572(40), 578(40), 586(57), 587(40,56,57), 588(56), 589(56), 595(40), 596(56,57)
Kohne, D. E., 47(42), 70(42)
Kohno, M., 172(64), 183(64)
Koike, T., 337(29), 343(30), 344(30), 352(30), 356(30), 363(30), 369(29, 30)
Koler, R. D., 78(19), 93(19), 461(93), 475(93), 623(24), 624(24), 629(24), 632(24), 635(24)
Koller, P. C., 17(13), 20, 409(107), 416(107), 428(42), 448(42)
Kolodny, E. H., 546(31), 556(31)
Kopp, P., 208(2), 228(2)
Koprowski, H., 53(27), 54(27), 70 (27), 163(73), 165(73), 166(73), 184(73), 525(30), 528(86), 531(86), 533(36), 534(35,36), 535(36), 556 (30), 557(35,36), 559(86)
Koral, W., 573(50), 596(50)
Korein, J., 620(27), 635(27)
Korsak, E., 422(218)
Kosenow, W., 289(59), 308(59), 457 (118), 476(118)
Koss, L., 424(43), 448(43)
Kotler, S., 378(108), 385(124), 387 (108,124), 416(108), 417(124)
Koulischer, L., 458(112), 462(112, 119), 476(112,119)

Koulisher, I., 290(55), 308(55)
Koziorowska, J., 168(74), 184(74)
Krakowski, A., 320(17), 332(17)
Krauss, S., 457(120), 462(120,121), 476(120,121)
Kravchenko, G. P., 468(67), 474(67)
Kravitz, B., 479(188)
Kreizinger, J. D., 89(25), 93(25)
Kremer, W. B., 457(31), 472(31)
Krishan, A., 590(58), 596(58)
Kritchevsky, D., 525(30), 556(30)
Krompotic, E., 390(109), 409(109), 416(109)
Kronman, B. S., 26(52), 27(14,29,30), 36(14,29,30), 38(52)
Kucerova, M., 222(91), 233(91)
Kudynowski, J., 136(1), 149(1)
Kullander, S., 362(14), 390(90), 368 (14), 415(90)
Kumatori, T., 219(83), 232(83), 645 (25), 648(24,25,26,27), 661(27), 663(25), 669(27), 673(24,25,26,27)
Kunze-Muhl, E., 219(66), 231(66)
Kupffer, Carl von., 4
Kurimura, T., 527(78), 537(80), 559 (78,80)
Kurita, Y., 587(59), 596(59), 689(14, 15,42), 690(16), 691(14,15,16), 693 (42)
Kurori, Y., 170(75), 184(75)
Kurth, R., 461(177), 479(177)
Kusano, T., 524(87), 559(87)
Kvasnicka, A., 158(76), 184(76)
Kwan, H. C., 577(15), 594(15)

Lacour, J., 398(39), 413(39)
Lacour, L. F., 122(63), 130(63)
Lacroix, J. V., 337(31), 369(31)
Lafourcade, J., 288(62), 293(62), 308 (62)
Laghi, V., 293(93), 310(93)
Laguardia, A. M., 400(162), 419(162)
Lajtha, L. G., 384(150), 418(150)
Lamb, D., 398(110), 416(110)
Lampert, F., 114(64), 130(64), 467 (122), 476(122), 632(28), 635(28)
Lamy, M., 390(86,87), 415(86,87)
Lander, M. R., 527(112), 561(112)
Lane, G. R., 100(65), 105(65,66), 125 (65,66), 130(65,66)

Lange, B., 466(107), 476(107)
Langlands, A. O., 100(14,67), 123(67), 127(14), 130(67), 212(24), 213(24), 219(23), 220(24), 229(23,24), 648 (11), 663(11), 672(11)
Langley, F. A., 424(30), 447(30)
Langley, R., 106(19), 127(19)
Lanz, E., 167(181), 190(181)
Lanzkowsky, P., 615(18), 617(18)
Lapidus, P. H., 464(132), 477(132)
Lapointe, N., 620(4), 634(4)
Lappat, E. J., 289(9), 304(9), 305(9)
Laskina, A. V., 394(116,117), 417 (116,117)
Lastha, L. G., 289(72), 292(72), 295 (72), 309(72)
Laurent, M., 401(111,153), 416(111), 419(153)
Lawler, S. D., 273(20), 285(20), 293 (53,60,61,95), 295(53,83), 298(69), 308(53,60,61), 309(69,83), 310(95), 456(130), 459(109,123), 460(130), 462(125), 463(109,126,145,160), 465(124), 469(123), 476(109,123, 124,125,126), 477(130,145), 478 (160)
Lazar, P., 170(21), 181(21)
Lazarus, H., 567(23), 594(23)
Lea, D. E., 97(68), 99(68), 100(16, 68), 119(16), 123(68), 127(16), 130 (68,69), 201(92), 205(92,93), 233 (92,93)
Lee, B. I., 343(85), 359(85), 372(85)
Lee, C. S. N., 463(28), 472(28)
Lee, S. L., 461(3), 470(3)
Leeper, D. B., 217(44), 230(44)
Legrand, E., 385(112), 390(112), 400 (112), 401(112), 416(112)
Lehman, J., 63(57), 71(57)
Lehman, J. M., 57(73), 72(73), 164 (34,35), 167(77), 178(34), 182(34, 35), 184(77)
Lehmann, A. R., 66(74), 72(74)
Lehto, L., 240(15), 250(15), 251(15, 57), 252(15), 254(57), 255(15), 259 (57), 260(15), 262(57)
Lejeune, J., 256(16), 260(16), 288 (62), 289(85), 290(88), 293(62,64), 294(88), 302(65,66), 304(63,65), 308(62,63,64,65,66), 309(85), 310

(88), 390(114), 398(39), 402(115), 409(113), 413(39), 417(113,114, 115), 458(127), 476(127), 648(35), 673(35)
Lelikova, G. P., 394(116,117), 417 (116,117)
Lengerova, A., 550(88), 559(88)
Leonard, A., 226(95), 227(94), 233 (94,95)
Leonard, C., 137(18), 149(18)
Leonard, M. J., 677(45), 693(45)
Lerch, V., 240(44), 262(44)
Lett, J. T., 100(70), 130(70), 208(2, 96,97), 228(2), 233(96,97)
Lettre, H., 247(17), 257(17), 260(17)
Lettre, R., 247(17), 257(17), 260(17)
Leuchtenberger, C., 359(36,37), 369 (36,37), 428(44), 429(44), 448(44)
Levan, A., 24(35), 37(35), 78(26), 79(17), 92(17), 93(26), 108(61), 130(61), 137(12,13), 149(12,13), 168(115,116,118), 171(2), 172(117, 119), 174(56), 178(118), 180(2), 183(56), 186(115,116,117,118,119), 241(19), 244(19,22), 247(18,60), 248(11), 249(60), 254(19), 255(20), 256(19,21), 260(11,18,19,20,21), 261(22), 263(60), 274(23), 285(23), 304(67), 308(67), 336(25,33), 352 (32), 361(32), 362(32,33,34,35,86), 363(21), 364(34), 365(32), 366(17, 32), 368(17,21), 369(25,32,33,34, 35), 372(86), 384(119), 385(98, 118), 387(98,118), 399(67,68,69), 402(67,69), 407(134), 409(120), 410(67,68,69,132,133,149), 414(67, 68,69), 416(98), 417(118,119,120), 418(132,133,134,149), 433(45), 448(45), 455(65), 460(128), 463 (129), 466(64,65,134), 469(65), 473 (64,65), 477(128,129,134), 482(2), 483(2), 494(2), 507(3), 516(3), 516 (13), 517(13), 590(34), 595(34), 638(20,21), 673(20,21), 675(8), 677 (13,17,18,20), 679(19,21,44), 680 (21), 681(21), 682(35), 684(18), 689(35), 691(8,13,17,18,19,20), 692(21,35), 693(44)
Levan, G., 137(22), 150(22), 410 (149), 418(149), 498(10,11), 507

(10,11), 512(11), 513(10,11), 514
(11), 515(10,11), 516(13), *517*(10,
11,13), 682(35), 689(32,34,35), 690
(22,23), *692*(22,23,32,34,35)
Leveque, B., 620(29), *635*(29)
Lever, C. S., 384(30), *412*(30)
Levin, R. H., 463(70), *474*(70)
Levin, W. C., 274(22), *285*(22), 466
(100), *475*(100)
Levine, A. J., 56(75), *72*(75)
Levine, P. H., 578(60), *596*(60)
Levis, A. G., 247(23), *261*(23)
Levisohn, S. R., 547(89), *559*(89)
Levy, C. I., 648(35), *673*(35)
Levy, J. A., 570(61), 573(61), 596
(61)
Levy, M., 467(131), *477*(131)
Lewin, S. A., 542(54), *558*(54)
Lewis, A. M., 154(79), *184*(79)
Lewis, F. J. W., 399(121), *417*(121)
Lewis, I. C., 467(96), *475*(96)
Lewis, P. D., 407(122), *417*(122)
Lewis, R. T., 586(63), 587(63), 591
(63), *596*(63)
Lewis, W. H., 522(90), *559*(90)
Lewison, E. F., 390(205), *421*(205)
Libby, R. I., 220(170), *237*(170)
Libre, E. P., 292(68), *309*(68)
Lieberman, M., 64(76), *72*(76)
Lin, C. C., 543(56), *558*(56)
Lindberg, U., 47(77), *72*(77)
Linder, D., 24(31), *37*(31), 281(29), *285*(29), 315(16,26), 316(28), 317
(26,27,29), 318(26), 329(29,30), *332*(16), *333*(26,27,28,29,30)
Lindsley, D. L., 602(23), *617*(23)
Lindsten, J., 136(2), 137(3), *149*(2,3),
211(4), 217(4), 228(4), 289(7), *305*
(7), 323(6), *332*(6)
Lis, H., 546(118), 548(118), *561*
(118)
Lisco, H., 648(36), *673*(36)
Liskay, R. M., 329(21), *333*(21)
Lisker, R., 623(30), 624(30), 629(30), *635*(30)
List-young, B., 592(106), *599*(106)
Lithner, F., 171(128), *187*(128)
Littlefield, J. W., 84(27,28), 85(28), *93*(27,28), 523(91), 525(93), 536
(91), *560*(91,93)

Littlefield, L. G., 224(22), *229*(22)
Livingstone, C. S., 621(11), *634*(11)
Lobb, D., 295(83), *309*(83)
Lobb, D. S., 298(69), *309*(69), 456
(130), 460(130), 463(160), *477*
(130), *478*(160)
Loewenstein, W. R., 546(6), 549(5,6,
19), *555*(5,6), *556*(19)
Loftus, J., 162(72), *184*(72), 466
(117), *476*(117), 587(56), 588(56),
589(56), *596*(56)
Log, T., 531(114), *561*(114)
Lohman, P. H. M., 208(98), *233*(98)
Lohmann, H., 390(84), *415*(84)
Long, C., 524(87), *559*(87)
Looby, P. C., 212(24), 213(24), 220
(24), *229*(24)
Lopez, V. F., 620(51), *636*(51)
Lopez-Saez, J. F., 122(47), *129*(47)
Lorz, A., 244(24), *261*(24)
Louis-Bar, M., 619(31), *635*(31)
Louro, J. M., 462(69), *474*(69)
Love, R., 169(158), *188*(158)
Lovelace, E., 589(101), *598*(101)
Low, B., 463(129), *477*(129)
Lowy, D. R., 64(92), *73*(92)
Lubs, H. A., 378(108), 385(123,124),
387(108,123,124), 400(123), *416*
(108), *417*(123,124), 488(3), *494*(3)
Lucas, L. A., 401(202), *421*(202)
Lucas, L. S., 573(110), 592(62), *596*
(62), *599*(110)
Luce, J. K., 289(51), *308*(51)
Ludwig, A. S., 430(62), *449*(62)
Ludwig, A. S. Jr., 391(169), *419*(169)
Luippold, H. E., 100(125), 119(126),
124(127), 125(127), *133*(125,127),
194(176), 205(177), 213(19), 217
(175), 221(21), *229*(19,21), *237*
(175,176,177)
Luippold, H. J., 106(30), *128*(30)
Lungeanu, A., 167(109), *186*(109)
Luzzati, L., 330(42), *334*(42)
Lyon, M., 226(99), *233*(99)
Lyon, M. F., 23(32), *37*(32), 85(29), *93*(29), 226(100), 227(100), *233*
(100), 315(31), *333*(31)
Lytle, C. D., 66(1), *68*(1)

McAllister, R., 155(120), *186*(120)

AUTHOR INDEX

McAllister, R. M., 154(78,79,80), *184* (78,79,80)
McBeath, S., 162(148), *188*(148), 586 (99), 587(99), 591(99), *598*(99)
McBride, J. A., 463(12), *471*(12)
McCarthy, R. E., 568(23), *594*(23)
McClelland, G., 317(50), *334*(50)
McClintock, B., 62(78), *72*(78)
McCollum, R. W., 577(80), 578(21), *594*(21), *597*(80)
McCormick, D. P., 465(141), *477* (141)
McCulloch, E. A., 314(53), *334*(53)
McDougall, J. K., 51(79), 67(79), *72* (79), 155(81,82,83), 157(82,83), 158(83), 178(82), *184*(81,82,83)
McFarland, V. W., 546(31), *556*(31)
McFarland, W., 292(68), *309*(68)
McFarlane, E. S., 157(141), *187*(141)
MacFarlin, D. E., 621(32), 623(32), *635*(32)
McGee, B. J., 293(17), 295(18), 304 (18), *306*(17,18), 458(52), 460(53), *473*(52,53), 525(43), *557*(43)
McGee, C. L., 290(19), *306*(19)
McGill, M., 461(142), *477*(142)
McGrath, R. A., 208(107), *234*(107)
Machala, O., 530(92), *560*(92)
MacKay, I. R., 465(57), *473*(57)
McKenzie, A., 32(33), *37*(33)
McKhann, C. F., 30(34), *37*(34)
Makowitz, A. D., 623(45), *636*(45)
McKusick, V. A., 602(19), *617*(19), 619(33,34), 620(33), *635*(33,34)
McLean, A. S., 613(7), *616*(7)
McLean, E. P., 26(39), 27(8), *35*(8), *37*(39)
McLeish, J., 108(73), *130*(73)
McLelland, J., 100(67), 123(67), *130* (67)
MacMahon, B., 467(131), *477*(131), 578(65), *596*(65)
McMahon, N. G., 401(202), *421*(202)
McMichael, H., 163(85), 178(85), *184* (85), 269(5), *284*(5)
McNutt, N. S., 546(95), *560*(95)
MacPherson, I., 154(79), 184(79)
McPherson, I. A., 164(86), *184*(86)
MacTaggart, M., 399(121), *417* (121)

Macek, M., 586(63,64), 587(63), 591 (63), *596*(63,64)
Mach, O., 56(66), *72*(66)
Maciera-Coelho, A., 568(91,92), 598 (91,92)
Mackey, C. M., 363(9), *368*(9)
Mackinnon, E., 155(84), *184*(84)
Maeda, M., 399(125), 400(125), *417* (125)
Majumdar, S. K., 689(39), 690(39), *692*(39)
Makino, S., 78(30,32), 87(31,33), *93* (30,31,32,33), 160(10,87), 170(8,9, 75,87), *180*(8,9,10), *184*(75,87), 248(26), 250(25), 255(27), *261*(25, 26,27), 269(7), 274(24), *284*(7), *285*(24), 288(70), *309*(70), 336(40, 42,44,47), 343(41,46,61,71,81), 344 (39,40,44,61,81), 348(71,81), 352 (40,41,46,61,71,81), 360(60), 361 (38,40), 362(38,40,43,44,47,48), 363(61), 364(61), 365(41,81), 366 (40,43,61), *369*(38,39,40,41), *370* (42,43,44,45,46,47,48,60,61), *371* (71,81), 384(126,128), 385(128), 390(126,128), 398(126), 399(126, 127,129,156,157), 400(126,128, 129), *417*(126,127,128,129), *419* (156,157), 429(46), *448*(46), 638 (37), *674*(37)
Malan, L., 578(60), *596*(60)
Malucci, L., 532(3,4), *555*(3,4)
Malutina, T. S., 222(29), *229*(29)
Manaker, R. A., 573(110), 592(62), *596*(62), *599*(110)
Mancinelli, S., 467(133), *477*(133)
Mann, F. C., 337(27), 341(27), *369* (27)
Manna, G. K., 251(28), *261*(28), 428 (47,48), *448*(47,48)
Manning, M. D., 138(25), *150*(25)
Mannini, A., 390(72,130), 391(74), *414*(72), *415*(74), *417*(130)
Manocha, S. L., 430(74), *450*(74)
Manolov, G., 24(35), *37*(35), 136 (23), *150*(23), 162(88), *185*(88), 407(134), 410(131,132,133), *418* (131,132,133,134), 466(134,135), *477*(134,135), 587(78), *597*(78)
Manolova, Y., 136(23), *150*(23), 162

(88), *185*(88), 410(131,132,133), *418*(131,132,133), 466(134,135), *477*(134,135)
Mantle, D. J., 467(136), *477*(136)
Manton, I., 122(71), *130*(71)
Mantooth, L., 400(189), *420*(189)
Maounis, F., 168(94), 170(93), *185* (93,94)
Marcellino, L. R., 384(138), 385(138), 387(138), 395(139), 400(139), *418* (138,139)
Marchand, J. C., 304(42), *307*(42)
Marco, F., 461(157), *478*(157)
Margulis, M. I., 468(67), *474*(67)
Marie, J., 620(29), *635*(29)
Marin, G., 218(101), *233*(101), 525 (93), *560*(93)
Marinelli, L. D., 100(72), *130*(72)
Mark, J., 137(22), *150*(22), 172(89, 90,91), 179(89), *185*(89,90,91), 269 (4), 270(9), 276(9), 277(9), *284*(4, 9), 376(135), 401(82), 406(135), 410(149), *415*(82), *418*(135,149), 470(137,138), *477*(137,138), 481 (12), 483(6), 484(5,7), 485(8,10), 486(7), 488(7,8), 489(5,8), 490(5, 7), 491(4,5,11,12), 492(9,11), 493 (10), *494*(4,12), *495*(5,6,7,8,9,10, 11,12), 498(6,9,10,11), 499(6), 504 (6), 506(4), 507(5,6,7,8,10,11), 508 (4), 512(6,11), 513(10,11), 514(11), 515(10,11), *516*(4,5), 516(13), *517* (6,7,8,9,10,11,13), 675(4), 676(24), 677(24,26,27,33), 682(35), 684(24), 685(24,33), 688(25,27), 689(34,35), 690(22), *691*(4), *692*(22,24,25,26, 27,33,34,35)
Markham, R. L., 467(163), *478*(163)
Markovic, V., 458(164), 468(164), *478*(164)
Marks, J. F., 316(41), *334*(41)
Marquez-Monter, H., 169(92), *185* (92), 394(166), *419*(166)
Marsh, J. C., 399(203), *421*(203)
Martin, D. C., 220(170), *237*(170)
Martin, G. M., 66(123), 75(123), 89 (34), *93*(34), 148(34), *150*(34), 248 (29), *261*(29), 527(135), 550(94), *560*(94), *562*(135)
Martin, G. S., 53(80), *73*(80)

Martin, H., 463(113), *476*(113)
Martin, M. A., 47(42), *70*(42)
Martin, P. G., 122(119), *133*(119)
Martineau, M., 382(137), 394(38,136, 137), *413*(38), *418*(136,137), 470 (139), *477*(139)
Maruyama, T., 639(18), *673*(18)
Maruyama, Y., 253(30), *261*(30)
Mashiach, S., 391(180), *420*(180)
Mastroangelo, R., 304(108), *311*(108), 463(140), *477*(140)
Masukawa, T., 390(103,205), *416*(103), *421*(205)
Matano, K., *370*(48)
Mathe, G., 569(4), *593*(4)
Matsaniotis, N., 168(94), 170(93), *185* (93,94)
Matsubara, K., 343(30), 344(30), 352 (30), 356(30), 363(30), *369*(30)
Matsubara, S., 406(178), *420*(178)
Matsui, T., 648(2), 664(2), 665(2), *672*(2)
Matsui, Y., 337(49), 338(49), *370*(49)
Matsunaga, T., 337(95), 343(95), 344 (95), 352(95), *372*(95)
Matsuya, Y., 530(14), *556*(14)
Matthews, F., 271(15), *285*(15)
Mattinson, G., 379(22), 405(22), *412* (22), 433(11), *447*(11)
Mauer, B. A., 577(66), *596*(66)
Mauler, R., 169(95), *185*(95)
Maxwell, P. M., 467(163), *478*(163)
May, E., 167(96), *185*(96)
May, P., 167(96), *185*(96)
Mayer, E. G. M., 169(92), *185*(92)
Mazurowa, N., 168(74), *184*(74)
Mazzone, H. M., 159(97), *185*(97)
Medenis, R., 623(45), *636*(45)
Medrek, T. J., 573(50), *597*(50)
Meera Khan, P., *560*(97)
Melander, Y., 362(14), 368(14), 390 (90), *415*(90)
Melendez, L. V., 159(63,107), *183* (63), *186*(107)
Mellman, W. J., 162(72), *184*(72), 170 (98), *185*(98), 192(103), 211(103), *233*(103), 466(117), *476*(117), 586 (57), 587(56,57), 588(56), 589(56), *596*(56,57), 645(39), *674*(39)
Mellors, R. C., 428(49), *449*(49)

Melnick, J. L., 47(17), 48(17), 69(17), 167(110), *186*(110)
Mendes, F. T., 464(46), *473*(46)
Merigan, T. C., 64(76), *72*(76)
Merker, H., 292(8), *305*(8), 456(199), 458(33), 460(143), 472(33), *477* (143), *480*(199)
Merz, T., 213(123), *234*(123)
Messinetti, S., 384(138), 385(138), 387(138), 395(139), 400(139), *418* (138,139)
Metcalf, D., 461(144), 462(144), 468 (144), *477*(144)
Meuge, C., 390(140), *418*(140)
Meuwissen, H., 620(4), 621(11), *634* (4,11)
Meyer, G., *556*(17)
Meyer, G. F., 59(55), *71*(55)
Meyer, H. M., Jr., 569(100), *598*(100)
Mezger-Freed, L., 85(35), *93*(35)
Micklem, H. S., 631(14), *634*(14)
Migeon, B. R., 157(99), *185*(99), 525 (96), 542(96), *560*(96)
Mikuni, C., 399(156,157), *419*(156, 157)
Miles, C. P., 160(122), 162(100), *185* (100), *186*(122), 364(50), *370*(50), 385(143), 387(143), 390(142,143), 394(142), 395(143), 398(142), 399 (145), 400(144), 409(141), *418* (141,142,143,144,145), *449*(50), 581(84), 587(67,68), 592(68), *597* (67,68,84)
Millard, R. E., 273(20), *285*(20), 293 (53), 295(53), *308*(53), 399(146), *418*(146), 459(109), 463(109,126, 145), *476*(109,126), *477*(145)
Miller, C. S., 157(99), *185*(99)
Miller, D. A., 138(24), 139(24), *150* (24), 525(96), 539(2), 542(96), *555* (2), *560*(96)
Miller, D. G., 364(50), *370*(50), 465 (141), *477*(141)
Miller, G., 462(98), *475*(98)
Miller, G. F., 29(36), *37*(36)
Miller, H. H., 121(34), *128*(34), 217 (44), *230*(44)
Miller, M. E., 621(37), *635*(37)
Miller, M. M., 619(10), 620(10), *634* (10)

Miller, O. J., 53(52), 71(52), 83(15), 88(15), *92*(15), 138(24), 139(24), *150*(24), 525(96), 532(61), 534(61), 535(61), 536(61), 538(61), 539(2, 61), 542(96), 543(61), 544(61), 545 (61), *555*(2), *558*(61), *560*(96,97)
Miller, R. W., 78(19), *93*(19), 138(25), *150*(25), 461(93), *475*(93), 621(35), 622(35), 623(24), 624(24), 629(24), 632(24), *635*(24,35,36)
Millman, P. A., 551(74), *559*(74)
Milne, J. S., 457(11), 463(11), *471* (11)
Milstein, C. B., 281(30), *285*(30)
Milton, R. C., 640(38), 641(38), 643 (38), 649(38), *674*(38)
Minkler, J. L., 384(147), *418*(147)
Minna, J., 552(98), *560*(98)
Minouchi, O., 344(51), *370*(51)
Minowada, J., 162(61,103), *183*(61), *185*(103), 568(5), 586(46,47), 587 (46,104), 588(46,47), 589(46), *593* (5), *595*(46,47), *599*(104)
Mintz, B., 314(33), 315(34,35), 328 (8,34,35,36), 329(36), *332*(8), *333* (33,34,35,36)
Minzell, M., 159(101), *185*(101)
Mitani, M., 289(84), *309*(84), 463 (162), 467(161,162), *478*(162)
Mitchell, E. Z., 578(102), *598*(102)
Mitelman, F., 137(22), *150*(22), 170 (102), *185*(102), 270(11), 275(11), 276(11), 277(11), *284*(11), 410 (148,149), *418*(148,149), 498(10, 11), 506(12), 507(10,11), 512(11), 513(10,11), 514(11), 515(10,11), 516(13), *517*(10,11,12,13), 676(29), 677(28,29,31,33), 681(29), 682(35), 683(31), 684(28,29), 685(29,30,31, 33), 686(30,31), 688(29,30,31), 689 (32,34,35), 690(22,23), *692*(22,23, 28,29,30,31,32,33,34,35)
Mitra, S., 225(102), *233*(102)
Mitus, W. J., 290(44,56), 293(56,57), 295(56), *307*(44), *308*(56,57), 457 (116), 462(32,114,115,116), 463 (114,115), 466(202), *472*(32), *476* (114,115,116), *480*(202)
Miura, A., 66(81), *73*(81)
Miwa, T., 463(174), *479*(174)

Miyata, H., 224(137), *235*(137), 648 (45), 660(45), 664(45), *674*(45)
Mizutani, S., 28(47), *37*(47), 49(118), *75*(118), 80(54), *94*(54), 526(131), 530(131), *562*(131)
Modest, E. J., 136(1), *149*(1)
Mohr, U., 390(60), *414*(60)
Moldvannu, G., 364(50), *370*(50)
Moller, G., 30(25), *36*(25)
Moller, T., 137(13), 149(13), 455(65), 466(65), 469(65), *473*(65), 507(3), *516*(3)
Moloney, W. C., 290(47), *308*(47), 462(92,165), *475*(92), *478*(165)
Monesi, V., 119(74), *130*(74)
Monk-Jones, M. E., 467(163), *478* (163)
Monod, J., 81(22), *93*(22)
Monroe, J. H., 569(25), 570(28), 573 (25), 588(28), *595*(25,28)
Moore, A., 364(50), *370*(50)
Moore, F. B., 135(9), *149*(9)
Moore, G. E., 162(103), *185*(103), *369*(24), 385(96,176), 390(96), 398 (96), *416*(96), *420*(176), 568(5,70, 73), 569(30,69,70,71,73,74), 573 (69,71), 583(73), 584(174), 586 (108), 587(44,45,104), 592(70,72), *594*(5), *594*(30), *595*(44,45), *597* (69,70,71,72,73,74,75), *599*(104, 108)
Moore, M., 26(5), *35*(5)
Moorhead, P. S., 78(18), *92*(18), 139 (19), *149*(19), 162(72), 163(73), 165(48,73,105,106,137,169,170), 166(48,73,105), 168(104), 170(22, 26,98), 178(48), *181*(22,26), *182* (48), *184*(72,73), *185*(98), 186(104, 105,106), *187*(137), *189*(169,170), 192(103), 211(103), *233*(103), 240 (32), 248(32), 249(31), *261*(31,32), 277(26), *285*(26), 466(117), *476* (117), 514(15), *517*(15), 583(76), 587(56), 588(56), 589(56), *596*(56), *597*(76), 645(39), *674*(39)
Mora, P. T., 546(31), *556*(31)
Moretti, G., 295(6), *305*(6)
Morgan, D. G., 159(107), *186* (107)
Morgan, T. H., 10(14), *20*

Mori, M., 364(52,72), *370*(52), *371* (72)
Morley, A. A., 468(146), *477*(146)
Morris, H. P., 270(10), 274(10), 277 (10), *284*(10)
Morris, T., 226(99), *233*(99)
Morrison, J. M., 54(113), *74*(113)
Morton, D. L., 29(36), *37*(36)
Moscarini, M., 395(139), 400(139), *418*(139)
Moser, H. W., 623(59), *636*(59)
Moses, H. L., 569(32,77), 573(77), 578(77), *595*(32), *597*(77)
Mosier, H. D., 304(24), *306*(24)
Moulinier, J., 295(6), *305*(6)
Moulton, J. E., 366(53), *370*(53)
Mouriquand, C., 304(71), *309*(71)
Moy, R. A., 578(27), *594*(27)
Mozziconacci, P., 289(85), *309*(85)
Mukerjee, D., 633(38), *635*(38)
Mul, N. A. J., 23(26), *36*(26)
Muldal, S., 289(72), 292(72), 295(72), *309*(72), 384(150), 402(151), *418* (150,151), 433(51), *449*(51), 458 (147), *477*(147)
Mulder, M. P., 210(12), *228*(12)
Muller, H. J., 96(75), *130*(75), 192 (104), 213(105), *233*(104,105)
Muller, T., 399(69), 402(69), 410(69), *414*(69)
Muller-Hill, B., 91(36), *93*(36)
Mullner, T., 219(66), *231*(66)
Multanen, I., 251(33), *261*(33)
Muratmasu, S., 226(106), *234*(106)
Murayama, F., 534(99), *560*(99)
Murayama-Okabayoshi, F., 540(100), *560*(100)
Murray, N. E., 61(82), *73*(82)
Murray, R. F., 24(37), *37*(37), 324 (37), *333*(37)
Mustafina, A. N., 165(108), *186*(108)

Nachtigal, M., 167(109,110), *186*(109, 110)
Nachtigal, S., 167(109), *186*(109)
Nadkarni, J., 24(35), *37*(35), 407 (134), *418*(134)
Nadkarni, J. J., 573(52), 587(78), *596* (52), *597*(78)
Nadkarni, J. S., 24(35), *37*(35), 407

(134), *418*(134), 573(52), 587(78), *596*(52), *597*(78)
Nagabhushanam, G. N., 468(180), *479* (180)
Nagaoka, S., 639(18), *673*(18)
Nagata, H., 100(56), 106(56), *129*(56)
Nagayama, T., 170(75), *184*(75)
Nahas, S., *636*(54)
Nahmias, A., 170(167), *189*(167)
Nakahara, H., 248(26,34), *261*(26,34)
Nakajima, K., 157(69), *184*(69)
Nakamura, W., 226(106), *234*(106)
Namunes, P., 464(132), *477*(132)
Nance, W. E., 314(38), *333*(38)
Nanta, A., 337(54), *370*(54)
Naspitz, C. K., 620(39), 621(39), *635* (39)
Nava, C. de., 289(35,40), 290(32,73, 75), 292(30,73,75), 293(31,33,75), 294(31,33,75), 295(37,41), 298(41), 300(74,75,102), 302(38), 304(40, 42), *307*(30,31,32,33,35,37,38,40, 41,42), *309*(73,74,75), *310*(102), 390(87), 402(88), 404(152), 409 (85), *415*(85,87,88), *419*(152), 458 (39,41,42), 459(40), 463(43), *472* (39,40,41,42,43), 677(11), *691* (11)
Neary, G. J., 100(77), 106(76,77), *130*(76,77), 205(108,109), 207 (142), 208(109,110), 217(141), *234* (108,109,110), *235*(141), *236*(142)
Nebel, B. R., 100(72), 122(78), *130* (72,78)
Neiman, P. E., 314(12), 327(12), *332* (12)
Nelson, P., 552(98), *560*(98)
Nelson, R. L., 532(39), *557*(39)
Neriishi, S., 648(2,7,8,9), 649(7,8,9), 663(8,9), 664(2), 665(2), *672*(2,7,8, 9,10)
Nesbitt, M. N., 314(20,39), *333*(20, 39)
Neu, R. L., 399(100), *416*(100)
Neubauer, A., 290(4), *305*(4), 460 (16), *471*(16)
Newcombe, H. B., 99(79,80), 108(79, 80), *130*(79,80)
Nezelof, C., 401(111,153), *416*(111), *419*(153)

Nichols, W. W., 155(120), 168(111, 113,114,115,116,118), 171(111), 172(111,112,117,119), 174(6,56), 178(118), 179(111,112), *180*(6), *183*(56), *186*(111,112,113,114,115, 116,117,118,119,120), 248(11), *260*(11), 366(55,56), *370*(55,56), 444(33), *448*(33), 460(128), 463 (129), *477*(128,129), 583(79), *597* (79), 676(36,37,38), 677(13,36), 690(23,36,37), *691*(13), *692*(23,36, 37,38)
Nicholson, M. O., 154(79,80), *184*(79, 80)
Nicoletti, B., 602(23), *617*(23)
Nicolson, G., 546(101), *560*(101)
Niederhalt, G., 461(83), 466(107), *474*(83), *476*(107)
Niederman, J. C., 573(53), 577(80), 578(21), *594*(21), *596*(53), *597*(80)
Nikaido, O., 100(55,56), 106(56), *129* (55,56)
Nilsson, S. B., 463(129), *477*(129)
Nirenberg, M., 552(98), *560*(98)
Nishimura, E. T., 461(95), *475*(95)
Nishizuka, Y., 689(14,15,42), 690 (16), *691*(14,15,16), *693*(42)
Nitowsky, H. M., 315(9), *332*(9)
Niyama, M., 364(72), *371*(72)
Noback, C. R., 337(6), *368*(6)
Nomura, S., 578(102), *598*(102)
Nonoyama, M., 576(81), *597*(81)
Noonan, K. D., 548(24), *556*(24)
Norden, A., 460(128), 463(129), *477* (128,129)
Norman, A., 100(81,82), 125(82), *131* (81,82), 209(64), 212(138), 213 (111), 220(170), 224(112), *231*(64), *234*(111,112), *235*(138), *237*(170), 651(40), 670(41), *674*(40,41)
Norrby, E., 168(116,118), 174(56), *183*(56), *186*(116,118), 248(11), *260*(11), 444(33), *448*(33)
Nossal, G. J. V., 28(38), *37*(38)
Notterman, R. F., 289(11), 304(11), *305*(11), 464(44), *472*(44)
Nowell, P. C., 83(37), *93*(37), 163 (85), 178(85), *184*(85), 192(103), 211(103), *233*(103), 269(5,8), 270 (8,10), 271(12,14,15), 272(18,19),

273(8,19), 274(8,10), 276(8), 277
(10), 279(8,27), *284*(5,8,10,12), *285*
(14,15,18,19), 289(76,77), 293(50),
295(76), *308*(50), *309*(76,77), 323
(40), *334*(40), 343(96), 352(96),
356(92,96), 359(96), 362(58), 363
(96), *370*(57,58), *372*(96), 378
(155), 409(154), *419*(154,155), 456
(148,149,150), 460(152), 462(151),
463(153), 464(101), 465(152), *475*
(101), *477*(148), *478*(149,150,151,
152,153), 591(82), *597*(82), 645
(39), 671(42), *674*(39,42)
Nowinsky, M. A., 336(59), *370*(59)
Nusbacher, J., 170(121), *186*(121)
Nyhan, W. L., 316(41), *334*(41)

Oakberg, E. F., 225(113,114), *234*
(113,114)
Obara, Y., 399(156,157), *419*(156,
157)
Obrecht, P., 292(8), *305*(8), 458(33),
472(33)
Ockey, C. H., 121(83), *131*(83)
Oettgen, H. F., 26(39), 27(8), *35*(8),
37(39), 577(83), *597*(83)
Oftebro, R., 248(35), *261*(35), 522
(102), *560*(102)
Oftedal, P., 226(115), *234*(115)
O'Grady-Boveri, Marcella, 3, 5
Ohha, Y., 172(64), *183*(64)
Ohira, M., 169(58), *183*(58)
Ohno, S., 81(41), 83(40), 85(39), 89
(38), *93*(38,39), *94*(40,41), 248(36),
261(36), 271(16), 275(16), 276(16),
279(16), *285*(16), 289(100), *310*
(100), 458(193), *480*(193), 523
(103), *560*(103)
Ojimo, Y., 343(84), 344(84), 348(83),
350(84), 352(84), 360(60,82), *370*
(60), *371*(82,83), *372*(84)
Okada, H., 642(28), *673*(28)
Okada, S., 108(116), *132*(116), 209
(143), *236*(143)
Okada, T. A., 105(23,24,25), 113(23,
24,25,26,27), 114(23,24,25), 123
(26), *128*(22,23,24,25,26,27), 136
(8), *149*(8)
Okada, Y., 522(158), 524(104), 534
(99), 540(100), *560*(99,100,104),

563(158)
Oksala, T., 241(38,39), 244(37,40),
247(40), 252(40), 253(40), 254(40),
255(40), 256(40), 257(40), *261*(37,
38,39), *262*(40), 442(52), *449*(52)
Old, L. J., 26(39), 27(8), *35*(8), *37*
(39), 80(52), *94*(52), 303(78), *309*
(78), 577(83), *597*(83)
Olinici, C. D., 390(158,159), 402(158),
419(158,159)
Olivieri, G., 125(84), *131*(84), 218
(20), 221(21), *229*(20,21)
O'Neill, F., 162(100), *185*(100), 399
(145), *418*(145), 587(67,68), 592
(68), *597*(67,68)
O'Neill, F. J., 160(122,123), *186*(122,
123), 581(84), *597*(84)
Onesti, P., 457(200), 458(47), 460(47,
48), 462(201), *473*(47), *480*(200,
201)
Oni, S. B., 330(42), *334*(42)
Oppenheim, J. J., 465(154), *478*(154),
591(107), *599*(107), 621(32), 623
(32), *635*(32)
O'Riordan, M. L., 136(26), *150*(26),
162(148), *188*(148), 272(17), *285*
(17), 586(99), 587(99), 591(99),
598(99)
Osato, T., 587(59), *596*(59)
Osgood, E. E., 568(85), *597*(85)
Oshimura, M., 343(61), 344(61), 352
(61), 363(61), 364(61), 366(61),
370(61)
Ostergren, G., 122(85), *131*(85), 168
(115), *186*(115)
Osunkoya, B. O., 162(61), *183*(61),
330(42), *334*(42), 586(46), 587(46),
588(46), 589(46), *595*(46)
Ottoman, R. E., 224(112), *234*(112),
670(41), *674*(41)

Padeh, B., 391(180), *420*(180)
Padgett, B. L., 163(124), *187*(124)
Paff, G. H., 337(6), *368*(6)
Pagano, J. S., 576(81), *597*(81)
Painter, R. B., 209(116,117,118,125,
126), *234*(116,117,118), *235*(125,
126)
Palcic, B., 208(119), *234*(119)
Palme, G., 457(87), *475*(87)

Palmer, C. G., 163(125), *187*(125)
Papanicolaou, G. N., 428(49), *449* (49)
Papworth, D. G., 213(149,150), 226 (152), *236*(149,152)
Paris conference, Standardization in Human Cytogenetics, 196(120), 221 (120), *234*(120)
Paris Conference 1971, 135(27), 136 (27), *150*(27)
Parish, C. R., 28(40), *37*(40)
Park, I., 322(43), *334*(43)
Park, I. J., 400(221), *422*(221), 430 (53), 433(36), 434(53), 436(36), 438(36), 439(34,88), 440(36), 442 (34), *448*(34,36), *449*(53), *450*(88)
Park, S. D., 203(87), 221(87), *233*(87)
Parkinson, J. S., 42(25), *69*(25)
Parmentier, R., 428(54), 444(54), *449* (54)
Parr, R. M., 219(66), *231*(66)
Parrington, J. M., 210(121), *234*(121), 614(20), *617*(20)
Passano, K., 106(17), *127*(17), 203 (28), *229*(28)
Patau, K., 241(41), *262*(41), 433(55), *449*(55)
Paterson, W. G., 399(160), *419*(160)
Pathak, S., 139(31), *150*(31)
Paton, G. R., 174(3), *180*(3)
Patrono, C., 293(93), *310*(93)
Patten, S. F., 424(60), 428(56), 442 (56), *449*(56,60)
Paul, B., 390(165), 409(165), *419* (165), 507(14), 508(14), *517*(14)
Paul, G., 137(28), *150*(28)
Paul, J., 153(126), *187*(126)
Paulete-Vanrell, J., 395(161), 400 (162), *419*(161,162), *449*(57)
Pawlowitzki, I. H., 404(163), *419* (163)
Payne, B., 24(37), *37*(37), 324(37), *333*(37)
Payne, L. N., 80(42), *94*(42), 159(17), *181*(17)
Peacock, J., 552(98), *560*(98)
Peacock, W. J., 122(86), 131(86)
Peacocke, A. R., 208(110), *234*(110)
Pearson, F. G., 221(21), *229*(21)
Pearson, G., 573(52,53,54), *596* (52,53,54)
Pearson, P., 135(29), *150*(29)
Pearson, P. L., 539(2), *555*(2)
Pedersen, B., 293(80), 295(79), 303 (81), *309*(79,80,81), 458(155), *478* (155)
Pees, E., 61(83), 73(83)
Pegoraro, L., 290(82), *309*(82), 458 (156), 460(156), 461(73), *474*(73), *478*(156)
Pegrum, G., 457(37), *472*(37)
Pelc, S. R., 194(79), *232*(79)
Pellett, O. L., 401(202), *421*(202)
Pelling, C., 59(84), 73(84)
Peluse, M., 155(120), 172(119), *186* (119,120)
Penn, I., 30(41,43,46), *37*(41,43,46)
Penrose, L. S., 379(164), *419*(164)
Pentycross, C. R., 465(124), *476*(124)
Pera, F., 248(42), *262*(42), 390(84), *415*(84), 522(105), *560*(105)
Perrin, E. V., 30(7), *35*(7)
Perry, P., 292(14), *306*(14)
Perry, S., 468(49), *473*(49)
Pescetto, G., 439(58,59), *449*(58,59)
Peters, A., 290(97), *310*(97)
Peters, J. J., 247(43), *262*(43)
Peters, R. F., 119(12), *127*(12)
Peterson, R. D. A., 619(41), 620(41), 621(40,42), 622(40,42), *636*(40,41, 42)
Pettijohn, D., 209(122), *234*(122)
Pfeiffer, R. A., 289(59), *308*(59), 457 (118), 464(45), *472*(45), *476*(118), 612(21), *617*(21), 623(43), 624(43), 629(43), 630(43), *636*(43)
Pickren, J. W., 385(176), *420*(176)
Piguet, H., 290(32), *307*(32)
Pike, M. C., 586(86), *598*(86), 645 (12), *672*(12)
Pileri, A., 290(82), *309*(82), 458(156), 460(156), 461(73), *474*(73), *478* (156)
Pimm, M. V., 26(5), *35*(5)
Pink, J., 281(30), *285*(30)
Pinkerton, P. H., 459(29), 463(29), *472*(29)
Plit, M., 623(55), *636*(55)
Plotkin, S. A., 170(22,26,98), *181*(22, 26), *185*(98)

Poen, H., 23(26), *36*(26)
Pogosianz, H. E., 222(29), *229*(29)
Pogosyants, E. E., 394(116,117), *417* (116,117)
Polani, P. E., 136(16), 149(16)
Pollack, R., 547(107), 560(107)
Pollack, R. E., 53(85), 73(85), 548(106), *560*(106)
Pollister, A. W., *371*(62)
Pomerat, C. M., 344(19), *368*(19)
Pontecorvo, G., 59(86), 73(86)
Ponten, J., 171(127,128), *187*(127,128), 487(15), *495*(15), 590(93), *598*(93)
Ponten, J. A., 163(73), 165(73), 166(73), *184*(73)
Pope, J. H., 569(87), 570(88), 572(87), *598*(87,88)
Pope, J. L., 570(89), *598*(89)
Popovic, M., 172(144), *188*(144)
Porter, I. H., 137(28), *150*(28), 387(31), 390(31,165), 391(32), 398(31), 409(165), *412*(31,32), *419*(165), 507(14), 508(14), *516*(1), *517*(14), 569(74), 584(74), *597*(74)
Poste, G., 525(108), 550(108), *560*(108)
Pot, F., 210(12), *228*(12)
Pothier, L., 579(1), *593*(1)
Potter, V. R., 270(10), 274(10), 277(10), *284*(10)
Poulik, M. D., 623(49), *636*(49)
Pourfar, M., 461(3), *470*(3)
Pouwels, P. H., 527(136), 542(136), *562*(136)
Power, J., 329(30), *333*(30)
Prasad, K. N., 82(43), *94*(43)
Prehn, R. T., 27(42), *37*(42)
Prempree, T., 213(123), *234*(123)
Prescott, D. M., 218(70,101), *232*(70), *233*(101)
Preston, E., 398(105), 400(105), *416*(105), 433(36), 436(36), 438(36), 440(36), *448*(36)
Preston, R. J., 205(108), 207(142), 220(124), 224(22), *229*(22), *234*(108,124), *236*(142)
Price, W. H., 137(3), *149*(3)
Prieto, F., 461(157), *478*(157)
Prieur, A. M., 289(85), *309*(85)

Prince, J. E., 252(6), *260*(6)
Prunerias, M., 163(129), *187*(129)
Puck, T. T., 428(77), *450*(77), 524(75), *559*(75)
Puget, 337(54), *370*(54)
Pugh, W. E., 527(112), *561*(112)
Pulvertaft, R. J. V., 570(90), *598*(90)
Pump, K. K., 621(11), *634*(11)
Punnett, H. H., 462(98), *475*(98)
Purrot, R. J., 224(47), *230*(47)
Putnam, C. W., 30(46), *37*(46)
Pye, C. P., 467(136), *477*(136)

Quaglino, D., 290(45), *307*(45), 458(85), 460(85), *474*(85)
Quinn, L. A., 586(108), *599*(108)
Quiroz-Gutierrez, A., 394(166), 398(167), *419*(166,167)

Rabinowitz, Z., 53(87), 54(58), 65(58), 71(58), 73(87), 410(92), *415*(92), 547(67,68), *558*(67,68)
Rabson, A. S., 324(7), *332*(7)
Rackley, J. W., 289(9), 304(9), *305*(9)
Radsel-Medvescek, A., 168(130), *187*(130)
Ragbeer, M. S., 343(85), 359(85), *372*(85)
Ragni, N., 390(172), *420*(172)
Randolph, M. L., 106(30), 124(127), 125(127), *128*(30), *133*(127), 205(177), *237*(177)
Rapp, F., 158(40), 159(131), 160(123,150), 161(16), *181*(16), *182*(40), 186(123), *187*(131), *188*(150), 528(55), 530(55), *558*(55)
Rapp, H. J., 26(52,53), 27(14), *36*(14), *38*(52,53)
Rappaport, H., 453(158), 457(158), *478*(158)
Rashad, M. N., 391(168), *419*(168)
Rastrick, J. M., 458(159), 462(159), *478*(159)
Rasmussen, R. E., 209(125,126), *235*(125,126)
Raujini, R., 399(89), *415*(89)
Ravdin, R. G., 163(73), 165(73), 166(73), *184*(73)
Raychaudhuri, R., 590(58), *596*(58)

Read, J., 109(87), 126(88), *131*(87, 88), 213(163), *237*(163)
Reagan, J. W., 424(60), *449*(60)
Reagan, S. W., 424(19), *447*(19)
Recio, R. G., 400(221), *422*(221), 439(88), *450*(88)
Reddi, P. K., 549(109), *561*(109)
Reed, G., 154(80), *184*(80)
Rees, E. D., 689(39), 690(39), *692*(39)
Reeves, B. R., 295(83), 298(69), *309*(69,83), 456(130), 460(130), 463(160), 465(124), *476*(124), *477*(130), *478*(160)
Regan, J. D., 78(50), *94*(50), 210(156), *236*(156), 537(117), 542(117), *561*(117), 614(25), *617*(25)
Reisman, L. E., 289(84), *309*(84), 463(162), 467(161,162), *478*(161,162)
Reiter, M. B., 344(63), *371*(63)
Reno, D., 210(132), *235*(132)
Reno, D. L., 66(94), *73*(94)
Report of the United Nations Scientific Committee on the Effects of Atomic Radiation, 647(43), *674*(43)
Rethore, M-O., 256(16), *260*(16), 289(85), 302(66), *308*(66), *309*(85), 402(115), *417*(115)
Revell, S. H., 97(89), 100(89,90), 103(89), 104(89), 105(89,90), 106(90), 109(89), 112(89), 125(89,90), *131*(89,90), 206(127), 207(127,128, 129), *235*(127,128,129)
Reye, C., 620(44), *636*(44)
Rhim, J. S., 64(88), *73*(88)
Rhodes, K., 467(163), *478*(163)
Rich, M. A., 172(132,161), *187*(132), *189*(161)
Richards, B. M., 78(1), 87(1), *92*(1), 433(13), 442(13), *447*(12,13)
Richardson, B. J., 85(44), *94*(44)
Richardson, L. C., 121(83), *131*(83)
Richart, R. M., 240(44), *262*(44), 391(169), *419*(169), 428(61), 430(62, 63,87), 441(87), *449*(61,62,63), *450*(87)
Richert, N. J., 56(89), *73*(89)
Richter, C. B., 527(132), 531(132), *562*(132)
Richter, M., 620(39), 621(39), *635*(39)

Rick, C. M., 99(91), 100(91), 106(91), 108(91), *131*(91)
Rickers, H., 464(45), *472*(45)
Ridge, O., 461(80), 465(80), *474*(80)
Rierm, H., 554(125), *561*(125)
Rigas, D. A., 78(19), *93*(19), 461(93), *475*(93), 623(24), 624(24), 629(24), 632(24), *635*(24)
Rigby, C. C., 394(170), 398(171), *420*(170,171)
Rigo, S. J., 290(86), *309*(86)
Riley, H. P., 100(93), *131*(93)
Rimoin, D. L., 320(13), *332*(13)
Ris, H., *371*(62)
Riser, W. H., 337(31), *369*(31)
Ritzman, S. E., 274(22), *285*(22)
Roath, S., 464(132), *477*(132)
Robb, P., 620(25), *635*(25)
Roberts, F. L., 89(45), *94*(45)
Robertson, M. R., 30(49), *37*(49), 622(56), *636*(56)
Robinson, G. A., 82(46), *94*(46)
Robinson, H. L., 64(90), *73*(90)
Robinson, J. A., 136(26), *150*(26)
Robinson, W. S., 64(90), *73*(90)
Robles, I. N. H., 398(167), *419*(167)
Roidot, M., 304(43), *307*(43)
Roizman, B., *449*(64), 524(110), 550(110), *561*(110)
Rokutanda, H., 49(49), *71*(49)
Rokutanda, M., 49(49), *71*(49)
Rolevic, Z., 458(164), 468(164), *478*(164)
Rollag, M. D., 119(12), *127*(12)
Röntgen, W. C., 4(15), *20*
Rosen, W. C., 391(32), *412*(32)
Rosenbloom, F. M., 85(49), *94*(49)
Rosenblum, E. N., 576(26), 578(27), *594*(26,27)
Rosenfeld, C., 568(91,92), *598*(91,92)
Rosenthal, A. S., 569(77), 573(77), 578(77), *597*(77)
Rosenthal, I. R., 623(45), *636*(45)
Rosenthal, P. N., 64(90), *73*(90)
Rosner, F., 461(3), *470*(3)
Ross, A., *183*(53)
Ross, A. J., 78(1), 87(1), *92*(1), 442(14), *447*(14)
Ross, J. D., 462(165), *478*(165)
Ross, R., 330(44), 331(44), *334*(44)

Rossen, J. M., 63(91), 73(91)
Rossi, H. H., 325(45), 334(45)
Rothfels, K. H., 192(130), 235(130), 344(64), 371(64)
Rothschild, H., 542(111), 551(111), 561(111)
Rousseau, M.-F., 401(111,153), 416(111), 419(153)
Rovera, G., 290(82), 309(82), 458(156), 460(156), 478(156)
Rowe, W. P., 64(92), 73(92), 527(112), 561(112)
Ruiz, F., 552(50), 557(50)
Rowley, J. D., 136(30), 150(30), 293(87), 310(87), 460(167), 462(168), 463(166,168), 478(166,167), 479(168)
Ruddle, F. H., 140(4), 149(4)
Rudivic, R., 458(164), 468(164), 478(164)
Ruffie, J., 290(88,89), 294(88), 310(88,89)
Rugiati, S., 390(47,172), 395(47), 413(47), 420(172), 439(23), 447(23)
Rupp, W. D., 66(93,94), 73(93,94), 210(131,132), 235(131,132)
Rus, H., 602(22), 617(22)
Russell, L. B., 227(133,134,135), 235(133,134,135)
Russell, W. L., 227(134), 235(134)
Rust, J. H., 337(65), 371(65)
Rutishauser, A., 122(63), 130(63), 248(45), 262(45)
Ryabov, S. I., 461(169), 479(169)
Ryan, W. L., 549(65), 558(65)

Sahin, A. B., 154(133), 187(133)
Sablina, E., 159(134), 160(134), 187(134)
Sachs, L., 53(87), 54(58), 56(46,95), 65(10,15,58), 69(10,15), 71(46,58), 73(87,95), 410(92), 415(92), 523(53), 543(53), 547(67,68), 548(72), 558(53,67,68), 559(72)
Saeki, Y., 337(66), 371(66)
Sagebiel, R. W., 320(13), 332(13)
Saito, H., 399(125), 400(125), 417(125)
Saksela, E., 163(73), 165(73,105,106, 135,137), 166(73,105), 168(7,25), 169(25,136), 180(7), 181(25), 184(73), 186(105,106), 187(135,136, 137), 514(15), 517(15), 583(76), 590(93), 597(76), 598(93)
Sakurai, M., 490(13), 495(13)
Salerno, R. A., 64(96), 73(96)
Salimi, R., 400(221), 422(221), 433(36), 434(65), 436(36), 438(36), 439(34,88), 440(36), 442(34), 448(34,36), 449(65), 450(88)
Salles-Mourlan, A. M., 290(89), 310(89)
Salmon, J. H., 488(3), 494(3)
Salvi, M. L., 528(79), 559(79)
Sambrook, J., 47(97), 73(97), 528(113), 561(113)
Sandberg, A. A., 78(47), 94(47), 169(138), 174(68,138), 183(68), 187(138), 269(3), 271(3), 272(3), 273(3), 274(3), 279(3), 284(3), 289(90), 290(48,90,92), 293(91), 295(90), 308(48), 310(90,91,92), 360(67), 361(67,68), 362(67,68), 369(24), 371(67,68), 385(94,95,96,176, 222,223), 387(95,174), 390(94,95, 96,223), 398(96), 399(175), 400(223), 409(173), 416(94,95,96), 420(173,174,175,176), 422(222,223), 457(56,99,120), 459(171), 460(170), 461(56), 462(99,120,170), 463(171,172,174,175), 464(170), 466(173), 467(172), 473(56), 475(99), 476(120), 479(170,171,172, 173,174,175), 490(13), 495(13), 507(16), 517(16), 568(70), 569(70), 586(47), 587(104), 588(47), 591(94), 592(70), 595(47), 597(70), 598(94), 599(104), 638(44), 674(44), 679(21), 680(21), 681(21), 688(10), 691(10), 692(21)
Sandler, L., 602(23), 617(23)
Sandritter, W., 430(67), 449(67)
Sargentini, S., 293(93), 310(93)
Sarma, P. S., 531(114), 561(114)
Sartori, R., 390(55), 414(55)
Sasaki, M. S., 100(82), 125(82), 131(82), 212(136,138), 213(111,136), 214(136), 215(136), 220(170), 222(139), 223(139), 224(112,137), 234(111,112), 235(136,137,138,139),

237(170), 269(7), 274(21), 284(7), 285(21), 336(42,47), 343(61), 344(2,61), 352(61), 362(43,47), 363(61), 364(61), 366(43,61), 368(2), 370(42,43,45,47,61), 384(128,177), 385(128), 390(128), 399(127,129, 157), 400(129), 406(178), 417(127, 128,129), 419(157), 420(177,178), 648(45), 651(40), 660(45), 664(45), 670(41), 674(40,41,45)
Satge, P., 288(62), 293(62), 308(62)
Sauer, G., 48(50), 71(50)
Saraiva, L. G., 464(19), 471(19)
Savage, J. R. K., 104(98), 119(40), 124(94,95,96), 125(94,96,97), 128(40), 131(94,95,96,97,98), 194(60), 205(108,109), 207(142), 208(109), 217(141), 221(140), 231(60), 234(108,109), 235(140,141), 236(142)
Sawada, S., 209(143), 236(143)
Sawicki, W., 525(30), 556(30)
Sawitsky, A., 78(48), 94(48), 612(24), 617(24)
Sax, K., 96(99), 97(100), 100(100), 109(99), 118(101,102), 119(100), 122(103), 125(101), 132(99,100, 101,102,103), 203(144,145,146), 216(147), 236(144,145,146,147)
Saylors, C. L., 227(135), 235(135)
Sbernini, R., 390(47), 395(47), 413(47), 439(23), 447(23)
Scaletta, L. J., 523(115), 525(115, 148), 533(115), 543(115), 547(148), 561(115), 563(148)
Scapoli, G. L., 390(46), 413(46)
Scarpelli, D. G., 247(46), 252(46), 253(46), 258(46), 259(46), 262(46)
Scharff, M. D., 571(3), 593(3)
Scheid, W., 109(105,106), 132(105, 106)
Schenck, S. A., 30(43), 37(43)
Schendler, S., 172(139), 187(139)
Schenken, J. R., 634(9)
Schlesinger, R. W., 48(98), 74(98)
Schlom, J., 49(104,105), 74(104,105)
Schmid, W., 247(47), 262(47), 464(176), 479(176), 623(46), 624(46), 628(46), 630(46), 631(46), 636(46)
Schneider, G., 460(143), 477(143)
Schneider, J. A., 579(95), 598(95)

Schneider, U., 161(180), 190(180)
Schoefe, G. I., 524(63), 558(63)
Schoffling, K., 290(4), 305(4), 399(29), 412(29), 460(16), 466(17), 471(16,17)
Schongut, L., 621(47), 623(48), 624(48), 636(47,48)
Schrader, F., 359(36), 369(36)
Schroder, J., 137(10), 149(10)
Schroeder, F., 464(46), 473(46)
Schroeder, J., 467(178), 479(178)
Schroeder, T. M., 461(177), 479(177)
Schryver, A. de., 577(43,96), 595(43), 598(96)
Schubert, W. K., 462(194), 480(194)
Schuler, D., 168(140), 187(140), 621(47), 623(48), 624(48), 636(47,48)
Schulte-Holthausen, H., 161(180,182), 190(180,182), 576(114), 577(115), 599(114,115)
Schuster, J., 623(49), 636(49)
Schwarzacher, H. G., 248(42), 262(42), 522(105), 560(105)
Scolnick, E. M., 49(99), 74(99)
Scott, D., 105(107), 106(107), 107(107), 117(107), 125(107), 132(107), 203(148), 207(61,148), 210(62), 212(76,160), 213(149,150), 219(35), 221(61,148), 230(35), 231(61,62), 232(76), 236(148,149,150), 237(160)
Scott, I. D., 289(13), 306(13)
Scott, W., 570(88,89), 598(88,89)
Seabright, M., 221(151), 236(151)
Searle, A. G., 226(50,152,153,154), 227(67), 230(50), 231(67), 236(152,153,154)
Sedgwick, R. P., 619(7), 620(6,7), 621(6,7), 622(6), 634(5,6,7)
Sedita, B. A., 119(33), 128(33), 208(82), 218(45), 230(45), 232(82)
Seegmiller, J. E., 85(49), 94(49)
Seidel, E. H., 586(63), 587(63), 591(63), 596(63)
Seif, G. S. F., 399(179), 420(179), 466(179), 479(179)
Selezniova, T. G., 161(37), 182(37)
Seligmann, C. G., 337(69), 371(69)
Seman, G., 569(4), 593(4)
Senz, E. H., 620(27), 635(27)

AUTHOR INDEX 729

Serra, A., 293(93), *310*(93)
Serr, D. M., 391(180), *420*(180)
Setlow, R. B., 78(50), *94*(50), 209 (155), 210(156), *236*(155,156), 537 (117), 542(117), *561*(117), 614(25), *617*(25)
Sevenkayev, A. V., 212(157), 213 (157), 214(157), *236*(157)
Shadduck, R. K., 468(180), *479*(180)
Shaki, R., 391(180), *420*(180)
Sharman, G. B., 85(44), *94*(44)
Sharon, N., 546(118), 548(118), *561* (118)
Sharpe, H., 213(149,150), *236*(149, 150)
Sharpe, H. B. A., 212(158,160), 220 (159), 224(158), *236*(158,159), *237* (160)
Shaw, G. J., 157(141), *187*(141)
Shaw, M. W., 89(25), *93*(25), 108(18), 109(10), *127*(10,18), 135(9,32), 144 (5), *149*(5,9), *150*(32), 380(181), *420*(181), 494(14), *495*(14)
Shein, H. M., 163(142), 165(142,175), 166(142), *187*(142), *189*(175)
Shen, M. W., 59(64), *72*(64)
Sheppard, C. W., 106(30), *128*(30)
Sheppard, J. R., 82(51), *94*(51), 549 (22,119,120), *556*(22), *561*(119, 120)
Shevliaghyn, V. J., 163(143), *188* (143), 530(121), *561*(121)
Shigematsu, S., 398(182), *420*(182)
Shins, S., *560*(97)
Shiragai, A., 639(18), *673*(18)
Shirasu, Y., 337(70), 340(70), *371* (70)
Shively, J. A., 592(106), *599*(106)
Shohoji, T., 640(38), 641(38), 643 (38), 649(38), *674*(38)
Shuck, A., 689(39), 690(39), *692*(39)
Shullenberger, C. C., 592(106), *599* (106)
Shy, G. M., 619(58), *636*(58)
Siebner, H., 399(183), *420*(183), 462 (71), *474*(71)
Siegler, J., 623(48), 624(48), *636*(48)
Siegler, R., 172(132), *187*(132)
Siguier, F., 300(74), *309*(74), 404 (152), *419*(152)

Silagi, S., 534(122), 536(122), 543 (122), 552(122), *561*(122)
Silfversward, C., 439(32), *448*(32)
Siminovitch, L., 192(130), *235*(130), 314(53), *334*(53), 344(64), *371*(64)
Simkovic, D., 172(144), *188*(144)
Simonsson, E., 136(1), *149*(1)
Simpson-Gildemeister, V. F., 208 (110), *234*(110)
Sims, P., 65(22,61), *69*(22), *71*(61)
Simu, G., 390(159), *419*(159)
Singer, H., 376(224), *422*(224), 470 (203), *480*(203), 498(17,18), 507 (17,18,19), 513(17,18), 514(17), *517*(17,18,19)
Singh, S., 315(22), *333*(22)
Sinha, A. K., 139(31), *150*(31)
Siniscalco, M., *560*(97)
Sinks, L. F., 399(184), *420*(184), 466 (181), *479*(181)
Siracky, J., 390(185), *420*(185)
Sjogren, H. O., 164(54,55), *183*(54, 55), 544(124), 545(123), *561*(123, 124)
Skarsgard, L. D., 208(119), *234*(119)
Skinner, G. R. B., 158(145), *188*(145)
Slemmer, G., 328(34), *333*(34)
Slot, E., 384(187), 390(186,188), *420* (186,187,188)
Smalley, R. L., 399(89), *415*(89)
Smalley, R. V., 290(94), *310*(94), 460 (182), 464(183), *479*(182,183)
Smart, G. E., 399(160), *419*(160)
Smeby, B., *636*(50)
Smith, B. D., 226(100), 227(100), *233* (100)
Smith, C. H., 621(12), *634*(12)
Smith, G., 336(73), 338(73), *371*(73)
Smith, G. H., 602(27), *617*(27)
Smith, H. A., 336(74), 337(74), 338 (74), 363(74), *371*(74)
Smith, J. D., 51(101), *74*(101)
Smith, J. W., 322(46), *334*(46), 431 (66), *449*(66)
Smith, P. G., 100(14,67), 123(67), *127*(14), *130*(67), 212(24), 213(24), 220, 224(25), *229*(24,25)
Smith, P. M., 170(41), *182*(41)
Smith, R. T., 162(61), *183*(61), 586 (46), 587(46), 588(46),

589(46), *595*(46)
Smithers, D. W., 30(44), *37*(44)
Smithies, O., 25(45), *37*(45)
Smith-Keary, P. F., 62(26,102), *70* (26,102)
Sobel, J. S., 554(125), *561*(125)
Socolow, E. L., 400(189), *420*(189)
Sofuni, T., 343(46,71), 348(71), 352(46,71), *370*(46), *371*(71), 463(175), *479*(175), 591(94), *598*(94), 648(2), 664(2), 665(2), *672*(2)
Sokal, J. E., 457(56,120), 461(56), 462(120), *473*(56), *476*(120), 688(10), *691*(10)
Solitare, G. B., 620(51), *636*(51)
Somers, C. E., 160(59), *183*(59)
Somers, K., 56(70), 57(70), 58(70), *72*(70)
Sonoda, M., 364(72), *371*(72)
Sorieul, S., 522(12,13), 523(13,46), 533(13), *555*(12,13), *557*(46)
Sox, H. C., 568(22), *595*(22)
Spanedda, R., 390(46), *413*(46)
Sparkes, R. S., 293(107), *311*(107), 431(66), *449*(66)
Sparks, R., 329(21), *333*(21)
Sparks, R. S., 317(50), 322(46), *334*(46,50)
Sparrow, A. H., 108(108), 122(117, 118), *132*(108,117), *133*(118), 222(63), *231*(63)
Sparvoli, E., 257(48), *262*(48)
Speed, D. E., 293(95), *310*(95), 457(88), *475*(88)
Spencer, R. A., 523(126), *561*(126)
Spiegelman, S., 49(104,105), 51(103), *74*(103,104,105)
Spiers, A. S. D., 399(190,191,192), *420*(190), *421*(191,192), 457(10), 466(184,185), *471*(10), *479*(184, 185), 590(97), *598*(97)
Spooner, M. E., 378(193), 395(193), 398(193), *421*(193)
Sprague, C. A., 248(29), *261*(29), 550(94), *560*(94)
Sprenger, E., 441(80), *450*(80)
Spriggs, A. I., 384(195), 385(194), 387(198), 390(197,198), 395(194), 398(198), 399(179,196), 400(194), 401(197), 402(51), 406(195), *413*(51), *420*(179), *421*(194,195,196, 197,198), 425(71), 428(69,69), 429(21,70), 435(70,71,72), 436(20), 438(71), 439(20,68,71), 440(71), 441(71), 442(71), 443(71), 444(21), 445(71), *447*(20,21), *450*(68,69,70, 71,72), 466(179,186), *479*(179, 186), 487(1), *494*(1)
Srinivasan, P. R., 47(97), *73*(97), 528(113), *561*(113)
Stadler, L. J., 100(109), *132*(109), 204(161), *237*(161)
Staeffen, J., 295(6), *305*(6)
Stafford, J. L., 289(54), 308(54), 457(110), 458(110), 462(111), *476*(110,111)
Stafl, A., 439(35), *448*(35)
Stanbury, J. B., 400(189), *420*(189)
Standardization of Procedures for the Study of Glucose-6-Phosphate Dehydrogenase, 315(47), *334*(47)
Stanley, M. A., 390(41), 395(199), *413*(41), *421*(199), 433(40,41), 436(73), 437(40), 438(40), 439(41), 441(40), 444(41), *448*(40,41), *450*(73)
Stannard, R. K., 290(86), *309*(86)
Starzl, T. E., 30(41,46), *37*(41,46)
Stearns, M. W., 387(200), *421*(200)
Steel, C. M., 161(147), 162(146,147, 148), 163(147), 178(147), *188*(146, 147,148), 569(98), 586(99), 587(99), 591(99), *598*(98,99)
Steele, H. D., 430(74), *450*(74)
Steenis, H. van., 384(201), *421*(201)
Steinman, P. A., 620(27), *635*(27)
Stent, G. S., 62(106), *74*(106)
Steplewski, Z., 528(86), 529(85), *559*(85,86)
Stern, C., 59(107), *74*(107)
Stern, H., 61(60,63,108), *71*(60,63), *74*(108), 225(78), *232*(78)
Steusing, J., 222(40), *230*(40)
Stevens, D. P., 569(100), *598*(100)
Steward, D. L., 208(82), *232*(82)
Stewart, H. L., 336(75), *371*(75)
Stewart, S. E., 578(102), 589(101), *598*(101,102)
Stich, H. F., 67(23), *69*(23), 155(32, 84,151,153), 157(152,155), 158(31,

AUTHOR INDEX 731

149,153), 160(150), 161(16), 168 (154), 174(151), 178(151,155), *181* (16,31,32), *184*(84), *188*(149,150, 151,152,153,154,155), 269(6), 276 (6), *284*(6), 430(74), *450*(74)
Stich, W., 290(96), 295(96), *310*(96), 460(187), *479*(187)
Sticker, A., 336(76,77), 337(77), 338 (77), 340(77), 356(77), 363(76), *371*(76,77)
Stimson, C. W., 623(49), *636*(49)
Stjernsward, J., 573(51), *596*(51)
Stock, Ad., 136(33), *150*(33)
Stock, N. D., 532(29), *556*(29)
Stockert, E., 80(52), *94*(52)
Stoitchkovy, Y., 159(156), 160(156), 168(157), *188*(156,157)
Stoker, M., 53(109), 55(116), 56(109), 74(109), 75(116)
Stoltz, D. B., 157(155), 178(155), *188* (155)
Stonova, N. S., 161(37), *182*(37)
Straub, D. G., 401(202), *421*(202)
Strauli, P., 248(45), *262*(45)
Strauss, C., 390(60), *414*(60)
Streiff, F., 290(97), *310*(97)
Strife, A., 568(12), 591(12), *593*(12)
Strong, L. C., 138(21), 147(21), *150* (21), 320(25), *333*(25)
Stubb, E. L., 337(78), 340(78), 366 (78), *371*(78)
Stubblefield, E., 248(49), *262*(49)
Sturtevant, A. H., 314(48), *334*(48)
Suarez, H., 157(23), *181*(23)
Subak-Sharpe, J. H., 54(110,111,113), 55(110,111,112), 62(112), 74(110, 111,112,113)
Sugahara, T., 100(55,56), 106(56), *129*(55,56)
Sugiyama, T., 689(14,15,42), 690(16, 40,41), *691*(14,15,16), *692*(40), *693* (41,42)
Sumner, A. T., 136(11), *149*(11)
Summers, D. F., 529(127), *562*(127)
Surimura, T., 528(79), *559*(79)
Sutherland, E. W., 82(46), *94*(46)
Suzuki, F., 100(55), *129*(55)
Svec, J., 172(144), *188*(144)
Svoboda, J., 530(92,128), *560*(92), *562*(128)

Swanson, C. P., 106(111), 122(110), *132*(110,111)
Swift, H. H., 359(79), *371*(79)
Swift, M., 613(26), 615(26), *617*(26)
Swift, M. R., 66(114,121), 75(114, 121), *562*(133)
Swisher, S. N., 289(29), *306*(29), 457 (76,78), 460(77), *474*(76,77,78)
Sylim-Rapoport, I., 464(75), *474*(75)
Syllaba, L., 619(52), *636*(52)
Szathmary, J., 168(140), *187*(140)
Szeinberg, A., 320(17), *332*(17)
Szybalska, E. H., 524(129), *562*(129)
Szybalski, W., 524(129), *562*(129)

Tabata, T., 399(125), 400(125), *417* (125)
Tachibana, T., 53(52), *71*(52), 83(15), 88(15), *92*(15), 532(61), 534(61), 535(61), 536(61), 538(61), 539(61), 540(100), 543(61), 544(61), 545 (61), *558*(61), *560*(100)
Tachikawa, K., 649(3), *672*(3)
Tadjoedin, H. K., 619(53), 621(53), *636*(53)
Takagi, N., 385(223), 390(223), 400 (223), *422*(223), 463(175), *479* (175), 587(104), 591(94), *598*(94), *599*(104)
Takayama, S., 343(46,80,81,84), 344 (2,81,84), 345(80), 348(81,83), 350 (84), 352(46,81,84), 360(82), 365 (81), 368(2), *370*(46), *371*(80,81,82, 83), *372*(84)
Taleb, N., *636*(54)
Tanaka, E., 639(18), *673*(18)
Tanaka, T., 100(56), 106(56), *129*(56)
Tandy, R. K., 384(147), *418*(147)
Tanner, R., 289(54), *308*(54), 457 (110), 458(110), *476*(110), 462 (111), *476*(111)
Tanzer, J., 159(19,156), 160(156), 168(157), *181*(19), *188*(156,157), 464(27), *471*(27)
Tarkowski, A. K., 314(49), *334*(49)
Tassoni, E. M., *479*(188)
Tataranni, G., 384(56), *414*(56)
Taylor, C. W., 424(30), *447*(30)
Taylor, H. C., Jr., 391(25), *412* (25)

Taylor, J. H., 59(115), 75(115), 218 (162), 237(162)
Taylor, J. J., 458(147), 477(147)
Taylor, M. C., 293(1), 305(1), 378 (20), 412(20), 459(4), 470(4)
Taylor-Papadimitriou, J., 55(116), 75 (116)
Teich, N., 64(92), 73(92), 527(112), 561(112)
Teller, J. D., 303(78), 309(78)
Temin, H. M., 28(47), 37(47), 49(117, 118), 56(117), 75(117,118), 80(53, 54), 94(53,54), 526(130,131), 530 (131), 562(130,131)
Tennant, R. W., 527(132), 531(132), 562(132)
Teplitz, R. L., 401(202), 421(202)
Terasima, T., 208(165), 209(165), 211 (165), 237(165)
Ter Meulen, V., 169(158), 188(158)
Terner, J. Y., 430(87), 441(87), 450 (87)
Tevethia, S. S., 47(17), 48(17), 69(17)
Therman, E., 240(54), 241(50), 244 (50,51,59), 247(53,58), 248(53,54), 249(52,58), 250(53,54), 252(58), 254(59), 255(53,55,58), 262(50,51, 52,53,54,58,59), 442(52), 449(52)
Thiede, H. A., 462(98), 475(98)
Thoday, J. M., 100(16,112), 106(112, 113), 119(16), 127(16), 132(112, 113), 213(163), 237(163)
Thomas, B., 210(34), 229(34)
Thomas, C. A., 59(119), 75(119)
Thomas, E. D., 314(12), 327(12), 332 (12)
Thomas, M., 159(19), 181(19)
Thompson, E. B., 547(89), 559(89)
Thompson, J. N., 462(194), 480(194)
Thompson, R. I., 304(108), 311(108), 463(140), 477(140)
Thomson, A. S., 336(26), 369(26)
Thorburn, M. J., 343(85), 359(85), 372(85), 399(1), 411(1)
Thorngate, J. H., 639(1), 649(1), 660 (1), 672(1)
Thouless, M. E., 158(145), 188(145)
Thoyer, C., 290(73), 292(73), 300 (74), 309(73,74), 404(152), 419 (152)

Thron, R., 222(40), 230(40)
Thurston, O. G., 551(74), 559(74)
Tiepolo, L., 390(72,73,130), 391(74), 414(72), 415(73,74), 417(130)
Tikhonova, Z. N., 530(121), 561(121)
Till, J. E., 314(53), 334(53)
Till, M., 457(88), 475(88)
Timonen, S., 240(54,55), 244(59), 247(53,55,58), 248(54), 249(58), 250(53,54,55), 251(57), 252(55,58), 254(57,59), 255(55,58), 259(56,57), 262(53,54,55,56,57,58,59), 428(76), 450(76)
Tipton, R. E., 293(17), 306(17)
Tisdale, V., 78(19), 93(19), 461(93), 475(93), 623(24), 624(24), 629(24), 632(24), 635(24)
Titzmann, S. E., 466(100), 475(100)
Tjio, J. H., 247(60), 249(60), 263(60), 289(25,98,105), 292(14), 306(14, 25), 310(98,105), 362(86), 372(86), 399(203), 421(203), 428(77), 450 (77), 457(31,189,197), 458(196), 459(196), 460(189,197), 463(70), 468(49), 472(31), 473(49), 474(70), 480(189,196,197), 591(107), 592 (62), 596(62), 599(107), 679(44), 688(43), 693(43,44)
Todaro, G. J., 53(85), 57(120), 64(62, 122), 65(69), 66(121,123), 71(62), 72(69), 73(85), 75(120,121,122, 123), 80(21), 93(21), 148(34), 150 (34), 157(159), 189(159), 164(173), 165(1), 167(1), 180(1), 189(173), 526(70,134), 527(135), 529(150), 547(107), 558(70), 560(107), 562 (133,134,135), 563(150), 633(26), 635(26)
Toews, H. A., 390(104,204,205), 394 (204), 416(104), 421(204,205)
Tohme, S., 636(54)
Tolmach, J. L., 209(46), 230(46)
Tomiyasu, T., 642(28), 673(28)
Tomizawa, J., 66(81), 73(81)
Tomkins, G. A., 586(103), 587(103), 588(103), 591(103), 598(103)
Tomkins, G. M., 549(7), 555(7)
Tonomura, A., 222(139), 223(139), 235(139), 269(7), 284(7), 336(44, 47), 344(44), 345(87), 362(44,47),

370(44,47), *372*(87), 384(126,128, 206), 385(128), 390(126,128), 398 (126), 399(126), 400(126,128), 406 (178), *417*(126,128), *420*(178), *421* (206), 429(46,78), *448*(46), *450*(78)
Toplin, I., 159(101), *185*(101)
Topping, N. E., 590(13), *595*(13)
Torres, R. A. de., 537(80), *559*(80)
Tortajada Martinez, M., 390(71), *414* (71)
Tortora, M., 390(207), *421*(207), 434 (79), *450*(79)
Toshima, S., 587(104), 592(72), *597* (72), *599*(104)
Tough, I. M., 162(160), *189*(160), 272 (17), *285*(17), 289(99), *310*(99), 362(3), *368*(3), 456(9), 457(9,36, 190,192), 458(190,192), 461(191), 462(190), 463(12), 466(103), *471* (9,12), *472*(36), *475*(103), *480*(190, 191,192), 589(49,105), *596*(49), *599*(105), 647(46), *674*(46)
Towner, J. W., 137(35), *150*(35)
Townsend, D. E., 317(50), 322(46), *334*(46,50), 431(66), *449*(66)
Traneus, A., 439(32), *448*(32)
Traut, H., 109(105,106), *132*(105, 106)
Travnicek, M., 49(104,105), 74(104, 105)
Trebuchet, C., 300(102), *310*(102)
Trentin, J. J., 155(151), 158(43), 174 (151), 178(151), *182*(43), *188*(151)
Trippa, G., 602(23), *617*(23)
Trosko, J. E., 119(54), 122(114), *129* (54), *132*(114,115), 210(33,164), 229(33), *237*(164)
Troup, S. B., 460(77), *474*(77)
Trowell, H. R., 467(96), *475*(96)
Trujillo, J. M., 289(100), 300(101), *310*(100,101), 458(193), 467(89), *475*(89), *480*(193), 592(106), *599* (106), 633(38), *635*(38)
Tscherman-Woess, E., 244(61), *263* (61)
Tseng, P.-Y., 395(208), *421*(208), 433 (36), 436(36), 438(36), 439(34), 440(36), 442(34), *448*(34,36)
Tsirimbas, A., 290(96), 295(96), *310* (96), 460(187), *479*(187)

Tsuboi, A., 208(165), 209(165), 211 (165), *237*(165)
Tsuchida, R., 172(132,161), *187*(132), *189*(161)
Tsuchimoto, T., 661(32), 662(32), *673*(32)
Tuan, T. Q., 568(91), *598*(91)
Tuda, F., 384(23), *412*(23)
Tudela, V., 161(36), 168(36), 170 (36), *182*(36)
Turleau, C., 137(18), *149*(18), 295 (103), 298(103), 300(74,102), 304 (42), *307*(42), *309*(74), *310*(102, 103), 404(152), *419*(152)
Turpin, R., 288(62), 293(62,64), *308* (62,64), 458(127), *476*(127)
Tutie, Y., 337(88), *372*(88)

Uchino, H., 290(52), *308*(52), 661(32, 33), 662(32,33), 669(33), 671(33), *673*(32,33)
Ueda, A., 337(89), 338(89), *372*(89)
Ulrich, K., 568(70), 569(70), 592(70), *597*(70)
UN Report, 213(166), 219(166), 224 (166), *237*(166)
Urano, K., 337(92), 338(92), 356(92), 363(92), *372*(92)
Urano, Y., 401(153), *419*(153)
Usui, K., 337(15), *368*(15)
Utian, H. L., 623(55), *636*(55)
Uzman, B. G., 568(23), *594*(23)

Vaage, J., 29(48), *37*(48)
Valentova, N., 172(144), *188*(144)
Vallee, G., 390(86,87), *415*(86,87)
Vanderpool, E. A., 158(45), *182*(45)
Van Hoosier, G. L., 155(151), 174 (151), 178(151), *188*(151)
Van Steenis, H., 450(75)
Veatch, W., 108(116), *132*(116)
Veiga, A. A., 295(5), *305*(5), 465(26), *471*(26)
Veldhuisen, G., 527(136), 542(136), *562*(136)
Vennart, J., 219(14), *228*(14)
Venuat, A. M., 568(91), 569(4), *593* (4), *598*(91)
Vesco, C., 56(124), *75*(124)
Vessey, M. P., 467(136), *477*(136)

Vetter, H., 219(66), 231(66)
Vialatte, J., 288(62), 293(62), 308(62)
Vietti, T. J., 621(12), 634(12)
Vigier, P., 530(137), 562(137)
Visfeldt, J., 211(167), 237(167)
Vlahakis, G., 602(27), 617(27)
Vogt, M., 52(125), 75(125), 164(163), 189(163), 529(127), 562(127)
Vogt, P. K., 81(55), 94(55)
Volkova, M. A., 468(66,67), 473(66), 474(67)
Von Essen, C., 210(132), 235(132)
Von Haam, E., 247(46), 252(46), 253(46), 258(46), 259(46), 262(46)
von Hansemann, David., 12(11), 13, 16(11), 19
Vrba, M., 172(164,165), 179(165), 189(164,165)
Vuopio, P., 137(10), 149(10)

Wade, H., 340(90), 372(90)
Waggoner, D. E., 578(60), 596(60)
Wagh, U., 136(1), 149(1)
Wagner, D., 441(80), 450(80)
Wagner, J. E., 269(5), 284(5)
Wagner, P. C., 163(85), 178(85), 184(85)
Wakabayashi, M., 390(209), 421(209)
Wakonig, T., 122(85), 131(85)
Wakonig-Vaartaja, R., 172(166), 189(166), 395(210,212), 421(210), 422(211,212), 433(17,82), 434(17,86), 438(82), 439(84,85), 440(84), 447(17), 450(81,82,83,84,85,86)
Walbaum, R., 636(57)
Wald, N., 215(26), 229(26), 462(25), 471(25)
Walder, B. K., 30(49), 37(49), 622(56), 636(56)
Walker, D. L., 163(124), 187(124)
Walker, M., 170(167), 189(167)
Walker, S., 464(195), 480(195)
Wall, R., 47(126), 75(126), 528(138), 562(138)
Wallach, D. F. H., 25(50), 37(50)
Wang, R., 48(7), 69(7)
Warburg, Otto., 7
Warburton, D., 539(2), 555(2)
Warkaney, J., 462(194), 480(194)
Warot, P., 636(57)

Warren, L., 546(139), 562(139)
Wasastjerna, C., 293(10), 305(10)
Washbourn, J. W., 336(73), 338(73), 371(73)
Watanabe, F., 337(91,92,93,94,95), 338(92,94), 340(93,94), 341(91,94), 343(91,94,95), 344(95), 345(94), 352(95), 356(92), 363(92,93,94), 372(91,92,93,94,95)
Watkins, J. F., 47(128), 65(127), 75(127,128), 524(62,63), 525(141), 528(140,141,142,144), 529(142), 532(141), 537(26), 546(26,141), 553(26,143), 556(26), 558(62,63), 562(140,141,142,143,144)
Watson, G. E., 203(168), 212(168), 216(168), 237(168)
Watson, J. A., 215(26), 229(26)
Watson, J. D., 50(129), 75(129)
Watson, K., 49(104,105), 74(104,105)
Waubke, R., 159(168), 160(168), 189(168)
Weatherall, D. J., 464(195), 480(195)
Webb, M., 27(23), 36(23)
Weber, W., 279(27), 285(27)
Weber, W. T., 343(96), 352(96), 356(96), 359(96), 363(96), 372(96)
Wegmann, T. G., 316(51,52), 334(51,52)
Weil, R., 167(96), 185(96)
Weinfeld, H., 169(138), 174(138), 187(138)
Weinstein, A. W., 293(104), 310(104)
Weinstein, D., 165(48,169,170), 166(48), 178(48), 182(48), 189(169,170)
Weinstein, E. D., 293(104), 310(104)
Weise, W., 400(213), 422(213)
Weiss, M. C., 524(147), 525(28,145,148,149), 529(145,150), 535(45), 547(148), 552(146), 556(28), 557(45), 562(145), 563(146,147,148,149,150)
Weiss, R. A., 81(55), 94(55)
Wells, C. E., 619(58), 636(58)
Wennstrom, J., 225(169), 237(169), 293(10), 305(10)
Wepsic, H. T., 26(52,53), 27(30), 36(30), 38(52,53)
West, B. J., 226(50,153,154), 227(67),

230(50), 231(67), *236*(153,154)
Westergaard, M., 63(91), *73*(91)
Westermark, B., 487(15), *495*(15)
Westphal, H., 47(97), *73*(97), 528 (113), *561*(113)
Wever, G. H., 528(151), *563*(151)
Whang, J., 289(25,98), 292(14), *306* (14,25), *310*(98), 399(203), *421* (203), 457(31,189), 458(196), 459 (196), 460(189), 463(70), 465 (154), 468(49), *472*(31), *473*(49), *474*(70), *478*(154), *480*(189,196), 688(43), *693*(43)
Whang, J. J., 578(102), 589(101), *598* (101,102)
Whang, J. J. K., 592(62), *596*(62)
Whang-Peng, J., 26(53), *38*(53), 162 (171), *189*(171), 289(105,106), *310* (105), *311*(106), 324(5), *332*(5), 398(214), *422*(214), 457(197,198), 459(91), 460(197), 463(91,198), 466(198), *475*(91), *480*(197,198), 569(32), 587(28), 591(107), *594* (28), *595*(32), *599*(107)
Wheeler, A. H., 31(15), *36*(15)
Whisell, D., 207(77), *232*(77)
Whissel, D., 107(53), *129*(53)
White, L., 401(215), *422*(215)
Whitehouse, H. L. K., 50(130), 59 (134), 60(53,134), 61(131,134), 63 (133), *71*(53), 75(130,131,132,133), *76*(134), 526(152), *563*(152)
Whitelaw, D. M., 399(216), *422*(216)
Whitmire, C. E., 64(96), *73*(96)
Wicker, R., 157(23), *181*(23)
Wiedeman, H. R., 170(172), *189*(172)
Wiener, F., 53(135), 76(135), 248(62), *263*(62), 404(217), *422*(217), 532 (82,156), 536(82), 537(82,155), 538 (82,155), 539(82,159), 540(156), 542(154), 543(82), 545(82,155), 551(153), 552(153,154), 553(153), *559*(82), *563*(153,154,155,156)
Wiernick, P. H., 459(91), 463(91), *475* (91)
Wilbanks, G. D., 430(63,87), 441(87), *449*(63), *450*(87)
Wilbert, S. M., 577(66), *596*(66)
Wilde, C. E., 66(94), *73*(94)
Williams, E. H., 586(86), *598*(86)

Williams, R. W., 208(107), *234*(107)
Williams, T., 22(51), *38*(51)
Williamson, B., 577(83), *597*(83)
Williamson, E. R. D., 289(99), *310* (99), 457(192), 458(192), *480*(192)
Williamson, M. E., 159(63), *183*(63)
Wills, M. R., 399(121), *417*(121)
Wilson, D. B., 271(14), *285*(14)
Wilson, E. B., 4, 5, 8(15), *20*
Wilson, G. B., 122(117,118), *132* (117), *133*(118)
Wilson, M. G., 137(35), *150*(35)
Wilson, S., 402(21), *412*(21), 431(10), 434(10), *446*(10)
Wimberly, I., 586(63), 587(63), 591 (63), *596*(63)
Winkelstein, A., 220(170), *237*(170), 293(107), *311*(107), 468(180), *479* (180)
Winiwarter, H. von., 13(16), *20*
Winocour, E., 55(136), 56(46), *71*(46), *76*(136)
Wisniewski, L., *422*(218)
Witkin, E. M., 66(137), *76*(137)
Witkowski, R., 398(219,220), *422* (219,220), 464(75), *474*(75)
Wlodarski, K., 168(74), *184*(74)
Wolf, H., 161(180), *190*(180)
Wolf, I., 248(35), *261*(35)
Wolf, S. M., 569(6), *593*(6)
Wolf, U., 3, 292(8), *305*(8), 344(97), 361(97), *372*(97), 390(84), *415*(84), 456(199), 458(33), 460(143), *472* (33), *477*(143), *480*(199)
Wolfe, S. L., 122(119), *133*(119)
Wolfendale, M. R., 430(20), 439(20), *447*(20)
Wolff, S., 100(120,122,125), 119(124, 126), 120(123,124), 122(115), 123 (121), 124(121,122,123,127), 125 (45,121,122,127), *129*(45), *132* (115), *133*(120,121,122,123,124, 125,126,127), 192(173), 194(176), 205(177), 209(172), 217(173,175), 218(174), 220(171), *237*(171,172, 173,174,175,176,177)
Wolman, S. R., 164(173), *189*(173)
Wood, D. A., 29(36), *37*(36)
Woodbury, L. A., 461(95), *475*(95)
Woodcock, A. S., *424*(30), *447*(30)

AUTHOR INDEX

Woodcock, G. E., 100(67), 123(67), *130*(67), 212(24), 213(24), 219(14, 23), 220(24), *228*(14), *229*(23,24), 648(11), 663(11), *672*(11)
Woodliff, H. J., 289(13), 290(12), *306* (12,13), 457(200), 458(47), 460(47, 48), 462(201), *473*(47,48), *480*(200, 201)
Woodruff, J. D., 398(105), 400(105, 221), *416*(105), *422*(221), 439(34, 88), 442(34), *448*(34), *450*(88)
Woods, L. K., 586(108), *599*(108)
Woods, P. S., 218(162), *237*(162)
Worst, P., 53(52), *71*(52), 83(15), 88 (15), *92*(15), 532(61), 534(61), 535 (61), 536(61), 538(61), 539(61), 543(61), 544(61), 545(61), *558*(61)
Worth, A., 269(2), *284*(2), 430(18), 439(16), 440(16), *447*(16,18)
Wrba, H., 390(60), *414*(60)
Wright, B., 586(86), *598*(86)
Wright, D. A., 466(103), *475*(103)
Wright, D. H., *589*(49), *596*(49)
Wright, J. E., 89(20), *93*(20)
Wu, A. M., 314(53), *334*(53)
Wurster, D. A., *372*(98)
Wyandt, E., 136(8), *149*(8)
Wyatt, R. G., 621(12), *634*(12)
Wyke, J., 547(157), *563*(157)

Yam, L. T., 462(32), 466(202), *472* (32), *480*(202)
Yamada, K., 361(68), 362(68), *371* (68), 385(222,223), 387(174), 390 (223), 398(223), 400(223), *420* (174), *422*(222,223)
Yamada, M., 160(10), *180*(10)
Yamagiwa, K., 337(99), 338(99), *372* (99)
Yamaguchi, J., 575(109), *599*(109)
Yamamoto, K., 552(33), *557*(33)
Yamanaka, T., 522(158), *563*(158)
Yamasaki, M., 461(95), *475*(95)
Yerganian, G., 159(97), 164(174), 165 (175), *185*(97), *189*(174,175), 677 (45), 693(45)
Yohn, D. S., 67(23), *69*(23), 155(32, 84,153), 157(152,155), 158(31,149, 153), 168(154), 178(155), *181*(31, 32), *184*(84), *188*(149,152, 153,154,155)
Yoon, J. S., 153(24), *181*(24)
Yoshida, A., 24(16), *36*(16), 323(10), *332*(10), 458(58), *473*(58)
Yoshida, M. C., 53(27), 54(27), *70* (27), 533(36), 534(36), 535(36), 541(159), *557*(36), *563*(159), 648 (2), 664(2), 665(2), *672*(2)
Yoshida, T., 361(100), *372*(100)
Yosida, T. H., 255(27), *261*(27)
Young, D. E., *183*(53)
Young, R. R., 623(59), *636*(59)
Younger, J. S., 344(101), *372*(101)
Yuncken, C., 402(51), *413*(51), 487 (1), *494*(1)

Zabel, R., *422*(220)
Zajac, B., 577(19), *595*(19)
Zakharov, A. F., 394(116,117), *417* (116,117)
Zang, K. D., 137(36,37), *150*(36,37), 376(224,225), 410(225), *422*(224, 225), 470(203), *480*(203), 498(17, 18,21), 507(17,18,19,20,21), 513 (17,18), 514(17), 515(21), *517*(17, 18,19,20,21)
Zankl, H., 137(36,37), *150*(36,37), 376(225), 410(225), *422*(225), 498 (21), 507(19,20,21), 515(21), *517* (19,20,21)
Zanoio, L., 390(44), *413*(44)
Zara, C., 390(72,73,130), 391(74), *414*(72), *415*(73,74), *417*(130)
Zbar, B., 26(52,53), 27(30), *36*(30), *38*(52,53)
Zecchi, G., 316(19), *332*(19)
Zech, L., 136(1,2), 137(3), *149*(1,2,3), 289(7), *305*(7), 323(6), *332*(6)
Zeleny, V., 550(88), *559*(88)
Zeliner, J. M., 390(109), 409(109), *416*(109)
Zelli, G. P., 384(138), 385(138), 387 (138), 395(139), 400(139), *418* (138,139)
Zellweger, H., 619(60), 623(60), *636* (60)
Zelman, S., 399(89), *415*(89)
Zeve, V. H., 573(110), *599*(110)
Zeve, Y. H., 592(62), *596*(62)
Zinder, N. D., 57(13,14), *69*(13,14)

Ziprkowsky, L., 320(17), *332* (17)
Zipser, E., 210(132), *235*(132)
Zito-Bignami, R., 119(74), *130* (74)
Zittoun, R., 295(37), 302(38), *307* (37,38), 463(43), *472*(43)
Zuelzer, W. W., 289(84), 304(108), *309*(84), *311*(108), 463(140,162), 467(161,162), *477*(140), *478*(161, 162)

Zur Hausen, H., 155(176,178,179), 156(179), 159(168), 160(168), 161 (180,182), 162(177), 163(57), 167 (181), 176(179), 178(178), *183*(57), *189*(168,176,177), *190*(178,179, 180,181,182), 249(63), *263*(63), 570(40), 572(40), 576(112,114), 577(115), 578(40), 583(113), 587 (40), 590(111), *595*(40), *599*(111, 112,113,114,115)
Zurhein, G. M., 163(124), *187*(124)

SUBJECT INDEX

Aberration formation, classical theory of, 203-205
Aberrations, in biological dosimetry, 223
  chromatid-type, 198, 646
  chromosome-type, 195, 646
  "derived," 201
  distribution of, 220
  in germ cells, 225
  half-chromatid, 122
  see also Chromosome aberrations
Abnormal mitosis, in cervix uteri, 444
Acentric chromosome, 118
Actinomycin D., 541
  resistance, 554
Acute granulocytic leukemia, 662
Acute leukemia, 294, 591, 612; see also Leukemia
Acute lymphoblastic leukemia (ALL), 467, 591; see also Leukemia
Acute lymphocytic leukemia, 591
Acute myeloblastic leukemia, 591
Acute myeloid leukemia (AML), 463, 591
Adenine phosphoribosyl transferase, 524
Adenocarcinoma, gastric, 141
  mammary, 141
Adenoma, 482
  bronchus, 387
  pituitary, 483, 492
Adenoma malignum, 259
Adenovirus, 154, 155, 529
  chromosome damage by, 154
  SA7, 158
  SV 15, 158

Adenovirus (continued)
  SV 20, 158
  transformation, 154
  tumors induced by, 157
  type 2, 158
  type 4, 158
  type 7, 158
  type 12, 154, 157
  type 18, 158
  type 31, 158
Adenyl cyclase, 546
Adrenalin, 549
Adrenocortical carcinoma, 400
African lymphoma, 159
Agammaglobulinemia, see Immunodeficiency
Age, 304
  influence of, 222
  -specific incidence of cancer, 31
  lung cancer, 32
Alanosine, 524
Aleukemic leukemia, 459, 463
Alga, 305
Alkaline phosphatase, 136
  leukocyte alkaline phosphatase (LAP), 461
  LAP and mongolism, 461
Allogeneic hosts, 544
Amethopterin resistance, 554
Aminopterine, 524, 551
Amitosis, 255
Anaphase, bridge formation, 646
  lagging, 646
Anemia, Fanconi's, see Fanconi's anemia
  hereditary spherocytic, 139

Anemia, Fanconi's (continued)
  idiopathic siderocytic (ISA), 463
  megaloblastic, 615
  refractory, 463, 464
  sideroblastic, 295, 304-305
Aneuploid, 140, 587, 670
Aneuploidy, 77, 592
  in cervix uteri, 444
  in human cancer, origin of, 406
    significance of, 409, 410
  modal, see Modal aneuploidy
  sex chromosomal, 586
Animal tumors, 276
Aniridia, 138
Anoxia, 255, 259
Antibodies, complement-fixing, 578
  to EB virus, 576
Antibody diversity, origin of, 35
Antibody production, specific, 571, 572
Antigen, EB viral, 577
  expression in hybrids, 543
  FMR, 545
  H-2, 538, 543, 545
  H-2 histocompatibility, 523
  histocompatibility, 543, 544
  L cell, 545
  membrane, 573
  MLV, 543
  specific, 571
  suppression in hybrids, 543
  virus-induced, 543, 544
  weak, 545
Antimetabolites, 528
Api marker, 439
Arbovirus, 170, 549
Aryl hydrocarbon hydroxylase, 532
Ascaris, 9, 10
Ascaris megalocephala, 9
Ascites, 252, 253, 254, 256
  Ehrlich, 252, 254
  Landschütz, 253-254
Asparaginase, 303
Astrocytoma, see Glioma
Ataxia telangiectasia, 461, 619
  chromosome abnormalities, 623
  chromosome No. 14 in, 630
  clinical features, 620
  clone, 627
  cytogenetic studies on

Ataxia telangiectasia (continued)
  malignant cells, 632
  fibroblasts, 631
  immunological deficiencies, 620
  lymphocytes, 623
  marrow cells, 632
  occurrence of cancer, 621
Ataxia-telangiectasia syndrome, 614
Atom bomb survivors, 640, 648, 649, 661, 663, 669, 671
  in Adult Health Study, 649
  exposed in utero, 649
  in Life Span Study Sample (Extended), 639
Autoradiography, 530
Autosome, 141
  recessive, 619
Auxotrophs, 524
Avian sarcoma virus, 531
Axolotl, 247
8-Azaguanine, 523, 528, 541
  -resistance, 534

Banding, 139
  patterns, 136, 138, 140, 141
  techniques, 367, 583
Benign tumors, 147
Benzo (α) pyrene, 532
3,4-Benzpyrene, 252
BHK 21, cell line, 158
Biohazards of established lymphocyte lines, 579
Bladder, carcinoma, 395
  carcinoma in situ, 398
Blastic crisis (CML), 458, 460
Blastic transformation, 289, 290, 305
bl gene, 601
Bloom's syndrome, 67, 304, 461, 464, 601, 602, 632
  cancer in, 612
  chromosome breakage, 607
  clinical features, 603
  cytogenetics, 607
  gene frequency, 615
  genetics, 606
  leukemia, 612
  sex ratio, 606
  skin, 605
Bone, tumors of, 401
Bone marrow, 592, 671

SUBJECT INDEX 741

Bone marrow *(continued)*
  and leukemia, 660, 661, 662
Boveri, Theodor, 3-20
  applications of his theory, 14, 15
  biography, 4
  theses on the nature of tumors, 12
Bowen's disease of vagina, 429
Brain, *see* Tumors, nervous system
  metastases, 492
Breakage, 638, 645
  of chromosome, 601
Breakage-first hypotheses, 100
Breast, carcinoma, 390
  carcinoma in situ, 390
  fibroadenoma, 390
Bridges, 255
5-Bromodeoxyuridine, 523, 528
Bronchus, adenoma, 387
  carcinoma, 387
Burkitt cells, 573
Burkitt lines, 586, 588, 589
Burkitt lymphoma, 24, 136, 145, 162, 466, 541, 566, 570, 572, 573, 576, 577, 587
  biopsis, 572
  EB viral antibodies, 572
  electron microscopy, 572, 573, 576
  immunofluorescence, 573
  viral etiology, 572
Burkitt lymphoma cells, 573
Burkitt tumor, 162
Bursa-dependent lymphocytes (BL), 571

Cancer, 138, 140, 147, 638, 639, 644, 669
  age-specific incidence of, 31
    lung cancer, 32
  chromosomal hypothesis of, 152, 153, 304
  evolution of, 280
  fetal antigens in, 26
  genetics of, 615
  irreversibility of, 282-283
  monoclonal character of, 23
    Burkitt lymphoma, 24
    chronic myelocytic leukemia, 24
    common warts, 24
    multiple myeloma, 23
    uterine leiomyomas, 24

Cancer *(continued)*
  nonhuman canine, 335-367
  predisposition, 613
  primary, 240, 248
  single-cell origin of, 302
  transplantable, 240, 252
  unicellular origin of, 280-281
  see also Neoplasm; Tumor
Carcinogenesis, chromosomal theory of, 304
Carcinogenic factors, 324
  radiation-related tumors, 325
  virus-associated tumors, 324
Carcinogens, chemical, 25, 26, 147
Carcinoma, adrenocortical, 400
  breast, 390
  bronchus, 387
  cervical, 429
  of cervix, 161, 424
    clonal evolution in, 434
    dicentrics in, 434
    DNA measurements in, 433
    glucose-6-phosphate dehydrogenase (G6PD), 430
    history of cytogenetic studies, 428
    hypodiploids, 432
    invasive, 431
    karyotype in, 433
    marker chromosomes in, 434
    microcarcinoma, 429
    minute chromosomes, 434
    mitotic abnormalities, 428
    modal chromosome numbers in, 431
    rings in, 434
    two separate stemlines in, 434
  corpus uteri, 394-395
  epidermoid, 251
  in situ, *see* Carcinoma in situ
  invasive, 252, 258
  laryngeal, 251
  mammary, 256
  microinvasive, 434
  nasopharyngeal (NPC), 566, 572, 577
  squamous, 424
Carcinoma in situ, 252, 258, 259, 437
  breast, 390
  of cervix uteri, 429
  in culture, 430

742   SUBJECT INDEX

Carcinoma in situ (*continued*)
  definition, 424
  and dysplasia, chromosome numbers in, 437
    chromosome No. 2 in, stretched areas, 439
    clonal evolution, 439
    Feulgen microspectrophotometry in, 438
    karyotype in, 439
    marker chromosomes in, 439
    markers in, 439
    minutes in, 439
    modal regions or clusters in, 437
    multiple samples in, 439-440
    rings in, 439
  glucose-6-phosphate dehydrogenase (G6PD), 431
  vagina, 436
Cell, 665, 670
  adhesion, 549
  division, 649
  fusion, 522, 527, 530, 531, 550
    in cervix uteri, 442
    virus-induced, 524
    in vivo, 551
  germ, 638
  hybrid, Chinese hamster, 548
    rat/mouse, HL, 547
    somatic, 147
  hybridization, 526, 529, 531, 544
    spontaneous, 522, 523
    in vivo, 550
  line, 138, 155
    adult skin fibroblasts, 164
    African green monkey, CV-1, 529
    African green monkey kidney (AGMK), 528, 529
    BHK 21, 164
    bovine lung fibroblasts, 171
    Burkitt lymphoblastoid, P3J-HR-1, 530
    C-band, 139
    Chinese hamster, *see* Chinese hamster cell line
    CV-1, 528
    cytogenetics of, *see* Cytogenetics of cell lines
    established human, 159
    G-band, 139

Cell, line (*continued*)
    human amnion, 155
    human embryo lung, 160
    monkey kidney, 159
    mouse, 3T3, 529
    primary human-embryo kidney, 155
    Q-band, 138, 139
    rabbit kidney cells, 169
    Syrian hamster kidney (BHK), 529
  -mediated immunity, 571
  membrane, 546
    protein, 25
  somatic, 638
  strains, 139, 140, 141
    C-band, 140
    Q-band, 139, 140
  transformation, 547
Centrioles, 522
Centromere, 138, 646
Centrosome, 13, 15
Cervical carcinoma, 429, 470
Cervical smear, 424
Cervix, carcinoma, *see* Carcinoma, of cervix
  nomenclature of lesions, 424
Chediak-Higashi syndrome, 569
Chemical carcinogens, 25, 26, 147
Chemical oncogenesis, 148
Childhood tumors, 137, 147
Chinese hamster, 155, 157
Chinese hamster cell line, 159
  DC3F, 541, 548, 554
  DC3F/A3, 554
  DC3F/AD/Aza, 541
  DC3F/ADIV, 554
  DC3F/ADX/Aza, 548
Choriocarcinoma, 400
Chorioadenoma destruen, 400
Chromatid, 646
  breaks, 109
  -or chromosome-type breaks, 583
  gaps, 109
    evidence that gaps are not true breaks, 109
    fine structure of gaps, 109
  half-, 122
  -type aberrations, 198
Chromatin, 145
Chromosome, 176, 601, 660

Chromosome (*continued*)
  aberration, specific, 162; *see also* Chromosome aberrations
  abnormal (aberrant), 638, 645, 670
  acentric, 118
  break, 95, 532
    nonrandom, 583
  breakage, 154, 157, 159, 304, 525, 601
    breakage syndromes, 18
    in cervix uteri, 444
    specificity of, 160
  changes, 174, 176
    early after virus infection, 174
    late after virus infection, 176
  Christchurch (CH_1) chromosome, 465
  chromosome No. 2, in cervix uteri, 444
    stretched areas, in carcinoma in situ and dysplasia, 439
  chromosome No. 10, 588
  chromosome No. 14, in ataxia telangiectasia, 630
  chromosome No. 17, 295
  damage, 161, 170
    bone marrow cells, 170
    specific, 155, 157
  deletion, 516
  dicentric, 155, 156, 513
  doubling, in human tumors, 384, 402
  endoreduplication, 155
  etiological role in cancer, 361
  exchanges, chromatid, 155
  fragments, 155
  in germ cells, 639, 645
  group, 591
  group C, 516, 587
  group D, 516
  group G, 503, 514
  instability, 178, 601
  interchanges, chromatid, 156
  isochromatid breaks, 155
  lag, 302
  lesions, 583
    specific, 156, 164
  loss of, 444, 529, 532, 537, 538, 550, 553
    malignancy, 535

Chromosome (*continued*)
  Madison, 466
  marker, *see* Marker chromosome
  Melbourne, 466
  minute, in carcinoma of cervix, 434
  mutation, 147
    rate, 140
  numbers, in carcinoma in situ and dysplasia, 437
    in microcarconima, 434-435
    modal, *see* Modal chromosome number
  overcontraction of, 176
  Philadelphia, *see* Philadelphia chromosome
  pseudodiploidy, 162
  pulverization, 155
  rearrangements, 138, 140, 141, 145, 147
    C-band, 145
    see also Rearrangements
  representation, 482
  ring, 155, 513
  in somatic cells, 638, 639, 645
  strandedness, 119
  structure, 651
  supernumerary, 293
  theory of carcinogenesis, 152, 153, 304
Chromosome aberrations, 96, 639, 647, 671
  asymmetric, 646, 648
  breaks, 97
  chromatid lesions, 97
  effects of, culture time on, 645
    dose, 211, 649, 651, 660
    dose-rate, 211, 649, 651, 660
    exposure, 211
    oxygen, 213
    radiation, 211
    temperature, 213
  exchanges, 651, 670
    complete, 99
    incomplete, 99
    U, 98
    X, 98
  gaps, 97
  interchange, 99, 646
  intrachange, 99, 646
  isodiametric deletions, 99

744 SUBJECT INDEX

Chromosome aberrations (*continued*)
   radiation-induced, 639, 645, 647,
      648, 669, 670
     endogenous, 217
     endogenous, ^{3}H-thymidine, 218
     ^{226}Ra, 219
     thorotrast, 219
   residual, 648, 660
   specificity in, 166
   stable, 648, 651, 661, 662, 665, 670
   -type aberrations, 195, 646
   unstable, 647, 648, 651, 662, 665
Chronic granulocytic leukemia, 638,
     661, 671; *see also* Leukemia,
     granulocytic, chronic
Chronic lymphatic leukemia (CLL),
   464; *see also* Leukemia,
   lymphatic, chronic
Chronic lymphocytic leukemia, 591,
   592; *see also* Leukemia,
   lymphocytic, chronic
Chronic myelocytic leukemia (CML),
   see Leukemia, myelocytic,
   chronic
Chronic myelogenous leukemia (CML),
   see Leukemia, myelogenous,
   chronic
Chronic myeloid leukemia (CML), *see*
   Leukemia, myeloid, chronic
Classical theory of aberration-formation, 203-205
Clonal evolution, 144, 287, 288, 289,
   290, 298, 301, 302, 303
   in AML, 463
   in carcinoma, of cervix, 434
   in carcinoma in situ, and dysplasia,
     439
   in CML, 458, 459
Clonal growth, leukemias, 270
   and neoplasia, 330
   non-neoplastic, 271
   sequential development, 274
   solid tumors, 269
   temporal correlation, 268
Clonal origin, 14, 568
Clonal populations, 590
Clonal selection, 167
Clonal sublines, 581
Clone, 147, 148, 157, 162, 165, 178,
     305, 577, 581, 584, 590, 627,

Clone (*continued*)
     639, 663, 670, 671
   in bone marrow, 669
   in human tumors, 381, 406
   in lymphocyte, 663
   in microcarconima, 435
   time of appearance, 268
Cloning, 523, 533
Co-cultivation, 587
C marker, 588
C-mitosis, 174, 244, 247, 255, 258
Colchicine, 240, 244, 247, 248, 257
Color blindness, 14
Complementation, 543
Complement-fixation, 570
Complement-fixing antibodies, 578
Concanavalin-A, 548
Conditional lethals, 523
Consanguinity, 15
Constitutive heterochromatin, *see*
   Heterochromatin, constitutive
Constrictions, secondary, *see* Secondary constrictions
Contact inhibition, 140, 527, 546,
   547, 549
Corpus uteri, carcinoma, 394-395
Crossing-over, 58, 59, 61, 67
   somatic, 88
Croton oil, 252
C-type virus, 168, 171
   RNA virus, 170, 527, 530
Culture, conditions, 569, 589
   time in, 589, 590
Cyclic AMP, 546, 548
Cytogen, of LCL's, diploid karyotype,
     581
   hypodiploid, 581
   inversion, 581
   polyploidy, 581
Cytogenetic analysis, 585
   normal lymphocytic cell lines,
     580
Cytogenetics, of cell lines, abnormal
     donors, 583
   of leukemia, general, 454, 455
   of lymphocyte cell lines, karyotype
     of donors, 580
   normal donors, 580
Cytomegalovirus, 158
Cytosine arabinoside, 554

Daunomycin, 542
Death certificate, 639
Defective virus, 529, 531
Degree of ploidy, 589
Delayed fertilization, 16
Deletion, 136, 137, 138, 147, 503, 509
  terminal, 646, 651, 661, 669
3'-Deoxyadenosine, 257
Dermatomyositis, 31
2,6-Diaminopurine, 524
Dicentric chromosome, 155, 156, 513
Dicentrics, 645, 646, 647, 648, 651, 660, 661, 670
  in carcinoma of cervix, 434
Differentiation, 552
  suppression, 552
Di Guglielmo's syndrome, 462, 464
Dihydroxyphenylalanine oxidase, 552
7,12-Dimethylbenz(α)-anthracene (DMBA), 689
  chromosome breakage, 690
  isozyme studies, 690
  karyotypic pattern, 689
  leukemias, 689
  nonrandom chromosome variation, 689
  sarcomas, 689
Dimethylnitrosamine, 547
Diplochromosome, 244, 247
  in cervix uteri, 442
Diploid, 139, 140, 141, 587, 670
  in carcinoma in situ and dysplasia, 438
  in human tumors, significance of, 410
  lines, 586
  in mild dysplasia of cervix, 441
Diploidization, 550
Diploidy, 591
  to hyperdiploidy, 586
Disorder, hematological, 137, 583, 661
  myeloproliferative, 671
Divisions, multipolar, 241, 247, 248, 249, 250, 251, 252, 253, 254, 255, 256, 257, 258
  quadripolar, 248
  tripolar, 247, 248, 250
DMBA, see 7,12-Dimethylbenz(α)-

DMBA (*continued*)
  anthracene (DMBA)
DNA, content, 359
  -DNA hybridization, 576
  EB viral, 576, 577
  estimations, in mild dysplasia of cervix, 441
  measurements in carcinoma of cervix, 433
  polymerase, 526, 615
  repair, 66, 67, 527, 542
  strand, 638
  strand breakage, 208
  subfractions of host cell, 176
  synthesis, 55, 56, 57, 58, 61, 63, 68, 571, 646
  value, modal, in ovarian carcinoma, 391
    prognostic significance of, in human cancer, 407
D 98/AG line, 139
Dominant lethal effects, 227
Dose, aberration relation, 638
  absorbed, 647, 669
  estimate, 639, 640
  response, 640
  T-57, 640
  T-65, 640
Dosimetry, 639
  air dose, 639
  biological, 645, 646
  gamma component, 639
  neutron component, 639
Down's syndrome, 66, 136, 222, 461, 584
  trisomy *21*, 136
Dragonfly, 241
Drosophila, 17, 602
Drug resistance, 554
Duplication, 147, 290
Dysplasia, 252, 259, 431, 437
  definition, 424
  DNA estimations in, 441
  mild, 440
  modal chromosome numbers in, 441
  polyploid in, 441
  regression in, 442
Dysproteinemias, see Immunodeficiency syndromes

SUBJECT INDEX

EB viral antigen, 577
EB viral DNA, 576, 577
EB viral genome equivalents, 576
EB virus, see Epstein-Barr (EB) virus
Ectopic proteins, 31
Ehrlich ascites tumor (EAT), 553
Ehrlich ascites tumor cells, 534, 536, 543, 551
Endoanaphase, 244
Endometaphase, 244
Endometrium, 249, 250
  hyperplasia of, 406
Endomitosis, 244, 253, 254, 255, 257, 258, 442
  in cervix uteri, 442
Endonuclease, 63, 66
Endopolyploidy, 241, 241n, 244
Endoreduplication, 241, 244, 254, 258
  in cervix uteri, 442
  partial, 256
  selective, 301
  trisomy, 256
Endotelophase, 244
Eosinophilic leukemia, 462
Ependymoma, 483, 489, 528
Epigenetic change, 85
Epstein-Barr (EB) virus, 159, 530, 565, 570, 571, 572, 578, 587, 588
  antibodies to, 578
  co-carcinogen, 578
  immune response to, 577
  -negative, 570
  passenger virus, 578
  -positive, 570, 587, 588
Erythremic myelosis, 464
Erythroleukemia (EL), 459, 464
Esophagus, carcinoma, 400
Establishment, 571, 572
Euchromatin, 148
Evolution, 305
  of cancer, 280
  clonal, see Clonal evolution
  of species, 288, 305
  stemline, 178
Exchange, 646
  hypotheses, 100, 206
  see also Chromosome aberrations

Fallopian tube, 251

Fallopian tube (*continued*)
  carcinoma, 400
Fall-out, Japanese fishermen, 648, 661
  Marshallese, 648
Fanconi's anemia, 66, 67, 157, 304, 461, 464, 527, 614, 615, 632
Feeder layer, 568
Feline leukemia virus, 531
Fertilization, delayed, 16
Fetal antigens, in cancer, 26
  in tumors, 29
Fetal leukocytes, 570
Feulgen, 259
Feulgen microspectrophotometry, in carcinoma in situ, and dysplasia, 438
Fibroblast, 139, 140
  feeder layers, 570
Fibrosarcoma, 141
2-Fluoroadenine, 524
Folate reductase, 554
FMR antigen, 545
Follicular lymphoma, 466
Fragment, acentric, 645, 651

Gammopathies, see Immunodeficiency syndromes
Gangliosides, 546
Gastric adenocarcinoma, 141
G-band, 136, 137
bl Gene, 601
Genome, 141, 147
  equivalents, 576
  EB viral, 576
Genomic integration, of oncogenic viruses, 79
Glioblastoma, 484, 488
Glioma, 470, 483, 484, 486, 488
  optimal, 491
Glucocorticoids, 549
Glucose-6-phosphate dehydrogenase (G6PD), 24, 430
β-Glucuronidase, 523
Glycoprotein, 546, 548
Group chromosomes, see Chromosome
Group C marker, see Marker, group C
3-Group metaphases, 428, 444
Growth retardation, 601

Half-chromatid, 122

SUBJECT INDEX 747

Half-chromatid (*continued*)
  aberrations, 122
Hamster cell lines, BHK, 548
H-2 antigens, 538, 543, 545
HeLa, 138, 250, 253
HeLa cell, 530
Helper leukemia virus, 531
Hepatitis, infectious, 139, 569
  viral, 569
  virus, 170
Hepatoma, rat, 549
HE_p-2, 139
Hereditary spherocytic anemia, 139
Herpes simplex virus, 583
Herpes-type, 577, 578
Herpesvirus, 158, 573
  cytomegalovirus, 158, 161
  Epstein-Barr (EB) virus, 159, 161
  herpes simplex, 158
  herpes simplex virus, type 1, 158, 159
    type 2, 158, 160
  herpes zoster virus, 161
  Lucké virus, 159
  Marek's disease, 159
    virus, 161
  saimiri, 159
  zoster, 159
Heterochromatin, 148, 256
  breaks in, 108
  constitutive, 581
  C-band, 145
Heterochromatization, 256
Heterokaryons, 524, 528
  human/Chinese hamster, 542
Heterophile-positive infectious mononuclesis, 577
Heteroploid, 670
  lines, 584
Heteroploidy, 586, 589
Heteropycnotic, 256
Heterotransplantation, 579, 580
  antilymphocyte serum (ALS), 579
  of LCLs, 579
  neonatal hamsters, 579
H-2 Histocompatibility antigen, 523
Histocompatibility antigens, 543, 544
Hodgkin's disease, 137, 399, 466, 578, 590
Hodgkin's lymphoma, 572, 579

Homograft rejection, 27
Homologs, 141
Homoploid, 584
^{3}H-thymidine, 571
Human cell lines, Fl amnion, 549
  HeLa, 549
  HE_p-2, 549
  W98 VaE, 528
Human immunoglobulins, 579
Human ovarian teratoma, 329
Hybrids, antigenic expression in, 543
  antigen suppression in, 543
  Chinese hamster, 541, 554
  differentiated/undifferentiated, 552
  Ehrlich/A9, 536, 544
  Ehrlich/CBA, 537
  Ehrlich/L, 540
  human/mouse, 531
  HyEN, 538
  LM/3T3, 547
  man/mouse, 525, 541
  melanoma/A9, 534, 536
  mouse, 523
  mouse/mouse, 525
  mouse/rat, 524
  rat, 549
  selective systems, 523
  SEWA/A9, 535
  SEWA/CBA, 537, 545
  SEWA/CBA/T6T6, 535
  Syrian hamster/mouse, PYY/EAT, 553
  TA3/ACA, 537
  teratoma/A9, 535
  tumor/A9, 538
  tumor/tumor, 539
    MBA/YACIR, 540
    MSWBS/YACIR, 540
  YACIR-A9, 531, 544
  YACIR/MSWBS, 535
Hybrid cell, 530
  selection of, 523, 536
  YAC-A9, 531
Hybrid cell lines, 525
Hybrid mouse, M cells, 533, 535, 540
Hydatidiform mole, 399
Hyperchromasia, in cervix uteri, 442
Hyperdiploidy, 141, 591, 592
  diploidy to, 586
Hyperplasia, adenomatous, 259

SUBJECT INDEX

Hypertriploid, 660
Hypocenter, 648, 661
Hypodiploid, 137, 141
Hypogammaglobulinemia, 584
Hypotetraploid, 137, 141, 586
Hypotetraploidy, origin of, in cervix uteri, 443
Hypoxanthineguanine phosphoribosyl transferase, 523

Idiopathic siderocytic anemia (ISA), 463
Immune response, to EB virus, 577
Immunity, 42, 43, 45, 46, 47, 49, 50, 51, 52, 55, 68, 338
  cell-mediated, 571
Immunization, tumor, 554
Immunoblastomata, 579
Immunodeficiency syndromes, 453, 465, 571
  Bruton's type, 465
  Swiss type, 465
  Wiscott-Aldrich syndrome, 465
Immunofluorescence, 530, 531, 571, 573, 588
Immunoglobulin, 23
  human, 579
  production, 571, 572, 579
Immunological deficiencies, 620
Immunological surveillance, 30, 623
Immunologic deficiency, primary, 569
Immuno-oncogenesis, 148
Immunopathy, 603
Immunosuppression, 30
Immunosuppressive therapy, tumors in patients on, 30
Inborn metabolic errors, 569, 571
Incontinentia pigmenti, 304
Infectious hepatitis, 139, 569
Infectious mononucleosis, *see* Mononucleosis, infectious
Infectious mononucleosis lines, 569
Influenza virus, 169
Inosinic acid pyrophosphorylase, 523, 551
Integration, of virus genome, 171
Intercellular connections, 546, 549
Interchanges, 196
Intersexuality, 304
Intrachanges, 195

Invasive carcinoma, of cervix, 431
Inversion, 147, 669
  paracentric, 646
  pericentric, 646, 651
5-Iododeoxyuridine, 528, 530
Ionizing radiation, 613
Irreversibility, of cancer, 282-283
Isochromosome, 137, 295, 460, 466
Isoproteronol, 549

Japan, 337
Jews, 606
Jijoye cell line, 570, 587
Junctional coupling, 549

Kaiser Wilhelm Institute of Biology, 5
Karyotype, 140, 141, 144, 165, 569, 638, 663
  in carcinoma of cervix, 433
  in carcinoma in situ and dysplasia, 439
  instability of, 157, 162, 165, 172
  in microcarconima, 435
  Q-band, 141
Karyotype evolution, 363, 498, 507, 586, 590, 591
Karyotypic changes, 583
Kinetochores, 148

Laggards, 255
Large intestine, carcinoma, 385
  polyp, 387
L cell antigen, 545
Lectins, 546, 548
Letterer-Siwe's disease, 579
Leukemia, 16, 137, 287, 527, 569, 572, 590, 612, 639, 640, 644, 661, 662
  acute, 294, 591, 612
  aleukemic, 459, 463
  cytogenetics of, general, 454, 455
  diploid karyotype, 590
  granulocytic, acute, 662
    chronic, 638, 661, 671
  lymphatic, 579
    chronic, 464
  lymphoblastic, 591
    acute, 467, 591
  lymphocytic, 590, 591
    acute, 591

SUBJECT INDEX

Leukemia (*continued*)
  chronic, 591, 592
  monocytic, 590
  murine, *see* Murine leukemia
  myeloblastic, 590, 591
    acute, 591
  myelocytic, chronic (CML), 136, 137, 145, 287, 301, 592
  myelogenous, 287
    chronic, 290, 305, 323
  myeloid, acute, 463, 591
    chronic, 136, 137, 145, 240, 287, 456, 592
      associated conditions, 462
      blastic crisis, 456
    preleukemic, 661
Leukemia lines, 591
Leukemic mononucleosis lines, 569
Leukemoid reaction, 462
Leukocyte, peripheral blood, 645
Leukoplakia, 259
Leukoproliferative disease, 573
Locus, 157
  thymidine kinase, 157
Loss of chromosomes, *see* Chromosome, loss of
Louis-Bar syndrome, 614, 619; *see also* Ataxia telangiectasia
Lucké virus, 159
Lung cancer, 32
Lymphatic leukemia, 579
Lymph node, hyperplastic of, 406
Lymph node biopsy, 592
Lymphoblastic leukemia, 591
Lymphoblastic sarcoma, 137
Lymphoblastoid line, 139
Lymphoblasts, 568, 572
Lymphocyte, 645, 647, 661, 670
  in culture, 647
  peripheral blood, 648, 669
  precursor, 665
Lymphocyte lines, established, 565
Lymphocytic leukemia, 590, 591; *see also* Leukemia, lymphocytic
Lymphoid tumors, 137
Lymphoma, 399, 465, 569, 579
  African, 159
  malignant, 466
    other than Burkitt, 590
  moloney, 531

Lymphoma (*continued*)
  mouse, *see* Mouse lymphoma
Lymphoproliferative diseases, 454
Lymphosarcoma, 137, 579
Lyon phenomenon, 23
Lysate method, 571
Lysogeny, 528
Lysolecithin, 525
Lysosomal enzymes, 174
Lysosomes, 532
Lytic cycle, 173

Macrophages in LCLs, 568
Madison chromosome, 466
Malignancy, 140, 141, 147, 148, 638, 642, 671
  a dominant character, 533
  a graded character, 534
  as recessive trait, 83, 536
  suppression of, 536
    cell hybridization, 532
    in hybrid cells, 534
Malignant lymphoma, 137, 466
  other than Burkitt, 590
Malignant melanoma, 141, 398
Malignant transformation, 147, 670, 671
Malignant tumor, 140
Mammary adenocarcinoma, 141
Marek's disease, 159, 161
Marker chromosome, 137, 139, 141, 144, 145, 162, 164, 171, 295, 300, 348, 482, 497, 499, 513, 514, 586, 587, 589, 590, 662, 670
  in carcinoma of cervix, 434
  in carcinoma in situ and dysplasia, 439
  C-band, 141
  in cervix uteri, 444
  Group C, 587, 588, 589
  in human tumors, 381, 382, 402, 404-405
  in ovarian carcinoma, 390
  in microcarconima, 435
  Q-band, 141
Mastomys, 155
Measles, 168
Measles virus vaccine, 168

SUBJECT INDEX

Medium 1640 of the Roswell Park Memorial Institute (RPMI 1640), 569
Medulloblastoma, 483, 486, 490
Megaloblastic anemia, 615
Melanoma malignant, 141, 398
Melbourne chromosome, 466
Membrane antigen, 573
Membrane receptors, 547, 548
Mendelian laws, 11
Meningioma, 137, 145, 469, 470, 481, 483, 497, 498, 503, 507, 509, 513
  chromosome numbers, 497
  stemline karyotypes, 497
Metaphase, 646, 649
  3-group, 428, 444
Metaphase/prophase ratio (M/P ratio), 240
Metastases, 483
Metastatic tumors, 325
Methylcholanthrene, 540
3-Methylcholanthrene, 532
α-Methyl-mannopyranoside, 548
Mice, 252
Microcarcinoma, 429, 434
  chromosome numbers in, 434-435
  clone in, 435
  karyotype in, 435
  marker in, 435
  modal numbers in, 434-435
  ring marker in, 435-436
Microchromosome, 171
Microinvasive carcinoma, 434
Micronuclei, 247, 255, 256
Microspectrophotometric data on Feulgen-stained, 428
Mild dysplasia, see Dysplasia, mild
Minute chromosomes, in carcinoma of cervix, 434
Minutes, in carcinoma in situ and dysplasia, 439
Misrepair, 207
Mitomycin C, 528
Mitosis, 148, 645, 646
  abnormal, in cervix uteri, 444
  in cervix uteri, 424
  irregular, 550
  multipolar, 10, 11, 12, 15, 249, 253, 254, 522

Mitosis (continued)
  quantal, 62
  tetrapolar, 550
  tripolar, 259
Mitotic cycle, 241, 244
Mitotic index, 250, 253
MLV antigen, YACIR, 543
Modal aneuploidy, 463, 467
  in ALL, 463, 467
  in AML, 463, 467
Modal chromosome number, 137, 141
  in carcinoma of cervix, 431
  of human tumors, relation to modal DNA values, 379, 381-382
  hyperdiploid, 137
  hypodiploid, 137
  in mild dysplasia of cervix, 441
Modal numbers, in microcarconima, 434-435
Modal regions or clusters, in carcinoma in situ and dysplasia, 437
Molluscum contagiosum, 167
Moloney lymphoma, 531
Moloney virus, 531
Mongolism, see Down's syndrome
Monoclonal character of cancer, see Cancer, monoclonal character of
Monoclonal origin, 18
Monocytic leukemia, 590
Mononucleosis, infectious, 162, 566, 569, 572, 579, 581, 586
  heterophile-positive, 577
Monosomy, 137, 139, 506
Monosomy G, 504, 509
Mortality, 639
  cancer, 669
  leukemia, 669
  ratio, 642, 644
Mosaic, 569, 586
Mosaic donors, 585
Mosaicism, 581
Mosaic systems, 313
Mouse, 247, 248, 252, 254, 256, 259
  YACIR, 543
Mouse ascites, 254, 257
Mouse cell lines, A9, 531, 534, 535, 536, 543, 553
  A9(IMP$^-$), 524
  APRT$^-$, 524

SUBJECT INDEX

Mouse cell lines, A9, (continued)
  BALB/C 3T3, LM(TK⁻) clone 1D, 547
  B 82(TK⁻), 524
  L, 523, 536, 540, 548, 549
  LAG, 534
  LM(TK)-Cl-1-D, 541
  NCTC 2472 (or N1), 533
  NCTC 2472 line, 538
  NCTC 2555 (or N2), 533
  N 1, 535
  N1=NCTC 2472, 522
  N 2, 535
  N2-NCTC 2555, 522
  NCTC 2455 (N2), 523
  3T3, 529, 548, 552
Mouse L 5178Y lymphoblasts, 551
Mouse lines, A9, 545
Mouse lymphoma, YAC, 531, 543
  YACIR, 531, 535, 553
Mouse lymphosarcoma, 6C3HED, 255
Mouse melanoma cells, 534
Mouse sarcoma, MSWBS, 532, 535
  TA3/Ha, 544
  TA3/St, 544
Mouse sarcoma-180, 551
Mouse strains, A, 538
  ACA, 537, 544
  A/Sn, 538
  A.SW, 535
  BALB, 538
  BALB/c, 527
  CBA, 533, 535, 537, 551, 543, 545, 553
  CBA T6T6, 537
  C3H, 543, 551
  C3H/He, 523
  C57 BL, 543
  L, 543
  129, 535
  Swiss, 523, 543, 553
Mouse teratoma cells, 552
Mouse tumor lines, A9HT, 539, 551
  B82HT, 551
  Ehrlich, 540
  MBA, 540
  MSWBS, 538
  SEWA, 538, 545
  TA3, 537
  YAC, 538

Mouse tumor lines (continued)
  YACIR, 538
M/P ratio, 241, 248, 250, 251, 252, 253, 254, 255, 256, 257, 259
Multiple samples, in carcinoma in situ and dysplasia, 439-440
Multipolar mitosis, see Mitosis, multipolar
Multipolar spindles, origin, 13
Mumps, 169
Murine leukemia, 172
  Friend virus, 172
  moloney virus, 172
  Rauscher virus, 172
Murine leukemia virus (MLV), 543
Murine sarcoma virus, 172, 531
Mutations, 46, 54, 64, 65, 66, 67, 68, 138, 147, 614, 638, 670
  chromosomal, 147
  somatic, see Somatic mutation
Myeloblastic leukemia, 590, 591
Myelofibrosis, 569
Myeloma, 399
Myeloproliferative diseases, general, 452
  non-leukemic, 453
  transitions, 453
Myelosclerosis, 462
Myelosis, erythremic, 464
Myxoviruses, 169

Nasopharyngeal carcinoma(NPC), 566, 572, 577
Necrosis, 255, 259
Neoplasm, 638, 670, 671
  benign, 644
  malignant, 638
  see also Cancer; Tumor
Neoplastic disease, 583
  of lymphoid tissue, 571
Nephroblastoma, 400
Nervous system, see Tumors, nervous system
Neuroblastomas, 483, 486, 490, 552
Neurinoma, 482, 483, 491
Nondisjunction, 141, 258, 301, 302
Nonrandom chromosome breaks, 583
Nuclear membrane, 148
Nucleolar organizers, 148
Nucleus, 148

SUBJECT INDEX

Nullisomy, 141

O., method of, 10
Oligodendroglioma, 483, 488
Oncogene hypothesis, 526
Oncogenesis, 147
　chemical, 148
　viral, 148, 526
Oncogenicity of lines, 579
Oncogenic DNA viruses, 566
　transformed cells, 566
　viral transformants, 566
Oncogenic viruses, effects on DNA, 29
Oncornaviruses, 170
Onion, 257
Ovary, benign teratoma, 391
　carcinoma, 390
　cystadenoma, 391

Papilloma, 163
Papilloma virus, 163
Papovaviruses, 163
Paracentrotus, 9
Parainfluenza virus, 524
Paramyxoviruses, 168
Paraproteinemia, see Immunodeficiency
PHA, see Phytohemagglutinin (PHA)
PH¹, see Philadelphia chromosome
Philadelphia chromosome (PH¹), 17, 136, 137, 289, 295, 301, 456, 592, 638, 661, 671
　common stemcell, 458, 459
　duplication of, 290
　early appearance, 457
　identity, structure, 457
　occurrence, 457
　origin, 460
　Ph¹, LAP, and mongolism, 461
　properties of positive cells, 461
　stability of positive cells, 458
　two Ph¹ chromosome, 458
　variant Karyotypes, 459
　Y chromosome, loss, 459
Phosphodiesterase, 549
Phytohemagglutinin (PHA), 155, 465, 568, 569, 591, 645, 661, 662
Picornaviruses, 170
Pituitary adenomas, 483, 492
Plaque assay, 527

Ploidy, degree of, 589
　and genetic regulatory systems, 90
Podophyllin, 442
Poland, 607
Poliovirus, 170
Polycythemia vera, 295, 305, 459, 462, 463
Polymorphs, leukemic, 461
　association, 461
　colonies in agar, 461
　motility, 461
　segmentation, 461
Polyoma-transformed BHK-21 cells, 547
　mouse cell line, 543, 548
　N2/Pyl98, 543
　virus-transformed cells, 533
　Py-198-1, 523
Polyoma-transformed Syrian hamster cells (PYY), 553
Polyoma transplantation antigen, 535, 538
Polyoma virus-induced transplantation antigen, 544, 545
Polyomavirus, 163, 164, 529, 535, 547
Polyploid, 244, 247, 248, 254, 586
　in mild dysplasia of cervix, 441
Polyploidization, 254, 300
Polyploidy, 241, 583
　in cervix uteri, 442
Poxvirus, 167
　molluscum contagiosum, 167
　Shope fibroma virus, 167
　vaccinia virus, 168
　Yaba virus, 167
Precancerous lesions, of cervix, 424
Predisposition, to malignancy, 15, 18
　to nondisjunction, 15
Primary immunologic deficiency, 569
Primary tumors, 688
　chromosome pattern and etiologic agent, 690
　DMBA-induced leukemias, 689
　DMBA-induced sarcomas, 689
　karyotypic evolution, 690
　RSV-SR-induced sarcoma, 689
　virus-induced, 18
Primordial dwarfism, 605

Progressive multifocal leukoencephalopathy, 163
Prostaglandin E_1, 549
Provirus hypothesis, 526
Pseudodiploid, 139, 587, 591
Pseudodiploidy, 16, 18
Pseudotetraploid, 586, 592
Pulverization, 168, 169, 171, 174
　fusion of asynchronous cells, 169

Q-band, 136, 137
Quadriradial configuration (Qr), 607, 611
Quantal mitosis, 62
Quinacrine fluorescence, 136, 139
Quinacrine mustard, 140

Rabies virus, 169
Radiation, 147
　ionizing, 613
Radiation effect, 639
　acute symptom, 665
　carcinogenic, 638
　late, 639
Radiation exposure, accidental, 647, 648, 660
　atom bomb, 639, 662
　clinical, 647
　fall out, 648, 661
　nuclear-explosion, 647, 660
　occupational, 647
　X-ray, 647, 660
　whole body (or total body), 648, 660, 671
Raji line, 570, 576
Rat, 253
Rat hepatoma line, HTC, 547
Rearrangements, 144, 145, 645, 651, 661, 662, 663, 670
　chromosomal, 138, 140, 141, 145, 147
　C-band, 145
Receptor site, 548
Reciprocal translocation, 136, 138
Recombination, 57, 58, 59, 61, 63, 67, 68
　repair, 210
Recurrent tumors, 326
Refractory anemia, 463, 464
Regression, of cervical lesions, 427-428

Regression, of cervical lesions (continued)
　in mild dysplasia of cervix, 442
Regulatory switch, 81
Repair enzymes, 176
Repair replication, 209
Replicons, 59, 61, 62, 63, 68
Restitution, 247, 247n, 258
Reticulosarcoma, 466
Retinoblastoma, 33, 137, 138, 145, 147, 491
Re-union, 638
Reverse transcriptase, 49, 56, 526
Reversibility, in mild dysplasia of cervix, 442
Rhabdomyosarcoma, 401
Ring chromosome, 155, 513
Ring marker, in microcarcinoma, 435-436
Rings, 645, 646, 647, 648, 651, 660
　in carcinoma of cervix, 434
　in carcinoma in situ and dysplasia, 439
RNA-dependent DNA polymerase, 171
RNA virus, 168
　C-type, 170, 527, 530
Röntgen rays, 15
Rous sarcoma virus (RSV), 171, 530, 675
　carcinogenesis, 676
　different tumor regions, 684
　histopathology, 687
　isozyme studies, 690
　karyotypic evolution, 676
　latent period and age of tumor, 684
　in mammialian cell systems, 171
　metastatic sarcomas, 685
　nonrandom chromosome variation, 689
　primary sarcomas, 676
　sequential karyotypic changes, 683
　transplanted sarcomas, 683
RSV, see Rous sarcoma virus (RSV)
RSV-transformed hamster cell line, 530
Rubella virus, 170

Sarcoma, 553
　embryonic, 401

Sarcoma (*continued*)
  venereal, 336
Scleroderma, 304
Scrotum, 338
Sea urchin, 9
Sea urchin egg, 10, 11, 12
  method of O., 10
Secondary constriction, 160, 165, 590
  accentuation of, 160
  in cervix uteri, 444
  enhancement, 165
Segregation, 241
  chromatid, 258
  somatic, 248
Selection, 581
Selective advantage, 587
Selective endoreduplication, 301
Sendai virus, 169, 248, 525, 530, 550
Senescence, 139
Senile keratosis, 429
Sex chromatin, in human tumors, 407
Sex chromosomal aneuploidy, 586
Sex chromosome anomalies, 569
Shope fibroma virus, 167
Shope papilloma virus, 163
Sialyl transferase, 546
Sideline, 482, 499
Sideroblastic anemia, 295, 304-305
Simian vacuolating virus 40 (SV40), 163, 164, 527, 528, 529
Single-cell origin, of cancer, 302
Site hypothesis, 221
Site-limitation hypothesis, 123
Solid tumor, 140
Somatic cell hybrid, 147
Somatic crossing-over, 88
Somatic mutation, 21, 29, 34
  increased vulnerability, 22
  rate of, 35
  suspicious nature of, 85
  tumor production, 22
Somatic segregation, 89
Species, 305
  evolution of, 288, 305
Specific antibody production, 571, 572
Specific antigen, 571
Spectrophotometry, ultraviolet, 428
Spemann, Hans, 5

Spermatogonial cells, translocations induced in, 225
Spinach, 244
Stemlines, 140, 157, 164, 171, 240, 241, 254, 258, 288, 482, 499, 503, 508, 513
Stemline cells, 348, 638
Stemline concept, 361
Stemline evolution, 178
Stemline karyotype, 365, 497
Stemline progression, 171
Sternberg-Reed cells, 466
Sticker's tumor, 336
Sticky (*st*) gene, 602
Stomach, carcinoma, 385
  polyp, 384
Sub-chromatid-type aberrations, 200
Submetacentric, 139
Subtelocentric, 139
Subterminal secondary constriction, 587
Superficial dyskaryosis, 441
Supernumerary chromosome, 293
Suppression of malignancy, *see* Malignancy, suppression of
SV 40, *see* Simian vacuolating virus 40
SV 40 (PARA), 167
SV40-transformed hamster cells, 551
SV40-transformed human cell, 583
SV40-transformed human cell line, WI-18-VA-2, 529
SV40 virus, 148, 528, 633
Syncytia, 169, 248
Syngeneic mice, 544
Syrian hamster, 155, 157

T antigen, 167, 527, 529, 534
Tapetum, 244
Telangiectasia, 605
Telomeres, 148
Telophase, in cervix uteri, 442
Temperature-sensitive mutant, 529, 548
Teratoma, human ovarian, 329
  sacrococcygeal, 401
  of testis, 394
Testicular tumors, 470
Testis, malignant tumors, 394
Tetraploid, 141, 244, 645
  in mild dysplasia of cervix, 441

SUBJECT INDEX

Tetraploid cells, 550
Tetraploidy, 583
  in cervix uteri, 442
Tetrasomy, 141, 145
Theophylline, 549
Therapy, effect on karyotype, 279
  immunosuppressive, tumors in patients on, 30
  individualized, 283
Thorotrast, 648, 663
³H-thymidine, 571
Thymidine kinase, 523, 542, 551
Thymine dimers, 542
Thymus-dependent lymphocytes (TL), 571
Thyroid, carcinoma, 400
Thrombocythemia, 462
Time in culture, 589, 590
T6 marker, 535, 537, 550, 551
Transformation, 163, 178, 179, 571, 572
  malignant, 147, 670, 671
  by SV40, 165
  viral, 527, 542
Transforming agent, 571
Translocation, 145, 147, 258
  induced in spermatogonial cells, 225
  reciprocal, 136, 138, 646, 649, 651, 661, 669
Trisomic line, 584
Trisomy, 141, 586
  double, 584
  group C, in carcinoma of the corpus uteri, 395
  in Hashimoto's disease, 406
  in large intestine polyps, 387
Trisomy 8, 137
Trisomy 21, 18, 288, 304, 527, 586
Trypsin, 548
Trypan blue exclusion, 569
TSV-5 transformed hamster cell line, 528
Tumor, 140, 141, 145, 147, 148
  adenovirus 12-induced, 158
  animal, 276
  benign, 147, 498, 513
  Burkitt, 162
  childhood, 137, 147
  Ehrlich, 253, 254, 257
  fetal antigens in, 29

Tumor (*continued*)
  Landschütz, 254, 257
  lymphoid, 137
  malignant, 140
  metastatic, 325
  nervous system, astrocytic malignant gliomas, 481
    ependymomas, 481
    medulloblastomas and neuroblastomas, 481
    metastases, 481
    neurinomas, 481
    oligodendrogliomas, 481
    pituitary adenomas, 481
    retinoblastomas and optical gliomas, 481
  in patients on immunosuppressive therapy, 30
  squamous cell carcinomas, 30
  primary, *see* Primary tumors
  recurrent, 326
  Sticker's, 336
  SV40, 148, 528, 633
  testicular, 470
  venereal, 365
  virus-induced primary, 18
  Wilms', 138, 147
  see also Cancer; Neoplasm
Tumor antigenicity, 26, 30
  antigenic individuality of tumors, 27
Tumor cell, 136
Tumor cell autonomy, 327
Tumor-forming dose 50% (TFD_{50}), 540
Tumor initiation, 317
  double-phenotype tumors, 320
  single-phenotype tumors, 317
Tumor progression, 322
  cervical cancer, 322
  chronic myelogenous leukemia, 323
Turner's syndrome, 467, 584
Twins, 289
Tyrosine aminotransferase, 549, 552

Ukraine, 607
Ultraviolet spectrophotometry, 428
Unicellular origin of cancer, 280-281
UV-irradiation, 542
UV radiation, 614

SUBJECT INDEX

Vaccination, 168
Vaccinia virus, 168
Vagina, 338
  carcinoma in situ, 400
Variant commun, 302
Venereal lymphosarcoma cells, of the dog, 87
Venereal tumor, 365
Vertical laminar flow hood, 580
Viable yellow, A^{vy}, 602
Vinblastine, 247
Vinblastin sulfate, 554
Viral hepatitis, 569
Viral infection, 139
Viral multiplication, 528
Viral oncogenesis, 148, 526
Viral rescue, 527
Viral transformation, 527, 542
Virus, 147, 366
  abortive infection, 173
  avian leukemia, 170
  avian sarcoma, 170
  cell interactions, 153
  chromosome breakage, 154
  C-type, 168, 171
  defective, 529, 531
  DNA synthesis, 155, 159
  DNA viruses, 154
  EB, *see* Epstein-Barr (EB) virus
  helper leukemia, 531
  hepatitis, 170
  herpes simplex, 583
  influenza, 169
  integration, 153, 157
  measles, 248
  moloney, 531
  multiplicity of infection, 153
  murine leukemia (MLV), 543
  murine sarcoma, 172, 531
  as mutagens, 153
  oncogenic DNA, *see* Oncogenic DNA virus
  papilloma, 163
  productive infection, 153
  rabies, 169
  RNA viruses, 168

RNA viruses (*continued*)
  C-type, 170, 527, 530
  rubella, 170
  rous sarcoma, *see* Rous sarcoma virus
  Sendai, 169, 248, 525, 530, 550
  Shope fibroma, 167
  Shope papilloma, 163
  transformation of cells, 153
  UV-irradiated, 155, 159
  vaccinia, 168
  wart, 163
  Yellow fever, 170
Virus genome, persistence of, 161, 162, 173, 178
Virus genome integration, 174
  productive infections, 174
Virus-incuced antigen, 543, 544
Virus-induced primary tumors, 18
Virus infection, 172
  early events, 173
  late events, 173
  lytic cycle, 173
Virus integration, 528
Virus-negative lines, 577
Vulva, carcinoma, 400

Waldenström's macroglobulinemia, 295, 305, 465, 466
Wart, 163
Wart virus, 163
Weak antigens, 545
Wheat germ agglutinin, 548
WI-Lz, 139
Wilms' tumor, 138, 147
WI-38, 139

Xeroderma pigmentosum, 66, 210, 527, 542, 614, 615
  de Sanctis-Cacchione variant, 543
X-irradiation, 544
X-rays, 248

Yaba virus, 167
Y chromosome, 139, 293
  loss of, 459
Yellow fever virus, 170

THE LIBRARY
UNIVERSITY OF CALIFORNIA
San Francisco
(415) 476-2335

THIS BOOK IS DUE ON THE LAST DATE STAMPED BELOW
Books not returned on time are subject to fines according to the Library Lending Code. A renewal may be made on certain materials. For details consult Lending Code.

14 DAY		
AUG 24 1988 RETURNED AUG 26 1988		